TECHNIQUES OF CHEMISTRY

ARNOLD WEISSBERGER, *Editor*

VOLUME I

PHYSICAL METHODS OF CHEMISTRY

PART IIIA

Interferometry, Light Scattering, Microscopy, Microwave, and Magnetic Resonance Spectroscopy

TECHNIQUES OF CHEMISTRY

ARNOLD WEISSBERGER, *Editor*

VOLUME I

PHYSICAL METHODS OF CHEMISTRY, in Five Parts
(INCORPORATING FOURTH COMPLETELY REVISED AND AUGMENTED EDITION OF PHYSICAL METHODS OF ORGANIC CHEMISTRY)
Edited by Arnold Weissberger and Bryant W. Rossiter

VOLUME II

ORGANIC SOLVENTS. Third Edition
John A. Riddick and William S. Bunger

VOLUME III

PHOTOCHROMISM
Edited by Glenn H. Brown

TECHNIQUES OF CHEMISTRY

VOLUME I

PHYSICAL METHODS OF CHEMISTRY

INCORPORATING FOURTH COMPLETELY REVISED AND AUGMENTED EDITION OF TECHNIQUE OF ORGANIC CHEMISTRY, VOLUME I, PHYSICAL METHODS OF ORGANIC CHEMISTRY

Edited by

ARNOLD WEISSBERGER

AND

BRYANT W. ROSSITER

Research Laboratories
Eastman Kodak Company
Rochester, New York

PART III

Optical, Spectroscopic, and Radioactivity Methods

PART IIIA

Interferometry, Light Scattering, Microscopy, Microwave, and Magnetic Resonance Spectroscopy

WILEY-INTERSCIENCE

A DIVISION OF JOHN WILEY & SONS, INC.

New York · London · Sydney · Toronto

Library of Congress Catalog Card Number: 45-8533

ISBN 0-471-92729-5

Printed in the United States of America.

10 9 8 7 6 5 4 3 2 1

PLAN FOR

PHYSICAL METHODS OF CHEMISTRY

PART I

Components of Scientific Instruments, Automatic Recording and Control, Computers in Chemical Research

PART II

Electrochemical Methods

PART III

Optical, Spectroscopic, and Radioactivity Methods

PART IV

Determination of Mass, Transport, and Electrical-Magnetic Properties

PART V

Determination of Thermodynamic and Surface Properties

AUTHORS OF PART IIIA

GEORGE G. COCKS
Department of Chemical Engineering, Cornell University, Ithaca, New York
RAYMOND V. ELY
Abingdon, Berkshire, England
W. H. FLYGARE
Noyes Chemical Laboratory, University of Illinois, Urbana, Illinois
EDWIN E. JELLEY, *Deceased*
A. N. MCKEE
Coates & Welter Instrument Corporation, Sunnyvale, California
NORBERT MULLER
Department of Chemistry, Purdue University, Lafayette, Indiana
ROBERT N. O'BRIEN
Department of Chemistry, University of Victoria, Victoria, British Columbia, Canada
GERALD OSTER
Mount Sinai School of Medicine of the City University of New York, New York
PHILIP H. RIEGER
Department of Chemistry, Brown University, Providence, Rhode Island
ROBERT G. SCOTT
Pioneering Research Division, E. I. du Pont de Nemours and Company, Experimental Station, Wilmington, Delaware

NEW BOOKS AND NEW EDITIONS OF BOOKS OF THE TECHNIQUE OF ORGANIC CHEMISTRY WILL NOW APPEAR IN TECHNIQUES OF CHEMISTRY. A LIST OF PRESENTLY PUBLISHED VOLUMES IS GIVEN BELOW

TECHNIQUE OF ORGANIC CHEMISTRY

ARNOLD WEISSBERGER, *Editor*

Volume I:	Physical Methods of Organic Chemistry *Third Edition—in Four Parts*
Volume II:	Catalytic, Photochemical, and Electrolytic Reactions *Second Edition*
Volume III:	Part I. Separation and Purification Part II. Laboratory Engineering *Second Edition*
Volume IV:	Distillation *Second Edition*
Volume V:	Adsorption and Chromatography
Volume VI:	Micro and Semimicro Methods
Volume VII:	Organic Solvents *Second Edition*
Volume VIII:	Investigation of Rates and Mechanisms of Reactions *Second Edition—in Two Parts*
Volume IX:	Chemical Applications of Spectroscopy *Second Edition—in Two Parts*
Volume X:	Fundamentals of Chromatography
Volume XI:	Elucidation of Structures by Physical and Chemical Methods *In Two Parts*
Volume XII:	Thin-Layer Chromatography
Volume XIII:	Gas Chromatography
Volume XIV:	Energy Transfer and Organic Photochemistry

INTRODUCTION TO THE SERIES

Techniques of Chemistry is the successor to the Technique of Organic Chemistry Series and its companion—Technique of Inorganic Chemistry. Because many of the methods are employed in all branches of chemical science, the division into techniques for organic and inorganic chemistry has become increasingly artificial. Accordingly, the new series reflects the wider application of techniques, and the component volumes for the most part provide complete treatments of the methods covered. Volumes in which limited areas of application are discussed can be easily recognized by their titles.

Like its predecessors, the series is devoted to a comprehensive presentation of the respective techniques. The authors give the theoretical background for an understanding of the various methods and operations and describe the techniques and tools, their modifications, their merits and limitations, and their handling. It is hoped that the series will contribute to a better understanding and a more rational and effective application of the respective techniques.

Authors and editors hope that readers will find the volumes in this series useful and will communicate to them any criticisms and suggestions for improvements.

Research Laboratories
Eastman Kodak Company
Rochester, New York

ARNOLD WEISSBERGER

PREFACE

Physical Methods of Chemistry succeeds, and incorporates the material of, three editions of *Physical Methods of Organic Chemistry* (1945, 1949, and 1959). It has been broadened in scope to include physical methods important in the study of all varieties of chemical compounds. Accordingly, it is published as Volume I of the new Techniques of Chemistry series.

Some of the methods described in *Physical Methods of Chemistry* are relatively simple laboratory procedures, such as weighing and the measurement of temperature, refractive index, and determination of melting and boiling points. Other techniques require very sophisticated apparatus and specialists to make the measurements and to interpret the data; x-ray diffraction, mass spectrometry, and nuclear magnetic resonance are examples of this class. Authors of chapters belonging to the first class of methods aim to provide all information that is necessary for the successful handling of the respective techniques. Alternatively, the aim of authors treating the more sophisticated methods is to provide the reader with a clear understanding of the basic theory and apparatus involved, together with an appreciation for the value, potential, and limitations of the respective techniques. Representative applications are included to illustrate these points, and liberal references to monographs and other scientific literature providing greater detail are given for readers who want to apply the techniques. Still other methods that are successfully used to solve chemical problems range between these examples in complexity and sophistication and are treated accordingly. All chapters are written by specialists. In many cases authors have acquired a profound knowledge of the respective methods by their own pioneering work in the use of these techniques.

In the earlier editions of *Physical Methods* an attempt was made to arrange the chapters in a logical sequence. In order to make the organization of the treatise lucid and helpful to the reader, a further step has been taken in the new edition—the treatise has been subdivided into technical families:

Part I Components of Scientific Instruments, Automatic Recording and Control, Computers in Chemical Research
Part II Electrochemical Methods
Part III Optical, Spectroscopic, and Radioactivity Methods

Part IV Determination of Mass, Transport, and Electrical-Magnetic Properties
Part V Determination of Thermodynamic and Surface Properties

The changes in subject matter from the Third Edition are too numerous to list in detail. We thank previous authors for their continuing cooperation and welcome the new authors to the series. New authors of Part IIIA are Professor George G. Cocks, Mr. Raymond V. Ely, Dr. A. N. McKee, Professor Robert N. O'Brien, and Mr. Robert G. Scott.

We are also grateful to the many colleagues who advised us in the selection of authors and helped in the evaluation of manuscripts. They are for Part IIIA: Mrs. Ardelle Kocher, Dr. Ronald G. Lawler, Dr. Louis D. Moore, Jr., Mr. Robert R. Parmerter, Dr. Anthony Pietrzykowski, Mrs. Donna S. Roets, Dr. Philip I. Rose, Dr. Norbert Stalica, and Dr. Don W. Vanas.

The senior editor expresses his gratitude to Bryant W. Rossiter for joining him in the work and taking on the very heavy burden with exceptional devotion and ability.

The experimental part of the chapter on "Refractometry" by S. Z. Lewin was not received in time to include it in Part IIIA as planned. Rather than delete this important subject, we have decided to relocate the entire chapter, including the theoretical section by K. Fajans, to a later part.

September 1971
Research Laboratories
Eastman Kodak Company
Rochester, New York

ARNOLD WEISSBERGER
BRYANT W. ROSSITER

CONTENTS

INTERFEROMETRY, LIGHT SCATTERING, MICROSCOPY, MICROWAVE, AND MAGNETIC RESONANCE SPECTROSCOPY

Chapter **I**

Interferometry
ROBERT N. O'BRIEN 1

Chapter **II**

Light Scattering
GERALD OSTER 75

Chapter **III**

Light Microscopy
GEORGE G. COCKS & EDWIN E. JELLEY 119

Chapter **IV**

Electron Microscopy
ROBERT G. SCOTT & A. N. MCKEE 291

Chapter **V**

Microwave Spectroscopy
W. H. FLYGARE 439

Chapter **VI**

Electron Spin Resonance
PHILIP H. RIEGER 499

Chapter **VII**

Nuclear Magnetic Resonance
NORBERT MULLER 599

Chapter **VIII**

X-Ray Microscopy
RAYMOND V. ELY 715

Subject Index 781

TECHNIQUES OF CHEMISTRY

ARNOLD WEISSBERGER, *Editor*

VOLUME I

PHYSICAL METHODS OF CHEMISTRY

PART IIIA

Interferometry, Light Scattering, Microscopy, Microwave, and Magnetic Resonance Spectroscopy

Chapter 1

INTERFEROMETRY

Robert N. O'Brien

---

1 **Introduction** **2**
2 **Theory of Optical (Visible) Interferometry** **4**
The Wave Nature of Light 4
Conditions for Interference 5
Light Sources 9
Coherence Theory 12
Nomenclature and Geometry of the General Interferometer 15
3 **Two-Beam Interferometers** **17**
Formation of Two Beams 17
The Rayleigh Interferometer 17
Uses 19
The Jamin Interferometer 20
The Mach-Zehnder Interferometer 21
The Michelson Interferometer 23
Uses 25
The Interference Microscope 27
The Polarizing Interferometer 28
Other Two-Beam Interferometers 32
4 **Multiple-Beam Interferometry and Interferometers** **32**
General Theory 32
The Fabry-Perot Interferometer 36
Fabry-Perot Interferometer with Wedge Fringes 38
Uses of Wedge Multiple-Beam Interferometry 41
Crystal Growth 41
Chemical Reactions 42
Thin and Thick Films 42
Dilatometry 44
Electrochemistry 44
Uncertainties in Electrochemical Applications 45
Electrodialysis 48
Thermal Conductivity of Transparent Liquids 48
Diffusion Measurements 48
Practical Details of Evaluation Calculations 49

5 Holographic Interferometry 51
General 51
Theory of Holographic Interferometry 52
6 Moiré Interferometers 59
7 Other Types of Interferometers Using Electromagnetic Radiation 62
X-Ray Interferometers 62
Microwave Interferometers 64
Radio Interferometers 65
8 Nonelectromagnetic Radiation Interferometers 65
Sonic Interferometers 65
Particle Interferometers 66
Neutron Interferometers 66
Electron Interferometers 66
9 Summary of Uses 66
References 67

1 INTRODUCTION

Although interferometers have been used for a long time, particularly in gas analysis, interferometry has only recently become of more general interest to chemists. Astronomers have used both optical and radio interferometry in measuring star sizes. Spectroscopists have used it for resolving multiplet spectral lines and for increased sensitivity in spectrometry. Surfaces including the surfaces of optical parts such as lenses have been investigated by optical physicists and engineers. Biologists have used interference microscopy as another advance in one of their basic tools. Interference methods are usual in converting the international standard of length into a practical scale. Interference refractometers have become common analytical tools, for example, in studies of diffusion. However, only since the laser has become available as a light source has optical interferometry become a reasonable alternative to more conventional methods in such areas as dilatometry and the study of crystal growth, film thickness, and chemical kinetics.

The basis for interferometry can probably best be stated by examining the Lorentz-Lorenz law. It can be expressed:

$$R = \frac{M}{\rho}\left(\frac{n^2 - 1}{n^2 + 2}\right)^{*}, \tag{1.1}$$

* Not strictly applicable to solutions. For the correct solution form, see chapter on refractometry in Part IIID.

where R is the molal refraction in cubic centimeters, M the molecular weight in grams, ρ the density in grams cm^{-3}, and n the refractive index, a pure number. Of all the parameters only R is considered constant (although it varies in solutions of high concentration). Hence any change in M such as a chemical reaction or an association affects n. Similarly, any change in temperature, concentration, or pressure causes a change in ρ and therefore in n. Moreover, changes in phase, crystal habit, and so on, affect n and are detectable.

From the time of Young's double-pinhole experiments which prompted Fresnel to postulate his wave theory of light, the study of interference has had a continuing influence on the development of physics. An indication of its increasing influence on chemistry is that in *Chemical Abstracts* of 1960 only 19 abstracts appeared under "interferometry" compared with 125 in 1966. In this interval lasers became commercially available, hence optical interferometry became much simpler. Interferometry has been included in physical optics treatises for some time, but monographs on interferometry only began to be written in 1930. Now several general references are available, listed in chronological order with laser interferometry last in the General References.

Interference occurs when radiation follows more than one path from its source to the point of detection. This description was adequate until the advent of the very stable unimodal output gas laser, but now interference has been observed between two beams from two different sources [1] and the statement should be altered to emphasize the path difference of the two interfering beams. Interference manifests itself as the departure of the resultant intensity from the law of additivity of intensities on superposition of beams. For two beams, as the point of observation of intensity is moved, the intensity oscillates about the sum of the intensities from each of the two paths. The character of this oscillation is affected by the differences in the two paths in such qualities as absorbance as well as optical path length. For more than two beams, the effect is more complicated. The types of interferometry now practiced include the whole range of the electromagnetic spectrum from x-rays to radio waves. In particle interferometers only electron and neutron interferometers have been constructed, but the use of all subatomic particles and a few of the lighter nuclei such as helium have been predicted. Sonic interferometers have been used since 1925 [2]. Recently, holographic interferometry has become possible and has begun to be employed by chemists. Moiré techniques using mechanically oriented grids in any kind of light do not involve interferometry in the true sense but they do when diffraction gratings are used. Moiré techniques are of use to chemists [3] and are included in this chapter. Interferometers of no specific use to chemists are only briefly mentioned.

2 THEORY OF OPTICAL (VISIBLE) INTERFEROMETRY

In order to understand interferometry, some knowledge of the wave nature of light is necessary. (See General References, Wave Nature of Light.)

The Wave Nature of Light

Visible electromagnetic radiation can be treated as two periodic transverse wave motions, the amplitude of each being the oscillating electric and magnetic vectors at 90° to each other, and the direction of propagation is as seen in Fig. 1.1. The equation describing such a motion is:

$$y = A \sin(\omega t + \epsilon), \tag{1.2}$$

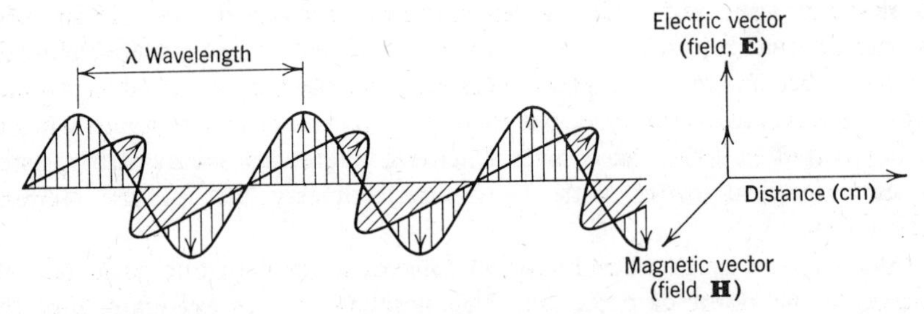

Fig. 1.1 Propagation of electromagnetic waves.

the well-known sine wave expression in which Y is the displacement (amplitude in our case) from zero at time t seconds, A is the maximum displacement amplitude, ω is the angular speed in radians per second, $\omega t + \epsilon$ is the phase angle and ϵ is the phase constant, that is, the phase angle at $t = 0$.

Classic electromagnetic theory states that acceleration of the motion of a charge leads to radiation. Quantum mechanics predicts radiation also when a dipole oscillates. We can use either of these models, but to simplify the model we hold one charge of the dipole steady and allow the other charge to oscillate, resulting in the equivalent of the one-charge system. Consider now that at either end of the oscillation (maximum and minimum distance of the two charges of a dipole) deceleration to a stop must occur. For simple harmonic motion, the instantaneous velocity is represented by:

$$v_{\text{inst}} = \omega\sqrt{A^2 - y^2}, \tag{1.3}$$

and an instantaneous acceleration at maximum displacement:

$$a_{\text{inst}} = -\omega^2 y. \tag{1.4}$$

At $Y = 0$, $v_{\text{inst}} = \omega A$, its maximum value, and all of the energy is present as kinetic energy (mv^2) and proportional to A^2 or the square of the amplitude.

This relationship holds at all times whether the energy is present totally in one form or the other at maximum and minimum displacements or apportioned between them at intermediate displacements. The time rate of flow of energy per unit area normal to the propagation direction, called the intensity, is therefore proportional to the square of the amplitude. The parameter "intensity" is the value normally recorded, although it frequently is convenient to use amplitude to describe energy flux. The fact that most optical effects including interferometry can also be explained in terms of particles is not discussed until particle interferometers are considered.

Conditions for Interference

From Fig. 1.1 it can be seen that if a wave that has the same amplitude, wavelength, and general propagation direction as another wave, but is 180° (π radians in $\omega t + \epsilon$) out of phase is superimposed on the latter, positive peaks of one ray will coincide with negative peaks of the other ray and total destruction of the wave amplitude will result. Perhaps the simplest example is interference in an unsupported film. If a monochromatic parallel beam strikes a thin film at normal incidence (90° to the plane of the film) and the film consists of a half-wavelength thickness (d) of transparent material, the light reflected from the second interface will have a path difference of $\Delta = n(AFB) - AD$ from the light reflected from the first surface, where n is the refractive index of the film as seen in Fig. 1.2. By using the construction AGF, another expression for the path difference is $\Delta = n(GB) - AD = n(GC + CB) - AD$ by similar triangles. Because AC is perpendicular to FB, AC and DB can represent two successive positions of a maximum in the sine wave function reflected from the second surface of the film. However, $n(CB) = AD$ because the general condition

$$c = \lambda\nu = n\lambda\nu \qquad \text{or} \qquad n\lambda = \frac{\nu}{c} \tag{1.5}$$

must be fulfilled. From this the path difference becomes:

$$\Delta = n(GC) = n(2d\cos\phi) \tag{1.6}$$

for the most general case of $\phi \neq 0$ or an angle of incidence not normal to the plane of the film. GC is equal to one wavelength of light, two for the next harmonic, three for the next, and so on. We can then make the general statement that:

$$2nd\cos\phi = N\lambda, \tag{1.7}$$

where N is the order of interference (which is discussed more extensively under multiple-beam interferometers) and, in the case considered, equal to one, and where $\cos\phi$ is also equal to one. Under these circumstances (1.6)

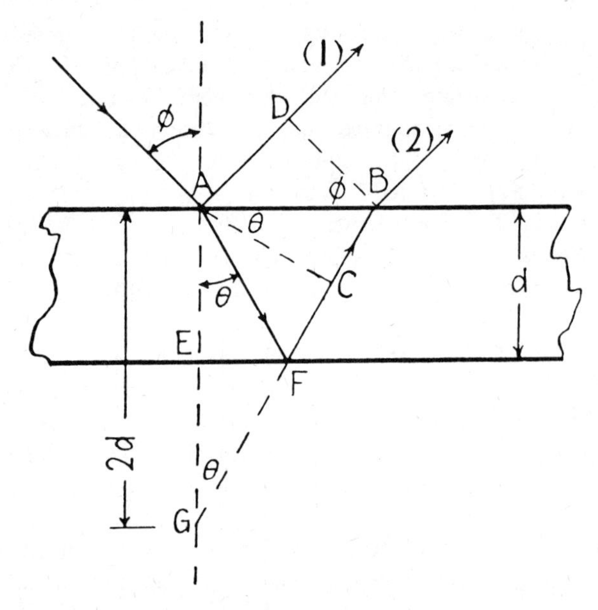

Fig. 1.2 Optical path difference between two consecutive rays in multiple reflection. (By permission, from *Fundamentals of Optics*, F. A. Jenkins and H. E. White, copyright 1950 2nd ed., McGraw-Hill, 1950.)

would be expected to give a maximum, but we have not taken into account the change of phase that occurs on reflection from a surface at which the change in refractive index is sharp and positive. Extensive treatments of this effect are given in the General References.

If N is one, the change of phase will cause a minimum to occur, that is, destructive interference so that no reflection occurs. At the other side of the film, the opposite situation obtains (less resistance to propagation, or lower refractive index) and the amplitudes add. The condition to obtain maximum transmission then is $2nd = [N + (1/2)]\lambda$. If the absorption of the film is low or negligible, almost all of the light will be transmitted. If the reflecting power of the plane parallel film in air is high, the film becomes an interference filter for the wavelength $\lambda = 2nd$ and all of that wavelength and only that wavelength is transmitted. This is interference by splitting amplitudes or intensities to produce two or more beams.

As an indication of how high the reflecting power must be to make a good interference filter, consider the simplified reflecting power expression:

$$R = \left(\frac{n_2 - n_1}{n_2 + n_1}\right)^2 \tag{1.8}$$

for the film in air ($n_1 \simeq 1$). If the film is ordinary glass $n_2 \simeq 1.5$ and $R = 0.04$

or the reflection is about 4%. Usually films of metal which are partially transparent are evaporated onto good glass flats, and a $\lambda/2$ thickness of a transparent dielectric such as MgF_2 is sandwiched between, usually all in one operation in a vacuum chamber. This system obviously provides successive reflections from both sides, that is, multiple-beam interferometry, which is extensively treated below. Pass bands can be as narrow as 20λ and have transmissions up to 40% of the selected wavelength.

In the classic experiment of Young, a source of monochromatic collimated light falls on a slit S (see Fig. 1.3), which then serves as the source for the optical system. Application of Huygens' principle—that all points on a wave

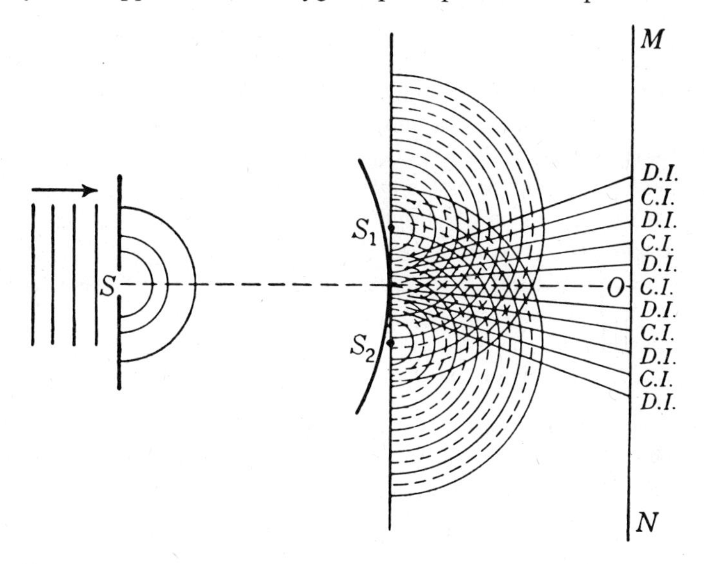

Fig. 1.3 Young's arrangement for interference from a double slit. *C.I.* is a region of constructive interference and *D.I.* a region of destructive interference. (By permission, from *Introduction to Geometrical and Physical Optics*, J. Morgan, copyright 1953, McGraw-Hill, 1953.)

front may be considered as centers of new disturbances, creating secondary wavelets which travel with a speed equal to that of the wave and that the envelope of these wavelets forms the new wave front at any time—leads to generation of wave fronts as shown in Fig. 1.3 at each of the slits S, S_1, and S_2. Huygens' principle is no longer the only method of explaining this phenomenon, a statistical theory is explained in Morgan (General References, Wave Nature of Light).

If the wave front from slit S is divided into wave fronts at S_1 and S_2, the arrangement is called two-beam interferometry by a divided wave front. Following S_1 and S_2, the crests (maximum positive amplitudes) and troughs

Fig. 1.4 Interference fringes from a double slit. (By permission, from *Fundamentals of Optics*, F. A. Jenkins and H. E. White, Copyright 1950, 2nd ed., McGraw-Hill, 1950.)

(maximum negative amplitudes) are shown as full and dashed lines, respectively. Where two crests meet, constructive interference occurs, and where a crest and a trough coincide, destructive interference results. Lines are drawn joining these points at successive positions of troughs and crests of the two wave fronts. Those for constructive interference have been labeled *C.I.*, and cause a bright fringe on the screen *MN*; those for destructive interference, *D.I.*, cause a dark fringe. The appearance of the fringe system, as photographed at *MN*, is shown in Fig. 1.4.

On the basis of this mechanism, we set down the conditions for interference between two beams of light:

1. The phase difference between the two beams must remain constant. This implies from (1.2) that $\omega_1 = \omega_2$, hence $\lambda_1 = \lambda_2$ (the subscripts referring to the two beams), that is, the two beams must have the same period and wavelength. If this is not the case, interference will only be visible with very short path differences. For white light, in which λ varies across the whole visible region, the two paths must be exactly equal. This condition is difficult to meet in condensed phases and when using multiple-beam interferometry.
2. The two beams, hence their wave fronts, must have a very small angle between their directions of propagation, such that the cosine of the angle is about equal to unity. Larger angles produce fringe systems but observation becomes more difficult as the angle increases, and the fringes become more closely spaced until the resolving power of most combinations of interferometer and microscope is exceeded. About 2 min of arc produce about two fringes per millimeter with gas laser light (6328 Å). The limits are discussed in the section on interference microscopy.
3. The intensities of the two waves must be reasonably close to equal, that is, in (1.2) $Y_1 \simeq Y_2$. If this condition is not met, fringe visibility may be poor, the fringes appearing only as a ripple on a large relative intensity. If one beam supplies only 10% of the total intensity, an intensity-time distribution of

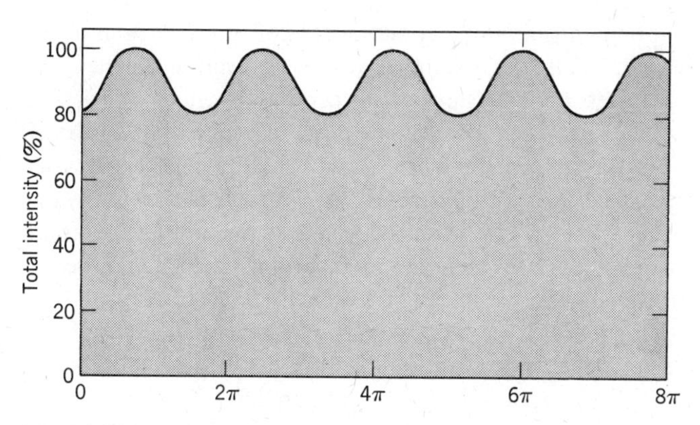

Fig. 1.5 Visibility of fringes formed by two beams of greatly differing amplitude.

light appears in general as in Fig. 1.5 and the dark fringes are not black. A correct description of conditions requires the summation of an infinite series. Contrast may be heightened by photographic recording under the right exposure conditions and with high-contrast film, but there is a limit to the effectiveness of this technique. Visual observation and mechanical manipulation of a low-contrast fringe system are difficult.

4. The two waves must have the same state of polarization. This condition is intuitively obvious from Fig. 1.1. If the oscillations are not in the same direction, the two waves cannot interfere. For further discussion, see the section on polarization interferometers.

Light Sources

The first light source used for interferometry was the sun. Interference fringes can be produced by white light but, except in the cases of fringes of equal chromatic order (a very seldom used system) or of the interference microscope, white light fringes are now used only as identifying markers in such a system as a Rayleigh interferometer. Sunlight passed through a monochromator is an inexpensive but not very convenient source of light.

Until lasers became commercially available, several kinds of metal vapor lamps were the principal light sources. The mercury vapor light was very widely used despite pressure broadening and the presence of isotopes which contributed extra lines. It is still useful for interferometry in which the fringe system is also to be observed visually. Soda glass shields, about $\frac{1}{8}$ in. thick, are needed to reduce the ultraviolet radiation to continuously tolerable levels. Filters are also required to remove all visible lines but the one desired. Unless very short path differences are to be used, the lines from high-pressure arcs are too broad.

Sodium vapor lamps are also good sources of monochromatic light; those commercially available, for example, for use in refractometers, require no filters or ultraviolet shields. They do not produce as much light as a medium-pressure mercury lamp, but after filtering the mercury light has been lowered to about the same intensity for the same level of coherence; coherence is defined here as the path difference allowable for distinct fringes to be obtained. The cadmium red line was used as a length standard and is a relatively weak source compared to mercury and sodium, which has been replaced as a standard by the emission line of krypton 86 at 606 nm.

Several other sources have been used, but lasers are likely to displace them all. For an evaluation of conventional low coherence sources, see Tolansky, 1955 (General References, Optical Interferometry). Only two years elapsed from the time the first laser was made to work in 1961 to the time when the first single-mode continuous laser became commercially available. Now, less than a decade later, such a large variety of lasers for visible and other wavelengths is available that it is difficult to imagine a need for other radiation sources in interferometry. For most interferometry important to chemists, a continuous source is most useful, and one with single-mode output is usually essential. The reflecting surfaces at each end of a laser tube, the oscillating chambers, cause the light to go back and forth a large number of times. Transmission percentages of the output end are typically $\sim 1\%$. If the end mirrors are strictly parallel, although one is usually curved, and exactly placed so that the resonant cavity extends over an even number of wavelengths, the TEM_{00} (transverse electromagnetic mode) is produced. This spatial mode corresponds to a constant distribution or phase of the electric field on the wave front and a Gaussian distribution of energy across the output spot (Fig. 1.6). Since the usual gas laser resonating tube represents about 10^6 wavelengths, a very small displacement of the end mirrors results in multimode operation. A mechanical shock to the laser head, such as rough handling, frequently gives this effect. For a more complete treatment, see General References, Patek or Françon.

For normal applications of interferometry to chemical problems, gas lasers in the milliwatt range are satisfactory; no protection of the eyes is necessary unless the beam is focused down. Usually beams are expanded. Helium-neon lasers of a nominal 1-mW output and a combination spatial filter and beam-expanding telescope—in which the beam emerging at a nominal 1.2-mm diameter was expanded to about 20 mm—gave slightly overexposed negatives at 16 frames/sec (exposure time $\sim\frac{1}{30}$ sec) with 16-mm motion picture film with ASA about 50, and no eye damage after seven years.

Because lasers produce highly coherent light, fringe systems are produced from all discontinuities of refractive index in an optical system which result

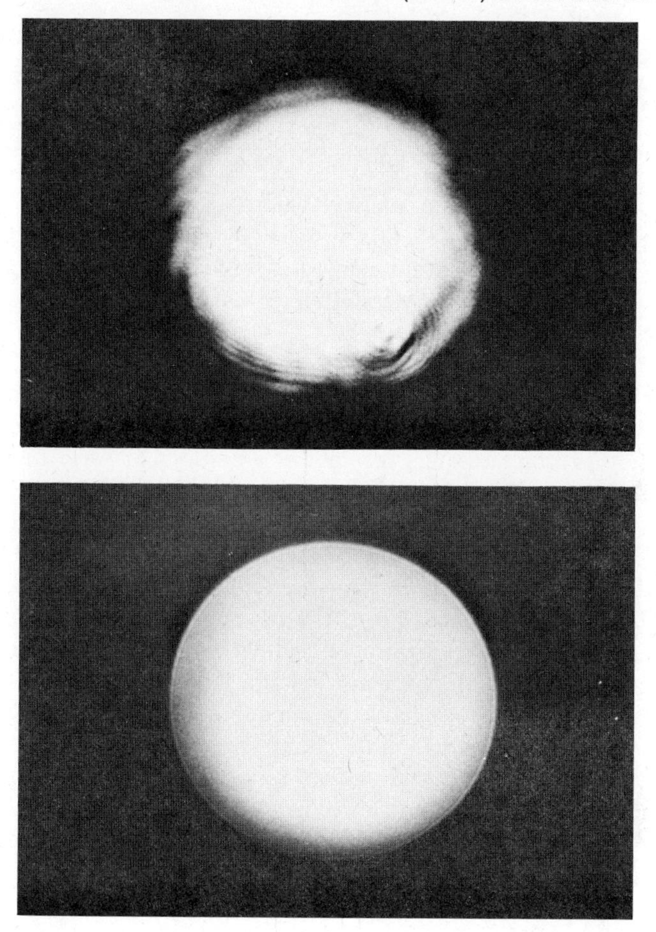

Fig. 1.6 The effect of inserting a spatial filter in a beam-expanding telescope on the output of a gas laser. (*a*) Shows the unfiltered output; (*b*) shows the filtered output which has been analysed and shown to be a smooth gaussian distribution of intensities across the diameter. (Photographs by courtesy of Spectra-Physics, Mountainview, California.)

in reflections [see (1.8)]. It is important to use as few optical surfaces as possible. All lenses should have antireflection coatings, and mirrors and beam splitters should have wedge angles of about 1° on the surfaces from which reflection is undesirable so that it will be outside the gathering angle of the instrument. If such an angle cannot be tolerated, antireflection coatings should be used.

In another method for reducing any background fringe system, a specially prepared optical system is used such as the Spectra Physics Model 333, 332 collimator and spatial filter combination. Figure 1.6 shows the smooth Gaussian energy distribution across the wave front obtained with this system, which does not introduce any spurious fringes. Modern still and motion picture cameras can usually be obtained with surfaces properly treated for laser work. Scattering of the background noise by a diffuser is discussed under multiple-beam interferometers.

The number of types of lasers and wavelengths available and the power of lasers continue to grow. A recent catalog of frequencies and types available is given by Schawlow (see General References), listing wavelengths from $\sim 2 \times 10^2$ to 10^6 nm.

Coherence Theory

Tolansky has defined coherent light beams as those radiated from one point on a source. Françon considers the case of vibrations or wave trains originating at a single atom and concludes that unless their frequencies are the same they will not interfere; their amplitudes do not combine, but their intensities do. This is another way of stating conditions one and two for interference, which stipulate both temporal (constant frequency) and spatial (good collimation) coherence. From the point of view of incoherence, it can be said that light waves from two thermal as opposed to laser sources are incoherent because no systematic relationship exists between the two beams even though they may have the same frequency. Interference occurs of course, but a fringe pattern lasts about the length of an average wave train, that is, it is of very short duration; hence there is no way of detecting this interference. To obtain completely incoherent sources is about as easy as obtaining any completely random quantity. The formation of recognizable interference requires careful choice of waves, usually from the same source even with laser light. In general, the coherence obtained with lasers is so good that they can be used in interferometers with path differences in the range of meters, whereas for conventional sources path differences of a few millimeters produce fuzzy fringes.

Coherence theory has made rapid advances in the last decade. A summary which is useful for understanding general interferometry theory is given below, but for those who only wish to use an interferometer the above remarks should suffice. Most modern monographs on interferometry and books on optics treat coherence. The discussion below is based on Steel (see General References).

Coherence theory should be viewed as the statistical correlation functions applied to radiation. Interference was the phenomenon that fostered the evolution of the theory. In any interferometer there must be a path difference between two beams, or all of them if it is a multiple-beam instrument. This can also be thought of as a time difference or delay for one beam, $\tau = l/c$, where τ is the time difference or delay, l is the path length difference, and c is the speed of light. If the interference is between two beams satisfying the interference conditions set out above, including a single frequency $\nu = c/\lambda$, the intensity varies with distance according to $I + \cos 2\pi\nu\tau$. (See for instance Figs. 1.5 and 1.13.) No radiation, however, contains only one frequency, since the uncertainty principle requires at least the natural broadening. When the radiation has a bandwidth $\Delta\nu$, a mean frequency ν_0, and a profile $g(\nu - \nu_0)$, the intensity becomes the integral over all these frequencies:

$$I_{(\tau)} = G_c(0) + G_c(\tau) \cos 2\pi\nu_0\tau, \tag{1.9}$$

where G_c is the Fourier cosine transform (1.17).

Inspection of this expression shows that the modulation is still cosinusoidal, but as τ increases the amplitude decreases or the fringe system fades out. The interpretation is then that the time difference for which fringe observation is possible is limited to the function $G_c(\tau)$. Using the uncertainty principle in the appropriate form:

$$\Delta\tau\, \Delta\nu \sim 1, \tag{1.10}$$

we can find $\Delta\tau$ or the coherence time of the radiation. This can be transformed into the coherence length by:

$$\Delta l = c\, \Delta\tau, \tag{1.11}$$

which is a measure of the temporal coherence of the radiation. The bandwidth of the mercury green line in a high-pressure arc can be 0.1 mm. Its coherence length then is 3 mm. For laser light the broadening is much less. In gas lasers only Doppler broadening is important, apart from mechanical vibrations and thermal expansions which can be controlled. For neon lasers the line width is commonly 0.03 Å or less and 0.007 Å for xenon lasers, being proportional to the reciprocal of the molecular weight to the one-half power ($\alpha M^{-1/2}$) (see General References, Wave Nature of Light, Patek, p. 125). The laser which for these purposes is treated as an interferometer with one bright fringe (TEM_{00} mode) as the output has a coherence length Δl of 13.35 cm. This is obtained by calculating $\Delta\nu$ from $\nu = \dfrac{c}{\lambda + \Delta\lambda}$, using $\lambda = 6.328 \times 10^{-5}$, $\Delta\lambda = 3 \times 10^{-10}$ cm, $\nu_0 = \dfrac{c}{\lambda}$, $\Delta\nu = \nu_0 - \nu$, the standard value of $c = 2.999774 \times 10^{10}$ cm sec^{-1} (see General References, Wave Nature of Light, Morgan, p. 195), and $\Delta\nu = 2.2474 \times 10^{9}$. Inserting this into (1.10) gives $\Delta\tau = 4.449586 \times 10^{10}$. If this is put into (1.11), Δl has the value 13.3478 cm. The same result can be obtained by assuming a doublet with discrete wavelengths at the two sides of the band and calculating the order of interference N, defined in (1.7), at which the two interference patterns exactly cancel each other. For the sodium *D*-line doublet 5890 and 5896 Å, with a 6-Å separation, the two interference patterns are exactly in phase and additive at 5893/6 or about 1000 wavelengths of path difference in a two-beam interferometer (see section on two-beam interferometry). At about 500 wavelengths path difference, $N = 500$, they are exactly out of phase and no fringes occur if their amplitudes are exactly equal. The coherence length is then about 500 × 5890 Å, that is, 294,5000 Å or about 0.03 cm. It has been our experience that sodium vapor light can be used and good fringes obtained with $N \simeq 10{,}000$ or $\Delta\lambda$ of the order of 0.5 Å. The doublet produces a recurring beat only. In the case of the laser, a similar calculation involves (6328/0.03) × 6328 = 13.35 cm. It has been stated above that path length differences of the order of meters are possible. Effective orders of interference of about 10^6 have been used in multiple-beam interferometry with good fringe definition, or coherence length of about 60 cm. This means that very little of the energy of the beam is distributed very far from ν_0, the center of the emission line. The spatial filter mentioned above also helps to increase the coherence length. The argument shows that as the bandwidth decreases the temporal coherence increases. Although the preceding discussion does not produce information unattainable by simpler methods, the concepts involved are important for an understanding of interference.

Spatial coherence in lasers is usually stated in terms of milliradians of beam divergence, a fraction of a milliradian or 10 to 50 sec of arc is a common value for lasers. It can be more formally stated, as Françon does, by considering an extended source (see Fig. 1.7). Consider radiating atoms at the opposite ends of the diameter of a disk source A_1 and A_2. A_1 radiates to P_1 and P_2 and, on arrival, the wave trains have a phase difference $n_1\pi$; A_2 radiates to P_1 and P_2 with a phase difference of $n_2\pi$ between the waves. If the differences in path lengths, $A_1P_1 - A_1P_2$ and $A_2P_1 - A_2P_2$, are smaller than the wavelength, the phase differences will be small and interference will

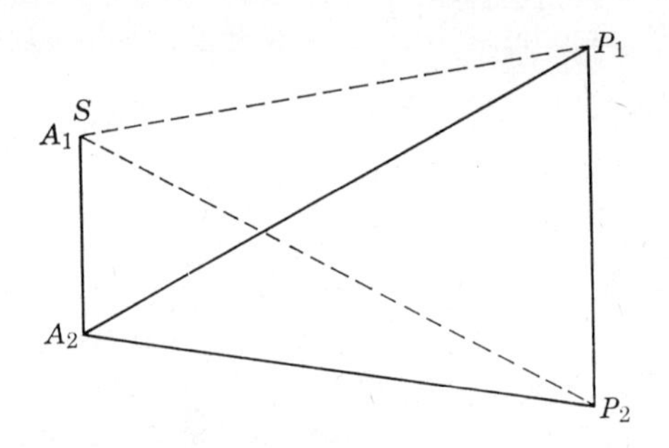

Fig. 1.7 Criteria for spatial coherence in an extended, radiating source.

occur. If P_1 and P_2 are two pinholes, interference fringes will be observed on a screen behind the holes. If P_1 and P_2 are moved further apart, at some point interference no longer will occur. At the separation at which fringes are about to disappear, the separation P_1P_2 is the diameter of a disk whose area can be called the area of coherence. If this is multiplied by the coherence length, the coherence volume is obtained.

These remarks on coherence apply to all light sources, but for thermal sources such a rigorous selection of waves must be made by slit or pinhole that the amount of usable light is a very small percentage of that emitted. In lasers, even with a spatial filter, much more than half of the light available is usable.

Nomenclature and Geometry of the General Interferometer

Recently, authors of monographs on interferometry (Françon and Steel) have adopted certain terms to describe the essentials of interferometry. As noted in Fig. 1.8, each interferometer has a source plane S which is shown as a laser source and a plane of observation. The instrument shown is a Michelson interferometer with collimation, a moving mirror M_2, a cell for the material whose refractive index is to be determined, a beam splitter of finite thickness S_1, a compensator C to compensate for the finite thickness of the beam splitter M, a fixed mirror and a combination beam expander and spatial filter. A quantity called by Steel *étendue:*

$$E = n^2 A\theta \tag{1.12}$$

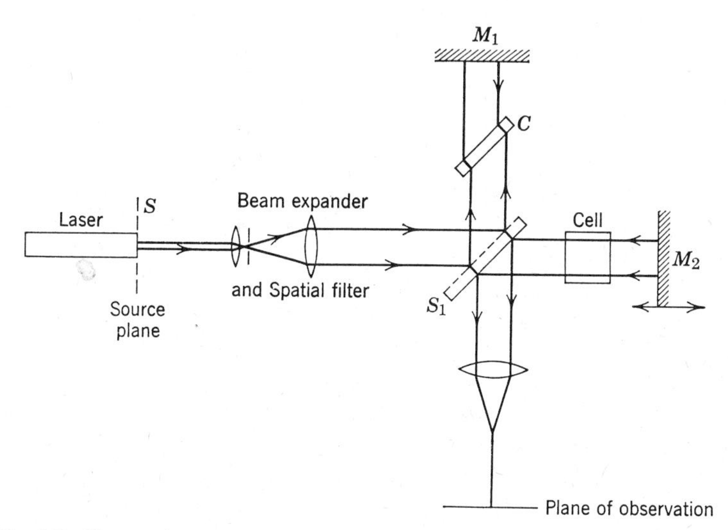

Fig. 1.8 Nomenclature and geometry of a typical (Michelson) interferometer.

of the instrument, and defined from the two planes mentioned above (but also called through-put, acceptance, or luminosity), is equal to the square of the refractive index (usually in air and taken equal to one) multiplied by the area of one plane times the solid angle subtended at it by the other. This quantity is related to the numerical aperture $n \sin \theta$.

Referring to Fig. 1.9, and following Steel, let the plane of observation have an origin at O. Viewing this plane from the source side of the interferometer, two images are seen, one from each beam (assuming a two-beam interferometer as in Fig. 1.8) and the point O appears in these images at O_1 and O_2 in two planes. These planes are tilted toward one another as shown in the figure,

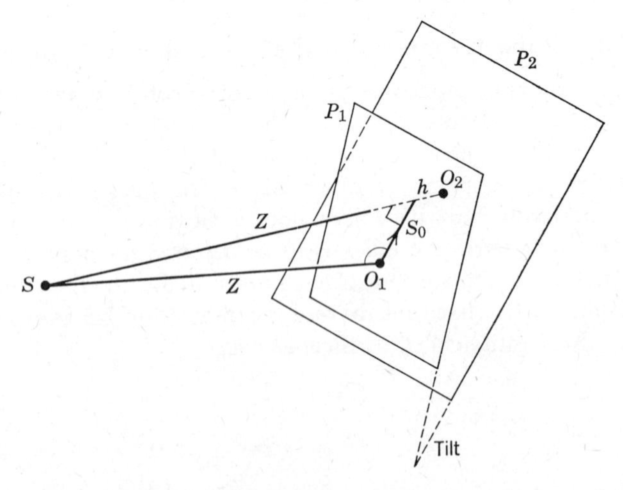

Fig. 1.9 Schematic drawing of the optical geometry of a two-beam interferometer illustrating some of the characterizing parameters.

by the "angle of tilt." O_2 is behind, that is, farther from S than O, by a distance $(Z + h) - Z = h$ which is called the "shift"—not to be confused with the term "fringe shift" which is used later. The lateral separation vector $\mathbf{S}_0$ is called the "shear" and of course the "angle of shear" is $\mathbf{S}_0/Z$. Shear can be a purely lateral displacement known as "lateral shear," a rotation of axis of images, "rotational shear" or, if there is a difference in magnification (the case of spherical mirrors at M_1 and M_2 in Fig. 1.8) as "radial shear." Other possibilities exist but are not usually met in chemical interferometers. Another parametric term often met is $\tau = (\Delta l)/c$, which is a form of (1.12) relating the path difference of two beams to the "delay" of one by dividing the path difference by c, the speed of light. From Δl, we obtain the "order of interference," $N = (\Delta l)/\lambda$. Δl may be the consequence of placing the cell, containing a material with $n \neq 1$, in one arm of the Michelson interferometer shown in Fig. 1.8.

3 TWO-BEAM INTERFEROMETERS

Formation of Two Beams

Although it is now possible to produce the two beams in two-beam interferometry by employing two lasers, it is still difficult to obtain usable interference effects in this way. In general, two beams are produced from the same source in either of two ways: by division of wave front as shown in Fig. 1.3, or by division of amplitude as shown in Fig. 1.2. Division of amplitude can also be effected in other ways. Beam splitters of various types can be used; they employ generally either partially silvered glass flats or prisms (see Fig. 1.12). The function of half-silvered glass flats is explained in detail in the section on the Michelson interferometer. Prisms can split the amplitude of a wave in two ways. In the first, reflection from two surfaces is utilized. If the prism is thick, two well-separated beams are produced. This system is used in the Jamin interferometer (see below). In the second method, a birefringent prism is used. In such a prism two rays are formed from the incident ray, the ordinary and the extraordinary ray (see Fig. 1.21). The phenomenon is fully treated in the section on polarizing interferometers.

The Rayleigh Interferometer

The Rayleigh interferometer produces two beams by means of division of wave front as shown in Fig. 1.10. The only other division of wave front–type interferometer is the grating or moiré interferometer (see below). The light from a point source, in reality as intense and small a source as possible, passes through a collimating lens L_1 and the center portion strikes a double-slitted stop B, allowing two beams to pass through two sample tubes T_1 and T_2 and two glass compensators C_1 and C_2. A lens L_2 now superimposes the two beams to give interference fringes. Because the separation of beams is

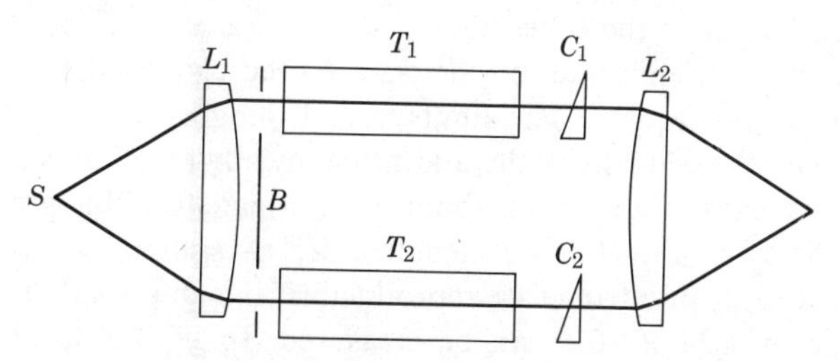

Fig. 1.10 Rayleigh interferometer with division of wave front.

great and the angle between the two superimposed beams is large, the fringes are closely spaced and are usually viewed with a cylindrical lens. With monochromatic light it is impossible to tell when sufficient optical path length compensation has been used. Therefore a second identical system may be combined in the instrument but using white light so that the fringes produced can be used as a scale to measure small fringe shifts (Fig. 1.11). Usually both tubes are evacuated. Then, as the gas to be tested is

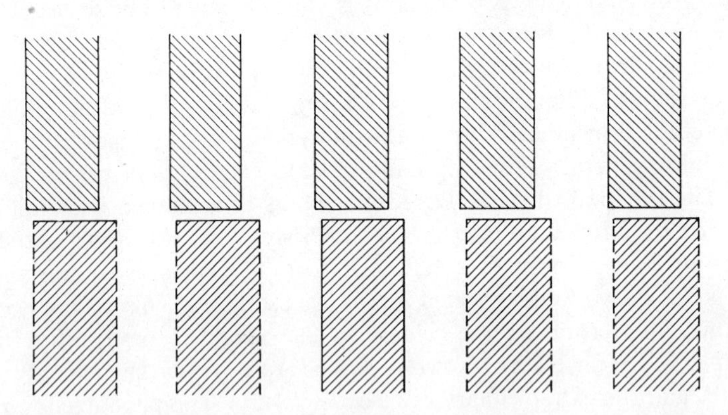

Fig. 1.11 The measurement of part-fringe displacements against white reference (bottom) fringes.

allowed to enter slowly, the number of fringes passing the fiduciary white fringes can be counted. Finally, the compensators (thin glass wedges of $n > 1$ which introduce an extra optical path length) can be used to determine any fraction of a fringe to be added to the number of fringes passed. In some instruments, however, as many as 100 fringes can be compensated for, so that compensations can be applied as the gas enters. The refractive index is calculated from:

$$(n - 1)l = N\lambda, \tag{1.12}$$

where l is the length of the tube cell.

Traditionally, Rayleigh interferometers are used to measure the refractive index of gases and liquids. The limits of cell lengths are dictated by the ability to isolate the cells thermally and mechanically from the surroundings. In gases this means a length of about 1 m, which for blue-green light at 500 nm (5000 Å) is 2×10^6 wavelengths. If, as some researchers claim, measurements to $\frac{1}{40}$ of a fringe are reproducible, this implies that a refractive index change of 1.24×10^{-8} can be measured. In liquids local heating is more serious since greater temperature differences and greater local refractive index changes can occur before relaxation through convection takes place. When light passes through nonhomogeneous media, it is at best displaced, at worst scattered. Transient fringe shifts and fuzzy fringes develop which prevent reliable work. Liquid tube cells are usually limited to 10 cm and must be well insulated thermally. A gas interferometer with 3-m-long cells now exists [4].

Recently, Svensson [5] has reported what amounts to a multiple-beam Rayleigh interferometer. The interference fringes from pairs of slits are brought into register so that a continuous set of fringes can be generated with about 150 fringes showing. The source is also modified by using a diffraction grating over a high-pressure mercury lamp. Results using such a system for sedimentation studies in an ultracentrifuge have been reported by Billick and Bowen [6].

Uses

Rayleigh-type interferometers have been used to determine the refractive index of gases and liquids, and of crystals, as described in the chapters on refractometry Chapter IA Pt III 4th Ed. and the ultracentrifuge Chapter XVII Pt II 3rd Ed. Many modifications of the Rayleigh interferometer have been made. The Haber-Löwe instrument and the Williams instrument and several more modern instruments embodying minor modifications are discussed in the chapter on refractometry Chapter IA Pt III 4th Ed. as is also the detailed handling of the instruments in refractometric determinations.

These instruments have been used to detect CH_4 in mine atmospheres (Baker et al. [7]) and both CH_4 and CO_2 (Subbotin and Zharikov [8]). Analyses of respiration gases [9], of the formation of ozone [10], as well as the evolution of CO_2 in the analysis of carbonate [11], are other examples for the use of Rayleigh interferometers. The concentration of D_2O in H_2O has been determined [12] and thermal diffusion in CCl_4–cyclohexane [13]. A Rayleigh interferometer using a microwave source to investigate magnetoplasmas has been described by Furdyna [14].

It is an advantage of the Rayleigh interferometer that it can be used with white light. Moreover, with careful isolation from the environment, very long optical paths can be used and consequently very high dilutions can be measured. As noted above, refractive index differences in the range 1.25×10^{-8} can be determined; with 3-m cells this limit is lowered to $\sim 5 \times 10^{-9}$. The apparatus is optically uncomplicated, but is bulky and requires a large sample. As noted below, multiple-beam interferometers can be built that produce long path lengths but remain compact and are more easily mounted as monitoring devices for refractive index change.

The Jamin Interferometer

This interferometer, devised by Jamin as a refractometer before the Rayleigh instrument existed, is still used. As shown in Fig. 1.12, the light from the source S has its amplitude split at the first surface of prism P_1, a thick slab of glass with parallel faces. One part is reflected and goes through a tube cell T_1 and one part is reflected from the silvered back surface and emerges to traverse tube cell T_2. The ray through T_1 is refracted and reflected in P_2, and the second ray is reflected at the first surface of P_2 so that superposition occurs and a fringe system is visible at D. In order to obtain a wide separation of beams, very thick prisms are needed, and light is lost by reflection as the beam through T_2 emerges from P_1 and again as the other beam enters P_2. As in the Rayleigh case, fine fringes are obtained and again strong magnification is necessary for observation. Thermal isolation of this system, which is quite similar to the Rayleigh interferometer, is more difficult since the masses of glass must be kept at constant temperature also.

This interferometer is used less since the Rayleigh interferometer has become common, but perhaps its real successor is a modification attributable to Mach and Zehnder which is discussed below. The Jamin interferometer has recently been used by Soskin [15] to find the refractive index of crystals, and a classic electrochemical investigation was performed with it by Ibl and Muller [16] in 1955. The diffusion layer at a working electrode was displayed as shown in Fig. 1.13. Along each fringe the refractive index times d the thickness of the cells is constant since [see (1.7)] $2nd = N\lambda$ and d is a constant for the parallel-sided cells used. Because the refractive index is directly proportional ($C = kn$) to concentration at the values used, the conversion to concentration gradient is simple and accurate. The degree of separation of beams possible precluded a simultaneous investigation of the diffusion gradient at both electrodes with a cell of the size required for the experiment.

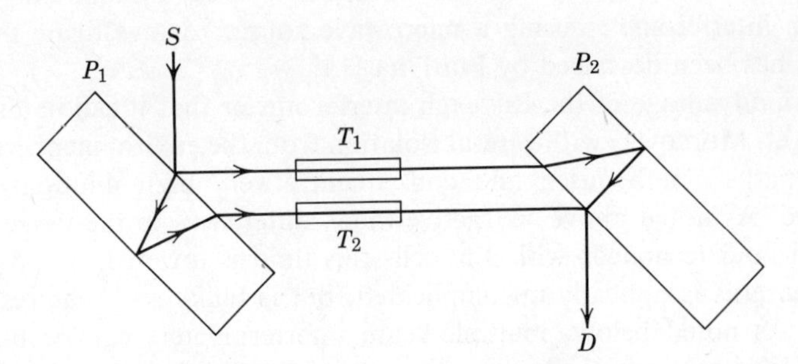

Fig. 1.12 The Jamin interferometer.

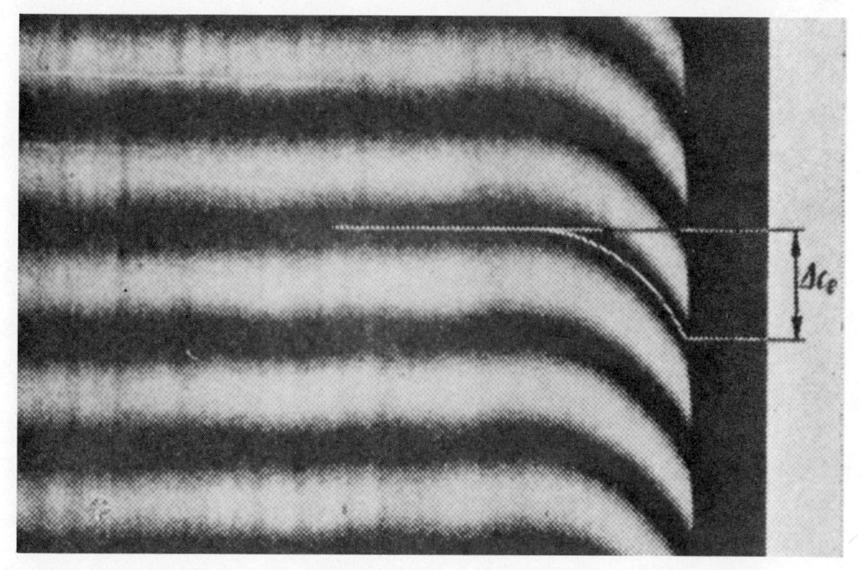

Fig. 1.13 Interference fringes from a Jamin interferometer showing the diffusion layer at a working vertical electrode after Ibl [98].

The Mach-Zehnder Interferometer

The Mach-Zehnder is a development of the Jamin interferometer in which the two massive glass blocks are replaced by a beam splitter, two mirrors, and another beam splitter to give beam separation. It is shown schematically in Fig. 1.14. Collimated monochromatic light from the source S is split into two beams at beam splitter B_1. On traveling counterclockwise through the cell C_1 and being reflected from the mirror M_1 to the second beamsplitter B_2, it combines with the other beam. The other beam, which originates at B_1, travels clockwise to be reflected at the mirror M_2 through the cell C_2 to the beam splitter B_2 to combine with the first beam. About one-half of the clockwise beam is lost by reflection from the back of B_2, and about one-half of the counterclockwise beam passes through B_2 and also out of the system. The light loss problem by absorption and reflection has recently been reviewed by Rienitz (see General References, Optical Interferometry). The cells C_1 and C_2 can contain a reference substance (liquid, gas, solid, or solution) and the substance whose refractive index is to be determined by difference, or only one cell may be used with a compensator in the other beam. As in the Rayleigh interferometer, a white light fringe is necessary to establish the zero-order fringe.

The uses of the Mach-Zehnder interferometer are generally related to the

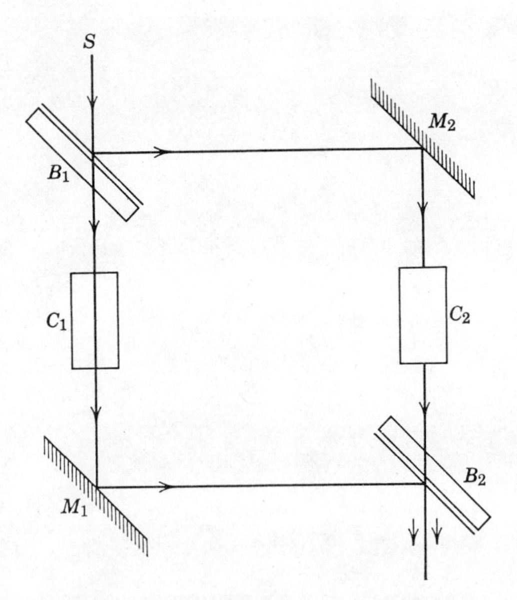

Fig. 1.14 The Mach-Zehnder interferometer.

wide separation of beams possible. This allows experimental setups to be inserted into the beams separately. The instrument is a major tool in aircraft and aerodynamic research. Large, square wind-tunnel sections can be accommodated in the arms so that no disturbance occurs in one beam while the disturbed flow around, for example, an airfoil section, occurs in the other. Since the refractive index of air is pressure dependent according to:

$$n = 1 + 0.0003P, \tag{1.14}$$

where P is the pressure in atmospheres [17], the flow lines around the solid object are outlined by interference fringes which actually show pressure changes [18].

This instrument has also recently become a common means for investigating plasmas and their electron densities. A general method for relating changes of refractive index to the interference pattern has been given by Howes and Buchele [19]. Electron densities in pinched [20] and shocked [21] plasmas have been measured, and interferometric measurements of electron density are compared to spectroscopic measurements in an electromagnetic shock tube by McLean and Ramsden [22]. Sodium atom concentration distribution in an ac carbon arc was measured by Dalgov and Petrov [23]. The instrument has also been used to measure strains in amorphous polymers [24] and convective motions through uneven heating in fluids [25]. Lin et al.

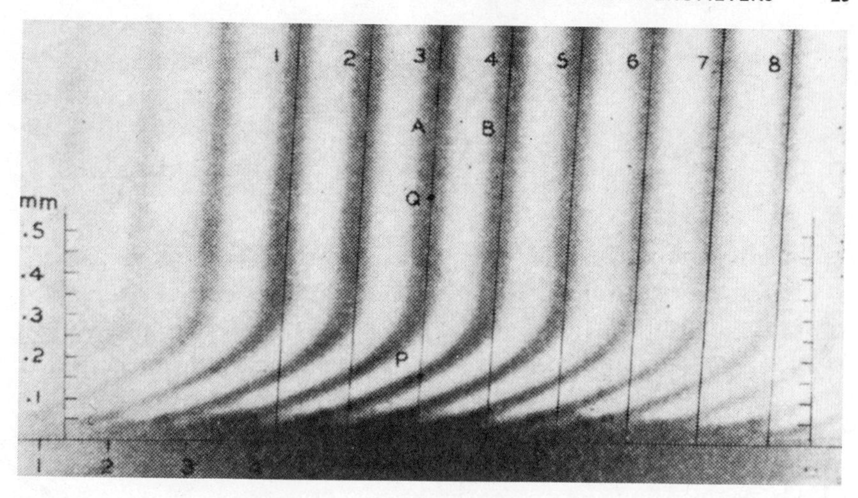

Fig. 1.15 Mach-Zehnder fringes showing the diffusion layer at a working electrode with the electrolyte flowing [26].

[26] made elegant use of the instrument to produce the fringes shown in Fig. 1.15. The diffusion layer next to a working electrode caused the bending in the fringes as the concentration and the refractive index changed. The authors measured the effect of forced convection on this layer by noting the degree of fringe bend with velocity of flow of electrolyte.

The Michelson Interferometer

This instrument, invented in 1881, and its modifications nave had a profound influence on the science of optics and its applications. A typical instrumental setup is given in Fig. 1.8. The instrument has also been used with thermal sources, without collimation, with equal and unequal arms, and with white and monochromatic light. Next to the Fabry-Perot instrument, it is probably still the most used interferometer.

Depending on the internal adjustments, ring or straight fringes can be obtained (see Figs. 1.16 and 1.17) if an extended source is used. An equivalent optical diagram can be drawn (Fig. 1.18). It shows two sources which corresponds to the fact that from the plane of observation the image in M_1 is seen directly at S_1, while that of M_2 is seen at S_2 after reflection from the beam splitter (S_1 in Fig. 1.8).

The arrangement shown results in circular fringes (Fig. 1.16) with equal inclination (Fig. 1.19). If the two mirrors M_1 and M_2 in Fig. (1.18) or P_1 and P_2 in Fig. 1.19 are not parallel, but tilted toward one another (see section on nomenclature and geometry), the fringes are not circles. If the two mirrors

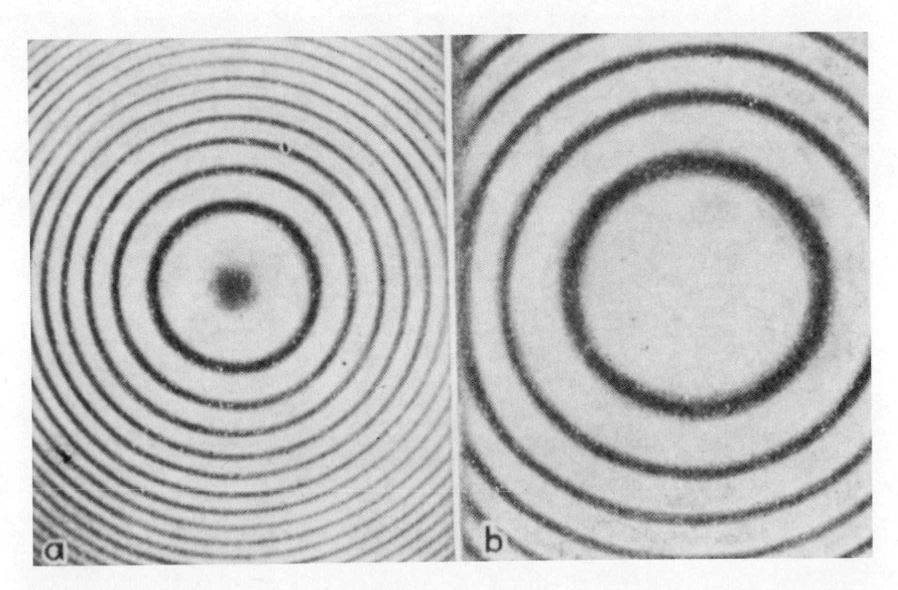

Fig. 1.16 Michelson interferometer fringes with no tilt using sodium vapor light with the distance d in Fig. 1.18, 5.1 mm in (*a*) and 2.1 mm in (*b*).

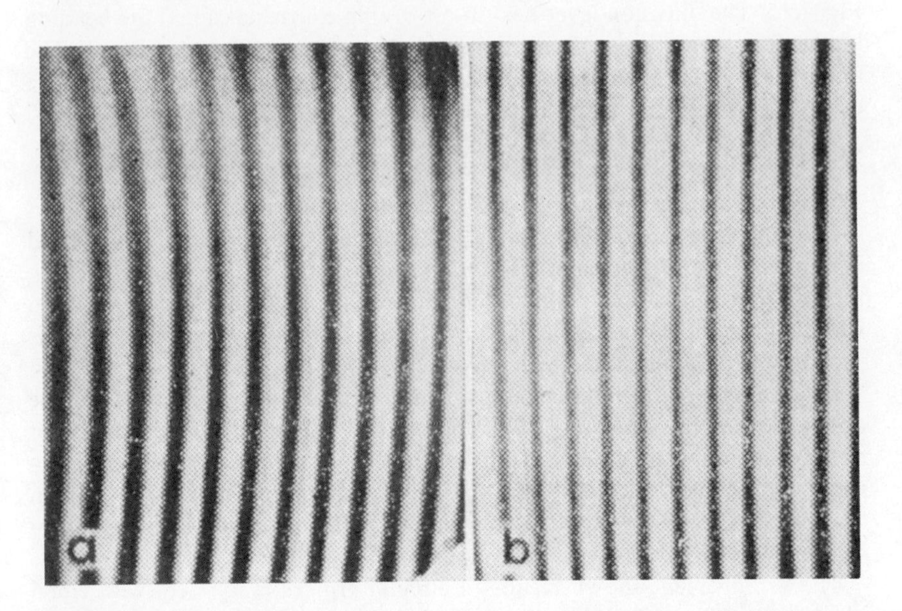

Fig. 1.17 Localized Michelson interferometer fringes. In (*a*) the wedge between the mirrors is truncated or M_1 and M_2 do not meet at the wedge apex (see Fig. 1.18; assume mirrors no longer parallel); in (*b*) the mirrors form the apex of a wedge.

P_1 and P_2 meet to complete a wedge apex giving zero or very small path difference, then the fringes are straight as in Fig. 1.7 and are called wedge or Fizeau fringes. If the mirrors form an open-ended wedge, then the fringes are curved and are sections of hyperbolas. Tolansky, 1955 (General References, Optical Interferometry) has given a full chapter to types of interferometers that are modifications of or derived from the Michelson instrument. One of these is the Twyman-Green modification.

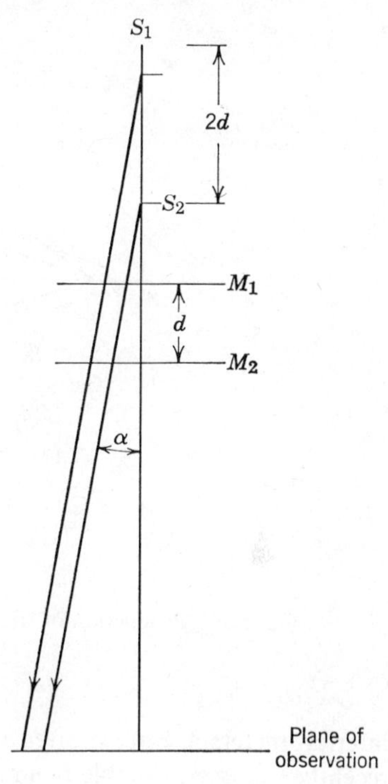

Fig. 1.18 Equivalent optical diagram of the Michelson interferometer.

The Twyman-Green modification uses a point source rather than an extended source as in the original Michelson instrument and requires very good collimation. Figure 1.8 corresponds to the Twyman-Green modification using a laser. This system does not have a range of incident angles producing circular fringes of equal inclination (see Fig. 1.19) as the Michelson instrument with an extended source and no collimation does. If the two mirrors P_1 and P_2 in Fig. 1.19 are parallel, then interference appears to occur between two beams of equal inclination originating at S and S'. This can also be seen from Fig. 1.18 although perhaps not as easily.

Uses

Michelson carried out fundamentally important experiments on the speed of light, to detect ether drift, to measure the meter in wavelengths of light, to measure the rotation of the earth and the effect of the moon on tides, and so on. (See General References, Françon, Morgan, and Steel.) He also used the instrument to investigate the fine structure of spectral lines. If the movable mirror is slowly moved back, the visibility of the circular fringes changes and from this can be calculated the widths of the spectral lines in the fine structure. The method of conversion from the visibility curve to the character of the original source is one of Fourier analysis. This work was the beginning of Fourier or interference spectroscopy. Fabry-Perot interferometers are now more commonly used for spectral line analysis, but Michelson

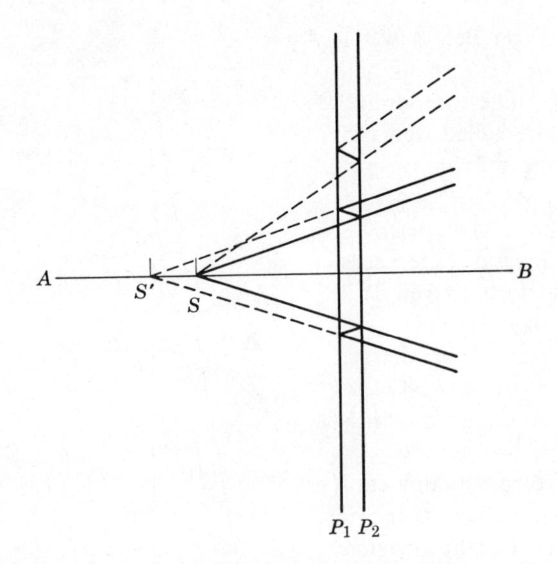

Fig. 1.19 Production of fringes of equal inclination by parallel plates.

interferometers are more often used in interference spectrometry; they have recently become available from several manufacturers.

In interference spectrometry the sample whose spectrum is to be analyzed is traversed by both beams of the interferometer, which takes the place of the monochromator slit. When one arm is slowly lengthened by moving the movable mirror back, the fringes change in intensity and the Fourier transform of the spectrum is traced out. Scanning can also be accomplished by changing the pressure of a gas in the spectrometer. For further treatment, see the chapters on spectrometry Chapter IVA Pt III 4th Ed.

The Michelson interferometer has been used to determine the refractive index of gases [27], liquids [28], and solids, and of inhomogeneities in solids [29] (see chapter on refractometry Chapter IA Pt III 4th Ed.). The instrument has also been used in dilatometry. Stevenson [30] has shown how it can be used to detect liquid level changes of less than 10^{-5} cm, and Herskovitz [31] has described a manometer with part of the gas-containing volume in one arm of the interferometer, which allows pressures in the range of 10 torr and above to be measured to ± 1.00 torr. The Michelson interferometer has also been adapted to microwave radiation (see below). The Twyman-Green interferometer has recently been used for measurement of small wedge angles [32] and to study ionization behind shock waves [33].

The Interference Microscope

An important adaptation of the Michelson interferometer is the Linnik interference microscope shown in Fig. 1.20. The main points of difference are that a double right-angled prism is used as a beam splitter to give path length compensation by insertion of the prism, and the eyepiece combination L_5 is added to constitute a microscope with L_2. The fringe visibility is good under the best circumstances, that is, with the microscope objectives L_2 and L_3 being about 4-mm lenses, as nearly identical in thickness and magnification as possible, and the reflectivity of the object to be examined and of the mirror M close to equal. The fringe dispersion and position can be controlled by tilting the compensating plate C. As a result, while the object O is in view as in a normal microscope, a system of fringes is superimposed, giving a contour pattern in the normal mapping sense, with reference to the optically flat mirror M_1. Usually, two to four different mirrors are supplied reflecting from 50 to 90%. There are now simple instruments in which the interference head can be screwed onto the nosepiece of any standard microscope. The Linnik instrument and its modifications are designed for the study of reflecting specimens. A modification of the Mach-Zehnder apparatus is used

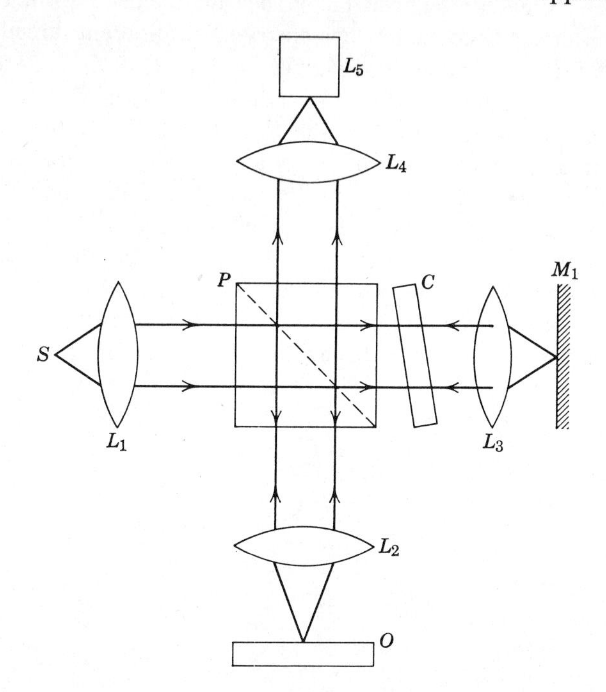

Fig. 1.20 Schematic optical diagram of the Linnik interference microscope.

for transmitting specimens. The adaptation of these instruments to microscopical use is very similar to that of the Michelson interferometer, except that microscope objectives must be inserted in each beam and lens matching is needed in four places. A multiple-beam modification has also been used and is described in the section on multiple-beam interferometers.

The adaptation of the Michelson interferometer for viewing surfaces can give contours accurate to 0.0002 mm [34] and has been used in preference to the phase-contrast microscope to determine the quality of electropolishing [35]. A birefringent crystal (Wollaston) has been used to displace the beams instead of a beam splitter [36], for example, for the determination of refractive indices in biological systems [37, 38]. For further treatment, see the chapter on optical microscopy Chapter IIA Pt III 4th Ed. Research in this area is rapidly expanding; recent ramifications are reported in detail in the translation by Dickson of Krug, Rienitz, and Schultz (General References, Optical Interferometry).

The Polarizing Interferometer

There are many types of polarizing interferometer (see General References, Francon, p. 137); all use the principle of shearing the beam from a source by a birefringent crystal into two beams emerging in different directions. Only the simplest form is discussed in detail. In Fig. 1.21 a Wollaston prism separates the beam from source S into two beams with rectangular polarization diverging at an angle θ. Since this form of interferometer is frequently used to determine refractive indices, especially in sedimentation experiments, the beams are shown traversing two tube cells T_1 and T_2 similar to the Rayleigh or Jamin arrangement. The lens L reunites the beams in W_2, the second Wollaston prism, where the beams are converted to the same polarization, see condition 4, p. 9 and passed into the viewing or recording optics after one ray has gone through compensator C. This instrument cannot be used with gas lasers containing Brewster angle mirrors which cause plane polarization of the light.

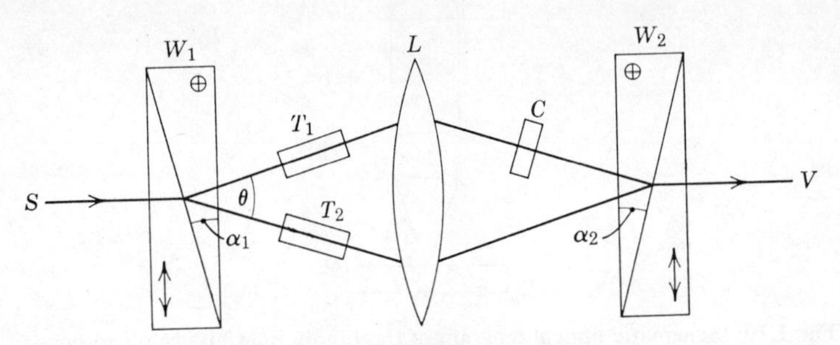

Fig. 1.21 Polarizing interferometer.

Sufficient separation of the beams has been accomplished so that the type described can be used in place of the Mach-Zehnder instrument for compressible boundary layers, that is, aerodynamic characteristics of shapes [39]. It has also been used in sedimentation studies in the ultracentrifuge [40], in diffusion experiments [41], and for the investigation of surfaces, optical [42] and otherwise [43]. Follenius [44a] and Follonius and Goldsztaub [44b] have studied the concentration changes around a crystal growing from solution. Recently, the instrument has been used in interference spectrometry and is stated to give a larger acceptance angle than a Michelson interferometer. Steel [45] disagrees with this and gives mathematical reasons.

More recently, a polarizing interferometer that splits the beam after passing through the cell has been introduced by Bryngdahl and Ljunggren [45a]. In this interferometer either one or two Savart plates are used. The two parts of the beam are displaced, hence if a nonlinear refractive index gradient exists there is a path length difference and the two parts interfere. The earlier form used one plate but the most recent with two has extra convenience.

The single-Savart-plate shearing interferometer has been adequately treated by Tyrrell [45b], especially in relation to diffusion studies. Figures 1.22*a* and *b* show how the two rays are displaced so that they overlap one another. The nonlinear refractive index gradient shown in the cell produces a fringe system. The less the displacement, the less the path difference between the two rays and the wider the spacing of the fringes (at zero displacement there are no fringes), or the less the sensitivity to detect Δn the differential refractive index. The number of fringes is proportional to Δn. Standardization with known refractive index changes is required.

The double-Savart-plate shearing interferometer [45a] with altered Savart plates always gives fringes. The method depends upon use of the modified

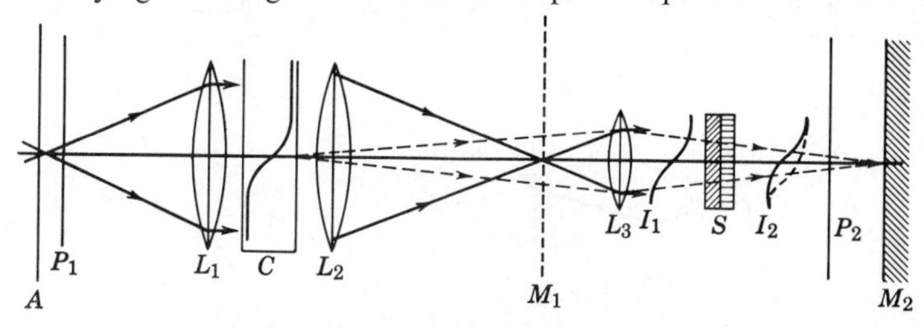

Fig. 1.22*a* Single Savart plate. Shearing interferometer applied to diffusion measurements. *A*, Horizontal slit; P_1 and P_2, polar screens; L_1, collimator; *C*, diffusion cell showing refractive index profile; L_2, lens forming Gouy fringe pattern at M_1, the primary image plane; L_3, convex lens having focal length equal to distance M_1L_3, which images diffusion cell in plane M_2; *S*, Savart plate; M_2, secondary image plane; I_1, shows single refractive index profile entering *S*; at I_2 this is shown split into two. (From Tyrrell, *Diffusion and Heat Flow in Liquids*, Butterworths, 1961.)

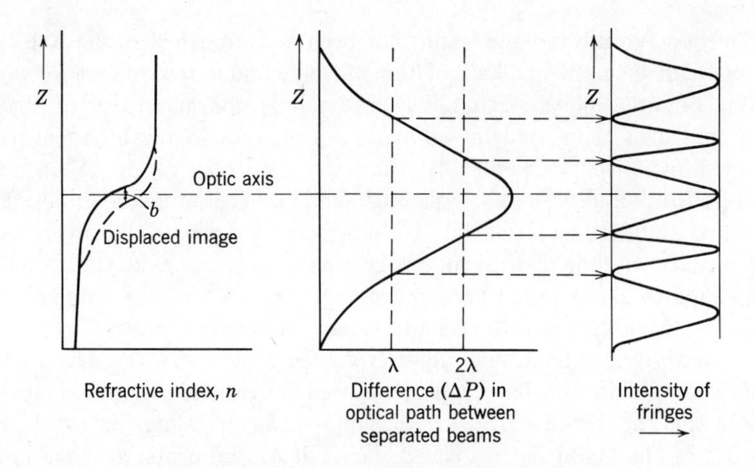

Fig. 1.22*b* Principle of shearing interferometer.

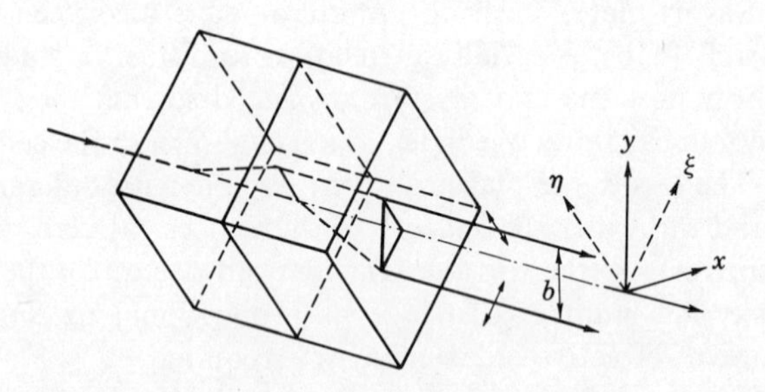

Fig. 1.23*a* Unmodified Savart plate. Arrows crossing the rays show polarization directions. Planes of oscillation are at 90°. η and ξ arrows indicate normal and extraordinary ray polarization directions.

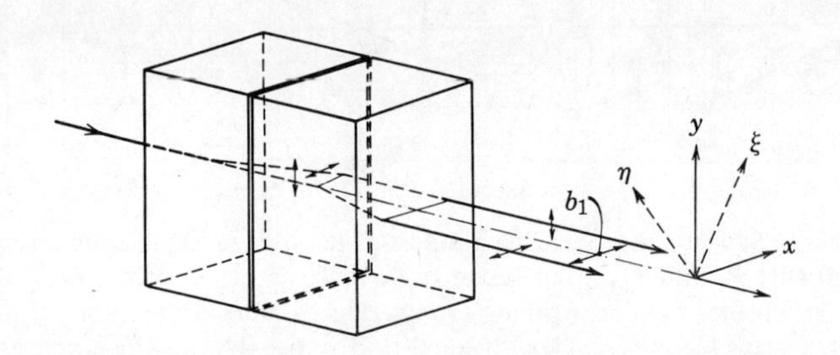

Fig. 1.23*b* Modified Savart plate. Insertion between the two halves (subplates) of a half-wave plate turns the oscillation planes 90° so that they are in the same plane for both rays.

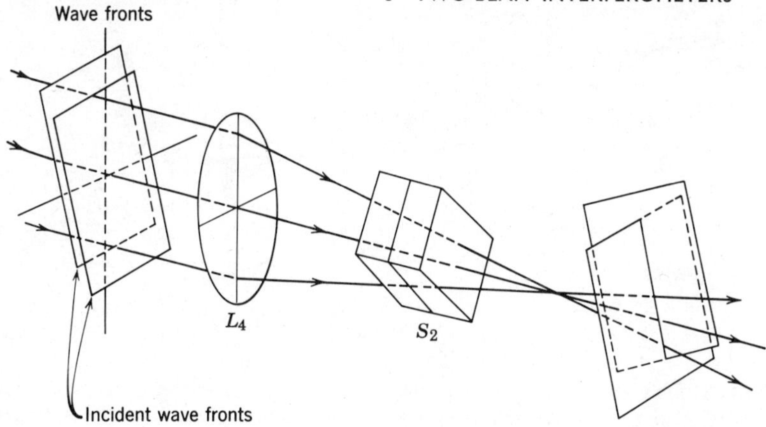

Fig. 1.23*c* Converging light traversing a modified Savart plate. The half-wave plate is not shown. For nonparallel light an angle is introduced between the emerging wave planes which will give fringes resembling Fizeau or wedge fringes. (From the Ph.D. thesis of M. W. Mitchell, London University, 1969.)

Savart plate and the difference in behavior of parallel and nonparallel light in the modified version. Figures 1.23*a* and *b* show the difference in behavior of parallel light traversing an unmodified and a modified Savart plate (often referred to as a beam splitter in the literature). The modification consists of insertion of a half-wave plate to turn the oscillation planes 90°, which places the entering and leaving rays in the same plane. Comparison of Figs. 1.23*b* and *c* shows the difference that occurs if the light traversing a modified Savart plate is not parallel. In the nonparallel case a slight angle is introduced (viewed from above), which results in a series of fringes of constant separation because the angle is constant. Although these fringes superficially resemble Fizeau or wedge-type fringes, Bryngdahl [45a] shows that they are hyperbolas whose common origin is normally so distant that the curvature is negligible.

A suitable optical arrangement following Bryngdahl is shown in Fig. 1.24. The light from the source after passing in parallel rays through the cell is polarized horizontally by polarizer P and then passes through the first modified Savart plate which is oriented so that the two emergent rays are of equal intensity. The rays are then passed while converging through S_2 and turned at 90° to S_1, which introduces a desired angle between the two wave planes. P_2, the analyzer that is perpendicular to P_1, allows the two wave fronts to interfere constructively and destructively as the path difference becomes a $\frac{1}{2}$ integral number of wavelengths in the area of overlap of the two beams.

In summary, this elegant method of measuring differential refractive indices seems likely to become widely used in sedimentation, diffusion, and

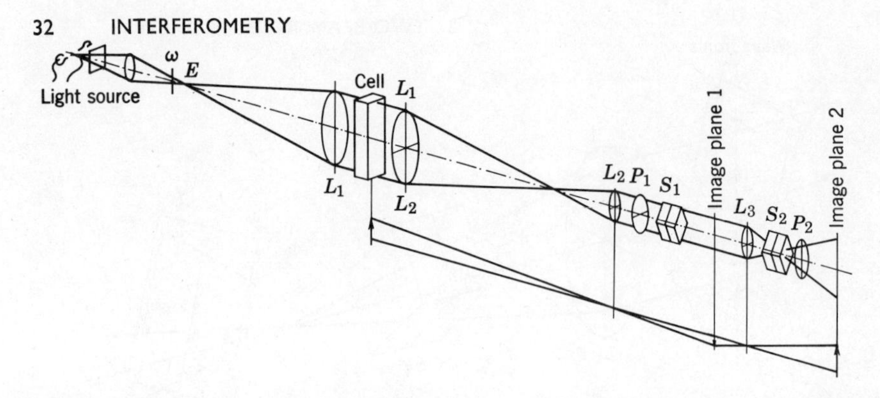

Fig. 1.24 Double Savart plate shearing interferometer. Light from the source is collimated, passed through the cell, recollimated to a smaller diameter, passed through the first polarizer P, and the first modified Savart plate S_1, set at an angle so that the ordinary and extraordinary ray will have the same intensity, then passed converging through the second Savart plate S_2, and the second polarizer P_2 to the camera. (From Reference 47.)

in general refractometry. Using a polarized continuous wave (CW) laser source the optics can be simpler. For experiments in which boundaries of a cell must be clearly observed, its application is difficult, but for a sheared liquid boundary layer where traversing beams can overlap it should make the study of slow diffusion less difficult, Bierlein [45c] has published results for thermal diffusion of $CdSO_4$, $AgNO_3$, and KCl with a clear explanation of optics and theory.

Other Two-Beam Interferometers

A number of historically important but now little-used interferometers have not been mentioned. Most of them have been important for electromagnetic radiation theory rather than in chemical applications; they are mentioned in the General References. Holographic interferometry, a very recent development, becoming practical with the availability of uniphase output lasers, is really a two-beam form of interferometry but has important differences. It is therefore considered at the end of this chapter together with other forms of interferometry difficult to classify. Likewise, moiré interferometry, which is also technically a two-beam system, is considered separately.

4 MULTIPLE-BEAM INTERFEROMETRY AND INTERFEROMETERS

General Theory

A start in discussing the theory of multiple-beam interferometers was made in the section on conditions for interference and the effects of reflecting surfaces of Fig. 1.2. In multiple-beam interferometers the reflectivities used for splitting the wave front are generally high, so that the light makes multiple traverses between the boundaries of the interferometer, usually glass mirrors or partially reflecting coated glass flats, and multiple beams are involved.

The amplitude of successive beams and how they interact to form a fringe system was first calculated by Airy. This calculation has become known as the Airy formula, or function. The derivation appears in both of Tolansky's books, and in Steel and Françon (see General References). An intuitive approach should begin by assuming that there is negligible absorption, which is by no means common if chemical solutions are used between the flats, and the definition:

$$T + R = 1, \tag{1.15}$$

where T is the fraction of incident light transmitted and R is the fraction reflected. The simplest optical arrangement consists of a parallel-sided plate of high refractive index (recall (1.8), $[(n_2 - n_1)/(n_2 + n_1)]^2 = R$) or possibly glass, coated with multiple quarter-wavelengths of dielectric materials. Suppose now that $R = 0.9$, then $T = 0.1$. Referring to Fig. 1.25, if ϕ is small such that $\cos \phi \simeq 1$, which must be so to obtain good multiple-beam fringes, then the assumption can be made that the light is at normal incidence. At R_1, 90% of the light is reflected and 10% transmitted, the beam intensities being R and T, respectively. At T_1, again 90% of the light is reflected back to R_2, having an intensity of RT or 0.09 of the incident light, and the transmitted beam has an intensity of T^2 or 0.01 of the incident light. At R_2 a beam is transmitted back toward the source having an intensity RT^2 or 0.009 of the

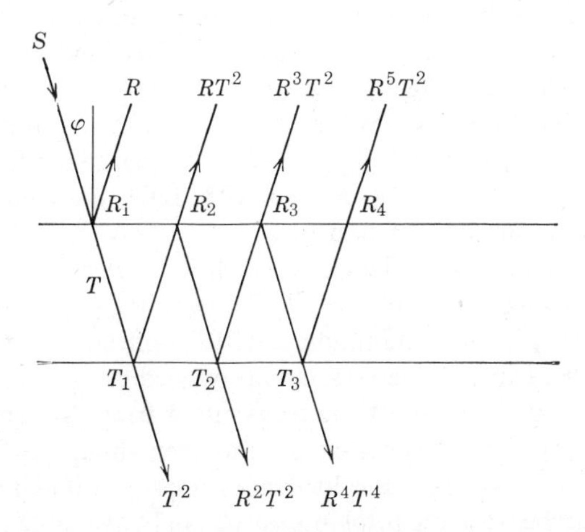

Fig. 1.25 Amplitude of successive reflections in multiple-beam interferometry.

incident light, and a ray having intensity R^2T or 0.081 of the incident ray is reflected to T_2. At T_2 a ray is reflected back to R_3 with intensity $R^3T = 0.0729$ and another ray is transmitted having intensity $R^2T^2 = 0.0081$. Obviously, the sum of an infinite series results both for transmitted and reflected light. It should be noted, however, that to satisfy condition 3, p. 8, the rays should be of nearly equal amplitude. The transmitted rays are falling off slowly 0.01 to 0.0081, and the next intensity would be $R^4T^2 = 0.00656$ as compared to the reflected rays 0.9 to 0.009 to $R^3T^2 = 0.00729$. The transmitted rays obviously meet our criterion, but in reflection the first ray predominates by two orders over the rest. It is obvious that multiple-beam interferometry by reflection requires lower reflectivity flats in order to make the series converge much faster. The consequences concerning fringe thickness are shown later.

Airy's formula for the intensity of reflected light is:

$$I_{\text{ref}} = \frac{4R\sin^2\left(\dfrac{2\pi nd}{\lambda}\right)}{(1-R)^2 + 4R\sin^2\left(\dfrac{2\pi nd}{\lambda}\right)}, \qquad (1.16)$$

assuming $\cos\phi = 1$. If (1.15) holds, then by difference:

$$I_{\text{trans}} = \frac{T^2}{(1-R)^2} \times \frac{1}{1 + \left(\dfrac{4R}{(1-R)^2}\right)\sin^2\left(\dfrac{2\pi nd}{\lambda}\right)}. \qquad (1.17)$$

At $\sin^2(2\pi nd/\lambda) = 0$, the transmitted intensity is $(T^2)/(1-R)^2 = 1$, which is the maximum and shows that under conditions of no absorption the intensity of the transmitted fringe is equal to the intensity of the transmitted light. It does not depend upon the reflectivity. The condition for the location of a maximum on a wedge was given previously in (1.7); the maximum occurs at $2nd\cos\phi = N\lambda$. When $\sin^2(2\pi nd/\lambda) = 1$, then $I_{\text{trans}} = (T^2)/(1-R)^2$ at $2nd = [N + (1/2)]\lambda$. To assess the depth of the minima, transform to $I_{\text{trans}} = (1-R)^2/(1+R)$. For 90% reflecting flats $I_{\text{trans}} = 0.00277$ or a very dark minimum is produced.

For reflected fringes and the same conditions, that is, $\sin^2(2\pi nd/\lambda) = 1$, the reflected fringe maximum is $I_{\text{ref}} = (4R)/[(1-R)^2 + 4R]$, which is 3.6/3.61 or a very shallow minimum; at 10% reflecting the minimum is 0.4 or a reasonably visible fringe system is produced.

The intensity at the peak of the transmitted fringe system is equal to the incident intensity if no absorption occurs, even though 90% of the light is rejected at the first surface. Bright fringes produced under these conditions must be thinner because the total intensity of light available is smaller and the dark fringes are broader.

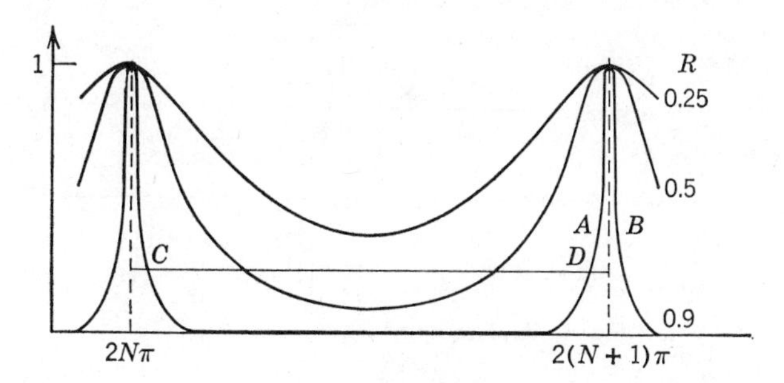

Fig. 1.26 The Airy formula. Intensity for multiple-beam fringes.

Figure 1.26 shows a plot of the Airy formula for various values of R. The influence of reflectivity on the minimum is great. Equation (1.17) can be rewritten:

$$I_{\text{trans}} = \frac{1}{1 + F \sin^2 \left(\dfrac{2\pi n d}{\lambda} \right)}, \tag{1.18}$$

and

$$F = \frac{4R}{(1 - R)^2}, \tag{1.19}$$

called the coefficient of finesse. The finesse is the ratio of the half-height thickness of the fringe to the fringe separation or distance between fringes. The width of the bright fringe is:

$$W = \frac{1 - R}{\pi \sqrt{R}} = \frac{2}{\pi \sqrt{F}}. \tag{1.20}$$

For a 90% reflecting flat, W is about 0.034 or the bright fringe should be about $\frac{1}{30}$ the width of the dark fringe. Using high-contrast film, these fringes can be further sharpened when recorded photographically.

Absorption has been ignored thus far, but may be a real problem when chemical solutions are interposed between the reflecting flats. For chemical experiments the most rugged "dichroic" coatings are usually required. They are now easily obtained in most sizes because they are used in lasers. If silver is used instead, absorption can be serious even if the substance between the flats does not absorb at the wavelength used. The fringe shape and general light distribution are not affected by absorption, but the peaks can be much reduced. Other problems encountered in multiple-beam interferometry are discussed with the applications.

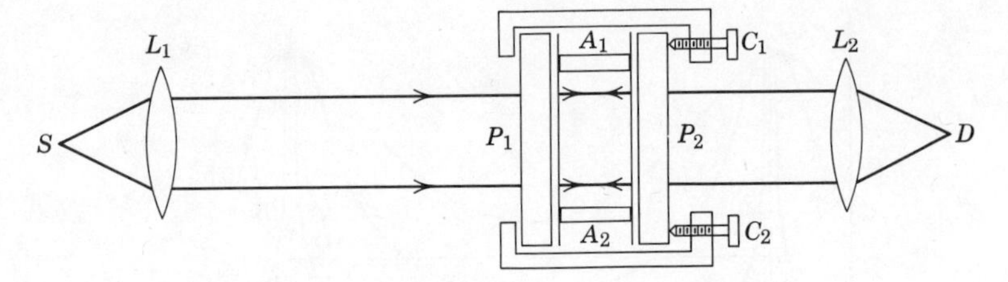

Fig. 1.27 The Fabry-Perot interferometer.

The Fabry-Perot Interferometer

The Fabry-Perot interferometer, next to the Michelson, has probably been the best known and now is the most used. With only one slight modification, it and its modifications are the only commonly used multiple-beam interferometers, and this situation is likely to continue. The relationship between the Fabry-Perot (Fig. 1.27) and Michelson interferometers is not obvious, although the fringes are similar (Fig. 1.28). The relation can be seen, however, if the etalon is considered; a device Michelson invented for use in measuring the standard meter. Figure 1.29 shows an etalon used in a series of substandards, measuring the meter in terms of the cadmium red line. The distance moved by the movable mirror in the Michelson interferometer (Fig. 1.8) is given by the number of fringes that vanish at the center of the field times one-half a wavelength. To measure a standard, the etalon is set in

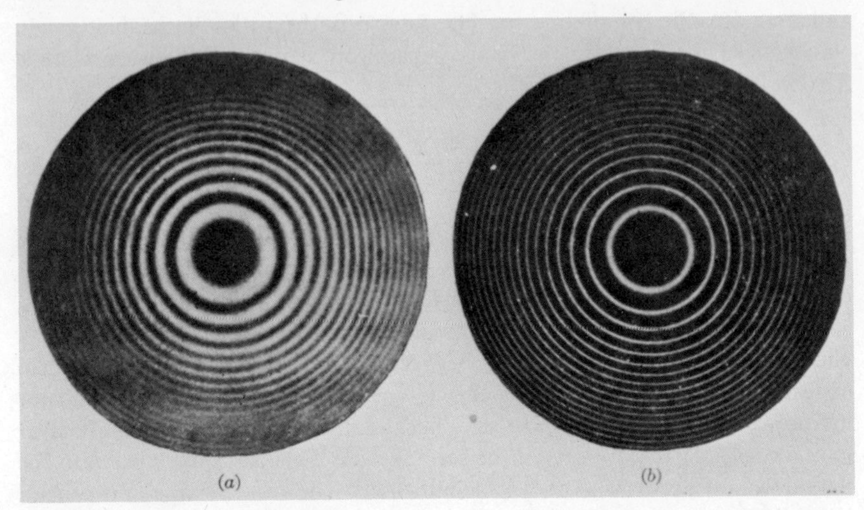

Fig. 1.28 Comparison of Michelson fringes (*a*) with Fabry-Perot fringes. (*b*) (By permission from *Fundamentals of Optics*, by F. A. Jenkins and H. E. White, 2nd ed., McGraw-Hill, 1950.)

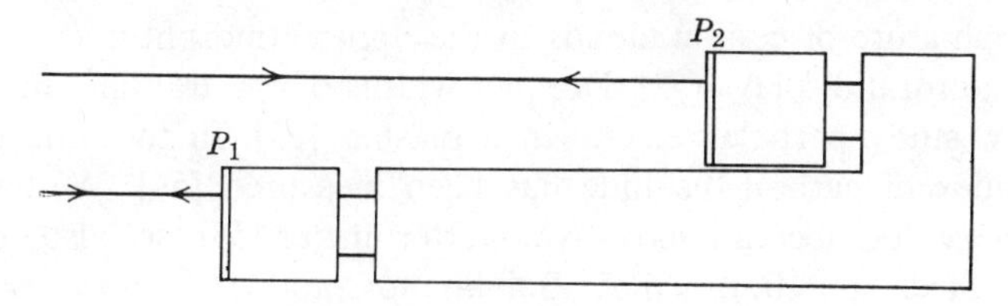

Fig. 1.29 Michelson's etalon.

place of the fixed mirror after a white light standardizing of the two arms to the movable mirror and to the mirror M_2 in Fig. 1.18. The number of red fringes disappearing is counted as the movable mirror is moved back until white light shows the two arms to be equal again with the second arm now extended to P_2. By using several etalons, each twice the size of the next, the meter is measured. The Fabry-Perot interferometer was made by setting the reflecting coatings of the two mirrors P_1 and P_2 opposite one another. When the two inner reflecting coatings are exactly parallel, that is, when the spacers A_1 and A_2 (Fig. 1.24) are the same size, this is also called an etalon, and when the distance is small it becomes an interference filter as discussed in the section on general theory of optical interferometry. The Fabry-Perot is the multiple-beam equivalent of the Michelson interferometer.

The instrument is used in the two modifications noted. The original instrument had no "tilt," that is, the reflecting inside surfaces in Fig. 1.27 were exactly parallel to form an etalon. The fringe pattern consisted of ring fringes as in Fig. 1.28 because the fringes (sometimes called Haidinger fringes) were of equal inclination (as explained in the section on the Michelson interferometer).

In a second form tilt is introduced by placing the reflecting surfaces at a slight angle to one another. As in the Michelson interferometer, straight, wedge-type Fizeau fringes are formed with monochromatic light.

Circular fringes are most often used, for example, to investigate the fine structure of spectral lines [46] of both Gaussian and Lorentzian [47] types because of their higher resolving powers. In fact, Michelson characterized the H_2 emission line as a doublet, whereas it is now known that there are at least three components (see General References, Tolansky, 1955, p. 90). Absorption-line profiles can be measured [48], but high finesse (fringe thickness at the half-height of the fringe compared to fringe separation) is required. Cooper and Grieg [49] described a time-resolved instrument in which a piezoelectric crystal accomplishes scans in 10^{-7} sec by mirror displacement.

The instrument has been operated over wide ranges of frequencies. A forbidden mercury emission line at 2656 Å in the ultraviolet has been analyzed [50]. In the visible region Doppler line widths have been used to find the temperature of cesium clouds in the upper atmosphere [51] and in airglow and auroral displays [52]. Doppler widths of spectral lines have also been used to study particle velocity in a plasma [53]. In the infrared the refractive index of carbon disulfide has been measured [54]. Westermann and Maier have described a microwave interferometer [55] (see also General References, Steel, p. 107), which Bolakhanov and Striganow used for plasma diagnostics [56]. Metal mesh interference filters have been used in the infrared region [57], and multislit metallic mirrors as reflectors have been shown to give excellent resolution in the 8-mm range [58].

The instrument has also been used to investigate the number and power of oscillating modes in a gas laser [59]. The density of sodium vapor has been determined [60] and isotope abundance in natural lithium [61]. High-resolution scanning interferometers [62] are being designed for use in the visible and infrared [63]. For very faint lines a fiber optic fringe splitter has been used [64]. Raeseler has shown that with the provision of limited wave-length ranges, an unambiguous interpretation of a spectrum can be made and that the instrument can be used within limited ranges in Fourier spec-troscopy [65].

Fabry-Perot Interferometer with Wedge Fringes

This interferometer, the second modification of the Fabry-Perot, has the inside surfaces of the glass flats inclined to one another, as described above, and of course some provision for removing the reflections from the outside surfaces of these flats. This can be done by a 1° wedge angle so that the reflections fall outside the gathering angle of the optical system. A better solution is the use of nonreflecting coatings; a quarter-wavelength thickness of MgF_2 reduces the amplitude of reflections to the point of negligibility. For best results it should be a quarter-wavelength of the light for which the instrument is intended. This is particularly important with laser light. Reflections giving unwanted interference patterns arise from all surfaces encountered. This effect is seen in the apparent "twinkling" when a laser beam is reflected by any material, producing interference patterns in standing waves. On reflection from surfaces of liquids with reflectory suspended particles, there should be no twinkling because the Brownian motion of the particles should result in reflection patterns that change too fast for the eye to record. Milk does not give the twinkling effect.

The introduction of a wedge angle or "tilt" can be accomplished by the arrangement shown in Fig. 1.27, except that the spacers A_1 and A_2 must be a ring and compressible by the clamps C_1 and C_2 to form a seal and a wedge. O'Brien has given details of such a cell [66]. Very thin cells have been constructed [67] consisting of microscope slides and used under a microscope to observe diffusion. Tolansky has used thin air cells to great advantage in observing and mapping surfaces [68]. Usually, silver is coated onto the surface by vacuum evaporation. Then a highly silvered glass flat ($R > 90\%$) is placed on it and collimated monochromatic light is used for illumination. Tolansky [68] claims that a step of 3 Å can be detected provided it extends over a linear distance of more than a fringe separation and gives examples of steps in mica, pits in diamonds, roughness of razor blade edges, directional resistance of the cutting edge of a diamond, spark erosion pits, indentations of single crystals, distortions of piezoelectric crystals, roughness of electro-deposits, and so on. The method has become one of the standard techniques—similar to true interferometric microscopy (see General References, Optical Interferometry, Krug, Rienitz, and Schultz) to which it bears strong resemblance—for observing junctions in semiconductor diodes and solid-state microelectronic components generally.

In wedge fringes the fringes are located on the wedge from (1.7), where again $\cos\phi \simeq 1$. Figure 1.30 shows a wedge, assumed to be of glass, perfectly planar on all surfaces and partially reflecting, for example, 90%, on the top and bottom. Plane parallel monochromatic light passing through the apex of the wedge does not interfere with itself. Proceeding up the wedge until its thickness is $\lambda/4$, if we assume no phase changes on reflection a ray having path length AB and passing straight through at this point differs in path length from a ray that goes through to A. It is reflected back to B and from B back to A and then emerges. The difference is $AB - 3(AB)$ or twice AB. If $AB = \lambda/4$, the two beams are 180° out of phase, destructive interference should occur, and the center of a dark fringe be observed. At $\lambda/2$, the path difference is λ, the twin beams are in phase, and a bright fringe

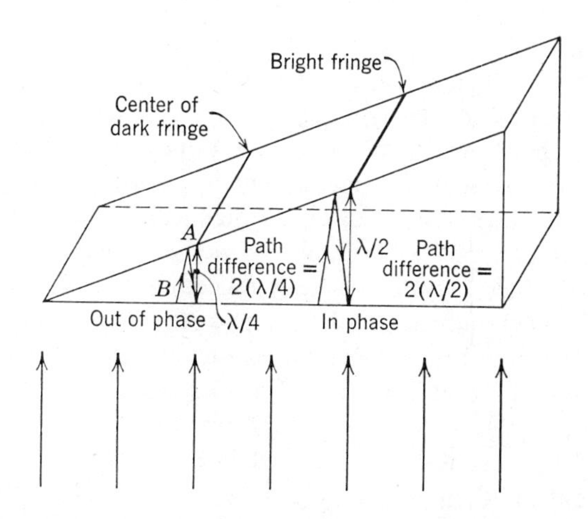

Fig. 1.30 The location of multiple-beam fringes on a wedge.

forms. Actually, quite a number of beams, in fact about 10, are involved. The Airy formula is needed to evaluate this situation completely (see introductory part of this section).

In the glass wedge in Fig. 1.30, of the parameters in (1.7) $2nd = N\lambda$, n the refractive index, and λ the wavelength stay constant, and N is constant along any fringe. The fringe is then a thickness (d) contour. This is the property Tolansky has so successfully used as illustrated in Fig. 1.31. The simplest way to illustrate that d is a constant along any contour is to slice a wedge off

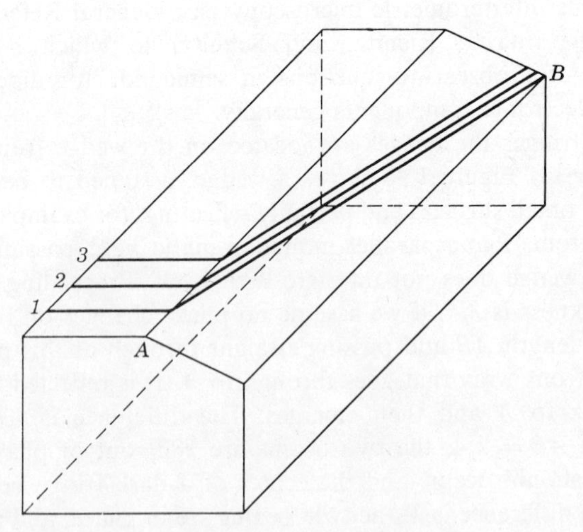

Fig. 1.31 Fizeau fringes on a beveled wedge.

the wedge. Only three fringes have been drawn. Note that they are equispaced and parallel on each face. The general wedge increases in thickness into the paper such that the sliced-off wedge has the same thickness at A as it has at B. Fringe 1 bends near A so that d remains constant all the way to B. Tolansky [68] calls these fringes of equal thickness, since he usually refers to the thickness of an air gap rather than a refractive index charge. If the wedge is a solution contained between partially reflecting glass flats and the refractive index is changed anisotropically in one of the two directions perpendicular to the direction of light propagation, a bending of the fringes occurs in the same manner as that shown in Fig. 1.31. An actual fringe system, reproduced from a thesis [69], demonstrating the change in refractive index when one side of the cell is cooled is shown in Fig. 1.32.

If the concentration, pressure, degree of association, or temperature is changed in particular sections of the cell, the wedge fringes in this section

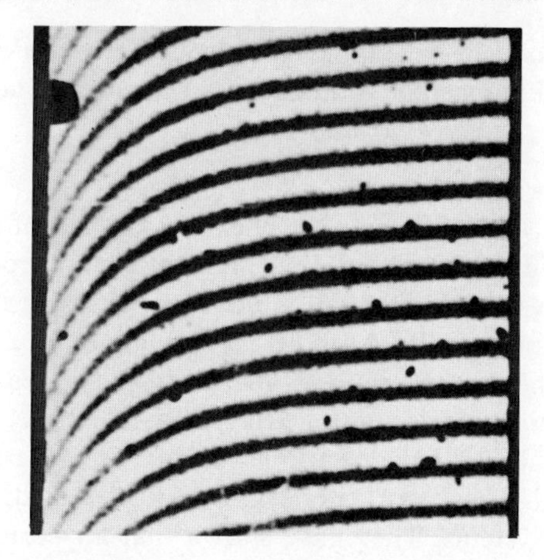

Fig. 1.32 Refractive index change on cooling. The left side-metal strip was drilled and cooling solution pumped through [69].

will record this perturbation by bending or shifting. The wedge thickness remains nearly constant and the fringes are a refractive index contour. If only one of the properties of a solution is known to vary, then the fringe is a contour of that property. The method of evaluation and minor errors involved are discussed below. It is this property of the fringes that O'Brien and co-workers have used to indicate changes of concentration, temperature, density, or molecular weight.

A series of cells has been designed by O'Brien et al. for electrochemical investigations and for nonelectrochemical applications [70]. The cells for electrodeposition studies were: one with luggin capillaries for studying relative corrosion rates of different crystal planes of copper similar to that previously described by O'Brien [66]; one for flowing electrolyte studies similar to that of Beach [71]; a cell for the study of transport phenomena at the dropping mercury electrode similar to that reported by O'Brien and Leja [72]; and a cell containing a rotating-disk electrode [73]. Among the cells for use for other than electrode position studies are cells for electrodialysis, for studying the diffusion of gases into liquids, and for studying thermal conductivity of transparent liquids.

Uses of Wedge Multiple-Beam Interferometry

Crystal Growth

Perhaps the first use of multiple-beam interferometry in thin cells was attributable to Berg [74] and Bunn [75], who both studied the concentration contours around crystals growing or dissolving in solutions of the crystal material. More modern observations of crystal growth from solution have been made with Rochelle salt single crystals [76] by observing the concentration gradients, including some estimation of the effect of convection currents [77].

Chemical Reactions

O'Brien and Glasel [78] studied the chemical kinetics of the dissociation of ethyl acetate in hydrochloric acid. Since reactants and products do not have the same collective or individual molecular refractions (see chapter on refractometry Chapter IA Pt III 4th Ed.) [78], the rate of the refractive index change is that of the disappearance of reactants and appearance of products, and can be determined by counting fringes that disappear from the wedge face. Vajagand et al. [79] have performed titrations in water solutions in the interferometric cell. These workers have also titrated acids and bases as well as more complex systems. They have shown that any reaction amenable to titration not involving solid or highly colored reactants or products can be measured. If refractive indices are plotted against volumes added, two inclined straight lines should be obtained which intersect at the equivalence point.

Thin and Thick Films

Tolansky's massive work on crystal surfaces is reviewed in Ref. 68. The largest amount of research using wedge multiple-beam interferometers has been applied to the study of surfaces. As mentioned above, multiple beam interferometry of the type Tolansky developed has become almost a standard technique for investigating etched surfaces in metallurgy and in semiconductor technology. Large numbers of workers have contributed in this area, and only a few typical examples and unusual work are mentioned. A typical example of an etched surface on which growth lines are shown by changes in impurity levels in a single crystal from the work of Johnston et al. [80] is shown in Fig. 1.33. Thermal etch patterns of potassium chloride cleavages have been investigated in this way and circularly terraced and spiral etch pits observed [81]. Patel and Bahl have studied the cleavage faces of graphite and the result of etching [82]. Thin films such as oxides on metals have been extensively investigated and thicknesses measured. Young's book [83] sets forth the theory of the determination of film thickness from interference colors, which has been largely replaced by elipsometry. Elipsometry [84] is primarily concerned with thin films, but some thin-film work is still done by ordinary Tolansky interferometry. Optical film densities of metals from 200 to 5000 Å have been measured by Walter interferometrically [85] in the visible, and Bates and Bradley have measured film transparencies of aluminum and Al-MgF_2 in the ultraviolet [86]. Uyeda [87] has described a jig for thickness measurements which allows adjustment of the wedge angle for proper fringes between the reflecting flat and the substrate by the turning of one nut.

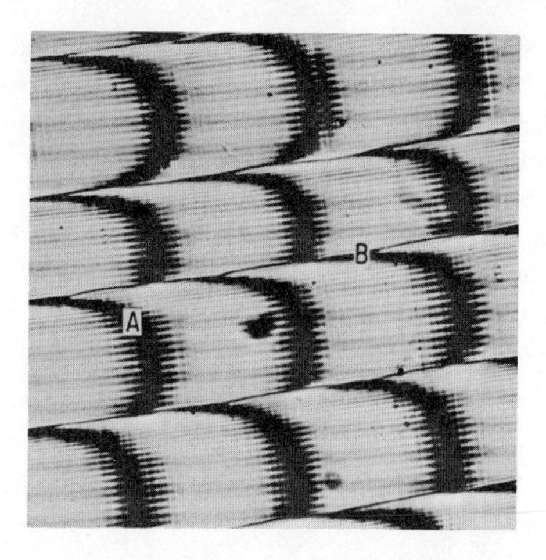

Fig. 1.33 Surface topology of InSb after etching outlined by multiple-beam fringes.

Thicker films can also be measured. A system described by Lockyer uses two fringe systems obtained with two monochromatic sources [88]. The thickness and refractive index of thin (500 to 1000 Å) slices of coal and graphite have been measured by interferometry [89]. Thick films that are transparent and have high refractive indices can give fringe patterns that are not easily interpreted. An example of work with thick films is given in the experimental arrangement shown in Fig. 1.34. Before the film is formed, the distances between the coated glass flat G and the substrate S, which may be a

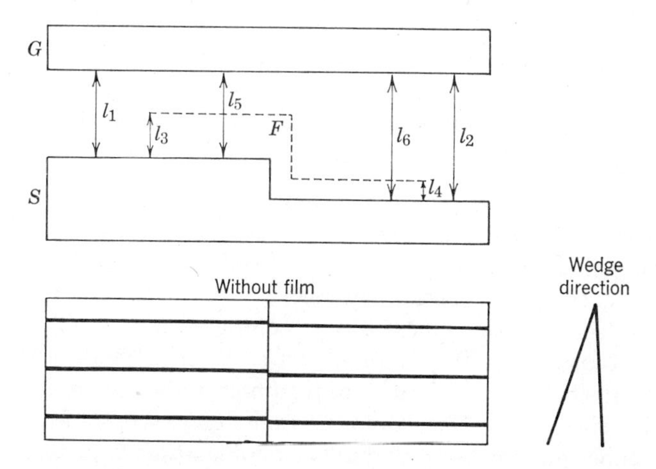

Fig. 1.34 Measurement of thick transparent films by multiple-beam fringes.

vacuum-evaporated metal film laid down over another highly polished metal surface, are l_1 and l_2. If the film forms more rapidly on one metal than on the other, $l_3 = 2l_4$ and further, if we assume that $n = 1$ in air and $n = 2$ in the film, then the resulting path lengths can be formulated and compared. If the wedge angle in the edge view (at the top) has its apex into the paper and in the plane view (at the bottom) has its apex up as shown at the side, then without the film the fringes would shift as shown at the step where the metal film began. The shift is about $\frac{1}{10}$ of a fringe and, using 6328-Å light, represents a step of about 600 Å. At this thickness all metals become opaque. After the film has been formed, if $l_3 = 2l_4$, then:

$$l_5 = n_{\text{air}}l_1 - n_{\text{air}}l_3 + n_{\text{film}}l_3,$$

and

$$l_6 = n_{\text{air}}l_2 - n_{\text{air}}l_4 + n_{\text{film}}l_4,$$

and

$$l_5 - l_6 = n_{\text{air}}l_1 - n_{\text{air}}l_2 - n_{\text{air}}l_4 + n_{\text{film}}l_4. \tag{1.21}$$

When $n_{\text{air}}l_1 - n_{\text{air}}l_2 = n_{\text{film}}l_4 - n_{\text{air}}l_4$, which might occur if l_1 were about equal to l_4, the fringes are no longer displaced at the step between the two metals, and at greater film thicknesses the fringes will be displaced in the opposite direction or downwedge. This condition can easily occur when the coated glass flat is simply laid over the film and l_1 and l_2 are therefore small.

Dilatometry

Interferometers have been used as dilatometers before [90], but a unique use by O'Brien [73] suggests that the instrument might also be used as a flowmeter. A rotating disk that fulfilled the Riddiford hydrodynamic criterion—that is, the flow to the disk was at 90° to its surface and the flow from the disk after impinging on the disk was parallel to the disk's surface—was found to cause dilation because of the shear imparted to the solution near the disk. Theoretical considerations of viscosity measurement based upon this shear dilation have been discussed by Flood [91].

Electrochemistry

The changes in concentration in a working electrolysis cell have been inferred by not totally unambiguous potential measurements, and at only a few points. Interferometry, as used by Ibl and Muller, gave a fair indication of the changes in the diffusion layer next to the electrode as shown in Fig. 1.13 [16]. The complete concentration contour from electrode to electrode was first recorded by O'Brien and Axon [92]. The technique was improved to the point that motion pictures of the onset of the concentration gradient could be made [93]. The complete transient concentration polarization phenomenon in a nonconvecting cell as outlined by interference fringes [94] was shown to agree with Rosebrugh and Miller's earlier calculation [95].

The effect of natural convection in several cell configurations has been investigated [96]. Concentration changes at ion-exchange membranes during electrodialysis have been photographed [97].

Uncertainties in Electrochemical Applications

In measurements of electrochemical concentration gradients by interferometry, two errors arise because of light displacement by the constantly changing refractive index near the electrode. Measurement of concentration depends on the assumption that the concentration gradients at the anode and the cathode are attributable only to change in the concentration of the electroactive species. Both theory and practice suggest that it is a good assumption. In two-beam interferometry, the change in the fringe direction, as shown in Fig. 1.31 (in which a decrease in concentration is simulated by a decrease in prism thickness) is measured in fringe shifts, and one fringe shift has occurred when the end of the first fringe has bent so that it is opposite the unbent portion of the second fringe. This is shown in Fig. 1.35, in which fringe no. *2* is produced by a dotted line to intersect fringe no. *1*. The fringe shift shown is about two.

As fringe no. *1* bends up the wedge in the *C*- or cathode volume (the area of decreasing concentration and refractive index), not only is the refractive index changing, but so is the thickness. When it reaches the dotted extension of fringe no. *2*, d has increased by $\lambda/2$ as previously noted from Fig. 1.27. Thus, not all of the fringe shift is attributable to refractive index change; d has changed to $N + 1$, where N is the order of interference (the multiple of $\lambda/2$ that the wedge thickness has attained at that point). Most electrochemical studies use water solutions with refractive indices of about 1.3, and the cells are usually at least 3 mm thick, which means that N is usually greater than $(2 \times 3 \times 1.3)/\lambda$ ($\sim 10^4$) therefore the error is about one part in 2×10^4, which is usually much smaller than other measuring errors. The second error is more serious and arises because of steep refractive index gradients simulated by thickness changes on the prism shown as added or subtracted wedges. As

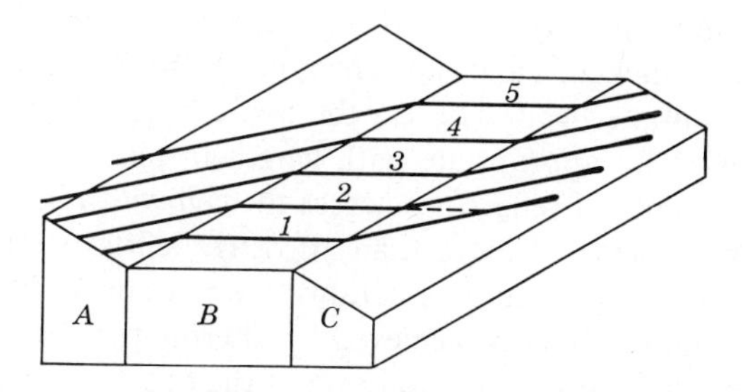

Fig. 1.35 Optical anomalies in multiple fringes.

fringe no. *1* is followed further out in the cathode region, it thickens, then terminates. This is the result of the light being at 90° incidence to the bulk or *B*-region, hence inclined to the *C*-surface (simulating the concentration gradient and having an equivalent optical effect) as shown, causing it to walk inward on successive reflections.

The opposite effect occurs in the *A*- or anode region and the fringes extend or appear to extend out past the solid, opaque anode. Ibl first treated this effect [98] and it has now been treated very thoroughly by Beach [71]. As the concentration gradient and the depth of the electrolyte increase, the effect becomes more serious. Ibl showed that the displacement of the fringe gives an error in concentration:

$$\Delta C = \frac{b\lambda}{aed} - \frac{k^2 d^2}{6an_1}, \tag{1.22}$$

where b is the fringe displacement, e is the fringe separation, a is the proportionality factor, $c = an$ for the electroactive substance, c is the concentration, d is the depth of the cell, k is related to the refractive index gradient, and n_1 is the refractive index in a particular region. He has shown that with natural convection occurring at vertical electrodes at a current density of 1 mA/cm^2 and a cell depth of 0.5 cm the displacement is 10^{-3} cm, which is near the lower limit of measurement, hence detection. The concentration measurement error likely under the same conditions is about 1%.

Beach [71] carefully considered the effects of the focusing of the bent rays to obtain minimum displacement. He calculated a table of values for fringe displacement away from the electrode and concentration errors inherent in displacement along the electrode. In general, errors are decreased by focusing inside the cell. For thin cells, in the range of 0.25 cm thick, the error is reduced by about one-half. At 2-cm depth, the error is about the same for focusing inside the cell and on the front flat of the cell. The errors predicted by Beach are found by an iterative computer program and are generally smaller than Ibl's prediction. Both workers considered a system in which convection was present and used two-beam interferometry. Figure 1.35 shows the case for multiple-beam interferometry. In two-beam interferometry the cathode end of the fringe is not thickened and shortened, nor is the anode side thinned and extended. The multiple beam case can be avoided [although errors should be calculated using an appropriate path length in place of d, for example, nd, where n = number of reflections across the cell width in (1.22)] by never attempting a fringe shift of more than 10 fringe shifts per millimeter perpendicular to the electrode and preferably fewer than 4 fringe shifts altogether. These conditions can be achieved by altering the current density and thickness of the cell. Considerable warning of the onset of conditions leading to appreciable error is given by the thickening of the fringes, and when laser

Fig. 1.36 Multiple-beam split fringes [99].

light is used, splitting of the fringes. Splitting occurs when the several beams making up the fringe are no longer lying upon one another and several more widely spread narrow fringes occur in the critical region in place of the solid fringe.

Fringe splitting can also be caused purposely by inclining the cell windows at a slight angle to the wave front of a laser beam. A typical example from the polarographic or dropping mercury electrode investigations of O'Brien and Dieken [99] is shown in Fig. 1.36. The fringes here are split into three fine dark lines, and since the limitation of accuracy of measurement is the precision of the location of the center of a fringe, the accuracy should be improved. O'Brien [66] has quoted $\pm\frac{1}{20}$ of a photographically recorded line in a fringe pattern as the limit. Since in this case the thickness of a dark line is about $\frac{1}{10}$ the fringe separation (from the center of a thick white line to the center of a thick white line), the limit should be $\pm 1/200$ of a fringe shift. Typically, in a $CuSO_4$ solution, one fringe represents about 0.4 g/liter or about 5×10^{-3} moles/liter in a 3-mm-thick cell. Then, about 2×10^{-5} M is the limit of accuracy. Obviously, with thicker cells and lower concentration gradients (and with photographic conditions correct) this figure could be lowered, but it also may be considerably exceeded under unfavorable conditions. Beach [71] has quoted Müller as calculating that in a two-beam interferogram an accuracy of 1/100 of a fringe separation can be attained by using high-contrast film and measuring along the edge of fringe rather than attempting to find the center of a fringe. This seems to be a reasonable estimate.

In the foregoing discussion it has not been considered necessary to mention that projection of a photograph of fringes or other optical magnification magnifies the fringe and the uncertainty of its center equally. Plotting table techniques using densitometer heads have been used [100] and give reproducibilities of about 0.1%

Electrodialysis

The concentration changes around an ion-exchange membrane at which electrodialysis is occurring have been followed interferometrically [97]. If the interferograms are obtained with a motion picture camera and the cell design and current density are correct, the onset of maximum concentration can be observed and related to conventional experimental data such as voltage impressed on the cell and current density.

Thermal Conductivity of Transparent Liquids

If massive blocks of highly conducting metal, drilled to accommodate a flow of thermostating liquid, are placed between the flats of a multiple-beam interferometer and a transparent liquid or solution is injected, the relaxation of the thermal gradient can be followed interferometrically [101]. Analysis of some data so obtained shows good agreement with published values of the thermal conductivity of water. Figure 1.29 is an interferogram [69] illustrating the thermal effect when one metal block is cooled with a solid carbon dioxide–ethanol mixture.

Diffusion Measurements

Diffusion measurements of high precision have long been carried out by the Gouy interferometric method as reported in the chapter on diffusion Chapter XVI Pt II 3rd Ed. Recently, a gradient produced by electrolysis in a wedge interferometer has been allowed to relax and the diffusion constant of the electrolyte has been determined by a computer program based on an 11-point grid superimposed on the interferogram [102]. The results for copper sulfate are in good agreement with literature values. The system has the advantage of short diffusion paths and rapid production of usefully accurate data. Typical experiments are finished in a few minutes, hence isolation of the cell from the surroundings need not be a major concern.

The diffusion of gases into liquids has been observed by wedge fringe laser interferometry by using the pressure jump technique [70, 103, 104]. The initial entry of all the gases used (O_2, N_2, H_2, He, Ar, CO_2, CH_4, C_3H_8, and iso-C_4H_{10}) resulted in a lowering of the refractive index of the solution, which has been interpreted as a decrease in density. In a cell with N, the order of interference, of about 50,000, fringe shifts of about one-half fringe per atmosphere of O_2 were observed corresponding to refractive index changes of -0.00002 per atmosphere or about -0.000004/mmole. Diffusion constants can be calculated from the relaxing gradients. In the case of the reactive gas

CO_2, a densification or at least an increase in refractive index of the solution began after the initial dilation about a second after the pressure jump. The interpretation of this observation as caused by the reaction $CO_2 + H_2O \rightarrow HCO_3^- + H^+$ through a solvated gas intermediate is supported by the behavior of CO_2 in 2 N NaOH, in which again almost a second elapsed before there was evidence of formation of carbonate anion. The evidence was a change in refractive index and the onset of natural convection as the first layers at the gas-water interface became more dense, presumably because of the formation of carbonate, and began to sink.

Propane and isobutane form clathrates with water at room temperature [105]. When either was dissolved in water at appropriate pressure; about 8 atm for propane and 2 atm for isobutane, a reversal of refractive index gradients occurred beginning at the gas-water interface. At 9°C both propane and isobutane caused seawater to partially freeze, rejecting the salt to the remaining liquid. The process can be followed by refractive index changes causing fringe bends in interferograms obtained with a motion picture camera.

Practical Details of Evaluation Calculations

In order to evaluate an interferogram [66] or the data obtained by counting fringes, only one variable that affects the refractive index in the Lorentz-Lorenz expression [see (1.1)] should change. If two or more change, the additional variables must change in a known and preferably uniform way. For each variable to be analyzed, data are required relating refractive index to that variable. Plots or tables of refractive index at the wavelength to be used versus the variable (such as refractive index versus concentration) are needed. The refractive index of a solution of O_2 in water is shown in Fig. 1.37

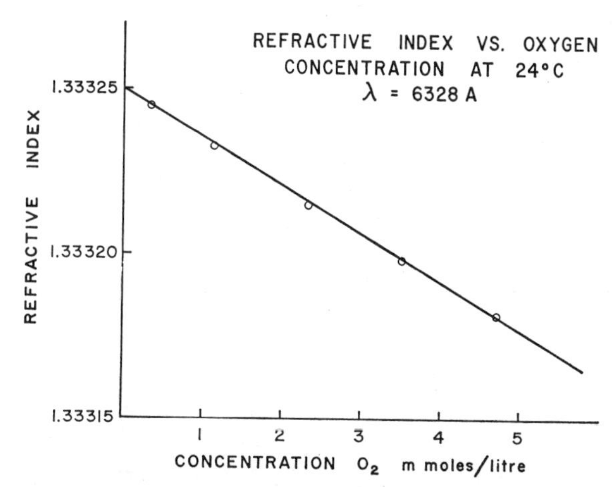

Fig. 1.37 The refractive index of oxygen dissolved in water.

obtained by combining refractive index–pressure data from a multiple-beam interferometer and conventional pressure-concentration data on the common axis. Refractive indices at laser frequencies are becoming available [107].

The cell width d must be measured as accurately as possible. One part in 10^{-4} cm can be measured by good machine tools. Interferometrically, the thickness can be determined to a much greater accuracy. Preliminary experiments show [108] that if pure water—whose laser frequency refractive index change with temperature has been measured [107]—is used to fill the cell and the temperature is known and well-controlled the fringes displaced can be counted. If the thermal expansion of the cell is known over the temperature range considered, the expansion with raised temperature can be equated to the number of fringes passed and subtracted from the total counted. Each net fringe counted passing a reference point, which should be on or near the optical center of the apparatus, represents a change of one-half the wavelength of the light in optical path length $N\lambda$. By referring to (1.7), this can be written $\Delta N\lambda/2 = \Delta nd = n_1 d - n_2 d$, where n_1 is the refractive index at T_1, n_2 is the refractive index at T_2, and ΔN is the number of fringes counted or the change in order of interference. Since ΔN, λ, and Δn are known, d can be calculated to as close a fraction of a fringe shift as can be measured. This may be $\pm 1/200$ at ideal conditions, but at least $\pm 1/20$. These errors represent either ± 15 or ± 150 Å, respectively.

To calculate the concentration change in oxygen after a pressure jump has caused the fringes to bend, the amount of bend is referred to the unperturbed portion of the fringe system. This part is straight and equispaced, hence eminently suitable as a measuring scale. Suppose the bend is two fringe shifts as shown in Fig. 1.38. The widely spaced lines at the top represent multiple-beam wedge fringes in oxygen, those at the bottom fringes in water.

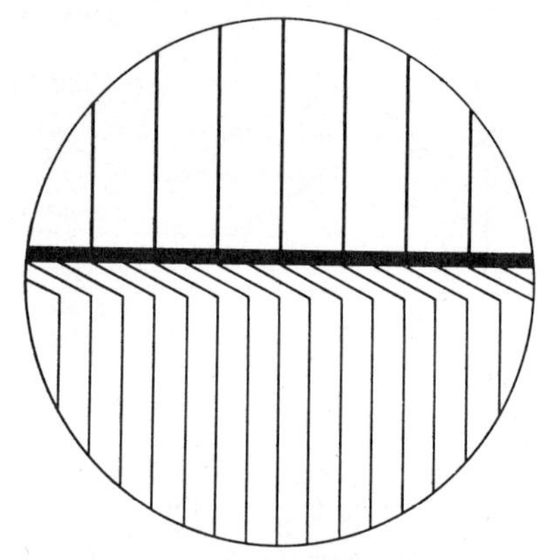

Fig. 1.38 Schematic diagram of gas and liquid fringes after a pressure jump.

The black horizontal line in the center results from total reflection at the meniscus, and the bends in the water fringes are caused by oxygen diffusing into the water.

If the cell is 5 cm thick, the refractive index of pure water at this temperature and wavelength is 1.333250, and λ is 6328 Å, N is calculated from (1.7) to be 210,690.6. It is assumed all other measurements are known to comparable precision. At the surface or meniscus the fringe shift is two, that is, each fringe bends so that at its end at the meniscus it is opposite the unbent portion of the next fringe but one. We know that dilation occurs (we also would know from setting up the apparatus which way the apex of the wedge pointed, in this case to the right) so at the meniscus N has decreased to $N - 2$. We can now recalculate n by inserting $N = 210{,}688.6$ in the equation. The refractive index is 1.3332374 or $\Delta n = 0.000013$. From Fig. 1.34 this can be read as about 0.6 mM. Undoubtedly, the meniscus hides some of the fringe, but it appears from analysis to be very little and the problem is being investigated.

All interference data are gathered by fringe counting or fringe-bend measuring, so all can be processed by the two principal means explained. In cases in which a cell is filled with one liquid, then refilled with another, and the number of fringes in a constant field of view is counted and the fringe separation for the two liquids is found, the relation is:

$$\frac{\text{Number of fringes in 1st liquid}}{\text{Number of fringes in 2nd liquid}} = \frac{n_2}{n_1}.$$

This is a variation of controlled-change fringe counting, but is not as accurate since only a small proportion of the total fringes on the wedge appear in the calculation.

5 HOLOGRAPHIC INTERFEROMETRY

General

Holography itself does not at the moment appear to have any applications in chemistry except in the investigation of aerosols and suspensions. Figure 1.39 is a print of a hologram of a suspension. If the reconstructed image is scanned with a microscope, the particles can be counted and sized in layers equal to the depth of focus of the microscope, provided the droplets or particles are not too dense. This technique has achieved commercial significance since it was first demonstrated by Silverman et al. [106]. A holographic interferogram has some advantages over an ordinary interferogram. The cell used requires less rigorous geometrical accuracy and background effects and optical artifacts are eliminated, but the actual technique is more difficult and can be adapted to motion picture recording only in certain cases.

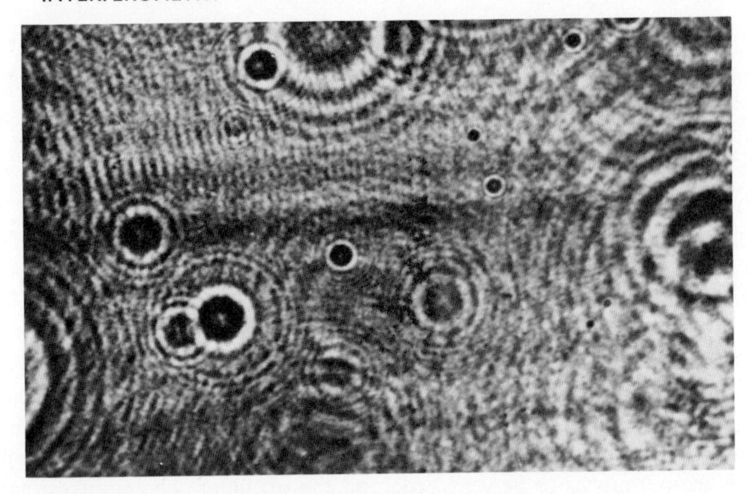

Fig. 1.39 Hologram of water droplets. This hologram can be reconstructed and droplets at any depth measured and positioned with a television raster scan. Particles or droplets of 4 to 200 μ can be counted and measured. Taken with a pulsed ruby laser, on Kodak SO 243 film at 20-nsec exposure time. (Reproduced by courtesy of Stat Volt Company, Santa Barbara, California.)

Theory of Holographic Interferometry

The original Gabor-type holograms [109] were formed by photographing the Fresnel (circular wave front) diffraction pattern produced by an object. Circular wave-front diffraction occurs if spatial coherence or collimation is not good or the light source is thermal and close to the object. To reconstruct the object, the hologram, which does not resemble the object but contains the amplitude information to reconstruct it, is illuminated with collimated monochromatic light. The density variations recorded in the hologram cause the inverse of the original diffraction effects that reconstructs the object. The original holograms did not contain phase information, hence only simple objects could be reconstructed, the usual choice being black, opaque lettering on a transparent background.

To register phase as well, Lieth and Upatnieks [110] added a uniform and coherent background. Inhomogeneities of diffraction now interact with the background or reference beam and the changes in phase angle $s(\omega t + \epsilon)$ [see (1.2)] in the diffraction pattern from the object are now converted to intensity according to:

$$I = \cos^2 \frac{s(\omega t + \epsilon)}{2} \tag{1.23}$$

and appear in the developed hologram as variations in photographic density. Such a hologram now contains sufficient information to construct a correct image.

The steps in forming such a hologram are identical to those of the Gabor type except that a part of the beam is folded around or allowed to pass the object and at a desired distance and angle to intersect the other part so that both beams are recorded. A possible arrangement is shown in Fig. 1.40. The object O is shown as a glass plate with a design painted on it. P is a prism which deflects part of the parallel monochromatic beam to AB in such a way

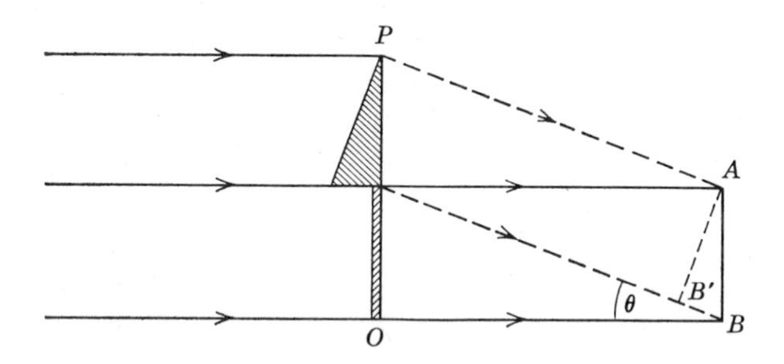

Fig. 1.40 Optical arrangement for making holographic interferograms.

that the wave front is at an angle θ to the general (but perturbed) wave front from O. AB is a photographic plate on which the Fresnel diffraction pattern from the optical inhomogeneities from O are recorded along with the uniform coherent background advancing from AB'. If it is remembered that at any point Z, Y in the photographic plate that extends into the paper the reference plane wave front has a complex amplitude:

$$U_R(Z, Y) = A_0 \exp\left[\frac{j2\pi\theta Z}{\lambda}\right], \tag{1.24}$$

and the amplitude of the perturbed wave front is:

$$U(Z, Y) = A(Z, Y) \exp[j\Phi(Z, Y)], \tag{1.25}$$

where $A(Z, Y)$ is the amplitude, $\Phi(Z, Y)$ the phase, θ the angle between the two wave fronts, A_0 the amplitude of the reference beam, and λ the wavelength of the light. The combined complex amplitude is now:

$$U_T(Z, Y) = A_0 \exp\left[\frac{j2\pi\theta Z}{\lambda}\right] + A, \tag{1.26}$$

giving an intensity at the photographic plate of:

$$I = \left\{A_0 \exp\left[j\frac{2\pi}{\lambda}\theta Z\right] + A\right\}\left\{A_0 \exp\left[-j\frac{2\pi}{\lambda}\theta Z\right] + A\right\}$$

$$= A_0^2 + A^2 + 2A_0A \cos\frac{2\pi}{\lambda}\theta Z, \tag{1.27}$$

a result foreseen from the argument developed around (1.3) and (1.4) and carried through to (1.6). These equations give the amplitude and intensity of a system of two-beam interference fringes. The arguments are entirely analogous and show that a hologram is really a two beam interferogram, but of a special kind. The difference is that a hologram, when illuminated by the same monochromatic light after development, gives a reconstructed object in the form of equal images of which the first- and zero-order are important (shown in Fig. 1.41) and virtual images of which only the first-order is shown.

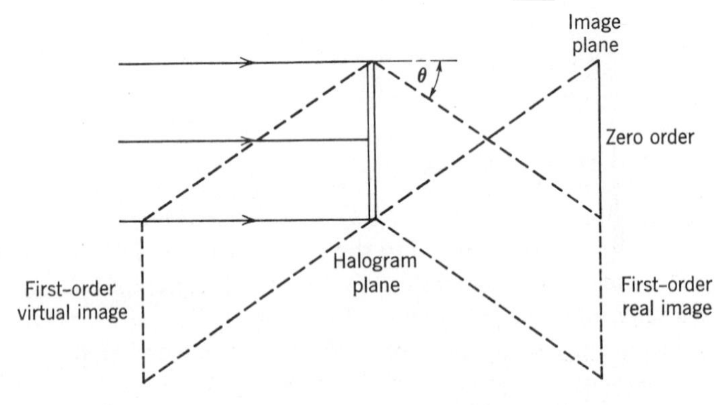

Fig. 1.41 Reconstruction of the image from a hologram.

The zero-order real image is difficult to photograph so one of the first-order images is usually used.

The hologram is a photograph of interference fringes produced by two wave fronts at an angle to each other. The reference wave front is plane, and if the Fresnel diffraction wave front were also plane, the hologram would be a true diffraction grating. Because the diffracted beam is perturbed, the grating is an imperfect one, and when illuminated, the direct or zero-order beam results and the first-order real and virtual images appear on either side of the zero as shown in Fig. 1.41.

There are three main ways to produce a holographic interferogram as opposed to a hologram. These are called double exposure, time-average, and real-time holography. In the first method the apparatus or cell is registered holographically on the photographic plate, then the perturbation of interest in the apparatus or cell is introduced or begun, and the plate is exposed a second time; both images are then developed. Providing the plate or film has sufficient resolving power, both holograms should be recorded, and when the hologram is illuminated for reconstruction, an interferogram should result in which only the differences in path length of light between the two diffracted beams appear.

Time-average holographic interferometry is used to study periodic perturbations such as deflections in piezoelectric crystals. The hologram is made over an interval long compared to the period of the vibration. This system is already being used by mechanical engineers in machinery testing.

Real-time holographic interferometry appears to be the most versatile and most suitable to chemical applications. It has already been used to observe the concentration changes around working electrodes [111] in much the same way Ibl and Mueller used two-beam interferometry. A holographic interferogram of an electrodeposition cell is shown in Fig. 1.42. A typical setup used by O'Brien and Glasel is shown in Fig. 1.43. Light from the laser L which has passed through a beam-expanding telescope T containing a pinhole which acts as a spatial filter. The beam splitter S folds about two-thirds of the beam around the cell by reflecting it toward the mirror M and through the lens L_2 to the photographic plate P. The other one-third goes through the lens L, the diffusing plate D, and the electrochemical cell C and onto the photographic plate. The one-third/two-thirds ratio has recently been tested by Caroll [111], who found maximum fringe visibility at 5:1 reference-to-object beam ratios. He also suggests that if the original hologram is recorded

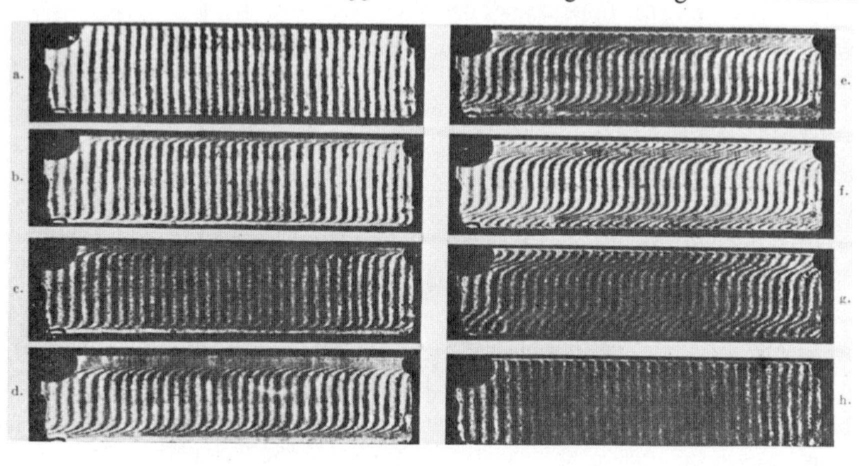

Fig. 1.42 A series of holographic interferograms in a $Cu/CuSO_4$ 0.1 M/Cu cell at 0.5 mA/cm^2 current density. (a) to (d) are interferograms of convectionless cells at elapsed electrolysis times up to 20 min, and (e) to (h) are interferograms of cells with natural convection occurring after the current in (a) to (d) has been reversed for various times up to 20 min [111].

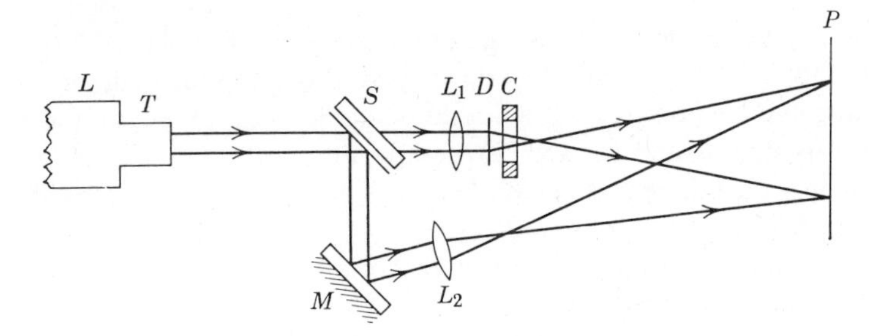

Fig. 1.43 Optical arrangements for real time holographic interferometry of an electrochemical cell.

with any beam ratio, fringe visibility can be improved by changing the beam ratio in the recording, that is, the reconstruction process. If the developed hologram taken of the zero or boundary condition in the cell is replaced in exactly the correct original position, then an infinite series of fringes develops. To obtain fringes that resemble wedge fringes, the hologram must be turned through a very small angle from the original position. Then, in the electrochemical application, the current is switched on and the appropriate first-order image is now recognizable as an interferogram when the reconstructing beam follows the original pathways except for the slight "tilt" introduced at the hologram.

The problems peculiar to holographic interferometry are generally physical and photographic, the optical ones being less onerous. The physical problems are mainly avoiding vibration or physical displacement effects. Vibrations that cause relative movement between two components of the optical setup of one-eighth of a wavelength blur the fringes appreciably, but up to one-half a wavelength movement over the extended exposures usually needed for the original hologram (which are of the order of minutes) generally produces a recognizable fringe system on reconstruction in real-time holographic interferometry. The replacing of the hologram after development in the real-time method has the same theoretical sensitivity as relative motion of optical components. If the replacement is $\lambda/2$ out of position, then a fringe system at infinity exists before a perturbation is applied, just as turning the hologram produces a fringe system which appears to be on the image. The fringe system can be seen without a lens system. Any perturbations to the cell or contents now appear as bends or shifts in the fringe system.

The role of the diffuser is to scatter light so that the light coming through the cell gives the interferogram a three-dimensional quality. If the imperfections, usually scratches on the diffusing plate, are too coarse, they introduce optical noise into the system. It is possible to predict whether this will be so or not by observing the image after a hologram has been formed and the reconstruction process is attempted. The image will appear grainy if the diffusing imperfections are too coarse. It has been found that using a good grade of float glass and grinding two surfaces together after they have been roughened will, when both are used, roughened surfaces in contact, make a satisfactory diffusing element. Another means of overcoming the coarseness is to move the diffuser back from the cell until only the light from the finer imperfections strikes the cell, the other light from the coarse parts leaving the optical system. This, however, has the effect of wasting light and can only be used to overcome small amounts of graininess.

The amount of three-dimensional character in the final interferogram also depends on the angle between the diffracted and background beam. About a 30° angle is the most usually attempted and this usually allows, under otherwise good optical conditions, a range of angles of viewing of about 30°. This means that if a holographic interferogram were taken of an electrochemical system in which a solid electrode had a physical feature 15° out of the optical axis of the diffracted beam, it might be just possible to see it in the interferogram if it were illuminated by laser light of the same frequency and the angle of observation were varied 15° to the plane of the hologram.

Such holographic interferograms as this, in which three-dimensional features are incorporated, require photographic plates of spectroscopic quality, usually spectroscopic plate 649F which has a resolving power of 2000 lines per millimeter and an ASA rating of 0.003. It must be exposed for about 1 min for a transparent electrochemical cell. For opaque objects in which the beam is scattered forward toward the source with poor reflection (and of course the reference beam has to be reflected forward as well), the exposure may require nearly $\frac{1}{2}$ hr and vibration becomes a serious problem. Exposure times may be lowered by a factor of about 5 by sensitizing the plates with a solution of triethanolamine [112]. If higher-speed plates are used, the angle between beams must be reduced because of the optical noise introduced by the graininess of the film, hence the three-dimensional properties are sharply reduced. Tri-X Pan film (ASA 400) has been used [113], but the angle between the beams was very small.

Exposure times can be cut by increasing the light flux. This can be accomplished by using a higher powered laser or reducing the magnification produced by the lenses L_1 and L_2. In the setup shown, the lenses were a matched pair of 88-mm focal length. Increased laser power would be preferred since the magnification achieved is real, that is, the image can be magnified over a reasonable range without magnifying the noise. This is possible because the optical noise level is very low until the first photographic recording, provided that the noise introduced by the diffusing element can be controlled, and this has been successfully achieved. Most noise in the original hologram in the real-time system can be removed by careful matching of optical components and then should be almost totally removed in the common path reconstruction. Stroke (General Reference, Holographic Interferometry) illustrates this. He also shows the further effect of using 1-Å x-ray construction in a two-step imaging process to 6328-Å reconstruction which gives 10^6 magnification. Since the beam expands while traversing the cell, the resulting interferogram is an average over an annular cone. For thin cells, this is not important. For thick cells, especially if there is optical anisotropy as would occur in electrochemical cells, it could be serious. There is also the possibility of reflection from the electrode surfaces which would confuse results. Stroke believes that resolutions of 1 Å are possible [114] using x-rays and visible light reconstruction.

Vibration elimination systems for holographic work in which exposures of about 1 min are contemplated are not difficult to arrange. Kits are now available. The system shown in Fig. 1.43 has consistently yielded holograms. It consists of a sturdy laboratory table in a third-floor laboratory with 8 in. of flexible urethane foam topped by a 300-lb sheet of $\frac{3}{4}$-in. steel. It has been suggested that a heavy wooden bench with legs set in buckets of sand might be sufficient [115].

The apparatus shown in Fig. 1.11 has been used with an argon ion laser and a filter to give a two-frequency output. The intention was to obtain split fringes as in multiple-beam interferometry (Fig. 1.33). The object is to increase the accuracy with which a concentration gradient can be measured this way.

The evaluation of holographic fringe shifts is identical to that for wedge fringes except that (1.7) must be modified to $nd = N\lambda$ because the effective path length difference between the two beams is the thickness of the cell and not twice this dimension. Fringe-counting evaluations are analogous, that is, the passage of one bright fringe means the optical path has increased or decreased by λ not $\lambda/2$. Obviously, although the same detailed control of the geometry as in interferometry, that is, optically flat parallel or near-parallel walls to the cell, is not necessary for accurate measurement, the geometry must be accurately known.

Because the same optical changes occur in an electrolysis cell under investigation as occur in ordinary interferometry, the errors as discussed by Ibl [98] and Muller and Beach [71] and mentioned in the multiple-beam interferometry section occur. The second beam (the reconstruction stage) does not coincide with the first beam (the hologram-forming stage). The effects are much less than for multiple-beam interferometry and half those for two-beam interferometry. Therefore displacement of the fringes, or of the shadow of the electrode to use Beach's concept, presents no serious problem. This holds until the concentration gradient is twice that at which the displacement would occur in a two-beam instrument or, alternatively, up to twice the thickness with the same concentration gradient.

As mentioned above, holographic interferometry has the possibility of a three-dimensional effect if the original zero-state hologram is made with a fine-grain emulsion such as 649F. There is also the possibility of two real magnifications. Real magnification indicates that the signal is increased without a proportionate noise increase. The imposition of a short-focal-length lens as shown in Fig. 1.11 and the change to a longer wavelength in a two-step reconstruction process is described by Stroke [114].

Most applications of holographic interferometry have been to surfaces. The detection of strain in evaporated films has been studied by Magill and Young [116], distortion in thermoelectric device surfaces by Wolfe and Doherty [117], and surface vibration by Powell and Stetson [118]. Apart from the electrochemical application of Knox et al. [111], the only published chemical study is that of a laser spark [119], in which electron densities were measured in a way analogous to those mentioned in the sections on the Michelson and Mach-Zehnder interferometers. The interferometer used in holographic interferometry has been called the Stroke interferometer.

Although the amount of published work is small, several laboratories are known to be investigating the possibilities of holographic interferometry. Since motion picture recording is possible under carefully controlled real-time circumstances, it appears that this system will be used in future in all chemical areas opened by interferometry, that is, diffusion, dilation, fluid flow, thermal conductivity, chemical kinetics, refractometry, titration, and plasmas.

6 MOIRÉ INTERFEROMETERS

Moiré patterns occur whenever two periodic structures overlap and are viewed by transmission. Familiar experiences are looking through two window screens in succession or the folds of a blowing nylon curtain. It is in fact a type of Vernier effect. The term comes from the name for a glossy silk cloth which has a pronounced parallel "weave" which has been folded over itself and pressed. The imprint from the pressing is slightly out of alignment, that is, not quite parallel with the original weave pattern. A shimmering fringe pattern results because of the superposition of slightly misaligned parallel systems of lines (Fig. 1.44). Diffraction gratings are systems of parallel lines and they are frequently used to produce moiré fringes. Coarse gratings such as the screens used by printers in half-tone reproduction which have pitches of about 0.5 mm (two opaque lines per millimeter) give fringes which can be observed in white light and with an extended source; printers must

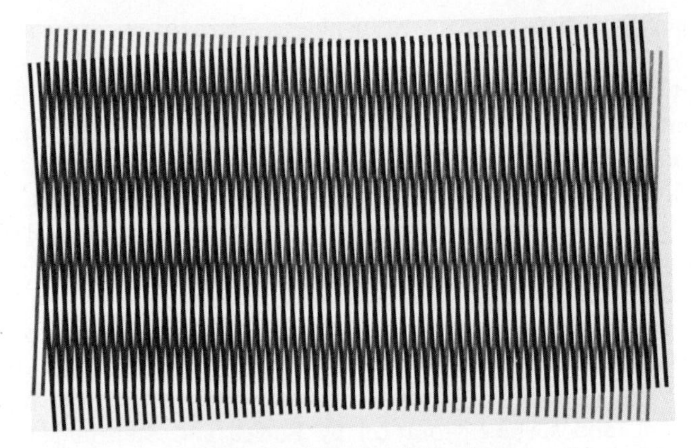

Fig. 1.44 Simple moiré pattern.

continually guard against them. With fine pitch gratings, collimated monochromatic light is used to obtain moiré fringes. Figure 1.45 shows that the fringes can really be wedge fringes. The source S_1 is a slit from which monochromatic, thermal-source light emerges which is collimated at L_1 and passes through the gratings G_1 and G_2 which are at a small angle (a few minutes of arc) and closely spaced. The decollimating lens L_2 is focused on the second slit S_2. Observing optics are placed on the other side of the grating.

The fringes in coarse gratings in which the opaque line occupies one-half the pitch distance can be understood by what Guild calls the "obstruction theory" or other writers the "mechanical interference theory." For fine-pitch gratings, true interference occurs and very complicated theory is required to explain the variety of results obtainable by changing conditions such as pitch, groove form, and order viewed.

Figure 1.46 is an arrangement adapted from Oster and Nishijima (General References, Moiré Interference). L is a laser and F is a spatial filter and beam-expanding telescope. The cell has walls composed of two identical, coarse-pitched gratings, G_1 and G_2. A lump of sugar is dissolving in the water contained in the cell. Moiré fringes should outline concentration changes

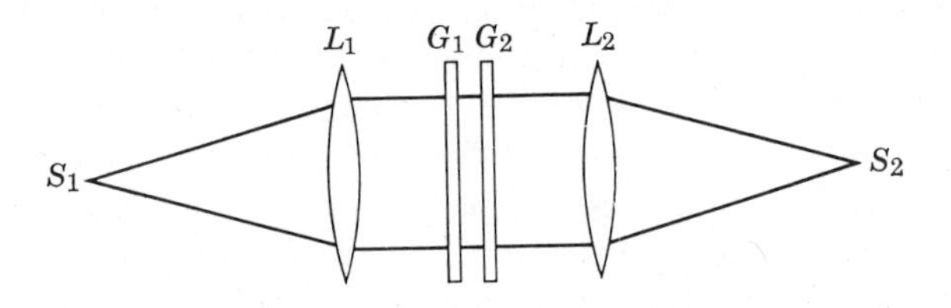

Fig. 1.45 Moiré interferometer.

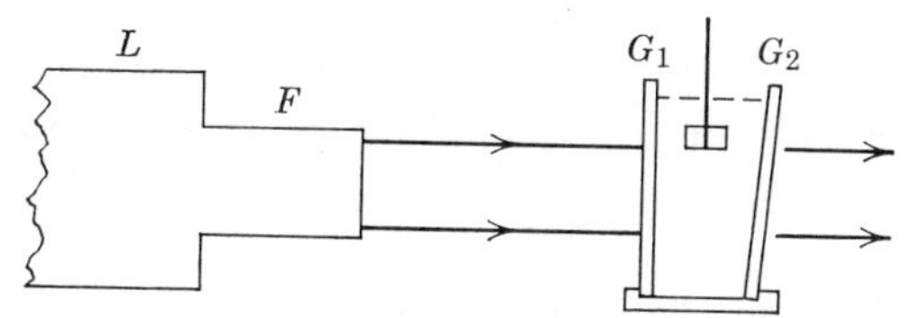

Fig. 1.46 Dissolution and diffusion in a solution contained in a moiré interferometer.

occurring below the sugar. Since sugar has a density greater than water, natural convection produces a complicated pattern. The cell walls could be parallel instead of inclined as shown, and made of ordinary clear glass. The gratings could be set off at a reasonable distance (a few centimeters) with an air wedge between them and the same fringe patterns would be obtained.

The evaluation of moiré fringes is the same as for holographic interference fringes. The modification of (1.7) for a single passage of light is used, that is, $nd = N\lambda$. For white light, colored fringes would result which would be more difficult to measure and to assign values of constants in (1.7).

According to Oster and Nishijima (General References, Moiré Interference), very large magnifications are possible. These occur when the angle between identical gratings is small and the two opaque lines in a coarse grating fall beside one another, giving a dark fringe and a light fringe, for example, every 100 spacings. If two crystals with slightly different-sized lattices but the same crystal habit are stacked over the object to be viewed with an electron microscope and the power to penetrate the crystals is available, a moiré fringe contour outline of the target surface should be seen. The authors suggest that dislocations of atomic size should be visible because the magnification in the moiré pattern may be as high as 100. The chemical uses they suggest are refractometry in the ultracentrifuge, diffusion and electrophoresis, x-ray crystallography, and in fact any of the experiments now performed with interferometry.

Surface roughness and the trueness of optical components such as lenses and gratings are the main field of application reported by Guild, who also gives an account of some very sophisticated fringe-counting devices.

Recently, the moiré device has been called a mock interferometer by Ring and Selby [120]. They used it with a spectrophotometer to simulate the output from a Michelson interferometer, making the spectrometer a multislit instrument with much greater input. The spectrum obtained is again the Fourier transform of the true spectrum. Dowling and co-workers have called this arrangement a lamellar grating interferometer and used it successfully to observe the pure rotational spectrum of H_2O, establishing three new lines in the H_2O vapor spectrum at 216, 202, and 199 μ [121].

The moiré system appears to have merit for electrochemical and chemical kinetics applications. The extremely simple optics, the possibility of having the gratings detached from the cell, and the magnification opportunities are attractive. The magnification, however, means that the values obtained are averaged over an annular cone of the cell because the beam expands within the cell. This is a minor defect for thin cells but could be a major problem with thick inhomogeneous cells as outlined for similar conditions in the section on holographic interferometry.

7 OTHER TYPES OF INTERFEROMETERS USING ELECTROMAGNETIC RADIATION

As has been mentioned, interferometers have been developed for all ranges of the electromagnetic spectrum as well as for sound and de Broglie particles. The instruments for the electromagnetic spectrum are considered, starting with the shortest wavelengths.

X-Ray Interferometers

The first x-ray interferometer was a von Laue-type interferometer described by Bonse and Hart [122], in which twisting or heating of a crystal produced interference fringes easily photographed. A simpler interferometer, using Bragg-type beam splitting and beam combination has been devised by the same authors [123]. The experimental setup shown in Fig. 1.47 is essentially an x-ray version of the Mach-Zehnder interferometer. A collimated beam of monochromatic x-rays enters at the bottom right of Fig. 1.47*a* and strikes the beam splitter *B* where some is transmitted to *M* and some reflected to M_2. M_1 and M_2 are mirrors which reflect the beams back to the beam splitter where they are reflected or refracted to recombine in the emergent beam. The cell *C* can be placed in either beam before recombination. The general theory of the x-ray applications can be found in the chapter on x-ray crystallography Chapter VIA Pt III 4th Ed.

Because of the nature of x-rays with wavelengths of, for example, 1 to 2 Å, conditions of alignment and surface smoothness are stringent. The cell shown was machined out of a nearly perfect single crystal of silicon. The beam splitter, much exaggerated in thickness in the drawing, probably represents the most difficult part of the construction considering the machining properties of silicon. The machining angles must be properly oriented to the crystal planes of the block as shown and so that the beam splitter is at a proper angle to the (220) planes. Since the depth of penetration is great, the smoothness of surfaces is not as critical as might be supposed. The angles to crystal planes and between surfaces, however, are very critical. This makes machining from one block necessary. The apparatus does not need to be

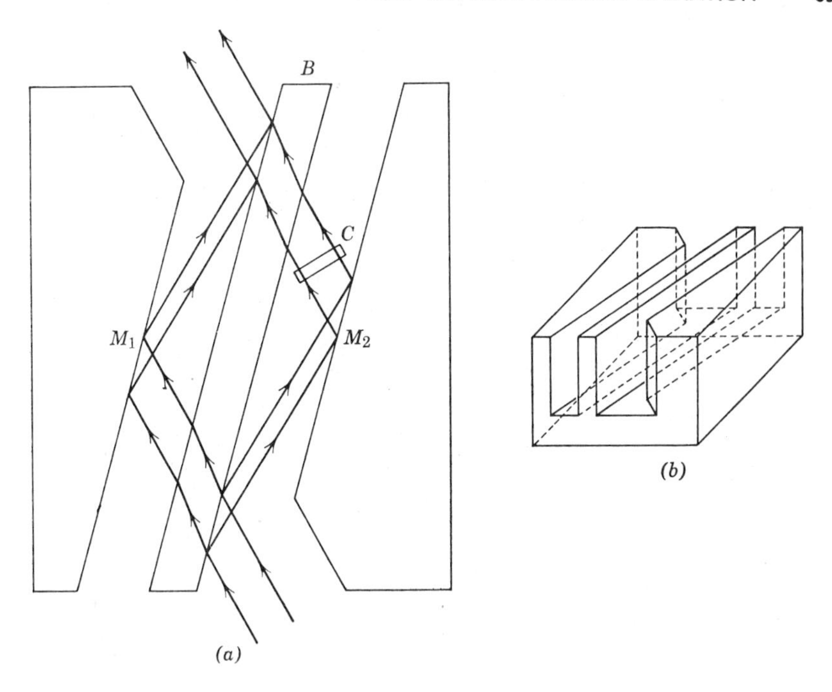

Fig. 1.47 A Bragg-Case x-ray interferometer [123].

thermostated, but must be protected from draughts. Bonse and Hart have also presented a device with combined Laue-and-Bragg case beams [124]. Interferograms of lattice imperfections, thermal gradients and of a piece of plastic are given.

The general theory of x-ray interferometers is presented by Schuelke [125]. Defects in germanium single crystals have been investigated with the aid of dynamic effects, and the field has been reviewed by Zielinska [126]. Polytypism has been detected in vapor-deposited crystals [127], and a one-sided Ca^{2+} chelate has been inferred in tetracycline-modified rat bones [128]. It is suggested that the electrical double layer in electrochemistry could be investigated with an x-ray interferometer with the cell oriented and the electrodes staggered so that the x-ray beam traverses the solution next to the electrode, parallel to the working face of the electrode. If a strong solution of CsOH were electrolyzed at high current density, the Cs^+ should cause scattering in such a way as to indicate either structure or lack of it in the double layer. It should also be possible to measure quantitatively the steps in the layer-type electrocrystallization observed, for example, in the electrodeposition of silver.

Microwave Interferometers

Microwave interferometers are used for probing plasmas, finding the dielectric constant of solutions and fluids generally, and in spectrometry. They provide the advantage of access to all the energy output of the radiation source in Fourier spectroscopy. Two general arrangements are possible, those in which the cell containing the probe is in a waveguide and those in which it is not.

Steel (General References, Optical Interferometry) gives a short historical treatment of the development of microwave interferometers. An early simple apparatus is shown in Fig. 1.48. The klystron K generates the required microwave signal which the horn H_1 introduces into the cell c, and the horn H_2

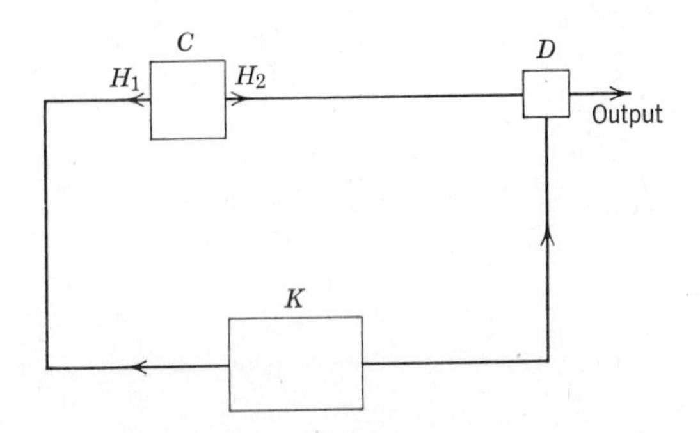

Fig. 1.48 Microwave interferometer.

collects the transmitted signal to the crystal detector D. This signal is now combined with the reference signal, arriving counterclockwise in the diagram, and the output contains the "fringes" which take the form of variations in signal intensity analogous to light intensity in visible fringes. The output usually goes to an oscilloscope in which the display can be analyzed in conventional ways, such as delaying the reference wave to find the null point and counting "fringes."

Although optical interferometers are used to examine the density of plasmas [129], microwave interferometers are more common for this purpose. Special optics of Teflon are required according to the design of Michaelis [130]. An interferometer has been designed to measure phase angles in an ionized gas as small as 1.2° [131] and another to measure electron density behind a shock wave [132]. Hufnagel and Klages [132a] have used a waveguide interferometer to measure dielectric constants in dilute solutions of phenyl chloride, benzophenone, and chloronaphthalene. They all showed a linear dependence of ϵ'' on concentration, ϵ'' being defined as the imaginary part of the complex dielectric from $\epsilon = \epsilon' - i\epsilon''$ where:

$$\epsilon'' = \frac{(\epsilon_0 - \epsilon_\infty)\omega\tau_D}{1 + \omega^2\tau_D}, \tag{1.28}$$

where ϵ_0 is the low frequency, ϵ_∞ the high frequency or optical value, τ_D the relaxation time, and $\omega = 2\pi f$, the circular frequency. A comparison of interferometric and standing-wave methods by Rao showed good agreement [133].

Radio Interferometers

Radio interferometry has been practiced by astronomers for about 10 years. The theory and practice are quite well developed and are discussed by Steel (General References, Optical Interferometry), but the system is unlikely to become important in chemistry.

8 NONELECTROMAGNETIC RADIATION INTERFEROMETERS

The idea of interferometry is quite general and may be used wherever a wave representation fits the physical phenomenon involved and rates of propagation can be compared in terms of path difference in two or more pathways. Some nonelectromagnetic wave phenomena that can be examined interferometrically are presented below. The treatment is neither intensively or extensively exhaustive; these interferometers have limited current use in chemistry and this is unlikely to change dramatically in the next decade.

Sonic Interferometers

Sonic interferometers have been used since 1925 [2] and have been very effective in investigating the mode of sound propagation. According to Herzfeld and Litovitz [134], however, the sonic interferometer has been rather unsuccessful in the liquid phase up till 1959. The great body of chemical research performed sonically consisted mostly of absorption and dispersion measurements. Despite this, a recent review gives 36 references to sonic interferometry [135]. New interferometers are continually being designed [136], including one using frequency modulation [137].

These devices usually have an input signal delivered to an immersed piezoelectric crystal and a suitably situated receiving crystal. The signal from the receiving crystal and the input signal to the input piezoelectric crystal are compared on an oscilloscope. A noise problem appears to be caused if the receiving piezoelectric crystal is placed so that the effects of diffraction are noticeable, which are caused by the finite size of the input crystal; by excitation of Rayleigh modes resulting from unwanted transverse vibrations of the input crystal; and by coherent or incoherent reflections. A recent paper gives a method of empirically determining required diffraction corrections [138]. Use of the instrument has also been reported for control of chemical processes [139], for research into the dissolved state of surface-active substances [140], and for a study of the hydration numbers of a series of chlorides [141].

Particle Interferometers

Any particle can be used to produce interferograms if it is small enough to be rationally treated by the de Broglie relationship $\lambda = h/(mv) = h/p$, where h is Planck's constant, λ the wavelength, m the mass of the particle, v its velocity, and p its momentum.

Neutron Interferometers

Both neutron [142] and electron interferometers have been designed and used for investigation in the field of physics. The neutron interferometer employs two glass prisms to separate the beams and glass lenses. The instrument has been used to find the dimensions of domains in ferromagnetic structures [143], but no purely chemical application appears to have been made.

Electron Interferometers

These instruments have been reviewed by Faget [144] and new developments in the electron optics reported by Kerschbaumer [145], but the only application related to chemistry has been the measurement of internal potentials of metals [145].

9 SUMMARY OF USES

Diffusion. Many types of interferometer have been used to study diffusion, including the Rayleigh, Mach-Zehnder, Jamin, multiple-beam wedge, Michelson, and polarization instruments and some of their modifications.

Dilation. Michelson, multiple-beam, Mach-Zehnder and Fabry-Perot interferometers have been used to measure length, and dilation attributable to flow or shear and for shock wave measurements. The Mach-Zehnder is most often used for flow and shock investigations.

Thermal gradients. The Mach-Zehnder instrument has most often been used for flow and heat flow measurements, especially in gases; the multiple-beam wedge and the polarization-type interferometers have been used to measure flow and heat flow in liquids.

Sedimentation. Sedimentation in the ultracentrifuge has been followed by the Rayleigh, polarization, Mach-Zehnder, and Jamin interferometers.

Raman Fourier spectrometry. A Michelson-type interferometer has been used recently with a solid-state exciting laser [146].

Fourier spectrometry. Almost all interferometers used with spectrometers are of the Michelson type. Over short spectral ranges the Fabry-Perot can be used. Grating or moiré interferometers are also used, especially in the microwave regions.

Dielectric constant. Sonic and microwave interferometers are most used.

Convection and mass transport. The Mach-Zehnder, Jamin, and multiple-beam wedge interferometers are generally used.

Electrochemistry. The multiple-beam wedge, the Mach-Zehnder, the Jamin, and the holographic interferometer have been most commonly used.

Chemical kinetics. Only the multiple-beam-wedge-type has been used, but others interferometers that produce parallel fringes could be employed.

Surface chemistry and topography. Multiple-beam and interference microscopes are most commonly used.

Fluorescence. A Mach-Zehnder interferometer has recently been used to measure decay times of the order of 10^{-11} sec using a helium-neon laser in tellurium-doped GaAs [147].

Crystal growth. The interference microscope and the wedge multiple-beam interferometer are most often used.

References

1. G. Magyar and L. Mandel, *Nature*, **198,** 255 (1963).
2. G. W. Pierce, *Proc. Am. Acad. Arts Sci.*, **60,** 271 (1925).
3. G. Oster, *Endeavour*, **27,** 60 (1968).
4. W. Nebe and S. Molinski, *Chemik* (*Gliwice*), **17,** 243 (1964).
5. H. Svensson, *Acta Chem. Scand.*, **5,** 1301 (1951); H. Svensson, *Opt. Acta*, **1,** 25 (1954).
6. I. H. Billick and R. J. Bowen, *J. Phys. Chem.*, **69,** 4024 (1965).
7. A. R. Baker, F. J. Hartwell, and D. A. Windle, *Min. Power* (*Gt. Brit.*) *Safety Mines Res. Establ.*, *Res. Rept.* **172** (1959).
8. G. K. Subbotin and I. J. Zharikov, USSR Pat. Nos. 131967 and 131969, Sept. 20, 1960.
9. R. Apfelthaler, *Rostl. Vyroba*, **12,** 9 (1966).
10. J. Pollo, B. Witkowska, and A. Narog, *Zeszyty Nauk. Politech. Slask. Chem.*, **35,** 97 (1967).
11. M. Vagner, *Rostl. Vyroba*, **13,** 775 (1967).
12. B. Versino, *Comm. Naz. Ric. Nucl.*, *CNI-48*, 14 (1960).
13. F. H. Horne and R. J. Bearman, *J. Chem. Phys.*, **37,** 2857.
14. J. K. Furdyna, *Intern. Conf. Microwave Behavior Ferrimagnetic Plasmas, London, 1965*, 36-1-36-3.

15. M. S. Soskin, *Opt. i Spektroskopiya*, **11**, 768 (1961).
16. N. Ibl and R. H. Muller, *Z. Elektrochem.*, **59**, 671 (1955).
17. P. M. Duffieux, *L'Integrals de Fourier et ses Applications a l'Optique*, Faculté des Sciences de Besançon, 1946, p. 11.
18. H. Lien and J. Eckershan, *AIAA J.*, **4**, 1988 (1966); J. H. Spark and D. D. Shear, *Phys. Fluids*, **8**, 1913 (1965).
19. W. L. Howes and D. R. Buchele, *J. Opt. Soc. Am.*, **56**, 1517 (1966).
20. D. Combecher, AEC Accession No. 13906, Rept. No. IPP-1/4Q (1965); F. P. Kuepfer, *Z. Naturforsch.*, **18a**, 895 (1963).
21. G. B. Wheeler and E. A. Dangor, *Conf. Proc.*, **2**, 517 (1965); S. Byron, *Agard Rept.*, **328** (1959); J. K. Wright, R. D. Medford, A. G. Hunt, and J. D. Herbert, *Proc. Phys. Soc. (London)*, **78**, 1439 (1961); S. Nakai and C. Yamanaka, *Kakuyugo Kenyu*, **15**, 203 (1965).
22. E. A. McLean and S. A. Ramsden, *Phys. Rev.*, **140**(4A), 1122 (1965).
23. G. G. Dalgov and G. D. Petrov, *Fiz. Sb. L'vovsk. Univ.*, **4**, 68 (1958).
24. K. Azuma and Y. Salzima, *Japan J. Appl. Phys.*, **6**, 909 (1967); *J. Appl. Phys.*, **6**, 915 (1967).
25. Y. Oshima, *Natur. Sci. Rept. Achanomizu Univ., Tokyo*, **17**, 17 (1966).
26. C. S. Lin, R. W. Moulton, and G. L. Putnam, *Ind. Eng. Chem.*, **45**, 636 (1953).
27. M. A. Jeppeson, *Am. J. Phys.*, **35**, 435 (1967).
28. E. M. Thorndike and A. P. Marion, *Anal. Chem.*, **40**, 236 (1968); W. Kinder, *Optik*, **24**, 324 (1967).
29. V. S. Doladugina and E. Berezina, *Rost Kristallov, Akad. Nauk, SSSR Inst. Kristallogr.*, **5**, 402 (1965).
30. W. H. Stevenson, *Rev. Sci. Instr.*, **36**, 704 (1965).
31. S. B. Herskovitz, *Rev. Sci. Instr.*, **37**, 452 (1966).
32. V. Met, *Appl. Opt.*, **5**, 1242 (1960).
33. A. L. Besse and J. G. Kelly, *Rev. Sci. Instru.*, **37**, 1497 (1966).
34. J. Montello, *Anales Real Soc. Espan. fis. quim. (Madrid)*, **56A**, 105 (1960).
35. H. J. Spies, *Freiberger Forschungsh.*, **B111**, 87 (1965).
36. G. Sabatier, *Bull. Soc. Franc. Mineral. Crist.*, **88**, 404 (1965).
37. C. Pellegrino and P. Ricci, *Sperimentale*, **110**, 48 (1960).
38. G. Bencke, *Introduction Quant. Cytochem.*, **1966**, 63.
39. G. Smeets, *Sci. Tech. Aerospace Rept.*, **4**, 2589 (1965).
40. V. N. Tsvethov, V. S. Skazka, and N. A. Nikitin, *Opt. i Spektroskopiya*, **17**, 119 (1964); V. N. Tsvetkov, *Vysokomolekul. Soedin., Ser. A*, **9**, 1249 (1967).
41. A. V. Podalinskii, *Zh. Fiz. Khim.*, **37**, 1189 (1963).
42. T. Tsurutu, *Japan J. Appl. Phys., Suppl. 1*, **4**, 172 (1965).
43. H. Eckstein, *Radex Rundschau*, **1967**(3–4), 629.
44a. M. Follonius, *Bull. Soc. Franc. Mineral. Crist.*, **82**, 343 (1959).
44b. M. Follonius and S. Goldsztaub, *Bull. Microscop. Appl.*, **7**, 8 (1957).
45. W. H. Steel, *Opt. i Spektroskopiya*, **20**, 910 (1966).
45a. O. Bryngdahl and S. Ljunggren, *J. Phys. Chem.*, **64**, 1264 (1960). O. Bryngdahl, *J. Opt. Soc. Am.*, **53**, 571 (1963).
45b. H. J. V. Tyrrell, *Diffusion and Heat Flow in Liquids*, Butterworths, London, 1961, p. 92.

45c. S. E. Gustafsson, J. G. Becsey, and J. A. Bierlein, *J. Phys. Chem.*, **69,** 1016 (1965).
46. R. Berg, *U.S. Atomic Energy Comm. UCRL-9512*, 1960.
47. T. Tako and M. Oi, *Bull. Natl. Res. Lab.*, *Metral* (*Tokyo*), **12,** 73 (1966).
48. R. J. Hull and L. C. Bradley, *J. Opt. Soc. Am.*, **57,** 11 (1967).
49. J. Ćooper and J. R. Grieg, *J. Sci. Instr.*, **40,** 433 (1963).
50. M. C. Bigeon, J. L. Cojan, and P. Giacomo, *Appl. Opt.*, **4,** 907 (1965).
51. C. D. Cooper, *U.S. Dept. Comm. Office Tech. Serv.*, AD273, 763, 27 (1961).
52. J. A. Nilson and G. G. Shepherd, *Planetary Space Sci.*, **5,** 299 (1961).
53. G. A. Odintsova and E. P. Vaulin, *Zh. Prikl. Spektrosk.*, *Akad. Nauk. Belarussk*, *SSR*, **3,** 105 (1965).
54. B. Wilhelmi, *Infrared Phys.*, **8,** 157 (1968).
55. F. Westermann and W. Maier, *Z. Physik.*, **179,** 244 (1964).
56. V. Y. Balakhanov and A. R. Striganov, *Zh. Prikl. Spektrosk.*, *Akad. Nauk. Belorussk*, *SSR*, **3,** 311 (1965).
57. R. D. Rawcliff and C. N. Randall, *Appl. Opt.*, **6,** 1353 (1967).
58. F. A. Korolev and V. I. Gridnev, *Opt. i Spectroskopiya*, **16,** 335 (1964).
59. T. G. Polanyi and W. R. Warren, *J. Opt. Sci. Am.*, **54,** 449 (1964).
60. S. M. Jarrett and P. A. Franken, *J. Opt. Sci. Am.*, **55,** 1603 (1965).
61. Y. Urano, S. Nakajima, Y. Ueda, K. Kosawa, Y. Maruyama, S. Katsube, and M. Iwata, *Osaka Kogyo Gijutsu Shikensho Kiho*, **14,** 260 (1963).
62. F. R. Raesler and J. E. Mack, *J. Phys.* (*Paris*) *Colloq.*, **28,** C_2-313-20 (1967).
63. R. Beer and J. Ring, *Infrared Phys.*, **1,** 94 (1961).
64. M. J. Forrest, *J. Sci. Instr.*, **44,** 26 (1967).
65. A. Roeseler, *Optik* (*Stuttgart*), **24,** 606 (1966–67).
66. R. N. O'Brien, *Rev. Sci. Instr.*, **35,** 803 (1964).
67. Y. Nishijima and G. Oster, *J. Chem. Educ.*, **38,** 114 (1961).
68. S. Tolansky, *Surface Microtopography*, Interscience, New York, 1960; S. Tolansky, *Microstructure of Surfaces Using Interferometry*, Edward Arnold, London, 1968.
69. F. W. Yakymyshyn, M. Sc. Thesis, University of Alberta, Edmonton, Alberta, Canada, April 1962.
70. R. N. O'Brien, E. A. Beer, K. W. Beach, and J. Leja, University of California, Lawrence Radiation Laboratory, UCRL-1700, Preprint, Part I; R. N. O'Brien, S. Weiner, and K. S. Spiegler, UCRL-1700, Preprint, Part II, July 1966.
71. K. W. Beach, M.S. Thesis, University of California, Lawrence Radiation Lab., UCRL-18037, Berkeley, California, 1968.
72. R. N. O'Brien and J. Leja, *Nature*, **210,** 1217 (1966).
73. R. N. O'Brien, *J. Electrochem. Soc.*, **114,** 710 (1967).
74. W. F. Berg, *Proc. Roy. Soc.* (*London*), **A164,** 79 (1938).
75. C. W. Bunn, *Discussions Farady Soc.*, **5,** 132 (1949).
76. V. M. Varikast, Z. Solc, and F. Myl, *Vestsi Akad. Navuk Belarusk. SSR. Ser. Fiz-Tekh. Navuk*, **1967,** 3, 129.
77. Roland Itti, *Comm. Energie At.* (*France*) *Rappt*, **1967** (3085).
78. R. N. O'Brien and A. Glasel, *Can. J. Chem.*, **47,** 223 (1969).

79. V. J. Vajagand, T. J. Pastor, F. F. Gaal, and M. Todorovic, *Proc. Conf. Appl. Phys. Chem., Methods Chem. Anal., Budapest*, **1,** 152 (1966).
80. D. C. Johnston, A. F. Witt, and H. C. Gatos, *J. Electrochem. Soc.*, **115,** 438 (1968).
81. A. R. Patel, O. P. Bahl, and A. S. Vagh, *Acta Cryst.*, **20,** 914 (1966).
82. A. R. Patel and O. P. Bahl, *Acta Cryst.*, **19,** 627 (1965).
83. L. Young, *Anodic Oxide Films*, Academic, New York, 1961.
84. E. Passaglia, R. R. Strombert, and J. Kruger, *Ellipsometry in the Measurement of Surfaces and Thin Films*, Symposium Proceedings, Washington, D.C., 1963, Natl. Bur. Stds. Publ. 256, issued 1964.
85. A. R. Walter, *J. Appl. Phys.*, **36,** 2377 (1965).
86. B. Bates and D. J. Bradley, *J. Opt. Soc. Am.*, **57,** 481 (1967).
87. O. Uyeda, *Rev. Sci. Instr.*, **38,** 277 (1967).
88. C. Lockyer, *J. Sci. Instr.*, **44,** 393 (1967).
89. H. T. McCartney and S. Ergun, *J. Opt. Soc. Am.*, **52,** 197 (1962).
90. H. L. Johnston, N. W. Altman, and T. Rubin, *J. Chem. Eng. Data*, **10,** 241 (1965).
91. E. A. Flood and R. F. Bartholomew, *Can. J. Chem.*, **46,** 249 (1968).
92. R. N. O'Brien and H. J. Axon, *Trans. Inst. Metal Finishing*, **34,** 41 (1957).
93. R. N. O'Brien, *J. Electrochem. Soc.*, **111,** 1300 (1964).
94. R. N. O'Brien, *J. Electrochem. Soc.*, **113,** 389 (1966).
95. T. R. Rosebrugh and L. Miller, *J. Phys. Chem.*, **14,** 816 (1910).
96. R. N. O'Brien, W. F. Yakymyshyn, and J. Leja, *J. Electrochem. Soc.*, **110,** 820 (1963); R. N. O'Brien and K. Kinashita, *J. Electrochem. Soc.*, **112,** 951 (1965).
97. R. N. O'Brien, K. S. Spiegler, and K. Beach, "Mass Transport at Ionic Exchange Membrane—Solution Interfaces," to be published in Proceedings of the 17th Meeting CITCE, Tokyo (1965).
98. N. Ibl, *Proceedings of the 7th Meeting, CITCE, Lindau, 1955*, Butterworths, London (1957).
99. R. N. O'Brien and F. P. Dieken, "The Polarographic Maximum by Laser Interferometry," *Can. J. Chem. 48*, 2651 (1970).
100. Unpublished work of R. N. O'Brien and D. Quon.
101. Unpublished work of R. N. O'Brien and D. Quon.
102. R. N. O'Brien, D. Quon, and C. R. Darsi, *J. Electrochem. Soc., 117*, 888 (1970).
103. C. W. Tobias and R. N. O'Brien, University of California, Lawrence Radiation Laboratory, Rept. No. UCRL, 16658, p. 53.
104. R. N. O'Brien, W. F. Hyslop, and C. W. Tobias, "The Refractive Index of Some Gases Dissolved in Water by Laser Interferometry Using the Pressure Jump Technique," to be published in *J. Electrochem. Soc.*
105. M. Hagan, *Clathrate Inclusion Compounds*, Van Nostrand Reinhold, New York, 1962, p. 60.
106. B. A. Silverman, B. J. Thompson, and J. H. Ward, *J. Appl. Meteorol.*, **3,** 792 (1964).

107. R. N. O'Brien, *J. Chem. Eng. Data*, **13,** 1 (1968); R. N. O'Brien and D. Quon, *J. Chem. Eng. Data*, **13,** 517 (1968).
108. R. N. O'Brien and W. F. Hyslop, to be published in *Can. J. Chem.*
109. D. Gabor, *Proc. Roy. Soc.* (*London*), **A197,** 454 (1949); D. Gabor, *Proc. Roy. Soc.* (*London*), **B64,** 449 (1951).
110. E. N. Leith and J. Upatnieks, *J. Opt. Soc. Am.*, **52,** 1123 (1962).
111. C. Knox, R. R. Såyano, E. T. Seo, and H. P. Silverman, *J. Phys. Chem.*, **71,** 3102 (1967); James P. Carroll, *Appl. Opt.*, **7,** 1642 (1968).
112. L. H. Lin and C. V. LaBianco, *Appl. Opt.*, **6,** 1255 (1967).
113. R. E. Brooks, *Appl. Opt.*, **6,** 1418 (1967).
114. G. W. Stroke and D. G. Falconer, *Phys. Letters*, **13,** 306 (1964).
115. J. T. Carcel, A. N. Rodemann, F. Florman, and S. Domeshek, *Appl. Opt.*, **5,** 1199 (1966).
116. P. J. Magill and T. Young, *J. Vacuum Sci. Technol.*, **4,** 47 (1967).
117. R. Wolfe and E. J. Doherty, *J. Appl. Phys.*, **37,** 5008 (1966).
118. R. L. Powell and Karl A. Stetson, *J. Opt. Soc. Am.*, **55,** 1593 (1965).
119. A. Kakus, G. V. Astrovskaya, Y. I. Ostrovskii, and A. N. Zaidel, *Phys. Letters*, **23,** 8103 (1966).
120. J. Ring and M. J. Selby, *Infrared Phys.*, **6,** 33 (1966); M. J. Selby, *Infrared Phys.*, **6,** 21 (1966).
121. J. M. Dowling and R. T. Hall, *J. Opt. Soc. Am.*, **57,** 269 (1967); J. M. Dowling and R. T. Hall, *J. Mol. Spectry.*, **19,** 108 (1965); R. T. Hall, D. Vrabec, and J. M. Dowling, *Appl. Opt.*, **5,** 1147 (1966).
122. U. Bonse and M. Hart, *Z. Phys.*, **190,** 455 (1966).
123. U. Bonse and M. Hart, *Z. Phys.*, **194,** 1 (1966).
124. U. Bonse and M. Hart, *Acta Cryst. Sect. A*, **24** (Pt. 1), 240 (1968).
125. W. Schuelke, *Ann. Physik.*, **17,** 207 (1966).
126. E. Zielinska, *Rotozenska*, **16,** 105 (1965).
127. G. K. Chadha and G. C. Trigunayat, *J. Phys. Chem. Solids* (*India*) *Suppl. No.* **1,** 31 (1967).
128. H. Kammerer, G. Gattaw, and W. Eager, *Acta Histochem.*, **28,** 112 (1967).
129. R. A. Alpher and D. R. White, *Plasma Diagnostic Tech.*, **431** (1965); *Nucl. Sci. Abstr.*, **20,** 1029 (1966).
130. M. Michaelis, AEC Accession No. 42408, Rept. No. IPP-2/49 (1966).
131. E. Hotson and M. Seidle, *J. Sci. Instr.*, **42,** 225 (1965).
132. W. Makias, AEC Accession No. 17510, Rept. No. IPP-3/25 (1965).
132a. F. Hufnagel and G. Klages, *Z. angew. Phys.*, **12,** 202 (1960).
133. V. M. Rao, *Can. J. Phys.*, **41,** 1679 (1963).
134. K. F. Herzfeld and T. A. Litovitz, *Absorption and Dispersion of Ultrasonic Waves*, Academic, New York, 1959, p. 366.
135. Y. Maiyahara, *Karaido To Kaiman—Kasseizai*, **2,** 609 (1961).
136. M. W. Kaulgud, *Akust. Beih*, **15,** 377 (1965); Z. Kozlawki and J. Wehr, *Polska Akad. Nauk Zeszyty Prob. Nauki Polskiej*, **26,** 272 (1965); P. Roy-Chowdhury, *Indian J. Pure Appl. Phys.*, **5,** 123 (1967); R. W. Leonard and H. Seguin, *J. Acoust. Soc. Am.*, **40,** 1467 (1966); I. I. Pervushin and L. P. Tilippov, *Akust. Zh.*, **7,** 385 (1961).

137. F. I. Gucker, C. L. Chernick, and P. Roy-Chowdhury, *Proc. Natl. Acad. Sci., U.S.*, **55,** 12 (1966).
138. O. Kubilliuniene and V. Ilgunas, *Lietuvos Fiz. Rinkinys, Lietuvos TSR Mokslu Akad. Lietuvos TSR Aukstosios Mokyklas*, **4,** 115 (1964).
139. E. Jaronis and K. Barsauskas, *Kauno Politech. Inst. Darbai*, **5,** 173 (1957).
140. K. Shigehara, *Bull. Chem. Soc. Japan*, **39,** 2332 (1966).
141. T. Isemura and S. Gato, *Bull. Chem. Soc. Japan*, **37,** 1690 (1964).
142. H. Maier-Leibnitz and T. Springer, *Z. Physik.*, **167,** 386 (1962).
143. P. Korpium, *Z. Physik.*, **195,** 146 (1966).
144. J. Faget, *Rev. Opt.*, **40,** 347 (1961).
145. E. Kerschbaumer, *Z. Phys.*, **201,** 200 (1967).
146. G. W. Chantey, H. A. Gebbie, and C. Hilsum, *Nature*, **203,** 1052 (1964).
147. R. J. Carbone and P. R. Longaker, *Appl. Phys. Letters*, **4,** 32 (1964).

General

WAVE NATURE OF LIGHT

Bloom, A. L., *Gas Lasers*, Wiley, New York, 1966.

Born, M., and E. Wolf, *Principles of Optics*, 2nd ed., Pergamon, Oxford, 1964.

Heard, H. G., *Laser Parameter Measurements Handbook*, Wiley, New York, 1966.

Heriot, D. H., "Applications of Laser Light," *Sci. Am.*, **219,** 140 (Sept. 1968).

Jenkins, F. A., and H. E. White, *Fundamentals of Physical Optics*, McGraw-Hill, New York, 1937.

Longhurst, R. S., *Geometrical and Physical Optics*, Longmans, London, 1957.

Morgan, J., *Introduction to Geometrical and Physical Optics*, McGraw-Hill, New York, 1953.

Patek, K., *Lasers*, H. Arend, *Transl.*, Chemical Rubber Co., Cleveland, 1966.

Schawlow, Ar. L., "Laser Light," *Sci. Am.*, **219,** 120 (Sept. 1968).

OPTICAL INTERFEROMETRY

Candler, C., *Modern Interferometers*, Hilger and Watts, London, 1951.

Connes, P., "How Light is Analyzed," *Sci. Am.*, **219,** 72 (Sept. 1968).

Davison, P. S., "Interferometry: A Review of the Sources of Published Information," *Spectrosc. Mol.*, **16,** 38 (1967).

Degenhard, W. E., "Interferometry," *Encycl. Ind. Chem. Anal.*, **2,** 334 (1966).

Françon, M., *Optical Interferometry*, I. Wilmanns, Transl., Academic, New York, 1966.

Heavens, O. S., "Recent Application of Lasers," *Brit. J. Appl. Phys.*, **17,** 287 (1966).

Krug, W., J. Rienitz, and G. Schulz, *Contributions to Interference Microscopy*, J. H. Dickson, Transl., Hilger and Watts, London, 1964.

Rienitz, J., "Effect of Partially Permeable Layers on the Efficiency of Double Beam Interferometers," *Abhandl. Deut. Akad. Wiss. Berlin Kl. Math. Physik. Tech.*, **1961**(2).

Steel, W. H., *Interferometry*, Cambridge Univ. Press, Cambridge, 1967.

Tolansky, S., *Multiple Beam Interferometry of Surfaces and Films*, Oxford Univ. Press, Oxford, 1948.

Tolansky, S., *An Introduction to Interferometry*, Longmans, London, 1955.

Williams, W. E., *Applications of Interferometry*, Metheun, London, 1930, reprinted 1950.

HOLOGRAPHIC INTERFEROMETRY

DeVelis, J. B., and G. O. Reynolds, *Theory and Applications of Holography*, Addison-Wesley, Menlo Park, California, 1967.

Ennos, A. E., "Holography and its Uses," *Contemp. Phys.*, **8,** 153 (1967).

Lieth, E. N., and J. Upatnieks, "Photography by Laser," *Sci. Am.*, **212,** 24 (1965).

Martienssen, W., *Physikertag Deut. Oesterr. Phys. Ges. Plenorvorts. Munich*, **1,** 92 (1966).

Stroke, G. W., *An Introduction to Coherent Optics and Holography*, Academic, New York, 1966.

Tanner, R. H., *J. Sci. Instr.*, **43,** 878 (1966).

Tsujiuchi, J., and T. Tsuruta, "Holographic Interferometry Using Diffusely Reflecting Surfaces" (review), *Oyo Butsuri*, **36,** 232 (1967).

MOIRÉ INTERFERENCE

Guild, J., *The Interference Systems of Crossed Diffraction Gratings, Theory of Moiré Fringes*, Clarendon, Oxford, 1956.

Guild, J., *Diffraction Gratings as Measuring Scales*, Oxford Univ. Press, London, 1960.

Oster, G., and Y. Nishijina, *Sci. Am.*, **208,** 54 (1963).

Chapter II

LIGHT SCATTERING

Gerald Oster

1 Introduction **75**

2 Outline of Theory **78**

Rayleigh Scattering 78
Debye Scattering 81
Mie Scattering 88
Nonindependent Particles 93
Fine Structure of Scattering 100
Multiple Scattering 101

3 Instrumentation **102**

Survey of Instrumentation 102
Angular Scattering Photometer 103
Absolute Turbidity 107
Forward-Angle Photometers 108
Fine-Structure Resolution 110

References **112**

1 INTRODUCTION

When a beam of light falls upon matter, the electric field associated with the light induces periodic oscillations of the electron clouds of the atoms of the material. The atoms then serve as secondary sources of light and radiate light in the form of scattered radiation. There is no clear-cut distinction between scattering and the diffraction of light, although the former term is usually applied to small particles arranged more or less randomly in space, whereas the latter is confined to large obstacles. The intensity, polarization, angular distribution, and fine structure of the scattered light is determined by the size, shape, optical constants, and interactions of the molecules in the scattering material. Conversely, from the light-scattering properties of a given system one should be able, with the aid of the electromagnetic theory of radiation and the kinetic theory of matter, to obtain a detailed molecular picture of that system.

The basic ideas of light scattering were laid down more than a half-century ago by Rayleigh, Mie, Smoluchowski, and Debye. It is convenient to divide light scattering by independent particles into three classes, which we shall refer to as (1) Rayleigh scattering,(2) Debye scattering, and(3) Mie scattering. The class into which a given case falls is determined by two parameters of the scattering particles, namely, their size relative to the wavelength of the incident light and the index of refraction of the particles relative to that of the medium. If the size parameter is very small, then the scattering is of the Rayleigh type. In practice, this type of scattering applies to particles that have their largest dimension less than about one-tenth of the wavelength of the scattered light, that is, about 50 nm (1 nanometer, nm, equals 1 millimicron, mμ, or 10^{-7} cm). Under this condition the scatterer acts as a point dipole oscillator [1]. If the size of the particle is not small relative to the wavelength of the light and if its refractive index is close to that of the medium, then the scattering is of the Debye type. Under these conditions there are fixed phase relations between wavelets scattered from different parts of the same particle, each scattering element of the particle acting as a point scatterer [2, 3]. For large particles of high relative refractive index not only are there fixed spatial relations between the scattering elements, but there are also distortions of the electric field of the incident and scattered light. This formidable theoretical problem was solved for spherical particles by Mie [4] and independently by Debye[5]. This type of scattering is referred to as Mie scattering. A more precise delineation of these three types of scattering, particularly for spherical particles, has been given (Ref. 6, Chapter 10).

The scattering by particles at finite concentrations is complicated by the fact that the particles can no longer be considered as being isolated scattering entities. The interferences between wavelets from different particles must also be taken into account. The problem of scattering from a system of particles whose positions are correlated (governed by a radial distribution function) was solved by Zernicke and Prins [7] in connection with the x-ray diffraction of liquids. It should be pointed out that the theory governing the scattering of matter by x-rays or by electrons is identical with that for light scattering in which only phase relations of the scattered wavelets are considered. The angular distribution of scattered intensity is governed entirely by these phase relations. Indeed, Guinier [8, 9] has shown that the size, shape, and interactions of macromolecules in solution can be determined by x-ray scattering. Here, because of the large size of the particles relative to the wavelength of the radiation, the phase relations are most pronounced in the forward direction, that is, observations must be made at very small angles (for recent developments, see Ref. 10).

A totally different approach to light scattering of condensed systems is attributable to Smoluchowski [11]. He considered scattering by a liquid as arising from the granularity caused by thermal fluctuations in local density. This fluctuation is inversely proportional to the compressibility of the fluid, hence the intensity of light scattering should be infinite at the critical temperature. At or very close to this temperature, the scattering is indeed very large (critical opalescence). As Ornstein and Zernicke showed theoretically [12], the character of scattering near the critical temperature is different from that at ordinary temperatures because of the correlations in fluctuation of neighboring portions of the liquid which set in as the critical temperature is approached [12].

For solutions, superposed on the scattering by the solvent is a scattering resulting from the fluctuations in concentration of the solute. Einstein [13] extended Smoluchowski's theory to the case of solutions and showed that the scattering is related to the variation of the thermodynamic activity (hence the osmotic pressure) with concentration of the solute. Thus from the concentration dependence of the intensity of scattering, one obtains thermodynamic values of the solute which by using modern statistical mechanical theories of solutions enables one to ascertain the nature of the interactions of the solute molecules.

The overwhelming amount of the light scattered from a solution illuminated by monochromatic light about the same wavelength, and in this chapter we neglect the small amount of light emitted with altered wavelength when the molecules are raised to higher vibrational states by the incident light, that is, the Raman effect. Examination under high resolution of the light scattered from liquids reveals, nevertheless, a small shift in spectra which can be distinguished from the Raman effect. This fine structure of the scattered radiation was anticipated by Brillouin [14], who showed that because of the presence of thermal (Debye) waves there should be a Doppler shift of the scattered light to give two bands, one of longer wavelength and the other of shorter wavelength than that of the incident monochromatic light. This fine structure of scattered light was first observed by Gross [15], who also observed a central, or undisplaced, band in addition to the symmetrically disposed Brillouin lines. Landau and Placzek [16] ascribed the central band as arising from entropy fluctuations, which vary relatively slowly with time and the side bands that arise from pressure fluctuations. Their theory predicts that the intensity of the central band relative to the intensities of the side bands is governed by the ratio of the specific heats at constant pressure and at constant volume. In many cases their relation is not obeyed. With the advent of the new highly monochromatic and intense light source, the laser, and its employment with the Fabry-Perot interferometer, new data on the fine structure of the Rayleigh line has appeared [17]. An interpretation of such data will, no doubt, influence our current concepts of the structure of actual liquids. An extension of the technique to solutions would certainly be welcomed by chemists.

Coupled with the examination of the spectral shifts is a study of the depolarization of the scattered light. Depolarization arises mainly from the optical anisotropy of the scattering molecules [18]. Depolarization contributes a relatively minor correction to light-scattering formulas, hence has until very recently received little attention. The center line of the fine structure of the Rayleigh line of liquids of anisotropic molecules is depolarized and reveals the relaxation time of the molecules. Indeed, it has been proposed that such a study can be used for the kinetics of fast reactions [19]. Laser light scattered by a suspension of particles exhibits a minute shift in spectra because of the Doppler effect produced by the thermally agitated particles which act as moving sources of light [20]. This Doppler shift effect therefore provides a means of determining the translational diffusion constants for the particles.

A review of all the light-scattering literature up to 1948 has been given [21]. Since that time the field has burgeoned and a number of reviews and monographs have appeared (see General References). A continuing bibliography on light scattering applied to chemistry is available [22]. Of particular interest to chemists is the use of light scattering to characterize the size and shape of synthetic high polymers in solution, work initiated by Debye in 1945 [23]. Much of the theory and methods outlined in this chapter is concerned with light scattering by solutions of macromolecules.

2 OUTLINE OF THEORY

Rayleigh Scattering

There are certain generalities in light-scattering theory that apply to any collection of particles. The intensity of scattering is of course proportional to I_0, the intensity of the incident beam of light. Since light is a transverse disturbance, the intensity of light scattered at an angle θ (the angle made with respect to the transmitted beam) is proportional to the polarization factor $(1 + \cos^2 \theta)/2$. For incident light polarized in the vertical plane, this factor becomes unity independent of angle, and for incident light polarized in the horizontal plane it becomes $\cos^2 \theta$. The distance r between the scattering particle and the point of observation is very large (of the order of a million or more times greater than the size of the source of the scattered light), hence the intensity falls off as the inverse square of the distance. The Rayleigh ratio R_θ is defined as $i_\theta r^2/I_0$ where i_θ is the intensity of scattering at the angle θ. Integration of R_θ over the solid angle yields the turbidity τ or extinction coefficient attributable to scattering, that is,

$$\tau = 2\pi \int_0^\pi R_\theta \sin \theta \, d\theta, \tag{2.1}$$

where τ is defined as

$$I_t = I_0 e^{-\tau l}, \tag{2.2}$$

I_t being the intensity of light transmitted by a sample of path length l.

Rayleigh scattering applies to independent point scatterers of light. For isotropic particles small compared with the wavelength of light λ' (λ' is the wavelength of light falling on the system divided by the refractive index of the medium n_0), then, as Rayleigh first pointed out [1], the light-struck particles may be regarded as Hertzian dipole oscillators whose radiant energy is proportional to the square of their polarizability α and varies inversely as the fourth power of λ'. The precise expression for Rayleigh scattering from ν-independent particles per unit volume is:

$$i_\theta = \frac{8\pi^2}{r^2\lambda'^4}\,\nu\alpha^2 I_0(1 + \cos^2\theta). \tag{2.3}$$

This equation together with (2.1) gives for the turbidity of a Rayleigh scattering system:

$$\tau = \frac{16\pi}{3}\,R_{90} = \frac{8\pi}{3}\,R_0. \tag{2.4}$$

The effective cross section, that is, the power removed per unit intensity from the incident beam, is extremely small for Rayleigh scattering and for an atom is of the order of 10^{-28} cm^2. For a free electron (Thomson x-ray scattering), the cross section is of the order of 10^{-24} cm^2 and for a highly absorbing molecule it is in the neighborhood of 10^{-16} cm^2, that is, nearly the area of the molecule itself. In other words, light scattering by an atom corresponds to electron displacements only one-millionth the diameter of the atom.

Combining the expression from electrostatics relating the polarizability α and the refractive index of the solution n and of the pure solvent n_0 (see Ref. 21, Section II, A, 1) and recalling that the number of particles per unit volume is $\nu = N_0 c/M$, where N_0 is Avogadro's number, c is the weight concentration, and M is the molecular weight, we obtain for Rayleigh scattering:

$$\tau = HcM \qquad \text{where} \qquad H = \frac{32\pi^3 n_0^2}{3N_0\lambda^4}\left(\frac{n - n_0}{c}\right)^2. \tag{2.5}$$

The refractive index increment $(n - n_0)/c$ is practically independent of concentration, hence the turbidity is proportional to the product of the weight concentration and the molecular weight. The turbidity of a polydispersed system of small independent particles is determined by the weight-average molecular weight. Hence in light scattering the higher-molecular-weight components of a polydispersed system predominate.

For green light H is of the order of 10^{-6} (c expressed in grams per milliliter) so that for a 1% solution of molecules of molecular weight 10^4 the optical density of a 1-cm path length (2.303τ) is only 10^{-4}. The optical density is measureable at this concentration for molecules of molecular weight 10^6, but then in order to insure that the particles are point scatterers one must measure the optical density at various wavelengths and extrapolate to infinite wavelength [24] (Fig. 2.1). It is obviously more practical to measure R_{90}, the

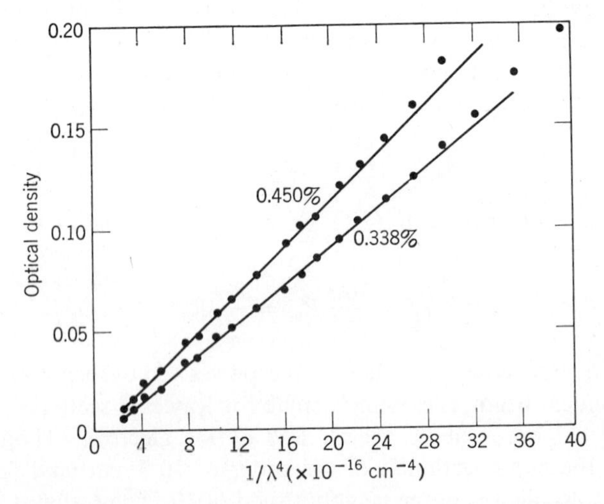

Fig. 2.1 Optical density as a function of the inverse fourth power of the wavelength for bushy stunt virus in water [24].

Rayleigh ratio at 90°. The optical density could of course be increased by increasing the concentration, but then the solution becomes nonideal since the particles are then nonindependent scatterers and the relation (2.5) no longer applies. Still further, at higher concentrations multiple scattering (see below) sets in and the above equations are no longer valid.

Light scattered at 90° from a system of small, isotropic particles is perfectly plane-polarized in the vertical direction since no horizontally polarized component is produced when oscillations take place in the direction of observation. For anisotropic particles, however, in which the polarizabilities along various directions in the particle are not equal, the direction of the electric field associated with the incident light may not coincide with the induced movement of the electron cloud. As a consequence, the light scattered at 90° is not perfectly plane-polarized perpendicular to the plane of the incident beam and the direction of the observation and exhibits a weak component in the horizontal direction.

If we denote the horizontal and vertical components of the intensity of light scattered at 90° by H and V, respectively, and the subscripts u, h, and v to indicate unpolarized, horizontally, and vertically polarized incident light, respectively, then the depolarizations are defined by:*

$$\Delta_v = \frac{H_v}{V_v}, \qquad \Delta_h = \frac{V_h}{H_h}, \qquad \text{and} \qquad \Delta_u = \frac{H_h + H_v}{V_h + V_v} = \frac{H_u}{V_u} \tag{2.6}$$

As a result of the reciprocity theorem of optics (the interchange of source and observer leaves the system unaffected), then $H_v = V_h$, whereupon we obtain the Krishnan relation [25]:

$$\Delta_u = \frac{1 + \dfrac{1}{\Delta_h}}{1 + \dfrac{1}{\Delta_v}}. \tag{2.7}$$

Thus if two of the depolarizations are known, the third can be calculated. This is particularly important in the determination of Δ_h, which is often difficult to measure directly. Values of Δ_u range from 10^{-3} for the rare gases to 0.10 for gaseous carbon disulfide and from 0.01 to 0.05 for low molecular-weight organic molecules in the gaseous state (see list of values in Ref. 17, Appendix III). High polymers are generally isotropic (DNA is a notable exception), but the depolarization of the solvent is usually high, hence renders the value for the polymer experimentally uncertain.

The expressions for the intensity and the turbidity (2.3) and (2.5), respectively, must be corrected for the depolarization by multiplying the relations on the right-hand side by the Cabannes depolarization factors:

$$\frac{6 + 6\Delta_u}{6 - 7\Delta_u} \qquad \text{and} \qquad \frac{6 + 3\Delta_u}{6 - 7\Delta_u}, \tag{2.8}$$

respectively [18].

Debye Scattering

Particles that have a linear dimension comparable to or greater than the wavelength of the incident radiation scatter more light in the forward than in the backward direction (Fig. 2.2). This asymmetry arises from the fact that in the forward directions ($\theta < 90°$) the wavelets scattered from two or more points in the particle are closer in phase than in the backward direction ($\theta > 90°$), resulting in enhancement of intensity in the forward directions and

* Some authors designate Δ_h by the reciprocal of the equation given here. Often the depolarization is designated by ρ.

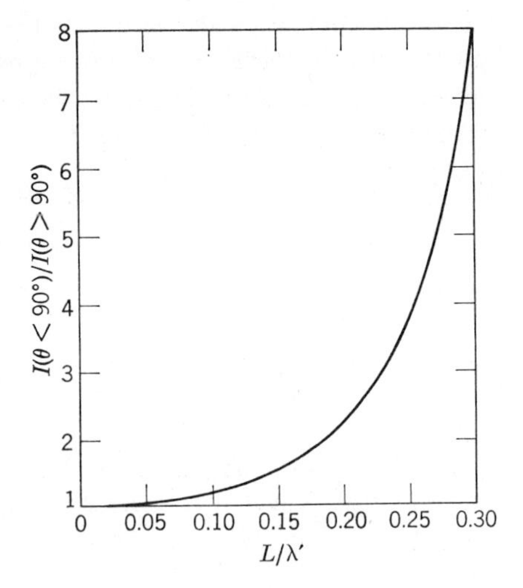

Fig. 2.2 Ratio of intensity of light scattered in the forward directions $I(\theta < 90^\circ)$ to that in the backward directions $I(\theta > 90^\circ)$ for spheres (of radius L) exhibiting Debye scattering.

diminution of intensity in the backward directions. In the limiting case in which the refractive index of the particle is equal to that of the medium, a plane wave falling on the particle is undistorted in its passage through the particle. In this case the mutual interferences of the wavelets can be computed from the relative dispositions of the scattering points. If the refractive index of the particles is exactly that of the medium, no scattering will result of course. For practical purposes, however, if the refractive index of the particle is within 10% of that of the medium, the theory is reasonably valid (see Ref. 6, p. 85).

As Debye has shown [2], the angular dependence of scattering from a large particle is determined by the interferences between paired scattering centers (P and Q of Fig. 2.3). The first problem is to determine the path difference between wavelets scattered from points P and Q at some distance

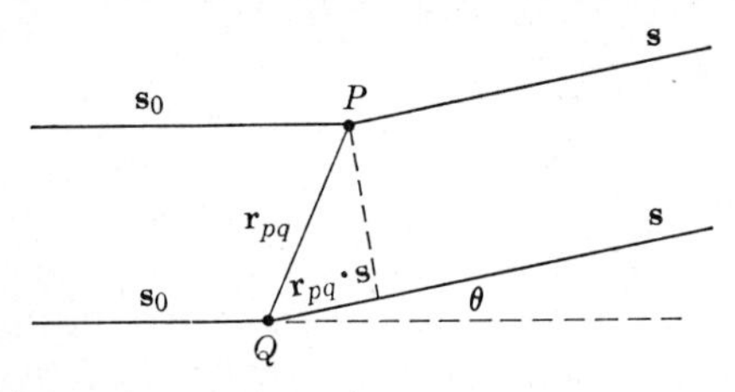

Fig. 2.3 Scattering of a parallel beam of light by two scattering centers P and Q. The resultant path difference is $\mathbf{r}_{pq} \cdot \mathbf{S}$ where $\mathbf{S}$ is the resultant vector of the incident and scattered beams.

large compared with their distance of separation r_{pq}, the phase difference being $2\pi/\lambda'$ times the path difference. If we denote $\mathbf{s}_0$ and $\mathbf{s}$ as the unit vectors defining the directions of the incident and scattered beams, respectively, and $\mathbf{r}_{pq}$ the vector defining the direction of the line joining P and Q, then the path difference is $\mathbf{r}_{pq} \cdot \mathbf{s} - \mathbf{r}_{pq} \cdot \mathbf{s}_0 = \mathbf{r}_{pq} \cdot \mathbf{S}$, where $\mathbf{S}$ is the resultant vector of the two beams hence $|S| = 2 \sin (\theta/2)$. The amplitude of scattering in complex notation is proportional to $\exp [i(2\pi/\lambda)\mathbf{S} \cdot \mathbf{r}_{pq}]$ and the intensity is the amplitude times its complex conjugate. We must consider all N pairs of scattering centers, so that the normalized scattering factor $P(\theta)$ is given by

$$P(\theta) = \frac{1}{N^2} \sum_p \sum_q \exp \left[-\frac{2\pi}{\lambda} \mathbf{S} \cdot \mathbf{r}_{pq} \right], \tag{2.9}$$

where N^2 is the normalization factor. If α is the angle between $\mathbf{S}$ and $\mathbf{r}_{pq}$, then the phase difference is

$$\frac{2\pi}{\lambda} \mathbf{S} \cdot \mathbf{r}_{pq} = \left(\frac{4\pi}{\lambda}\right) \sin (\alpha/2) r_{pq} \cos \alpha = k r_{pq} \cos \alpha,$$

where $k = (4\pi/\lambda') \sin (\theta/2)$. Since the vector $\mathbf{r}_{pq}$ can assume all directions, we integrate over the solid angle. Then, since

$$\frac{1}{4\pi} \int_0^{\pi} \exp (ikr_{pq} \cos \alpha) 2\pi \sin \alpha \, d\alpha = \frac{\sin kr_{pq}}{kr_{pq}}, \tag{2.10}$$

the particle-scattering factor becomes:

$$P(\theta) = \frac{1}{N^2} \sum_p \sum_q \frac{\sin kr_{pq}}{kr_{pq}}. \tag{2.11}$$

For a continuous series of points on a line of length L, the number of pairs of scattering points is proportioned to $L - r$ so (2.9) becomes

$$P(\theta) = \frac{2}{L^2} \int_0^L (L - r) \frac{\sin kr}{kr} \, dr \tag{2.12}$$

$$P(0) = \frac{1}{u} \int_0^{2u} \frac{\sin w}{w} \, dw - \left(\frac{\sin u}{u}\right)^2, \tag{2.13}$$

where $u = kL/2 = 2\pi(L/\lambda') \sin (\theta/2)$. This is the scattering factor for a stiff rod whose thickness is small compared with the wavelength of light [26]. In a similar way the scattering factor for a thin disk of diameter L can be shown to be [27a]

$$P(\theta) = \frac{2}{u^2}\left[1 - \frac{J_1(2u)}{u}\right], \tag{2.14}$$

where $u = 2\pi(L/\lambda') \sin (\theta/2)$ and $J_1(x)$ is the Bessel function of order one.

For a random Kuhn coil of $N - 1 \simeq N$ links the coil may be considered continuous and

$$P(\theta) = \frac{2}{N^2} \int_0^N (N - x) \exp\left(-\frac{\beta k^2}{4}\right) dx, \tag{2.15}$$

where the root mean square end-to-end distance is $L = (3\beta N/2)^{1/2}$. The final result is [23b]:

$$P(\theta) = \frac{2}{u^2} [\exp(-u^2) - 1 + u^2], \tag{2.16}$$

where $u = (2/3)^{1/2} 2\pi(L/\lambda') \sin(\theta/2)$. For the case of a wormlike (Kratky-Porod) particle, $P(\theta)$ is a power series involving a function of the type of (2.16) which reduces to (2.16) for zero persistence length and to (2.13) for infinite persistence length (i.e., a fully stretched coil) [27a, 27b].

For centrosymmetric particles (2.11) becomes:

$$P(\theta) = \left[\frac{1}{N} \sum_i \frac{\sin kr_i}{kr_i}\right]^2, \tag{2.17}$$

where r_i is the distance between the ith scattering element and the center of the particle. If there is a continuous distribution of scattering elements, $P(\theta)$ is given by the square of

$$\frac{\int_0^\infty 4\pi r^2 G(r) \frac{\sin kr}{kr} dr}{\int_0^\infty 4\pi r^2 G(r)\, dr}, \tag{2.18}$$

where $G(r)$ is the radial distribution function of scattering elements in the particle. Thus for a solid sphere of diameter L, $G(r) = 1$ for $0 < r \leq L/2$ and $G(r) = 0$ for $r > L/2$. Then from (2.9) the particle scattering factor is given by

$$P(\theta) = \left[\frac{3}{u^3} (\sin u - 3 \cos u)\right]^2, \tag{2.19}$$

where $u = 2\pi(L/\lambda') \sin(\theta/u)$. Using (2.18) one can readily calculate the particle-scattering factor for spherical particles with internal radial structures [28]. Particle-scattering factors have been evaluated numerically for circular cylinders [29] and for ellipsoids of revolution [29, 30].

In Fig. 2.4 is illustrated the scattering factor as a function of $2\pi(L/\lambda') \times \sin(\theta/2)$. As seen, the more attenuated the particle the slower $P(\theta)$ falls off with angle. A measure of the extent of falling off of intensity with angle of observation is the dissymmetry of scattering which is the ratio of the intensity at some angle θ and its supplement. If the shape of the particle is known, its

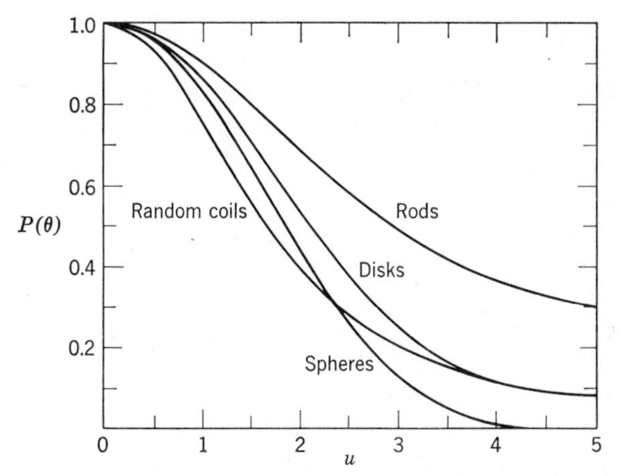

Fig. 2.4 Particle-scattering factor as a function of the size parameter u for various shapes.

dimension is readily obtained from the observed dissymmetry. As seen from Fig. 2.5, the dissymmetry for $\theta = 45°$ is very sensitive to particle size for spheres but is very much less so for rods. Indeed, for rods the dissymmetry becomes independent of particle size for large values of L/λ' and for $\theta = 45°$ approaches the value 2.4 in the limit. More generally, if the observation is made at angles θ and $180° - \theta$, the dissymmetry approaches cot $(\theta/2)$ for rods as seen from (2.13) for large values of u (whereupon the integral

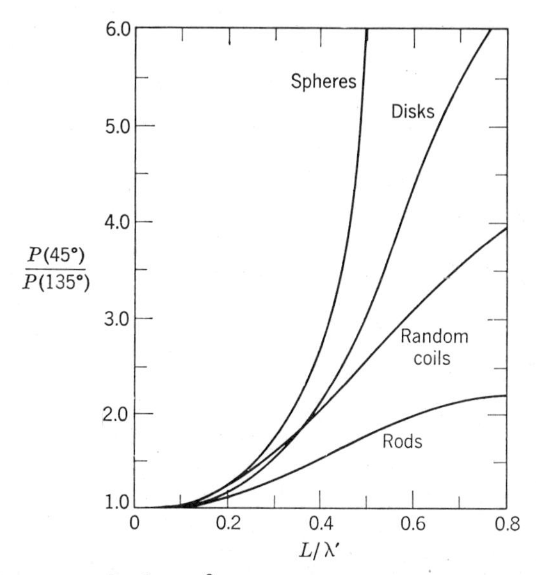

Fig. 2.5 Dissymmetry for $\theta = 45°$ as a function of the size (relative to the wavelength of light) for various shapes.

approaches $\pi/2$ and the other term approaches zero). Similarly for disks and for random coils the dissymmetry approaches $\cot^2(\theta/2)$ and for spheres it oscillates about $\cot^4(\theta/2)$. The shape of the particle can also be determined by measuring the dissymmetry as a function of wavelength. In practice it is simpler to measure the entire scattering envelope at one wavelength and determine which $P(\theta)$ curve is followed.

The actual shape of the particle may not be one of the idealized forms discussed above. If $kL/2$ is smaller than unity we can expand (2.11) in a power series $[\sin(w/w) = 1 - (w^2/6) + (w^4/120) + \cdots]$ to give:

$$P(\theta) = \frac{1}{N^2}\left(N^2 - \frac{k^2}{6}\sum_p\sum_q r_{pq}^2 + \frac{k^4}{120}\sum_p\sum_q r_{pq}^4 - \cdots\right), \tag{2.20}$$

or

$$P(\theta) = 1 - \frac{k^2\rho_2^{\;2}}{3} + \frac{k^4\rho_4^{\;4}}{60} - \cdots, \tag{2.21}$$

where ρ_2 is the second moment of the mass averaged over all directions (i.e., the radius of gyration about its center of gravity, the factor of 2 being obtained from the double summation), ρ_4 is the fourth moment, and so on. Thus the limiting slope of $P(\theta)$ versus k^2 is determined by the radius of gyration of the particle independent of its shape. Here we approximate the power series (2.20) by two terms, namely,

$$P(\theta) = 1 - \frac{(4\pi)^2}{3}\frac{\rho_2}{\lambda'}\sin^2(\theta/2), \tag{2.22}$$

so to this level of approximation the dissymmetry at angles of observation and its supplement becomes

$$\frac{P(\theta)}{P(180^\circ - \theta)} = 1 + \left(\frac{4\pi}{3}\right)^2\left(\frac{\rho_2}{\lambda'}\right)^2 \cos\theta \tag{2.23}$$

In Table 2.1 are given a list of radii of gyration for various-shaped particles.

Table 2.1 Square of the Radius of Gyration for Various Shapes

Sphere of radius R	$\frac{3}{5}R^2$
Spherical shell of outer radius R and of inner radius cR ($c < 1$)	$\frac{3}{5}\left(c^2 + \frac{c+1}{c^2+c+1}\right)R^2$
Infinitesimally thin spherical shell of radius R	R^2
Elongated right circular cylinder of length $2R$ and axial ratio p	$(4 + 3/p^2)R^2/12$
Infinitesimally thin rod of length $2R$	$R^2/3$
Infinitesimally thin circular disk of radius R	$R^2/2$
Ellipsoid of revolution of major axis $2R$ and axial ratio p	$(2 + p^2)R^2/5$
Random coil of root-mean-square end-to-end distance R	$R^2/6$
Gaussian distribution e^{-ar^2}	$3/2a$

For a polydispersed system of random coils, the radius of gyration is the z-average (third moment in weight distribution) radius of gyration [31]. More complicated (mixed) average radii of gyration are obtained for polydispersed systems of particles of other shapes [32]. Graphical methods have been devised for determining the various averages of sizes of polydispersed systems of particles of a single shape [32].

Optical anisotropy of the particles of finite size can have an influence on the scattering envelope. Thus a thin rod of negative birefringence behaves optically like a rod of finite thickness and thereby shows a stronger dependence of scattering on angle than an isotropic thin rod of the same length would. The theory for scattering from thin rods has been extended to include two polarizabilities, one along and one perpendicular to the axis of the rod [33]. The depolarization for anisotropic rods is dependent on angle and at 90° the value of the H_h component, for example, depends both on the length of the rod and on the ratio of the two polarizabilities. Random coils containing anisotropic links are not, however, overall anisotropic, hence do not exhibit a depolarization of scattering [34].

Because of the asymmetry of scattering by particles of finite size, the intensity of scattering at 90° is less than that for a Rayleigh scatterer of the same molecular weight. Hence the scattering envelope (or the dissymmetry, if the shape of the particle is known) must be ascertained together with the turbidity in order to determine the molecular weight. For a dilute solution of large particles, the equation for the turbidity, replacing (2.5) for point scatterers, becomes

$$\tau = \tfrac{3}{8}HcM\int_0^{\pi} P(\theta)(1 + \cos^2 \theta) \sin \theta \, d\theta. \tag{2.24}$$

The Rayleigh ratio may be expressed in terms of the particle-scattering factor as

$$R_\theta = KcMP(\theta)\left(\frac{1 + \cos^2 \theta}{2}\right), \tag{2.25}$$

where $K = (16\pi/3)H$, H being defined in (2.5).

Scattering at large angles is also of some interest. In the case of spheres [see (2.19)], the particle-scattering factor falls off as $1/u^4$ for large values of z, hence as $(k/L)^4$. Rayleigh scattering is proportional to the square of the polarizability (hence proportional to L^6) and to the inverse fourth power of the wavelength, so the intensity of scattering for large spheres is proportional to L^2. Therefore the surface area of spheres can be determined from large-angle scattering [35, 36]. For a polydispersed system of random coils, $P(\theta)$ at large angles is determined by the number-average molecular weight of the particles [31], so that combining the small-angle data with that of the large-angle data one obtains information regarding the size distribution of the particles in the system.

The turbidity of very large particles does not obey the inverse fourth power of wavelength relationship (see Fig. 2.1). In general, the turbidity is proportional to $\lambda^{-4+\beta}$, where β equals zero for a Rayleigh scatterer ($u \to 0$). For very large particles (u large), β equals 1.0, 1.74, and 2.0 for rods, coils, and spheres, respectively. These results may be shown analytically by solving (2.24) when $P(\theta)$ is given by (2.14), (2.16), and (2.19), respectively, for large values of u [36]. Values of β for intermediate values of u for these three types of particles have been determined numerically [37].

Mie Scattering

The scattering of large particles of any value of the relative refractive index is of considerable practical importance in the field of atmospheric visibility and in the field of microwave radio transmission through fogs. It has been deduced from the light scattered by "black clouds" in nebulas that much of the matter of the universe is in the form of colloidal particles [38]. The theory of scattering by spheres initiated more than a half-century ago by Mie [4] and Debye [5] has remained unchanged because of its completeness but has since been applied by great numbers of investigators [6, 39–41]. Unfortunately, the exact solution cannot be expressed in closed form, and the terms of the series expressions are themselves complicated functions. The radius R of the sphere relative to the wave length λ' of light in the medium is expressed as the parameter $x = 2\pi R/\lambda'$. The refractive index of the particle divided by the refractive index of the medium is written as m. The intensity of scattering is a function of x, m, and θ. The intensities of the vertically and the horizontally polarized components of the scattered light, i_1 and i_2, respectively, are given by [42]:

$$\begin{aligned} i_1 &= \left(\frac{\lambda'}{2\pi r}\right)^2 \left| \sum_{n=1}^{\infty} [A_n \pi_n(\cos\theta) + B_n \tau_n(\cos\theta)]^2 \right| \\ i_2 &= \left(\frac{\lambda'}{2\pi r}\right)^2 \left| \sum_{n=1}^{\infty} A_n \tau_n(\cos\theta) + B_n \pi_n(\cos\theta) \right|, \end{aligned} \tag{2.26}$$

where A_n and B_n, the scattering coefficients, are complex numbers and are involved functions of x, m, and fractional [order $n + (1/2)$] Bessel functions of x and mx. The angular function π_n is the first derivative, with respect to $-\cos\theta$, of the nth-order Legendre polynomial, and τ_n is an algebraic function of $-\cos\theta$ and the first and second derivatives of the nth-order Legendre polynomial. The intensities are the squares of the complex amplitudes, which differ in phase depending on the value of θ. For very small values of x (i.e., Rayleigh scattering) (2.26) reduces to:

$$i_1 = \left(\frac{\lambda'}{2\pi r}\right)^2 \left(\frac{m^2 - 1}{m^2 + 2}\right) x^6$$

and

$$i_2 = \left(\frac{\lambda'}{2\pi r}\right)^2 \left(\frac{m^2 - 1}{m^2 + 2}\right)^2 x^6 \cos^2 \theta, \tag{2.27}$$

which for $i_1 + i_2$ is the Rayleigh expression [see (2.1)] since the polarizability of a sphere of relative index of refraction is:

$$\alpha = \frac{m^2 - 1}{m^2 + 2} R^3. \tag{2.28}$$

It is convenient to define the term extinction efficiency Q, which is the turbidity per particle cross section. Q is calculated by inserting (2.26) into (2.2) and dividing by the cross-sectional area of the spheres, namely, πR^2. For very small values of x, the extinction efficiency becomes:

$$Q = \frac{8}{3} x^4 \left(\frac{m^2 - 1}{m^2 + 2}\right)^2, \tag{2.29}$$

which is proportional to the turbidity for Rayleigh scattering. When x is large but m is close enough to unity so that $2x(m - 1)$, the change of phase of a light ray passing through the sphere along a diameter, is very small (i.e., Debye scattering), then the extinction coefficient is given by [43]:

$$Q = 2x^2(m - 1)^2. \tag{2.30}$$

Hence for Debye scattering by spheres the turbidity is proportional to the fourth power of the radius and the inverse square of the wavelength, as mentioned earlier. For extremely large values of x and arbitrary values of m (the geometrical optics region), $Q \rightarrow 2$, a surprising result since it means that a very large particle removes from the incident beam twice the amount of light it can intercept. This result is not strange, however, when it is recalled that all the scattered light including that at small angles is counted as removed from the beam and that the observation is made at a very great distance beyond which a shadow can be distinguished (see Ref. 6, Chapter 8). The transition from physical to geometric optics of scattering by an object of rotational symmetry has been reexamined in recent years, notably by Keller [44], who uses the Young interference approach coupled with the newer asymptotic expansion techniques. Incidentally, an elementary demonstration of the connection between physical and geometrical optics is vividly carried out using moiré patterns [45a].

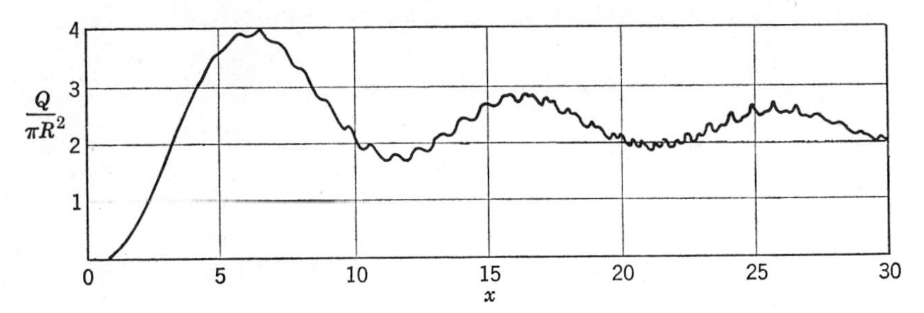

Fig. 2.6 The extinction coefficient (per unit cross-sectional area) for spheres of refractive index 1.33 as a function of the size parameter x. (Redrawn from Ref. 45b.)

Intermediate values of the phase shift $2x(m - 1)$ provide the most troublesome cases for the computation of Q, and many terms of (2.26) must be taken into account. These stupendous computations lead to results such as those shown in Fig. 2.6. With increasing x, Q starts as proportional to x^4 and undergoes oscillations (the small subsidiary maxima are real) about the value 2. The power of x likewise undergoes oscillations (roughly the derivative curve of the Q-versus-x curve) and is even negative for certain values of x. For example, it is -2 in the neighborhood of $x = 5$ for the case given in Fig. 2.5, which means that the scattering actually increases with increasing wavelength, a feature which no doubt contributes to the bizarre colors that sometimes occur in sunsets and are readily seen in monodispersed sulfur sols [39, 46]. The situation is further complicated if the refractive index of the medium is varied. Thus in Fig. 2.7 a system of spheres with size parameter $x = 13$ exhibits more turbidity when the refractive index of the medium is closer to that of the particles, although with the smaller size range (x less than 8) the reverse is the case.

The angular scattering pattern as computed from (2.26) becomes increasingly more complicated the greater the value of the phase shift $2x(m - 1)$. The convergence of the series may be slow, and as many as 20 terms (for a given value of x, m, and Q) must be taken into account. Obviously, such computations are best carried out with high-speed computers. Tables of these computations [47], although extensive, may still not be sufficiently detailed for some particular values of m, x, and Q. Computer programs for these calculations are now available [48]. It is often difficult to interpolate between computed values because most scattering patterns show wild intensity fluctuations. Phases as well as intensities are contained in this formulation since the Mie equation [see (2.26)] is given in terms of the complex amplitude. Fortunately, precisely in the size range of particular interest to most colloid chemists, namely, $1 < x < 10$, interpolations using the phases can safely be made [49].

The Mie theory also predicts that spherical particles, although isotropic,

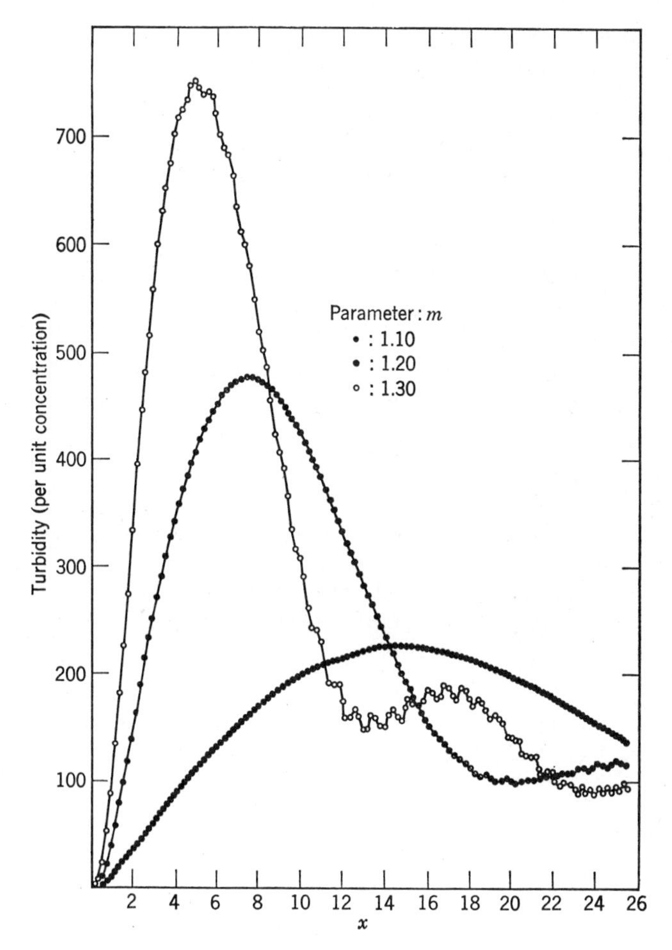

Fig. 2.7 Variation of turbidity per unit concentration for spheres of various relative refractive index as a function of the size parameter [40].

will exhibit a depolarization at 90°. In addition to V_v differing from zero, H_h and V_h will both be zero. Hence Δ_v and Δ_h will be zero but Δ_u will differ from zero [50]. Indeed, the depolarization at 90° is a sensitive measure of the size of the spheres if the particle parameter x is greater than about 0.5 and less than about 2 for the case of sulfur sols ($m = 1.44$) [51], although for $m = 1.20$, x must be at least 2 for the depolarization to be appreciable [52].

A beam of uniform spherical particles exhibits a corona about the central beam such as that seen about a lamplight on a misty evening. The size of the ring is smaller the greater the size of the particle. For very large particles ($x > 10$), when the corona appears at small angles ($\theta_{max} < 10°$), the theory is complicated by anomalous interference effects (see Ref. 6, Section 13.41).

For smaller particles (x about 2 to 6), it can be shown from the Mie theory that the size parameter x is related to the angle θ_{max} of the corona by [53]

$$x \sin \theta_{max} = p, \tag{2.31}$$

where $p = 3.45$ for $m = 1.40$–1.60 and $p = 3.92$ for $m = 1.33$. From classic diffraction theory, an opaque disk would exhibit its first bright ring (Airy disk) according to (2.31) with $p = 5.25$. The formula predicts that a fine mist of water particles of diameter L would show a corona at $\theta_{max} = \sin^{-1}(1.25\lambda/D)$. Uniform latex particles (of polyvinyltoluene) dried down from extremely dilute aqueous suspensions to give a slight haze on a glass slide exhibit coronas obeying (2.31) [53] (here $n_0 = m = 1.58$), where now $\theta_{max} = \sin^{-1}(1.10\lambda/D)$. Samples of latex particles dried down from more concentrated suspensions exhibit an iridescence attributable to diffraction from two-dimensional [54] or even three-dimensional [55] structures which these uniform size particles readily form. Likewise the iridescence exhibited by solutions of tobacco mosaic virus [56], and crystals of this virus [57] or some other viruses [58], is attributable to Bragg diffraction of visible light since the lattice spacings are comparable with the wavelength of light.

In his original paper [4], Mie considered the case in which m is large and complex. This is applicable to metallic particles that are electrically conducting or absorbing. Debye [5] showed that when m is infinite the problem reduces to one of modes of oscillation in a conducting sphere, a result of great interest in microwave theory [42]. If the particles are considered to have the same optical characteristics as a macroscopic gold surface, then the

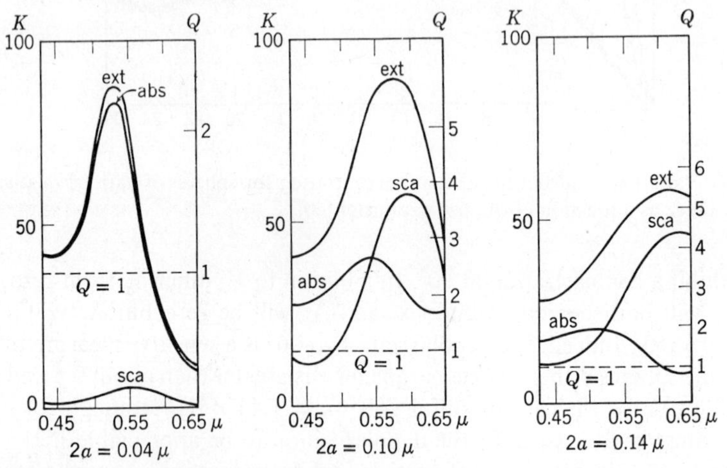

Fig. 2.8 Extinction coefficients according to the Mie theory for spherical gold particles of radius 20 nm (red sol), 50 nm (purple sol), and 70 nm (blue sol) as a function of wavelength (in microns). The total extinction coefficient Q_{ext} is the sum of that for absorption Q_{abs} and that for scattering Q_{sca}. K is Mie's notation and equals Q times $\frac{3}{4}$ divided by the radius [6].

extinction coefficients as a function of the wavelength of incident radiation are as shown in Fig. 2.8. The resultant transmission color of the gold sol when illuminated with white light varies with particle size. As the particle size increases, the color shifts from red ($2a = 0.04\ \mu$), to purple ($2a = 0.10\ \mu$), and then to blue ($2a = 0.14\ \mu$). The Mie theory applied to gold sols has been completely verified experimentally [59].

Nonindependent Particles

The scattering per molecule for a substance in the liquid state is considerably less than that in the gaseous state. Similarly, for a nonideal solution, the scattering per solute particle is less than that calculated for an ideal solution of the same concentration. As a result of intraparticle interferences, the intensity of scattering of independent particles is reduced by the Zernicke-Prins factor [7]:

$$1 - \nu \int_0^\infty 4\pi r^2 [1 - g(r)] \frac{\sin kr}{kr}\, dr, \tag{2.32}$$

where $k = 4\pi/\lambda' \sin(\theta/2)$, ν is the number of particles per unit volume, and $g(r)$ is the radial distribution function, the particles being considered spherically symmetrically distributed in space. This introduces an angular dependence of the scattering even for point scatterers. For particles of finite size, the angular dependence of the scattering is governed by the product of the particle-scattering factor $P(\theta)$ and the Zernicke-Prins factor. Because of interparticle correlations, the scattering is reduced in the forward direction, the effect becoming more pronounced as the concentration is increased (Fig. 2.9). For dilute solutions this produces a lower dissymmetry than that expected for infinite dilution. If only first nearest neighbors of spheres are considered, then the dissymmetry will be given by [36]

$$\frac{I(\theta)}{I(180° - \theta)} = \left[\frac{I(\theta)}{I(180° - \theta)}\right]_{\nu=0} \left[1 - \nu \frac{(4\pi)^3}{30} \frac{L^{*5}}{(\lambda')^2} \cos\theta\right], \tag{2.33}$$

where L^* is the effective diameter of the spheres. The effective diameter of the particles may be greater than the actual diameter L in the case of ionically charged colloidal particles. We note that for hard spheres of diameter L the van der Waals covolume $b = 2\pi L^3/3$ and for elongated cylinders of length L and diameter d, $b = \pi d L^2/4$. Since for spheres the ratio of b to the volume of the particles is independent of the size of the particles, whereas for elongated cylinders the ratio is L/d, we expect interparticle correlation to set in at lower concentrations for rods than for spheres [36, 56].

The thermodynamic properties of a pure liquid or of a solution are governed by the radial distribution function $g(r)$ and the interparticle forces. The theory of fluctuations enables one to directly relate the thermodynamic properties of the system and the turbidity without recourse to a knowledge of the molecular arrangements or the nature of the intermolecular forces. The

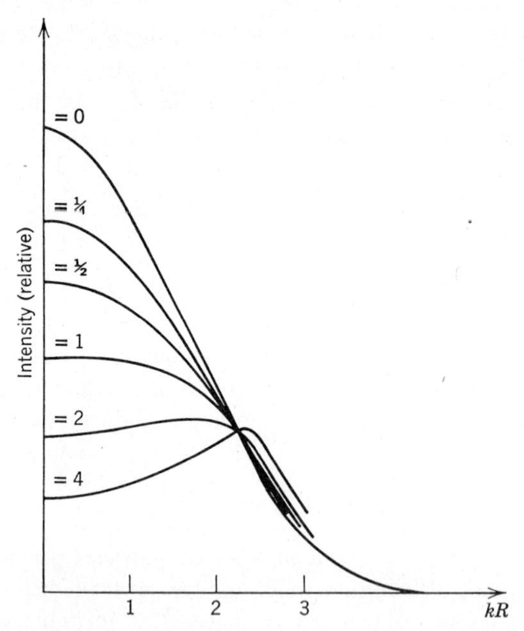

Fig. 2.9 Debye scattering of spheres (of radius R) as a function of kR for various values of the relative volume parameter (proportional to the concentration of spheres) [60].

molecules in a fluid are not distributed uniformly, but rather in any small volume the number of particles varies from instant to instant as a consequence of their thermal motions. This local fluctuation in density of the particles gives local inhomogeneities in the index of refraction and scatters light. The elementary theory of fluctuations (see, for example, Ref. 21) shows that the average of the square of the deviation of some thermodynamic quantity x from its average value x is given by:

$$\overline{(x - \bar{x})^2} = \frac{kT}{\left(\dfrac{\partial^2 A}{\partial x^2}\right)_{x=\bar{x}}}, \tag{2.34}$$

where k is Boltzmann's constant and A is the Helmholtz free energy. Still further, one can show that the average value of the product of deviations of two independent quantities is zero. The fluctuation in refractive index of a pure liquid is proportional to the fluctuation in density (the temperature dependence being negligible), which in turn is proportional to the compressibility [11]. The Smoluchowski theory gives for the intensity of scattering (relative to the incident intensity I_0) for a liquid at temperature T, density ρ, and compressibility β

$$\frac{i_\theta}{I_\theta} = \frac{\pi^2}{2\lambda^4 r^2}\left(\rho\,\frac{\partial\epsilon}{\partial\rho}\right) kT\beta(1 + \cos^2\theta), \tag{2.35}$$

where ϵ is the dielectric constant ($\epsilon = n^2$) and k is Boltzmann's constant. Taking the compressibility for an ideal gas one obtains the Rayleigh expression for the intensity, namely (2.3). For liquids there is some question about the relation of the refractive index to the density. The electric field associated with the light is not that applied locally to the molecules. Experimentally [62], it is found that in order to account for the absolute turbidity of liquids using the Smoluchowski formula relating turbidity and compressibility [11] the local field lies between the Clausius-Mossotti and Onsager theories. Water is somewhat anomalous and this may be a result of local correlations in dipole orientation which are known to be particularly great in the case of liquid water [63].

For solutions the fluctuations in concentration likewise contribute to the turbidity (in excess of the pure solvent) as Einstein showed [13]. The second derivative with respect to concentration of the Helmholtz free energy is simply related to the first derivative with respect to the osmotic pressure P of the solution (Ref. 21, Section III, 2) to give for the turbidity:

$$\tau = H \frac{RT_c}{\left(\dfrac{\partial P}{\partial c}\right)_T} . \tag{2.36}$$

Equation (2.36) becomes identical with (2.5) in the case of an ideal solution. The osmotic pressure of a nonideal solution may be expressed in terms of a power series in concentration. For sufficiently dilute solutions only the first two terms may be considered:

$$P = \frac{c}{M} RT + Bc^2, \tag{2.37}$$

where R is the gas constant and B is a measure of the deviation from ideality. B is proportional to the so-called second virial coefficient, hence to the van der Waals covolume b. To this approximation (2.36) becomes on rearrangement:

$$H \frac{c}{\tau} = \frac{1}{M} + \frac{2Bc}{RT} . \tag{2.38}$$

A plot of Hc/τ versus c gives a straight line whose slope is determined by B and whose intercept is the reciprocal of the molecular weight (weight-average molecular weight) of the solute. For large particles at finite concentrations [but neglecting concentration terms of higher order than c^2 as in (2.37)], one obtains an equation analogous to (2.38), namely,

$$\frac{Kc}{R_\theta} = \frac{1}{MP(\theta)} + 2Bc. \tag{2.39}$$

Extrapolation of Kc/R_θ to both zero angle and zero concentration gives M^{-1}.

A general procedure attributable to Zimm [64a] is to plot Kc/R_θ against $\sin^2(\theta/2) + k'c$, where k' is an arbitrary constant chosen to space the data conveniently and to facilitate the extrapolations to zero angle and concentration. Then, as seen from (2.39), the data falls in a grid of parallel lines (Fig. 2.10) since the concentration dependence and angular dependence of Kc/R_θ are independent of each other (see Ref. 32 for further details).

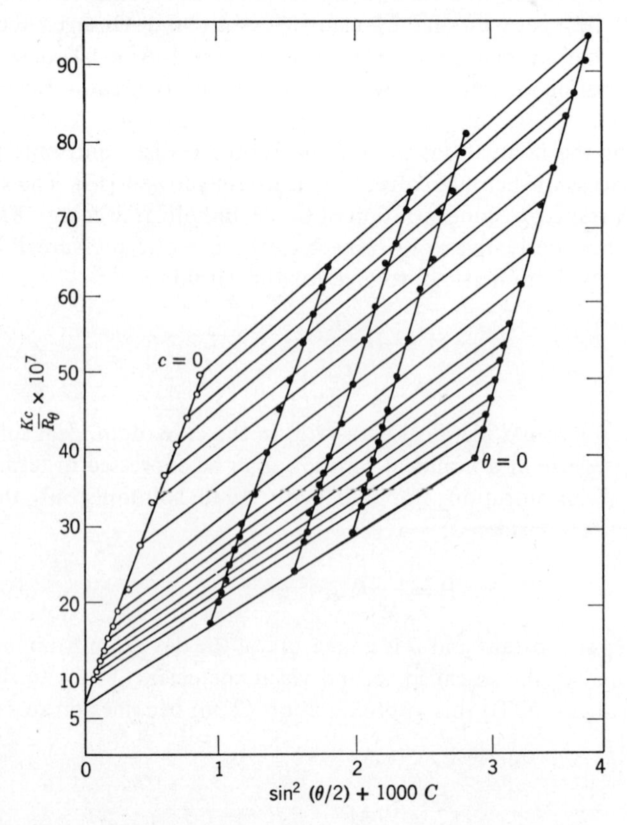

Fig. 2.10 Zimm plot for a solution of a fraction of cellulose nitrate in acetone [64b]. The left-hand edge of the grid gives the values on extrapolating to zero concentration, and the lower edge on extrapolating to zero angle.

Fluctuation theory has also been applied to the scattering of light at an interface [65]. The surface of a liquid has a microscopic roughness because of fluctuations in position of the molecules perpendicular to the surface. The fluctuations, hence the intensity of scattering, is inversely proportional to the surface tension. For two immiscible liquids in contact, the intensity of scattering is inversely proportional to the interfacial tension, and indeed light scattering provides a means of determining this quantity. For this two-dimensional structure, the intensity of scattering is inversely proportional to the square of the wavelength of incident light and this accounts for the whitish appearance of the interface between two immiscible liquids. In the detailed theory [66] extrapolation to smooth surfaces yields the classical Fresnel relations for reflection and refraction at interfaces. The theory also accounts for polarization effects [66]. Agreement with experiment and theory is quite good [67]. Interest in the scattering from free surfaces has been quite limited since surface tension measurements of liquids can be made by more convenient means. However, scattering by soap films [68] can be quite revealing, as should be studies of the interfaces between immiscible liquids since they provide models for biological membranes. The scattering from an interface as the critical temperature is approached increases owing to a lowering of the surface tension, and this has been observed for liquid carbon dioxide [69] and with liquid mixtures [70].

Most of the interest in critical phenomena in liquids is concerned with bulk properties. The basic theory in this area is attributable to Ornstein and Zernicke [71], who reasoned that close to the critical temperature there would be a correlation in fluctuation of density in neighboring regions of the liquid. The correlation falls off with distance and, if it is assumed to fall off exponentially, we define a correlation distance a which is some measure of the extent of intermolecular force. The scattering intensity is now given by a modification of the Smoluchowski [see (2.35)], where now it is inversely proportional to $1/\beta$ plus a term proportional to the square of $2\pi a/\lambda \sin(\theta/2)$. Close to the critical temperature, $1/\beta$ is small, so that the intensity of scattering is inversely proportional to the square of the wavelength of light. This accounts for the milky appearance of fluids close to the critical temperature. Debye [72] has analyzed the data [73] on the angular dependence of scattering of fluid mixtures near the critical temperature and showed that a plot of i_θ versus $\sin^2(\theta/2)$ gives a straight line whose slope should be determined by the distance a of intermolecular force. Critical solutions of polymers can be treated in a similar manner, but now the particle-scattering factor $P(\theta)$ must be included in the formula [74]. Several attempts have been made to test the Ornstein-Zernicke theory very close to or at the critical temperature, but the measurements are fraught with practical difficulties including the instability because of gravity, slow changes over days, and local heating by the light source [75].

The turbid appearance of many plastics is attributable to inhomogeneities in density of the crystalline and amorphous regions. The density correlations of the frozen-in inhomogeneities may be determined from the angular dependence of the light scattering [76] and have been the subject of detailed investigations by Stein [77]. The refractive index difference between amorphous and crystalline regions is small, so that Debye scattering may be assumed to apply. If, as an approximation, the crystalline regions are taken as spheres, then from (2.19) $P(\theta)$ is zero for $u = 4.49$ when the scattering pattern falls to zero at an angle θ_{min} governed by the equation $4\pi R\lambda' \sin(\theta_{\text{min}}/2) = 4.49$, where R is the radius of the crystallite. This formula serves only to give a rough estimate of crystallite size. As a next approximation, the position of the spherical domains may be considered in relation to their first nearest neighbors [see (2.33)] or, more generally [78], in terms of an interparticle radial distribution function [see (2.32)]. Microscopical examination of many highly crystalline plastics reveals that the crystallites consist of spherulites whose density varies as one goes out from the center. Hence the variable intraparticle radial distribution function [see (2.18)] should be used to calculate the scattering of isolated spherulites [78]. The spherulites consist of radially symmetric sheaths of anisotropic material and, as a consequence, exhibit interesting patterns under polarized light. Indeed, such patterns are readily seen about the transmitted beam of a helium-neon laser beam passing through the plastic and using a polarizing dichroic filter. The horizontally polarized component of the incident light is accelerated in two diagonally apposed quadrants and retarded in the other two quadrants. As a consequence, the H_v scattering pattern should exhibit a cloverleaf pattern [79] (Fig. 2.11). The technique has been used to study

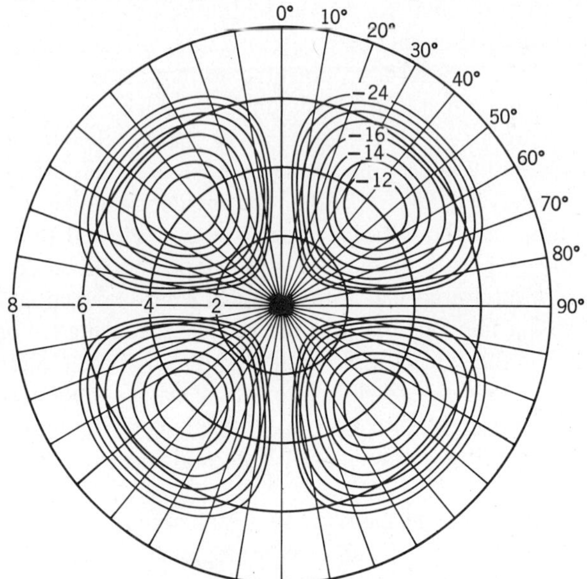

Fig. 2.11 Polar contour plot of calculated *Hv* scattering from anisotropic spheres where the tangential polarizability is greater than the radial polarizability [80].

deformations under stress which give the cloverleaf pattern distorted by an applied affine transformation [80]. The growth of crystallites from polymer melts can also be followed by light scattering [81].

The time course of crystallite formation in the case of, for example, the reaction of halide ions with soluble silver salts has often been followed by light scattering using a simple turbidimeter but not, unfortunately, by angular scattering measurements. The kinetics of coagulation and of polymerization can be analyzed from light-scattering data obtained as a function of time. In the comparatively rare cases in which each particle grows uniformly, so that the system remains monodispersed, the methods described earlier can be employed. Mie scattering of the uniform slow growth of colloidal sulfur particles has been studied [39]. For random coagulation, wherein any particle can coalesce with any other particle, the problem is more complicated. If a coagulating system exhibits Rayleigh scattering throughout the time of examination, then the turbidity is given by:

$$\tau = A \sum_j j^2 \nu_j, \tag{2.40}$$

where ν_j is the number of particles having a size j times the volume of the starting particles and A is a constant independent of time. Equation (2.40), which merely states that for small, independent particles the turbidity is proportional to the number of particles and the square of their volume, has been solved for typical coagulation and polymerization processes by using the appropriate time dependence of ν_j [82]. For rapid coagulation or for condensation polymerization, the turbidity increases linearly with time and the slope is proportional to the elementary rate constant times the square of the initial concentration of the reactants. For addition polymerization and for nucleation coagulation, the turbidity varies quadratically with time. In random depolymerization the turbidity decreases with the hyperbolic cotangent of the time. When the aggregates grow so large that the turbidity is determined only by the cross-sectional area of the particles, then, as can be shown with the use of (2.40), the turbidity for a randomly aggregating sample is inversely proportional to the time. The effect of the increase in cross section is offset by the decrease in number of particles, the latter having a stronger time dependence than the former. As a consequence, starting from small aggregating particles, the turbidity first increases with time and then decreases.

Changes in dissymmetry provide a convenient means of following coagulation. This method was employed in studying antibody-antigen reactions [83]. Serological tests, which are of considerable importance in clinical medicine, can be rendered far more sensitive by using uniform latex particles as a carrier [84]. The light-scattering effect is particularly noticeable on viewing at angles close to the transmitted beam.

Fine Structure of Scattering

In the Debye theory of specific heat of solids [85], the thermal vibrations of the atoms are considered as sound waves. These thermal elastic waves are related to the density fluctuations of the Smoluchowski theory of scattering [86]. Brillouin [14, 86] predicted a shift in wavelength of scattered light attributable to diffraction of the incident light by the thermal waves. For a liquid the scattered light consists of a doublet whose components are shifted from the wavelength λ of the incident light by $\Delta\lambda$, the Brillouin doublets, where

$$\Delta\lambda = \pm 2\lambda_0 \frac{v}{c} \sin(\theta/2), \tag{2.41}$$

where v and c are the velocity of sound and of light, respectively, in the medium. As seen from (2.41), the spectral shift $\Delta\lambda$ depends on the angle of observation. Indeed, to determine $\Delta\lambda$ to within $\pm 1\%$ at 90° the angle θ must be set within $\pm 1°$. Sound velocities of liquids are in the neighborhood of 1×10^5 cm/sec and light velocities are about 10^5 times as great, so that for red light the wavelength shift is only a few angstroms (frequency shift of about 0.1 cm^{-1}). Obviously, to observe the Brillouin doublets a highly monochromatic light source is required, such as now is readily available in the form of the helium-neon laser. The Brillouin components are not infinitely sharp because of damping of the elastic waves. The lines have a Lorentzian profile with a half-width proportional to the amplitude damping coefficient of the sound wave [87]. With extremely high light intensities, such as achieved with gigawatt pulse lasers, the fine structure of scattering has many components. This has been attributed to the shock waves produced by the electrostriction of the liquid caused by large electric fields associated with such intense illumination (for review see Ref. 17, Chapter 10).

So far, we have considered scattering of liquids as arising only from density fluctuations. In addition, there is a contribution attributable to entropic fluctuations. These fluctuations manifest themselves as a central (undisplaced) line in the fine structure [15, 16]. In the theory of Landau and Placzek, the ratio of the intensity I_C of the central line to that of the two Brillouin lines $2I_B$ is determined by the ratios of the isothermal and the adiabatic compressibilities or, equivalently, by the ratios of the heat capacities C_p and C_v, namely,

$$\frac{I_C}{2I_B} = \frac{C_p - C_v}{C_v}. \tag{2.42}$$

The central component for many liquids is more intense than that predicted from (2.42) [88, 89]. The Landau-Placzek theory neglects the dispersion of the thermodynamic properties. The Brillouin components are generated by high-frequency sound waves, while the central component results from relatively slow disturbances. Liquids such as carbon disulfide show a large dispersion in the high-frequency range. This dispersion factor [90] explains the relatively high values of the central band observed for this liquid (reviewed in Ref. 17, Chapter 6). Despite this correction the observed values of the ratio for acetic acid and for benzene are still too high (see Ref. 91 for discussion).

Radiating atoms in the gaseous state exhibit a line broadening as a result of the Doppler shift, which is two orders of magnitude greater than the natural broadening. In solution the scattering by the solute molecules in thermal motion should exhibit a Doppler line broadening. The effect should be very small, however, because of collisions of the solute molecules with the solvent molecules. The elementary theory of the Doppler scattering effect for solutions [92, 93] shows that the broadening is proportional to the linear diffusion constant of the solute molecules times the square of sin $(\theta/2)$. To observe this broadening for macromolecules in solution, an optical resolving power of about 10^{14} is required, that is, a resolution of 10 cps. This has been achieved only in recent years using a laser in conjunction with optical heterodyne detection techniques (see below). By this means the translational diffusion constant of uniform latex particles was measured [20]. The technique has an advantage over other light-scattering methods in that it is not sensitive to the presence of dust and other slowly diffusing particles which may be present [93]. The method can be extended to the determination of rotational diffusion constants of elongated particles [92, 93].

Multiple Scattering

If a scattering system is highly dense (optical density for a 1-cm path length greater than about 0.1, corresponding to τ about 0.2), then a new complication sets in which had not previously been considered. Now, the light scattered from one particle illuminates another particle further in the sample (secondary scattering), which in turn serves as the light source for still another particle, and so on (multiple scattering). The net effect is that more light is scattered in the back direction regardless of the optical characteristics of the individual particles. That is, the angular scattering of a system exhibiting multiple scattering has no relation to that expected for a highly transmitting system and all the theories described earlier are not applicable. For example, a dense collection of uniform spheres, owing to multiple scattering, exhibits lower dissymmetry than that expected for isolated spheres of the same size. Furthermore, maxima expected from the Mie theory for such particles are washed out [94].

The polarization relations are upset even if the slightest amount of secondary scattering occurs (see Ref. 21, Section II.B.4). This effect renders polarization observations from macromolecular systems difficult to interpret in terms of the characteristics of the individual particles unless the measurements are carried out in extremely dilute solution, but then the measurements are subject to error. The blue of the sky is clearly interpretable in terms of Rayleigh scattering of isolated particles, but the polarization distribution of the sky is explainable in terms of multiple scattering (see Ref. 95, Chapter 10).

The scattering pattern of a multiple-scattering system is also dependent on the size l of the scattering system. Obviously, the smaller the path length the less the chance for scattering from one portion of the sample to act as a light source for another portion. One test for multiple scattering is that the angular scattering characteristics of the system decrease with decreasing size of the sample. Another indication of multiple scattering is that the scattering becomes more pronounced in the forward direction as the sample is diluted, but this could also arise from interparticle interference as discussed earlier. For dilute systems the concentration dependence of multiple scattering is greater than that for interparticle correlation effects unless one is dealing with critical solution phenomena.

The mathematical difficulties associated with multiple scattering are formidable. One approach is to treat the scattering as a problem of diffusion in which the scattered photons are undergoing a random walk. There has been a considerable effort to solve the mathematical problem of multiple scattering [95–97], but the results so far are in too complicated a form for ready application to cases of interest for chemists. Perhaps the best practice in the laboratory is to avoid the phenomenon. This can be achieved by diluting the sample and/or reducing the size of the illuminated volume. When faced with the problem of particle size determination in, for example, the atmosphere of Venus or a biological tissue, the only correct approach is to study the light-scattering characteristics of a small sample.

3 INSTRUMENTATION

Survey of Instrumentation

All light-scattering instruments contain as their basic components a light source, a sample holder, and a light detector. The relative disposition of these elements and the use of accessories (collimators, diffusers, and so on) requires careful design if the measurements are to be of any significance.

Light-scattering instruments in common use fall into three categories. There are angular light-scattering photometers that measure relative intensities at various angles; these instruments are also capable of measuring absolute intensities of scattering. Transmission ($\theta = 0°$) measurements can be carried out in these instruments, but spectrophotometers with long path length have frequently been used for this purpose. Then there are photometers that measure scattering mainly in the forward directions and therefore are particularly useful for large particles. Individual large colloid particles can be counted under dark-field illumination and automatically recorded by electronic devices. Finally, there are the miscellaneous varieties of turbidimeters. Perhaps the most poorly designed instruments fall into this last category. In a special class are instruments used to determine the fine structure of the scattered light. As mentioned earlier, this involves the use of the helium-neon laser and of spectral analysis instruments capable of extremely high resolution.

There are instruments designed specifically for the measurement of depolarization. Such measurements can, however, be made with commercially available angular scattering photometers which are supplied with holders for dichroic polarizing filters. There are several other accessories which can be added to an angular light-scattering instrument, such as holders for flat samples and thermostats for the samples. Because of the adaptability of the angular scattering photometers to nearly all light-scattering problems, they have been widely adopted in laboratories. However, for some particular problem such as smoke control, less universal but more convenient instruments may be preferred.

An essential parameter in the light scattering by solutions is the relative refractive index of the particles or its equivalent, the refractive index increment. This subject is treated in considerable detail elsewhere in this volume (Chapter XVIII of the Third Edition of Part II of Vol. I of this series).

Angular Scattering Photometer

The intensity of light scattered by the sample can be measured by rotating a phototube about the plane of the incident beam and the detector with the sample at the axis. This is the arrangement preferred by most instrument designers, although two instruments [98, 99] have a stationary detector and the direction of the incident beam is varied with the use of mirrors. All the angular light-scattering photometers have high-intensity light sources and sensitive light detectors. This is necessary since the scattered light may be one-millionth or less intense than the incident beam. The most common monochromatic light source for the visible region is a high-intensity mercury lamp of the H-3 or H-4 type used in conjunction with colored glass combinations or interference filters to isolate the strongest lines of the mercury spectrum, particularly the 436- and 546-mm lines. Perhaps the most convenient detectors are phototubes of the multiplier type, although single-stage phototubes with amplifiers have been used in some cases [98, 100].

The three most commonly used commercially available angular scattering photometers are the Brice-Phoenix [101], the Aminco [102], and the Sofica

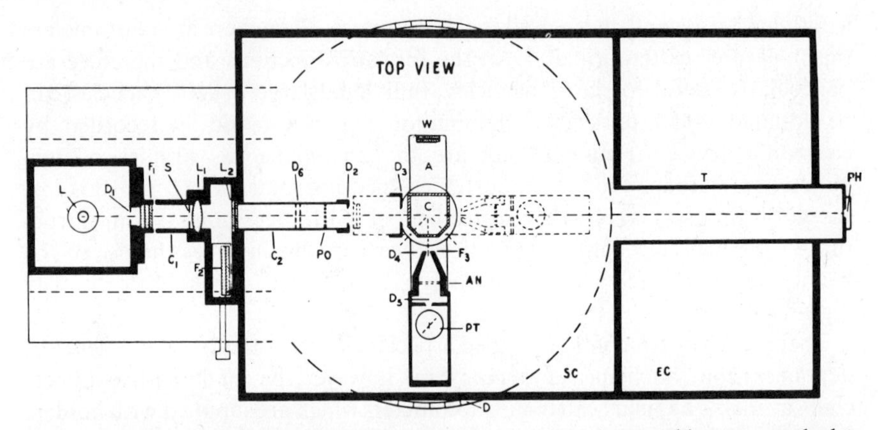

Fig. 2.12 The Brice-Phoenix light-scattering photometer. *A*, rotatable arm attached to disk; *AN*, analyzer; *C*, semioctagonal tube; C_1, shutter collimating tube; C_2, primary beam collimating tube; *D*, graduated disk; G_1, lamp diaphragm; D_2, removeable collimating tube diaphragm; D_3, cell table diaphragm; D_4, nosepiece diaphragm; D_5, cathode diaphragm; D_6, collimating tube diaphragm; *EC*, electrical compartment; F_1, monochromatic filter; F_2, neutral filter; F_3, location of filters used for correcting for flourescence; *L*, mercury lamp; L_1, achromatic lens; L_2, planocylindrical lens; *PH*, covered peephole; *PT*, photomultiplier tube; *PO*, demountable polarizer; *S*, photographic shutter; *SC*, scattering compartment; *T*, light trap tube; *W*, working standard.

[103] instruments. The Phoenix instrument is based on the original design of Brice et al. [104] illustrated in Fig. 2.12, but the electronics have been modified to include photomultiplier detectors and ratio recording circuitry. The Aminco instrument illustrated in Fig. 2.13 was designed by Kremer [105] with the photomultiplier electronic unit described by Oster [106]. The Sofica

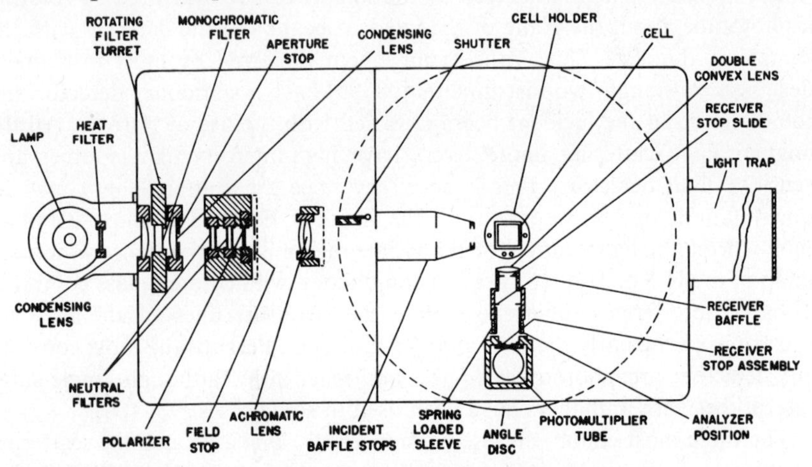

Fig. 2.13 The Aminco absolute light-scattering photometer.

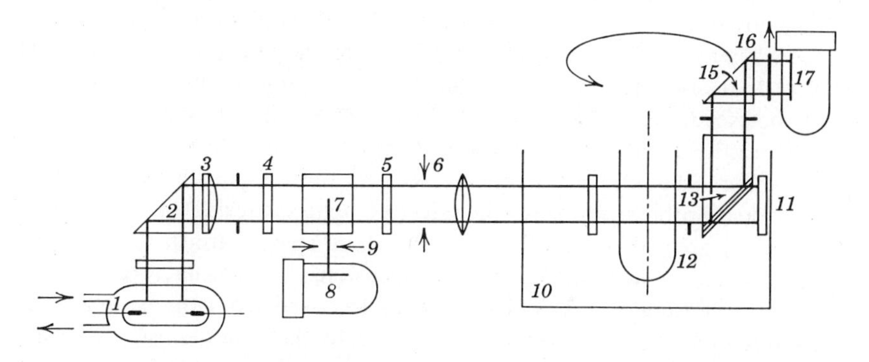

Fig. 2.14 The Sofica light-scattering photometer. *1*, High-pressure water-cooled mercury lamp; *2*, total reflection prism; *3*, achromatic condensing lens; *4*, filters; *5*, polarizer; *6*, variable slit; *7*, glass diffuser; *8*, reference photomultiplier; *9*, iris diaphragm; *10*, constant-temperature vat; *11*, light trap; *12*, entrance slit; *13*, air blade total-reflection prism; *14*, exit slit; *15*, total-reflection prism; *16*, manual shutter; *17*, photomultiplier.

instrument illustrated in Fig. 2.14 and designed by Wippler and Scheibling [107] is of a more radical design than either the Brice-Phoenix or the Aminco instrument. For all three instruments the detector, a photomultiplier tube, is rotated externally. In the Sofica instrument the rotation is controlled by a motor, whereas in the other instruments the angle is achieved by hand. The photomultiplier used in all three instruments is an RCA type IP28 selected to have a low noise level.

The Sofica instrument is also unique in that the sample cell and the observation prism are immersed in a bath which serves not only as a thermostat but also eliminates unwanted refraction and reflection effects. Of particular importance in light scattering is the problem of reflection at the air-glass and liquid-glass interfaces. Reflections can be reduced somewhat by blackening the further wall of the sample vessel. A still further correction, namely, the refraction effect, must be made when the emerging scattered beam is not perpendicular to the wall of the vessel. Some cells have been constructed in which the faces are at 45 and 135° (as well as 90°) with respect to the transmitted beam, but such cells cannot be used for observations at other angles. With cylindrical cells whose diameters are not large compared with the width of the incident beam, there is also a refraction correction associated with an effective converging effect of the cylinder. These troublesome effects are essentially eliminated in the Sofica instrument in which the immersion bath has a refractive index nearly that of the glass and in which the observations are made within the bath. In the Sofica instrument the cell is in the form of a test tube about 25 mm in diameter.

Other corrections are required when the sample shows appreciable fluorescence when excited by the incident beam. In this case the apparent turbidity is larger than expected. The contribution attributable to fluorescence can usually be ascertained by exciting with blue light and inserting a yellow glass filter before the detector [108]. Still further corrections must be made if the sample absorbs light. This effectively reduces the dissymmetry [109, 110]. The converse problem, namely, the effect of light scattering on absorption spectra, is of great concern when the absorbing species are in the form of large aggregates, as with biological cell suspensions. In the usual arrangement in a spectrophotometer, the detector is far removed from the sample and detects not only the transmitted beam but also the small-angle scattering contributions. The scattering contribution can be minimized by placing the detector close to the sample cell or, equivalently, by simply inserting a diffusing plate just behind the sample; the same should be done for the blank cell, with the detector remaining at long distance, as in the conventional spectrophotometer [111–113].

A good light source for light scattering are the newer point source mercury-xenon Osram lamps used in conjunction with interference filters. The Phoenix, Aminco, and Sofica instruments, however, use extended-source mercury lamps. The Sofica instrument uses a water-cooled high-intensity mercury lamp and the fluctuations in light intensity are automatically compensated for by using a reference photomultiplier cell. One means of producing more light at the sample cell is to focus the light toward the center of the cell. This complicates the geometry of the system since now the scattering angles are undefined and the phototube "sees" different volumes of the scattering sample at different angles [114]. Furthermore, there is a depolarization correction which is proportional to the square of the convergence angle [115]. The incident beam can be rendered approximately parallel by means of lenses (preferably achromatic) together with slits or simply by slits alone. For point source lamps the achievable parallelism is of course greater than for the extended sources now employed in the three commercial instruments described. The best parallel beam is achieved with a laser. The recent commercially available [116] infrared nyodinium lasers used in conjunction with quantum-doubling substances, such as barium sodium niobate (emission at 530 mm), are very well suited for light scattering.* Here, filters, lenses and most apertures for the incident beam can be eliminated.

The incident beam should be sufficiently large so that only light from within

* Perhaps more convenient sources are the newer commercially-available lasers. For example, the helium-cadmium laser available from Spectra Physics (Mountain View, Calif. 94040) produces a strong line at 441.6 nm. Also available (from Coherent Radiation Laboratories, Palo Alto, California 94303) is the argon ion laser with strong lines at 476.5, 488.0, and 514.5 nm and the krypton ion laser with strong lines at 530.8, 568.2, and 547.1 nm.

the scattering volume falls upon the phototube. At the same time the beam should be sufficiently narrow that measurements at small angles can be made without picking up the transmitted beam. Furthermore, multiple scattering is minimized by using a smaller beam. Apertures in the phototube housing limit the angle of light falling on the phototube and ensure that the edges of the illuminated region of the sample are not viewed. When collimation is achieved with only two slits, the angular resolution is given by the sum of the widths of the slits divided by their distance apart. However, if the light from the source is focussed on a slit in front of the phototube, very high resolution can be achieved. Thus in one instrument [117] now commercially available [101] a resolution of 0.02° has been achieved in this manner and the diffraction image of the first slit is removed by the slit at the phototube. Another high-resolution light-scattering photometer has been designed [118] and is being manufactured [119]. One criterion of good resolution is that the volume viewed by the receiver is inversely proportional to $\sin \theta$. For the Sofica instrument this relation holds to within 1%.

All the commercially available instruments have provisions for the insertion of polarizing dichroic filters (e.g., Polaroids) in the incident beam and in the receiver nosepiece. The plane of the polarizers should be checked by reflection from a glass surface (the light is mainly horizontally polarized) to insure that the plane is exactly perpendicular or parallel to the edge (or other marking) of the holder. In one commercially available [119] instrument an accessory is supplied that allows the coupled rotation of the two polarizing elements, a feature of particular importance in the study of the scattering by plastic films. For incident unpolarized light the vertical component of the scattered light is, for many liquids and solutions, of the order of a hundred times greater than the horizontal component. One approach to the measurement of the feeble horizontal component is to amplify the signal from the chopped beam and display it on an oscilloscope [114].

Absolute Turbidity

If a solution exhibits a symmetrical scattering envelope (i.e., the particles are Rayleigh scatterers), and yet is a sufficiently intense scatterer so that its optical density is measurable, then it may be used to standardize the light-scattering photometer. Aqueous solutions of tomato bushy stunt virus satisfy this criterion [24, 120], as does the more accessible material, Ludox [121]. The latter material is a fairly stable silica suspension in which the particles have a molecular weight of four million [122]. A 3% solution of Ludox exhibits an optical density inversely proportional to the fourth power of the wavelength. When this solution is measured in a light-scattering photometer, such as the older Aminco instrument [106] which has a narrow incident beam, it exhibits an angular scattering envelope which is symmetrical about

90° and exhibits a negligible depolarization at 90°. For wider-beam instruments such as the Brice-Phoenix, the Ludox solutions exhibit depolarization [123] indicative of multiple scattering. With such instruments the readings should be extrapolated to infinite dilution of the Ludox. The commercially available light-scattering photometers are supplied with a standard opal glass diffuser which has a correction factor associated with it to account for its nonideality as a diffuser.

Using an opal glass diffuser standard, the Rayleigh ratios for freshly distilled benzene at 25°C are 47×10^{-6} and 17×10^{-6} cm^{-1} for 436 and 546 mm, respectively, and the corresponding values of Δ_u are 0.45 and 0.43. Water is a less suitable standard because of its much lower turbidity (about $\frac{1}{20}$ that of benzene) but, more importantly, because of the difficulty of rendering water free of motes. Extraneous particles are readily seen by eye on observing the sample close to the transmitted beam using a narrow pencil of light for illumination. Such particles, presumably dust or, in the case of biomacromolecules, denatured protein, are often difficult to remove especially if the solution is highly viscous. Low-speed centrifugation seems to be the most practical procedure for their removal.

If the geometry of the photometer is precisely defined, as with the Aminco instrument, then there is no need for turbidity standards. In particular, R_{90} is proportional to the ratio of the photocurrent readings at 90 and 0° with a proportionality constant involving the size of the slits at the photocell nosepiece and their distance apart [105].

Forward-Angle Photometers

There has been a considerable effort in recent years to develop photometers capable of measuring particle concentration in samples exhibiting feeble Mie scattering. In some cases the samples, for example, dilute aerosols, may have a turbidity of 10^{-10} cm^{-1} or even less. Transmission measurements could not be performed in the laboratory because the length of the sample would have to be inordinately long. Right-angle observation of the sample in a strongly convergent beam, as in the ultramicroscope, is one means of studying the particles. Because of the preponderant forward-angle scattering by large particles, however, it is more effective to make observations close to the transmitted beam as in dark-field microscopy.

A forward-angle scattering photometer that is particularly useful for continuously measuring the relative concentration of aerosol particles of constant size is illustrated in Fig. 2.15 [124] and is commercially available [125]. The sample is illuminated with a cone of light whose center has been blocked out by a black disk, and only the scattered light falls on the phototube. Another forward-angle photometer has the stop in a position closer to

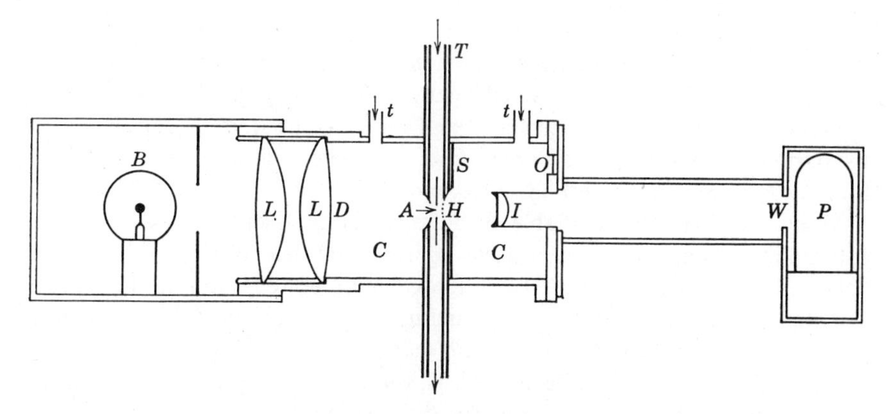

Fig. 2.15 The Sinclair-Phoenix forward-angle photometer. Light from the tungsten lamp *B* is focused at the center *A* of the sample by means of the lenses *L*. An opaque disk *D* removes the center of the converging light cone. The forward-scattered light is made to fall in the phototube *P* by means of the lens *I*.

the light source [126], while in another version [127] the stop is at the surface of the detector (a silicon photovoltaic cell in this case).

The advantage of forward-scattering instruments in collecting the more intense forward-scattered light from the sample is offset somewhat by the difficulty of reducing background light from the source in the forward direction. Optimum design conditions are governed by the scattering diagram of the particles to be measured. An analysis of the geometry of forward-angle photometers has been made [128] and it is concluded that the instrument must be designed around the particle size range to be studied. The problem is further complicated if, as is usually the case with aerosols, the sample contains a wide size range of particles. For these reasons, many workers have resorted to single-particle-counting techniques and to pulse-height analysis techniques to determine the size distribution of the particles [126]. The scattered light produced by individual particles as they enter the light field falls on the phototube, and the electrical impulses are amplified to actuate a counter. An electrical discriminator circuit allows the recording to be limited to particles down to a lower value determined by the noise level of the phototube. Such an instrument can be calibrated with aerosol particles of known size and may be sensitive enough to be calibrated with pure air [129]. The range of particle size applicable with this instrument may be shifted somewhat by altering the color of the incident light [126].

Another approach to the measurement of particle size and of particle concentrations involves photography of the sample under dark-field illumination [130]. More recently, holography has been used to determine particle sizes in an aerosol [131]. In particular, Fraunhofer, that is, far-field, holographs are made by the interference of the diffraction patterns produced by the particles with the undiffracted beam passing through the system. This holographic technique is particularly well suited to the study of fogs in which the particles may be of the order of 10 μ in diameter and when large depth of focus, a feature of this method, is desired.

Many commercially available turbidimeters are essentially adaptions of colorimeters using rather insensitive photovoltaic detectors sometimes in a balanced circuit arrangement. The photocell may be very close to the sample and the optics are generally poorly defined. These instruments detect only very large concentrations of scattering material. Since the detector is at 90° with respect to the incident beam there is, for large turbidities, an attenuation in the entering beam, and consequently the photoelectric response may actually decrease with increasing scattering. Also, no recognition is made of the fact that the scattering by many dilute systems is predominately in the forward direction. Hence the response of the turbidimeter may be greater for small particles than for larger particles of the same weight concentration. The problem is further complicated in that most turbidimeters are supplied only with broad-band colored filters and so the wavelength of the incident light is likewise poorly defined.

Fine-Structure Resolution

Although much of the original work on the fine structure of the Rayleigh line was carried out with conventional light sources and monochrometers it does not match the finesse that can be achieved with present-day components, most of which are commercially available. In particular, all work in this area is now carried out using lasers of the CW or of the pulsed type. For the study of Brillouin splitting, many workers use a Fabry-Perot interferometer. This device consists of two thick glass plates whose separation is adjusted. The opposing faces are highly reflecting. The interference rings may be observed with a telescope focused at infinity. The resolving power is determined by the number of reflections, hence by the reflectivity of the surfaces (see, for example, Ref. 132, Chapter 6). With present coating techniques resolutions of as high as 10^5 are achievable.

In one study of the fine structure of scattering, the Fabry-Perot interferometer was arranged as illustrated in Fig. 2.16. The overall line width was 0.05 cm^{-1}. Instead of moving the photomultiplier across the line, the detector remains stationary and the line is scanned by varying the air pressure in the interferometer. Other high-resolution techniques might also be applicable to the study of fine structures. Thus in Fourier spectroscopy [133] the Michelson interferometer is employed wherein one of the mirrors is translated at a slow uniform rate and the output of the detector is recorded as a function of this

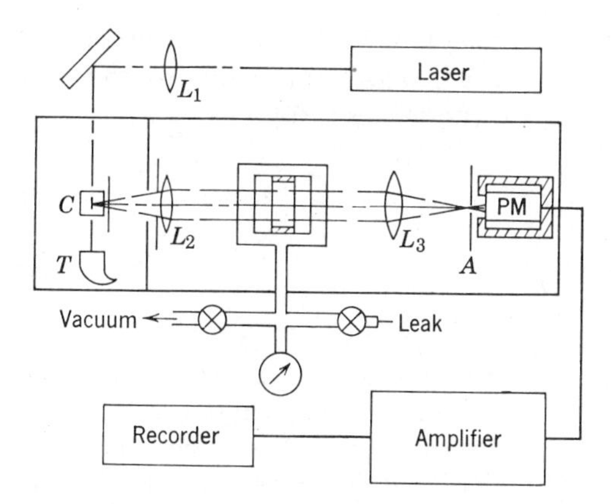

Fig. 2.16 A high resolution spectrometer. Laser light is focused through the lens L, onto the sample cell C with light trap T. The scattered light is rendered parallel with lens L_2, which passes through the Fabry-Perot etalon and is then focused on the photomultiplier PM via the slit A. The spectrum is scanned by varying the air pressure in the etalon [90].

placement. The resultant curve, called an interferogram, is the Fourier transform (carried out by a computer routine) of the detected radiation.

In the case of Doppler shifts from the scattering by moving particles, the resolution of the instrument must be enormous (of the order of 10^{14}) and one must resort to heterodyne beat techniques [20, 93]. In particular, in one method [20] the laser light is split at a half-silver mirror, one beam passing into the sample and the other going to a photocell where it combines with the scattered light from the sample. Each of the two beams at the beam splitter are modulated (by acoustical-Bragg reflection modulators, for example) at two different frequencies and the resultant beat is analyzed by electronic methods. By this means line widths of only a few cycles per second can be measured. A helium-neon laser radiates in a series of modes which are separated in frequency in the megahertz region. This property is utilized in the self-beat method wherein the scattered light is compared with one of the modes of oscillation in an audiofrequency spectrum analyzer [134]. The Doppler broadening is obtained from the Fourier transform of the observed power spectrum [93].

The determination of the distribution of the relative velocities of pairs of scattering centers can be carried out, using a coherent light source, by obtaining the spectrum of intensity fluctuations [93a]. One procedure for extracting spectral information, called digital autocorrelation, involves the measurement and analysis of the fluctuations in intensity caused by the beating together of the various Doppler-shifted frequencies [93b]. By this latter method it was possible to simultaneously determine the diffusion constants of two macromolecular species in a mixture where the relative concentration was adjusted to give approximately the same light scattering contribution of the two substances [93b].

References

1. Lord Rayleigh, *Phil. Mag.*, **12,** 81 (1881).
2. P. Debye, *Ann. Physik*, **46,** 809 (1915). For an English translation, see *The Collected Papers of Peter J. W. Debye*, Interscience, New York, 1954, p. 40.
3. The special case of spherical particles of low relative refractive index was treated by Lord Rayleigh, *Proc. Roy. Soc. London*, **A90,** 219 (1914), and by R. Gans, *Ann. Physik*, **76,** 29 (1925). Debye scattering is also called Rayleigh-Gans scattering.
4. G. Mie, *Ann. Physik*, **25,** 377 (1908).
5. P. Debye, *Ann. Physik*, **30,** 59 (1908).
6. H. C. Van der Hulst, *Light Scattering by Small Particles*, Wiley, New York, 1957.
7. F. Zernicke and J. Prins, *Z. Physik*, **41,** 184 (1927).
8. A. Guinier, *Ann. Phys.*, **12,** 161 (1939).
9. A. Guinier and G. Fournet, *Small-Angle Scattering of X-Rays*, Wiley, New York, 1955.
10. H. Brumberger, Ed., *Small-Angle X-Ray Scattering*, Gordon and Breach, New York, 1967.
11. M. Smoluchowski, *Ann. Physik*, **25,** 205 (1908); *Phil. Mag.*, **23,** 165 (1912).
12. L. S. Ornstein and F. Zernicke, *Physik. Z.*, **27,** 761 (1926); see also V. Rocard, *J. Phys. Radium*, **4,** 165 (1933).
13. A. Einstein, *Ann. Physik*, **33,** 1275 (1910).
14. L. Brillouin, *Ann. Phys.*, **17,** 88 (1922).
15. E. Gross, *Acta Physiocochim. URSS*, **20,** 459 (1945).
16. L. Landau and G. Placzek, *Sowjet. Phys.*, **5,** 172 (1934).
17. I. L. Fabelinskii, *Molecular Scattering of Light*, Plenum, New York, 1968.
18. J. Cabannes and Y. Rocard, *La Diffusion Moléculaire de la Lumière*, Les Presses Universitaires de France, Paris, 1929.
19. B. J. Berne and H. L. Frisch, *J. Chem. Phys.*, **47,** 3675 (1967); L. Blum and Z. W. Salsberg, *J. Chem. Phys.*, **48,** 2292 (1968).
20. H. Z. Cummins, N. Knable, and Y. Yeh, *Phys. Rev. Letters*, **12,** 150 (1964).
21. G. Oster, *Chem. Revs.*, **43,** 319 (1948).
22. *Selected Bibliography on Light Scattering*, Phoenix Precision Instrument Co., Philadelphia, Pennsylvania.

23a. P. Debye, *J. Appl. Phys.*, **15,** 338 (1944).
23b. P. Debye, *J. Phys. Colloid Chem.*, **51,** 18 (1947).

24. G. Oster, *Science*, **103,** 306 (1946).
25. R. S. Krishnan, *Proc. Indian Acad. Sci.*, **A3,** 126 (1936).

26. T. Neugebauer, *Ann. Phys.*, **42,** 509 (1943).
27a. O. Kratky and G. Porod, *J. Colloid Sci.*, **4,** 35 (1949).
27b. A. Peterlin, *J. Polymer Sci.*, **10,** 425 (1953); A. Peterlin, in *Electromagnetic Scattering*, M. Kerker, Ed., Pergamon, Oxford, 1963.
28. G. Oster and D. P. Riley, *Acta Cryst.*, **5,** 1 (1952).
29. N. Saito and Y. Ikeda, *J. Phys. Soc. Japan*, **6,** 305 (1951).
30. C. C. Shull and L. C. Roess, *J. Appl. Phys.*, **18,** 295 (1947).
31. H. Benoit, *J. Polymer Sci.*, **11,** 507 (1953). See also H. Benoit, in *Electromagnetic Scattering*, M. Kerker, Ed., Pergamon, Oxford, 1963.
32. P. Geiduschek and A. Holtzer, *Advan. Biol. Med. Phys.*, **6,** 323 (1958).
33. P. Horn, H. Benoit, and G. Oster, *J. Chim. Phys.*, **48,** 1 (1951).
34. H. Benoit and G. Weill, *Coll. Czech Chem. Commun.*, **22,** (Special Issue), 35 (1957).
35. G. Porod, *Kolloid-Z.*, **124,** 83 (1951). See also P. Debye, H. R. Anderson, and H. Brumberger, *J. Appl. Phys.*, **28,** 679 (1957).
36. G. Oster, *Rec. Trav. Chim.*, **68,** 1123 (1949).
37. P. Doty and R. F. Steiner, *J. Chem. Phys.*, **18,** 1211 (1950).
38. H. C. Van der Hulst, *The Solid Particles in Interstellar Space*, Schotanus and Jens, Utrecht, 1949. See also L. Spitzer, Jr., *Diffuse Matter in Space*, Interscience, New York, 1968.
39. D. Sinclair and V. K. La Mer, *Chem. Rev.*, **44,** 245 (1949).
40. W. Heller, in *Electromagnetic Scattering*, M. Kerker, Ed., Pergamon, Oxford, 1963.
41. D. Diermendjian, *Electromagnetic Scattering on Spherical Polydispersions*, Elsevier, Amsterdam, 1969.
42. For detailed derivation see J. A. Stratton, *Electromagnetic Theory*, McGraw-Hill, New York, 1941, Chapters 7 and 9, or M. Born and E. Wolf, *Principles of Optics*, Pergamon, Oxford, 1959, Chapter 13.
43. R. W. Hart and E. W. Montroll, *J. Appl. Phys.*, **22,** 376 (1951) obtain the scattering cross section in closed form when $m < 1.1$ and x is any value.
44. J. Keller, *J. Opt. Soc. Am.*, **52,** 116 (1962); J. Keller and B. R. Levy, in *Electromagnetic Scattering*, M. Kerker, Ed., Pergamon, Oxford, 1963. See also M. Kline and I. W. Kay, *Electromagnetic Theory and Geometrical Optics*, Interscience, New York, 1965.
45a. G. Oster, in *Modern Optics*, J. Fox, Ed., distributed through Interscience, New York, 1967; G. Oster, *The Science of Moiré Patterns*, Second ed., Edmund Scientific Co., Barrington, New Jersey, 1969, Chapter 7.
45b. B. Goldberg, *J. Opt. Soc. Am.*, **43,** 1221 (1953).
46. V. K. La Mer and M. D. Barnes, *J. Colloid Sci.*, **1,** 71 (1946).
47. For example, A. N. Lowan, *Tables of Scattering Functions for Spherical Particles*, Natl. Bur. Standards (U.S.), Appl. Math. Ser. 4, Gov't. Printing Office, Washington, D.C., 1948; R. O. Gumprecht and C. M. Sliepevich, *Light Scattering Functions for Spherical Particles*, Univ. of Michigan Press, Ann Arbor, 1951; W. J. Pangonis and W. Heller, *Angular Scattering Functions for Spherical Particles*, Wayne State Univ. Press, Detroit, 1960; F. T. Gucker, R. L. Rowell, and G. Chien, *Proc. Conf. Aerosols*, Czech. Acad. Sci., 1964.

48. For example, M. Kerker of the Clarkson College of Technology has a large collection of punched cards of the Mie functions.
49. H. C. Van der Hulst, *J. Colloid Sci.*, **4,** 79 (1949).
50. H. Blumer, Z. Physik, **38,** 304 (1926).
51. M. Kerker and V. K. La Mer, *J. Am. Chem. Soc.*, **72,** 3516 (1950); M. Kerker and E. Matijevic, *J. Opt. Soc. Am.*, **50,** 722 (1960).
52. W. Heller, W. J. Pangonis, and N. A. Economon, *J. Chem. Phys.*, **34,** 971 (1961).
53. G. Oster and Y. Nishijima, *129th Ann. Meeting Am. Chem. Soc., Dallas, Amer. chem. Soc. Texas, April 1956.*
54. T. Alfrey, E. B. Bradford, and G. Oster, *J. Opt. Soc. Am.*, **44,** 603 (1954).
55. I. M. Kriegler and F. M. O'Neill, *J. Am. Chem. Soc.*, **90,** 3114 (1968).
56. G. Oster, *J. Gen. Physiol.*, **33,** 445 (1950).
57. M. H. F. Wilkins, A. R. Stokes, W. E. Seeds, and G. Oster, *Nature*, **166,** 127 (1950).
58. A. Klug, R. E. Franklin, and S. P. F. Humphreys-Owen, *Biochim. Biophys. Acta*, **32,** 203 (1959).
59. J. Turkevich, G. Garton, and P. C. Stevenson, *J. Colloid Sci.*, *Suppl.*, **I,** 26 (1954).
60. G. Fournet and A. Guinier, *Compt. Rend.*, **228,** 66 (1949).
61. L. Landau and E. Lifshitz, *Statistical Physics*, Oxford Univ. Press, London, 1938.
62. G. D. Parfitt and J. A. Wood, *Trans. Faraday Soc.*, **64,** 805 (1968).
63. G. Oster and J. G. Kirkwood, *J. Chem. Phys.*, **11,** 175 (1943).
64a. B. H. Zimm, *J. Chem. Phys.*, **16,** 1099 (1948).
64b. A. M. Holtzer, H. Benoit, and P. Doty, *J. Chem. Phys.*, **58,** 624 (1954).
65. L. I. Mandelshtam, *Ann. Physik*, **41,** 609 (1913).
66. A. A. Andronov and M. A. Leontovich, *Z. Physik*, **38,** 485 (1926).
67. C. V. Raman and L. A. Ramdas, *Proc. Roy. Soc.*, **A108,** 561 (1925); **A109,** 150, 272 (1925).
68. A. Vrij, *J. Colloid Sci.*, **19,** 1 (1964); A. Vrij, *Advan. Colloid Interface Sci.*, **2,** 39 (1968).
69. L. A. Ramdas, *Indian J. Phys.*, **1,** 199 (1927).
70. F. Barikhanskaya, *Zh. Teor. Eks. Fiz.*, **7,** 51 (1937).
71. L. S. Ornstein and F. Zernicke, *Proc. Acad. Sci. Amsterdam*, **17,** 793 (1914); *Physik. Z.*, **27,** 761 (1926).
72. P. Debye, *J. Chem. Phys.*, **31,** 680 (1959).
73. B. H. Zimm, *J. Phys. Colloid Chem.*, **54,** 1306 (1950).
74. P. Debye, D. Woermann, and B. Chu, *J. Chem. Phys.*, **36,** 1803 (1962).
75. P. Haller, *Rept. Prog. Phys.*, **30,** Part II, 731 (1967).
76. P. Debye and A. M. Bueche, *J. Appl. Phys.*, **20,** 518 (1949).
77. R. S. Stein and J. J. Keane, *J. Polymer Sci.*, **17,** 21 (1955).
78. A. E. Kiejers, J. J. van Aartsen, and W. Prins, *J. Am. Chem. Soc.*, **90,** 3107 (1968).
79. R. S. Stein and P. R. Wilson, *J. Appl. Phys.*, **33,** 1914 (1962).
80. R. S. Stein, P. Ehrhardt, J. J. van Aartsen, S. Clough, and M. Rhodes, *J. Polymer Sci. C*, **13,** 1 (1966).

81. R. S. Stein, in *Electromagnetic Scattering*, M. Kerker, Ed., Pergamon, Oxford, 1963.
82. G. Oster, *J. Colloid Sci.*, **2,** 291 (1947).
83. D. Gitlin and H. Edelhoch, *J. Immunol.*, **66,** 77 (1951).
84. J. M. Singer and C. M. Plotz, *Am. J. Med.*, **21,** 888 (1956); J. M. Singer, *Am. J. Med.*, **31,** 766 (1961).
85. P. Debye, *Ann. Physik*, **39,** 789 (1912).
86. L. I. Mandelshtam, *Zh. Russ. Fiz. Khim.*, **58,** 381 (1926).
87. M. A. Leontovich, *Z. Phys.*, **72,** 247 (1931).
88. C. S. Venkateswaren, *Proc. Indian Acad. Sci.*, **15,** 322 (1942).
89. I. L. Fabelinskii, *Soviet Phys. Dokl.*, **1,** 115 (1956).
90. H. Z. Cummins and R. W. Gammon, *J. Chem. Phys.*, **45,** 4438 (1966).
91. R. D. Mountain, *Rev. Mod. Phys.*, **38,** 205 (1966).
92. R. Pecora, *J. Chem. Phys.*, **43,** 1562 (1965).
93. M. J. French, J. C. Angus, and A. G. Watson, *Science*, **163,** 345 (1969).
93a. S. B. Dubin, J. H. Lunacek, and G. B. Benedek, *Proc. US Nat. Acad. Sci.*, **57,** 1164 (1967).
93b. R. Foord, E. Jakeman, C. J. Oliver, E. A. Pike, R. J. Blagrove, E. Wood, and A. R. Peacocke, *Nature* (London), **227,** 242 (1970).
94. C. Smart, R. Jacobson, M. Kerker, J. P. Kratohvil, and E. Matijevic, *J. Opt. Soc. Am.*, **55,** 947 (1965).
95. S. Chandrasekhar, *Radiative Transfer*, Oxford Univ. Press, Oxford, 1950.
96. V. Kourganoff, *Basic Methods in Transfer Problems*, Oxford Univ. Press, Oxford, 1952.
97. V. V. Sobolev, *Radiative Transfer*, van Nostrand Reinhold, New York, 1963.
98. P. P. Debye, *J. Appl. Phys.*, **17,** 392 (1946).
99. P. Bosworth, C. R. Masson, H. W. Melville, and F. W. Peaker, *J. Polymer Sci.*, **9,** 565 (1952).
100. R. Speiser and B. A. Brice, *J. Opt. Soc. Am.*, **36,** 364A (1946).
101. Available from the Phoenix Precision Instrument Co., Philadelphia, Pennsylvania.
102. Available from the American Instrument Co., Silver Spring, Maryland.
103. Available from the Société Francaise d'Instruments de Controle et d'Analyses, Le Mensil Saint-Denis (S.-et-O.), France, and distributed in the United States by Bausch and Lomb Company, Rochester, New York.
104. B. A. Brice, M. Halwer, and R. Speiser, *J. Opt. Soc. Am.*, **40,** 768 (1950).
105. J. Kremer and J. J. Shapiro, *J. Opt. Soc. Am.*, **44,** 500 (1954).
106. G. Oster, *Anal. Chem.*, **25,** 1165 (1953).
107. C. Wippler and G. Scheibling, *J. Chim. Phys.*, **51,** 201 (1954).
108. Y. Tominatsu, K. J. Palmer, and W. H. Ward, *J. Polymer Sci.*, **36,** 129 (1959).
109. B. A. Brice, R. Nutting, and M. Halwer, *J. Am. Chem. Soc.*, **75,** 824 (1953).
110. H. P. Frank and R. Ullman, *J. Opt. Soc. Am.*, **45,** 471 (1955).
111. K. Shibata, A. A. Benson, and M. Calvin, *Biochim. Biophys. Acta*, **15,** 461 (1954).
112. P. Latimer, *Science*, **127,** 29 (1958).
113. J. Amesz, L. M. N. Duysens, and D. C. Brandt, *J. Theoret. Biol.*, **1,** 59 (1961.)
114. S. Guinard and J. Tonnelat, *J. Chim. Phys.*, **51,** 276 (1954).

115. R. Gans, *Phys. Z.*, **28,** 661 (1927).
116. Available from, for example, TRG Laser Products, Melville, New York.
117. W. H. Aughey and F. J. Baum, *J. Opt. Soc. Am.*, **44,** 833 (1954).
118. A. E. M. Keijgers, J. J. van Aartsen, and W. Prins, *J. Appl. Phys.*, **36,** 2874 (1965).
119. Available from Nederlandsche Optieken Instrumentenfabrick, Zeist, Holland.
120. For review of light-scattering studies on viruses see Section III of H. K. Schachman and R. C. Williams, in *The Viruses*, F. M. Burnet and W. M. Stanley, Eds., Vol. I, Academic, New York, 1959.
121. Ludox is available in concentrations of approximately 30% by weight of solids from Grasselli Chemical Dept., E. I. duPont de Nemours and Co., Wilmington, Delaware.
122. G. Oster, *J. Polymer Sci.*, **6,** 525 (1952).
123. W. F. H. M. Mommaerts, *J. Colloid Sci.*, **7,** 71 (1952); S. H. Maron and R. L. H. Lou, *J. Polymer Sci.*, **14,** 29 (1954); J. Kraut and W. B. Dandliker, *J. Polymer Sci.*, **18,** 563 (1955).
124. D. Sinclair, *Air Repair*, **3,** 51 (1953)
125. Phoenix Precision Instrument Co., Philadelphia, Pennsylvania. A somewhat simpler version is also manufactured by T. D. Associates, Baltimore, Maryland.
126. F. T. Gucker and D. G. Rose, *Brit. J. Appl. Phys. Suppl.*, **3,** 138 (1954).
127. C. R. Weston, W. Vishniac, and D. Buchendahl, *Ann. N. Y. Acad. Sci.*, **157,** 149 (1969).
128. J. R. Hodkinson and J. R. Greenfield, *Appl. Opt.*, **4,** 1463 (1965).
129. F. T. Gucker, Jr., and A. H. Patterson, *J. Colloid Sci.*, **10,** 12 (1955).
130. P. Lee and V. K. La Mer, *Rev. Sci. Instr.*, **24,** 1004 (1954).
131. B. J. Thompson, J. H. Ward, and W. R. Zinky, *Appl. Opt.*, **6,** 519 (1967).
132. M. Françon, *Optical Interferometry*, Academic, New York, 1966.
133. J. Connes and J. Connes, *J. Opt. Soc. Am.*, **56,** 896 (1966).
134. J. B. Lastovka and G. B. Benedek, *Phys. Rev. Letters*, **17,** 1039 (1966).

General

MONOGRAPHS AND REVIEWS

Bhagavantum, S., *Scattering of Light and the Raman Effect*, Chemical Publishing Co., Brooklyn, New York, 1942.

Boedtker, H., in *Methods in Enzymology*, S. P. Colowick and N. O. Kaplan, Eds., Vol. 13B, Academic, New York, 1968 (mainly RNA).

Cabannes, J., and Y. Rocard, *La Diffusion Moléculaire de la Lumière*, Les Presses Universitaires de France, Paris, 1929.

Chu, B., *Ann. Rev. Physical Chem.*, **21** (1970) (laser light scattering).

Doty, P., and J. T. Edsall, *Advan. Protein Chem.*, **6,** 37 (1951).

Fabelinskii, I. I., *Molecular Scattering of Light*, Plenum, New York, 1968.

Fishman, M. M., *Light Scattering by Colloid Systems: An Annotated Bibliography*, Technical Services Laboratories, River Edge, New Jersey, 1958.

Geiduschek, P., and A. Holtzer, *Advan. Biol. Med. Phys.*, **6,** 323 (1958) (mainly DNA and muscle proteins).

Guinier, A., and G. Fournet, *Small-Angle Scattering of X-Rays*, Wiley, New York, 1955.

Kerker, M., *The Scattering of Light*, Academic, New York, 1969.

Kratky, O., and G. Porod, *J. Colloid Sci.*, **4,** 35 (1949).

Oster, G., *Chem. Rev.*, **43,** 319 (1948).

Peterlin, A., *Prog. Biophys. Biophys. Chem.*, **9,** 176 (1959).

Riley, D. P., and G. Oster, *Discussions Faraday Soc.*, **11,** 107 (1951).

Stacey, K. A., *Light Scattering in Physical Chemistry*, Academic, New York, 1956.

Tonnelat, J., *J. Chim. Phys.*, **47,** 821 (1950).

Van der Hulst, H. C., *Light Scattering by Particles*, Wiley, New York, 1957.

SYMPOSIA AND COLLECTIONS OF PAPERS

Blumberger, H., Ed., *Small Angle X-Ray Scattering*, Gordon and Breach, New York, 1967.

Green, M., and J. V. Senges, Eds., *Critical Phenomena*, Misc. Publ. No. 273, Natl. Bur. Standards, Washington, D.C., 1966.

Kerker, M., Ed., *Electromagnetic Scattering*, Pergamon, Oxford, 1963.

McIntyre, D., and F. Gornick, Eds., *Light Scattering from Dilute Polymer Solutions*, Gordon and Breach, New York, 1965.

Rowell, R. W., and R. S. Stein, Eds., *Electromagnetic Theory*, Gordon and Breach, New York, 1967.

Chapter **III**

LIGHT MICROSCOPY

George G. Cocks
Edwin E. Jelley*

1 Introduction 121

2 The Optics of the Microscope 122
Characteristics of the Microscope 122
Magnification 122
Resolving Power 125
Image Intensity 129
Image Contrast 130
Microscope Lenses 131
Aberrations of Lenses 131
Objectives 133
Eyepieces and Amplifying Lenses 134
Illuminating Systems 135
Microscope Illuminators 136
Complete Optical Systems 139
Stops and Diaphragms 139
Field Stops 139
Aperture Stops 141

3 The Mechanical Parts of the Microscope 143

4 Specimen Preparation 146
Preparation of Particulate Materials 147
Mounting and Mounting Media 147
Separations 149
Preparation of Crystals 149
Crystallization from Solvents 150
Crystallization by Sublimation 151
Crystallization from the Melt 151
Preparation of Materials for the Examination of Surfaces 154
Replication 154
Shadowing 154

* Deceased

Sectioning Techniques 155
Fractured Sections 155
Plane Sectioning 156
Thin Sectioning 156
Staining and Etching 157

5 Microscopes and Microscopical Techniques 158

Microscopy with Transmitted Bright-Field Illumination 158
Bright-Field Microscopes and Illumination 158
Qualitative Chemical Analysis Based on Morphology 160
Quantitative Chemical Analysis 161
Polarized Light Microscopy 166
The Polarizing Microscope 166
Optics of Transparent Crystals 171
Crystal Systems 171
Habit, Distortion, and Twinning 171
Polarized Light 172
Optically Isotropic Crystals 173
Optically Anisotropic Crystals 173
Interference Colors 179
Optical Dispersion 183
Molecular Structure and Optical Properties 188
Microscopy of Transparent Crystals 189
Orthoscopic Examination 189
Conoscopic Examination 191
Determination of Refractive Indices 199
Determination of Birefringence and Its Dispersion 210
Determination of Optic Axial Angles and Their Dispersion 216
Microscopy of Colored Crystals 219
Theoretical 219
Preparation of Colored Crystals 234
Orthoscopic Examination 237
Conoscopic Interference Figures 237
Conoscopic Absorption Figures 238
Microspectrograph 238
Vertical Illumination 238
Thermal and High-Pressure Microscopy 239
Hot and Cold Stages 239
High-Pressure Stages 242
Chemical Applications of Thermal Microscopy 242
Properties Measurable with the Aid of a Hot Stage 242
Phase Studies on One-Component Systems 245

Phase Studies on Binary Systems 250
Phase Studies in Ternary Systems 254
Identification of Compounds Using Thermal Data 254
Dispersion Staining 255
Theory 255
Mounting Media for Dispersion Staining 257
Microscopical Equipment for Dispersion Staining 257
Applications of Dispersion Staining 258
Dark-Field Microscopy 260
Microscopy with Reflected Light 263
Reflected Bright-Field Illumination 263
Reflected Dark-Field Illumination 264
Ultraviolet and Infrared Microscopy 264
Microspectrophotometry and Microphotometry 266
Fluorescence Microscopy 268
Interference Microscopy and Interferometry 270
Principles of Interference Microscopy 270
The Polarizing Microscope 270
Interference Microscope Construction 271
Classes of Interference Microscopes 271
Reflected Light Interference Microscopes 280
The Applications of Interference Microscopy 281
6 Photomicrography 282

I INTRODUCTION

This chapter deals with chemical microscopy using the light microscope. In it an attempt has been made to describe various microscopical techniques which can be used to solve chemical problems. Because the intelligent application of these techniques requires a knowledge of the optical principles of the instruments, there is considerable discussion of various kinds of microscopes. There is also an extensive discussion of crystal optics because chemical microscopy is heavily dependent on determining or understanding the optical properties of crystalline materials.

Clearly, it is impossible to give in this chapter such complete and adequate instruction on the methods of chemical microscopy that a chemist who wishes to apply microscopical methods needs no other reference. It is also impossible to refer to all of the important works that are of value to a chemical microscopist. Therefore an attempt has been made to provide more complete bibliographical material for the techniques that are treated briefly and to provide some references for all of the techniques discussed. In writing this chapter the author has made extensive use of books on microscopy written by Chamot and Mason, Françon, Martin, Kingslake, and McCrone.

This chapter is the successor to chapters on the same subject written by E. E. Jelley for the earlier editions of *Physical Methods of Organic Chemistry*. Although most of the material has been rewritten and revised, the sections on polarized light microscopy and optical crystallography have been changed very little.

2 THE OPTICS OF THE MICROSCOPE

Characteristics of the Microscope

The compound microscope is a two-stage magnifier. A lens, the objective, forms a real image of an illuminated or self-luminous object. This real image is further enlarged by another lens, the eyepiece. The microscope is basically an instrument that extends human vision to objects too small to be seen by the unaided eye. To do this the output of the microscope must have certain characteristics: (1) there must be sufficient magnification so that the fine structures can be observed visually or photographed; (2) there must be sufficient resolution so that the fine structures of the object are distinguishable in the image; (3) the rays in the various parts of the image must be intense enough to stimulate the retina; (4) there must be sufficient contrast between the various parts of the image so that the fine structures can be distinguished from their surroundings; and (5) the output rays must be in such a state that they can form an image on the retina. This means that for a normal eye the rays emanating from a single point on the object must be parallel or slightly diverging as they leave the microscope.

Magnification

The magnification (M) of a microscope is equal to the magnification of the objective (M_o) times the magnification of the eyepiece (M_e). Some of the important qualities of the objective can be brought out by considering the image formed by means of a pinhole (Fig. 3.1). Every point on an illuminated or self-luminous object gives off rays of light in many directions. Consider only two points on the object in Fig. 3.1, located on the head and the tail of the arrow. Only those rays that pass through the pinhole contribute to the image. The magnification is by definition:

$$\frac{\text{Size of the image}}{\text{Size of the object}} = \frac{S_i}{S_o}, \tag{3.1}$$

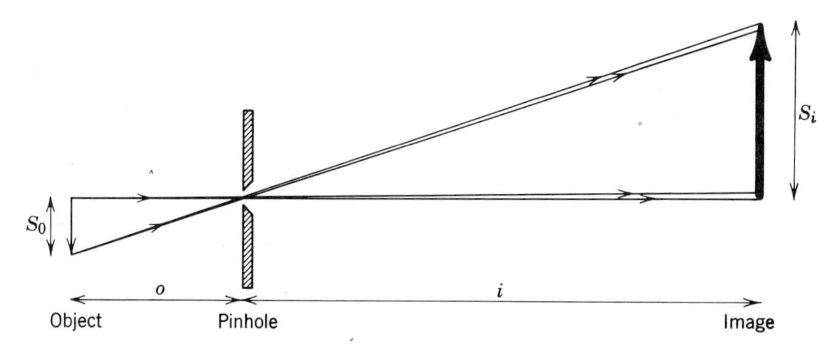

Fig. 3.1 An image formed by a pinhole. Physical optical effects are neglected.

and by simple geometry:

$$\frac{S_i}{S_o} = \frac{i}{o}. \tag{3.2}$$

The magnification can be varied infinitely by varying i and o, which are independent variables.

The way in which a lens forms an image of an object is indicated schematically in Fig. 3.2. As there is always a ray from a point on the object that does not change direction after having passed through the lens on its way to the corresponding image point, the magnification of a lens is essentially the same as it is for a pinhole, i/o. The magnification of a lens can be varied infinitely by varying i and o, however, i and o are no longer independent variables but

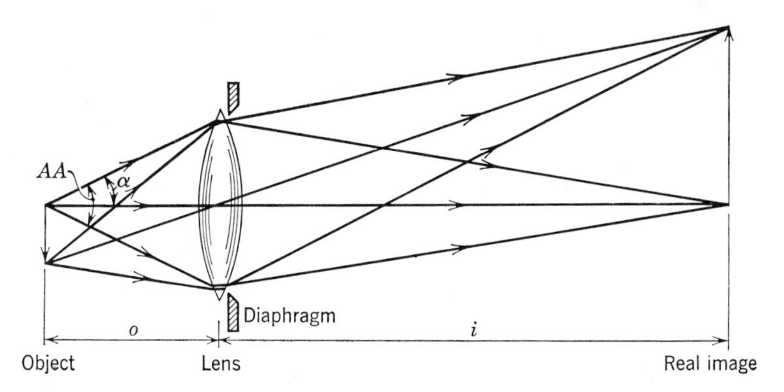

Fig. 3.2 An image formed by a lens. Only the cones of rays from points on the object that pass through the diaphragm can contribute to the formation of the image. These cones are represented by three rays from each of two points on the object. The apex angle of the axial cone is AA.

are related by the equation:

$$\frac{1}{i} + \frac{1}{o} = \frac{1}{f}, \tag{3.3}$$

where f is the focal length of the lens.

Although a single lens can in theory give any magnification desired, and although early microscopists, especially Leeuwenhoek, used single lenses very successfully, they are seldom used nowadays except as low-power magnifiers. The use of a single lens for high-power microscopy is inconvenient because a suitable lens must have a very short focal length and must be held inconveniently close to the eye and to the specimen. These difficulties are overcome if an objective lens is used to form a real image at relatively low magnification and an eyepiece is used to further magnify this real image. The eyepiece is designed so that the eye can be held at a comfortable distance from it. The eyepiece acting in conjunction with the cornea and the lens of the eye forms a real image on the retina. This mode of operation is sufficiently different from image formation as discussed above so that the eyepiece is said to be functioning as a "simple magnifier." Figure 3.3 illustrates this type of functioning. To an observer the object appears enlarged and located somewhere beyond the true position of the object. This subjective image is called a virtual image. The observer cannot judge either the location or the size of the virtual image, therefore for purposes of calculating the magnification it is arbitrarily specified that $i = 250$ mm. For a normal human eye an image can be formed only if the rays entering the eye from an object point are parallel or slightly diverging. The most divergent rays that the average normal eye can focus comfortably are those arising from an object point 250 mm from the eye. To put it in another way, the eye can accommodate for virtual images lying anywhere between $i = 250$ mm and $i = \infty$.

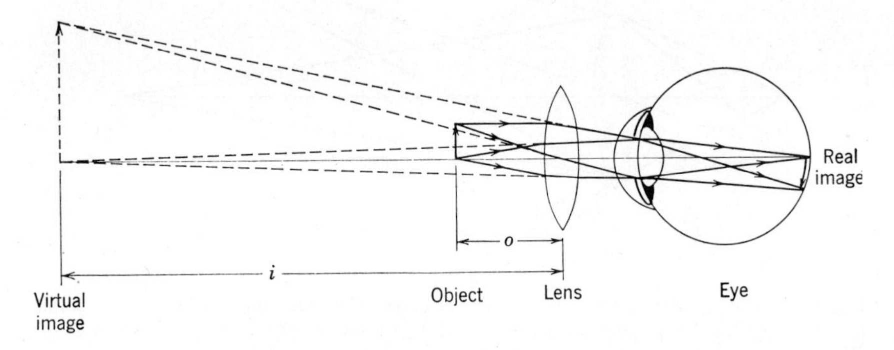

Fig. 3.3 A lens used as a simple magnifier.

Typical objective magnifications (M_o) may vary from 5 times (5×) to 100 times (100×) at $i = 160$ mm, where i is determined by the tube length of the microscope. Typical eyepiece magnifications (M_e) may vary from 5 to 20×. Thus M, which equals M_o times M_e, may vary from 25 to 2000×. Magnifications greater than 2000× are of little use because of limitations on resolving power which will be discussed below.

Resolving Power

The resolving power of a microscope is a measure of its ability to reveal closely adjacent structural details as actually separate and distinct. Resolving power is a property of the microscope and can be defined mathematically in various ways. Often it is defined as the smallest distance between two points in the object such that the points can be distinguished in the image. Considered from a purely geometrical optical viewpoint, it can be seen that the resolving power of a pinhole is limited by the size of the pinhole (Fig. 3.1). A cone of light rays defined by the size of the object point, by the object distance o, and by the size of the pinhole produces the conjugate image point. Object points that are further apart than the diameter of the pinhole in general produce recognizably distinct image points. From the same viewpoint there is no limit on the resolving power of an ideal lens. The size of an image point depends only on the size of the object point and on the magnification, and the image points formed by distinct object points never overlap.

However, neither the resolving power of microscopes, nor image formation in general, can be considered solely from the viewpoint of geometrical optics. Light is diffracted or bent when it passes close to something such as an element of the object structure or the edge of the pinhole. These diffracted rays can interfere among themselves or with the undeviated rays. As a result a pinhole illuminated by a point source of light produces at an appropriate image distance a diffraction pattern in the image consisting of a bright central disk surrounded by alternating dark and light fringes. This central disk is considerably brighter than the surrounding fringes and is smaller than would be predicted from geometrical optics. Effectively, a pinhole has a focal length, and if this focal length is taken into account when setting o and i, the resolution is better than would be calculated on the basis of geometrical optics. For a given o and i, an optimum pinhole size can be calculated using diffraction theory.

If diffraction theory is applied to the formation of images by lenses, it is found that the image of a point source also consists of a bright disk surrounded by dark and light fringes. This interference pattern is called "Airy's disk." The size of the central bright spot of Airy's disk formed by an ideal lens is always larger than the geometrical size of the image of a point source. The size of the disk depends upon the angular aperture of the light used to form the image. Rayleigh's criterion for resolution states that two points in the object are resolved if the center of their conjugate Airy disks in the image are separated by a distance equal to or greater than the diameter of the central bright spot. Based on this criterion, the resolving power of a microscope (d) is:

$$d = \frac{1.2\lambda}{2n \sin \frac{AA}{2}} = \frac{1.2\lambda}{2n \sin \alpha}, \tag{3.4}$$

where d, the resolving power, is the minimum distance between two points on the object that can be distinguished in the image, n is the index of refraction of the medium in which the object and the front of the objective lens are immersed, λ is the wavelength, in vacuo, of the light used to form the image, and $\alpha = AA/2$ is defined in Fig. 3.2. The effective wavelength of the light that interacts with the specimen is λ/n.

An increased understanding of resolving power and the factors that affect it can be obtained by considering the theory developed by Abbe. If a diffraction grating is used as a test specimen for microscopical examination, the factors that affect resolution can be strikingly demonstrated. A narrow beam of light striking a grating normally is diffracted according to the following formula: $a(\lambda/n) = d \sin \theta_a$ (see Fig. 3.4*a*), where a is the order of the diffracted beam. If such a grating is observed microscopically, Abbe's theory states that the grating lines can be resolved only if at least two diffracted beams enter the objective of the microscope. The greater the number of diffracted beams entering the objective, the better the quality of the image formed. Figure 3.4*b* shows three diffracted beams entering the objective. The zero-order beam and the two first-order beams recombine to form an image of the grating. If the grating spacing d is slightly smaller, θ_1 will be slightly larger, the two first-order beams will not enter the objective lens, and no image of the grating will be formed. Therefore, according to Abbe's theory, the grating will be resolved if the angular aperture α of the objective lens is equal to or greater than θ_1. The resolving power of the objective can be specified by substituting α for θ_1 in the grating formula, $a(\lambda/n) = d \sin \alpha$. As only the first-order beam need enter the lens, $a = 1$. The formula can be rearranged to give $d = \lambda/(n \sin \alpha)$ and d, the minimum grating spacing that can be resolved, becomes a direct measure of resolving power.

The above formula for resolving power was derived for a narrow beam of light striking the grating at 0° incidence. This beam of light was also assumed to be coincident with the optic axis of the objective lens. As Abbe's theory specifies that resolution can be obtained if only two beams diffracted by the specimen enter the objective lens, it is clear that tilting the illuminating beam improves resolution. Figure 3.4*c* illustrates this. The resolving power d can

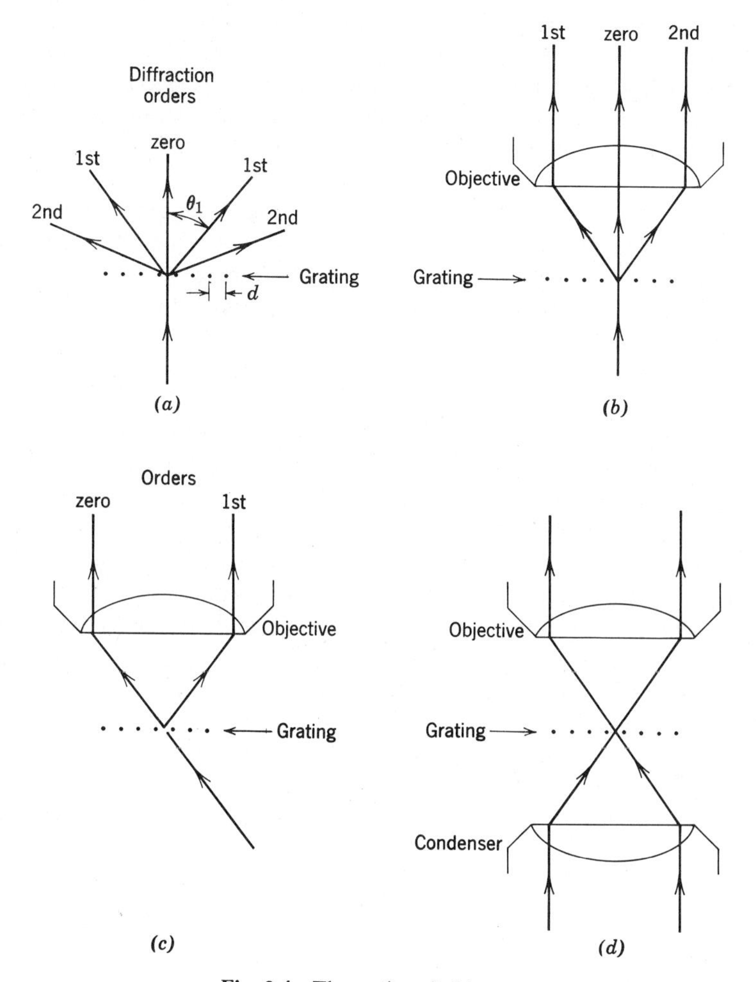

Fig. 3.4 The optics of object space.

be smaller by a factor of 2 if the incident beam is tilted so that the zero-order beam just enters one edge of the objective. The formula for resolving power then becomes $d = \lambda/(2n \sin \alpha)$. Because it is not generally desirable to have the resolving power varying with the azimuth, oblique light is ordinarily supplied in all azimuths by means of a condenser lens as shown in Fig. 3.4*d*. For maximum resolution the condenser should be of high quality, and its angular aperture should match that of the objective.

Allowing for the fact that Rayleigh's theory of resolution considers point-to-point resolution, while Abbe's theory as derived above considers line resolution, the two theories are in close agreement. Modern light microscopes attain resolving powers predicted by these theories, a fact that indicates that well-made microscope objectives can closely approach perfection at least for objects near the optic axis. More extensive discussion of resolution theory can be found in many texts on optics and in books concerned with microscopy [1, 2, 3].

It is common practice for lens manufacturers to describe objective lenses by giving the focal length or the magnification, and a characteristic called the numerical aperture (N.A.). The numerical aperture of a lens is $n \sin \alpha$. Therefore the resolution formula is often stated in the following form:

$$d = \frac{1.2\lambda}{2\text{N.A.}} \quad \text{or} \quad d = \frac{0.6\lambda}{\text{N.A.}}. \tag{3.5}$$

The resolving power of a microscope is limited primarily by the numerical aperture of the objective lens. However, the numerical aperture of the condenser should match that of the objective, and if the objective is an immersion lens the immersion medium must be placed between the condenser and the slide as well as between the specimen and the objective lens.

Attempts to improve resolving power must involve increasing the numerical aperture or decreasing λ. The numerical aperture of light microscope objectives, $n \sin \alpha$, is limited by space considerations and by the refractive index of the immersion medium. The angular aperture cannot exceed 180°, and in any case diffracted beams having an angle $\theta > 180°$ do not occur. As stated above, the use of an immersion medium having $n > 1$ shortens the wavelength, resulting in a reduction of the angle θ and an improvement in resolving power. To be effective the immersion media must surround the specimen and fill the space between the condenser and the objective. The higher the refractive index of the immersion medium the better the resolving power, but this method of improving resolving power is limited by the availability of suitable materials for lenses and for immersion media.

Decreasing the wavelength (λ) of the radiation used to form an image is the most effective way of improving resolving power. With present materials λ can be decreased by a factor of less than 2 through the use of immersion media. Changing the type of radiation used can be much more effective. For visual microscopy blue light ($\lambda \simeq 4300$ Å) produces the best resolving power and red light ($\lambda \simeq 7000$ Å) the worst. Microscopy can be extended to the ultraviolet region if the eye is replaced by photographic film or other types of receptors. The limiting wavelength is then about 2000 Å because below 2000 Å air absorbs the radiation. The vacuum region of the spectrum has not been utilized for microscopy, but microscopes have been built using x-radiation ($\lambda \simeq 0.1$ to 10 Å). The resolving power of x-ray microscopes has been limited to approximately the same value as light microscopes because, although the wavelengths are shorter, suitable lenses have not yet been developed. The most satisfying improvements in resolving power have come from the development of electron microscopes. Electron beams have associated with them electromagnetic radiation with wavelengths varying from about 0.01 to 0.1 Å and, although electron lenses are poor (N.A. about 0.02) compared with light lenses (N.A. up to 1.4), resolving powers of 1 to 2 Å have been attained.

Image Intensity

If it is assumed that the magnification and resolving power are adequate to reveal the fine structures to be examined, the next requirement is to provide sufficient illumination (illuminance) in the image so that the structures can be observed. A pinhole permits only a small cone of light to pass through it and therefore the illumination of the image is low. The illumination can be greatly improved by replacing the pinhole with a lens as is shown schematically in Fig. 3.2. The cone of rays accepted by the lens (AA) is greatly enlarged over that accepted by the pinhole, and because these rays are concentrated on a corresponding image point the illumination of the image point is much greater even though the luminance (brightness) of the object point remains constant. If the magnification is held constant, the illumination of the image increases as the square of the angular aperture of the lens.

The resolving power also increases as the angular aperture increases. Because the output characteristic of the microscope must match the input characteristic of the eye, this increased resolving power can be utilized only if the magnification is adequate. As a general rule for light microscopes the magnification should be about 1000 times the numerical aperture of the objective. The area of the image increases as the square of the magnification, and therefore if the luminance of the object is constant, the illumination of the image decreases as the square of the magnification. These factors are combined, and therefore, the illumination of the microscopical image should remain approximately constant as the magnification increases. In practice this is not the case, primarily because increasing the angular aperture of a lens requires increasing the number of elements in the lens, and there is an unavoidable light loss caused by reflection at the surfaces of these elements and by absorption of light within the elements. Antireflection coatings can minimize reflection losses and the elements of well-made modern lenses are so coated. Absorption losses also increase with the number of lens elements, but these losses are less important than reflection losses.

If it is assumed that the microscope and its illuminating system are designed correctly and aligned properly, the effects of low image illumination can be overcome in only two ways: increasing the luminance of the source or replacing the eye with a more sensitive receptor such as photographic film or an image intensifier. Consider the optics of the illuminating system using critical illumination, that is, with the light source imaged in the specimen plane. The light source is generally demagnified by the illuminating system. The size of the source must be such that its demagnified image fills the field of the objective lens being used. If reflection and absorption losses are neglected, the illumination of an image (B') depends only on the luminance of the source (B), the angular aperture of the lens nearest the source (α), and the convergence angle of the light emerging from the final lens α':

$$B\alpha = B'\alpha'.$$

Luminance (brightness) refers to light emitted by a surface and is the flux per unit area per unit solid angle. Illumination (also called illuminance) has the same units but refers to light striking a surface. The above equation assumes a perfect optical system. In actual practice there are inevitable losses in the optical system because of absorption and reflection, and there may be additional losses because of poor optical design, poor execution of the design, or poor alignment.

Image Contrast

To be recognized an image of a structure must be different from the image of its surroundings. The eye can detect differences in luminance or differences in color. Thus there are two types of contrast: luminance contrast and color contrast. A definition of luminance contrast (γ) is

$$\gamma = (L_1 - L_2)/L_1, \tag{3.6}$$

where L_2 is the luminance of the image of the structure and L_1 is the luminance of its surroundings. The human eye can detect a γ of about 0.04 for a dark structure on a bright field. Color contrast is more difficult to define, but the eye is quite sensitive to small changes in hue or saturation.

Contrast arises from the interaction of radiation with the material of the specimen. When radiation strikes a specimen it may be transmitted, absorbed, or scattered. Microscopes are designed to form an image by exploiting one or more of these interactions. Most specimens interact with radiation in all three ways, but each effect can be considered separately. Light may be transmitted without appreciable absorption or scattering, however, for nearly all specimens, light travels at different speeds in the different parts of the specimen with resulting phase shifts. The eye cannot detect phase shifts, but the polarizing microscope, the interference microscope, and the phase microscope are designed to transform these phase shifts into luminance or color contrast in the image.

Absorption of radiation produces changes in luminance contrast directly and is a major factor utilized in bright-field microscopy. Absorption also produces color contrast if the absorption varies with wavelength and if polychromatic light is used for illumination. Light energy absorbed by the specimen may cause chemical changes but more often it is transformed into heat energy which may be reradiated or dissipated by conduction or convection. However, some absorbed energy may be reradiated at wavelengths in the visible region of the specimen, thus producing fluorescence or phosphorescence. Fluorescence microscopy utilizes this phenomenon to produce image contrast. Heat (infrared) radiation can also be used to form microscopical images and, incidentally, to determine the temperature of various portions of a specimen. Of course the heat in specimens examined in such an infrared microscope may have been caused by phenomena other than absorption of an illuminating beam.

The scattering of radiation results from both reflection and refraction. Bright-field microscopes, whether transmission or reflection types, derive luminance image contrast from scattering effects. Bright-field microscopes also make use of absorption effects to produce image contrast. Absorption may modify both luminance and color contrast in dark-field microscopy. Phase microscopes mentioned above as depending on phase effects for production of image contrast also depend on scattering and on absorption. Certain microscopical techniques, such as dispersion staining and Rheinberg illumination [4], also depend on scattering of radiation by the specimen.

One of the primary tasks of the microscopist is the preparation of specimens, and one of the primary purposes of specimen preparation is the control of interaction of the specimen with light or other radiation so as to produce the desired contrast in the image. Specimens often must be made thin to prevent excessive scattering or absorption and to minimize confusion resulting from the overlapping of structures within the specimen. Cover glasses and mounting media are used to control scattering of light, staining is used to enhance absorption selectively, etching is used to enhance the scattering of polished materials to be examined by reflected light, and so on. It is clear that the method of specimen preparation must be appropriate for the type of microscopical examination to be carried out; for example, staining is used for transparent specimens that are to be examined using a transmission microscope.

Microscope Lenses

Aberrations of Lenses

The foregoing discussion assumes that lenses are ideal. The diagrams show thin, single-element lenses with spherical surfaces having relatively large radii of curvature. Such thin lenses do approach ideality for object points lying on the optic axis if monochromatic light is used and if the angular aperture is very small. Practical optical instruments such as microscopes must be able to form images of objects that do not lie on the optic axis, and the angular aperture must be large if the instrument is to have a useful resolving power. Monochromatic illumination can be used in many cases but color is an important element in most microscopical studies. Single-element glass lenses are unsuitable for microscope objectives because of their aberrations. Aberrations are present even though the lenses themselves are perfect in their manufacture and are manufactured of perfect material. Aberrations are not defects but rather are inherent properties of lenses. Aberrations can be classified in several ways. One classification distinguishes between axial aberrations and field aberrations. Aberrations can also be classified as monochromatic or chromatic aberrations. Axial aberrations, that is, aberrations that result in the failure of a lens to form a point image of an object point that lies on the optic axis, are of two types: spherical and chromatic. Spherical aberration is present when monochromatic light is used. It is related to the size of the angular aperture of the lens. Paraxial rays, that is, those lying very near the optic axis, focus at a certain distance from the lens. Rays that are not paraxial, that is, rays that pass through outer zones of the lens, are focused at a distance different from the paraxial rays. This aberration is measured by determining the difference in distance between the focal point of the axial rays and the focal point of the rays passing through the outer zones of the lens. Because the index refraction of glasses varies with the wavelength of the light, it follows that rays of different wavelengths also focus at different points. This variation of focal length with the wavelength of radiation is called chromatic aberration.

In addition to axial aberrations, there are a number of other aberrations, so-called oblique or field aberrations. These include distortion, astigmatism, coma, curvature of field, and oblique spherical aberration. These aberrations must be corrected if useful objective lenses having high resolving power are to be manufactured. Aberrations that result from the fact that most lenses have spherical surfaces, for example, spherical aberration, can be corrected by producing lenses that have aspherical surfaces. This has not proved to be a practical solution for microscope objectives since production of an aspherical surface is difficult and expensive. Spherical aberration and other aberrations are corrected by designing lenses that have multiple elements, that is, a number of individual lenses are properly spaced and mounted together and act as a single lens. These different elements are made of glasses having different physical properties. In this way many of the aberrations can be corrected or in any case a lens can be designed in which the aberrations are minimized. The aberrations of lenses and the means by which these aberrations are corrected are discussed in many textbooks on optics [5–7].

Objectives

Microscope objectives are classified according to the degree to which their chromatic aberrations have been corrected. Achromatic objectives are constructed of crown glass converging lenses and flint glass diverging ones and are able to bring light of two widely separated wavelengths to the same focus. Because they lack chromatic correction for intermediate or extreme wavelengths, a so-called secondary spectrum can generally be observed in images formed by these lenses. Achromats are spherically corrected for one wavelength only. Semiapochromatic objectives, often called fluorite objectives contain, as their name suggests, one or more fluorite lenses. These lenses give a less pronounced secondary spectrum and an overall improvement in the correction of spherical aberration. Apochromatic objectives are corrected chromatically for three wavelengths and spherically for two wavelengths. Apochromatic objectives also contain lenses made of the mineral fluorite. For general microscopy there can be no doubt of the great superiority of apochromats, but few apochromats can be used for polarized light work because of the anomalous birefringence of most naturally occurring fluorite. Because of the difficulty in obtaining suitable material for lenses and because of the difficult computations involved in computing an objective lens, it has not been possible to correct for all aberrations simultaneously over a usable field. In recent years new materials for lenses have become available and the computer has made computation of lenses much easier. As a result, many new lenses have been computed. Older lenses usually exhibited considerable curvature of field. This was not objectionable for most visual work, and for photographic work it was possible to correct for curvature field using so-called negative eyepieces. However, in recent years it has been possible to compute flat-field objectives with adequate correction for the other aberrations over the entire field. Thus objectives known as planachromats or planapochromats have become available.

Objectives used for transmitted light microscopy are designed to be used with a cover glass over the specimen. Omitting the cover glass from a specimen results in a poor image. Objectives that can be adjusted for cover glass thickness can be obtained, but ordinarily the lens is corrected for one thickness only. Commonly, the thickness chosen is 0.18 mm. Cover glasses designated number $1\frac{1}{2}$ are on the average the correct thickness. Number $1\frac{1}{2}$ or number 1 cover glasses should be used whenever possible. With microscopes having variable tube length, it is possible to correct for cover glass thickness. Objectives used for reflected light microscopy are corrected for use without a cover glass. Other types of objectives for special purposes, that is, immersion objectives, metallography objectives, phase objectives, interference objectives, and so on, are described by Benford [6] and others [2, 8, 9]. Some of these are discussed briefly in the following sections.

Eyepieces and Amplifying Lenses

The eyepieces ordinarily encountered by the microscopist are of two general types: Huygensian (negative) and Ramsden (positive). They are shown in Fig. 3.5*a*. Eyepieces are usually designed to correct, insofar as possible, lateral chromatic aberration, astigmatism, and distortion. Coma is partially corrected. Axial chromatic and spherical aberrations are accepted, as it is relatively easy to design an objective to compensate for these.

The most common eyepiece is the Huygensian, consisting of two simple planoconvex lenses with a field stop located between them. Although they

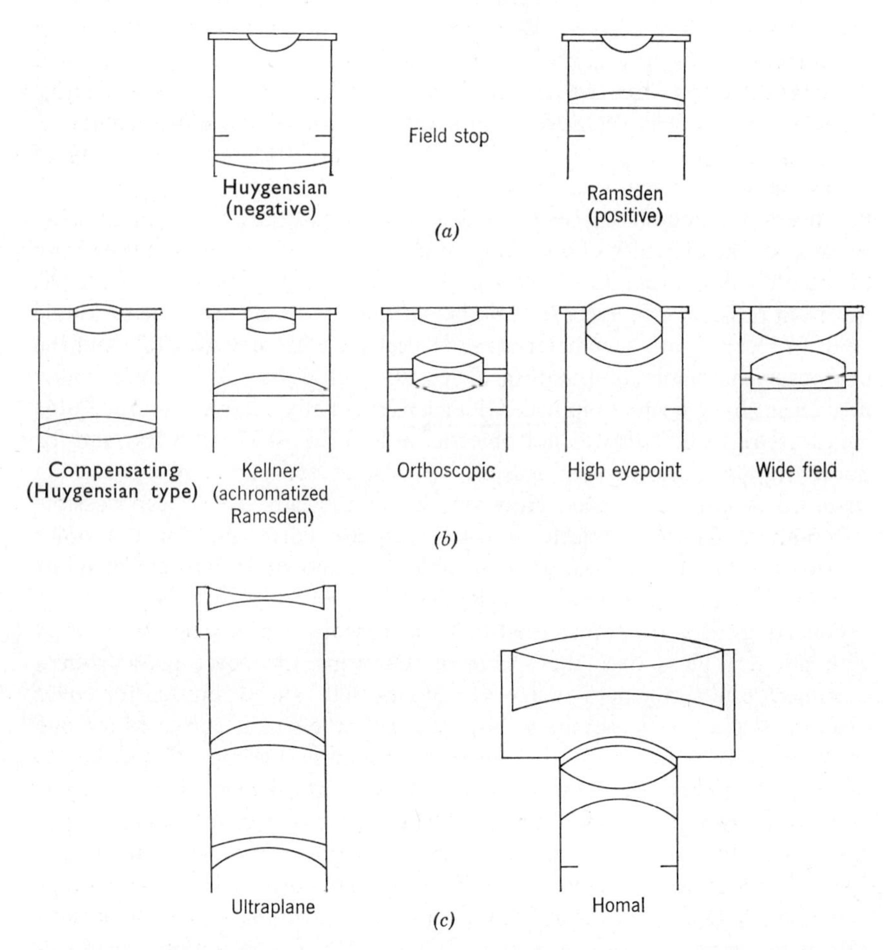

Fig. 3.5 Schematic diagrams of eyepieces and projection lenses. (*a*) General types; (*b*) some common modifications of the general types; (*c*) projection lenses (negative or amplifying lenses).

are simple and inexpensive, the Huygensian eyepieces are well corrected for lateral chromatic aberration and they work well with low- and medium-power objectives. Huygensian eyepieces manufactured by various companies are so nearly alike that they generally can be used with objectives of a different make. High-power objectives are more difficult to correct for aberrations, and it is advantageous to design eyepieces that compensate for the residual aberrations of specific objectives. Therefore high-power objectives, especially apochromats, semiapochromats, and flat-field objectives should be used with eyepieces designed especially for them. These eyepieces are generally of a more complex design than the simple Huygensian or Ramsden type (see Fig. 3.5*b*).

Projection-type eyepieces are designed to give a real image at some finite distance. They are used in photomicrography and in microprojection. Often provision is made for adjusting the separation of the components so that the focal length can be constant while the magnification is varied. Although these lenses are not designed for visual observation, a visual image can be obtained by looking through them.

Another type of projection lens is shown in Fig. 3.5*c*. These lenses are negative lenses and are called by various names such as amplifiers, negative projection lenses, or by trade names such as Homal or Ultraplane. They are used to correct for curvature of field in high-power objectives and are similar to compensating lenses with regard to the correction of lateral chromatic aberration. These lenses cannot be used for visual observation.

Illuminating Systems

To provide adequate image intensity, it is generally necessary to provide intense illumination for the specimen. Specimens that are essentially transparent are nearly always illuminated by means of a substage condenser, and the image is formed by light rays transmitted through the specimen. Images of opaque specimens are formed by light rays reflected from the surface of the specimen. For opaque specimens the illuminating system is frequently a "vertical" illuminator which uses the objective lens as a condenser, but other means such as ring illuminators, special condensers surrounding the objective, spot lamps, or reflectors are common, particularly at medium or low magnification.

A condenser should supply to the specimen plane a uniformly illuminated spot which just fills the field of view. The numerical aperture of the condenser should be as large as that of the objective. In fact, the optimum design of a condenser is identical with that of the objective being used. However, it is generally desirable to use the same condenser for a wide range of objectives so that the numerical aperture of a condenser is usually at least as high as that of the highest power objective. This leads to difficulty in filling the field of low-power objectives with uniform illumination. To overcome this difficulty provision is often made to reduce the numerical aperture of the condenser by removing or changing the top lens, by inserting another lens into the optical system, or by changing the spacing between the lens elements.

Although the optimum design of a condenser is identical with that of an objective, most condensers have much poorer correction than objectives. Reasonably good results can be obtained with poorly corrected condensers such as the Abbe type shown in Fig. 3.6, but when the best results are required

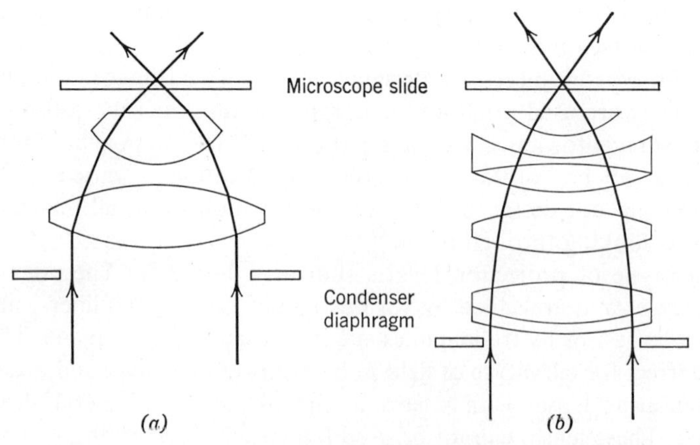

Fig. 3.6 Types of substage condensers. *(a)* Abbe; *(b)* achromatic.

achromatic condensers should be employed. Achromatic condensers are corrected for both spherical and chromatic aberration. Condensers corrected for spherical aberration only, called aplanatic, are also available.

The general principles of illumination of transparent objects also apply to opaque objects. However, in "vertical" illuminators designed for bright-field work the objective lens serves as the condenser. Dark-field and bright-field vertical illuminators for reflected light microscopy are shown in Fig. 3.41. Other illuminating systems are generally modifications of those mentioned above. These specialized illuminating systems are discussed in later sections.

Microscope Illuminators

The light source is an important part of the illuminating system. The important characteristics of the light source are its size, its luminance (brightness), its uniformity, and its spectral quality. The condenser forms a demagnified image of the source in the specimen plane. Therefore the source must be large enough so that its demagnified image fills the field of the objective. Broad, diffuse light sources such as the sky, an ordinary frosted lamp bulb, or a piece of ground glass illuminated by a lamp bulb may be adequate for visual work.

A light source can be too broad. If its image more than fills the field of the objective, the excess illuminated area is not only useless, but it is harmful to the final microscopical image because some of the excess light finds its way into and through the microscope resulting in loss of image contrast. This deterioration of the image can be minimized by limiting the size of the light source by means of a field stop such as a diaphragm or an opaque shield with a hole of the proper size. The size of the hole varies with the objective being used.

Broad diffuse sources are inefficient and usually have low luminance. Highly luminous sources such as carbon arcs, mercury or other gaseous discharge arcs, or tungsten filament lamps are used when the specimen must be strongly illuminated. Highly luminous light sources are usually small if they are to be uniform and if the power needed to operate them is kept to a reasonable value. Such sources, if focused by the condenser alone, generally produce images too small to fill the field of the objective. Therefore they are used with an auxiliary condenser, a lens placed in front of them, which serves to magnify the image of the source and to collect the light from the source more efficiently.

Auxiliary condenser lenses should have high angular apertures, that is, they should be fairly large in diameter and have a short focal length. They should also be corrected for spherical aberration but they need not be highly perfect lenses. Aspheric lenses are often used as condenser lenses. They are generally made of low-expansion glass such as Pyrex because of the heat given off by highly luminous sources. The light source and the auxiliary condenser are mounted in a housing that provides for ventilation while preventing stray light from getting into the observer's eyes. Provision is made for focusing the light by moving either the light source or the auxiliary condenser lens. The better illuminators also have a field stop (a diaphragm) closely adjacent to the auxiliary condenser lens. A good microscope illuminator is a necessity for microscopical work requiring high image intensity, for example, photomicrography, and dark-field microscopy. The luminance of these lamps can be reduced for visual microscopy by introducing neutral density filters, or by reducing the power to the light source.

If the light source is to be focused in the plane of the specimen as it is with critical illumination (see Fig. 3.7), it must be uniformly luminous over an area large enough to fill the field of the objective. If it is assumed that the objective has been selected and that the uniform area of the light source itself is fixed by physical considerations, the only factor that can be varied is the magnification of the condenser–auxiliary condenser system. The ratio of the focal lengths of these two components controls the magnification. If the

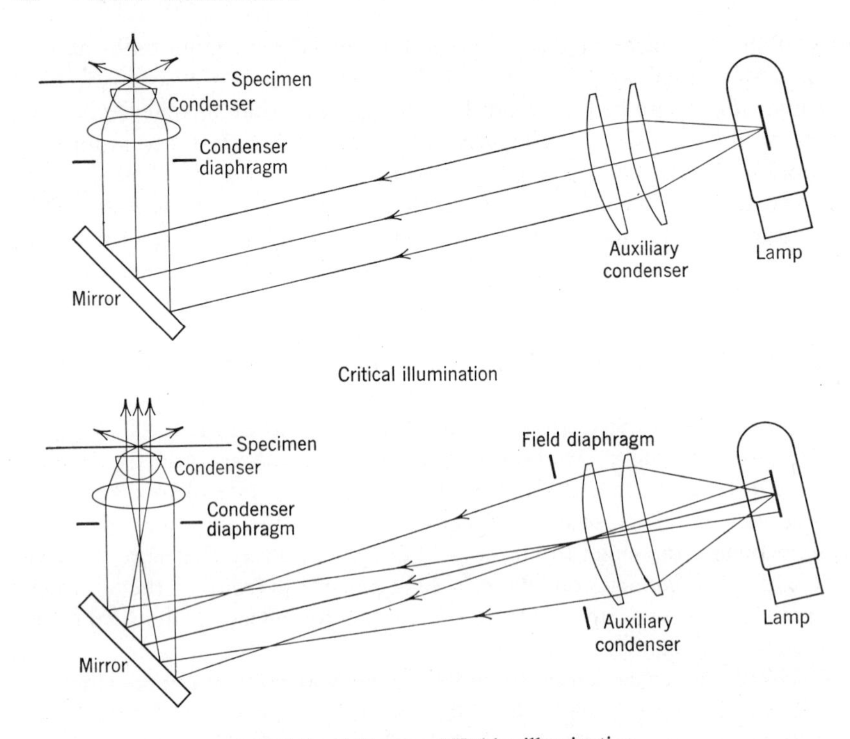

Fig. 3.7 Critical and Köhler illumination.

focal length of the condenser is larger than that of the auxiliary condenser, the image of the light source is magnified. The practical magnification of the illuminating system is limited by physical considerations such as space required, the complexity of the auxiliary condenser lens system, and the light losses suffered when lens systems become more complex. Unfortunately, the higher the luminance of a source the smaller it tends to be.

Sources such as coiled tungsten filaments are not uniformly luminous. They do, however, have inherent advantages in that they can be manufactured easily and inexpensively, and by proper arrangement of the coils they can be made to cover a relatively large area with some degree of uniformity. Because of their large size they can be used with auxiliary condensers having long focal lengths and relatively simple design. Such sources cannot be used for critical illumination, but they are quite satisfactory when Köhler illumination [10] (see Fig. 3.7) is used. With Köhler illumination the auxiliary condenser lens itself is focused in the plane of the specimen, and the filament is focused in the lower focal plane of the condenser. A diaphragm or stop placed near the auxiliary condenser lens serves as a field stop. Köhler illumination is widely used in microscopical work because it provides good control of the illumination and because it can be used with readily available coiled filament lamps.

Complete Optical Systems

Figure 3.8 shows the complete optical system of a compound microscope arranged for transmission microscopy using Köhler illumination. To simplify the diagram only two rays from each of two points on the source are traced through the system. The light source (A_1) is imaged at locations lettered A_2, A_3, and A_4. The object located at F_2 is imaged at F_3 and F_4. The diaphragm F_1 is imaged in the plane of the object F_2 and of course at F_3 and F_4. In Figure 3.8 the final image is formed on the retina of the eye, however, the eye can be replaced by a camera.

When the microscope is focused properly, the rays from any point in the image at F_3 are parallel after leaving the eyepiece. This makes it possible to focus the image on the retina while the eye is relaxed, as it is when viewing a distant object. The microscope can of course be focused so that the rays reaching it from a point in the image at F_3 are slightly diverging. In the latter case the eye must accommodate for near vision. Microscopists suffering from myopia or hyperopia need not wear glasses when using a microscope as it is possible to accommodate for these visual defects by changing the focus of the microscope.

Stops and Diaphragms

The magnification of a microscope is controlled by selecting the objective and the eyepiece. Once the objective has been selected, the resolving power of the microscope is determined. However, if it is assumed that the magnification is adequate, the resolution attained depends on the skill of the microscopist. He must of course prepare the specimen properly, but he must also understand the effects of the stops and diaphragms on the image and he must be able to manipulate these stops and diaphragms to obtain the effect he desires.

Stops are openings of a fixed size; diaphragms are openings that are variable in size. The edge of a lens may act as a stop. For purposes of this discussion, the terms stop and diaphragm are equivalent. Stops should absorb all of the light rays they intercept, and for this reason they are nearly always flat black. There are two classes of stops: field stops and aperture stops. Field stops control the size of the field, and aperture stops control the angular aperture of the lenses.

Field Stops

The primary purpose of a field stop is to prevent the loss of contrast that results when stray light enters the microscope and arrives at the image. In

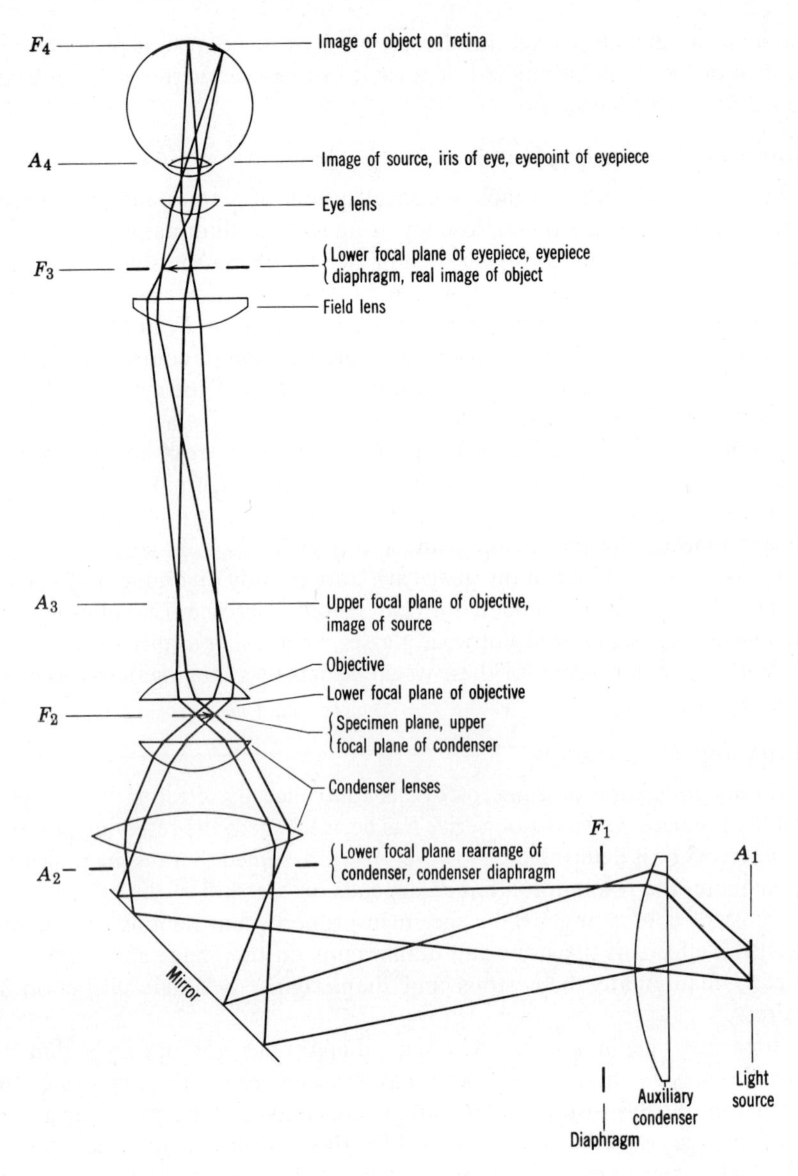

Fig. 3.8 The optical system of the microscope arranged for critical illumination.

the section on microscope illuminators, it was pointed out that illuminating parts of the specimen that are outside the field of view introduces stray light into the microscope. However, stray light may also result from the reflection of image-forming rays at the surfaces of lens elements or from the scattering of light by dirt on the surfaces or within the lens elements. Coating the lens elements reduces reflection and cleanliness reduces scattering caused by dirt, but stray light from these sources cannot be completely eliminated. The proper use of field stops prevents much of the stray light from reaching the image.

In Fig. 3.8 the possible locations of field stops are lettered *F*. It is often impossible, inconvenient, or unnecessary to place physical stops at all of the lettered locations. The placing of a stop in the plane of the specimen (F_2) is difficult and inconvenient. A stop at F_4, at the retina of the eye, is unnecessary as well as impossible, however, if the eye is replaced by a camera, the film holder usually acts as the limiting field stop.

All of the field stops are at image positions conjugate with the specimen, and therefore all field stops can be imaged on the retina. Ideally, their images on the retina should all be the same size. In most microscopes the stop in the eyepiece (F_3) is a fixed stop and the stop in front of the auxiliary condenser (F_1), which is usually a diaphragm, is adjusted to match F_3.

Insofar as field stops are concerned, a microscope is properly adjusted when all of the field stops that can be manipulated are in focus and the same size. In visual microscopy this means that F_1 should be focused in the plane of the specimen and should coincide with the field of view as limited by F_3. In photomicrography F_1 is adjusted to match F_4. F_1 is imaged in the plane of the specimen F_2 by focusing the condenser.

Aperture Stops

The primary purpose of an aperture stop is to control the angular aperture of a lens. Controlling the angular aperture of the objective lens sets a limit on the resolving power and also affects the contrast in the final image. These effects were discussed on p. 122. The possible locations of aperture stops are lettered *A* in Fig. 3.8. There are aperture stops in all of these locations because even in the absence of a separate stop the edge of the lens serves as a stop.

The effective angular aperture of a lens is reduced if the beam of light passing through it is too small in diameter or not sufficiently divergent. For example, if the size of the source (A_1) is too small, then its first image at A_2 will not fill the condenser lens with light, with the result that the angular aperture of the rays emerging from the condenser will be reduced and, consequently, the aperture of the objective may not be filled with light.

Microscopes are designed so that no part of the optical system prevents the angular aperture of the objective from being filled with light.

The numerical aperture of the objective limits the resolving power of the microscope, and therefore the objective chosen for a given task depends on the size of the structures that are to be examined. The objective chosen should have the lowest numerical aperture capable of giving the required resolving power. The use of an objective having a numerical aperture higher than necessary for the task results in loss of contrast and loss of depth of focus. The necessary magnification can be obtained by choosing the proper eyepiece. Thus A_3 in Fig. 3.8 is determined by choosing the appropriate objective.

As pointed out on p. 131, the condenser has a maximum numerical aperture at least equal to any objective that may be chosen. It was also shown on p. 125 that in order to obtain the highest possible resolution the numerical aperture of the condenser must match that of the objective. The numerical aperture of the condenser is controlled by a diaphragm A_2 in Fig. 3.8. When the microscope is focused, the diaphragm A_2 is imaged at A_3 and A_3 can be observed by removing the eyepiece of the microscope and observing the back focal plane of the objective. The numerical apertures of the objective and condenser are matched when the image of A_2 and the back of the objective lens coincide. Ordinarily, the numerical aperture of the condenser is made slightly smaller than that of the objective so as to reduce the loss of contrast resulting from stray light.

Aperture stop A_4 is located at the eyepoint of the eyepiece. The physical stop is the iris of the eye. For some designs of eyepieces and depending on the intensity of the light, the iris may contract enough to limit the aperture of the microscope system. However, this is not common as eyepieces are usually designed so that the diameter of the beam at the eyepoint (the Ramsden disk) is smaller than the iris of the eye. The eye should be located so that the eyepoint of the eyepiece is coincident with the iris.

The size of the light source can limit the effective numerical aperture of the objective. When Köhler illumination is used, the image of the light source A_1 must fill the condenser aperture A_2. This depends on the physical size of the light source and on the focal length of the auxiliary condenser and on the distance from F_1 to A_1 (see p. 135). With "built-in" illuminating systems, the requirements for proper illumination must be met by the designer of the illuminator. When using a separate microscope illuminator, the microscopist must select an illuminator that fulfills the requirements. The adequacy of the illuminator can also be checked by observing the back focal plane of the objective.

Just as the field stops can be checked by observing the image formed by a microscope, the aperture stops are checked by observing the back focal plane of the objective after removing the eyepiece. If the microscope is properly adjusted and aligned, the back of the objective will be uniformly filled with light and the condenser diaphragm will be just visible in and concentric with the back lens of the objective. To check for the concentricity of the objective and condenser diaphragm, it is necessary that the eye be centered at the top of the body tube. A pinhole eyepiece can be used to ensure that the eye is centered. To aid in focusing on the back of the objective, the pinhole eyepiece can be fitted with a +6 diopter lens. Another device which is very useful for observing the back aperture of the objective is the special eyepiece microscope usually used for aligning the optical system of a phase microscope. The Bertrand lens in petrographic microscopes can also be used for this purpose. The condenser should be made concentric with the objective by means of centering screws in the condenser mount.

3 THE MECHANICAL PARTS OF THE MICROSCOPE

The mechanical system of a microscope has several important functions: it supports the imaging system, the specimen, and all or part of the illuminating system; it provides means of focusing the imaging and illuminating systems on the specimen; it provides means for aligning these systems; and it often supports accessory equipment such as a camera.

From the latter part of the nineteenth century until fairly recently, the most popular microscope stands have been similar to that shown in Fig. 3.9. The base consists of a horseshoe-shaped foot with a pillar rising from it. The arm, to which is attached the body tube, the stage, the condenser, and the mirror, is attached to the pillar by means of an inclination joint. The mirror is mounted in a fork which is inserted into the lower end of the arm. The mirror can be tilted and rotated so as to direct a beam of light from the illuminator into the condenser lens. The condenser is mounted on a dovetail slide and can be focused on the specimen by means of a rack and pinion. The stage that supports the specimen is usually rigidly attached to the arm. The body tube, which carries the imaging system, is mounted at the upper end of the arm on a dovetail slide. The imaging system can be focused on the specimen by means of a rack and pinion, the pinion being driven by a coarse-adjustment knob, and although this motion is suitable for focusing low-power lenses on the specimen it is too coarse for focusing medium- or high-power objectives. Therefore the dovetail slide of the coarse-focusing mechanism is in turn mounted on a second dovetail slide. The body tube plus the coarse-focusing mechanism on the second dovetail slide is moved by means of a fine adjustment. The fine adjustment is an arrangement of screws and levers, cams, inclined planes, or the like, which imparts a very slow movement to

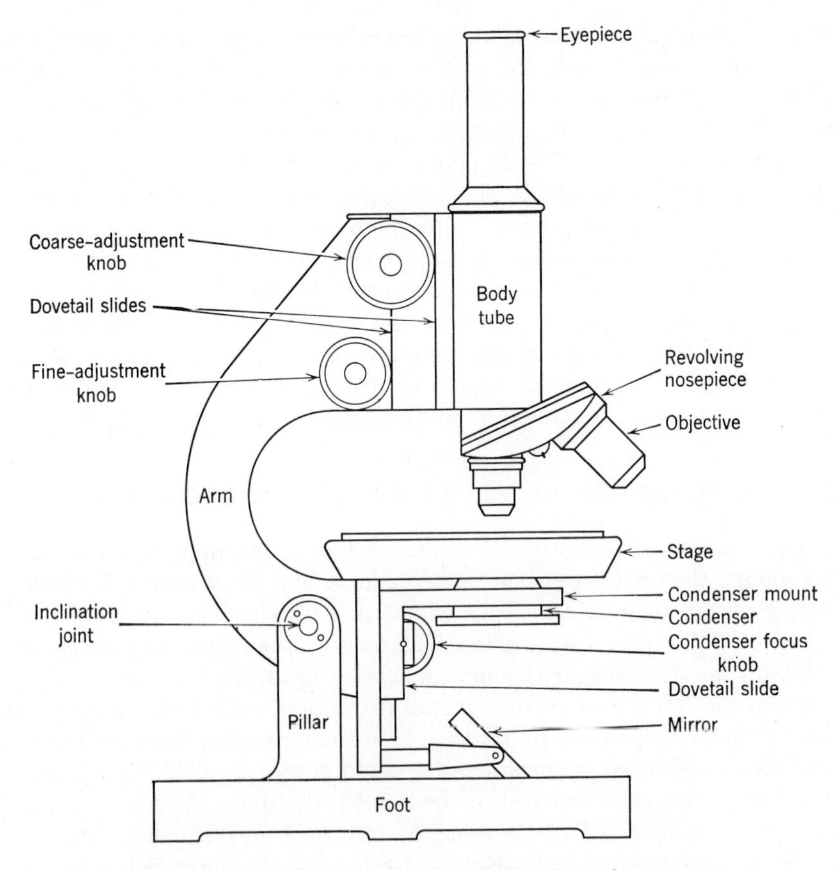

Fig. 3.9 Drawing of a microscope showing the various parts.

the body tube. Commonly, a complete turn of the fine adjustment knob moves the body tube about 0.1 mm. Usually, the knob of the fine adjustment is graduated to serve as a micrometer for vertical measurement.

Objectives are commonly held in a revolving nosepiece at the lower end of the body tube. The revolving nosepiece, which may hold from two to five objectives, makes it possible to change objectives quickly and safely and provides storage for objectives when they are not being used. The eyepiece is inserted into the top of the body tube. Objectives are designed to be used at a designated "tube length." The tube length so designated is the mechanical tube length: the distance from the surface of the nosepiece against which the shoulder of the objective fits to the top of the body tube. Older microscopes are often fitted with a draw tube which allows the mechanical tube length to be adjusted, but in recent years this adjustment has usually been eliminated.

Fig. 3.10 A modern microscope. The Orthoplan, manufactured by E. Leitz, Inc. A photomicrographic camera is attached.

Microscope stands must be extremely rigid and stable and the adjustments, particularly the coarse and fine adjustments, must be precise and essentially free from sideward motion. Sideward motions of the specimen relative to the objective may be induced by vibrations in the laboratory or by pressure applied by the microscopist as he operates the microscope. Such motions should be kept below the resolving power of the highest power objective (0.0002 mm). This requires rigid stands and slideways which are fitted to very close tolerances. The focusing mechanisms must be capable of positioning the body tube within the depth of focus of the highest power objective, approximately 0.0002 mm. There should be no lag in the response when the fine adjustment is changed, and there should be no sideward motion when the fine adjustment is started or reversed. The revolving nosepiece should be constructed so that when objectives are changed their optic axes are accurately centered with respect to the condenser. Microscope stands should also be constructed so that they can be used without causing unnecessary operator fatigue. Accessories and fittings should be easily attached and adjusted.

Figure 3.10 shows a contemporary microscope which exemplifies some of the recent trends in microscope design. The base and the arm are one large hollow member providing great rigidity. The inclination joint has been eliminated, but the body tube is inclined for comfortable viewing. Binocular body tubes have become very common, and "trinocular" tubes are available for photomicrography. The focusing adjustments move the stage rather than the body tube and objectives. The illuminating system is built into the hollow base and arm, and it is possible to use either transmitted illumination or vertical illumination separately, or both types of illumination simultaneously.

4 SPECIMEN PREPARATION

Samples of materials for microscopical examination are seldom received in a suitable condition for examination. In addition to being able to choose the appropriate type of microscope, to set it up, and to use it, the microscopist must be able to prepare the specimen properly, and finally he must be able to interpret what he sees. The interpretation of microscopical images and the validity of the scientific conclusions drawn depend on the microscopist's knowledge of the sample and on his skill in, and understanding of, the techniques by which he prepares the specimen. It is of course his background knowledge that distinguishes the chemical microscopist from the petrographer, the metallographer, the biological microscopist, or other similar microscopists. However, to be fully effective a microscopist must be informed about and adept in all types of specimen preparation, not just those commonly used in his specialty.

Preparing a specimen that is truly representative of the material to be examined is difficult. The sampling problems encountered in almost all chemical analyses are increased because the volume of specimen that can be examined in practical microscopical studies is quite small. These problems can usually be solved by using the lowest magnifications possible, by examining many fields, and by following all steps in the preparation of the specimen with low-power microscopes or magnifiers. The examination of a specimen at high magnifications without the thorough understanding of the relationship between the specimen and the original sample is very likely to result in misinterpretation of data and erroneous conclusions.

The operations involved in preparing a specimen from a sample often cause changes in the sample material. For example, ductile materials, such as metals and polymers, may be distorted by cutting or polishing operations, parts of a sample may be dissolved or altered if treated with a solvent or other chemical reagent, or changes may occur if the sample is heated. Although it is often impossible to avoid these changes, it is possible to avoid misinterpretations of data by thoughtful selection of specimen preparation techniques, by the use of control specimens, and by following the steps in the preparation, carefully using low-power microscopes and magnifiers.

Preparation of Particulate Materials

Mounting and Mounting Media

Particulate materials may be prepared for examination at low powers and by reflected light simply by spreading them on a microscope slide. However, even these simple objects generally must be mounted in some medium and covered with a cover glass if they are to be examined at medium or high powers using transmitted light. The mounting medium controls the effective transparency of the preparation by reducing refraction and reflection. If the refractive index of the mounting medium is the same as that of a colorless particle, the particle will become invisible, that is, the contrast will disappear. The refractive index of the mounting medium generally should differ from that of a colorless object by at least 0.1, the greater the difference the greater the contrast. Colored objects or objects for which inner structures are to be examined can be mounted in media of matching refractive index. This tends to make surface structure disappear and enables the microscopist to see inner structures more clearly. If surface structures are to be examined, the refractive index of the mounting medium should differ appreciably from that of the object. Water or organic liquids are often used for temporary mounts, while resins or polymers are common as media for more permanent mounts.

Media for permanent mounts must fulfill three criteria: they must not dissolve or react with the material to be mounted; they must have a suitable refractive index; and they should be nonvolatile and as viscous as possible at room temperature. Most mounting media are organic materials and therefore little difficulty arises when inorganic specimens are to be mounted, however, the first condition is usually difficult to fulfill if organic compounds are to be mounted. Of two mountants that have a low solvent action for many classes of organic compounds, liquid petrolatum (USP, heavy grade) is too fluid and glycerol jelly has too low a refractive index in most cases. An alternative is saturating a medium of suitable viscosity and refractive index with the substance under study, a convenient way being to dissolve some of the substance in a drop of hot mountant on an object slide, cool it to about 50°C, place some well-formed crystals in it, and cover with a no. 1 cover glass. The slide is then placed under gentle pressure and allowed to cool to room temperature. Many commercial synthetic resins are now available which, either alone or in a mixture, are suitable for making semipermanent mounts. The following list of mountants is given by way of illustration:

Polyvinyl alcohol mixtures, $n \simeq 1.38$ [11]
Glycerol jelly, $n \simeq 1.40$
Hard Canada balsam in xylene or fluid Canada balsam, $n = 1.53$
Fifteen grams of hard Canada balsam in 10 ml of α-bromonaphthalene, $n = 1.59$
Ten grams of hard Canada balsam in 30 ml of Aroclor no. 2354, $n = 1.61$
Twenty grams of polystyrene in α-bromonaphthalene, $n = 1.64$
Aroclor no. 1262, $n = 1.65$

Glycerol jelly is made by soaking dry gelatin in 6 parts of water for 2 hr; the solution is mixed with an equal weight of glycerin, 2% phenol added, and the mixture warmed. To prepare the polystyrene medium, 40 ml of α-bromonaphthalene are poured over 20 g of polystyrene resin contained in a tall-form 200-ml beaker. A short stirring rod is placed inside the beaker, which is then covered with a watch glass to prevent evaporation of the solvent. The beaker is maintained at 70°C for about 1 week, with occasional stirring until the mixture is homogeneous. It is then allowed to remain undisturbed in a hot-air oven for a further 3 or 4 days to allow undissolved solid material to settle and is finally decanted into a clean, dry, wide-mouth bottle. This medium has a low solvent power for organic materials and is viscous at room temperature. However, it hardens slowly by evaporation of the α-bromonaphthalene and its index falls from 1.64 to 1.61.

Excellent permanent mounts can be made using catalyzed monomers that subsequently harden by polymerization. Epoxy resins work very well and a wide variety of such resin systems are available. Those used for embedding biological specimens for ultramicrotomy are convenient and easily obtained [12, 13]. Polymethacrylates, polystyrene, polyesters, and other polymers are also useful.

Specimens of particulate materials are usually mounted by placing them in a drop of the mounting medium on the microscope slide. Resinous media and glycerol jelly are heated to reduce their viscosity. Care must be taken to eliminate air bubbles which are often entrained in the medium when the specimen is added. A cover glass is carefully placed on the mixture and is pressed down while the mounting medium hardens. For permanent mounts it is usually desirable to seal the edges of the cover glass with varnish, gold size, shellac, marine glue, asphalt varnish, or similar materials. Diluted liquid glue or casein cement is often useful if it is found that other varnishes diffuse into or dissolve the preparation.

Separations

Often it is advantageous when working with mixtures of particulate materials to separate the various constituents. This can be done in many ways such as selective solution, heavy liquid separation, centrifugation, sublimation, or ashing. However, the microscope is unique in permitting separations based on the morphology of the constituents. In many cases physical separations are unnecessary for microscopical analyses. For example, paper fibers spread on a slide may be recognized by their morphology or by selective staining reactions. When physical separations are desirable, they can often be accomplished by sorting out various constituents. Such separations can be made using tweezers, probes, probes with adhesive on the tips or other similar instruments. The instruments can be hand held or attached to a mechanical micromanipulator of some kind.

A type of microscope that is useful for sorting and for general low-power work in the laboratory is the two-objective stereoscopic binocular, often referred to as the Greenough binocular. This type of instrument uses matched pairs of objectives, Porro erecting prisms in the body tube, and matched pairs of high-power, wide-field eyepieces. The long working distance, erect image, and true stereoscopic vision are particularly useful for low-power viewing of manipulations with uncovered and unmounted objects and in mounting specimens for high-power optical and electron microscopy.

Preparation of Crystals

The chemical microscopist is often called upon to examine crystalline materials. Frequently, these crystals lack well-formed recognizable faces because they have been ground to reduce their size or abraded during handling. If they are soluble, specimens of these materials can be prepared by crystallization. Crystallization can be carried out on a macroscopical scale, for example, in laboratory vessels or in chemical processing equipment. Alternatively, samples may be crystallized on a microscopical scale while the process is observed under the microscope. Both macro- and microscopical techniques have characteristic advantages and disadvantages, and usually it is desirable to use both techniques.

Macroscopical techniques allow more exact control of concentrations, temperature, pressure, and similar variables. Extreme temperatures, potentially damaging to a microscope, can be used. The possibility of simulating actual chemical processes is an important advantage. However, in large-scale equipment it is usually impossible to observe crystal growth continuously, and as a result important growth phenomena may not be noticed.

Crystallization on a microscopical scale, that is, in a drop of solvent on a microscope slide or in a microchamber has the great advantage that growth phenomena can be followed continuously. It is also a process that can be carried out quickly and easily. However, growth conditions cannot be exactly controlled. This lack of control may in some cases be an advantage, as a range of conditions can be observed in the same specimen. Crystals grown on a microscope slide are likely to be small but more perfectly formed and with fewer faces than those produced in larger batches. Techniques for crystallization are described briefly below, but a more extensive description can be found in the *Handbook of Chemical Microscopy* [1].

Crystallization from Solvents

Among the various ways of crystallizing a substance from a solvent, the following are the most useful:

1. Slow evaporation of the solvent is useful for crystallization of organic compounds on microscope slides. The rate of evaporation of very volatile solvents can be slowed down by covering the preparation with a watch glass. Care should be taken to prevent the absorption of water vapor from the air when working with acetone or methanol. An effective way of doing this is to allow the evaporation to take place in a desiccator containing a drying agent and an evaporating basin half filled with an absorbent such as butyl phthalate. Activated charcoal can also be employed for this purpose. Crystallization on microscope slides is a rapid means of checking the purity of an organic substance when the impurity has different optical properties. When the object is to obtain well-formed crystals, it is sometimes advantageous to perform the crystallization in a small beaker.

Crystallization of water-soluble inorganic compounds on a microscope slide is usually easier than crystallization of organic compounds. A drop of water about 7 to 10 mm across is saturated with the compound. As evaporation proceeds, crystals form at the edge of the drop. These crystals are poorly formed, and it is best to push them into the center of the drop with a stirring rod. When crystals start to grow in the interior of the drop, they are usually well formed and suitable for study. Continued evaporation can be slowed by covering the drop with a cover glass.

2. Cooling a hot, saturated solution usually works well when crystallizing on a macroscopical scale. It can also be used to advantage for some microscopical preparations, especially with water-soluble inorganic materials.

3. Dilution of a solution of a substance with a nonsolvent; for example, the addition of water to alcohol solutions, or of ligroin to benzene solutions, often produces usable crystals. Inorganic salts can often be precipitated by addition of alcohol or acetone to an aqueous solution.

4. Salting out substances from their solutions allows sulfonates to be thrown down as sodium salts by the addition of a large quantity of sodium chloride or sodium sulfate to their solutions. From the microscopic point of view, sodium chloride has the advantage that it forms easily recognizable

isotropic crystals of square or rectangular outline, so that crystals of the organic substance are readily distinguished under polarized light.

5. Crystals may be formed by precipitation reactions of various types. Such reactions can be carried out on a macroscopical scale or on a microscopical scale. For example, lead iodide can be formed by allowing a drop of potassium iodide solution to flow into a drop of lead nitrate solution on a microscope slide. Crystals of free organic acids or bases may sometimes be obtained by addition of a mineral acid to aqueous solutions of salts or organic acids, or by addition of strong alkali to aqueous solutions of organic bases.

Crystallization by Sublimation

Substances that have appreciable vapor pressures at temperatures below their melting points are often readily crystallized by sublimation. This method is useful with substances that do not readily crystallize from solvents and particularly as a means of separating a small quantity of a volatile substance from a large bulk of a relatively nonvolatile one. The rate of sublimation of a substance depends not only on its vapor pressure at the operating temperature but also on the temperature of and distance from the condensing surface and on the atmospheric pressure. Much can be accomplished by an experienced worker without the aid of elaborate apparatus; material can be sublimed from one object slide to another, or from a small watch glass to an object slide. With such simple apparatus there are three important rules for successful operation: (1) avoid drafts; (2) use a small source of heat such as a microburner or a small coil of electrically heated Nichrome wire; (3) heat the specimen very slowly. It is sometimes desirable with a substance of low melting point or great volatility to cool the object slide that receives the sublimate by means of a small block of aluminum or a small vessel containing water. However, condensation on too cold a surface may cause the crystals to be too small for useful observation. Mixtures of two volatile substances may be fractionally sublimed if their volatilities differ appreciably, but care must be taken to heat the mixture very slowly and evenly, the sublimate being received on a series of object slides laid down sublimate side up in sequence as obtained. The progress of sublimation can be watched under the microscope, and the temperature range of sublimation observed with the aid of a hot stage; applications of this method are discussed in Kofler and Kofler [14, 15]. Various types of apparatus have been devised to facilitate sublimation under reduced pressure; a list of references is given in Chamot and Mason [1].

Crystallization from the Melt

The preparation of crystal films of organic compounds by cooling the fused substance as a thin film under a cover glass has some very useful applications. The crystals may be made as thin as desired, and quite often a variety of optical orientations are obtained in the same slide so that the interference figures can be observed to the best advantage. The usual way of preparing microscope slides by fusion is to place a few milligrams of substance on an object slide and cover it with a no. 1 cover glass. It is then heated very slowly, the temperature being raised up to the melting point and held there until fusion is complete. The slide is cooled slowly. It is sometimes necessary to seed the edge of the preparation by touching it with a trace of the solid substance on the point of a needle. After the substance has solidified, it may be recrystallized by cautiously melting the substance under about three-quarters of the cover glass. The slide is then allowed to cool slowly. The solid substance seeds the melt and much larger uniform areas of crystal are formed. A metal bar heated at one end and cooled at the other forms a very convenient hot plate for fusion work. It is easily calibrated for temperature gradient, so that it becomes a simple matter to crystallize a melt at any desired temperature. Under certain conditions it is an advantage to solidify the fusion preparation very suddenly; this is the case with substances that normally crystallize to yield a single optical orientation as revealed by covergent polarized light. Very rapid crystallization often results in the formation of some crystals with a different orientation, after which the preparation may be partly remelted and then slowly cooled to give larger crystals of new orientation.

The fusion method is particularly valuable as a rapid means of comparing a known with an unknown substance. Very small quantities of the two substances are melted at opposite edges of a cover glass and allowed to flow under the cover glass so that they meet somewhere near the center (Fig. 3.11). The slide is slowly cooled, preferably on a heated bar, and the known substance is seeded. The course of crystallization may then be watched with a hand magnifier or a low-power bench microscope in order to observe whether or not the crystals continue their growth with undiminished velocity at the boundary between the known and unknown substance. If the two substances are identical, the crystals will continue past the boundary, and if the slide is then heated from one end of the boundary, so as to establish a temperature

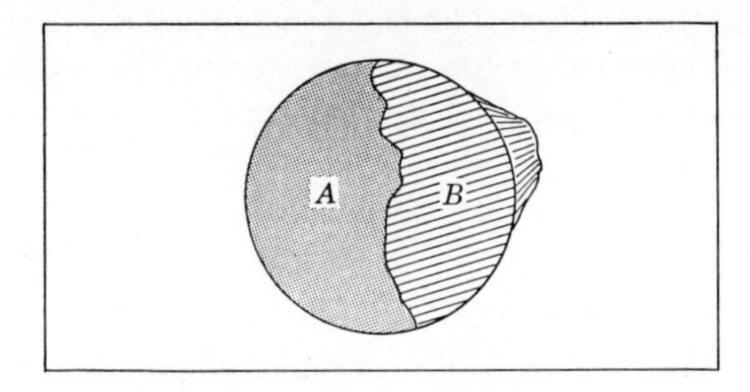

Fig. 3.11 Preparation of gradient fusion slide. Substance *A* is melted under the cover glass and allowed to freeze so that part of the space is filled. Substance *B* is placed at the edge of the cover glass, melted, and allowed to fill the rest of the space.

gradient across the slide from A to B, the two substances will melt simultaneously. The optical properties of the crystals that are determinable, such as retardation and optical orientation, do not exhibit any discontinuity at the boundary if the two substances are identical. Likewise, polymorphic transitions are not checked at the boundary. It should be remembered that a small proportion of an impurity in either the known or unknown substance may differentially slow up the crystallization and diminish the degree of perfection of the crystals obtained. If the two substances are dissimilar, various differences may occur. When the substances are dissimilar and not isomorphous, crystals grow through the known substance and stop at the boundary. Even if the crystallization of the known creates sufficient disturbance at the boundary to start crystallization of the unknown, the rate of growth and optical properties of the crystals will present a discontinuity at the boundary. On melting back the substance, differences on either side of the boundary will be apparent, and also in the narrow boundary region where the two substances have mixed, for there is the possibility that eutectics or even definite addition compounds may be formed. In comparison slides in which the known and unknown are isomorphous, a slight slowing up of crystallization at the boundary may often be observed, and the crystals that grow in the narrow region of mixing are usually not as well formed as those of the pure known and unknown. Some pairs of substances are not easily differentiated by fusion comparison methods: for example, *p*-dichlorobenzene and *p*-dibromobenzene, 2-4-dichloro-1-naphthol and 2-4-dibromo-1-naphthol.

Jelley [16] used an interferometric method employing half-silvered slides and cover slips as a rapid means of comparing the refractive indices at the boundary of a crystal that had grown from a known into an unknown substance by studying the interference fringes given in plane-polarized monochromatic light. Both the object slide and the cover glass must have fairly good optical surfaces since the best results are obtained when the two half-silvered surfaces are parallel. Particularly good results are obtained from slides and covers constructed from optically worked pieces of fused quartz that have been given a light coating of platinum by cathodic sputtering. The platinum as first deposited is very easily removed by rubbing, but after the coated slides have been heated to 700°C and slowly cooled in a muffle furnace the coat becomes very resistant. When a single substance such as β-bromonaphthalene is fused between the two platinized surfaces and is then allowed to crystallize, the individual crystals show only very broad interference bands when examined with a low-power microscope in polarized monochromatic light. The same interference pattern is observed with β-chloronaphthalene. If, however, a mixed slide is prepared with β-chloronaphthalene on one side and β-bromonaphthalene on the other, and if some mixing has occurred at the boundary, crystals growing through the boundary suffer a change of refractive index. This change is shown in polarized monochromatic light, such as sodium ($\lambda = 589$ mμ) or mercury ($\lambda = 546$), when the crystal is in the extinction position and the analyzer has been withdrawn. Under these conditions the crystal in the boundary zone shows some closely spaced interference bands. The position, and sometimes the spacing, of the bands changes when either the crystal or the polarizer is rotated through 90°. If the two refractive indices presented by the "known" part of the crystal are n_1 and n_2, and those of the "unknown" part are n_1' and n_2', and if the crystal has a thickness of d mμ, then the number of bands in the boundary zone is $2d(n_1' - n_1)/\lambda$ and $2d(n_2' - n_2)/\lambda$, respectively (n_1' and n_2' being taken as greater than n_1 and n_2). With a preparation 0.03 mm thick, a refractive index difference of 0.01 gives a single interference band. Under favorable conditions fractions of a band can be estimated.

Preparation of Materials for Examination of the Surfaces

Replication

The surface of a sample may be examined directly by using techniques described in Section 5, p. 263. However, in many cases the surface may be inaccessible for direct viewing, or interpretation of the image may be complicated by structures beneath the surface. One of the most valuable techniques that can be applied to such samples is replication. Replication has been developed extensively for electron microscopy [12], but it has also been used for light microscopy [1, 17]. The simplest way to prepare a replica is to coat the surface to be examined with a viscous lacquer, for example, collodion, allow the solvent to evaporate, and strip the dried collodion film from the surface. Another alternative is to soften a piece of sheet polymer, for example, cellulose acetate, in a solvent, and place it on the surface to dry. Still other varieties of replicas can be used, but those described above are usually satisfactory for light microscopy. Replicas can be removed from a surface by mechanical stripping or by dissolving the specimen. Loose material on the surface of the specimen often adheres to the mechanically stripped replica. The chemical microscopist can take advantage of this in cleaning a surface or in removing deposits, for example, corrosion products, from a specimen for further analysis. If the replica is stripped by chemical means, a separation of phases may be accomplished. For example, steels containing carbides can be coated with the replicating lacquer and the iron dissolved in acid leaving the carbide particles clinging to the replica [12]. Such replicas are called extraction replicas.

Shadowing

Metal shadowing [12, 17, 18] is another technique, developed primarily for electron microscopy, that can be useful in light microscopy. Shadowing enhances the contrast of surface structures that are in relief. We have used a thin layer of gold, deposited by vacuum evaporation, for this purpose. A layer of gold of the proper thickness, a few hundred angstroms, is transparent and gives a green color which is excellent for light microscopy. Shadowing has been used to study the surface structure of paper, crystals, fibers, and many other specimens having inherently low contrast when examined with high-power objectives.

Sectioning Techniques

Many samples consist of pieces of material too large to be treated as particles. Although grinding or dissolving and recrystallizing can render these samples particulate, such methods destroy the internal structures and relationships. Sectioning provides a means for examining interior structure. Depending on its properties, a piece of material may be sectioned by fracturing it, by cutting or grinding a plane surface on it, or by cutting a thin, transparent slice from it. Thin, transparent slices are examined by transmitted light, while samples prepared by the other two methods are examined either by reflected light or by replica techniques. Except for fractured sections an ideal section is essentially a mathematical plane passed through the sample. If such planes are chosen in strategic directions or if a series of sections are cut parallel to each other, the three-dimensional interior structure of the sample can be revealed.

Fractured Sections

Hard, brittle materials can be fractured with little damage to the internal structural relationships. If ductile or plastic materials are fractured, the structural relationships of the material in the fractured surface are destroyed or at least badly distorted. However, many ductile or soft materials can be rendered hard and brittle by freezing; for example, many polymers can be sectioned using this method.

Fracture surfaces are often rough and may be difficult to examine microscopically because of the limited depth of field of objective lenses. The Greenough-type microscope is useful for low-power examination. Examination at high power is restricted to small areas. Electron microscopes, particularly scanning electron microscopes, have relatively great depth of field and are very useful for studying fractured surfaces. Replicas are also useful for fractography.

Fractographic studies reveal the manner of failure of a material. The fracture follows the path of least resistance, through cavities and weak constituents, leaving these structures exposed to view. The weak constituents may cover a large proportion of the fractured surface even though they may constitute only a small portion of the sample. The interpretation of fractured surfaces is complicated by the ever-present possibility that tearing and distortion of the surface may have occurred.

Plane Sectioning

Flat surfaces, approaching the ideal plane, may be produced on hard, plastic materials by grinding and polishing with abrasives. Metals are perhaps the principal example of this type of material, and the principles and practices of grinding and polishing metal surfaces are described in a great many publications [1, 19–21].

Samples are usually cut with a hacksaw or cutoff wheel, then ground on successively finer abrasive papers, and finally polished with still finer abrasives on rotating laps usually covered with cloth. Sometimes wax, plastic, or soft metal laps are used instead of cloth-covered laps. The abrasives are usually SiC, Al_2O_3, or diamond. At each stage in the grinding and polishing, the material deformed by the previous step must be removed. The final deformed layer remaining after the last polishing step is removed by etching, usually with chemical reagents. Etching also serves to reveal the structure of the specimen because the etchant is chosen to etch some parts of the structure more rapidly than others. There are variations in this general procedure; for example, polishing and etching may be done electrolytically, or by ion bombardment of the specimen in a vacuum.

Resins and polymers can also be prepared using metallographic techniques and, in fact, resinography has become a specialized branch of science analogous to metallography and petrography [22–24].

Hard, brittle materials can be prepared in much the same way as metals, that is, by grinding on finer and finer abrasives. The techniques used are somewhat different in that the laps are usually of cast iron and the abrasive is loose rather than fixed as it is in metallography. The final polishing may be done on cloth-covered laps as in metallography. Etching to remove deformed materials is not necessary, but it may be used to reveal structure. Plane sections of hard, brittle materials are used primarily for opaque minerals, ores, and ceramics [25, 26]. The methods for preparing plane sections are almost identical with those used to prepare thin sections of hard, brittle materials.

Thin-Sectioning

Most hard, brittle materials, such as rocks, minerals, and ceramics, are composed of transparent constituents. Thick pieces of such materials appear opaque because they scatter and absorb light. However, thin sections can be examined by transmitted light. Even coal can be sectioned thin enough to be transparent. The method of preparing thin sections begins with the preparation of a plane section as described above. The plane surface is then cemented to a microscope slide, and most of the piece is cut and ground away, leaving a thin slice of the sample cemented to the microscope slide. A cover glass is then cemented over the slice. Because the cement tends to reduce the contrast of the surface, it is usually unnecessary to polish the surfaces of the thin section. Techniques for preparing thin sections of rocks and minerals are given by Reed [27] and by Jones and Hawes [28]. Similar techniques for preparing ceramics, nonmetallics, and refractories are described by Rigby [29] and by Insley and Fréchette [30]. Porous friable materials or loose particulate materials can be sectioned by these techniques if they are first embedded in polymers or resins [31].

Thin sections of soft materials can often be cut by hand using a knife or razor blade. Such sections are quickly and easily prepared and may give much valuable information. The knife should be very sharp to avoid distorting the specimen. Specimens that are too soft and flexible to support themselves during cutting can be sandwiched between pieces of cork, pitch, soap, or similar materials.

More accurate control of the sectioning process can be obtained by using a microtome. There are many kinds of microtomes available ranging from simple hand-held types with a micrometer screw feed to very precise models for cutting sections a few hundred angstroms thick. A variety of knives, made of steel, glass, or diamond enable the microtomist to section almost any material, including metals and ceramics as well as polymers, fibers, and other soft materials. We have cut thin cross sections of anodized aluminum sheet in order to study the structure of the Al_2O_3 film and its relationship to the metal substrate. Discussions of microtomy are usually concerned with biological tissues [32], but techniques for sectioning textiles [33, 34] and paper [34] are well established. Techniques for sectioning many types of materials can be found in the literature. Ultramicrotomy (sectioning for electron microscopy) is of special interest to the chemical microscopist because the techniques that have been developed are applicable to many technological materials [12, 13].

Sections, whether fractured, ground, or cut can be used by the chemical microscopist in a variety of ways. Ordinarily, they are used to study morphology, structural relationships, and optical properties, but the surfaces of sections can also be etched, stained, or treated with reagents to identify the phases present. It is possible to subject such surfaces to selected environments to study corrosion reactions, oxidation, and similar chemical reactions while observing them microscopically.

Staining and Etching

Staining is a specimen preparative technique that can be applied to each of the types of specimens mentioned in the preceding paragraphs. Particulate materials (including crystals), fractured sections, plane sections, and thin sections can at times be stained to improve image contrast or to identify certain constituents. Biological microscopists, particularly histochemists and cytochemists, have made extensive use of staining for identification of various tissues and cell constituents. Stains are of two types: general and selective or specific. General stains stain all tissue, while selective stains color only certain constituents of the specimen. Techniques for staining biological tissues can be found in standard works on biological microscopy and in the journal *Stain Technology*. Techniques for staining textiles [33, 35], paper [34], and starch [36] are also well developed. Metals, rocks and minerals, polymers, and other technological materials can also be stained.

Etching is a technique that produces surface relief on the specimen. This surface relief can be used to enhance contrast, to identify various constituents, to delineate crystal and grain boundaries, and to reveal the orientation of crystals. The technique has been developed most extensively for metallographic studies [19–21], but numerous nonmetallic materials such as ores and minerals can be etched and stained selectively by appropriate reagents [37–39].

5 MICROSCOPES AND MICROSCOPICAL TECHNIQUES

Microscopy with Transmitted Bright-Field Illumination

Most microscopical observations are made by using transmitted bright-field illumination. It is principally this form of illumination that is used in biological microscopy. It is also used extensively in petrography and in chemical microscopy. Specimens must be essentially transparent, and thin enough so that confusion does not occur because of superposed structures. Image contrast is either color contrast or contrast arising from reflection or refraction. The contrast arising from refraction and reflection can be controlled to some extent by proper mounting of the specimen (see Section 4, p. 147). However, mounting cannot affect the contrast of structures that lie entirely within a thin section, and often these structures are invisible, especially when objectives of high angular aperture are used. Many of the microscopes and microscopical techniques described later in this section were developed to improve the contrast of internal structures, thus making them visible. Staining techniques can also be used for this purpose.

Bright-Field Microscopes and Illumination

The description of microscopes in Section 2 refers primarily to bright-field transmission microscopy, so further description of the microscope itself is unnecessary, however, a brief discussion of illuminating systems and how to arrange them properly may be helpful. For work with low- and medium-power objectives, diffuse sources of illumination, such as a frosted incandescent lamp, may be used. Even with such a simple light source it is necessary to adhere to the correct principles of illumination if the best results are to be obtained. The diffuse source should be imaged in the plane of the specimen and if possible the size of the source should be adjusted, by means of a simple stop, so that its image just fills the field of the objective. This arrangement is essentially critical illumination using the diffuser as the source.

For microscopy at high magnification, Köhler illumination is preferable. A ribbon filament lamp is an excellent source. The lamp should be mounted in a well-ventilated housing provided with a focusing auxiliary condenser, an iris diaphragm, a filter holder, and a water cell for cooling the beam. Other types of sources such as coiled tungsten filaments, quartz iodide lamps with coiled tungsten filaments, gas discharge lamps (usually xenon or mercury arc lamps), zirconium arc lamps, or carbon arc lamps are also used, but the lamp housings should be similar to that described for the ribbon filament illuminator.

The procedure for setting up the microscope for Köhler illumination is as follows (see Fig. 3.7).

The microscope lamp is placed so that its auxiliary lens is about 25 cm from the microscope mirror and the lamp is adjusted so that the beam of light falls on the center of the mirror. A piece of paper held in front of the mirror helps in locating the beam of light.

The mirror is tilted and rotated to direct the reflected beam upward through the substage condenser.

The iris diaphragm of the substage condenser is nearly closed, and the image of the light source is focused on the diaphragm by adjusting the auxiliary condenser.

The microscope equipped with a low-power objective, for example, 10×, is focused on a specimen. The diaphragm on the illuminator is closed to about 5 mm in diameter, and the image of this diaphragm is brought into the center of the field by adjusting the mirror. The image of the lamp diaphragm is focused in the plane of the specimen by moving the substage condenser up or down as necessary.

The low-power objective is replaced by the objective to be used, and the microscope is focused on the specimen. The lamp diaphragm, which is now a field diaphragm, is closed so that it matches the field diaphragm in the eyepiece. This reduces stray light to a minimum and improves contrast. If the object does not fill the field of view, the lamp diaphragm can be closed still further.

Finally, the angular aperture of the condenser is adjusted, by means of the condenser diaphragm, until it matches or is slightly smaller than that of the objective. This should be done by observing the back aperture of the objective as discussed in Section 2, p. 139. If the illumination has been set up properly, the back aperture of the objective is completely and uniformly filled with light and the condenser diaphragm is just visible in and concentric with the back aperture of the objective. For many types of specimens, good objectives give their best image when the angular aperture of the condenser is between 0.7 and 0.9 of the full aperture of the objective. With experience the proper adjustment of the condenser aperture can be judged by the disappearance of flare from the image.

At this stage the light is probably much too bright for comfortable observation. If so, its intensity should not be reduced by closing the condenser diaphragm or defocusing the condenser. Some lamps are provided with a rheostat or variable transformer so that the intensity of the illumination can be reduced. For visual work this is usually satisfactory, but it does change the color temperature of the light source, thus changing the apparent color of the specimen. This is especially undesirable if color photomicrographs are to be taken. Neutral density filters are excellent for reducing the intensity of illumination.

Köhler illumination is used for nearly all the microscopes discussed in the remainder of this section. There are of course modifications in the setting-up procedure because of variations in the illuminating systems.

Transmitted bright-field illumination is used for many observations of interest to the chemical microscopist. Such observations can be carried out using a microscope equipped for biological or medical work, however, bright-field observations can also be carried out using a polarizing microscope equipped as described on p. 166. A polarizing microscope is indispensable to the chemical microscopist. A biological microscope can often be converted to a makeshift polarizing microscope by inserting polarizing filters into the system, however, this is generally unsatisfactory for a number of reasons concerned with the ways in which polarizing microscopy is used.

Qualitative Chemical Analysis Based on Morphology

Transmitted bright-field microscopy is best suited for examining the morphology and color of a specimen. Many materials have a characteristic shape and color, and many samples can be analyzed on this basis alone. Ordinarily, the microscopist uses a combination of properties, such as refractive index, behavior in polarized light, and staining character or chemical reactivity, as well as morphology and color, to identify materials. The microscopist learns to recognize materials by their morphology either by examining known materials or by studying atlases containing micrographs of the materials. There are atlases of paper fibers [34, 40], textile fibers [35, 41], cement clinkers [42], cement clinkers, refractory materials and slags [43], and many other materials. An interesting atlas of particulate materials often found as atmospheric pollutants has been published by McCrone et al. [44]. This book is not only an atlas but also discusses the microscopical methods and techniques used for the analysis of particles.

Another method of chemical microscopical analysis that depends primarily on the recognition of characteristic morphology is that described in Volume II of the *Handbook of Chemical Microscopy* [45]. In this method precipitation reactions are carried out under the microscope, and identification of unknown materials depends on the recognition of the crystals of the precipitate formed. This technique has several advantages. A reagent usually forms precipitates with relatively few ions or organic functional groups, so that only a few characteristic precipitates need be recognized. Highly selective reagents are not necessary because the various precipitates have characteristic form and color and can readily be distinguished from other precipitates formed by the same reagent. This kind of test is quick and does not ordinarily require extensive separations and manipulations. When separations or manipulations are necessary, they can be carried out rapidly because of the small amounts of materials involved. Finally, this type of test can be quite sensitive.

Reactions are carried out on a microscope slide in one of several ways, commonly a drop of reagent is allowed to flow gently into a drop containing the unknown in solution, or the solid reagent is added to the drop. The reagent may be contained in some medium such as a fiber [45–47] or a layer of gelatin gel [48]. Another technique is to use electrodeposition to precipitate metallic ions [49, 50]. Rachelle [51] has developed a technique that is extremely sensitive. He places tiny drops of reagent and unknown solution in a drop of oil. When the reagent and unknown droplets are put in contact they coalesce, allowing the reaction to occur.

Microchemical tests as described above can be used for qualitative analysis on a systematic basis if the amount of unknown material available is reasonably large. For very tiny amounts of unknown, it is better to use other methods of analysis [44]. However, if the purpose of the test is to determine whether or not a specific ion is present, the sensitivity of the method may allow detection of 10^{-15} g [51]. The method can be adapted to electron microscopy [50, 52] and the sensitivity improved.

The microchemical tests described by Chamot and Mason [45] are for identifying inorganic ions. The method is also applicable to the identification of organic compounds. Behrens and Kley published a fourth edition of their *Organische Mikrochemische Analyse* in 1921. Recently, this book was translated by Stevens [53]. Drugs and pharmaceuticals are frequently identified by microanalytical tests as well as by their crystalline structure [54]. Biennial reviews of chemical microscopy have been published in *Analytical Chemistry* [55], and a bibliography is included in *The Particle Atlas* [44].

Quantitative Chemical Analysis

Microscopical quantitative analyses are based on measurements made using the microscope. In this section analysis based on the measurement of length, area, and volume are discussed, although it is possible to base quantitative and qualitative chemical analyses on measurements of temperature, optical absorption, and other properties.

Linear measurements can be made in several ways. Estimates of length, or area, can be made if the magnification of the microscope is known. The apparent size of the object is estimated and the estimated value is divided by the magnification. Another method of estimating is based on the measured diameter of the field of view. Methods of measuring this diameter are given below, but it can be accomplished simply by focusing on a ruled scale and reading the diameter of the field directly. A ruled scale made especially for microscopical measurements is called a stage micrometer and is available from microscope supply houses. Objects can of course be placed directly on a stage micrometer and measured, however, the scale and the object are in different planes and cannot be focused simultaneously. A more convenient and accurate method of measuring involves placing a ruled scale (called a reticule) in the focal plane of the eyepiece, position F_3 in Fig. 3.8. An eyepiece so equipped is called a micrometer eyepiece. It is necessary to calibrate micrometer eyepieces against a standard scale, such as a stage micrometer, and a calibration must be made for each objective used. Varying the tube length of the microscope changes the calibration, so when measurements are made the tube length must be set as it was when the eyepiece was calibrated.

A ruled scale can also be placed at F_1 in Fig. 3.8. The image of this scale appears at the specimen plane F_2 as well as at F_3 and on the retina of the eye. A scale placed at F_1 can be calibrated against a stage micrometer or, if the demagnification of the condenser is known, the scale can be calibrated with an ordinary ruler. If a photomicrograph is taken or if an image of the specimen is projected on a screen, measurements on the image can also be made with an ordinary ruler provided the magnification of the image is known. Ordinarily, the magnification of projected images or photomicrographs is obtained by using a stage micrometer. Other methods and more details regarding the methods mentioned above can be found in various books on microscopy [1, 44, 56].

The methods of measurement discussed above apply only to measurements made in the plane of the image. Measurement of the dimension parallel to the optic axis of the microscope requires a different technique. The fine-focus adjustment is usually graduated, and thus it can be used for measurements. The objective is focused on the top and on the bottom of the object and the difference in the plane of focus is read from the graduated fine adjustment. For accurate measurements the fine-adjustment scale should be calibrated against some standard such as a micrometer. Thickness measurements can also be made using shadowing techniques (Section 4, p. 154) if the shadow angle is known. A still more accurate method is interferometry, discussed on p. 281.

The accuracy of the measurements made depends on the resolving power of the microscope, on the accuracy of the standard scale used for calibration, on the precision of the mechanical parts of the microscope, and on the skill of the microscopist. Measurements along the optic axis made by focusing the microscope also depend on the depth of focus of the objective being used. The depth of focus is approximately 0.0001 mm for an objective of N.A. = 1.30, ±0.0006 for an objective of N.A. = 0.65, and ±0.004 when N.A. = 0.25. By using interferometry it is possible to measure distances parallel to the optic axis to less than a few tens of angstroms. The resolving power of the microscope can be calculated using the formulas given in Section 2, p. 125. Stage micrometers are sufficiently accurate and precise for calibration for ordinary work. If extraordinary accuracy and precision are needed, the stage micrometer can be sent to the Bureau of Standards for testing.

Particle size analysis is probably the most important and widely used quantitative determination carried out microscopically. Other methods of particle size analysis generally rely on the microscopical method as the ultimate standard. Particle size analyses are usually carried out either by counting the total number of particles in a known weight of sample, or by measuring the size of a representative number of particles. The first method is often used for particles too small to be resolved but large enough to be seen and counted by dark-field or ultramicroscopy. If the sample is a dry powder, a known weight of it can be dispersed in a liquid and all of the particles in an aliquot of the dispersion counted. If the powder is already dispersed, a method for determining the weight of particles per unit volume of the mixture must be devised. The average particle size $\bar{D}$ of particles of density ρ is calculated:

$$\bar{D} = \sqrt[3]{\text{Wt. of particles counted}/\rho n},$$

where n is the number of particles. This formula assumes the particles are cubical.

The more common method is to measure the size of a representative number of particles. To do this one must decide how the diameter is to be defined. If the particles are of regular shape, for example, spheres or cubes, a single diameter gives complete information. If the particles are nonequidimensional, measuring a meaningful diameter becomes more difficult. A relatively accurate method is to measure the length, width, and thickness of each particle. However, it is inconvenient to measure thickness so two measurements, length and width, are taken. In the calculations various arithmetic or geometric mean diameters can be derived from the measurements.

It is still more convenient to measure some sort of statistical diameter directly. The most common statistical diameter is Martin's diameter [57], the length of a line bisecting the projected area of the particle. The line is taken in the same direction for all particles, usually parallel to a crosshair in the eyepiece. In this way a useful statistical diameter can be obtained provided the particles are randomly oriented in the mounting medium. Another statistical diameter is Feret's diameter, the average of two perpendicular diameters measured between tangents of the particle, always in the same direction. A third commonly used statistical diameter, the maximum horizontal intercept, is the maximum diameter of the particle, again measured in a given direction.

Statistical particle diameters can also be obtained by matching the projected area of each particle with the area of a circle of known diameter. A special eyepiece reticle having a series of circles of known diameter engraved upon it is used.

Data from particle size measurements can be treated statistically in a variety of ways. Commonly, the particles are grouped into size classes and the number in each class is plotted against the average diameter of the class to form a histogram. The histogram is a size distribution, and the shape of the distribution may be of great significance. However, a single numerical "average particle" size $\bar{d}$ is usually calculated: $\bar{d}_1 = \sum nd/\sum n$, where n is the number of particles and d is the particle diameter.

In chemical work other types of averages may be more significant. If surface area is important, an average particle size with respect to surface ($\bar{d}_3$) can be calculated: $\bar{d}_3 = \sum nd^3/\sum nd^2$. If volume or weight are more important, an average particle size with respect to volume ($\bar{d}_4$) can be calculated: $\bar{d}_4 = \sum nd^4/\sum nd^3$. Curves showing surface distribution and volume distribution can be obtained by plotting nd^2 versus d and nd^3 versus d, respectively. Cumulative-type curves are often useful.

Particle size analysis has been discussed much more completely by a number of authors [1, 58–63].

Quantitative analysis of mixtures of particulate materials can be carried out microscopically if the different constituents can be recognized. This method is rapid compared with wet methods of analysis and has the great advantage that the phases present are determined and not just the amounts of elements.

In some samples the phases present can be recognized by their morphology and color. If such easy recognition is impossible, differential stains or etchants may be used to label the particles, or they may be mounted in a medium of proper refractive index, thus causing the constituents to appear in varying contrast. Dispersion staining (p. 255) may also be used to aid in identifying the particles.

Once the specimen has been prepared so that the constituents can be recognized at a glance, the analysis may be carried out in a number of ways.

The most straightforward and quickest method is estimation. If the constituents are of the same fineness, as would be the case if a sieve fraction is being analyzed, estimations of composition can usually be made to $\pm 10\%$ for mixtures of two or three components. Experience and or comparison with known mixtures improves the accuracy of estimation.

Still more accurate results can be obtained by counting. If the particles are of the same fineness, the number percent is equal to the volume percent. If the constituents are of different degrees of fineness, it may be necessary to make particle size distributions to determine an average particle size for the various constituents.

For some types of analysis, it is advantageous to count all the particles in a known volume of suspension. One way of doing this is to place a known volume of suspension under a cover glass and count all the particles in the preparation. However, counting cells, constructed so that the distance between the cover glass and the slide is accurately known, can be used to count known volumes without the necessity of counting all the particles in the preparation. A good example of such a cell is the hemacytometer ordinarily used for counting the number of blood cells per unit volume of blood.

Areal analysis can be made on plane sections or thin sections of samples of aggregated heterogeneous materials. If the constituents of such a material are randomly oriented and distributed, the area of each constituent exposed in the plane of the section is proportional to its volume in the aggregate. Thus if the densities of the constituents are known, the weight percent of each constituent can be calculated from the ratio of its area exposed in the surface of the section to the area of the section [64, 65]. Samples of loose particles can be formed into aggregates for purposes of this type of analysis by embedding them in some solid medium.

Relative areas of the constituents can be determined microscopically in a number of ways. Photomicrographs can be taken or drawings of the specimen made with a drawing camera and the areas of the various constituents measured with a planimeter or cut out and weighed. A simple way of measuring is to trace the outlines of the constituents on coordinate paper and count the squares occupied by each constituent.

The relative areas in the section can also be measured by statistical methods such as lineal or point counts. If lines are drawn across the specimen (or an image of the specimen), the total of the intercepts of each constituent along the line is proportional to the area, hence to the volume of the constituents. Point counting is a statistical method of making a lineal analysis. If an array of regularly spaced points is made along a line, the proportion of points falling in each constituent is the same as the proportion of lineal intercepts, which is in turn proportional to the area and volume of that constituent in the sample. Integrating stages and integrating eyepieces are available to aid in lineal analysis directly on the specimen. Special accessories for point counting directly on the specimen are also available.

A thorough discussion of methods of quantitative analysis with many references is given by Chamot and Mason [1].

Polarized Light Microscopy

The Polarizing Microscope

In order to convert a microscope to a polarizing microscope, it is necessary to polarize the light entering the substage and to insert another polarizer, known as an analyzer, above the objective. The analyzer may be arranged to slide into the body tube above the objective or be mounted in a cap which fits over the eyepiece. An ordinary type of substage condenser cannot be used with a polarizing prism since the prism would have to be very large in order not to obscure part of the back lens of the condenser; consequently, a substage condenser of a modified design is often employed. Other necessary features of a polarizing microscope are a graduated rotating stage and crosshairs in the eyepiece, which are accurately aligned with the vibration directions of the polarizer and analyzer.

The most elaborate polarizing microscopes are those designed for petrographic research. The determination of the optical properties of crystals, such as those occurring in thin sections of minerals, necessitates very frequent changes from orthoscopic to conoscopic methods of observation. The orthoscopic observation of a crystal employs the ordinary optical system of the microscope with the addition of a polarizer and an analyzer; a low-power objective (4 to 10×) is used in conjunction with a low-power condenser, which provides a narrow cone of illumination over a relatively large area. The conoscopic method of examination involves the study of the interference figure formed at the back focal plane of a high-power objective of large numerical aperture and requires the use of a condenser of equal or greater aperture. The back focal plane of the objective is imaged in the eyepiece with the aid of an auxiliary lens, a Bertrand lens, which is inserted between the objective and the eyepiece. Petrographic microscopes are so constructed that the change from one method of observation to the other is easily carried out.

A typical example of a modern petrographic microscope is the Bausch & Lomb LC model shown in Fig. 3.12. It has a graduated stage which rotates on ball bearings and which is fitted to take a mechanical stage and other equipment. The analyzer prism slides in and out of the body tube between a parallelizing lens system of the type first described by Becher [66]. This lens system ensures freedom from astigmatism and permits the analyzer to be inserted and withdrawn without change of focus, magnification, or location

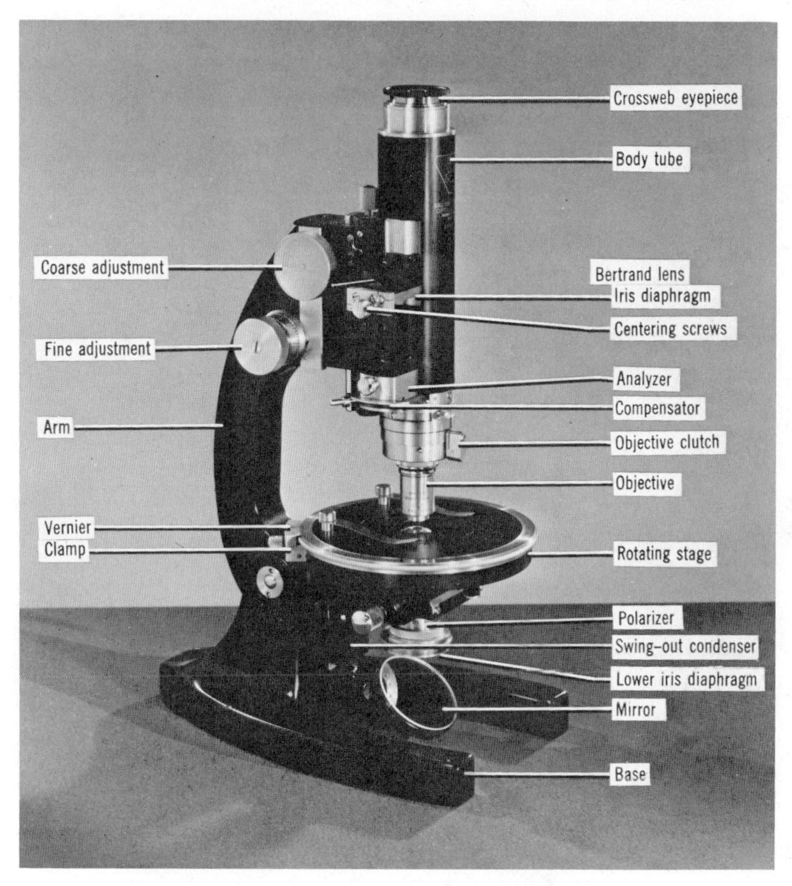

Fig. 3.12 Bausch & Lomb LC petrographic microscope.

of the image. The change from a low-power to a high-power condensing system is effected by turning a knurled head which swings the high-power condensing system into place. This mechanism is shown in Fig. 3.13. Each objective is carried in its own centering collar, which is readily attached to a special nosepiece, so that any number of precentered objectives may be readily interchanged. Under normal working conditions the objectives remain centered indefinitely. The Bertrand lens slides into the body tube in a focusing sleeve. It is equipped with centering screws and an iris diaphragm for the optical isolation of the interference figures of small crystals. A slot, provided with a dust cover, is placed just beneath the analyzer for the accommodation of a quartz wedge, first-order red selenite, quarter-wave mica, or other form of compensator.

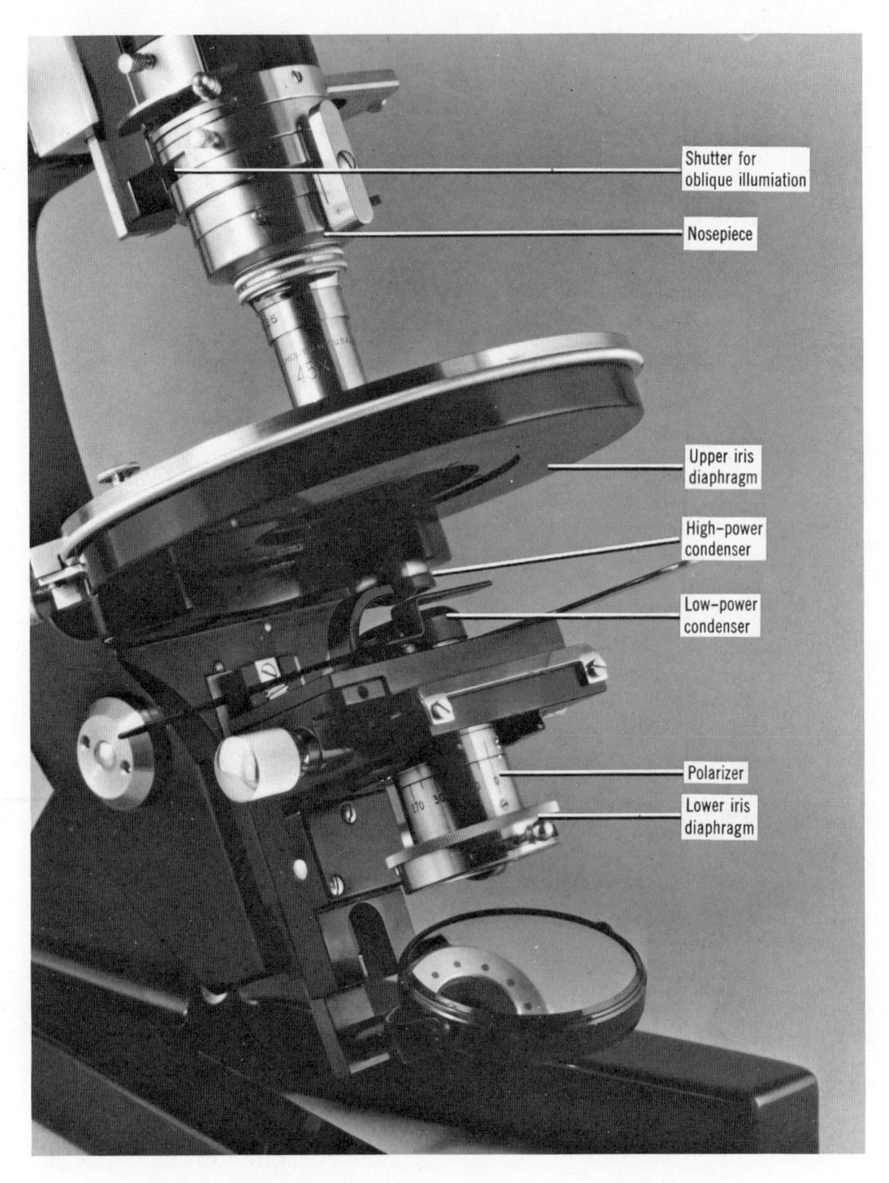

Fig. 3.13 Swing-out condenser on Bausch & Lomb microscope.

Chemical microscopes are polarizing microscopes of simple and sturdy design according to specifications originally laid down by Chamot and Mason [1]. They differ principally from the more elaborate petrographic instruments described above in that they are not equipped with a built-in Bertrand lens. The best models are equipped with a swing-out high-aperture condenser system which, as already explained, greatly facilitates the study of interference figures. Most makers of chemical microscopes are able to supply either a cap analyzer or an analyzer that slides into the body tube. The latter is decidedly preferable for workers who wear glasses. In place of the Bertrand lens, a pinhole diaphragm or an auxiliary lens above the eyepiece, known as a Becke lens, is used. The self-contained conoscopic apparatus of the type discussed on p. 191 is worthy of special consideration since the microscopy of organic compounds depends a great deal on the study of interference figures given by small crystals.

At the present time nearly all chemical microscopes are equipped with polarizing filters in place of the polarizing prisms formerly used. These filters are of excellent optical quality. Their use makes the design of the optical system simpler because they are free from astigmatism and do not require parallelizing lenses. Large polarizing prisms are expensive and difficult to obtain and substage condensers must be specially designed to accommodate them, but polarizing filters are obtainable in large sizes and this, plus their thinness, makes the design of the condenser much simpler.

Chamot and Mason recommend the use of an accurately centered revolving nosepiece, or one with individually centerable openings, for three objectives. This is a sound practice, since most chemical microscopy, apart from more-or-less fundamental research on the optical properties of crystals, can be carried out with three objectives.

The choice of accessories for chemical microscopy is naturally governed by the extent of the optical investigations to be undertaken. The minimum requirements for useful work in the organic laboratory are: a chemical microscope equipped with a swing-out high-power condenser and apparatus for the observation of conoscopic interference figures of small crystals; an accessory for determining optical sign, such as a quarter-wave mica plate, a first-order red selenite plate, a quartz wedge, or a Johannsen or Berek compensator; a triple nosepiece equipped with three parfocal and centered strain-free objectives and two or three eyepieces with crosshairs and focusing eye lenses. The optical characteristics of a selection of Bausch & Lomb strain-free achromatic objectives are given in Table 3.1 by way of illustration. A similar range of objectives is made by other manufacturers of polarizing microscopes.

When only three objectives are to be employed in a nosepiece, our preference is for a 6, 21, and 45× (N.A. = 0.85). It is important to note that

Table 3.1 Objectives for Polarizing Microscope

Initial Magnification	*Focus* (mm)	N.A.	*Working Distance* (mm)	*Type*
4×	32	0.10	38	Dry
6×	22.7	0.17	15.5	Dry
10×	16	0.25	7.0	Dry
21×	8	0.50	1.6	Dry
45×	4	0.85	0.3	Dry
or 36×	4	0.95	0.25	Dry
97×	1.8	1.25	0.14	Oil immersion

different manufacturers employ different tube length corrections for their objectives. Low-power objectives up to 10× initial magnification are not at all critical with respect to tube length correction, but objectives of over 30× initial magnification are very critical. It is not a sound policy to equip a laboratory with an odd collection of microscope accessories.

Focusing crosshaired eyepieces are supplied with magnifications, calculated on a 250-mm-image distance, of 5, 7.5 or 8, 10, 12.5×, and sometimes higher. When only two eyepieces are to be chosen, our preference is for a 5 and 10×. A very useful addition is a 7.5× eyepiece with a focusing eye lens which takes a micrometer disk. This is used for estimating the optic axial angle in air ($2E$) of biaxial crystals (see p. 216).

A mechanical stage is a very great convenience. It enables one to remove a slide from the microscope, apply some physical or chemical treatment to it, and return it to the same position on the microscope stage. This particular application is useful when studying the action of heat on extremely small particles. The slide is allowed to cool before being returned to the microscope. More obvious uses of a mechanical stage are to aid in the systematic exploration of a large number of minute crystals and to render easier the task of centering them on the crosshairs for conoscopic examination. A mechanical stage with a very fine adjustment over a limited travel is particularly advantageous for the study of very small crystals. An adjustment of this type is provided by the built-in mechanical stage of the Bausch & Lomb LD petrographic microscope designed by F. E. Wright.

For work on colored crystals, a dichroscope eyepiece of the Wollaston prism-type (p. 221) is convenient, although not absolutely necessary. It is essential to have a good microscope lamp equipped with a water cooling cell and a set of Wratten M filters. A source of monochromatic light, such as a sodium vapor lamp for $\lambda = 589$ mμ, or a mercury vapor lamp equipped with Wratten filters nos. 77A and 58 for $\lambda = 546$ mμ, is very useful for the preliminary conoscopic examination of absorbing crystals and for studying optic normal, obtuse bisectrix, and uncentered interference figures [67]. A monochromator with a wavelength scale is invaluable for determination of dispersion of birefringence and the dispersion of extinction commonly met with in monoclinic and triclinic crystals of colored organic compounds (p. 237).

Optics of Transparent Crystals

CRYSTAL SYSTEMS

Although a knowledge of geometric crystallography is a necessary foundation for the study of crystal optics and for success in chemical microscopy, an adequate presentation of the subject would require far more space than is available here. The reader is therefore referred to the standard works on crystallography and mineralogy cited in the references at the end of this chapter. It is sufficient for our present purposes to classify crystals in seven systems arranged in order of descending symmetry, an order that is logical in the study of the optical properties of crystals, as these properties become more complex as the symmetry becomes lower.

1. Cubic, having three equal axes mutually at right angles ($a_1 = a_2 = a_3$, $\alpha = \beta = \gamma = 90°$)
2. Rhombohedral or trigonal, having three equal crystallographic axes equally inclined at an angle other than a right angle. This system, which is characterized by a trigonal axis of symmetry, is often included in the hexagonal system by mineralogists and crystallographers ($a_1 = a_2 = a_3$, $\alpha = \beta = \gamma \neq 90°$)
3. Hexagonal, having three equal axes mutually at 120° in a plane, with a longer or shorter axis perpendicular to this plane
4. Tetragonal, having two equal axes and a third axis longer or shorter than these two. The three axes are mutually at right angles ($a_1 = a_2 \neq c$, $\alpha = \beta = \gamma = 90°$)
5. Orthorhombic, having three unequal axes mutually at right angles ($c < a < b$ or $a < b < c$, $\alpha = \beta = \gamma = 90°$)
6. Monoclinic, having three unequal axes; a and c make an obtuse angle β in a plane that is perpendicular to the b axis ($a \neq b \neq c_1$, $\beta > 90°$, $\alpha = \gamma = 90°$)
7. Triclinic, having three unequal axes inclined at unequal angles

HABIT, DISTORTION, AND TWINNING

It is to be emphasized that crystallographic symmetry is determined by the lattice structure of the crystal; the external geometric shapes of a number of crystals of the same substance may differ greatly among themselves, although the angles between corresponding faces on the different crystals are constant for a given substance. Some factors that change the external shape, are discussed below.

Variations in the development of different forms may completely change the shape of crystals of a given compound. A striking example, familiar to students of mineralogy, is that of calcite, which occurs naturally in acute and obtuse rhombs, hexagonal prisms, and many types of scalenohedra. All these geometric forms are built from the unit rhombohedron of calcite, as is readily shown by their identical cleavage and optical properties and by x-ray diffraction methods. Variation in the forms developed on crystals of a given compound are often a result of variations in the solvent from which they were crystallized, or even of the presence of a foreign substance in the mother liquor. Three well-known examples are those of sodium chloride, which normally develops only cubic (100) faces, but develops octahedral (111) faces when grown from solutions containing a high concentration of urea; potash alum, normally octahedral, which develops (100) faces when grown from solutions to which some sodium hydroxide has been added; and the normally octahedral lead nitrate which develops (100) faces when grown from solutions saturated with methylene blue. Crystal habit is also greatly influenced by the rate of growth of the crystals. Rapid growth from supersaturated solutions tends to make crystals grow fastest at their edges, so that relatively few faces form.

Different faces of the same form may develop unequally and so give rise to "distorted" crystals. This is usually the case when crystals grow in contact with a surface, such as the bottom of a beaker or crystallizing dish.

Under certain conditions it is not easy to recognize multiple twinning. The twinned crystals may simulate symmetry higher than that of the untwinned crystals. Two main types of multiple twinning may occur: multiple twinning on a pinacoid is known as polysynthetic twinning; multiple twinning on a prism produces cyclic twins.

The general shape of a crystal, as governed by the relative development of certain forms, or of particular faces of one form, is referred to as the crystal habit.

POLARIZED LIGHT

In studying the optical properties of crystals, we are concerned with the wave nature of light. Light consists of transverse electromagnetic waves; by virtue of the transverse nature of these waves, they possess a vector property known as polarization. As ordinarily emitted by the sun or a lamp, light consists of a mixture of waves having every possible vibration plane: it is however, possible to polarize light so that every ray has the same vibration direction. In former times, the plane of polarization was defined as the plane of incidence in which the polarized light was most copiously reflected. This plane is at right angles to the direction of the electrical vibrations, which are now recognized as being responsible for the chemical and photographic effects of light; consequently, the vibration plane of polarized light is defined as the direction of the electric vector. Wood [68] uses the term "plane of polarization" as synonymous with the vibration plane, a very desirable procedure which is not yet followed in all works on mineralogy and petrography.

OPTICALLY ISOTROPIC CRYSTALS

Crystals of the cubic system in general are optically isotropic, that is, they have a single refractive index regardless of the direction of light passing through them, and of its plane of vibration. They share this property with true liquids and unstrained glasses. A very few cubic crystals are known that possess optical rotatory power—the best known example is sodium chlorate which rotates the plane of polarization of sodium light 3° 1′ per millimeter of thickness, some crystals being levo- and some dextrorotatory. Apart from these rare exceptions, it may be stated that a flat plate of an isotropic substance does not affect the plane of polarization of light passing perpendicularly through it, so that a flat plate of an isotropic substance does not brighten the microscope field with crossed polarizers in any position of rotation of the stage. This is true no matter what the orientation of the plate is with respect to the crystal axes. However, steeply inclined interfaces can rotate the vibration plane to an extent dependent on the vibration azimuth of the incident light, so that such interfaces may appear gray or white between crossed polarizers.

OPTICALLY ANISOTROPIC CRYSTALS

Crystals of the remaining six classes of crystal symmetry possess the property of double refraction, whereby light passing through such crystals is in general resolved into two beams, polarized at right angles, which travel through the crystal at different velocities. The velocity of light in any medium is inversely proportional to the refractive index of that medium. If n is the refractive index of that medium, V the velocity of light in it, and V_a the velocity of light in air, then:

$$n = V_a/V. \tag{3.7}$$

Consequently, it may be stated that in general an optically anisotropic crystal has two refractive indices for each and every ray direction other than along an optic axis.

Optically anisotropic crystals form two principal groups, known as uniaxial and biaxial. There are various ways of representing the optical

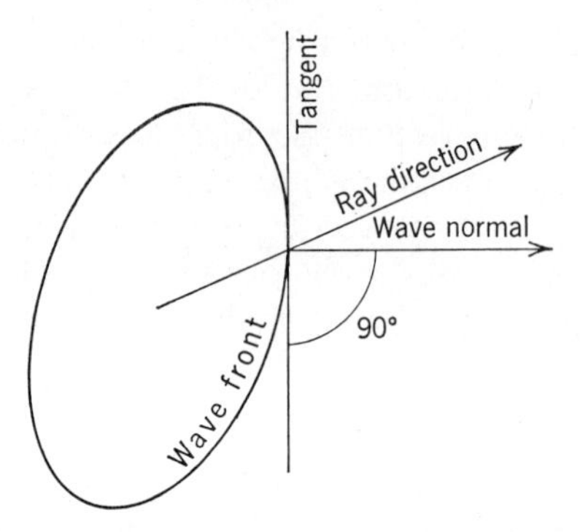

Fig. 3.14 Corresponding ray and wave-normal directions.

properties of these two classes of crystals. The best known geometric construction is Fletcher's [69] optical indicatrix, in which the ray and wave-normal surfaces are derived from an ellipsoid of revolution for uniaxial crystals and a triaxial ellipsoid for biaxial crystals. The ray-velocity surface is a three-dimensional representation of the two velocities corresponding to each ray direction; the wave normal–velocity surface represents the pair of velocities for each wave normal. The logical construction for the purposes of the crystallographer and chemical microscopist, however, is the surface derived by plotting the pair of refractive indices for each direction of the wave normal, since the direction of light traveling through a crystal is that of the wave normal. The relation between a corresponding ray and wave-normal direction is shown diagrammatically in Fig. 3.14. We now consider the wave normal–refractive index diagrams for the two groups of optically anisotropic crystals.

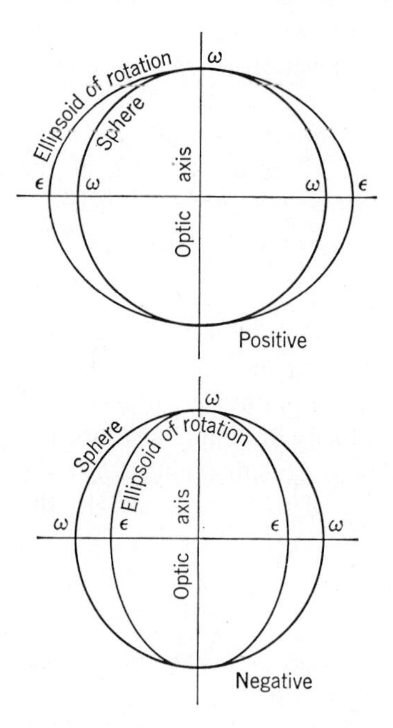

Fig. 3.15 Wave normal–refractive index surface of positive and negative uniaxial crystals.

Uniaxial Crystals. With crystals of the tetragonal, trigonal, and hexagonal systems, one of the sheets of the wave normal–refractive index surface is a sphere, as in an optically isotropic medium; the other is an ellipsoid of rotation which touches the sphere at the poles of its axis. Its equatorial diameter may be greater or less than that of the sphere, in which cases the crystals are said to be positively or negatively uniaxial, respectively (Fig. 3.15). The wave with the spherical front is known as the ordinary ray, since it obeys the ordinary (Snell's) law of refraction; consequently, in our wave normal–refractive index diagram the radius of the sphere is known as the ordinary refractive index and is usually represented by the symbol n_ω or its abbreviation ω. The wave with the ellipsoidal front is known as the extraordinary ray, since it does not in general obey Snell's law because of the difference between the ray and wave-normal directions. The refractive index for the extraordinary ray varies from ω along the optic axis to its maximum or minimum value of n_ϵ or ϵ at the equator of the ellipsoid. The intermediate values of the refractive index are sometimes designated $n_{\epsilon'}$ or ϵ'. If the wave normal of an intermediate ray makes an angle θ with the optic axis of the crystal, the refractive index of the crystal for the extraordinary ray is given by

$$n_\epsilon = \sqrt{\epsilon^2\omega^2/(\epsilon^2 \cos^2 \phi + \omega^2 \sin^2 \theta)}. \tag{3.8}$$

Since there is but one refractive index along the optic axis, light is transmitted in this direction through a crystal without modification.

In optically active uniaxial crystals, plane-polarized light traveling along the optic axis is resolved into two circularly polarized (right-handed and left-handed) beams which traverse the crystal with slightly different velocities. They recombine on emerging from the crystal to give plane-polarized light which, however, has had its plane of polarization rotated. The phenomenon of optical rotation does not modify the interference effects studied with the polarizing microscope, and therefore, is not considered further in this chapter. Optical rotatory power is dealt with in Part IIIC, Chapter 3.

A very important and easily memorized rule is that the extraordinary ray vibrates in the plane of the optic axis, which in turn coincides with the crystallographic *c*-axis of trigonal, tetragonal, or hexagonal symmetry.

Biaxial Crystals. The wave normal–refractive index surface of a biaxial crystal is more complicated than that of a uniaxial crystal. It is shown diagrammatically in Fig. 3.16, which should be traced on tracing paper, cut and folded, as indicated on the diagram, to yield one octant of a three-dimensional representation. The folded diagram should resemble Fig. 3.17. In this diagram the three coordinate axes X, Y, and Z, are the principal vibration directions corresponding to the smallest, intermediate, and highest refractive indices, which are designated α, β, and γ, respectively. (Winchell uses the symbols N_p, N_m, and N_g.) The abbreviations α, β, and γ should not be used unless the context makes it clear that the angles between the *a*-, *b*-, and *c*-axes of a triclinic crystal are not being referred to. The most important principal plane of our refractive index diagram is the X-Z plane, shown in

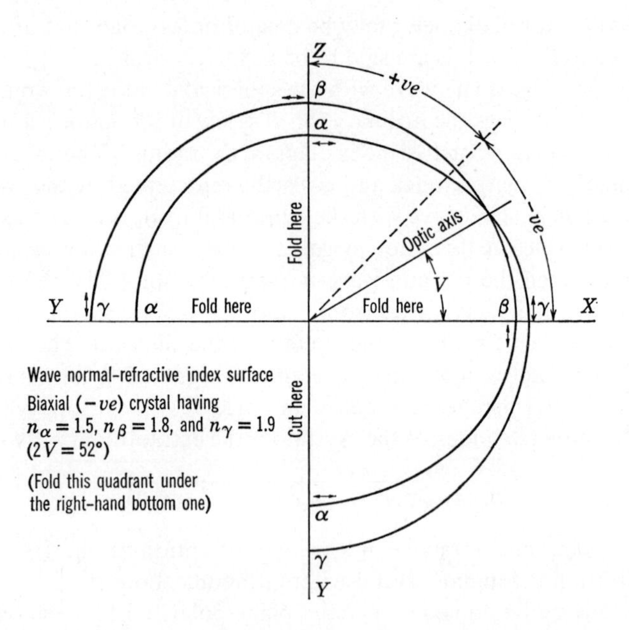

Fig. 3.16 Wave normal–refractive index surface of biaxial crystals.

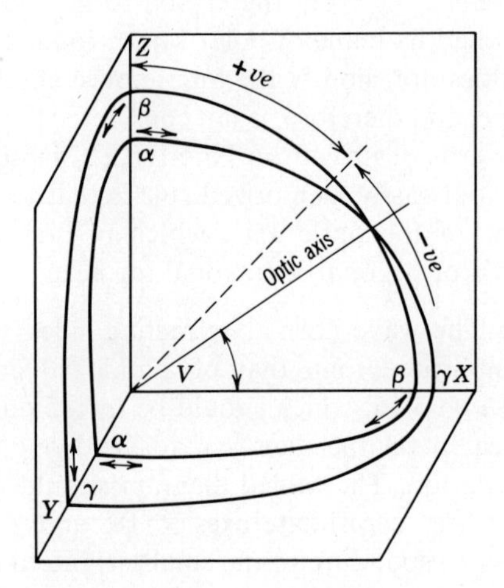

Fig. 3.17 Wave normal–refractive index surface of biaxial crystals (folded diagram).

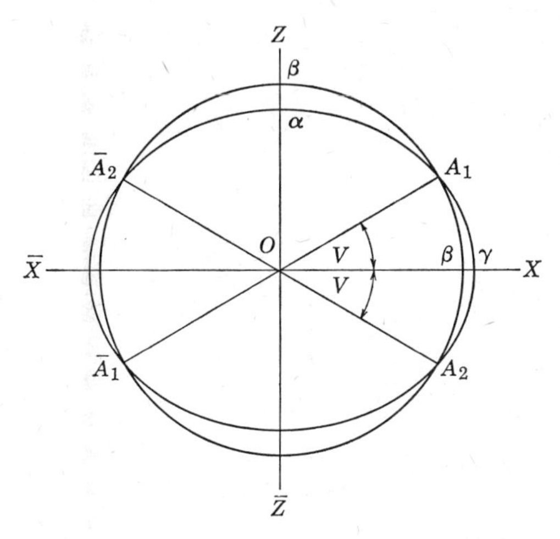

Fig. 3.18 *X-Z* Plane of wave normal–refractive index surfaces showing the primary optic axes of a biaxial crystal.

Fig. 3.18. It consists of a circle of radius β which cuts an ellipse which has a semimajor axis of γ and a semiminor axis of α at points A_1, A_2, $\bar{A}_1$, and $\bar{A}_2$. The directions A_1-$\bar{A}_1$ and A_2-$\bar{A}_2$ are the primary optic axes, usually referred to as the optic axes. Light traveling along the optic axes of a colorless biaxial crystal does not change its state of polarization. The refractive index for waves traveling along an optic axis is obviously β.

If, instead of employing a wave normal–refractive index diagram, we show the velocity of the waves in the *X-Z* plane as a function of the ray direction, we then have a circle of radius $1/\beta$, which cuts an ellipse having a semimajor axis of $1/\alpha$ and a semiminor axis of $1/\gamma$ at points S_1, S_2, $\bar{S}_1$, and $\bar{S}_2$ (Fig. 3.19). The directions S_1-$\bar{S}_1$ and S_2-$\bar{S}_2$ are known as the secondary optic axes. Rays that travel through a crystal along a secondary optic axis deviate according to their plane of polarization; the rays for all possible planes of polarization form a hollow cone of a small angle which diverges from the point of emergence. This effect is known as external conical refraction. In a solid model of the ray-velocity surface, it can be seen that the four points of emergence of the secondary optic axes are in conical depressions, "dimples," in the surface. A plane surface tangent to the ray surface in these locations makes contact everywhere in a ring; this means that all the rays within a hollow cone have the same wave normal. This effect is known as internal conical refraction. The angle of the cones is very small in most minerals and inorganic crystals but may be large with crystals of aromatic compounds having extremely large values of $(\gamma - \alpha)$ and a large value of $2V$. The phenomena of

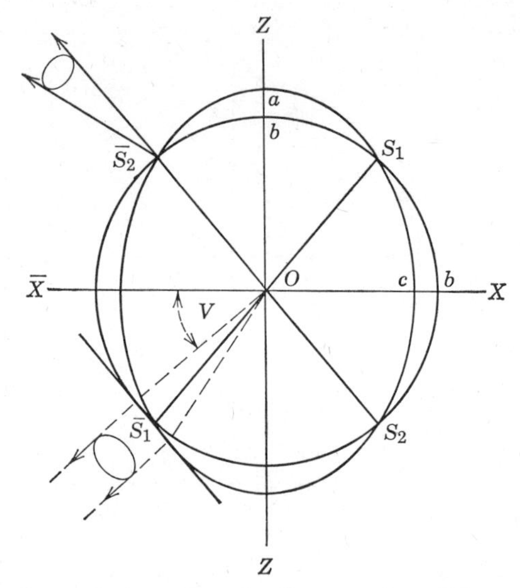

Fig. 3.19 *X-Z* Plane of ray-velocity surface showing secondary optic axes, external (upper left) and internal (lower left) conical refraction.

external and internal conical refraction are of no practical significance in petrography and chemical microscopy.

The optical axial angle is expressed as $2V$, where V is the angle between an optic axis and the principal vibration direction X or Z, whichever is nearer. It is nearer to X for negative biaxial crystals and nearer to Z for positive ones; this is equivalent to stating that for negative biaxial crystals $(\gamma - \beta)$ is less than $(\beta - \alpha)$ and for positive ones, greater.

The principal vibration directions in the plane of the optic axes are known as the bisectrics or median lines. The acute bisectrix Bx_a, or first median line, makes an acute angle with each optic axis; it is X for negative and Z for positive crystals. The obtuse bisectrix Bx_o, or second median line, is Z for negative and X for positive crystals. In orthorhombic crystals the principal vibration directions, X, Y, and Z coincide with the crystallographic axes. In monoclinic crystals one of the principal vibration directions coincides with the crystallographic orthoaxis b, whereas in triclinic crystals the principal vibration directions are without necessary relationship to the directions of the crystallographic axes.

The relation between V and the principal refractive indices is readily derived from the wave normal–refractive index diagram in Fig. 3.18, in which the elliptical curve has the following equation in polar coordinates:

$$\frac{\cos^2 \theta}{\gamma^2} + \frac{\sin^2 \theta}{\alpha^2} = \frac{1}{r^2}. \tag{3.9}$$

For a negative crystal V is the angle XOA. For a positive crystal it is the angle ZOA. If we call these angles V_n and V_p, respectively, we have:

$$\frac{\cos^2 V_n}{\gamma^2} + \frac{\sin^2 V_n}{\alpha^2} = \frac{1}{\beta^2} = \frac{\sin^2 V_p}{\gamma^2} + \frac{\cos^2 V_p}{\alpha^2}. \tag{3.10}$$

Solving for V_n and V_p,

$$\sin^2 V_n = \cos^2 V_p = \alpha^2(\gamma^2 - \beta^2)/\beta^2(\gamma^2 - \alpha^2) = (\gamma^2/\beta^2 - 1)/(\gamma^2/\alpha^2 - 1). \tag{3.11}$$

When γ is less than 0.1 greater than α, we may assume that $\alpha^2(\gamma + \beta) \cong \beta(\gamma + \alpha)$ and obtain the simple approximate formula

$$\sin^2 V_n = \cos^2 V_p \cong (\gamma - \beta)/(\gamma - \alpha). \tag{3.12}$$

INTERFERENCE COLORS

An important property of doubly refracting crystals is that they exhibit interference effects in polarized light when three conditions are satisfied: the light entering the crystal must be polarized; the crystal must be so oriented that it splits the light into two beams, vibrating in planes at right angles, which traverse the crystal with different velocities; the emerging beams must be brought into the same plane of vibration by means of another polarizing prism, which is termed the analyzer when used for this purpose. The interference is easily understood by means of vector diagrams. In Fig. 3.20*a* the amplitude and vibration plane of a polarized light wave are represented by the length and direction of vector **A**. This wave is resolved into two waves

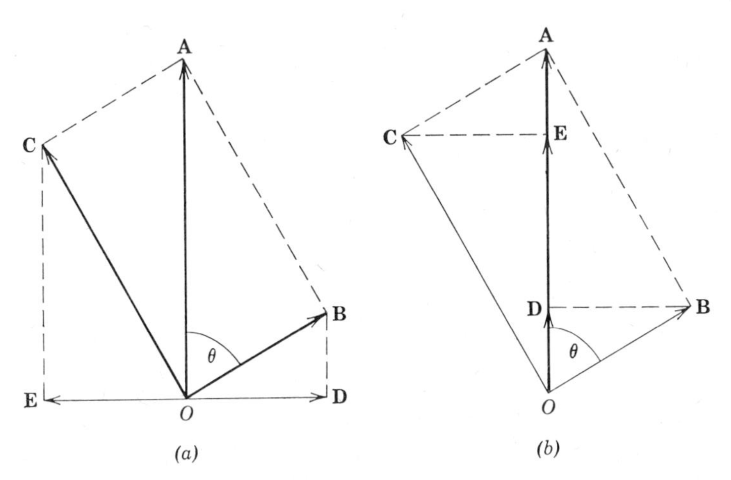

Fig. 3.20 (*a*) Vector diagram for crossed polarizers; (*b*) vector diagram for parallel polarizers.

(represented by vectors **B** and **C**) vibrating at right angles by the crystal. Vectors **B** and **C** are resolved in the vibration plane of the analyzer, which in this diagram is taken as being perpendicular to that of **A**, and we obtain vectors **D** and **E**. In Fig. 3.20*b* the analyzer plane coincides with that of **A**. The former case is that of crossed polarizers and the latter that of parallel polarizers. For crossed polarizers the waves have the same amplitude, since $OD = OE = OA \sin\theta \cos\theta$, while with parallel polarizers $OD = OA \cos^2\theta$ and $OA \sin^2\theta$, which are only the same when $\theta = 45°$. Moreover, with crossed polarizers the vectors are at 180°, which means that a phase difference of $\lambda/2$ has been introduced between the two waves.

We must now consider the effect of the difference in length of the optical paths traversed by the two waves within the crystal. If the thickness of the crystal is l and its two refractive indices are n_1 and n_2, the path difference of the two waves is $l(n_1 - n_2)$. For our present purpose we are interested in the phase difference δ, expressed in degrees,

$$\delta = 360l(n_2 - n_1)/\lambda, \tag{3.13}$$

where λ is the wavelength of the light. In order to determine the amplitude of the light emerging from the analyzer, we make use of another simple vector diagram. The case of crossed polarizers is shown in Fig. 3.21. One of the vectors is retarded by an angle δ to give vector **E**′ (Fig. 3.21*a*). The vector sum of **E**′ and **D** is **F**, whose scalar, the distance OF, is the required amplitude:

$$OF = 2OA \sin\theta \cos\theta \sin(\delta/2). \tag{3.14}$$

If θ is varied with δ remaining constant, OF is a maximum for $\theta = 45°$. This position of the crystal is known as the 45° or diagonal position. In this position $OF = OA \sin(\delta/2)$. Since the intensity of light is proportional to

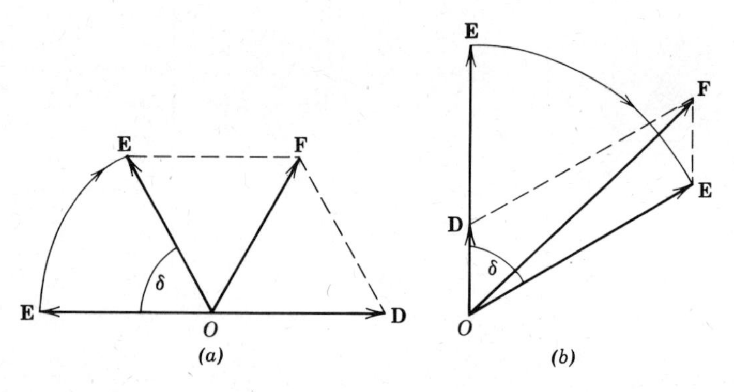

Fig. 3.21 (*a*) Vector sum for crossed polarizers; (*b*) vector sum for parallel polarizers.

the square of the amplitude, we may write:

$$I = I_0 \sin^2 (\delta/2) \equiv I_0(1 - \cos \delta)/2, \tag{3.15}$$

where I_0 is the original intensity and I that transmitted by a crystal having a phase retardation δ between crossed polarizers. It is noted that OF becomes zero for all values of δ when $\theta = 0$, 90, 180, and 270°; hence when a crystal is rotated between crossed polarizers through 360°, there are four positions in which light is not transmitted. These are known as the extinction positions. We now consider the effect of varying δ with θ remaining constant. When δ is 0 or 360°, $\sin (\delta/2)$ is zero, so that OF is zero. The maximum value of OF occurs for $\delta = 180°$, or for any number of complete revolutions plus 180°; this may be expressed as $\delta = (2\eta + 1)\pi$ radians for maximum transmission, and $\delta = 2\eta\pi$ for darkness.

The case of parallel polarizers is treated in Fig. 3.21*b*. As before, vector **E** is retarded by an angle δ to give vector **E**′, which when added to **D** gives **F**, whose scalar, the distance OF, is the amplitude of the transmitted wave.

$$OF = \sqrt{\sin^4 \theta + \cos^4 \theta + 2 \sin^2 \theta \cos^2 \theta \cos \delta}. \tag{3.16}$$

This expression can only equal zero when two conditions are simultaneously fulfilled, namely, $\cos \delta = -1$ and $\theta = 45°$. For $\theta = 0$, 90, 180, and 270°, $OF = 1$, which means that the maximum light is transmitted. We obtain the following expression for the light transmission of a crystal in the 45° position between parallel polarizers:

$$I = I_0(1 + \cos \delta)/2 \equiv I_0 \cos^2 (\delta/2). \tag{3.17}$$

It is noted that the sum of the intensities transmitted by crossed and by parallel polarizers is I_0. Since this is true for all values of δ, it follows that the interference colors obtained in the two cases are exactly complementary. As a practical example of the use of the above formulas, a comparison is given in Table 3.2 of the transmissions of optic normal sections of benzil and quartz, 30 μ thick, between crossed polarizers.

Table 3.2 Transmission of Benzil and Quartz between Crossed Nicols

	Benzil			*Quartz*		
λ (mμ)	$\epsilon - \omega$	δ (degrees)	I/I_0	$\epsilon - \omega$	δ (degrees)	I/I_0
420	0.0000	0	0.000	0.0094	242	0.745
450	0.0074	178	1.000	0.0094	226	0.847
500	0.0138	298	0.265	0.0093	201	0.967
550	0.0175	344	0.019	0.0092	187	0.996
600	0.0201	362	0.000	0.0091	164	0.981
650	0.0216	359	0.000	0.0090	150	0.933

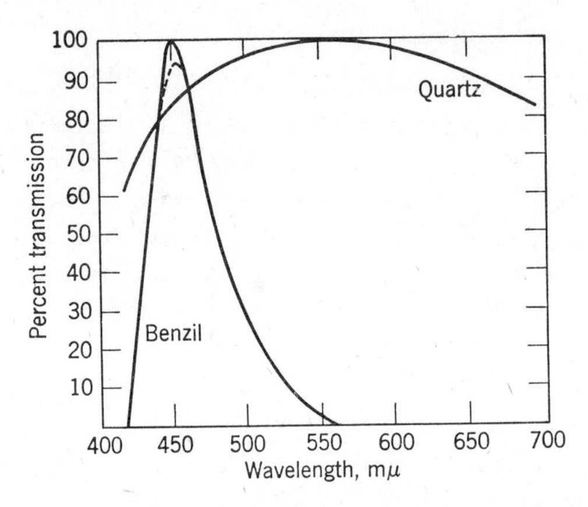

Fig. 3.22 Transmission of 30-μ optic normal sections of quartz and benzil between crossed polarizers.

The results in Table 3.2 are plotted in Fig. 3.22, which clearly shows that benzil transmits practically only the blue, while a quartz section of the same thickness transmits most of the spectrum and appears a slightly yellowish white. The transmission maximum of benzil is not 100% because of absorption of the extraordinary ray; this is indicated by the dotted line in the figure.

When a wedge of a doubly refracting crystal is examined between crossed polarizers with monochromatic light of wavelength λ, it extinguishes the light at positions where the retardation is an integral number of wavelengths. If the angle of the wedge is θ, the refractive indices are n_1 and n_2; when the spacing d of the fringes that cross the wedge is given by:

$$d = \lambda/(n_1 - n_2) \tan \theta, \tag{3.18}$$

if $(n_1 - n_2)$ were the same for all values of λ, d would be directly proportional to λ. Such a wedge, when examined between crossed polarizers with white light presents Newton's color scale, which is the order of the interference colors shown by reflection from an air gap between two plane glass surfaces in contact at one edge and separated by about 1 μ at the opposite edge. The interference colors shown by a trace of lubricating oil on a wet road very nearly follow Newton's color scale; the very slight departure is caused by the dispersion of the oil. With crystals, the dispersion of birefringence may be very large and the departure from Newton's color scale considerable. When d is greater for shorter wavelengths than for longer ones, the interference colors are said to be anomalous, as in the case of benzil discussed above. This subject is discussed further under the heading of dispersion (p. 183).

OPTICAL DISPERSION

Dispersion of Refractive Index. The refractive index of any substance depends on the wavelength of the transmitted radiation. This property is known as the dispersion of refraction and is possessed by all substances in varying degree. Dispersion of refraction is often expressed as the inverse relationship:

$$\nu = (n_D - 1)/(n_F - n_C), \tag{3.19}$$

where n_D, n_F, and n_C are the refractive indices for the sodium D- ($\lambda =$ 5893 Å), hydrogen F- ($\lambda = 4861$ Å), and hydrogen C- ($\lambda = 6463$ Å) lines. However, the constant ν does not take into account the shape of the dispersion curve; two substances might have the same values of n_D and ν and yet have different refractive indices for the extreme red and violet ends of the spectrum. The principal refractive indices of uniaxial and biaxial crystals are usually subject to differing degrees of dispersion.

Dispersion of Birefringence. The birefringence of a crystal may be expressed as the difference between the two refractive indices for light traveling in any given direction through it. This means that the birefringence of a crystal is a function of the ray direction. In the case of uniaxial crystals, the dispersion of birefringence is very nearly independent of ray direction and for all practical purposes may be considered as constant. Theoretically, a slight deviation is possible with crystals possessing widely different values of ω and ϵ, together with widely different dispersions of the two principal refractive indices. With biaxial crystals both the birefringence and the dispersion of birefringence are dependent on ray direction. This is particularly true of the dispersion of birefringence of crystals viewed in a direction close to that of an optic axis, as the directions of the optic axes of biaxial crystals may vary considerably with change of wavelength. For this reason, a measurement of the birefringence of a biaxial crystal is useless unless the orientation of the direction of viewing with respect to the wave surface is also recorded.

Dispersion of Optic Axes. The directions of the two optic axes of a biaxial crystal are usually different for different wavelengths. The nature and extent of this dispersion of the axes is of considerable importance in identifying organic compounds.

Two different effects may be combined in the dispersion of the optic axes. The first of these effects is a change of optic axial angle with change of wavelength and is best referred to as "differential dispersion of indices" since the optic axial angle is governed by the ratio of the three principal refractive indices. If the dispersion of one of the indices differs markedly from the other two, the crystal may be uniaxial at a specific wavelength, whereby crossed axial plane dispersion must result. Furthermore, it is possible for a biaxial crystal to change sign. This occurs when $2V$ passes through 90°. The second effect, which occurs only with monoclinic and triclinic crystals, is a dispersion of the principal vibration directions of the wave surface.

Orthorhombic dispersion. With orthorhombic crystals, the dispersion of the optic axes is attributable solely to differential dispersion of the principal indices of refraction, as the principal vibration directions are constant for all wavelengths. All three of the principal vibration directions of orthorhombic crystals coincide with crystallographic axes; consequently, orthorhombic dispersion is characterized by the fact that the optic axial plane does not change and the two axes converge or diverge equally with change of wavelength. There are two special cases of orthorhombic dispersion that deserve attention. The first of these, crossed axial plane dispersion, occurs when the optic axial angle decreases to zero at some point in the spectrum. At this point, the Y-vibration direction interchanges with X or Z and the optic axial angle then increases in a plane at right angles to its former one. In the second special case, the optic axial angle is 90° at some point in the spectrum, so that X is the acute bisectrix for one end of the spectrum and Z is for the other end. This is equivalent to saying that the sign of birefringence is changed at the particular wavelength.

Crossed axial plane dispersion is shown by a number of substances, among which may be mentioned *p*-nitrobenzaldehyde [70], picrates of various amines [71], the high-temperature polymorphs of *o*-dinitrobenzene and 1,8-dinitronaphthalene, and the low-temperature form of dibenzoylmethane [72].

Monoclinic dispersion. This is characterized by the fact that only one of the principal vibration directions must necessarily coincide with a crystallographic axis for all wavelengths. The particular crystallographic axis is that which is at right angles to the plane of the other two and is called the ortho or *b*-axis. Any one of the three principal vibration directions of the wave surface may be coincident with the crystallographic *b*-axis, so that three principal types of dispersion result.

1. *Crossed dispersion.* This type of dispersion results when the acute bisectrix (X of negative and Z of positive crystals) is fixed, whereby the plane of the optic axes rotates about the acute bisectrix. Mitchell and Bryant [71] recently found a number of amine picrates that exhibit this property in marked degree.

2. *Horizontal dispersion.* This type of dispersion results when the obtuse bisectrix (X of positive and Z of negative crystals) is fixed. This results in the plane of the optic axes being displaced horizontally in the acute interference figure. It should be noted that a substance that shows horizontal dispersion of the axes in the acute figure must show crossed dispersion in the obtuse figure, and vice versa, so that in the special case in which axial angle $2V$ is

90° for some particular wavelength, the birefringence changes sign and the two acute interference figures show horizontal and inclined dispersion, respectively. A good example of this effect is provided by stilbene [76], which shows horizontal dispersion in the visible spectrum down to 4070 Å, and crossed dispersion in the ultraviolet.

3. *Inclined dispersion.* Inclined dispersion results when the normal to the optic plane (*Y*) is fixed. Both acute and obtuse interference figures show displacement in the plane of the optic axes. Good examples of inclined dispersion are provided by *o*-nitroacetanilide [73] and hexamethylenetetramine picrate [71].

In addition to the dispersion of the principal vibration directions, the optic axial angle may also vary with change of wavelength, so that the interference figure may be quite complex when viewed with white light. It is theoretically possible for the optic axial angle to become zero at some point in the spectrum and then to increase in a direction at right angles. This would result in a modified form of crossed axial plane dispersion in which inclined dispersion would become horizontal, and vice versa. However, crossed dispersion would remain crossed dispersion under these conditions. Perhaps the finest example of monoclinic crossed axial plane dispersion is α,α'-dipyridyl, which has been studied by Bryant [74]. This substance is uniaxial at about 4150 Å and exhibits horizontal dispersion for shorter wavelengths and inclined dispersion for longer wavelengths. The various types of monoclinic dispersion are summarized in Table 3.3.

Triclinic dispersion. With triclinic crystals, all three principal vibration directions of the wave surface may vary with change of wavelength. Since the axes of the wave surface bear no relationship to the crystallographic axes, the dispersion of the principal vibration directions has no element of symmetry, so that the acute interference figure could present the appearance of a

Table 3.3 Types of Monoclinic Dispersion

b =	*Sign*	*Acute*	*Obtuse*	*Acute**	*Obtuse**
X	−	Crossed	Horizontal	Crossed	Horizontal
X	+	Horizontal	Crossed	Inclined	Inclined
Y	−	Inclined	Inclined	Horizontal	Crossed
Y	+	Inclined	Inclined	Horizontal	Crossed
Z	−	Horizontal	Crossed	Inclined	Inclined
Z	+	Crossed	Horizontal	Crossed	Horizontal

* After crossed axial plane dispersion has taken place. The obtuse bisectrix changes place with *Y*.

mixture of horizontal and crossed, horizontal and inclined, and so on, monoclinic dispersion. Dispersion of the optic axial angle also occurs, and crossed axial plane dispersion is theoretically possible when the axial angle becomes zero for some particular wavelength. The dispersion of the optic axes would possess the characteristic asymmetry of the triclinic system at wavelengths both longer and shorter than the point of uniaxiality. As far as we are aware, triclinic crossed axial plane dispersion has not yet been observed, but it must be remembered that a small measure of monoclinic or triclinic dispersion combined with crossed axial plane dispersion can only be detected by careful measurements in monochromatic light of various wavelengths.

Change of sign in uniaxial crystals. In uniaxial crystals the direction of the optic axis is constant for all wavelengths, so that in a centered uniaxial interference figure differential dispersion of the ordinary and extraordinary rays gives rise to a departure from the Newton's rings series of colors. It is possible for the dispersion of the lesser refracted ray to be so much greater than that of the more highly refracted one that at some specific wavelength the two refractive indices become equal and then change places. This means that the sign of birefringence changes on passing through the wavelength for which the crystal is isotropic. This effect is shown by benzil [75, 76], which is isotropic at 4205 Å, optically positive for longer, and optically negative for shorter wavelengths.

Anomalous interference colors of thin crystals. It has already been mentioned that the scale of interference colors given by increasing thicknesses of a crystalline organic compound between crossed polarizers may deviate widely from Newton's color scale. When the sequence of colors is reversed, the scale is said to be anomalous. The effect is most striking in the first order of interference colors, brilliant blues and greens often being exhibited. This effect may result from one or more of the following causes:

1. The birefringence of the crystals for the blue end of the spectrum is less than one-half that of the red end. Benzil is an example of a uniaxial substance with this type of dispersion of birefringence that has been dealt with above. The transmission curve of a plate of benzil showing an anomalous blue is shown in Fig. 3.22.

2. The birefringence of the crystal for the blue end of the spectrum is very much higher than that of the red end (at least twice as much). Such dispersion is sometimes observed in substances having strong adsorption in the violet or near ultraviolet. A crystal of such a substance just thick enough to have a half-wave retardation in the blue would have such a low retardation for green and red that it would appear a fairly bright blue between crossed polarizers.

3. Biaxial substances having crossed axial plane dispersion with uniaxiality in the yellow or green usually show a very much more rapid divergence of the optic axial angle on going to shorter wavelengths than on going

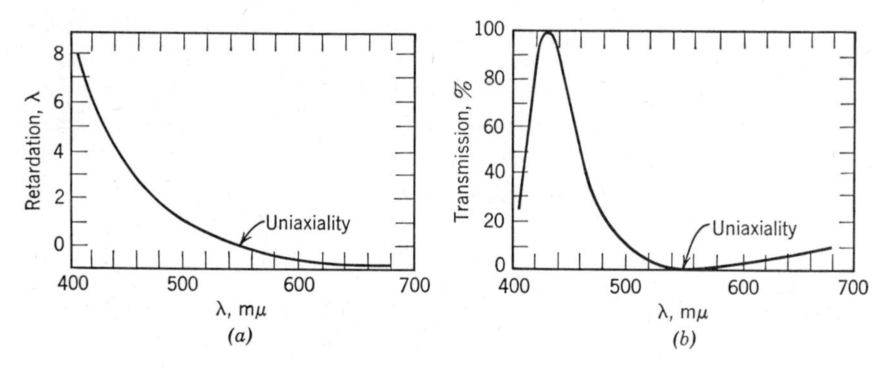

Fig. 3.23 (*a*) Retardation of 70-μ Bx_a plate of brookite; (*b*) transmission of 7-μ Bx_a plate of brookite.

to longer ones. As a consequence, thin crystals of such a substance viewed along their acute bisectrix exhibit brilliant blues and purples because of their high birefringence for the blue end of the spectrum. By way of an actual example let us consider brookite, the naturally occurring orthorhombic form of titanium dioxide. Figure 3.23*a* gives the retardation (in wavelengths) plotted against the wavelength for a 70-μ acute bisectrix plate. These values were obtained from a microspectrogram [75]. From thc values in this graph, the spectral transmission of a 7-μ plate of brookite was calculated, reflection and absorption losses being ignored. The transmission curve is shown in Fig. 3.23*b*, which represents a pure blue, as saturated as that of benzil. The actual hue given by a thin plate of brookite, or other substance of this optical type, is very sensitive to changes of thickness and of orientation, whereas that of benzil is very much less sensitive to either.

4. The fourth type of anomalous interference color is shown by crystals having strong monoclinic crossed dispersion. It is attributable to dispersion of extinction. Many of the amine picrates studied by Mitchell and Bryant [71] show strong monoclinic crossed dispersion. The values in Table 3.4 are taken from their paper.

Table 3.4 Monoclinic Crossed Dispersion in Amine Picrates

	Dispersion (degrees) at λ =				
Picrate	6438 Å	5780 Å	5461 Å	5086 Å	4800 Å
n-Propylamine	2	6	13	—	72
Isopropylamine	5	28	49	65	69
n-Butylamine	0	1	3	9	54
Pyridine (I)	1	5	9	20	35

The extinction angle is rotated through the above angles in crystals viewed along, or near to, the acute bisectrix. It should be noted that the dispersion increases rapidly for the blue end of the spectrum, so that if a very thin crystal of one of these substances is at extinction for red or green light it appears a brilliant blue or violet. This effect is independent of the thickness of the crystal, provided it is thin enough to have a retardation not exceeding about 200 mμ. Unlike the other types of anomalous interference color, the hue changes when the microscope stage is rotated.

As types 3 and 4 anomalous interference color are given by acute bisectrix orientations, they can only be observed with a narrow cone of illumination. Conoscopic observation, when this can be made, readily differentiates between the four types.

MOLECULAR STRUCTURE AND OPTICAL PROPERTIES

Since a crystal of an organic compound consists of a regularly arranged group of identical molecules, the shape of the molecule and its polar characteristics play a large part in determining the optical properties of the crystal. Rod-shaped molecules, such as those of long-chain hydrocarbons and their simple derivatives, have a greater refractive index for light vibrating along their length than for transverse vibrations. If such molecules are approximately parallel to one another in the crystal, the crystal is positive biaxial or positive uniaxial. However, considerable inclination of molecules in one crystal layer relative to those in an adjacent layer changes the sign of the birefringence. The refractive indices of crystals built from rod-shaped molecules are not very high unless the molecule contains highly refractive substituents, such as bromine, iodine, or sulfur. As with the refractive index of liquids, a conjugated double-bond linkage causes an increase in the indices.

Planar molecules, particularly those with conjugated chain ring systems such as benzene and naphthalene derivatives, confer two high-principal refractive indices (β and γ) on their crystals so that in general crystals of carbocyclic substances possess large values for their maximum birefringence ($\gamma - \alpha$). Exceptions to this general rule may result when molecules in different planes are staggered with respect to each other. When this occurs, γ and β are lower and α is higher than is usual for this class of compound. The effect of substituents is of considerable interest inasmuch as the dispersion of refractive indices and of principal vibration directions is often greatly modified. Heavy atomic substituents such as bromine, iodine, and selenium increase both the refractive indices and their dispersion, but polar substituents such as amino, nitro, carbonyl, and thiocarbonyl often affect the dispersion of the optical properties so strongly that striking abnormalities are exhibited by the crystals. The optical properties of salts of a colored aromatic compound may differ markedly among themselves, a property often of importance in establishing the identity of a colorless acid or base which itself is not well characterized optically but which yields a crystalline salt with a colored base or acid, respectively.

Microscopy of Transparent Crystals

ORTHOSCOPIC EXAMINATION

Orthoscopic examination refers to the examination of the image of the crystal in ordinary light, in polarized light, or between crossed polarizers. When the crystal is of sufficient size, a low-power objective and very narrow cone of illumination should be used. The crystalline form, habit, and cleavage should be observed with the analyzer withdrawn; it is not necessary to remove the polarizer for these observations. If the crystal is mounted in a medium of medium refractive index, its relief may vary considerably when either the microscope stage or the polarizer is rotated; this effect is dealt with under the heading of refractive index determination (p. 199).

Geometry of the Crystals. The observation of the form of a crystal is usually the first concern of the microscopist. If the geometrical shape of a crystal is understood, the crystal can usually be assigned to one of the crystal systems without further data. Well-formed crystals should be selected and drawings made of them. As crystals may lie on various faces and so present different appearances, the microscopist should study and record the shape of all of the crystals that seem to be different. The shapes of the top and bottom faces are relatively easily determined, the side faces are more difficult. It is helpful to section the crystal optically by focusing at different levels and recording the outline of the crystal at these levels. In this way the direction of slope of the faces and their general shape can be determined. All significant angles should be measured. It is also helpful to draw possible orthographic projections of other views of the crystal by using the techniques of projective geometry. By combining the information from the various crystals drawn, it should be possible to deduce the symmetry of the crystal and its ideal undistorted shape. It is also helpful to record optical data, obtained as described, for each crystal drawn, as this information always helps in understanding the symmetry.

In some cases it may be difficult or impossible to obtain certain views of a crystal because it does not lie in the proper orientation. This may interfere with the determination of optical properties as well as the geometrical symmetry. Crystals can usually be rolled over in any direction and any view exposed by mounting them in a viscous medium such as Aroclor 1260 or the polystyrene–bromonaphthalene medium described in Section 4, p. 147 and sliding the cover glass. The shearing action in the viscous medium causes the crystal to roll in the general direction the cover glass is moving. Other methods of orienting crystals are described on p. 201.

Extinction Angles. When a crystal is rotated on the microscope stage between crossed polarizers, it passes through an extinction position every 90°, as explained on p. 181. Extinction is classified under three headings, parallel, symmetrical, and oblique, which are illustrated in Fig. 3.24*a*, *b*, and *c*, respectively. Parallel and symmetrical extinction are characteristic of all possible orientations of crystals of the hexagonal, trigonal, tetragonal, and orthorhombic systems. Symmetrical extinction is symmetrical in that the vibration planes of polarizer or analyzer bisect the edge angles presented by the crystal; irregular development or distortion may cause the crystal to deviate markedly from the ideal form, but the bisection of edge angles still holds. Parallel and, more rarely, symmetrical, extinction also occurs in monoclinic crystals viewed exactly at right angles to the crystallographic ortho *b*-axis, but its occurrence in other orientations of monoclinic crystals, or in any orientation of triclinic crystals and in all orientations of monoclinic crystals other than those in which the ortho *b*-axis is at right angles to the line of vision, occurrence of parallel extinction being fortuitous. In the case of oblique extinction, the extinction angle is the angle, less than 45°, between the most prominent or characteristic crystal edge of the crystal at extinction and the nearest vibration plane of analyzer or polarizer. An important aspect of extinction angles in the study of organic crystals is that of dispersed extinction. Dispersed extinction is most clearly seen with thick crystals when

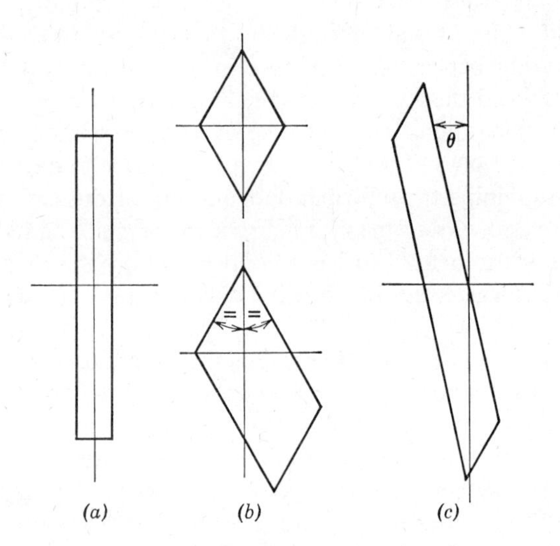

Fig. 3.24 Extinction: (*a*) parallel; (*b*) symmetrical; (*c*) oblique.

a very narrow cone of illumination and a powerful light source are used. When the dispersion is slight, the color of crystal changes from brownish to bluish on passing through the extinction position, whereas brilliant colors are seen when the dispersion is large. Dispersed extinction can occur in triclinic crystals and in monoclinic crystals not in their position of parallel extinction. Under no circumstances can dispersed extinction occur in crystals of symmetry higher than monoclinic.

Sign of Elongation. In elongated crystals, such as those of columnar or acicular habit, the character of the elongation is regarded as positive or negative according to whether it corresponds to the higher (slow ray) or lower (fast ray) refractive index, respectively. In uniaxial crystals the direction of elongation is of necessity the crystallographic c-axis, since the tetragonal or hexagonal axes are equal. We have already noted that vibrations parallel to the c-axis are those of the extraordinary ray, consequently, positive and negative uniaxial crystals show positive and negative elongation, respectively. With orthorhombic crystals the elongation is negative if it takes place in the principal X-vibration direction, positive for the Z-direction and both positive and negative if the elongation is along Y, depending on the orientation of the crystal. Crystals of this type, when rotated through 360° along a horizontal axis, pass alternatively through two positive and negative elongations separated by four extinction positions which occur when the visual axis coincides with an optic axis. With crystals of the monoclinic and triclinic systems, the elongation is regarded as positive or negative according to whether it most nearly agrees with the direction of higher or lower refractive index, respectively; when the direction of elongation most nearly corresponds to Y, the crystals exhibit both positive and negative elongation, depending on their orientation.

CONOSCOPIC EXAMINATION

The Conoscope. The conoscope is a device for determining the nature of the wave surface of a crystal. Its principle is shown diagrammatically in Fig. 3.25, in which a roughly parallel beam of light is polarized and is then passed through a converging lens which brings it to a focus within the crystal, which we will suppose to be a plate of calcite cut perpendicular to the optical axis. The light diverging from the crystal is rendered parallel by a collecting lens and passes through an analyzer, crossed with respect to the polarizer, which causes interference to occur. As seen from above, the collecting lens resembles a flat disk. The center of the disk is illuminated by light which has passed perpendicularly through the crystal; its circumference is illuminated by light which has passed obliquely through the crystal. This illuminated disk is therefore an orthographic projection of a segment of the wave surface as refracted into air.

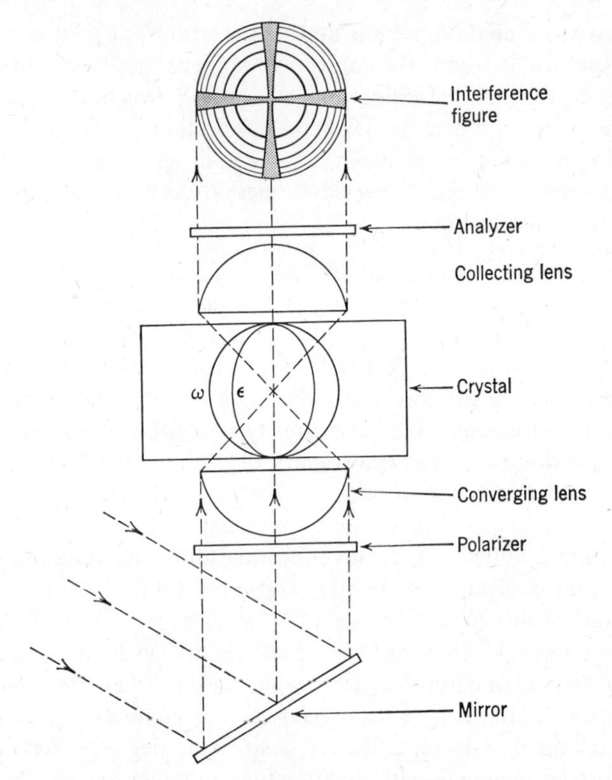

Fig. 3.25 Formation of interference figure.

If we bear in mind that the vibration direction of the extraordinary ray is in the optic axial plane, we can readily understand the formation of the interference figure of our optic axial section of calcite since all radii of the circular disk represent vibration directions of the extraordinary rays having directions through the crystal indicated by their location on the disk of the interference figure. The center of the figure appears dark because it is made up of rays that have traveled along the optic axis of the crystal. The radii of the figure that have the same, or very nearly the same, directions as the vibration planes as the polarizer and analyzer must appear dark, since along these directions the crystal is at extinction. If monochromatic light is used, a series of concentric dark rings will appear. These occur for directions in which the retardation of the crystal is an integral number of wavelengths as explained on p. 182. The ratio of the diameters of the 1st, 2nd, 3rd . . . rings is roughly $1:\sqrt{2}:\sqrt{3}\cdots$. In white light the rings form an ascending series of interference colors.

In actual practice single lenses are not favored as convergers and collecting lenses because of their bad spherical aberration and relatively small angular aperture. It is very desirable, particularly when working with small crystals, to be able to view them orthoscopically and then, with the minimum of manipulation, to view their conoscopic interference figures. We therefore consider some of the ways in which a polarizing microscope can be used to serve as a conoscope.

The simplest method, which, however, works only with crystal plates of large area, consists in using a high-power dry objective, such as 4-mm achromat in conjunction with a condensing system capable of filling its back lens with polarized light. The crystal is brought into focus, the eyepiece is removed, and the back lens is viewed through an analyzer. A small, but distinct, interference figure is seen. The method may be applied to smaller crystals by using a pinhole diaphragm over the drawtube of the microscope; this ensures that only light passing through a particular crystal reaches the eye. In place of a pinhole diaphragm, a Wright crossed-slit diaphragm may be used [77].

Petrographic microscopes are usually fitted with a Bertrand lens which slides into the body tube of the microscope below the eyepiece. In conjunction with the eyepiece, the Bertrand lens forms a low-power microscope which is focused on the back lens of the objective, an enlarged image of which is seen. The more elaborate instruments have an iris diaphragm fitted immediately above or below the Bertrand lens, which serves to isolate the light from a small crystal. It is necessary to focus the image of the crystal on the Bertrand lens diaphragm. This operation may require the use of an auxiliary magnifier over the eye lens until some experience has been gained, after which it is easy to judge how far the objective must be raised by observing the edge of the interference figure.

The interference figures of small crystals may be observed on an ordinary "biological" microscope by equipping it with Polaroid polarizer and analyzer and a Johannsen lens [78]. A Johannsen lens is made by drawing out a heated glass rod to hairlike thinness and breaking the glass filament into pieces 3 or 4 cm in length. The ends of these filaments are fused by brief heating on the edge of a Bunsen flame in order to form spherical globules. On examination under the microscope, some spherules will be found that are free from bubbles and are less than 0.1 mm diameter. The crystal is first focused under a 16 or 8-mm objective and a 5 or 10× eyepiece and is brought into the center of the field. The small spherical lens is then placed in contact with the cover glass, directly over the crystal. On racking up the objective, a plane of focus is reached in which an interference figure is seen. The angle of the cone of rays making up the interference figure is, of course, the same as that of the cone of illumination given by the substage condenser; consequently, the diaphragm on the Abbe condenser should be fully opened. The maximum angle of the cone of rays visible in a Johannsen lens is approximately 90°, which corresponds to an N.A. of 0.7.

Another way of observing the interference figures of crystals is to magnify the Ramsden disk given by a low-power (5×) eyepiece by means of a high-power (20×) lens usually known as a Becke-Klein magnifier. Jelley [72] found it better to employ two 5× eyepieces equipped as in Fig. 3.26. Each eyepiece has a field lens *A* and an eye lens *B*. One of the eyepieces (shown at

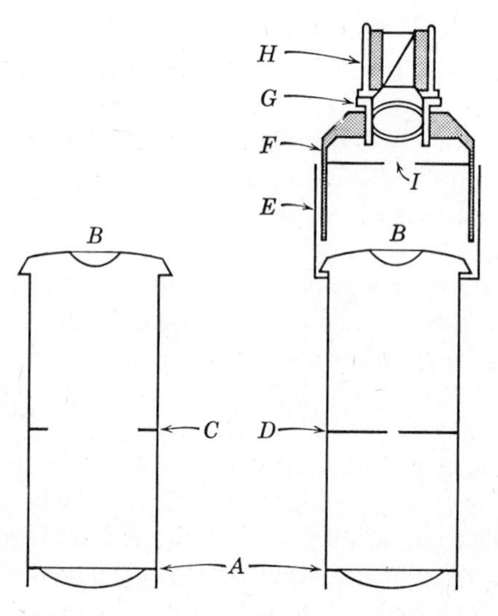

Fig. 3.26 Conoscopic apparatus for use with student-type microscope.

the left of the figure) is equipped with crosshairs in the field diaphragm *C*; this eyepiece is used for orthoscopic observations in conjunction with a cap analyzer. The other eyepiece is provided with a pinhole aperture in the center of the diaphragm *D* which replaces the usual field diaphragm. This eyepiece is converted to a conoscopic apparatus by inserting it in sleeve *E*, into which slides fitting *F*, which carries a 20 or a 30× magnifier *G* and analyzing prism *H*. A reticle with a micrometer or coordinate scale may be fitted in a diaphragm *I* at the focus of *G*. In using the apparatus the image of the crystal under study is first centered and focused on the crosshairs of the orthoscopic eyepiece, which is then withdrawn from the microscope body tube. The conoscopic apparatus is then inserted and is rotated to an extinction position with respect to the substage polarizer. The interference figure is focused by sliding fitting *F* into sleeve *E*.

The angle θ of the cone of light within the crystal that is represented by the diameter of the conoscopic interference figure is related to the numerical aperture of the objective, and the mean refractive index n presented by the particular orientation of the crystal by the approximate relationship:

$$\theta = 2 \sin^{-1} (\text{N.A.}/n). \tag{3.20}$$

It is obviously advantageous to use an objective with as high a numerical aperture as possible for the purpose of studying the general nature of the wave surface of a crystal, such as a 4-mm objective of N.A. = 0.85 or 0.95. A wider conoscopic angle may be obtained by the use of a 1.8-mm oil-immersion objective of N.A. = 1.25 or 1.30, but this entails the use of an oil-immersed condenser, which is rather a messy operation since in this case a return from conoscopic to orthoscopic observation necessitates the removal of the condenser and cleansing of the slide. An alternative procedure is to use distilled water as an immersion fluid for the condenser.

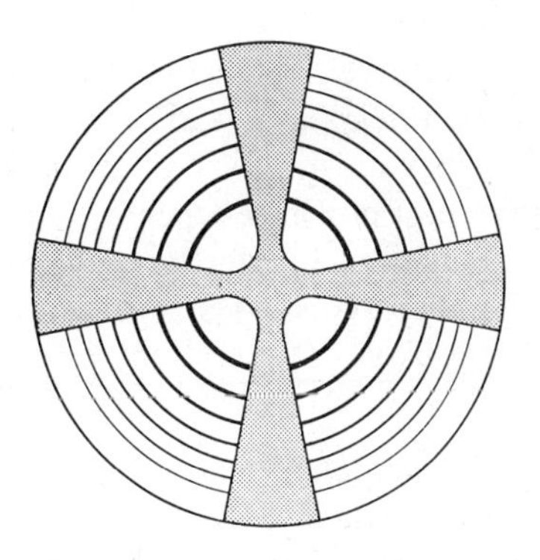

Fig. 3.27 Interference figure of uniaxial crystal.

Uniaxial Interference Figures. Two types of centered uniaxial figures are possible: the optic axial figure (see Fig. 3.27), the formation of which has been described above, and the optic normal figure (see Fig. 3.28). The optic axial figure is very easily recognizable because it consists of a black cross and a number of concentric circles of rising order of interference. The appearance of the uniaxial figure in monochromatic light is shown in Fig. 3.27. The number of circles in the figure increases with increase of thickness and of birefringence of the crystal, and with increase of numerical aperture of the objective used. The determination of the sign of a uniaxial crystal is quickly made by inserting a birefringent plate of known fast and slow vibration directions in the 45° position between the objective and analyzer of the conoscopic system when compensation or extinction takes place in a spot in each of two opposite quadrants. These are of course the quadrants in which the fast and slow vibrations of the crystal correspond to the slow and fast vibrations, respectively, of the compensator. Since we know that the vibration direction of the extraordinary rays are radii in the figure, we can at once tell whether ϵ is fast (uniaxial negative) or slow (uniaxial positive); when the crystal is positive, the pair of quadrants in which compensation occurs is at right angles to the pair of quadrants in the slow direction of the compensator;

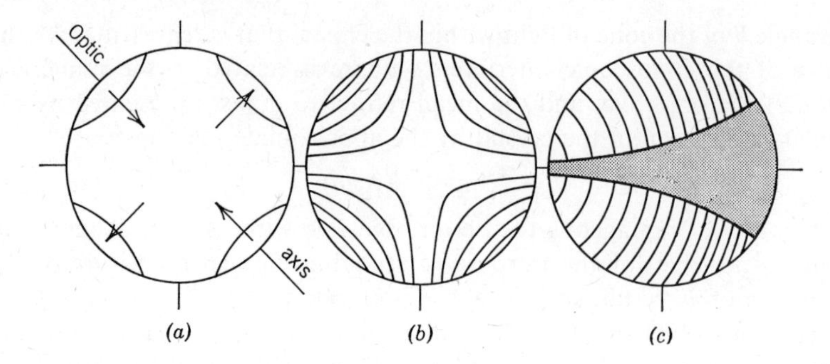

Fig. 3.28 Interference figures of uniaxial crystals: (*a*) optic normal, white light; (*b*) optic normal, monochromatic light; (*c*) uncentered figure, monochromatic light.

for negative crystals the compensation occurs in the same pair of quadrants. With thick basal sections of uniaxial crystals possessing marked optical rotatory power, the center of the uniaxial cross is replaced by a bright field, the color of which depends on the thickness and optical rotatory power of the substance. A perfectly centered optic axial figure does not change when the microscope stage is rotated; if the optic axis is not quite perpendicular to the plane of the section, the center of the figure describes a circle on rotating the stage, but the uniaxial cross does not rotate.

Biaxial Interference Figures. Three types of centered biaxial figures are possible: the acute bisectrix (Bx_a), obtuse bisectrix (Bx_o), and optic normal figures. The optic normal ("flash") figure is indistinguishable from the uniaxial optic normal figure when $2V$ is small. When $2V$ is in the region of 90°, the edges of all four quadrants show the same slight increase of retardation, so that the figure presents a symmetrical appearance. The acute bisectrix figure is the most important of those presented by biaxial crystals since it permits the estimation of the optic axial angle with a fair degree of accuracy.

The appearance of an acute bisectrix figure for a crystal in the 45° position is shown in Fig. 3.29*a*. In the extinction position the appearance changes to that shown in Fig. 3.29*b*. The points of emergence of the optic axes are marked by the "eyes" of the figure, which are points of zero retardation and must therefore appear dark for any position of rotation of the microscope stage. Johannsen [78] has termed these points in the interference figure melatopes. Surrounding the melatopes are bands of equal retardation, which correspond successively to retardations of λ, 2λ, 3λ, ... In monochromatic light these bands are black; in white light they form a rising series of interference colors. The apparent distance apart of the melatopes in a centered acute bisectrix figure depends on (1) the numerical aperture of the objective

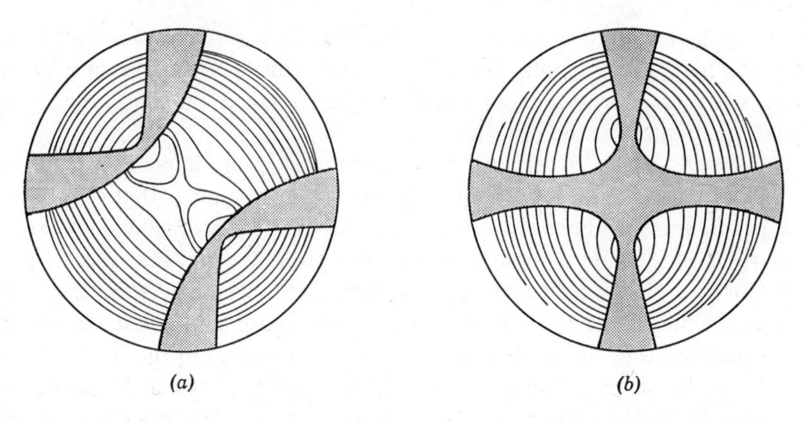

Fig. 3.29 Biaxial interference figures: (*a*) 45° position and (*b*) extinction position.

used; (2) the optic axial angle $2V$ of the crystal; and (3) its refractive index β along the optic axis. The curves in Fig. 3.30 give the maximum observable values of $2V$ for objectives of various numerical apertures as a function of β. Two isogyres pass through the melatopes; with the crystal in the 45° position they form two hyperbolic brushes with their convex side toward the acute bisectrix. The smaller $2V$, the greater the curvature of the isogyres; for $2V = 90°$ the isogyres are straight. When the stage is rotated, the isogyres rotate in the opposite direction. If, instead of rotating the stage, the polarizers are rotated synchronously, then the isogyres rotate in the same direction as the polarizers but twice as fast. It follows therefore that the isogyres of a

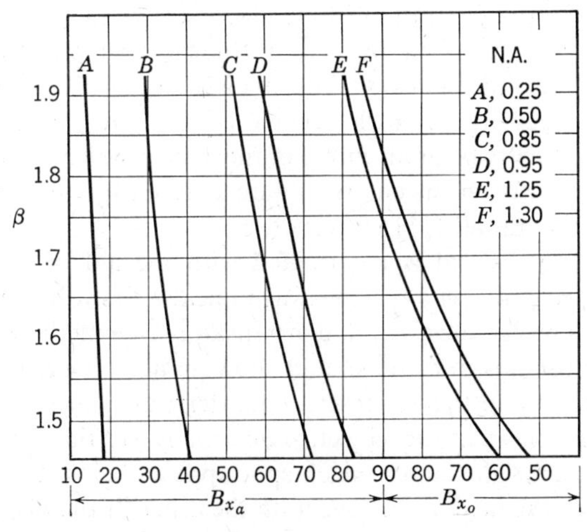

Fig. 3.30 Conoscopic field of various microscope objectives.

biaxial crystal, unlike those of a uniaxial one, do not remain parallel to the vibration planes of the polarizer and analyzer. This fact often permits one to judge whether a crystal is uniaxial or biaxial when only an uncentered conoscopic figure can be obtained.

The appearance of a centered obtuse bisectrix figure depends very much on the optic axial angle. When $2V$ is nearly 90°, it resembles that of an acute bisectrix figure with the melatopes just outside the conoscopic field. When $2V$ is nearly 0°, the figure is practically indistinguishable from that given by the optic normal. The appearance may of course be anywhere between these two extremes. One other type of biaxial conoscopic figure is of particular importance, namely, that in which a single optic axis emerges in the field of view. The single brush establishes beyond doubt that the crystal is biaxial; its curvature when the crystal is in the 45° position renders possible a rough estimation of $2V$, and from its movement when a compensator is inserted between the objective and the analyzer, the sign of the crystal can be determined by one of the methods given below.

It is convenient to classify conoscopic interference figures of biaxial crystals into the groups centered, displaced, and uncentered, according to whether two, one, or no symmetry planes pass through the center of the figure. Using this classification, we find that the following are essential relationships between the type of biaxial figure and the face upon which a crystal rests.

Centered figures are given by {100}, {010}, and {001} plates of orthorhombic crystals and by {010} plates of monoclinic crystals. Centered figures by other monoclinic orientations are fortuitous.

Displaced figures are given by $\{hk0\}$, $\{h0l\}$, and $\{0kl\}$ orthorhombic plates and by {100}, {001}, and $\{h0l\}$ monoclinic plates.

Uncentered figures are given by $\{hkl\}$ orthorhombic plates; by $\{hk0\}$, $\{0kl\}$, and $\{hkl\}$ monoclinic plates; and by all orientations of triclinic crystals. Centered and displaced figures with triclinic crystals are fortuitous. In the above classification any two, or all three, of the Miller indices, denoted by h, k, l, may be alike; thus an uncentered figure is given by an orthorhombic crystal resting on a pyramid {111} face.

The sign of biaxial crystal is determined with the aid of some sort of compensating device, such as a quarter-wave mica, a first-order red selenite (gypsum), a quartz wedge, or a Berek compensator. One method is to rotate the crystal, which for this purpose should show an acute bisectrix figure, to its extinction position and then to insert a quarter-wave mica of known fast and slow directions between the objective and analyzer; the melatopes and isogyres now appear bright, and a black spot appears below the optic axial plane on one side of the field and above it on the other. If the black spots are in the same pair of quadrants as those in the slow direction of the mica, the crystal is negative, whereas it appears in the adjacent pair of quadrants for a positive crystal. A convenient way of using a first-order red selenite (gypsum) plate is by rotating the crystal to that 45° position in which its optic axial plane is in the same direction as that of the slow component of the selenite. The isogyre, or isogyres, appears red, which shades to yellow on its convex side and blue on its concave side for positive crystals, and to blue on its convex side and yellow on its concave side for negative crystals. This method can obviously be used when only a single isogyre appears in the field provided, however, that $2V$ is not so nearly equal to 90° that the isogyre is devoid of curvature.

When a quartz wedge of the Wright or Johannsen type, or a Berek compensator (described on p. 211) is used to determine the sign of a crystal, it is well to adopt the routine of placing the crystal in the 45° position for which its optical axial plane coincides with the slow vibration direction of the wedge or compensator. Then, on inserting the wedge or rotating the Berek compensator to either side of its normal (30° graduation) position, the isogyres appear to move toward Z. It follows therefrom that when the isogyre moves in the direction of its convex side the crystal is positive and that movement in the concave direction shows that the crystal is negative. This method is equally applicable to figures presenting a single isogyre. When an undoubted obtuse bisectrix figure is studied by this means, the unseen concave sides of the isogyres face the center of the figure. On sliding in the wedge, or on rotating the Berek compensator, the color fringes move along the optic axial plane toward the center of the figure if the crystal is negative, and out from the center toward the optic axes if the crystal is positive.

It is advisable, when working with unfamiliar polarizing microscopical equipment, to check the above techniques on known crystals. A basal section of calcite, or a fusion slide of sodium nitrate, provides a convenient example of a negative uniaxial substance. A cleavage plate of muscovite ("ordinary") mica, or a fusion slide of piperonal, provides a convenient example of a negative biaxial crystal. The piperonal preparation presents acute and obtuse bisectrix figures and optic normal flash figures.

DETERMINATION OF REFRACTIVE INDICES

Crystal Orientation. Refractive index determinations with the microscope are usually made by the immersion method, in which a very narrow axial cone of illumination is used. The orientation of the crystal is therefore given by the center of the interference figure. In the great majority of polarizing microscopes, the vibration direction of the polarizer is in the N-S direction (to and from the observer) or, in the case of microscopes with synchronous rotation of the polarizers, it is easily set in this position. This position of the vibration plane of the polarizer is therefore assumed in the following discussion. Isotropic crystals, having a single refractive index, do not require special orientation.

Uniaxial crystals have two principal indices, ω and ϵ. Optic normal orientations present both. For ω set the crystal at extinction with its crystallographic *c*-axis (direction of the optic axis) E-W and withdraw the analyzer. For ϵ set *c* in the N-S position. Centered optic axis figures (basal plates) give ω only for all settings.

For biaxial crystals centered acute bisectrix, obtuse bisectrix, and optic normal figures indicate that two optical symmetry planes intersect in the center of the figure. Consequently, crystals giving these conoscopic figures present two of the three principal refractive indices. Displaced figures present only one principal index and uncentered figures present none.

The particular refractive index for plane-polarized light transmitted through a biaxial crystal in an extinction position is readily deduced by considering the type of interference figure in conjunction with the following rules.

1. When the convergent polarized light interference figure shows that axial rays are transmitted along the acute or obtuse bisectrix, the crystal is at extinction when the plane of the optic axes is in the N-S or E-W position. The vibration at right angles to the plane of the optic axes is β regardless of whether the figure is that of the acute or obtuse bisectrix, and the vibration in the plane of the optic axes must therefore be for α or γ, depending on whether it corresponds to an index lower or a higher than β, respectively. It follows therefore that when a crystal showing a centered bisectrix figure is rotated to an extinction position, and the analyzer prism and converger are withdrawn, light is transmitted through the crystal with a velocity corresponding to α, β, or γ. The various possibilities are given in Table 3.5.

2. When convergent polarized light gives a centered optic normal or flash figure, one of the extinction positions presents the lowest principal index α, and the other presents the highest index γ.

3. One principal index is presented by crystals that give a displaced interference figure, since an optical symmetry plane passes through the center of the figure. Possible cases are obtuse and acute bisectrix figures, either positive or negative in sign, which when displaced in the plane of the optic axes present β when the direction of the optic axial plane is E-W and the transmitted light is vibrating N-S. Obtuse and acute bisectrix figures when displaced laterally present α or γ when the optic axial plane is N-S (see Table 3.5). In the case of crystals giving an optic normal flash figure displaced along a principal plane, a very simple rule gives the principal index presented. The crystal is placed

Table 3.5 Optical Orientation

Interference Figure	*Sign*	*Direction of Optical Axial Plane*	*Index*
Acute bisectrix	Positive	N–S	α
Acute bisectrix	Positive	E–W	β
Acute bisectrix	Negative	N–S	γ
Acute bisectrix	Negative	E–W	β
Obtuse bisectrix	Positive	N–S	γ
Obtuse bisectrix	Positive	E–W	β
Obtuse bisectrix	Negative	N–S	α
Obtuse bisectrix	Negative	E–W	β

at extinction with the optic normal E or W of the microscope axis; the N-S vibration direction then corresponds to α or γ depending on whether this is the "fast" or "slow" direction.

Methods of Changing Crystal Orientation. Frequently, crystals do not lie in a favorable orientation on the microscope slide, making it difficult to observe the interference figure in the desired direction. Such crystals can be rolled as described on p. 189. However, with this arrangement it is not possible to measure refractive indices as described later in this section. Hartshorne and Stuart [70] have described rotation stages used for orienting crystals for viewing their interference figures and for measuring refractive indices. A more complex apparatus, the universal stage [70, 77, 78, 107, 108], can also be used to measure optical properties such as optic axial angle and refractive indices. However, the universal stage is a complex instrument, and its limited usefulness does not justify the high cost of the stage and the necessary accessories for routine chemical microscopy.

Immersion Methods. The immersion method of determining the refractive indices of small crystals may be said to date from the work of Maschke [79], who described the appearance of crystals immersed in liquids of differing refractive indices. Becke [80] first described the movement of fringes of light seen around immersed crystals when the microscope objective was raised or lowered, an effect now known as the Becke line. Shroeder van der Kolk [81] did much to popularize the immersion method of determining the refractive index of crystalline substances. Stated simply, the immersion method consists in matching a refractive index of a crystal with that of a liquid whose refractive index is either known or can be measured; and the many variations in present-day procedure depend on alternative methods of ascertaining when the index has been matched and of measuring the refractive index of the liquid. There are two principal methods of comparing the refractive index of a transparent solid with that of a liquid in which it is immersed: the central illumination method, which makes use of the Becke line effect and the method of oblique illumination.

Method of central illumination. This method depends on the fact that a crystal or crystal grain having a prismatic or lenticular cross section does not deviate axial light rays when immersed in a liquid having an identical refractive index, whereas the rays are deviated if the crystal and immersion liquid differ in refractive index. The nature of this deviation is readily understood from the diagram in Fig. 3.31, which shows that rays passing through

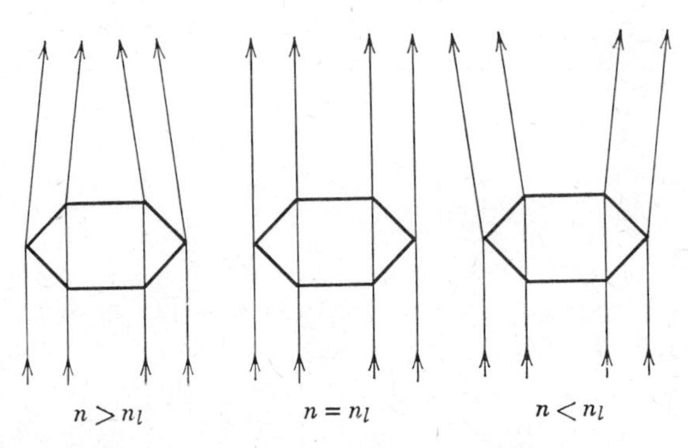

Fig. 3.31 Becke line of prism with index n in liquid with index n_l.

a prismatic (or lenticular) edge of a crystal are bent toward the crystal if it has the higher index. Consequently, when the microscope system is racked upward, the band of light from the crystal edges appears to move toward the substance with the higher refractive index. These Becke line phenomena are best observed by using a low-power objective (a 6 or 10× objective not exceeding 0.25 N.A.) and a high-power ocular (10 to 20×). An important feature of the Becke line procedure as described above is that the refractive index of the immersion liquid is compared with that of the crystal for rays traveling along the microscope axis and therefore corresponds to the center of the interference figure when the Becke line is not displaced on racking up the objective.

Method of oblique illumination. This method depends on the fact that no shading is visible on immersed crystals observed in oblique light when the index of the crystal and liquid are identical. The usual method of obtaining oblique illumination is by inserting a card below a high-aperture condenser so that only rays from one side of the condenser pass through the immersed crystal. A low-power objective is used for observing the shading on the crystal, which appears on the side facing the direction in which the card is inserted if the crystal has the higher refractive index, and on the opposite side if it has the lower index. The substage condenser should be racked up fully. This method of oblique illumination is open to the objection that the rays used in the test do not pass axially through the crystal; although the error so introduced is small with crystals of low birefringence, such as are common among aromatic compounds. Saylor [82] has discussed the sources of error both in the Becke line method of central illumination and the method of oblique illumination. He found the latter to be unreliable with strongly birefringent crystals and described a much improved method employing two diaphragms—one inserted below the converger element of the condenser from the right-hand side so as to obscure about half the aperture, and the other a stop covering the right-hand half of the back lens of a 3.2× objective. Saylor recommends that the vibration plane of the beam of polarized light be horizontal (E-W) rather than vertical (N-S) as is commonly used with Becke line methods.

Immersion Liquids. Adjusting the refractive index of the immersion liquid. The Becke line effect and the shadows in oblique light indicate when the refractive index of a liquid is higher or lower than that of the immersed solid. In fact, with some experience, the difference between them can be estimated to 0.1. The problem of the immersion method is, therefore, to adjust the refractive index of the liquid so that it substantially matches that of the immersed solid. The following is a list of methods that can be used for this purpose.

1. *Changing the immersion liquid.* This is usually accomplished by having a set of liquids of known refractive index from about 1.360 to 1.780 in steps of about 0.005. The most expeditious way of working with such a set of liquids is to start with one of $n \simeq 1.55$ and use it as a mountant for crystals held between an object slide and cover glass. The interference figures of the crystals should be studied in order to make sure that principal indices are being determined. It is then noted, by examination in central or oblique illumination, whether the principal indices of the crystal are higher or lower than the index of the liquid. Liquids nearer to the estimated indices are then tried on another preparation of the crystals; this operation is repeated until finally each of the indices is matched to within 0.005. If an index of the crystal falls between those of two liquids, it is usually possible to judge the index to 0.002. It should be remembered, however, that inaccurate orientation of the crystal and temperature changes during the manipulations can introduce an error of greater magnitude than 0.002. When only a very small quantity of crystalline material is available, it is advantageous to use volatile immersion media

that have been redistilled, since in this case one liquid may be allowed to evaporate before the next is applied to the same preparation. The high-index melts used by Larsen and Berman [83] for determining the refractive index of minerals are unsuitable for refractive index work on organic compounds.

2. *Varying the temperature of the immersion liquid.* Of the various methods that have been developed to permit the use of fewer liquid standards with a given specimen, that of changing the refractive index of the liquid by varying its temperature has possibly found most favor among petrographers. The thermal variation method was originated by Gaubert [84] and is based on the fact that the temperature coefficient of refractive index, dn/dT, of a liquid is considerably greater than that of most minerals and inorganic compounds. In principle, the transparent or translucent substance is immersed in a liquid of slightly higher refractive index, and the temperature of the system is then raised slowly to the point at which a central or oblique illumination test shows that the indices of liquid and solid are equal. Both Emmons [85] and Saylor [86] have described heating cells for this purpose. That of Saylor has the special advantage of thinness, so that it permits the use of a conoscopic observation of an immersed crystal in order to determine if one or both of its vibration directions are on optical symmetry planes and therefore correspond to principal indices.

3. *Varying the wavelength of the light used for observation.* Posnjak and Merwin [87] originated the method of adjusting the refractive index to that of the immersed solid by using a monochromator as a light source and by varying the wavelength until the central or oblique illumination test shows that a match has been obtained. By working with an overlapping series of liquids of known refractive index and dispersion, it is possible to obtain values for the refractive index of a given solid at various temperatures.

The last two methods (2 and 3) of varying the refractive index of an immersion liquid have been combined in the double-variation method of Emmonns [88], who has also devised a universal stage technique which employs a special lower segment with a water circulation device for regulating the temperature. The universal stage is used for orienting the crystals or crystal fragments held between a pair of cover glasses "oiled" to the segments with an appropriate immersion fluid. In the absence of a specially equipped universal stage, the necessary determination of the optical orientation can be carried out in convergent polarized light with a Bertrand lens if a Saylor heating cell is used, as this cell is of course equally usable for the double-variation method.

4. By mixing two nonvolatile liquids on the microscope slide. Jelley [89] proposed the system of mixing two substantially nonvolatile liquids of widely separated refractive indices in a cavity on a microscope object glass, varying

their proportion until the refractive index of an immersed solid has been matched. Some of the liquid is then transferred to the prism of a microrefractometer, and its refractive index is read. Provided that the immersion liquids, microscope, and microrefractometer have reached room temperature to within ±2°C, no temperature control is necessary. For work with crystals of inorganic compounds and organic compounds of low solubility, the liquids originally chosen are given in Table 3.6. However, since this work was

Table 3.6 Liquids for Determination of Refractive Index

Substance	*Boiling Point* (°C)	d_4^{20}	n_D^{20}	dn/dT
Ethyl oxalate	73 at 10 mm	1.0793	1.410	0.00049
Ethyl citrate	180 at 11 mm	1.1369	1.443	0.00036
n-Butyl phthalate	155 at 10 mm	1.0388	1.492	0.00032
α-Bromonaphthalene	148 at 16 mm	1.4876	1.658	0.00048
α-Iodonaphthalene	160 at 14 mm	1.7344*	1.702	0.00047
Methylene iodide	180 at 760 mm	3.325	1.742	0.00068

* d_4^{15}.

published, Jelley has substituted di-*n*-butyl carbonate, $n_D^{20} = 1.411$, for diethyl oxalate and tri-*n*-butyl citrate, $n_D^{20} = 1.445$, for triethyl citrate. There are now available a large number of commercial plasticizers which serve for the lower refractive index range.

The mixing of the liquids is readily carried out by stirring with a glass rod drawn out to about a 1-mm diameter at the stirring end. The liquids can be handled conveniently in small hypodermic syringes, which are well adapted for transferring small amounts of liquid to the slide excavation in which the mixing is performed [90]. Since the refractive index of the liquid is determined after a match has been obtained, the liquids used need not be pure—the only necessary condition is that their refractive indices shall not change within a few minutes exposure to the air.

The refractive index of a mixture of two liquids of different volatilities and refractive indices may be regulated by differential evaporation. This method is particularly applicable to Clerici's method of determining refractive indices by means of the apparatus described below. Promising results have been obtained with this method by using the following mixtures [91]:

1. Two volumes of diethyl carbonate and 1 volume of α-iodonaphthalene has $n_D^{20} = 1.491$, which rises on evaporation of the ethyl carbonate to 1.702.

2. Two volumes of *n*-butyl acetate and 1 volume of α-iodonaphthalene has $n_D^{20} = 1.497$, which rises on evaporation to 1.702.

3. Two volumes of *n*-heptane and 1 volume of α-iodonaphthalene has $n_D^{20} = 1.492$, which rises on evaporation to 1.702.

4. By substituting *n*-butylphthalate for α-iodonaphthalene in the above formulas, liquids are obtained having a refractive index of about 1.42, which rises on evaporation to 1.49.

Similarly, the refractive index of mixtures of methylene iodide and Merwin's solution [92] may be adjusted by allowing the methylene iodide to evaporate, whereby the index is raised. Merwin's solution consists of methylene iodide saturated with sulfur, iodoform, stannic iodide, arsenic triiodide, and antimony triiodide. A rather deep cell was used for the "variation-by-evaporation" technique, which worked well with moderately large crystals. Once the index corresponding to a known vibration direction had been obtained, some of the liquid was transferred by means of a micropipet (constructed from a medicine dropper) to a modified prism on Jelley's microrefractometer described below.

The solubility of many crystalline organic compounds in organic immersion media—especially methylene iodide—frequently imposes a restriction on the number of methods of matching the refractive index of the crystal with that of the liquid. Thermal variation methods can rarely be used, partly because the solvent action of the media is much more rapid at higher temperatures, and partly because dn/dT of the crystal may be appreciable. The wavelength variation method sometimes fails because $dn/d\lambda$ is of the same order of magnitude as that of organic immersion liquids. When sets of known liquids are used, they may be saturated with the compound under study and recalibrated with an Abbe refractometer [93]. One particular advantage of Jelley's technique of varying the refractive index by mixing two liquids, or by differential evaporation, is that any moderate degree of dissolution of the solid is automatically compensated for, since the refractive index is measured after the match has been obtained. Certain aqueous and glycerol solutions of highly refractive inorganic salts, particularly of potassium mercuric iodide, are useful immersion media. Sonstadt's or Thoulet's solution [78] is prepared by dissolving 230 g of potassium iodide and 270 g of mercuric iodide in 80 ml of water and then cautiously evaporating on a steam bath until a crystalline film forms on the surface of the liquid. The clear portion is decanted. It has a refractive index of about 1.72. It should be noted that this solution contains 2.33 times as much potassium iodide as that required to form potassium mercuric iodide ($KHgI_3$). If an excess of mercuric iodide is used, a solution of potassium mercuric iodide is obtained that has a considerably lower refractive index. Jelley prepared aqueous solutions of various alkali mercuric iodides of the general formula $M \cdot HgI_3$, saturated at 20°C. These had the following refractive indices:

$M =$	Li	Na	K
$n_D^{20} =$	1.620	1.621	1.615

Dunningham [94] studied the system $KI–HgI_2–H_2O$ and found that the solid phase, K_2HgI_4, does exist, contrary to the statements in many textbooks. Jelley confirmed, by means of a microscopical study of the crystals that separated from Sonstadt's solution on evaporation, that only KI and $KHgI_3$ are found at 30°C. The principal objection to Sonstadt's solution is its very poisonous and corrosive nature.

Nonprincipal refractive indices. In chemical work, particularly with monoclinic and triclinic crystals, it is often useful to measure the refractive indices corresponding to the two vibration directions presented by some well-defined and easily recognizable crystallographic orientation of the crystal. This is especially true of crystals having a lamellar or tabular habit, since such crystals usually present the same orientation, and measurements of the refractive indices presented are of value in identification work. For purposes of publication such indices should be accompanied by an accurate sketch showing the crystallographic and, if possible, the optical orientation.

Measuring the Refractive Index of an Immersion Liquid. When the refractive index of the immersion liquid has been adjusted by either the method of mixtures or the method of evaporation, it becomes necessary to determine its refractive index. The quantity of liquid used is usually so small that the index must be measured on a single drop. The choice of method employed is usually governed by the optical apparatus available.

Wright [95] gave an account of known methods and of several original methods. The first three of his methods depend on the use of easily constructed hollow prisms which are used on a spectrometer or a goniometer. The fourth method uses an Abbe-Pulfrich crystal refractometer. A drop of liquid is placed on the horizontal surface of the hemisphere and is covered with a piece of matt tinfoil which is made by pressing the smooth foil onto ground glass by means of a pencil eraser. Its function is to prevent the appearance of troublesome interference bands which are present when the thin film of liquid is bounded by a substantially plane liquid-air interface. "Dull" and "medium" grades of aluminum foil work equally well. We have tried Wright's fourth method on a Bausch & Lomb Abbe refractometer with excellent results. However, the dividing line is not easily seen when reflecting matt foil is used, but it is readily seen if black or dark foil is used, such as oxidized copper foil or sulfided silver or lead foil. For nonaqueous immersion liquids a piece of gelatin-coated black paper serves very well, provided that it has been pressed flat. Advantages of this technique are that the lower prism of the Abbe can be closed and the temperature control used and that the dispersion of the liquid is compensated for by the Amici prisms, the degree of rotation of which can be used in computing the dispersion ν. It should be remembered, however, that the prisms of the Abbe refractometer are constructed of heavy flint glass which is rather easily scratched, hence the foil, or coated black paper, should be kept free from dust.

Wright's fifth method is an ingenious application of the critical-angle method of determining refractive indices. The apparatus, shown diagrammatically in Fig. 3.32*a*, is in effect a miniature pair of Abbe prisms. The prisms have a refractive index of 1.92 and an angle of 60°. The lower prism has a ground face and the upper prism has a polished one. A drop of the immersion liquid is placed in the gap between the two prisms. A low-power objective is focused on the top surface of the cell, and the Bertrand lens of the microscope is introduced. The field is then seen to be divided into light and dark portions separated by a color fringe if white light is used. In monochromatic light, which today is conveniently obtained from sodium vapor and mercury vapor lamps, the dividing line is sharp. Its position depends on the refractive index of the immersion liquid and is recorded by means of a filar micrometer. The apparatus is calibrated with known liquids. Wright's sixth, seventh, eight, and ninth methods are alternative forms of critical-angle apparatus for use with the petrographic microscope and do not appear to have gained much popularity.

Clerici's method [96] employs a cell containing a prism mounted over an index line, as shown diagrammatically in Fig. 3.32*b*. When the cell is filled with immersion liquid, the compound glass liquid prism displaces the image of the index line by an amount that is a function of the refractive index of the

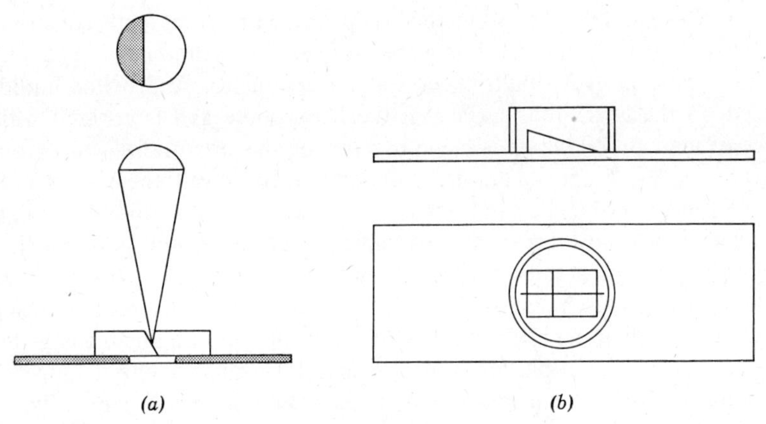

Fig. 3.32 (*a*) Wright microrefractometer; (*b*) Clerici stage refractometer.

liquid. The displacement is measured at moderate magnification with a filar micrometer eyepiece. The setup can be calibrated by means of known liquids. Viola [97] improved Clerici's method by using a double prism with its apex vertically over the index line. Nichols [98] uses two prisms cemented side by side and sloping in opposite directions. Both Viola's and Nichols' methods have the advantage that the displaced lines are at the same focus, whereas in Clerici's original design the greater the displacement the greater is the difference in focus. Wright [95] found that Clerici's method was accurate to only ± 0.005 and that Viola's modification was somewhat more accurate. Nichols claims an accuracy of ± 0.001.

Jelley [89] described a new type of microrefractometer, which is a self-contained, direct-reading instrument. Its optical principle is shown in Fig. 3.33. Light from a fixed slit *A* passes through a narrow slot in a scale *B*, which is graduated in refractive indices. The slit is viewed through aperture *C* in stage plate *D*. Over the aperture is placed a plate of glass *E* about 1 mm thick, which supports microprism *F*. The microprism, which is obtained by beveling one edge of an optically worked cover glass at 45°, is placed with the bevel facing the glass plate. The lower (obtuse) edge of the bevel is placed approximately midway over the aperture in the stage plate, and parallel with the slit. A drop of the immersion liquid is placed in the bevel so as to form a compound glass-liquid prism. On looking through the aperture, a deviated image of the slit is seen superimposed on an undeviated image of the scale. This image of the slit, which is obtained with white light, is spread out into a spectrum, the position of the yellow on the scale giving the approximate value of n_D. With sodium light, the principal image is bordered by a number of diffraction bands. A form of the apparatus is made and sold by E. Leitz of Wetzlar under the name of the Leitz-Jelley microrefractomer, which incorporates two improvements suggested by Jelley. The first of these is a didymium (neodymium plus praseodymium) glass filter over the slit for use

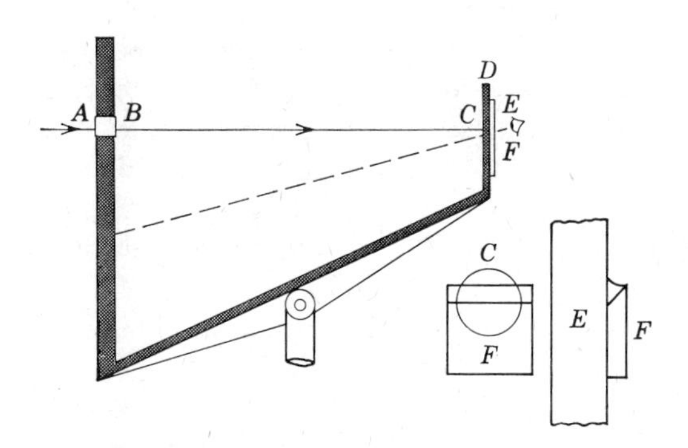

Fig. 3.33 Jelley microrefractometer.

with white light. When this is used, the spectrum image of the slit is seen to be crossed by two sharp bands, the one nearer the red giving the value of n_D to ± 0.001. The second improvement is a transparent index, which moves over the scale and is set to correspond with the didymium band. The refractive index may then be read and checked at leisure.

Two other modifications of the Jelley microrefractometer have appeared which, however, both employ the same optical principle. They are the modification of Edwards and Otto [99], which is calibrated by known liquids instead of the more accurate but less convenient method of optical computation, and that made and sold in the United States under the name of Fisher refractometer.

In its original form, the microrefractometer was used to measure the refractive index of about 0.1 ml, or 0.0001 ml. The Leitz-Jelley instrument has a more sturdy prism and needs about 0.0005 ml of liquid. In the original publication the author suggested two special uses of the microrefractometer in chemical microscopy: checking the purity of a microquantity of liquid by measuring its refractive index as it evaporates and characterizing organic compounds by determining n_D at their melting points. A modification of the Fisher refractometer with a heating stage for carrying out this measurement has been described by Frediani [100].

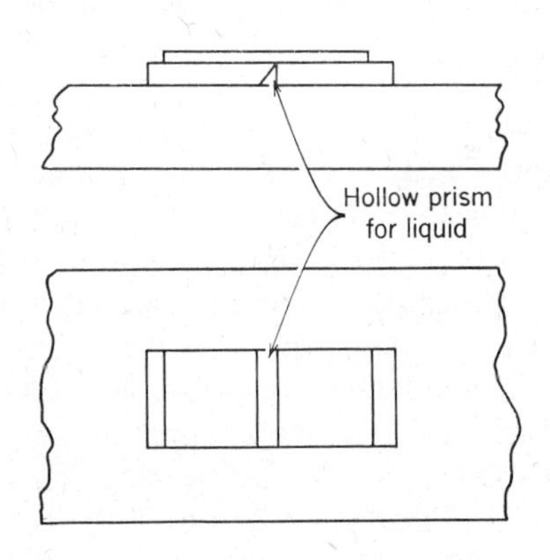

Fig. 3.34 Microprism for volatile liquids.

A modification of the microprism, shown in Fig. 3.34, permits the use of the microrefractometer with volatile liquids. This enclosed prism, which has a side 0.1 to 0.4 mm, holds the liquid in the vertical position and so can be used with a vertical slit and horizontal scale. A direct-reading, self-contained microrefractometer eases the work of determining refractive indices by the immersion method. There does not appear to be any good reason why the Wright prism (Fig. 3.32*a*) could not be built into a simplified Abbe-type refractometer without Amici prisms. This could be used with monochromatic light.

DETERMINATION OF BIREFRINGENCE AND ITS DISPERSION

Compensators. It is necessary to distinguish between the principal birefringences that are the differences between principal indices, for example, $\omega - \epsilon$ of uniaxial crystals, $\gamma - \alpha$, $\gamma - \beta$, and $\beta - \alpha$ of biaxial ones, and the nonprincipal birefringence presented by some known crystallographic orientation of a crystal. In organic chemical microscopy it is often possible to grow relatively large crystal plates by crystallization from a melt and quite frequently these crystal plates have a constant optical orientation. Consequently, measurements of birefringence on such plates can be of value in comparing an unknown substance with a known one. Principal birefringences are of value in estimating γ and sometimes β of aromatic substances that usually are beyond the range of measurement by the immersion technique.

It has been noted (p. 179) that polarized light entering a crystal not in the extinction position is resolved into two beams vibrating at right angles which travel with different velocities, so that one ray is retarded relative to the other. The distance by which the one ray is retarded on its emergence from the crystal is known as the retardation of the crystal for that particular direction of light transmission. Retardation is given by the expression:

$$R = t(n_2 - n_1), \tag{3.21}$$

where t is the thickness of the crystal, n_2 is the higher index, and n_1 the lower. R and t are of course expressed in the same units, such as millimeters, microns, or millimicrons. If n_2 and n_1 are principal indices, it follows that R/t is a principal birefringence. Measurements of retardation can usually be made without much difficulty on small crystals, but the measurement of thickness is another matter. At first sight it might appear to be possible to orient the crystal so that only rays corresponding to n (or the lower index of a uniaxial substance) are transmitted and then to note the distance traversed on the fine-adjustment drum on racking up the microscope from a focus on the bottom of the crystal to one on the top. However, in practice the method is found to be quite inaccurate for a variety of optical and mechanical reasons. Sometimes crystals prepared by the fusion method contain air bubbles, or holes, which extend from the object slide to the cover glass; in this case a somewhat less inaccurate measurement of the crystal thickness may be made.

The measurement of retardation is usually made with the aid of a compensator. A compensator is a means of introducing a known retardation in the faster beam emerging from a crystal. The amount of known retardation is adjusted so that it neutralizes or compensates the retardation of the crystal. The earliest type of compensator was the quartz wedge of Biot [78], which suffered from the disadvantage that the thin end still had an appreciable retardation. Wright [101] combined a quartz wedge with a parallel plate having a retardation corresponding to that of the middle of the wedge but with its fast vibration direction at right angles to that of the wedge. The Wright combination wedge has zero retardation in the center, with increasing retardation on either side—one side having its fast vibration parallel, and the other at right angles, to the length of the wedge. This wedge, which is usually engraved to give direct readings of retardation, is particularly useful when used with the Wright universal ocular [102] in which it slides in the focal plane of the ocular. The Johannsen [103] wedge is somewhat similar but has zero birefringence at one end. The Berek compensator [104], which is an improvement on the earlier one of Nikitin [105], consists essentially of a thin plate of calcite cut perpendicularly to the optical axis, which is rotated about a diameter by means of a graduated drum. This compensator is inserted in a slot over the microscope objective and so does not entail the use of an analyzer over the eyepiece.

Compensators do not work well with substances having a strong dispersion of birefringence, as the compensation point is often masked by color effects, particularly at higher retardations. The problem cannot be solved by using monochromatic light because compensation occurs at wavelength intervals in monochromatic light, and only the difference of the retardation from the nearest integral multiple of wavelengths can be determined. However, the strong dispersion of birefringence of many organic crystals makes measurements of this dispersion highly desirable as a means of characterization of organic compounds.

Microspectrographs. In order to overcome the difficulty of determining the compensation point at a number of wavelengths, Jelley [67, 75, 106] worked out three methods of determining the dispersion of birefringence of organic compounds that melt without decomposition, and one that is applicable to crystals that are not recrystallizable by fusion. These methods were designed primarily for use with the grating microspectrograph, three different designs of which have been described in the publications referred to above. In view of the increasing interest in optical methods of identifying organic compounds, it seems desirable to summarize the methods here:

Method 1. A square or rectangular piece of number 2 or number 3 cover glass is placed on an object slide, which is preferably but not necessarily made of optically worked fused quartz. Another square of cover is placed so that one edge rests on the first cover and the other on the slide. Some crystals of the substance to be studied are placed near the sloping cover slip, and the slide is then heated on a hot plate until the substance melts and is drawn under the sloping cover glass to form a wedge of liquid. The preparation is then cooled and, if necessary, seeded by touching the liquid with some solid substance on a needle. Usually, the preparation needs careful "melting back," until crystals remain only under one corner of the thin end of the wedge. The slide is then allowed to cool very slowly in order to obtain crystals of uniform orientation. Such a wedge preparation shows ascending orders of retardation when examined in polarized light (crossed polarizers) with a low-power objective. The image of the wedge is focused on the slit of the microspectrograph so that the slit cuts through the color bands at right angles. Spectrograms are then made on Panatomic-X 35-mm film. From these spectrograms the dispersion is readily obtained by one of two methods. The first is applicable when many interference bands are visible. A line is ruled through the spectrogram in a position corresponding to the thick end of the wedge crystal, as in Fig. 3.35. Irregularities in the crystal cause many lines to run through the length of the spectrum: one of these should be chosen as a guide for the ruled line, thereby avoiding errors resulting from aberrations of the optical system. A wavelength scale is then attached to the spectrogram, and the wavelengths λ_n, λ_{n+1}, λ_{n+2}, . . . at which the nth, $(n + 1)$th, $(n + 2)$th, . . . interference bands cross the ruled line are recorded. The reason for using a wedge is of course that the value of n is immediately given if the extreme tip of the wedge appears in the spectrogram. The value of the retardation for any wavelength λ_n at which an interference band of the nth order intersects the ruled line is $n\lambda_n$, and the corresponding birefringence is $n\lambda_n/t$, where t is the thickness of the wedge at the corresponding point. The values of $n\lambda_n$ are plotted against λ, if necessary, to obtain the value of the retardation for sodium light ($\lambda = 589.3$). The curve is then redrawn with the values of the retardation at various wavelengths being expressed relative to that of $\lambda = 589.3$, which is taken as unity. In this way, the need for measuring the thickness of the wedge is avoided. The manner of using a monochromator for this method is explained later in this chapter. It should be noted that the wedge crystal can be used for determining the value of the birefringence in

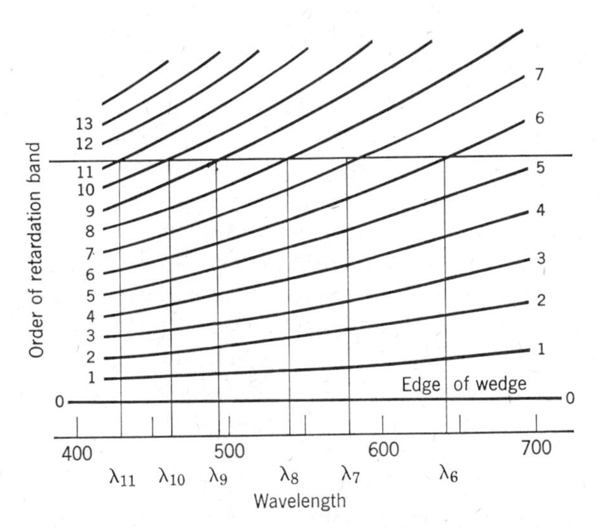

Fig. 3.35 Computation of dispersion of birefringence from spectrogram.

monochromatic light if the slope of the wedge is known: thus if the slope of the wedge is θ and the distance between two bands (measured by means of a micrometer eyepiece) is d for a wavelength λ, the birefringence for that wavelength is $\lambda/(d \tan \theta)$. Since the angle of the wedge is small, usually about 2°, it is necessary to measure it to within $\pm 1'$ if reasonable accuracy is desired.

Method 2. A small planoconvex lens, preferably of fused silica, and an optically worked object slide of fused silica are required. The microspectrograph is calibrated for any given microscope setup (usually the lowest power the microscope will give) by photographing the Newton's rings given in sodium light with the slit of the microspectrograph wide open. If transmitted light is used, the upper surface of the object slide and the convex surface of the lens must either be half-silvered or else coated with a thin dye film which strongly reflects sodium light. By far the best dye of those studied is 1,1'-diethyl-2,2'-cyanine chloride, which is used as a 1% solution in a mixture of 9 parts of methanol and 1 part of water. When a drop of this solution is applied to each surface, it spreads out and evaporates, leaving a film of dye having a very sharp absorption band at 578 mμ and, in consequence, a very high refractive index for sodium light. The Newton's rings produced by dye-treated surfaces are just as good as those obtained with half-silvering. In order to prepare lens crystals, the lens is placed convex side down on the object slide and touching a little of the substance under study. The slide is then heated in order to melt the substance and cause it to flow under the lens. The slide is then cooled and seeded, if necessary, to crystallize the melt. It is usually necessary to melt back the crystals until only a trace remains solid and then cool very slowly in order to obtain well-formed crystals. When such a preparation is examined with crossed polarizers in monochromatic light, a series of dark rings is seen, corresponding to retardations of λ, 2λ, 3λ, and so on, whose radii are proportional to 1, $\sqrt{2}$, $\sqrt{3}$, and so on. When more than one optical orientation is presented, different series of rings are apparent. Thus piperonal may give rings corresponding to $\beta - \alpha$ and rings of about four and a one-half times the diameter corresponding to $\gamma - \beta$. However, it may happen that the orientation does not correspond to an axis of the wave surface so that rings do not correspond to a principal birefringence. In this case the measurements of dispersion are only of value if the substance always presents the same orientation in a lens crystal; fortunately, this is often the case. The image of the lens crystal is thrown on the slit of the microspectrograph, care being taken to assure that the slit passes through the center of the figure. From a comparison of the Newton's rings and the interference spectrogram, the birefringence of the substance can be determined for all wavelengths from 400 to 660 mμ with panchromatic film, and in the infrared with suitably sensitized film or plates.

Method 3. This method employs "droplet crystals" and has the advantage of simplicity, but the disadvantage that it gives the dispersion of birefringence only. Fragments of the substance under study weighing between 0.1 and 10 mg are fused on a microscope slide. On cooling, the droplets crystallize to yield wedge shaped-crystals, which are examined as in method 1.

Method for Substances Not Recrystallizable by Fusion. The method of computing the retardation-λ graph and of deriving the dispersion of birefringence described under method 1 is readily applied to crystals that possess either a crystallographically or an artificially produced wedge-shaped edge. The crystals should be mounted in a medium having an index not too far removed from the means of the two indices presented. Some of the viscous Aroclors, mixtures of Aroclors with Canada balsam, and the other viscous media mentioned earlier in this section are very useful. If possible, the crystal is rolled under the cover glass until its interference figure shows that a principal birefringence is presented. Since the purpose of the wedged part of the crystal serves merely to indicate the order of the retardation bands, the wedge may be quite crude; even a rounded end of a crystal obtained by dissolving one end of the crystal, can serve.

Monochromators. A small monochromator, such as the Bausch & Lomb laboratory wavelength spectrometer, may be used for determining the dispersion of birefringence of crystals having a wedge-shaped edge. The microscope is illuminated with light from the exit slit of the monochromator. An appropriate crystal, whose optical orientation has previously been determined conoscopically with white light, is brought to the center of the field and is placed in the 45° position. The light from the monochromator is thrown on the crystal so that the wedge-shaped edge is illuminated, as well as a flat, thick part of the crystal that is imaged on the crosshairs. The wavelength drum is turned to the extreme red and is then brought slowly back until the part of the crystal on the crosshairs is at extinction. The order of the retardation band n and wavelength λ_n are noted. The wavelength of the light is shortened until the crystal is again dark and the wavelength λ_{n+1} for the $(n + 1)$th retardation band is noted. This operation is repeated for the entire spectrum. The computation of dispersion and its graphical expression are carried out as described above. The lens method can be used for determining the value of the birefringence for sodium 589 mμ or mercury 546 mμ in order to convert the values of dispersion of birefringence into birefringences at various wavelengths.

Approximate values of dispersion of birefringence may be obtained with the aid of a microspectroscope having a wavelength scale. A crystal with a wedge-shaped edge is used, as in the method described above, to determine the order of the retardation band; the wavelengths $\lambda_n, \lambda_{n+1}, \lambda_{n+2}, \ldots$ at which the nth, $(n + 1)$th, $(n + 2)$th, . . . bands across the spectrum are noted, and the necessary computations are made.

Rarely, crystals are examined in which the birefringence is greater in the red than in the blue. In these cases care must be exercised so that the right order is ascribed to the retardation bands. Benzil is a substance that falls into the category. Other examples are to be found in crystals showing crossed axial plane dispersion; the birefringence for the acute bisectrix of such crystals passes through zero at the wavelength of uniaxiality. Since the retardation bands on either side of this wavelength have the order 1, 2, 3, . . . , it is not necessary to have a wedge-shaped edge on the crystal in order to determine its dispersion of birefringence.

DETERMINATION OF OPTIC AXIAL ANGLES AND THEIR DISPERSION

Visual Determination. The precision with which optic axial angles can be measured varies within wide limits depending on the direction of the axes with respect to the morphology of the crystal. The most favorable condition is that in which both axes appear in the convergent polarized light interference figure, and in this case the measurement can be made with relatively simple apparatus. When neither, or only one, axis appears in the interference figure, a reasonably accurate determination is very time-consuming and requires a universal stage. Universal stage methods cannot be adequately described in this chapter; for a description of them, the reader is referred to books by Wright, Johannsen, Winchell, and Emmons and Emmons [70, 77, 78, 107, 108].

The optical axial angle is designated as $2V$. This is the acute angle between the primary optic axes within the crystal. The angle of the axes, as measured in air from a crystal plate cut normal to the acute bisectrix, is $2E$. When an oil-immersion condenser and objective are used, the angle is $2H$. The relations between these angles are:

$$\sin E = \beta \sin V \tag{3.22}$$

and, for an immersion oil with $n_D = 1.515$,

$$\sin H = 0.66 \sin E. \tag{3.23}$$

One of the most used methods is that of Mallard [109], which is based on the observation that the distance between the optic axes as measured on the interference figure with any particular microscope system is proportional to $\sin E$, the sine of one-half the optic axial angle in air. This relationship is usually expressed:

$$D = K \sin E, \tag{3.24}$$

where D is the measured distance and K is known as Mallard's constant, which is determined from a known crystal or by an apertometer. The accuracy of Mallard's method is usually greater with thicker crystals. Wright has shown that Mallard's method may be applied to uncentered acute bisectrix figures by the use of a double-screw micrometer eyepiece [110], or an eyepiece fitted with a coordinate micrometer scale [111].

Centered acute bisectrix figures having $2E$ so large that the axes fall outside the interference figure may be measured by Wright's modification [112] of Michel Lévy's method. In this method a circle of reference is introduced into the interference figure. This is effected by introducing a glass plate having a circle (or concentric circles) into the focal plane of the eyepiece, or at a conjugate focus, such as the back lens of the objective, or the upper iris diaphragm of the petrographic swing-out condenser. A crystal of known $2E$ is placed in the crossed (extinction) position, and the stage is rotated until the isogyres are tangential to the reference circle. The angle of rotation ϕ is noted. The constant C for the particular microscope system is given by:

$$C = \sqrt{\sin 2\phi \cdot \sin E}. \tag{3.25}$$

With C determined, the value of $2E$ for another crystal plate can be determined by noting the new value of ϕ and substituting in the equation:

$$\sin E = C/\sqrt{\sin 2\phi}. \tag{3.26}$$

Johannsen and Phemister [113] have described a more accurate method in which a plate of substance of known $2E$, such as a plate of muscovite mica, is fixed in the diagonal position upon the uppermost lens of the condensing system. An object slide with the crystal of unknown $2E$ is placed on the microscope stage so that its optic axial plane is parallel to that of the mica. The ease and accuracy of this operation are increased if the microscope is provided with synchronous rotation of the polarizers, since in this case the mica is brought to extinction by rotating the polarizers through 45°. The crystal of unknown $2E$, which has been roughly oriented with the aid of the Bertrand lens, is then rotated to exact extinction and the polarizers are brought back to the former direction. The Bertrand lens is again inserted, and the stage is rotated; the confused interference figure as first seen changes until, at the end point, the dark brushes come into view because the isogyres of the unknown substance have been superimposed on those of the mica. If ϕ is designated the angle through which the stage has been rotated, then:

$$\sin E \text{ of unknown} = \sin E \text{ of known}/\sqrt{\sin 2\phi}. \tag{3.27}$$

When a single optic axis appears in the interference figure, a rough estimate of $2V$ can be made from the curvature of the isogyre. Wright [77] has computed the trace of the isogyre for centered and uncentered optic axis figures for $\beta = 1.6$ and $2V = 15$, 30, 45, 60, 75, and 90°.

Accurate measurements of optic axial angles have been made by rotating crystals in a trough filled with a liquid having a refractive index of β. The axis of rotation must be normal to the optic axial plane, that is, about the Y-axis. Except in favorable cases, the orienting of the crystal is a difficult operation. A simple stage goniometer combined with a thin cell has been described by Wood and Ayliffe [114]. The apparatus has the advantage that it can be used with the high-power objective and condenser necessary for observing the convergent polarized light interference figure, so that the mounting of the crystal is not as critical. It should be possible, by using the Wood-Ayliffe goniometric apparatus on a microscope equipped with a calibrated micrometer eyepiece for measuring the direction of emergence of the optic axes in a plane normal to that of rotation, to compute $2V$ with fair accuracy on crystals rotated about an axis making a considerable angle with Y. Bryant [115] has used a modification of this apparatus for measuring optic axial angles.

Photographic Determination. Since, as already mentioned, the dispersion of the optic axes of crystals is usually much more marked with organic than with inorganic compounds, the measurement of this property is important. Bryant, who has done so much work in this field, has lately adopted a photographic method of recording the dispersion [116]. The principle is quite simple. The interference figure, which should of course show both optic axes, is photographed with various wavelengths of monochromatic light using a micrometer ocular in order to obtain a reference scale. The photographs are measured in order to determine: (1) the nature of dispersion of the axes of the wave surface; and (2) the variations of $2E$ or $2H$ as a function of λ.

In his publications dealing with the grating microscopectrograph, Jelley has drawn attention to the value of having a wide spectrum, in which lines corresponding to the various wavelengths are straight, for obtaining a photographic record of the dispersion of optic axes. Unfortunately, the accuracy of the method is offset by the disadvantage that it cannot be used for measuring monoclinic inclined or horizontal dispersion, or for triclinic dispersion. As mentioned earlier, crossed axial plane dispersion occurs in orthorhombic, monoclinic, and triclinic crystals when the dispersions of the principal refractive indices are so different that at some particular wavelength β changes places with α or γ. This wavelength, often referred to as the "wavelength of uniaxiality," may be determined by using a monochromator as a source of illumination for the convergent polarized light system, or by the use of the microspectrograph. The temperature coefficients of the principal refractive indices are usually different, which means that the λ of uniaxiality may vary with change of temperature.

Microscopy of Colored Crystals

THEORETICAL

Pleochroism and Dichroism. Colored crystals of other than cubic symmetry are usually, although not invariably, pleochroic, that is, the degree of absorption of light of any wavelength by unit thickness of the crystal is a single-valued function of the vibration direction with reference to the crystal orientation. The amplitude A of a ray that has traversed a thickness d of an absorbing medium is given by:

$$A + A_0 e^{-2\pi\kappa d/\lambda}, \tag{3.28}$$

where λ is the wavelength in the medium, A_0 is the incident amplitude, and κ is the absorption index. Since intensity I is proportional to the square of the amplitude, the absorption formula can be written:

$$I = I_0 e^{-4\pi\kappa n d/\lambda_\theta}, \tag{3.29}$$

where n is the refractive index of the medium and λ_θ is the wavelength in air of the incident light.

In the most general case, monochromatic light incident on a plate of an absorbing crystal is resolved into two elliptically polarized rays. The ellipses have the same ratio of major to minor axes and the same direction of rotation but have their major axes at right angles. However, light vibrating along a vibration plane of the crystal plate is transmitted as plane-polarized light and can therefore be extinguished by the analyzer in the crossed position. The crystal section has refractive indices of n_1 and n_2 and absorption indices of κ_1 and κ_2, respectively, for its two extinction directions. Differently oriented sections of the same crystal present different values of n_1, n_2, κ_1, and κ_2. These values vary with change of wavelength, the absorption indices in particular being liable to wide variations. Just as we can construct a wave normal–refractive index surface as an aid to visualizing the nature of double refraction in a crystal, so we can construct a wave normal–absorption index surface. For a perfectly transparent crystal the absorption surface is a geometrical point, while for an absorbing optically isotropic crystal it is a sphere of radius κ. With uniaxial crystals the absorption surface for the ordinary ray is a sphere of radius κ_ω, and for the extraordinary ray it is an ovaloid varying between κ_ω along the optic axis and κ_ϵ along the optic normal. Usually, crystals of positive birefringence have $\kappa_\epsilon > \kappa_\omega$, while those of negative birefringence have $\kappa_\omega > \kappa_\epsilon$. This is equivalent to stating that the higher absorption is usually associated with the higher refractive index. With biaxial orthorhombic substances the absorption surface resembles that of the wave normal–refractive index surface in that there are two directions for which the two absorption indices are the same. These directions do not coincide except by chance with the optic axis and have no particular significance. What is important is that the principal axes of the absorption surface of an orthorhombic crystal coincide with X, Y, and Z of the refractive index surface, which in turn coincides with the crystallographic symmetry. Only one of the principal absorption axes coincides with a principal vibration direction in a monoclinic crystal, and this is the one which in turn coincides with the crystallographic b-direction. The complete optical asymmetry of crystals of the triclinic system is also shown in the absorption surface, since the principal absorption axes do not except by chance coincide with the principal vibration axes.

It is convenient at this point to discuss briefly the meaning of the terms dichroism, trichroism, and pleochroism. In petrography and chemical microscopy, dichroism has two meanings. Any crystal or mineral section that changes in color when rotated on the microscope stage in polarized light (analyzer withdrawn) is said to be dichroic; the term is applied in a more specific sense to absorbing uniaxial crystals that exhibit two extreme colors with optic normal sections viewed in polarized light vibrating parallel to, and normal to, the plane of the optic axis, respectively. In the latter sense, an absorbing orthorhombic crystal may be trichroic when it exhibits three different absorptions corresponding to the X-, Y-, and Z-vibration directions. Absorbing monoclinic and triclinic crystals may be trichroic only if there is no appreciable dispersion of the principal axes of vibration and absorption. Usually, at least as far as crystals of dyes are concerned, dispersion of these axes is present. This dispersion causes the crystals to exhibit a great many colors on changing their orientation in plane-polarized light. Such crystals are said to be pleochroic. The term pleochroism is also used to describe the absorption phenomena of colored uniaxial and biaxial crystals in general.

Optical Phenomena of Pleochroic Crystals. Ordinary light. Most of the optical studies of strongly pleochroic crystals have hitherto been made on naturally occurring substances, especially the uniaxial minerals chlorite and tourmaline; the biaxial minerals cordierite, epidote, andalusite, and glaucophane; and a very curious group of highly colored crystals of the platinocyanides first studied by Haidinger [117, 118].

Plates of equal thickness cut at different orientations from a colored uniaxial crystal appear different when viewed by transmitted ordinary light. The maximum difference is shown by sections normal to and parallel to the optic axis. If, for example, a uniaxial crystal transmitted only red for the ordinary ray and only green for the extraordinary ray, the optic axis section would appear red. The optic normal section would appear yellow, since half of the transmitted light would be red and the other half green. A mixture of red and green light appears yellow.

Similarly, plates of equal thickness cut normal to the principal vibration axes of a trichroic orthorhombic crystal appear different when viewed by transmitted ordinary light. Thus half the light incident on a plate cut normal to X suffers an absorption corresponding to the Z-axis, whereas plates cut normal to Y have a mixture of absorptions corresponding to X and Z, and those normal to Z have a mixture of absorptions corresponding to X and Y. An example of pleochroism visible in ordinary light is afforded by a fusion slide of azobenzene.

Plane-polarized light. The light transmitted by a section of a pleochroic crystal is readily resolved into its two components, except in the case of monoclinic and triclinic crystals showing dispersed extinction, by successively rotating the section on the stage of the polarizing microscope to extinction positions in adjacent quadrants and removing either the polarizer or analyzer in order to observe the color of the transmitted light for each of the two vibration planes of the section. However, this technique, which is the one commonly adopted by the chemical microscopist, is not very good for showing slight degrees of dichroism. Haidinger [119, 78] introduced the dichroscope, which is substantially a prism of calcite interposed between a square aperture and focusing lens. The calcite splits the light from the aperture into two beams polarized at right angles, whereby two images of the square are seen in approximate contact. The plane of vibration of the ordinary image is parallel to the junction between the two squares, that of the extraordinary images is perpendicular to this junction. Many modifications of the dichroscope have been reported in the literature, which covers improvements such as rotating specimen holders with two mutually perpendicular axes of rotation for studying pleochroism. A similar device has been used as an eyepiece dichroscope, in which the square aperture is in the image plane of the microscope [120]. The usual construction in present-day eyepiece dichroscopes is to equip a Huygens eyepiece with a diaphragm having a square aperture, which is placed near the field lens, and a Wollaston prism which is placed just beneath an eye lens with focusing adjustment. The Wollaston prism splits a beam of light into two beams, polarized at right angles, which deviate equally from the microscope axis, so that when the eye lens is properly adjusted for focus, two squares are seen in approximate contact. The innermost side of each square is strongly fringed with yellow, and the outermost with blue, but this is not quite as objectionable as the appearance in the Haidinger type of dichroscope eyepiece, in which the extraordinary image is strongly fringed with yellow on the side nearest the ordinary image and fringed with blue on the other side, since in addition the focus is different from the two images. It is necessary to check the vibration planes of the two images of the Wollaston prism–type of dichroscope eyepiece; they may be parallel and perpendicular to the dividing line of the field, or at +45 and −45°. This check is readily made on the microscope with the polarizer in position.

The dichroscope eyepiece is used without polarizer or analyzer. Daylight is an unsatisfactory illuminant for a microscope employing a dichroscope eyepiece because it may have an appreciable plane-polarized component.

Convergent ordinary light. If a section of a pleochroic uniaxial crystal cut perpendicular to the optic axis is viewed conoscopically in convergent ordinary light, the direction of the optic axis is indicated by a small spot different in color from the rest of the field. The apparent diameter of the spot diminishes as the thickness of the section is increased, and its boundary becomes more sharply defined. Seen in monochromatic light, it is lighter than the surrounding field if the ordinary ray is less strongly absorbed than the extraordinary one, and conversely. It is possible for the optic axis to appear as a dark spot at one wavelength and as a light spot at another.

Optic axial sections of biaxial crystals exhibit a dark brush, perpendicular to the optic axial plane, which is intercepted by a bright spot on the axis.

Convergent plane-polarized light. The most important conoscopic observations of pleochroic crystals are made in convergent plane-polarized light. Absorption figures are obtained by the use of the polarizer (or analyzer) alone. Interference figures are obtained by the use of a crossed polarizers as they are with transparent crystals.

The absorption figure of a uniaxial crystal, as seen with light comprising a fairly narrow band of the spectrum, may present either of two appearances. If the absorption of the ordinary ray is less than that of the extraordinary ray, a brush is seen which is parallel to the plane of vibration of the polarized light; the brush is intercepted by a bright spot on the optic axis. With most polarizing microscopes the vibration plane of the polarizer is in the N-S position (vibrations to and from the observer), so that it is in this position that the brush appears. If, however, the absorption of the ordinary ray exceeds that of the extraordinary ray, the brush is perpendicular to the plane of vibration and is continuous through the optic axis. With a uniaxial crystal having different absorption spectra for the ordinary and extraordinary rays, the parallel (N-S) brush with the bright axial spot appears at some wavelengths, and the perpendicular (E-W) continuous brush at others. When examined with white light, the conoscopic absorption figure of such a crystal consists of a pair of differently colored brushes at right angles, that corresponding to the ordinary ray being perpendicular to the plane of polarization and continuous. Magnesium platinocyanide is an example of this type; its conoscopic absorption image consists of violet continuous brush perpendicular to the vibration plane of the polarizer and a red brush at right angles which fades out on the optic axis.

The interference figure of an optic axial section of a pleochroic uniaxial crystal consists of dark cross and some rings if sufficient light is transmitted for them to be visible. This is the case when the absorption of the ordinary ray is slight, and the number of rings that can be seen depends on the degree of absorption of the extraordinary ray. When the absorption is appreciable, the rings are indistinct because they are produced by the interference of two elliptically polarized rays of different intensities which cannot give extinction. When the ordinary ray is very strongly absorbed, the conoscopic interference figure is uniformly dark.

The absorption figures of pleochroic biaxial crystals are observed conoscopically in plane-polarized light. The following considerations apply to orthorhombic crystals and to monoclinic and triclinic crystals showing only slight dispersion of the principal vibration directions.

Optic axial absorption figures. In absorbing biaxial crystals, the absorption suffered by light traveling along an optic axis depends on its plane of vibration. The vibration planes of an optic axial section are in the plane of the optic axes (the *X-Z* plane) and perpendicular to it (the *Y*-axis), respectively. Let us call the corresponding absorption indices κ_p and κ_y. When the section is viewed conoscopically in monochromatic light vibrating in the plane of the optic axes, a continuous brush is seen in this plane if $\kappa_p > \kappa_y$, whereas the brush is perpendicular to this plane and is intercepted by a bright spot on the optic axes if $\kappa_p < \kappa_y$. When the section is viewed in light vibrating perpendicularly to the optic axial plane, a continuous brush is seen in the optic axial plane if $\kappa_p < \kappa_y$, while the perpendicular brush intercepted by the bright axial spot is seen if $\kappa_p < \kappa_y$. The figures seen with white light are colored; the colors are different from the two extinction positions.

Acute bisectrix absorption figures. When an acute bisectrix section is viewed conoscopically in plane-polarized white light vibrating perpendicularly to the optic axial plane, a band of uniform color, corresponding to the *Y*-absorption, goes straight across the figure along the optic axial plane. On either side of this band, the color changes since these parts of the conoscopic figure comprise vibrations approaching the Z-direction with negative crystals and the *X*-direction with positive ones, so that the color of these outer parts of the figure changes symmetrically from the center to the sides of the field, with abrupt changes in the region of the optic axes. When the same section is viewed in white light vibrating in the optic axial plane, a band of uniform color crosses the conoscopic figure through the acute bisectrix, perpendicular to the optic axial plane. In the case of optically negative crystals, this color corresponds to the *X*-absorption; with optically positive ones it corresponds to the Z-absorption. On either side of this band, the color changes toward that for the *Z*-direction with negative crystals, and that for the *X*-direction with positive ones. The changes in the region of the optic axes are abrupt. The colored band corresponding to the *Y*-direction, which is seen in white light vibrating perpendicular to the optic axial plane, is very much narrower than that corresponding to the *X*- or *Z*-band seen when either the crystal or the plane of polarization is rotated 90°.

Obtuse bisectric absorption figures. When an obtuse bisectrix section is viewed conoscopically with white light vibrating perpendicular to the optic axial plane, the narrow colored band, corresponding to the *Y*-absorption, crosses the figure along the optic axial plane. The rest of the figure shades toward the color corresponding to the *X*-absorption for optically negative crystals, and to the *Z*-absorption for optically positive ones. When either the crystal or the plane of polarization is rotated 90°, a broad band of color passes through the center of the figure, perpendicular to the optic axial plane. Its color corresponds to the *Z*-absorption for optically negative crystals and to the *X*-absorption for optically positive ones.

Uncentered figures presenting one axis show the colored band along the optic plane corresponding to the *Y*-absorption.

Optic normal sections. When an optic normal section is viewed conoscopically in plane-polarized light, the absorption figure is uniformly colored. The two extreme colors, corresponding to the *X*- and *Z*-absorptions, are obtained when the crystal is rotated to extinction positions perpendicular to each other.

The interference figures of pleochroic biaxial crystals in monochromatic light differ in certain respects from those of transparent crystals. For wavelengths at which the absorption is negligible, the interference figures are perfectly normal, whereas with moderate absorption the isogyres are intercepted by bright spots on the optic axes unless vibration directions of the crystal coincide with the vibration when the crystal is in the diagonal position, when the lemniscates nearest the optic axes also have moderately bright spots at their points of intersection with the optic axial (*X-Z*) plane. The bright spot effect is not usually visible in thick crystal plates having a very strong absorption in one region; such plates give a fairly normal interference figure by their transmitted light. It follows that the conoscopic interference figures of absorbing crystals are profoundly influenced by their thickness.

Dispersion of the optic axial angle causes dispersion of the optic axial brushes but does not affect the appearance of bands of uniform color seen in the conoscopic absorption figures along the *Y-Z*, *X-Z*, and *X-Y* plane for *X*-, *Y*-, and *Z*-vibrations. However, dispersion of the principal vibration directions, such as may occur in monoclinic and triclinic crystals, causes the absorption spectrum of the crystal to become a function of the vibration direction. Instead of there being three principal absorption spectra, the absorption maxima may vary continuously with change of vibration direction.

This type of dispersed pleochroism is strikingly shown by microscopical preparations of pinakryptol green, the photographic desensitizing dye. A few milligrams of pinakryptol green are dissolved in a drop of pyridine on an object slide. The slide is warmed to hasten the evaporation of the pyridine and to prevent its dilution by atmospheric moisture. The droplet is spread out with a glass rod so that the crystals that form are extremely thin. Many of the crystals have irregular contours (anhedral), but they all show the dispersed pleochroism. Viewed orthoscopically by transmitted ordinary white light, they appear a dull olive-green. By substituting plane-polarized light, the crystals show many colors on rotating the microscope stage, passing consecutively through dull red, orange, yellow, green, blue, and purple. As would be expected, these crystals show strong dispersion of extinction. Although crystals showing dispersed pleochroism must of necessity show dispersed extinction, the converse is not necessarily true.

Measurement of Absorption. The absorption index κ was defined on p. 219. It is the ratio of the amplitude of light at any instant to its amplitude 1 radian of a vibration earlier. Other constants expressing the absorption of light by a medium can be derived from the absorption index. Two that are common in crystal optics are the absorption coefficient k, which is numerically equal to $n\kappa$ when n is the refractive index, and the extinction modulus m, which is numerically equal to:

$$(4\pi n\kappa \log_{10} e)/\lambda_0, \tag{3.30}$$

where λ_0 is the wavelength in air.

In computing κ, k, or m from measurements of transmission, the loss of light by reflection must be taken into account. For an absorbing medium bonded by air, the transmission is given by the expression:

$$T = I/I_0 = [16n^2/(n + 1)^4]10^{-md}, \tag{3.31}$$

where n is the refractive index, m is the extinction modulus, and d is the thickness, from which we obtain

$$m = \{2 \log_{10} [4n/(n + 1)^2] - \log_{10} T\}/d. \tag{3.32}$$

When an absorbing medium is mounted between two pieces of glass with a mounting medium having refractive index n', the transmission T', is measured relative to that of two pieces of glass separated by the same mounting medium. If $n'' = n/n'$, we have:

$$m = \{2 \log_{10} [4n''/(n'' + 1)^2] - \log_{10} T'\}/d. \tag{3.33}$$

The extinction modulus, absorption index, or absorption coefficient of plates of weakly absorbing crystals, such as cobalt copper sulfate, cobalt potassium sulfate, and other substances of similar absorption, have been determined by cutting suitably oriented plates of the crystal and measuring the absorptions for the two vibration directions of each plate on a spectrophotometer. (For references, see Pockels [118].) Berek [121] has described a slit microphotometer which may be used with transmitted polarized light. This instrument gives satisfactory results on crystals about 1 mm across when a mercury vapor lamp is used in conjunction with the appropriate Wratten filters as a source of monochromatic light. What is needed, however, is a combined microphotometer and monochromator designed to function with the polarizing microscope.

Measurements on crystals with very strong selective absorption, such as organic dyes, are hardly ever possible because crystals of the required thickness (often not more than a few microns) cannot be grown of a sufficient area for use with low magnification. From what has been said above regarding conoscopic absorption figures, it is realized that the cone of illumination must not exceed a few degrees when the principal absorptions of an acute bisectrix section are being measured. It is essential that only light that has passed through the crystal be permitted to enter the microscope objective; this means that an image of a distant exit pupil of a monochromator must be thrown on the crystal. The ordinary plane mirror of the microscope gives a double image and should therefore be replaced by a surface aluminized one. The swing-out type of substage condenser cannot be used if the polarizing prism is equipped with cylindrical correcting lens, but an adequate polarizer can be constructed from any type of polarizing prism, over which is mounted a short-focus achromatic lens. The Bausch & Lomb 21 × achromat with an iris diaphragm is a good lens for this work, especially if the crystals are mounted between two microscope cover glasses so that the correction of the lens is disturbed as little as possible. The crystal preparation is fixed to a 1 × 1 in. slide of Bakelite with a $\frac{1}{2}$-in. hole in the center. As the light entering the condenser is very nearly parallel, an ordinary nicol prism is quite satisfactory as a polarizer. Since accurate measurements of crystal thickness are rarely possible, quite large errors in the absolute values of the extinction modulus are to be expected. However, the wavelength of the absorption maximum and the shape of the absorption curve are obtainable even when the thickness of the crystal is not known.

A considerable amount of work has been done on the dichroism of planar metallic complexes [122] and of benzene ring systems [123], in which images of the crystals were projected on the slit of an ultraviolet spectrophotometer. Both Yamada and Nakamoto refer to the earlier work of Tsuchida and Kobayashi [124].

Measurement of Dichroism. The degree of dichroism at any particular wavelength of a section of a pleochroic crystal is conveniently expressed by $m_2 - m_1$, the difference between the two extinction moduli. Let us suppose that the extinction moduli of the section are m_1 and m_2, the corresponding intensities of the transmitted beams are I_1 and I_2 for an incident intensity of I_0, and the thickness of the crystal is d. Since the original unpolarized light is resolved into two beams by the crystal we have:

$$I_1 = I_0 \cdot 10^{-m_1 d}/2 \tag{3.34}$$

$$I_2 = I_0 \cdot 10^{-m_2 d}/2, \tag{3.35}$$

so that

$$\log_{10} (I_1/I_2) = (m_2 - m_1)d. \tag{3.36}$$

Hence the dichroism can be determined by measuring the ratio of intensities of the two transmitted beams and the thickness of the crystal section.

The simplest optical method of determining dichroism on the microscope consists of the use of a dichroscopic eyepiece of the Wollaston prism type over which rotates an analyzing prism on a graduated circle. Neither the polarizer nor the body-tube analyzer of the microscope is used during the actual measurement, although a polarizer must be used for the purpose of orienting the crystal so that its vibration planes coincide with those of the dichroscope images. It is absolutely essential that only light that has passed through the crystal be allowed to enter the microscope in order to avoid the very serious errors that result from lens flare when the objective is flooded with light. The microscope is illuminated with monochromatic light which, because of the small field and narrow cone of illumination required, is conveniently provided by a Bausch & Lomb laboratory wavelength spectrometer. If the intensities of the two adjacent images comprising the dichroscope field are I_1 and I_2, and the vibration plane of the analyzer makes an angle θ with the vibration plane of the image with intensity I_1, then $I_1/I_2 = \tan^2 \theta$. Substituting in (3.36), we obtain:

$$m_2 - m_1 = (2 \log \tan \theta)/d. \tag{3.37}$$

The readings are repeated at a number of wavelengths in order to obtain the curve for dichroism. No correction has been applied for the loss of light by reflection at the two crystal faces in the above formula. Such a correction could be applied, but it would fail to take account of light reflected back from the microscope objective to the upper surface of the crystal and back again into the microscope system. The most practical solution is mounting the crystals in a medium having a refractive index about half-way between the two indices for the crystal section, when the reflection error becomes negligible.

As with other crystal optical measurements, measurements of dichroism are valueless unless the optical orientation of the crystal section is known. The principal dichroisms are those of optical normal sections that give $m_z - m_x$ and bisectrix sections that give $m_z - m_y$ (negative acute and positive obtuse) and $m_y - m_x$ (positive acute and negative obtuse). Obviously, if the dichroism is known for two principal sections, the third is obtained from the sum or difference of the other two. Dichroism curves are obtained by plotting dichroism against wavelength.

Metallic Reflection. Metallic reflection is reflection of the larger part of the incident light. The reflecting power R of a plane surface of a substance is defined as the ratio of the intensity of the reflected light to that of the incident light for normal incidence. If the substance has refractive index n and absorption index κ,

$$R = [(n - 1)^2 + n^2\kappa^2]/[(n + 1)^2 + n^2\kappa^2]. \qquad (3.38)$$

Obviously, R approaches unity as κ increases and as n increases above unity or decreases below it. Metals have high values of κ and consequently reflect a large part of the incident light, whereas the reflecting power of a transparent substance is only considerable if its refractive index is high. Thus for $\lambda =$ 589 mμ, silver and platinum have reflecting powers of 0.93 and 0.70, respectively, while those of water, window glass, and diamond are 0.023, 0.04, and 0.172, respectively, the high reflecting power of the diamond being attributable to its high refractive index (2.417). In general, any face of an anisotropic absorbing crystal has different reflecting powers for its two vibration directions.

Metallic reflection is of interest to the organic chemical microscopist because this type of reflection occurs with a great many dyes. Very strongly colored crystals of dyes have large value for one or more of the principal absorption indices with correspondingly high refractive indices on the long-wavelength side of the absorption band. Berek [121] has evolved microscopical methods of determining the reflecting power and double reflection of anisotropic absorbing crystals. One of these methods employs a slit microphotometer. Another uses a specially constructed elliptical analyzer. The success of these methods owes much to Berek's ingenious modification of the vertical illuminator. In place of a simple right-angled prism, he uses a compensating prism, which is constructed of glass having a refractive index of $\sqrt{3}$. Plane-polarized light entering the prism suffers three internal reflections and emerges as plane-polarized light for all azimuths of the plane of vibration. This work does not appear to have been extended to the study of dyes, possibly because of the difficulty of growing dye crystals with optically flat faces. However, a qualitative study of the phenomena exhibited by dye crystals under vertical illumination is of interest.

The double reflection of the different faces of an absorbing anisotropic crystal can be observed in three ways. First, the crystal can be examined by vertical illumination with plane-polarized light. The crystal is conveniently mounted on a simple rotation device, such as a dissecting needle pushed horizontally through a cork which is supported on a 1 × 3 in. microscope slide. An ordinary prism vertical illuminator gives fairly satisfactory results if the light entering it is polarized in the plane of incidence, or perpendicular thereto, by a Polaroid polarizing filter or by a nicol. A cover glass vertical illuminator may be used with polarized light vibrating perpendicular to the plane of incidence, that is, at right angles to the microscope body tube [125]. A low-power objective such as a 6× achromat should be used. The presence of double reflection is demonstrated by rotating the microscope stage. The second way is to use ordinary light with the vertical illuminator but to examine by means of a rotating analyzer, or to rotate the crystal and use a fixed analyzer. The third way is to use ordinary light for vertical illumination and to analyze the double reflection with a dichroscope eyepiece.

Doubly reflecting crystals can also be studied by vertical illumination with crossed polarizers. An ordinary prism or cover glass vertical illuminator can be used if the vibration plane of the polarizer is in the correct azimuth. With the cover glass reflector, the most intense reflected beam of plane polarized light is obtained when the vibration plane of the polarizer is perpendicular to the plane of incidence. Therefore, if the light enters the vertical illuminator along a path directly toward the observer (N-S), the analyzer is in the crossed position when its vibration plane is in this same (N-S) direction. This is 90° from the usual setting of the analyzer. The prism vertical illuminator may be used with the same polarizer setting, or with one parallel to the plane of incidence. The latter setting has the advantage that when the light enters the illuminator directly toward the observer (N-S) the analyzer is crossed in the customary E-W position. The setting of the polarizer can be checked by observing the darkness of the field when the microscope is focused on a clean microscope slide. The polarization effects of a face of a doubly reflecting crystal as seen by vertical illumination with crossed polarizers are as follows:

1. On rotating the microscope stage the crystal goes through four extinction positions, 90° apart, in one complete revolution, unless the crystal is monoclinic or triclinic with the particular face being studied showing appreciable dispersion of its vibration directions. Such dispersion of the vibration directions gives vivid and often characteristic colors in place of extinction. In the four diagonal positions, the image has maximum brightness.

2. If the reflecting power for one vibration direction is considerably greater than for the other, the crystal in the diagonal position appears the

same color as that observed with a polarizer or an analyzer alone for the vibration direction of the stronger reflection.

3. If the two reflections are both high and of not greatly different intensity, interference colors may be produced that are unrelated to either the absorption or the reflection of the crystal. The colors result from the interference of the resolved components of two elliptically polarized rays and are in consequence less saturated than the interference colors of the first and second orders obtained with transmitted light from colorless crystals between crossed polarizers. A good example of this type of interference is given by the mineral marcasite, the orthorhombic form of FeS_2. A polished specimen of this mineral exhibits curious pastel colors with crossed polarizer vertical illumination.

The reflecting power of a plane surface of an absorbing substance is changed to some degree by immersing it in a medium of refractive index n_m. The change is slight with metals having a high absorption index, this being the case for silver ($\kappa = 20.6$), chromium ($\kappa = 4.9$), and cadmium ($\kappa = 5.0$). An easily perceptible lowering of reflecting power occurs with metals having a somewhat lower absorption index such as nickel ($\kappa = 1.86$) and tungsten ($\kappa = 0.94$). (The above values of κ are for $\lambda = 589$ mμ.) The loss of reflecting power with many metals is greater at shorter wavelengths since κ is usually lower in this region. Crystals of dyes have considerably lower absorption indices than those of metals; even the highest value of κ rarely exceeds 0.1 and is usually much lower, so that the effect of immersion is great. The reflecting power for any vibration direction is lowered when the corresponding refractive index is greater than $\sqrt{n_m}$ and is a minimum when it is n_m. When, however, the refractive index is less than $\sqrt{n_m}$, the reflecting power is increased. Both of these effects can occur for one vibration direction of a single face of a crystal since the refractive index is abnormally high for the longwave side of an absorption band and abnormally low for the shortwave side. This lowering of reflecting power on the longwave side of an absorption band together with an increase on the shortwave side in the immediate vicinity of the absorption band gives rise to a color shift toward the blue as well as to an increase of color saturation. The effect is present with dyes having a narrow and intense absorption band in the middle of the visible spectrum. Examples occur in the cyanine class of photographic sensitizing dyes. This effect is also shown by amorphous films of some readily accessible dyes such as basic fuchsin, pararosaniline hydrochloride, and the various methyl violets; the films are prepared by rapidly drying thin layers of saturated alcohol solutions of the dyes on glass plates. It should be noted that with crystals embedded in a mounting medium between an object slide and cover glass approximately 4% of the vertically incident white light is reflected by the top surface of the cover glass when a dry objective is used. This white light is mixed with the light reflected from the crystal and reduces its color saturation; it can be eliminated when desired by using a prism of small angle (10 to 15°) as a cover.

Color and Constitution. Textbooks on petrography usually divide colored minerals into two classes: idiochromatic, in which the color is attributable to the absorption of the mineral itself; and allochromatic, in which the coloring matter is suspended in the mineral as minute inclusions. In chemical microscopy the colored crystals with which we are concerned are idiochromatic.

The absorption of light by a substance, crystalline or otherwise, is attributable to electron transitions for some frequency range (or ranges) of the electromagnetic waves that constitute light. The absorption of an individual molecule or ion is modified by its environment; the greatest modification usually occurs when the molecules or ions are packed in a crystal lattice. We first consider the principal types of absorbing ions and molecules and classify them as atomic, atomic-molecular, or molecular absorbers.

Atomic absorbers have an incomplete subshell of electrons which is shielded by completed subshells. This is of course the electronic structure of the rare earth elements, atomic numbers 58 to 71, inclusive. Salts of the rare earth elements are remarkable for the narrowness of their absorption bands. In crystals, even at room temperature, the absorption bands are very narrow, some of them appearing as lines comparable in width to the broader lines in the solar spectrum. Rawlins [126] has pointed out that the sharpness of these absorption bands is attributable to the shielding of the incomplete $4f$-group of electrons by the completed $5s$ and $5p$ of the triply ionized atoms. A good example of this type of absorption spectrum is afforded by neodymium sulfate octahydrate crystals [127, 128].

Atomic-molecular absorbers are present in molecules or ions having an atom with an unshielded incomplete subshell. Elements with this type of electronic structure are the metals with atomic numbers 22 to 29, 41 to 46, 73 to 79, and 92. Not only are transitions possible for electrons in lower levels to the incomplete subshell of such an atom, but transitions may also occur between this subshell and coordinated groups surrounding it. Consequently, the absorption spectrum of a compound of one of these elements also depends on the nature of the coordinated groups. The very wide range of colors displayed by bivalent and trivalent cobalt compounds illustrates this fact. In certain ions such as CrO_4^{2-} MnO_4^-, and UO_2^{2+}, there is an empty subshell immediately below the valency electron level, so that orbital transitions from lower levels to the otherwise empty subshell are possible. Brode [129] considers that the coordinated shell of oxygen atoms in the permanganate ion shields the third subshell and that this accounts for the well-defined structure of some seven components in its principal absorption band. However, the four oxygen atoms of the chromate and manganate ions do not afford any substantial degree of shielding.

Molecular absorbers are the type encountered in colored organic compounds. Such compounds are highly unsaturated structures in which resonance occurs between two or more electronic configurations of the molecules. A conjugated double-bond system is present in the great majority of colored organic compounds, either in the form of carbocyclic or heterocyclic nuclei, or as chains such as

$$\left(\begin{matrix}-\mathrm{C}=\!=\mathrm{C}\\ \quad\mathrm{H}\quad\;\mathrm{H}\end{matrix}\right)_n\!- \qquad \text{and} \qquad -\mathrm{N}\!=\!\mathrm{N}-.$$

Usually, both conjugated ring and chain systems occur in the same molecule. In general, the frequency at which conjugated chain and ring systems absorb is lower the greater the number of atoms in the conjugated system. The resonance frequency is also lowered by attaching polar groups such as $-NO_2$, $-CN$, $-NR_2$, $=O$, $-S$, $-NO$, and $-OH$ to the terminal atoms in conjugated systems. Ketenes represent a rare type of a resonating consecutive double-bond chain of atoms. Detailed discussion on resonance in its relation to the color and constitution of organic compounds has been given by Lewis and Calvin [130] and by Brooker [131, 132]. The color of free radicals results from transitions between an incomplete shell of a carbon or nitrogen atom and other atoms or groups attached to it; the absorbing system is therefore of the atomic-molecular type of the present classification.

It is of interest to consider the effect of environment on the absorption of molecules, with particular reference to the absorption of crystals. When the molecules of a light-absorbing substance are so far apart in space that there is no appreciable interaction between them, the absorption spectrum consists of many thousands of sharp lines arranged in bands which result from the superimposition of many quantized levels of rotational and vibrational energy on the energy of the various electron transitions. The energy absorbed by such isolated molecules is reemitted as resonance radiation [68].

At somewhat higher vapor pressures the resonance radiation ceases, for the potential energy of electrons in higher levels is translated into energy of vibration and rotation of other molecules with which the excited molecule collides, but the character of the absorption remains substantially unchanged. However, when the absorbing molecules are in a liquid or solid, molecular collisions result in a considerable broadening of the fine structure of the absorption bands. At room temperature the absorption bands may still show several moderately sharp maxima, but the fine structure entirely disappears. The influence of dipole moment, refractive index, and other factors on the molar extinction coefficient and position of the absorption maximum of dyes in solution is discussed by Sheppard and co-workers [133] in a paper on the effects of environment and aggregation on the absorption spectra of dyes.

In crystals of an absorbing compound, the molecules or ions are arranged in a regular manner, and the absorption may become quite complicated. The effect of packing an atomic absorbing system, such as the neodymium sulfate octahydrate mentioned above, in a crystal lattice is to make the absorption bands much narrower than they are for the same absorbing ion in aqueous solution or in a glass, such as neodymium glass. Spedding [134] has shown that the grouping of the absorption lines of a rare earth salt depends on the crystal symmetry of the substance. The absorption spectrum of atomic-molecular absorbers is not usually greatly changed in a crystal lattice unless strongly polar ions, such as NO_3^-, are also present. The platinocyanides, referred to on p. 220, are a marked exception to this rule; the $Pt(CN)_4^{2-}$ ion in aqueous solution and in anhydrous crystals is colorless, whereas crystals containing water of crystallization are very strongly colored and exhibit metallic reflection for certain wavelengths. The color is deeper the greater the number of moles of water of crystallization per $Pt(CN)_4^{2+}$ ion, as is shown in Table 3.7. Apparently, the water molecules and platinocyanide ions are involved in an electronic resonance mechanism which is a characteristic of the lattice structure of the crystals.

The absorption spectrum of molecular absorbers may be very greatly changed in a crystal lattice. When all the molecules in the unit cell of the crystal have the same orientation, there is considerable interaction between neighboring molecules, and the absorption spectrum for the *Z*-vibration direction suffers considerable broadening both to longer and to shorter

Table 3.7 Colors of Platinocyanide Crystals

Substance	H_2O *per* $Pt(CN)_4^{2-}$	*Color by Transmission*	*Metallic Reflection*
$Ag_2Pt(CN)_4$	0	Colorless	None
$HgPt(CN)_4$	0	Colorless	None
$(NH_4)_2Pt(CN)_4 \cdot H_2O$	1	Colorless	Blue-violet
$(NH_4)_2Pt(CN)_4 \cdot 2H_2O$	2	Yellow	Blue
$K_2Pt(CN)_4 \cdot 3H_2O$	3	Yellow	—
$BaPt(CN)_4 \cdot 4H_2O$	4	Green and yellow pleochroic	—
$MgPt(CN)_4 \cdot 5H_2O$	5	Deep yellow	—
$MgPt(CN)_4 \cdot 7H_2O$	7	Purple and red pleochroic	Blue and green
$Y_2[Pt(CN)_4]_3 \cdot 21H_2O$	7	Red, yellow, green, and colorless pleochroic	Blue and green

wavelengths. With this disposition of the molecules in the crystal lattice, both the birefringence and the pleochroism are extreme. Usually, however, the molecules are arranged so that they can be considered as belonging to two interpenetrating lattices in which they are differently inclined. This "staggering" of the molecules considerably affects the adsorption characteristics of the crystal. In general, the color of a crystal composed of a nonionized molecule, such as phenol blue or a merocyanine dye [135], is similar to that of the substances dissolved in a liquid with a high refractive index and large dipole moment. However, when the dye is a salt, the electromagnetic field of the ion attached to the dye may exert a very considerable influence on the light-absorbing properties of the crystal. This is especially true of basic dyes of the cyanine [136] class. Crystals of the chlorides of these dyes are not very different in color from their solutions, but the colors of crystals of the perchlorates of these dyes are often quite different from those of their solutions. This is particularly true of crystals whose principal refractive indices are very close together; not only is their absorption weak, but also the absorption maximum is often shifted toward the violet end of the spectrum, an effect clearly shown by different salts of 1,1′-diethyl-2,2′-cyanine, which vary from the deep red halides to the pale straw-colored trithionate [67].

PREPARATION OF COLORED CRYSTALS

The preparation of colored crystals for microscopic observation is complicated by the necessity for obtaining crystals thin enough for examination by transmitted light and yet of sufficient area to permit easy observation of their pleochroism and optical orientation. Crystals of lamellar habit are ideal for examination. The difficulty is of course that such crystals present only one optical orientation. The habit of crystals or organic compounds is often different from different solvents, but the possibility that the crystal may contain solvent of crystallization or exist in polymorphous forms must be kept in mind. Since much of the microscopy of colored organic crystals consists of comparing known with unknown crystals, it is important the same salts of acidic and basic dyes be compared after crystallization from the same solvent under identical conditions of evaporation or cooling.

Basic Dyes. The choice of anion depends on the solubility of the dye. By far the most generally useful salts are the perchlorates, which usually show strong pleochroism. The perchlorates are prepared by the addition of a large excess of sodium perchlorate solution. When the quantity of dye is very small, its perchlorate may be collected in a filter tube, rinsed with a drop or two of water and then recovered by extracting the filter with menthanol or other suitable solvent. With very soluble dyes, a larger anion is an advantage, such as that of Reinecke salt $[Cr(CNS)_4(NH_3)_2]^-$. Other well-characterized crystals of basic dyes are the picrates, in which the strong blue absorption of the picrate anion is little changed by the presence of the dye cations in the crystal, while the absorption of the cation may be completely changed.

Acidic Dyes. The alkali and alkaline earth salts of sulfonated dyes do not always form crystals suitable for microscopic study. In this case it is best to prepare the free acid and add an excess of a volatile alkali such as ammonia or morpholine. Well-characterized crystals can often be obtained by double decomposition with salts of bulky cations such as $[Co\ en_3]Cl_3$, $[Co(NH_3)_6]Cl_3$, $[Zn\ en_2]SO_4$, $[Cu\ en_2]Cl_2$, where en indicates ethylenediamine. An excess of the metallic coordination compound is used to precipitate the dye complex, which is then collected and recrystallized from water. Silver–thiourea complexes of acidic dyes are readily obtained by adding to the dye solution an excess of thiourea previously saturated with silver chloride [137].

Evaporation. As with colorless crystals, the evaporation of a solvent may yield crystals suitable for study. The simplest case is that of the slow evaporation of a pure solvent such as *n*-hexane, benzene, or water. Evaporation from a small beaker gives crystals more suitable for the study of crystal form than of optical properties, whereas evaporation of a film of solution on an object slide often yields crystals thin enough, and of sufficient area, for optical study. Both types of crystals are usually necessary in order to correlate the geometric and optical characteristics. Complicated organic compounds, such as dyes in general, crystallize very slowly, so that the rate of evaporation must be retarded lest the preparation dry before the crystals have time to grow to a satisfactory size. Object slide preparations should be covered with a watch glass and protected from drafts so that the evaporation takes from 15 to 60 min. Extremely concentrated solutions can be handled by the object slide technique; thus nitrobenzene and pyridine are useful solvents for cyanine dyes. Use can be made of the property of certain solvents of not wetting glass in order to collect traces of a dye left on an object slide by the evaporation of a wetting solvent such as acetone, alcohol, or water. A small globule of the nonwetting solvent is rolled over the object slide with the aid of a very thin glass rod until it has picked up sufficient dye and is then allowed to evaporate. Evaporation of mixed solvents, in which the more volatile constituent is also the best solvent, is a very useful method of growing crystals of dyes. For example, dyes insoluble in water but soluble in dilute alcohol can be crystallized by allowing their solution in dilute alcohol to evaporate. The percentage of alcohol in the mother liquor falls, and the dye crystallizes out from solution.

Cooling. When a substance is substantially more soluble in a given solvent at high than at low temperatures, it can be recrystallized by cooling its hot, saturated solution. This principle may often be applied to the micro-recrystallization of dyes on an object slide. The liquid chosen should be such that the solubility of the dye has a very high temperature coefficient, such as pyridine and quinoline for acidic dyes, α-bromonaphthalene for many basic dyes, and triethylbenzene for nonionized dyes of low solubility. When this method of crystallization, polymorphous varieties of both unsolvated and solvated crystals may occur. The cooling should in any case be slow—over a 1- or 2-hr period at least.

Dilution with a Nonsolvent. Addition of a nonsolvent to a solution of a dye throws it out of solution if there is an appreciable affinity between the molecules of the solvent and the nonsolvent. Typical systems of this nature are alcohol (solvent) diluted with water (nonsolvent) and pyridine or nitrobenzene (solvents) diluted with benzene or cyclohexane (nonsolvents). The nonsolvent must be added very slowly to the solvent, since too rapid precipitation yields very small and aggregated crystals. A convenient way of carrying out crystallization by dilution is to place the more-or-less saturated solution of the dye, contained in a microbeaker or a cavity in an object slide, in a desiccator containing a beaker of nonsolvent. Dilution from the vapor phase then takes place very slowly and evenly and good crystals are formed in a few days.

The method of Vesce [138] of growing dye crystals probably is a dilution method. He uses 93% sulfuric acid as a solvent for both sulfonated and unsulfonated azo dyes. Some of the powdered dye is stirred into a drop of acid on an object slide and the slide is then covered. Crystals form after several minutes and may subsequently change; the general appearance of the crystals serves for their characterization. The formation of the crystals and their subsequent modification could be attributable to dilution of the sulfuric acid with water from the atmosphere. The method does not, however, lend itself to a study of the optical properties of the crystals.

Diffusion Method. A modification of the dilution method has proved itself particularly useful for the preparation of large areas of strongly absorbing crystals of cyanine dyes [137]. A milligram or so of dye is dissolved in 1 drop of nitrobenzene and the concentrated solution of dye is transferred to a clean object slide. The droplet is covered with a cover glass 18 or 22 mm in diameter, where it spreads out to form a somewhat irregular film approximately 10 mm in diameter. Some xylol balsam, warmed if necessary, is run around the edge of the cover circle with the aid of a turntable, and the slide is stored for some days at 35 to 45°C, after which crystals usually form. The thickness of the crystals is limited by the distance between the cover and object glass and may be between 10 and 100 μ. This method, which has given particularly good results with perchlorates of cyanine dyes, has the advantage that the crystals are grown and mounted in a single operation.

ORTHOSCOPIC EXAMINATION

The morphology of the crystal should first be studied as far as possible and then observations on the nature of the extinction between crossed polarizers should be made. Organic dyes in crystalline form rarely have a symmetry higher than orthorhombic and consequently are quite likely to show one or more types of optical dispersion, often to an extreme extent. Most triclinic crystals, and monoclinic crystals viewed along other than a perpendicular to the ortho *b*-axis, show some dispersion of extinction. The nature and extent of this dispersion should be noted. Observations on pleochroism should be made on crystals in their extinction positions, when they possess them. It should be remembered that dye crystals often have an abnormally high reflectivity as a consequence of very strong selective absorption. This may completely change the apparent pleochroism of crystals viewed with a dry objective when the field is filled with light, since light is reflected from the front glass surface of the objective to the top surface of the crystal, thence back into the microscope. Crystals that should appear very deep blue or purple for one vibration direction often appear bright yellowish green. A striking example of this effect was shown by 1,1′-diethyl-2,2′-cyanine perchlorate crystallized from nitrobenzene by the diffusion method. When examined with a 10× objective in a narrow cone of polarized light, they presented colorless and pale brown pleochroism, whereas the true pleochroism was colorless and deep red, as was demonstrated by restricting the field of illumination until it fell within the crystal. The only certain way of studying the pleochroism of organic crystals with strong selective absorption is to restrict the area of illumination so that it falls entirely within the crystal being studied.

CONOSCOPIC INTERFERENCE FIGURES

The conoscopic examination of colored crystals between crossed polarizers follows the general lines of that of transparent crystals discussed on p. 191. Preliminary work is best carried out with monochromatic sodium ($\lambda =$ 589 mμ) or mercury ($\lambda = 546$ mμ) light if the crystals are sufficiently transparent in this region of the spectrum. Crystals transmitting the red end of the spectrum are best examined with tungsten light filtered through a Wratten no. 29(F), or a combination of no. 29(F) and no. 35(D). If one or both of the optic axes appear in the interference figure as seen with monochromatic light, it is then profitable to change over to white light. The interference figure appears more recognizable when the crystal is in the extinction position. Crossed axial plane dispersion is quite common, so that the interpretation of the interference figure is often impracticable. However, it is this complexity of the interference figures of dyes that makes them all the more useful as a qualitative basis of comparison of known with unknown substances for purposes of identification. The appearance of the interference figure of a crystal of a dye depends on the thickness of the crystal. With crystals thin enough to have only weak absorption, very brilliant color effects attributable to a combination of optical dispersion and absorption are often exhibited, while with thick crystals an almost monochromatic band of light is transmitted and the interference figure appears to be fairly normal.

CONOSCOPIC ABSORPTION FIGURES

Conoscopic absorption figures of small dye crystals are usually difficult to obtain but are a valuable aid in their study and optical characterization. Absorption figures are obtained by using either a polarizer or an analyzer in the conoscopic mode of examination. The principal practical difficulty is lens flare, which arises from the circumstance that the microscope objective is usually flooded with light that has passed around the crystal, whereas in the study of the corresponding interference figure this light is extinguished by the analyzer. Obviously, the difficulty is diminished by using an objective with as small a field of view as possible. A 2-mm oil-immersion objective and 1.40-N.A. oil-immersion condenser give the clearest absorption figures. If the illuminated area can be restricted to fall within the crystal under examination, lens flare is eliminated. However, this usually requires a polarizing substage of special construction.

Conoscopic absorption figures are of great value in determining whether or not a number of crystals presenting different dichroism when examined orthoscopically are different orientations of the same crystalline modification.

MICROSPECTROGRAPHS

Jelley's microspectrograph is well adapted to the study of pleochroism, both orthoscopically and conoscopically. Neither the polarizer nor the analyzer of the microscope is used for microspectrography. Instead, the image of the crystal or its conoscopic disk is projected on the slit of the microspectrograph, and a Wollaston prism is inserted in the collimated light beam to produce two adjacent spectra corresponding to the vibration planes of the crystal. In order to set a crystal so that its vibration planes agree with the vibration planes of the Wollaston prism, the microscope polarizer is temporarily inserted and rotated until it extinguishes one of the images given by the Wollaston, the crystal is rotated to extinction, and the polarizer is withdrawn.

VERTICAL ILLUMINATION

Large crystals of dyes having strong selective absorption are conveniently studied by vertical illumination with a 6 or 10× objective. Some form of simple rotation apparatus is essential for orienting the crystal; the reflected light is analyzed as explained on p. 219. The extreme colors of the crystal are sometimes quite characteristic, and different, for homologous dyes.

Thermal and High-Pressure Microscopy

A major branch of chemical microscopy is based on the study of phase equilibria using the microscope. The microscope has been and is one of the most important instruments available for determining phase diagrams for metals and alloys, rocks and minerals, and ceramics. Microscopy in these studies is to a large extent confined to the examination, at room temperature and atmospheric pressure, of solids of known composition and thermal history. Microscopical examination of metallic, mineral, or ceramic systems at temperatures where interesting phase transitions occur is usually very difficult because of the high temperatures involved. In spite of the difficulty, experimental observations of transitions at high temperature are being carried out by metallurgists, mineralogists, and ceramists. It is not possible to discuss these fields of application here, so microscopists interested in these fields should consult the appropriate books and journals.

Traditionally, those microscopists who classify themselves as chemists or chemical engineers have been interested in vapor-liquid equilibria or in the solid-liquid equilibria that occur in chemical solutions. Vapor-liquid systems can in some cases be studied microscopically, but systems involving liquids and solids are more generally amenable to microscopical examination. For example, solutes in solution can be identified by the methods described on p. 161, or solids that precipitate can be identified by the methods outlined on p. 160 or by the crystallographic methods described on p. 166. Many compounds that interest chemists melt at relatively low temperatures, and phase equilibria that exist between these materials and their melts or between polymorphs of the materials are readily observed microscopically.

Studies of phase equilibria can be used in two ways. They can be used to predict what will happen if a system of known composition is subjected to certain pressures and temperatures. This use implies that the phase diagram of the system is known or can be determined readily. Phase equilibrium data can also be used to identify chemical compounds. For example, melting points, boiling points, phase transitions, and similar data can serve as criteria for identification.

Hot and Cold Stages

The factors affecting phase equilibrium are temperature, pressure, and composition. Variations in composition are effected during preparation of the specimen. The pressures and temperatures are produced by special stages. In the past most of the work on phase equilibrium has been carried out at atmospheric pressure and only a hot or cold stage was needed. Many hot and cold stages have been described in the literature. The large number of stages described apparently results from the tendency of microscopists to build hot or cold stages adapted to their particular needs. Some of the factors that must be considered in designing or choosing a stage are: its heat capacity, its insulation, its temperature range, the accuracy with which temperatures can be measured, the magnitude of the temperature gradients in the viewing chamber, the ease and rapidity of changing the temperatures, and the convenience of the necessary manipulations.

A type of warm or cold stage that is easily constructed is one in which the heat is provided by circulation of a heat-exchange liquid. Cells of this type have been described by Saylor [139, 70] and McCrone et al. [140, 141]. They consist of thin cells with glass windows through which the thermostatically controlled heat-exchange liquid flows. The specimen is placed on the cell and viewed using transmitted illumination. Such cells usually have high heat capacity which gives thermal stability but makes it difficult to change the temperature rapidly. The temperature range is usually low, from about −50 to +100°C, depending on the heat-exchange liquid and the thermostat. The amount of insulation is usually small and the sample and cell are easily manipulated.

Higher temperatures can be reached with electrical heating. The heating element can be an electrically conductive coating on glass [142–144]. Such stages have low heat capacity, little insulation, and their temperature can be changed easily and quickly. Temperatures up to about 300°C can be attained. However, to assure accuracy and control of temperature the electrical controls may be complex.

A more common form of electrically heated stage consists of a block of metal, usually aluminum or copper, with high heat conductivity, heated by a resistance winding such as Nichrome wire. The characteristics of these stages varies widely depending on their construction. Chamot and Mason [1] and Grabar and McCrone [145, 140] have described relatively thick massive stages with high heat capacity and thick insulation. Temperatures up to 500°C can readily be obtained. Hartshorne and Stuart [70] have described two thin electrically heated stages. These stages are only about $\frac{1}{16}$ in. thick and thus do not raise the specimen much above the stage and the substage condenser, permitting conoscopic observation of the specimen. Matthews [146] designed a stage consisting of an electrically heated copper or brass block surrounded by a water jacket through which water at 25°C is circulated. The use of this arrangement protects the microscope from undue heating, allows the stage to be cooled rapidly, and provides highly controllable heating rates with a relatively simple electrical supply.

Hot stages for the range from 500 to 3000°C are more difficult to design and construct. The problem of protecting the microscope is much more difficult, and it is often necessary to surround the specimen and the heating element with an atmosphere of inert gas or a vacuum. Often the specimen can be placed directly on a ribbon of platinum or other filament material which is heated electrically. McCrone [140] suggests the use of a soldering gun as a power supply. A different approach was taken by Ordway [147], who used a very small heat source to minimize the total heat output. He placed the specimen to be examined in a small loop made of a thermocouple. The thermocouple was electrically heated and the specimen temperature measured with the thermocouple. The thermocouple hot stage has been modified and improved by a number of investigators [148–151]. Dodd [152] also used an electrically heated loop filament. The loop served as a resistance thermometer for measuring the temperature.

Cold stages are often of the circulating liquid type mentioned above [1], but lower temperatures can be attained by circulating gases cooled by liquified gases, for example, nitrogen or helium. Temperatures can be controlled by mixing warm and cold gas streams. Another method of obtaining cold gas is to allow it to expand just before it contacts the specimen, the temperature being regulated by regulating the amount of expansion of the gas. Thermoelectric cooling has also been used [153].

Although it is possible for the microscopist to construct hot or cold stages, such homemade stages are usually relatively crude, and the cost of refining them is high. Thus it is often advisable to purchase hot or cold stages. The Kofler hot stage is widely used for temperatures up to 350°C. It is manufactured by Reichert and can be obtained from A. H. Thomas. Reichert also manufactures other stages, some of which can attain high temperatures. E. Leitz also produces a variety of hot stages. The Mettler Instrument Corporation has recently produced a hot stage which has many interesting and valuable features. McCrone has written a pamphlet describing the stage and its use in chemical microscopy [158].

Temperatures can be measured in various ways using thermometers, thermocouples, optical pyrometers, resistance thermometers, and thermistors. With some types of stages it is possible to estimate temperatures from the current and voltage of the power supply. Temperature-measuring instruments are usually calibrated using the melting points of pure compounds as reference standards.

Microscopes used with hot or cold stages can be simpler than the polarizing microscope used for chemical microscopy. Hot and cold stage work usually must be done at low magnifications with long-working-distance objectives. The microscope should, however, be equipped for polarized light work. Repeated heating and cooling of objectives and condensers causes strain birefringence to develop in these lenses, therefore, the lenses should be kept as far as possible from the stage except during actual use. For high-temperature work an image transfer lens can sometimes be used to project a real image into the focal plane of the objective. Cold stages are often difficult to use because of the condensation of moisture from the atmosphere on the stage windows and on the objective lenses. This can usually be prevented by directing a stream of dry gas over the surfaces where condensation occurs.

This description of hot stages is necessarily very brief. Further information can be obtained from Chamot and Mason [1], Hartshorne and Stuart [70], McCrone [140], and from other articles and books [14, 15, 55, 154–157].

High-Pressure Stages

Phase equilibria in condensed systems, that is, systems consisting of liquids and solids only, are affected very little by moderate changes in pressure. As pointed out above, most phase diagrams for condensed systems are determined at atmospheric pressure. In recent years there has been considerable interest in high-pressure transformation in solids and liquids. While most of this work has been carried out using large presses [159], Van Valkenberg [160] has invented a high-pressure stage suitable for microscopical studies at very high pressures.

The apparatus consists of two anvils made of diamond, which can be pressed together by using a system of levers and a screw. The sample is placed between the anvils and observed through the anvils as pressure is applied. This apparatus makes it possible to study many phase transformations that were hitherto unobservable microscopically.

Chemical Applications of Thermal Microscopy

PROPERTIES MEASURABLE WITH THE AID OF A HOT STAGE

Material properties that can be measured with the help of a hot stage include melting point, boiling point, transformation temperature, critical solution temperature, refractive index, molecular weight, the growth rate of crystals, and the morphological and optical character of the crystals. Microscopical melting-point determinations are of much more value than capillary melting points because of the details observable. Melting is the transition from a crystalline lattice to a liquid. Except for isotropic crystals this transition is easily observable between crossed polarizers as it results in a loss of birefringence. Other changes that occur on heating such as dehydration, sublimation, polymorphic transformation, and decomposition are revealed. Pure materials melt sharply at one temperature while impure materials or mixtures melt over a temperature range, thus making it possible to judge the purity of a compound. Noncrystalline materials do not possess a true melting point, but their softening points may be observed with the microscope.

Melting points can be taken at equilibrium and, with a good hot stage, should be accurate to 0.1°C. Because microscopical melting points can be observed so readily, they tend to be somewhat lower than those measured by the ordinary methods.

Molecular weight determinations based on the lowering of the freezing point of a solvent when a solute is added [140, 164] can be carried out microscopically. Solvents suitable for this type of determination as given by McCrone [140] are listed in Table 3.8. A carefully weighed mixture of the solvent and compound whose molecular weight is to be determined are thoroughly mixed by grinding in a mortar. The mixture, containing about 10 parts solvent to 1 part solute, should be wetted with ether during mixing and grinding. A small portion of the mixture is placed on a slide covered with a cover glass and melted. As soon as it is melted it should be frozen rapidly

Table 3.8 Solvents for Use in the Microscopical Determination of Molecular Weight

Compound	*mp* (°C)	*Compound*	*mp* (°C)
Perylene	276	Tetrabromoethane	93
Borneol	204	2,4,6-Trinitrotoluene	81
Hexachlorethane	186	Cyclopentadecanone	65
Camphor	178	Camphene	49.5
Bornyl chloride	130		

by placing the preparation cover glass down on a cold metal cooling block. After freezing, the cover glass is sealed with Duco cement and the melting point determined very precisely, using a stopwatch as described by McCrone [140]. The melting point is taken as the disappearance of the last crystals in the solvent. It is advisable to determine the melting point a number of times using several preparations. The melting point and the molar depression constant must be determined for each sample of solvent since both depend on the purity of the solvent.

Boiling points and critical solution temperatures are determined using small capillary tubes mounted in the hot stage. To determine the boiling point, the end of the capillary is sealed and a small drop of the liquid is placed in the capillary. A small bubble of air should be left at the bottom of the capillary. Both the air and the liquid can be placed in the capillary using a drawn-down Pasteur pipet. As the specimen is heated, the junction between the air bubble and the liquid is watched. At the boiling point the liquid is suddenly blown out of the capillary.

The critical solution temperature is the temperature at which two liquids become completely miscible. To determine a critical solution temperature, the two materials are placed in contact in a sealed capillary tube and the meniscus is observed as the specimen is heated. The meniscus disappears at the critical solution temperature during heating and reappears on cooling to that temperature [140, 161–163].

The refractive index of a melt can be determined using heated refractometers, but it is simpler to use the Becke test as described on p. 199. However, the techniques used and the type of data obtained are different [14, 15, 140, 158]. The reference standards are a series of glasses of known refractive indices. Particles from one of these standards are placed in the melt, and the Becke test is used to determine whether the n of the melt is higher or lower than that of the reference standard. If the n of the liquid melt is higher than that of the standard, the melt is heated to a higher temperature. On heating, the n of the melt decreases and the temperature at which the two indices match is recorded. If the n of the melt at its melting point is lower than that of the standard, a new standard must be used. If the temperature of matching refractive index is measured using two standard glasses, the temperature coefficient of refractive index can be calculated. Standard glass powders can be obtained from Arthur H. Thomas Company or from William J. Hacker & Company, Inc.

The growth rates of crystals vary with the degree of undercooling, with the amount of impurities, with viscosity, and with the shape and complexity of the molecules, ions, or atoms that form the crystal lattice. In spite of this complexity, crystal growth rates can be used in several ways for analyzing materials. McCrone et al. [141] used crystal growth rate to analyze for *p,p'*-DDT in technical grade DDT. The analysis is based on the fact that the crystals grow more and more slowly (at a given temperature) as the proportion of *p,p'*-DDT increases. Other uses for crystal growth rate determinations are mentioned below.

The morphological and optical properties of crystals and their use in chemical microscopy are discussed on p. 158. However, the techniques described there are generally applied to more-or-less well-formed crystals grown from solvents. Crystals from a melt are less likely to show recognizable habit because the external shape of the crystals is determined largely by the growing together of the various crystals. Also, growth is often rapid, resulting in dendritic crystals. However, at the interface between the crystal and the melt, faceting may occur and at times some idea of crystal habit can be obtained. In spite of the lack of faceting, the overall structure of the frozen crystals, and the growing crystal front, may be very characteristic of the compound. In fact, if the microscopist is dealing with a limited number of compounds, he can often learn to recognize them from their characteristic growth pattern. If the number of compounds involved becomes large, then measured properties boiling point, melting point, molecular weight, and refractive index must be used. In addition, properties such as crystal angles, extinction, birefringences, and interference figures can be observed on crystals grown from a melt.

Often it is unnecessary to use a hot stage for operations in thermal microscopy. The specimen can be melted over a microburner, on a hot plate, or with an electrically heated wire loop or small soldering pencil. With these methods of heating, the specimen is usually observed as it cools. Many of the thermal phenomena can be observed in this way. A hot stage makes it possible to control the phenomena somewhat better, as the specimen can be heated, cooled, or held at a chosen temperature. In this way the microscopist can be more certain that an equilibrium state has been attained, and it is possible to grow larger crystals or crystals with a given orientation (see also p. 210).

PHASE STUDIES ON ONE-COMPONENT SYSTEMS

One-component systems may exhibit several phenomena on heating at constant pressure. As it is heated, a solid, crystalline compound may undergo one or more polymorphic transformations before melting. Further heating of the melt causes boiling. Instead of melting, the solid may sublime, passing directly from the solid to the vapor state; and there is always the possibility that it may decompose.

Decomposition. Decomposition may occur when the substance is a solid, a liquid, or a gas. Decomposition is usually easily recognized, but care should be taken not to confuse it with other phenomena such as phase transformations or dynamic isomerism. A special case of decomposition often encountered is the loss of solvent of crystallization. This type of decomposition is properly considered under the heading of two-component phase systems, but since it usually occurs on heating a pure compound it is discussed briefly here. The behavior on heating of a crystal containing solvent of crystallization depends on the boiling point of the solvent and on the forces binding it into the crystal lattice. There are several possibilities. First, the solvent may be driven off as a vapor. In this case the crystal usually becomes opaque. Although it may retain its original shape, the crystal lattice has changed and this results in a porous pseudomorph. The loss of solvent may also cause decretpitation. Second, the solvent may be liberated as water and then dissolve part of the unsolvated crystal. Third, the crystal may melt to form a solution of the substance in water. Finally, a part of the solvent may be liberated with the formation of a new crystal phase containing less solvent of crystallization.

The presence of a low-boiling solvent of crystallization may often be detected by heating some of the substance in a capillary tube, the upper part of which is kept cold. Condensed droplets may be tested by their solvent

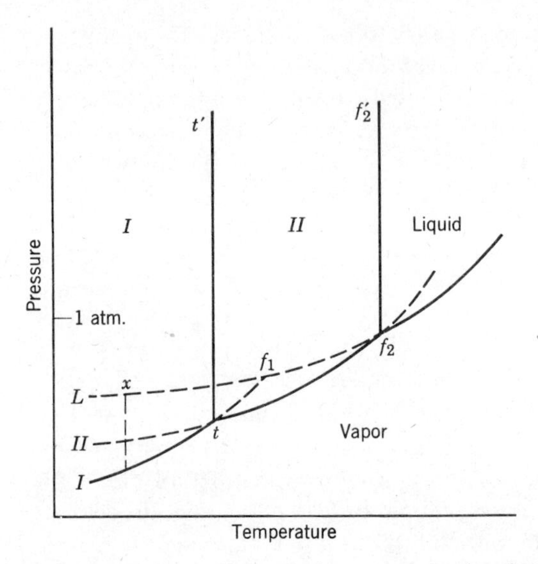

Fig. 3.36 Pressure-temperature diagram of an enantiotropic substance.

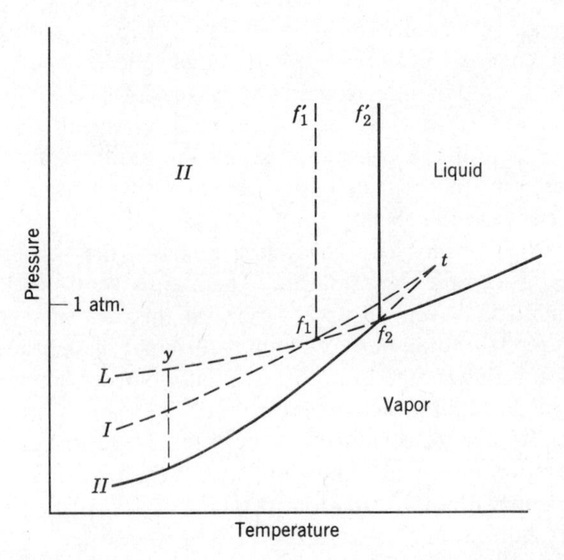

Fig. 3.37 Pressure-temperature diagram of a monotropic substance.

action on different highly colored materials. For example, crystals of potassium permanganate yield an intense purple solution with droplets of water, a weak purple with methanol, a faint purple with ethanol, and no color at all with benzene. Many simple unsulfonated azo dyes are insoluble in water but are very soluble in alcohols and may also be used as indicators. Another useful way of detecting solvent of crystallization is by heating the crystals under a heavy aliphatic hydrocarbon, such as petrolatum liquid USP XIII. As the solvent is driven off, it forms bubbles in the immersion liquid. If a substance crystallizes in the same form from a number of different solvents, it is almost a certainty that no solvent of crystallization is present.

The behavior of a substance that is thermally stable is best understood from a phase diagram such as those shown in Figs. 3.36 and 3.37. If the pressure (commonly atmospheric pressure) is above point f_2, heating will result in a phase transformation from polymorph *I* to polymorph *II*; then polymorph *II* will melt and finally the melt will boil. If the pressure lies below f_2, sublimation will occur. The behavior of materials undergoing polymorphic transformations is interesting and complex.

Polymorphism. Substances that exhibit more than one crystalline form are said to be polymorphous. Polymorphous crystals of the same substance differ one from another in their physical properties but yield identical liquids and vapors.

Polymorphs may be divided into two classes: those that are enantiotropic and those that are monotropic. Enantiotropic polymorphs interchange reversibly on changing the temperature, each form having its own stability range of temperature. Monotropic polymorphs, however, do not interchange reversibly on changing the temperature. One form is unstable and the other stable at all temperatures. In Fig. 3.36, curves for the polymorphs *I* and *II* and the liquid *L* of an enantiotropic system are shown. In this diagram the conditions under which form *I* is stable are represented by the area to the left of boundary Itt', form *II* is stable under the conditions denoted by area $t'tf_2f_2'$. On gradually heating crystal form *I* at atmospheric pressure, it remains unchanged, except for slight loss by vaporization, until temperature t is reached, when it changes over to crystal form *II*. On further heating, form *II* melts at temperature f_2. In the complete absence of supercooling, the liquid solidifies at f_2 to form *II*, and form *II* changes to form *I* at t. However, many organic substances supercool considerably. In the case under consideration, the vapor-pressure curve of the supercooled liquid would continue in the direction $f_2 \rightarrow f_1 \rightarrow x$. In the complete absence of form *II*, form *I* may persist above the transition temperature until its melting point f_1 is reached; similarly, form *II* may supercool below the transition temperature.

Monotropic polymorphs are characterized by the fact that one form is stable at all temperatures below its melting point and never exhibits transformation into the second form, while the second form has a lower melting point and is metastable at all temperatures below that. The pressure-temperature diagram of a monotropic substance is given in Fig. 3.37, in which the stable form *II* has melting point f_2, and the metastable form *I* has melting point f_1. Curves IIf_2 and If_1, if extrapolated, meet at a point t, representing the hypothetical transition point which cannot be attained experimentally because it lies above both melting points.

Ostwald [165] formulated a general rule, the law of successive reactions which, in connection with crystallization from supercooled liquids and from supersaturated solutions, implies that the form that crystallizes out is not the most stable one but the one that can be reached with the minimum loss of free energy. In a dimorphous system this, of necessity, is the metastable form. If in the pressure-temperature curve of an enantiotropic dimorphous substance given in Fig. 3.36 the liquid has supercooled to point x, the nearest state will be *II*, which is in its metastable region at x. Contact with stable form *I* causes the transformation of *II* into *I*. Similarly, in the case of the monotropic dimorphous system shown in Fig. 3.37, liquid supercooled to y crystallizes to the metastable form. On heating, this form melts at f_1. A good example of this phenomenon is supplied by benzophenone, as first reported by Zincke [166]. If the stable orthorhombic modification, which melts at 48.5°C, is heated to about 60°C and then cooled, it does not crystallize in the stable modification but remains liquid. Occasionally, a monoclinic modification crystallizes from the supercooled liquid; this metastable modification has a melting point of 26.5°C. Inocculation of the supercooled liquid with the metastable solid causes immediate crystallization to this modification.

The displacement of transition and melting points with increase of pressure is so small that it need not be considered in connection with microscopical work. When a substance possesses more than two polymorphous forms, they may be all enantiotropic, or all monotropic, or some may be enantiotropic and the rest monotropic. With many organic compounds the transformations of the different modifications take place readily unless the cooling has been too rapid and severe supercooling has taken place. Supercooling may be avoided by having the cooling liquid in contact with crystals of stable and metastable modifications. Thus Dippy and Hartshorne [167, 168] recommend a preliminary crystallization from a melt of a substance under a microscope cover glass. The central portion of the cover is then heated to remelt the crystals, either by means of a small gas flame or by contact with the heated end of a small metal rod. In preliminary observations the establishment of a temperature gradient across a small area of the slide allows different enantiotropic forms to exist together in the right temperature zone of the crystalline layer, whereas under similar conditions a stable monotropic form always grows at the expense of the unstable monotropic form over the whole temperature range.

Dynamic Isomerism. Care must be exercised not to confuse dynamic isomerism with polymorphism. Dynamic isomerism, or tautomerism, is the more-or-less ready change of one isomer to another, such as that of the α- and β-benzaldoximes and α- and β-acetaldoximes studied by Bancroft and his associates [169]. An isomer melting at 168 to 170°C was obtained by repeated crystallization from 95% ethanol; the other isomer, melting at 157°C, was obtained by sublimation. When the second modification came in contact with the first, liquefaction occurred through the lowering of the melting point. Crystallization from the melt gave an equilibrium mixture melting near 148°C. The thermal behavior of a dynamic isomer is that of two different substances mutually interconvertible. The rate of interconversion may be greatly modified by traces of catalysts, such as acids. This often accounts for apparently erratic behavior.

Glass Formation. When a melt is cooled quickly it may supercool to form a vitreous mass of "glass," which may remain in this state for hours, days, or even years. The physical properties of such a glass are those of a liquid of extremely high viscosity. It is optically isotropic unless subjected to mechanical strain. A glass does not possess a true melting point. Heating often induces crystallization or, if crystallization does not take place, the glass softens and finally regains the fluidity of the normal liquid. The phenomenon of crystallization on heating is very well shown by aceto-β-naphthalide, a substance that melts at 130 to 134°C and solidifies to a clear glass when cooled suddenly. On warming a microscope slide preparation of the vitreous substance, crystals radiate from a number of centers, a growth which, however, is instantly checked by recooling the slide.

Mesomorphism (*Liquid Crystals*). Certain substances when heated do not pass directly from the crystalline form to the liquid melt. Instead, these substances, called "liquid crystals," transform from the solid to a pasty or viscous liquidlike material which is often turbid and has some of the optical properties of a crystal. At a still higher temperature, these materials transform to a normal clear melt. The transition temperatures are definite temperatures which vary with pressure just as do melting points and polymorphic transformation. The mesomorphs can be either enantiotropic or monotropic.

Similar intermediate phases can be produced at constant temperature in some materials by mixing the material with a solvent. Such behavior is called lyotropic mesomorphism, while the temperature-dependent behavior is called thermotropic mesomorphism.

Thermotropic mesomorphs exist in two well-defined states, the smectic and the nematic. The smectic state may be considered to be solid in two dimensions and liquid in one, while the nematic state is solid in one dimension and liquid in two. Another type of mesomorph having very unusual optical properties is the cholesteric. This is a modification of the nematic-type mesomorph. Liquid crystals usually consist of elongated stiff molecules having strong dipoles and easily polarizable groups. They often have weak dipolar groups at the ends of the molecules. Such molecules tend to associate in various types of parallel arrangements, and because of this alignment liquid crystals are anisotropic. Within the temperature range between the true crystalline state and the true molten state, liquid crystals may show many transformations between the various types and between polymorphs of these types.

When working with liquid crystals, it is often advantageous to use the thermal gradient type of hot stage so that the full range of mesomorphs can exist in one preparation. The differentiation between nematic and smectic mesomorphs is not difficult after some experience has been gained by working with known substances. The following is a guide to the principal differences between the two states when observed under a microscope within the temperature range of stability. Nematic mesomorphs are much more fluid than smectic mesomorphs and are markedly oriented by solid surfaces, in contrast to the behavior of smectic mesomorphs which are not so oriented. When the nematic mesomorph is formed by cooling the true (isotropic) liquid, it first appears in anisotropic droplets of circular contour, while angular bodies (*bâtonnets*) are characteristic of the smectic state under similar conditions. The nematic state in bulk is often isotropic when quiescent, but movement causes alignment of the threadlike aggregates, with the production of strong anisotropy and, in the case of colored substances, dichroism. Uncovered layers of smectic mesomorphs show a terraced structure; thus thallous soaps exhibit the layer structure known as "Grandjean's terraces," and many other smectic substances exhibit a curious terraced structure of droplets known as *gouttes àgradinsa*. Smectic mesomorphs in thin layers between a microscope slide and cover glass often exhibit a characteristic focal conic structure. This behavior is very clearly shown by ethyl *p*-azoxybenzoate between 114 and 120°C.

There has been great interest in liquid crystals in recent years, and many books and articles have been published [172–178]. There is considerable information in Hartshorne and Stuart's book [70]. A journal entitled *Molecular Crystals and Liquid Crystals* is also available.

PHASE STUDIES ON BINARY SYSTEMS

As mentioned in Section 4, p. 151, it is possible to determine whether or not two compounds are identical by the mixed fusion test. A preparation of the two compounds is made by melting them together under a cover glass as shown in Fig. 3.11. If the compounds are identical, crystals growing from the mixed melt will not be affected when they cross the zone of mixing. Thus a mixed fusion test is very valuable in identification when the unknown is one of a limited number of compounds. If the two materials in a mixed fusion are not identical, then the zone of mixing contains varying composition going from 100% of one material to 100% of the other. If such a gradient mixture is cooled until it freezes, there are four general possibilities. The system may form a continuous series of solid solutions, it may form a eutectic, or it may form addition compounds with accompanying eutectic or peritectic reactions. There may of course be combinations of the above, for example, a eutectic with limited solid solubility. Also, the continuous solid solution may show maxima or a minima in their melting-point curves. The fourth possibility is that the melts are immiscible or partially immiscible. In these cases two liquid phases are visible in the melt.

Determination of the Type of Binary Phase Diagram. The type of phase diagram for binary mixtures can be determined very rapidly, usually within a few minutes. A hot stage is not needed, as such determinations can be made by melting the mixed zone of a mixed fusion slide and observing it while it solidifies. However, the use of a hot stage permits determination of the quantitative aspects of the phase diagram, and it is essential for determining whether an addition compound system is peritecic or not, whether a solid solution system shows a maximum, minimum, both, or neither, and whether a eutetic system shows limited solid solubility [158]. Some hot stages are designed to establish a temperature gradient. For example, if a wire is stretched over the cover glass and heated by passing a current through it, a stable temperature gradient can be established in the specimen. If the temperature gradient is arranged to be perpendicular to the concentration gradient, the various phases in the system will have an opportunity to stabilize so that they can be examined at leisure.

Figure 3.38 shows schematically how mixed fusion preparations might appear for various systems being held in a thermal gradient. These diagrams can aid in understanding what happens as a mixed fusion is cooled or heated on a hot stage. Consider what happens when a mixed fusion of two compounds that form a continuous series of solid solution is cooled from above T_1 (Fig. 3.38*a*). When T_1 is reached, crystals of B start to form on the right and grow toward the left. The melt is on the left and between the ends of the crystals. As cooling continues, the crystals continue to grow but at a reduced rate. The rate becomes slower as the solid solution that is freezing out approaches a 50-50 mixture. After reaching this composition the growth rate increases again until the entire mixture is frozen at T_2. The resulting

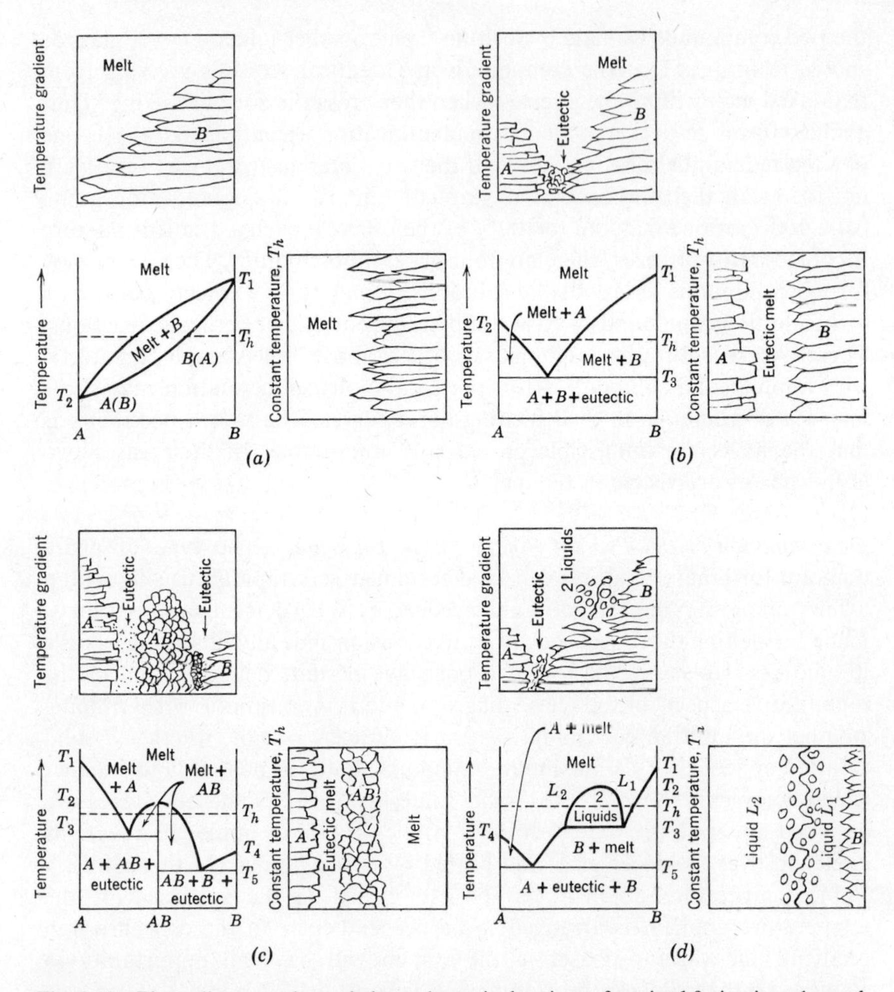

Fig. 3.38 Phase diagrams shown below schematic drawings of a mixed fusion in a thermal gradient. (*a*) Solid solution; (*b*) eutectic; (*c*) compound formation plus two eutectics; (*d*) eutectic with two liquid layers.

solid crystals may vary in some of their properties, for example, refractive index and lattice parameters, but they all have the same crystallographic structure. When melting, the reverse occurs and if a given temperature T_h is held the preparation should appear as illustrated in the drawing to the right of the phase diagram.

If the system shown in Fig. 3.38*b* is cooled from above T_1, crystals of B will first appear on the right, growing toward the left. At T_2 crystals of A appear at the left growing toward the right. These two sets of crystals approach each other until at temperature T_3 a eutectic grows in the remaining melt, filling the space between the two crystal fronts. The system shown in Fig. 3.38*c* acts in much the same way with crystals of A starting to freeze at T_1. However, at T_2 the compound AB starts to freeze in the center of the concentration gradient. At T_3 a eutectic forms between A and AB. At T_4 crystals of B start to form and finally at T_5 another eutectic forms. The system shown in Fig. 3.38*d* is similar to that shown in Fig. 3.38*b* except that at T_2 two immiscible liquids form and at T_3 one of these liquids disappears, having been converted into crystals of B plus liquid L_1.

Determination of Composition in Binary Systems. The compositions of the eutectics can be determined by estimating a possible composition and then weighing out the estimated composition, grinding it, and heating it to just above the eutectic temperature. If the estimate is incorrect, crystals of the compound in excess will be found in the melt and a new estimate can be made. Estimates can be based on the Clapeyron equation if the heats of fusion and the melting points are known. If only the melting points are known, the mole percent (X) of the lower-melting compound in the eutectic can be estimated from the following equation, which is based on the assumption that the slopes of the liquidus curves are the same [140].

$$X = \frac{100(T_1 - T_3)}{T_2 + T_1 - 2T_3}.$$

The temperatures in the equation are defined in Fig. 3.38*b*.

The composition of the eutectic can also be determined by measuring its temperature of refractive index match [179]. To do this a curve showing the temperature of refractive index match as a function of the composition of the binary mixture must be derived from experimental data. Generally, this curve is a straight line between the temperatures of refractive index match for the pure components, however, it is advisable to obtain at least one other experimental point near the center of the curve to be sure that the curve is actually a straight line. It may be that a single refractive index standard will not suffice to cover the entire range of compositions. In this case it is necessary to obtain more experimental points so as to construct a branch of the curve for each refractive index standard. Once the curve has been obtained, the composition of the eutectic can be determined by distributing the refractive index standard power throughout the mixed fusion preparation. On heating the preparation the eutectic is first to melt, and particles of the index standard lying in the eutectic are observed as heating is continued. The temperature of refractive index match is noted and the composition read from the curve.

The use of the temperature of refractive index match to determine composition is not confined to the determination of eutectic composition in mixed fusion preparations. It can be used to determine composition in any binary system that does not decompose on heating [140, 179–184].

Another method of determining the composition of binary systems is based on freezing-point depression. The methods are the same as those used for molecular weight determinations. Although estimates can be made based on a melting-point depression of 0.5°C for every weight percent of the solute, the estimates are not very reliable. To obtain good results using this method, it is necessary that each binary system be calibrated [140, 185].

Other methods of determining composition that can be used when appropriate are: crystallization rate as discussed on p. 244, counting of constituents prior to melting as outlined on p. 161 or, if the melts of the constituents are immiscible, areal analysis as described on p. 165.

Polymorphism in Binary Systems. The determination of phase diagrams can be complicated by the occurrence of polymorphism. The techniques used to study polymorphism in binary systems are similar to those used for one-component systems (see p. 247).

PHASE STUDIES IN TERNARY SYSTEMS

Ternary phase diagrams can be determined microscopically by using the same techniques described for binary diagrams [140]. In fact, the first step in the procedure is the determination of all of the binary diagrams involved. These three diagrams form the sides of the ternary diagram. The interior of the diagram is then filled in by making mixtures of the three components and checking the thermal properties of these mixtures. Ternary mixed fusion preparations can also be made. Addition compounds and polymorphism can complicate the determination [15, 140].

IDENTIFICATION OF COMPOUNDS USING THERMAL DATA

Although thermal analysis provides data that can be used to characterize materials, these data must be arranged in a useful form if they are to be of value in the identification of compounds. The Koflers have been most successful in compiling data so that it is useful. In their book [15] there are identification tables that present data on about 1200 compounds. The tables have the following column headings.

Melting point in degrees centigrade
Substance
Eutectic melting temperature with azobenzene
Eutectic melting temperature with benzil
Refractive index of the melt
  Standard refractive index powder
  Temperature of melt at matching index
Miscellaneous characteristics

The table is arranged in the order of increasing melting point. Eutectic melting temperatures are measured for mixtures of the unknown with reference materials such as azobenzene and benzil. Data on the refractive index of the melt are obtained as described on p. 244. Matching temperatures for more than one powder are often given. Under miscellaneous characteristics are given data on color, odor, habit, polymorphic forms, sublimation temperatures, and so on.

McCrone [140] has rearranged the data in the Koflers' table to make it more useful to the analyst. Table I lists the compounds alphabetically with their melting points. Table II lists the compounds in the order of increasing melting point. Table III gives the eutectic melting points using standard reference compounds. They are arranged in the order of increasing melting point under each standard compound. Table IV lists data on refractive index. The data are listed in the order of increasing temperature of match for each reference standard glass powder.

Any method of analysis that depends on previously determined properties of a material necessarily suffers in that tables recording such properties are always incomplete. This problem is particularly acute for microscopical analysis based on optical and geometrical properties because the determination of such properties requires knowledge and skill that are rather uncommon. Microscopical analysis based on thermal data such as are contained in the Koflers' tables can be carried out more easily by microscopists who are not optical crystallographers. Therefore although tables of thermal data are now very limited, it should be possible to extend these tables rather easily. It is obvious of course that the value of analytical methods that depend on previously determined data is not destroyed because tables of data for all substances are not available.

Dispersion Staining

Theory

Recently, a technique has been developed for identifying particulate materials which depends on the dispersion of the refractive index of the particle as well as its refractive index [44, 186–190]. More specifically, it depends on the difference in dispersion between the particle and the medium in which it is mounted. The dispersion curves of the medium and of the particle must cross somewhere in the visible spectrum in order for the method to work (see Fig. 3.39). Consider a particle with wedge-shaped edges mounted in a suitable medium and illuminated with a beam of parallel rays. The wedge-shaped edges of the particle act as prisms, spreading the beam into spectra. Light of wavelength λ_m, for which the refractive indices of the particle matches that of the mounting medium, passes through the prismatic edge without being deviated. Light beams of other wavelengths are deviated

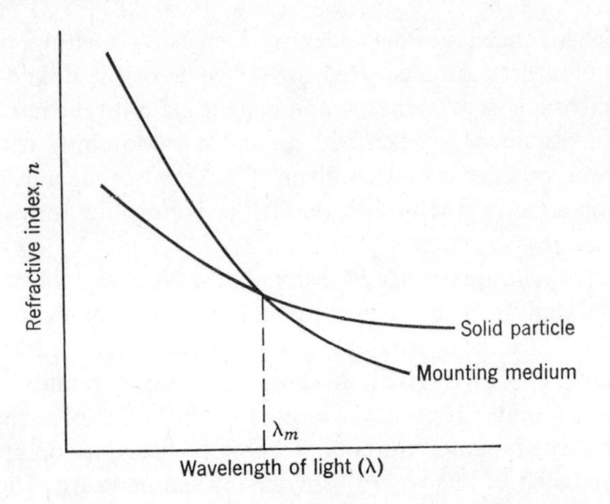

Fig. 3.39 Dispersion curves for particle and mounting medium.

at various angles on either side of the undeviated beam. If the undeviated beam is selected for formation of a microscopical image, the wedge-shaped edges of the particle will have the color of λ_m. However, if the undeviated beam is prevented from contributing to the image, while all deviated beams are allowed to contribute to the image, the particle edges will have a color complementary to that of λ_m.

From the colors exhibited by the edges of the particles, λ_m can be estimated or measured in various ways. With practice, estimates to ± 20 mμ can be made. By using a monochromator λ_m can be measured much more accurately;

Table 3.9 Dispersion Staining Colors

λ_m (mμ)	*Undeviated Beam*	*Deviated Beam*
<400	Black	White
430	Bluish violet	Yellow
450	Blue	Yellow
480	Bluish green	Golden magenta
510	Green	Reddish magenta
550	Yellowish green	Magenta
590	Yellow	Bluish magenta
610	Orange	Blue
640	Reddish orange	Bluish green
660	Brownish red	Pale bluish green
>680	Black	Bluish white

it is the wavelength at which the particle becomes invisible. Estimates are made by comparing the colors with charts showing the spectral colors and their complements, or by using a table such as Table 3.9 which lists the colors by name. Color comparisons can be made more accurate through the use of a comparison microscope [191].

Knowing λ_m is equivalent to knowing the refractive index of the particle if the dispersion curve of the mounting medium is known. By determining λ_m for several mounting media, the dispersion curve for the particle can be determined.

Mounting Media for Dispersion Staining

As stated above, the dispersion curve of the mounting medium must cross that of the particle to produce dispersion staining colors. Mounting media have been suggested by several authors [186, 52, 193]. The suggestion that the refractive index liquids manufactured by R. P. Cargille Laboratories (33 Village Part Road, Cedar Grove, New Jersey) be used seems most practical [193]. These liquids cover a wide range of refractive indices and in general their dispersions are satisfactory. Cargille has recently made available some high-dispersion liquids to supplement their regular liquids. The dispersion curves of these liquids can be derived from the value of dispersion (ν), given on each bottle, using (3.19). Forlini [194] discusses this method and two others, one based on published data [44] and one based on experimental determination using a solid with a known dispersion curve.

Microscopical Equipment for Dispersion Staining

As pointed out above, there are two ways of obtaining λ_m. In one the undeviated beam is used to form the image, and in the other the deviated beams are used. The first method makes use of a small aperture in the back focal plane of the objective. This method gives bright-field illumination and is termed annular screening [44]. The second method makes use of an opaque central screen at the back focal plane of the objective. This gives dark-field illumination and has certain advantages such as higher resolution. A dispersion staining objective that can be used with central screening, annular screening, or without screening is available from W. C. McCrone Associates (493 East 31st Street, Chicago, Illinois).

Crossman [186] uses conventional dark-field techniques for obtaining dispersion staining. Schmidt [192] uses phase-contrast techniques. Cheraksov [190] describes several arrangements which can easily be adapted to conventional microscopes. The equipment described by McCrone [44] and by Cheraksov [190] illuminate the specimen with parallel light, and this is desirable for examining anisotropic crystals.

A microscope can be adapted for central or annular screening by placing the appropriate spatial filter in the back aperture of the objective. For example, as opaque disk with a hole 2 to 3 mm in diameter may be placed on, or just above, the back lens of a 10× objective for annular screening. For central screening the filter can be made by putting a spot of India ink on a cover glass of the right size to fit into the back of the objective. If desired, the black spot can be placed directly on the back lens of the objective. For a 10× objective the spot should be about 4 mm in diameter. Some objectives have an iris diaphragm in their back focal plane which can be used for the annular screening arrangement.

The setup should be adjusted for optimum dispersion color. When observing the back aperture of the objective, the condenser diaphragm is closed until it is just smaller than the central screening filter spot or just larger than the opening of the annular screening filter. The condenser must of course be centered and properly focused.

To use dispersion staining a specimen is mounted in a selected medium, and observed by using either of the methods described. If the dispersion curve of the particle crosses that of the medium, a color will appear. If the curves do not cross, the edges of the particles will be either white on a dark-field for central screening, or black on a light-field for annular screening. If this happens, the Becke test is used to determine whether the n of the medium is too high or too low and on this basis another medium is selected. This process is continued until colors appear. When colors appear, the λ_m can be determined as described above.

Of course it is also possible to heat or cool the mounting medium to enter the dispersion color range. In practice, temperatures should probably be restricted to the range from −50 to +100°C. The n_D of the heated liquids must be calculated from the temperature coefficient given on the Cargille liquid, and the assumption must be made that the dispersion curves do not change with temperature.

Applications of Dispersion Staining

Dispersion staining has been used for two closely related purposes: for the identification of unknown materials and for differential staining for counting analyses described on p. 164.

Unknown substances can be identified by dispersion staining methods if the dispersion data for the substance is available in the literature. Compilations of dispersion data have been published by several authors [44, 194, 195]. The most recent compilation [194] is based on data published by Winchell and Winchell in their book [196], and on data determined by Forlini. Forlini's compilation gives dispersion data arranged in three tables. Table I is arranged alphabetically by chemical formula. Table II is arranged in the order of increasing n_D^{25}, and Table III is arranged in alphabetical order by common name. In the compilation in *The Particle Atlas* [44], there are tables similar to Forlini's Tables I and II, but there are also charts on which are plotted λ_m versus n_D^{25} of Cargille liquids for various compounds.

Isotropic particles have only one refractive index and show only one color in a particular mounting medium. Therefore dispersion color can be used as a rapid means of identifying isotropic materials. The examination of anisotropic materials is more complex, as the refractive index of the particle varies with its orientation and with the vibration direction of the light waves traveling through the crystal. Thus polarized light is used so that the direction of vibration of the light waves is known.

Uniaxial crystals oriented at random exhibit refractive indices varying from ω to ϵ. Also, uniaxial crystals exhibit ω no matter what their orientation. If the *c*-axis of the crystal lies in the plane of the specimen, the refractive index of the crystal will vary from ω to ϵ as the polarizer or stage is rotated. If the *c*-axis lies neither in nor perpendicular to the plane of the specimen, the index of refraction exhibited will vary from ω to some value between ω and ϵ, depending on the orientation of the crystal. Therefore if a preparation of randomly oriented crystals is observed while the plane of vibration of the polarized light is rotated, the dispersion colors will vary between definite limiting colors. One of these limiting colors will represent the dispersion color for ω and the other limiting color the dispersion color for ϵ. If now the individual crystals are observed, each crystal will exhibit at some orientation the limiting color for ω. Thus the microscopist can decide which color extreme represents ω and which represents ϵ. From these data λ_m for both ω and ϵ can be determined. If annular screening is used, the color extreme nearest the red will represent ω when the crystal is optically positive and ϵ when the crystal is optically negative.

Biaxial crystals exhibit a somewhat more complex behavior. If a specimen consisting of randomly oriented biaxial crystals is examined while the plane of vibration of polarized light is rotated, the crystals will exhibit eight possible types of behavior. If two of the three principal indices α and γ lie in the plane of the specimen, two dispersion color extremes will be observed in the crystal. The dispersion color nearest the red end of the spectrum will usually represent α and the color nearest the blue will represent γ. (If the dispersion of the solid is stronger than that of the mounting liquid the colors corresponding to α and γ will be reversed.) All of the colors between the extremes will also appear as the polarizer or stage is rotated. If α and β lie in the plane of the specimen, the color will vary from the red extreme to some intermediate color. If γ and β lie in the plane of the specimen, the color will vary from the blue extreme to the same intermediate color observed when α and β lie in the plane of the specimen. However, while one can recognize color extremes, it is impossible to decide that a given color represents β. If only α lies in the plane of the specimen, the color will again vary from the red extreme to some intermediate color, and if only γ lies in the plane of the specimen, the color will vary from the blue extreme to some intermediate color. If only β lies in the plane of the specimen, there is no way of being certain when one has isolated β. If no principal index lies in the plane of the specimen, the colors will vary without exhibiting the extremes. And, finally, if an optic axis of the crystal is parallel to the optic axis of the microscope, only one color will be exhibited and that color will represent β.

Observing the above behavior enables one to determine that a crystal is biaxial, that the red extreme color corresponds to α, and that the blue extreme corresponds to γ. It is unusual to find the optic axis view, but if it is found the color corresponding to β can be determined. Thus the λ_m-values of α and γ and possibly β can be determined.

From the information derived from dispersion staining experiments, it is possible to identify an unknown if the data for the unknown material is tabulated. The information can also be used to check for the suspected presence of a material. This is true whether or not data for the suspected material are tabulated, provided that a sample of the suspected material is available for comparison. Dispersion staining is an excellent method for determining whether or not two substances are identical.

Complete dispersion curves can be derived if the λ_m-values for a substance are determined for several mounting media, or if they are determined for one mounting medium at several temperatures. From these curves it is possible to calculate n_o, n_D, and n_f for a compound. Formulas for this calculation are given by McCrone [44].

Dark-Field Microscopy

For bright-field microscopy the illuminating system is adjusted so that the field of the microscope is uniformly illuminated and the aperture of the objective is also filled uniformly. Under these circumstances the structures in a specimen absorb or scatter light, removing it from the image-forming beam so that the structure appears darker than the background. However, if the structure scatters or absorbs light to a slight extent, the luminance contrast γ, which is equal to $(L_1 - L_2)/L_1$ (see Section 2, p. 130) is small because L_1, the background luminance, is large, while L_2, the luminance of the structure, is only slightly different from L_1.

For dark-field microscopy the illumination is arranged so that no light enters the objective unless it is scattered by the specimen structure. A structure so illuminated appears bright against a dark field. With this arrangement γ is high even though L_2 may be quite small because L_1 is essentially zero.

A structure need not scatter very much light to be visible against a dark field. Also, as the direct beam does not enter the eye, its luminosity can be made very high so that even structures that scatter very weakly are visible. Structures that are smaller than the resolving-power limit of the microscope are visible. Such submicroscopic structures have an apparent diameter equal to their actual diameters plus the diameter of the Airy disk $(1.2\ \lambda)/(2\ \text{N.A.})$.

Dark-field condensers are designed to produce a hollow cone of illumination, the minimum numerical aperture of which is greater than the numerical aperture of the objective. For objectives having numerical apertures below about 0.65, a condenser with a central stop (Fig. 3.40*a*) may be used for dark-field microscopy. Abbe-type condensers can be used, but aplanatic achromatic condensers give much better results. Either Köhler or critical

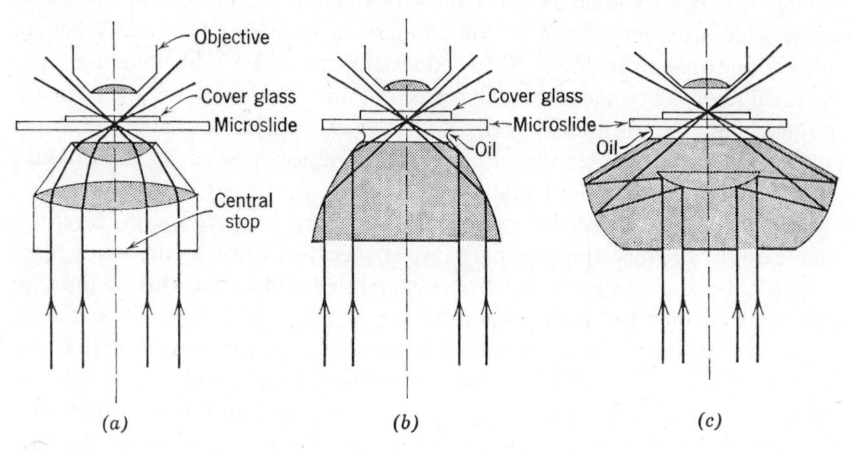

Fig. 3.40 Three common types of dark field illuminators. (*a*) Central Stop; (*b*) paraboloid; (*c*) cardioid.

illumination can be used. The condenser must be focused carefully so that an image of the field diaphragm (F_1, Fig. 3.8) or of the source lies in the plane of the specimen being examined. A central stop can be made of a disk of opaque material supported by a circle of glass which fits into the slot usually found just below the condenser diaphragm. The size of the opaque disk needed for any objective can be determined as follows. A properly illuminated specimen is brought into focus and the eyepiece is removed so that the back focal plane of the objective can be observed. A pinhole eyepiece or an eyepiece microscope (see Section 2, p. 143) can be used to observe the back focal plane. The condenser diaphragm is opened until its image is just larger than the back lens of the objective. The diameter of the opaque disk should then be the same as the opening in the condenser diaphragm.

Condensers designed especially for dark-field microscopy are usually reflecting condensers (Figs. 3.40*b* and *c*). Such condensers are inherently achromatic. The reflecting surface may be a section of a paraboloid, however, it is difficult to grind aspheric surfaces accurately so the quality of paraboloidal condensers is not as good as for condensers with spherical reflecting surfaces. Condensers made with spherical reflecting surfaces have two reflecting surfaces and are called bispheric, bicentric, or cardioid condensers. Descriptions of dark-field condensers can be found in several books [1, 197] and journal articles [198–200].

Dark-field condensers must be carefully focused and centered. They are usually designed for a certain thickness of microscope slide. Too thick a slide makes it impossible to focus on the specimen. Dark-field condensers having N.A. values greater than 1 must be oiled to the bottom of the microscope slide whether or not an oil immersion objective is used. When oil immersion objectives have N.A. values approaching 1.40, the minimum N.A. value of the condenser must exceed 1.4 and the immersion medium and mounting medium must have refractive indices greater than 1.45. Dark-field condensers can be centered and adjusted, using low-power objectives and a specimen that scatters light well.

Specimens for dark-field microscopy must be prepared very carefully. They should be very thin, as particles or structures not in the focal plane scatter light and reduce image contrast. Slides and cover glasses must be very clean or they too scatter light.

Ultramicroscopy is the study of particles or structures smaller than the resolving-power limit of the microscope. Dark-field techniques using condensers such as described above can be used, but Seidentopf and Zigsmondy [201] devised a slit ultramicroscope that illuminates the specimen by projecting an image of a brightly illuminated slit into the focal plane of the specimen. The length of the slit and the optical axis of the illuminating system are both perpendicular to the optic axis of the microscope, thus the portion of the specimen lying in the focal plane of the objective is illuminated by a ribbon of light. The thickness of the ribbon of light and its width can be set by adjusting the width and length of the slit. Knowing the thickness of the illuminated region makes it possible to calculate the volume of the sample examined and quantitative data on the number of particles per unit volume can be obtained. The diameter of ultramicroscopic particles can also be determined by interferometry [202–204].

The chemical microscopist uses dark-field microscopy principally for the study of colloids. Electron microscopy has in many cases replaced dark-field and ultramicroscopy, but when liquid suspensions must be examined the electron microscope is often unsuitable. Amelinckx [205] uses dark-field microscopy to observe dislocations in crystals. Heilman [206] claims that thermal vibrations of the atoms in very pure single crystals can be observed using ultramicroscopy.

Microscopy with Reflected Light

Many objects are best examined by using light reflected from their surfaces. For opaque objects this is the only method that can be used. This type of illumination is useful primarily for examining surface structures.

There are two general types of reflected light microscopy, one analogous to bright-field microscopy and one analogous to dark-field illumination. These names refer to the appearance of a flat reflecting surface; in bright-field illumination such a surface looks bright and in dark-field illumination it looks dark. A rough surface appears dark in bright-field and bright in dark-field reflected illumination.

Reflected Bright-Field Illumination

A common optical system for reflected bright-field illumination is shown in Fig. 3.41*a*. This type of illuminator is commonly called a vertical illuminator. A semitransparent reflector, often a thin piece of glass, is arranged at a 45° angle above the objective lens. A beam of light from the light source is reflected downward through the objective, illuminating the specimen. Light rays from the specimen pass back through the objective and through the reflector to form the image. Thus the objective acts as the condenser. The illuminating system may be arranged for Köhler or critical illumination.

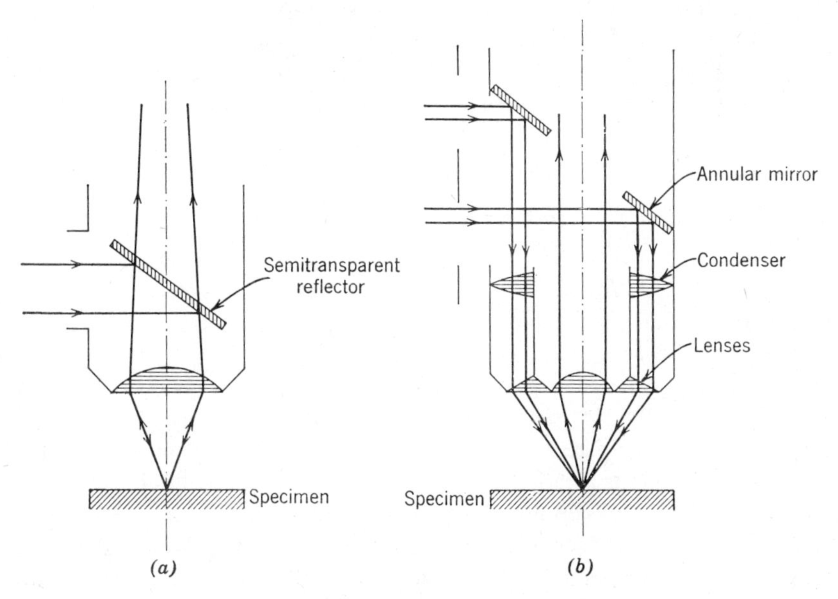

Fig. 3.41 Illuminators for reflected light microscopy. (*a*) Bright-field; (*b*) dark-field.

Often the entire illuminating system is an integral part of the vertical illuminator. Modern vertical illuminators have built-in aperture and field diaphragms and these should be adjusted just as they are for transmitted illumination while using a first surface mirror or a smoothly polished surface at the specimen position.

There are other designs of vertical illuminators, some serving special purposes such as polarizing microscopy, phase microscopy, and interference microscopy. The use of polarized light for reflected light microscopy has already been discussed on p. 238. Reflected polarized light has been used for the study of opaque minerals [25, 26, 121] and metals and alloys [207, 208]. Phase microscopy and interference microscopy are discussed briefly on pp. 278 and 270. A good discussion of the various types of vertical illuminators can be found in Françon's book [2] and in Chamot and Mason's *Handbook of Chemical Microscopy* [1].

Reflected Dark-Field Illumination

Figure 3.41*b* is a schematic diagram of an arrangement for producing reflected dark-field illumination. Light from the illuminator enters from the side and is reflected downward by an annular mirror through a condenser which focuses the illumination on the specimen. Structures that scatter light into the objective appear bright in the image, smooth areas which reflect the light specularly appear dark in the image.

The apparatus described above gives conical dark-field illumination Illumination from a single direction or from several directions simultaneously can be obtained by directing spot lamps at the surface of the specimen. This technique is very simple, but it is difficult to use with high-power objectives. It is the common way of illuminating specimens being examined with a Greenough or other stereomicroscope.

Ultraviolet and Infrared Microscopy

When microscopy using ultraviolet illumination was first investigated by Köhler [209], the purpose was to attain higher resolving power. The technical difficulties encountered apparently discouraged application of this technique. It was necessary to record the image photographically and to focus by trial and error. Later work with ultraviolet microscopy was primarily spectrophotometric [210, 211]. Materials that are colorless often absorb in the ultraviolet and can be identified and determined quantitatively using spectrometric measurements. These techniques have made it possible to photograph live, unstained tissues and to locate and determine the concentration of certain cellular constituents. Certain materials that are opaque in the visible range become transparent in the ultraviolet, so that their internal structure can be observed.

Infrared radiation, having a longer wavelength than visible radiation, has the disadvantage that the resolving power of infrared microscopes is relatively poor. Therefore, as with ultraviolet microscopy, the principal use of infrared microscopy is in the spectrometric identification and determination of compounds. Also, many opaque materials are transparent in the infrared and their internal structures can be investigated using infrared microscopy.

Ultraviolet radiation for early investigations was usually obtained from metal arc lamps, principally magnesium or cadmium arcs. Below 2000 Å a spark discharge between tin or zinc electrodes is the best source. At present, high- and medium-pressure enclosed gaseous arcs are used. Mercury arcs are common and they emit a number of strong lines. Hydrogen, xenon, and deuterium arcs provide a continuous spectrum. Filters made of glass or solutions of various compounds can be used to isolate certain lines or portions of the spectrum, but quartz prisms or grating monochromators are so convenient and easy to use that they have largely replaced filters.

Most infrared microscopy is carried out using tungsten lamps or globar sources used in conjunction with a monochromator. Highly intense infrared sources include carbon arcs, zirconium arcs, and tungsten glowers.

Objectives made of optical glasses can be used in the near ultraviolet and the near infrared portions of the spectrum. They transmit radiation with wavelengths from about 0.3 to about 1.5 μ. If lenses are made of quartz, fluorite, or lithium fluoride, the range of transparency can be extended, but below about 0.2 μ it is preferable to use reflecting objectives. At wavelengths below 0.2 μ it is necessary to have the entire system in a vacuum or a nitrogen atmosphere to prevent absorption of the radiation by the air. With reflecting objectives in a vacuum system, it is possible to extend ultraviolet microscopy to about 0.1 μ.

Reflecting objectives have limitations on their numerical apertures which can be overcome if both reflecting and refracting elements are used to construct the lenses. Thus lenses consisting of reflecting elements and lenses made of silver chloride or thalium bromide iodide have been constructed with N.A. values up to 1.5. These lenses are useful in the range from 2 to 20 μ.

Reflecting objectives are achromatic and can be focused using visible light. However, it is desirable to be able to convert ultraviolet or infrared images into images that can be observed visually. Fluorescent screens have been used for observing ultraviolet images, but the results are generally unsatisfactory. An image orthocon can be used with a television display tube and appropriate circuitry. A simpler arrangement uses an ultraviolet image converter built by RCA. Bausch & Lomb have built an apparatus using the RCA tube to focus the ultraviolet image for photomicrography. Image-converter tubes suitable for use in the infrared can be obtained from Bausch & Lomb.

A good brief description of lenses and image converters can be found in Françon's book [2]. Casperson [212] has given a brief review of ultraviolet microscopy. Ultraviolet and infrared microscopy have been applied primarily to identify substances in biological tissues and to determine the concentration and distributions of these substances. However, ultraviolet microscopy has also been used for examining resins [213], examining metals [214], and the observation of sintering at high temperatures [215]. Transmission ultraviolet microscopy has been used to study the internal structure of the alkali metals, silver, and silver-based alloys [216]. In a similar manner infrared microscopy has been used to examine the interior structure of silicon [217] and coal [218].

Microspectrophotometry and Microphotometry

In recent years microscopy has become increasingly more quantitative. Microspectrophotometry and microphotometry are examples of this trend. Microspectrophotometry provides absorption spectral data on very small objects, of the order of 1 μ in diameter. Spectral data supply information on the composition and concentration of substances in very localized areas, thus making it possible to correlate microscopical structure with composition. Microphotometry is a simpler quantitative technique for measuring absorptivity of a specimen for a selected spectral range.

Microspectrophotometers can cover the spectral ranges already discussed, namely, the ultraviolet, the visible, and the infrared. The optical systems employed are often composed of reflecting-type lenses because their achromatism permits scanning of the spectrum without varying the focus of the microscope, however, ordinary glass microscope lenses can be used if only the visible range is of interest.

In addition to the microscope itself, a microspectrophotometer must have components that make it possible to select and vary the wavelength of the radiation, to select the structure to be examined, to isolate the structure selected, and to measure and record the radiation transmitted or reflected by the structure selected. The wavelength is selected and varied by a monochromator, which may be placed between the radiation source and the microscope or between the microscope and the photometer. It is generally preferable to place the monochromator between the light source and the microscope because in this way the amount of stray light reaching the image is minimized.

A variety of apparatuses has been devised for selecting and isolating the structure to be examined. If the photoreceptor is photographic film, the structure and its surroundings are visible in the photographic image and all that is necessary is to measure the blackening of the film within the area of the structure. However, if the photoreceptor is a photocell of some kind, the image is usually observed with an eyepiece while a mask (field aperture) is inserted and adjusted so as to allow only the radiation passing through the structure to reach the photoreceptor.

Two types of photoreceptors are commonly used: photographic film and photoelectric cells. Photographic film has the advantage that the transmission of the entire area of the specimen in the field of the microscope is recorded, and the transmission of any structure can then be determined photometrically. However, the photograph must be taken with monochromatic radiation, and variations in transmission with wavelength can be obtained only by taking numerous photographs. Photographic photometry requires that the exposure and development of the film be done carefully using standardized techniques. It is also necessary to obtain sensitometric data for the film at each wavelength used.

Photocell receptors have the advantage that the wavelength of the radiation can be varied continuously while the transmission is measured and recorded. However, only a single selected area of the specimen can be examined at one time. The sensitivity of the photocell receptors also varies with wavelength, and this variation must be taken into account.

Microspectrophotometers can be classified into two general groups, single-beam and double-beam instruments. In single-beam instruments the radiation travels from the light source through the specimen and on to the photoreceptor. With this type of instrument, it is necessary that the light source and the photocell with its associated electronic equipment be highly stable. Fluctuations in the luminosity of the light source or in the sensitivity of the photoreceptor cause the instrument to be inaccurate. With a single-beam instrument, it is necessary to run a standardization spectral curve using no sample, or some standard sample. This standard curve is then compared with the curve obtained for the sample so that the absorption can be obtained. The instrument must be stable over the period of time required to run both the standardization curve and the sample curve.

Double-beam instruments have a reference beam which originates from the source and is conducted to the photoreceptor without passing through the specimen structure being examined. The reference beam may bypass the microscope entirely or it may pass through an area of the specimen adjacent to the structure being examined. The reference beam and the sample beam are compared by any number of methods so that the ratio of the intensities of the two beams is the value read out and recorded. Thus the two-beam instrument eliminates the problems arising from instabilities in the light source and photoreceptor, and at the same time automatically compensates for the spectral sensitivity of the photoreceptor.

Sources of errors in microspectrophotometry that are not eliminated by the two-beam arrangement are stray light, the numerical aperture of the microscope, aberrations in the lenses (particularly the objective lens) and, for small structures, the diffraction effects inherent in optical systems.

Microspectrophotometry is a technique that requires considerable knowledge and skill on the part of the operator. It also requires equipment which may be quite expensive. Although microspectrophotometric equipment is commercially available, many investigators construct their own instruments. Discussions of microspectrophotometry can be found in several recent articles and books [2, 219–222]. Many articles describing instruments and their applications are reviewed by Cocks [55].

It should be pointed out that microspectroscopy, that is, the visual observation of the spectral quality of structures in the microscopical image, can be carried out using spectroscopic eyepieces. The Jelley microspectrograph [1, 67, 73, 75] is an eyepiece instrument, and some of its uses are described on pp. 212 and 238. Microspectroscopes or microspectrophotometers can be used for reflected light as well as for transmitted light microscopy.

Fluorescence Microscopy

When a specimen absorbs radiant energy it may fluoresce, that is reradiate the energy at another wavelength. Many substances fluoresce, and this characteristic can be used to identify them and to determine their concentration. However, fluorescence is often greatly affected by the presence of small amounts of impurities or degradation products. This sensitivity to small amounts of material can be used to advantage in detecting and determining these materials, but considerable care must be taken in setting up and checking proposed analytical procedures.

Substances that do not fluoresce naturally may be made to do so by staining procedures. The stains used for this purpose are called fluorochromes. Fluorescent staining is widely used in biology and medicine because many fluorochromes are specific for certain tissues and cell components. Fluorochromes have also been used as vital stains because they can be used at very low concentrations.

Specimens to be examined for fluorescence are illuminated with the exciting radiation, which is usually the ultraviolet or the blue-violet portion of the spectrum. Often only a small part of the exciting radiation is converted into fluorescent radiation, so powerful light sources may be necessary. Many of the fluorescent stains excited by the blue-violet or near ultraviolet portion of the spectrum, and for these materials incandescent lamps are used. For this portion of the spectrum, glass lenses are sufficiently transparent to be used in the illuminating system. For materials that require exciting radiation of shorter wavelength, the lamps mentioned on p. 265 for ultraviolet microscopy are used in conjunction with quartz, fused silica, or reflecting-type condenser lenses.

Bright-field illumination may be used, however, many lamps give off light at other than the exciting wavelengths and the fluorescence of the specimens may be obscured by the general illumination. To prevent this difficulty the exciting radiation is usually isolated by the use of filters or a monochromator, and a barrier filter is used in the imaging system to prevent the exciting radiation from reaching the image. Under these conditions the fluorescent portions of the specimen appear as self-luminous against a dark field. A typical setup for excitation in the near ultraviolet includes a mercury arc lamp with a Corning 5860 excitation filter, an ordinary glass lens condenser, and a Wratten no. 22 barrier filter.

Dark-field illumination is particularly advantageous for fluorescence microscopy. Either transmission or reflection dark-field illumination may be used, depending on the type of specimen to be examined. The use of excitation and barrier filters is advisable and is often necessary.

Microspectrophotometry is a valuable tool for studying fluorescence. A monochromator between the light source and the specimen makes it possible to determine which wavelengths absorbed by the specimen cause fluorescence. A spectroscope or spectrometer can also be used to analyze the spectral quality of the fluorescent radiation.

Ordinary specimens are mounted on glass slides, although fused silica slides may be necessary for ultraviolet excitation. Mounting media and immersion oils should not be fluorescent. Paraffin oil, glycerin, and water are satisfactory. Fluorite and apochromatic objectives may fluoresce because of their fluorite lenses. Also, the fluid in the eye fluoresces when excited with near-ultraviolet radiation. Fluorescence in the microscope or the eye can be prevented by using cover glasses made of barrier filter material. A piece of Wratten no. 2B filter can be used for this purpose. Silver mirrors absorb in the ultraviolet so the mirror of the microscope should be replaced by a first surface aluminized mirror or by a silica prism.

Ultraviolet radiation is harmful to the eyes, so ultraviolet-absorbing filters should be used over the eyepieces of the microscope. The lamp housing and the illuminating system may allow ultraviolet to escape into the room. The presence of stray ultraviolet can be detected by using a fluorescent test paper made by soaking filter paper in a saturated solution of anthracene in benzene and allowing it to dry. The test paper is used in a darkened room to explore for stray radiation.

Fluorescence microscopy has been applied to the solution of many chemical problems, including studies of colloid systems [223], the structure of natural and artificial cellulose fibers [224], the identification of polyvinylpyrrolidone in tissues [225], fossils, rocks, and minerals [226, 227], and wood and fiber structure [228]. More general information along with further references to applications can be found in a number of books and articles [1, 229–231].

Interference Microscopy and Interferometry

Principles of Interference Microscopy

The eye and photographic film are sensitive to luminance and color contrast only. When light passes through a specimen, it interacts with the specimen in three ways (Section 2, p. 130). The light may be absorbed, reflected, or refracted. In ordinary bright-field microscopy, objects become visible because they refract or reflect light outside the aperture of the objective. Reflection and refraction effects can scatter enough light outside the aperture of the objective only when the refractive index of the mounting medium is quite different from that of the specimen. Colorless objects, whose refractive indices are close to those of the surrounding media, are invisible in ordinary bright-field illumination. Such colorless objects do, however, cause the rays that pass through them to be retarded or advanced in comparison to the rays passing through the surrounding media. In other words, if a coherent wave front passes through such a specimen, the portion of the wave passing through the object will be out of phase with that portion of the wave passing around the object. Such phase shifts are not detectable by the eye or by photographic film. However, if the out-of-phase rays are superimposed in an appropriate way, they can interfere with each other, producing luminance contrast in the image.

Microscopes that produce luminance or color contrast in the images of phase objects by causing the beams put out of phase by the specimen to interfere are called interference microscopes. There are many kinds of interference microscopes, but they all have several features in common. A coherent beam of light is divided into two beams. One of these beams, the sample beam, passes through the object; the other, the reference beam, generally but not invariably passes around the object. The two beams are then recombined so that they interfere.

THE POLARIZING MICROSCOPE

Polarizing microscopes and phase microscopes are interference-type microscopes, although for historical reasons, and because of their specialized usage, they are usually considered separately. In a polarizing microscope a beam of plane-polarized light is resolved by an anisotropic crystal into two beams whose vibration directions are perpendicular one to the other. One of these beams is retarded behind the other because the refractive indices of the crystal for the two vibration directions are different. Upon leaving the crystal the phase shift between the beams persists, but interference does not occur because the beams are not vibrating in the same plane. The analyzer of the polarizing microscope causes the two beams to vibrate in the same plane and interference occurs. As a result of this interference, the crystal exhibits either luminance contrast if the illumination is monochromatic, or color contrast if polychromatic illumination is used.

INTERFERENCE MICROSCOPE CONSTRUCTION

A coherent beam of light can be divided into two beams by any of four general methods: the use of semitransparent beam dividers, the use of diffraction, the use of double refraction by anisotropic crystals, and by simple refraction or reflection. The use of semitransparent beam dividers needs no further explanation. Diffraction beam dividers use a diffraction grating that produces diffracted beams of various orders. Any two orders are coherent and may be used, for example, the two first-order beams or a first- and a second-order beam. One method of using anisotropic crystals to produce two coherent beams has already been mentioned in connection with the polarizing microscope. In this instrument the specimen itself serves as the beam divider and the beams travel very nearly along the same path, although they are polarized in planes perpendicular to each other. However, doubly refractive crystals can be used to produce separate beams, for example, a plate of calcite or quartz cut at an angle to the optic axis produces two beams whose separation depends on the thickness of the plate and on the angle between the normal to the plate and the optic axis of the crystal. These beams are polarized in planes at right angles to each other. Other devices such as Savart plates or Wollaston prisms produce separate beams of polarized light. Beam separation can also be produced by simple refraction or reflection effects. Refraction and reflection in the specimen itself produce the beam separation in the phase microscope. However, the beams can be separated before or after reaching the specimen by means of a Fresnel biprism, Fresnel mirrors, or by a lens and apertures.

After passing the specimen the two beams are usually recombined so that they interfere. The devices used for recombining the beams are identical with those used for beam separation. The recombining device may be the same type as the beam divider or it may be different. A number of instruments dispense with beam-recombining devices if the beams are not separated very far and if the lateral coherence of the two beams is adequate.

CLASSES OF INTERFERENCE MICROSCOPES

The distance by which the two beams of an interference microscope are separated can serve as a basis for classification. Classification according to the separation of the beams is not trivial, as the optical and mechanical design, the appearance of the specimen image, and the use to which an interference microscope can be put all depend on the separation of the two beams. Figure 3.42 shows some of the features of the four classes of interference

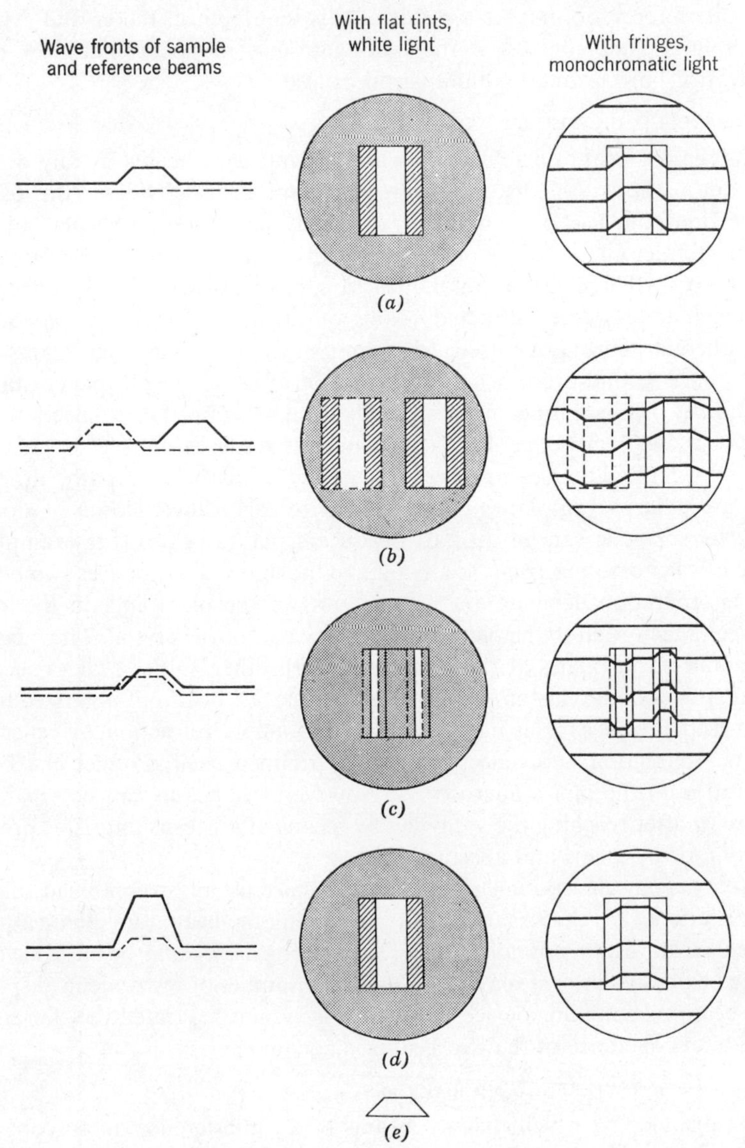

Fig. 3.42 Schematic diagrams of wave fronts and microscopical appearance of a sample for the various types of interference microscopes. (*a*) Complete separation of reference beam; (*b*) complete image duplication; (*c*) partial image duplication; (*d*) complete image superposition; (*e*) end view of object.

microscopes. There are many optical and mechanical designs for interference microscopes. Most of these designs are discussed in books by Krug et al. [9], Francon [2], and Martin [3].

Microscopes with Complete Separation of the Reference Beam. The interference microscope, having complete separation of the reference beam (Fig. 3.42*a*), is simple in principle, although the optical systems must be highly perfect. A coherent beam is divided into two beams, for example, by semitransparent reflectors as shown in Fig. 3.43. One of the beams passes through the specimen, the other through a reference similar to the specimen except for the absence of the object to be examined. There are objectives for both the reference and the sample beam. The beams from these objectives are recombined by a system of prisms and a semitransparent film. In the eyepiece images of the specimen and the reference are superimposed. If then the only difference in optical path length between the two beams is caused by

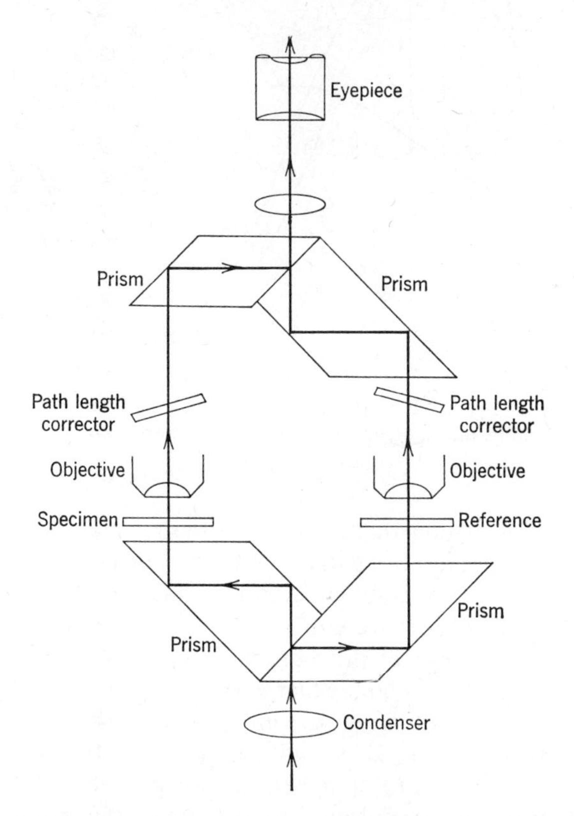

Fig. 3.43 An interference microscope with complete separation of reference beam. This design is used by E. Leitz in their interference microscope.

the object, the wave fronts for the two beams should be superimposed as shown in Fig. 3.42*a*. The refractive index of the object is shown as being lower than that of the surrounding mounting medium, and therefore the wave front that has passed through the object is advanced with respect to the wave front of the portion of the beam that passed through the surrounding medium. The reference beam wave front is shown as lagging slightly behind the wave front of that portion of the beam passing through the surrounding medium. The path length corrector plates shown in Fig. 3.43 can be used to make the path lengths exactly equal or to introduce any desired differences in path length.

Figure 3.44 shows the phase relationships that might exist at the image when the specimen is in focus and illuminated with monochromatic light.

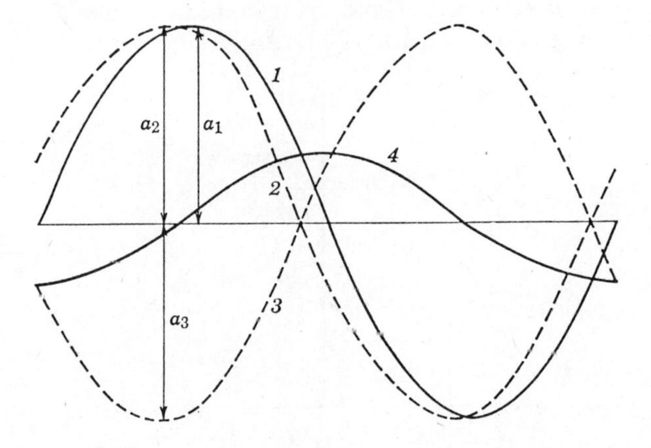

Fig. 3.44 Phase relationships in an interference microscope. *1*, wave passing through the object; *2*, wave passing through the surrounding mounting medium; *3*, reference wave retarded $\lambda/2$ behind wave *2*.

Consider first the situation that exists when the reference beam is cut off. Within the image of the object, the intensity of light is a_1^2, a_1 being the amplitude of wave 1. As the object is colorless and refraction and reflection effects are negligible, the intensity of light in the image of the object is the same as that in the image of the surrounding areas (a_2^2) and there is no contrast in the image. Now suppose that the reference beam is added and that it is in phase with the wave front passing around the object, wave 2. Being coherent, these waves can be added and the light intensity in the background around the object becomes $(a_2 + a_r)^2$. In the image of the object, the same kind of addition can be carried out, but because of the phase shift caused by the object $(a_1 + a_r)^2$ is slightly smaller than $(a_2 + a_r)^2$. Thus there is some image contrast, but the contrast $\gamma = [(a_2 + a_r)^2 - (a_1 + a_r)^2]/(a_2 + a_r)^2$ is low. However, the phase relationship between the sample beam and the reference beam can be adjusted using the path length correctors (Fig. 3.43). With proper adjustment of these correctors, the reference wave can be retarded exactly $\lambda/2$ behind the background wave (wave *2* in Fig. 3.44). Wave *3* in Fig. 3.44 represents the reference wave retarded by $\lambda/2$. Under these conditions, if the amplitude a_3 of this wave is equal to the amplitude of a_2 of the background wave, complete destructive interference occurs and the background intensity $(a_2 + a_3)^2$ is zero. In the image of the object, addition of waves *1* and *3* yields wave *4*. Thus the intensity of light in the image of the object is $a_4{}^2$, and the image contrast

$$\gamma = [(a_2 + a_3)^2 - (a_1 - a_3)^2]/(a_2 + a_3)^3 = a_4{}^2/{\sim}0$$

is high. The intensity of light in the background is not zero because of scattered light and imperfect optical systems.

If white light is used for illumination, Newton's series of colors will appear, as the retardation between the reference beam and the sample beam is increased from zero. With zero retardation the background illumination is white. When the retardation reaches $\lambda/2$ for violet light, violet light interferes destructively and the background appears yellow, and so on through Newton's series of colors. Since the retardation of the specimen differs from that of the background it is colored differently, producing color contrast in the image. The most sensitive background color is first-order red.

If the wave fronts of the specimen beam and reference beam can be tilted with respect to one another, interference fringes can be produced in the field of view. This can be done by inserting a glass wedge into the reference beam or the sample beam or both. Fringes can be produced using either monochromatic or white light. Fringes are often advantageous when measurements are being made, as the displacement of the fringes crossing the object is a direct measure of the optical path difference between the background and the object.

Complete Image Duplication Microscopes. The other classes of interference microscopes use only one objective lens. The separation of the reference and sample beams must be less than the diameter of the field of view of the microscope. Therefore the separation depends on the magnification of the objective. Complete image duplication as in Fig. 3.42*b* depends on the size of the object as well as the magnification. This places some serious restrictions on the type of object that can be examined. If accurate measurements of path length differences are to be made, the specimen must be an isolated object surrounded by the mounting medium or at least it must have an edge such that the reference beam can pass by the object. If the microscope is used only to produce image contrast and not to make measurements of path length difference, the type of specimen that can be examined is less restricted.

Complete image duplication microscopes also usually show an out-of-focus image of the object as indicated in Fig. 3.42*b*. The degree to which this secondary image is out of focus depends on the design of the particular microscope. If measurements are to be made, the out-of-focus image should not be superimposed on the images of other objects that are in focus.

Complete image duplication interference microscopes are usually polarizing-type interference microscope, that is, they use doubly refractive crystals as beam dividers. With this type of microscope the beams that are reunited spatially by the beam-recombining device still do not interfere until their planes of vibration are made identical by an analyzer. The resulting interference effects are therefore somewhat different than those described above for an instrument with complete separation of the reference beam. The background and reference beams are polarized at right angles to each other, and if the phase retardation is zero and the analyzer is crossed with respect to the polarizer, the background is dark rather than light. As the retardation between the background and reference beams is increased, the background becomes lighter until the retardation reaches $\lambda/2$. When white light is used, the regular series of polarization colors is observed. If the polarizers are parallel, a complementary series of colors starting with white is observed. With polarizing-type interference microscopes, regular compensators can be used to determine path length differences.

Partial Image Duplication Microscopes. If the reference and object beams are very close together, primary and secondary images are partially superimposed on one another as is shown in Fig. 3.42*c*. In all areas where the primary and secondary images overlap the color, or shading, is identical with that of the background. This shading or color depends on the phase shift (retardation) between the reference and the background wave fronts, and if the microscope is a polarizing type a zero retardation will give a black background. Whenever the specimen slopes in the direction of image duplication, the retardation changes and a difference in shading or color is observed. In the diagram in Fig. 3.42*c*, the edges slope in the direction of image duplication and thus the sloping right and left edges are shaded or colored differently from the background. However, the top and bottom edges of the object do not slope in the direction of image duplication and they are therefore the same shade or color as the background. The central part of the object has zero slope and it is also the same shade or color as the background. For a given retardation setting of the microscope, the angle of a slope in the direction of image duplication is indicated by the shade or color at that point.

In most partial image duplication microscopes, the separation of the sample and reference beam is quite small, commonly less than the resolving power of the microscope. In the instrument designed by Françon [2], beam separation occurs after the objective. In fact, the apparatus is built into an eyepiece. Another eyepiece interferometer designed by Johansson [232] uses the complete image duplication method.

Fringe systems can be developed in partial image duplication systems just as they are in the full image duplication system (Fig. 3.42*c*). However, the displacement of the fringes indicates the slope of the object surface rather than the optical thickness of the object.

Complete Image Superposition Microscopes. The most important examples of this class of interference microscopes are the phase microscope and the multiple-beam interference microscope. Objects having top and bottom surfaces nearly parallel exhibit interference fringes. Such fringes can be seen without a microscope as, for example, in slides and cover glasses. Interference effects arise between a portion of beam that traverses the specimen once and another portion of the same beam that is reflected from the top surface back to the bottom and then back out the top surface. Interference effects from this type of system are faint and difficult to see, particularly when the object is viewed in transmission. The transmission fringes are difficult to observe because the unreflected beam is so intense compared with the reflected beams.

Objects are seldom of such shape that their top and bottom surfaces are parallel enough to produce interference effects. Also, the reflectivity of the surfaces is usually unsatisfactory for optimum production of interference effects. To overcome these difficulties the object can be mounted between a slide and a cover glass that have semitransparent layers deposited on them as shown in Fig. 3.45. A single beam T will then be divided by reflection, as shown, into a number of beams T_0, T_1, T_2, T_3, and so on. Each succeeding beam has traveled farther than the preceding beam by approximately two times the optical thickness of the specimen in that spot. Thus these beams can

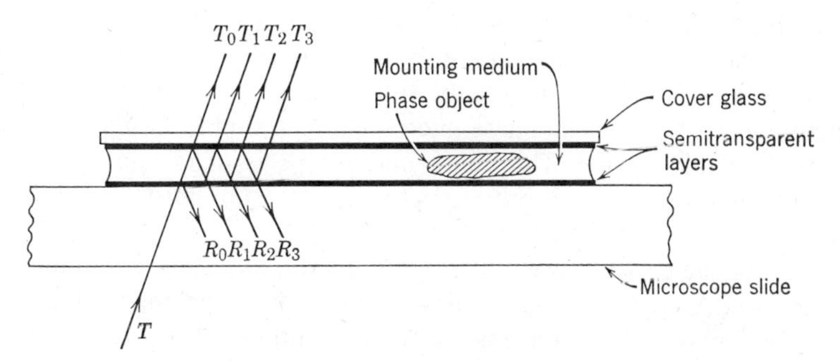

Fig. 3.45 Multiple-beam interference.

interfere. The reflected beams R_0, R_1, R_2, R_3, and so on, can also interfere. If the illuminating beam strikes the reflecting surfaces normally, the various transmitted and reflected beams are superimposed.

Fringes produced by multiple beams are much narrower than those produced by two-beam instruments. Narrow fringes are advantageous for making measurements. The greater the number of beams, the narrower the fringe, and in good multiple-beam instruments there may be 50 to 100 beams. To obtain large numbers of beams, the semitransparent layers must have high reflectance, for example, 95 to 98%, and very low absorption. The best films are multilayer films, but evaporated films of aluminum are also satisfactory.

A phase object in the mounted specimen causes the optical path length between the reflective layers to change and therefore the object appears in contrast in the image. Since the fringes are narrow, small retardations cause large changes in shading and the instrument can be quite sensitive.

Multiple-beam interference microscopes have a number of requirements which restrict their use considerably. The illumination must be highly monochromatic, the illuminating beams should be nearly normal to the reflecting surfaces, and the reflecting surfaces should be very close together. The second requirement demands the use of objectives with low numerical apertures, so multiple-beam interferometry is usually done at low magnification. Under the most favorable conditions, the multiple-beam interference microscope is sensitive to phase shifts of a few tens of angstroms.

Multiple-beam fringe systems can be produced simply by making the space between the reflecting surfaces wedge-shaped. In transmission multiple-beam interferometry, the fringes represent contours of the same optical path length in the specimen.

Phase Microscopes. The phase microscope deserves special attention because of its widespread use, particularly in biological microscopy. Living tissues are generally phase objects, and as already pointed out, the images of such objects formed by ordinary microscopes are completely lacking in contrast. Staining or other specimen preparative techniques are likely to harm living tissues, so the use of interference microscopes is extremely valuable to biologists. The phase microscope is a simple, easily used type of interference microscope. The phase microscope is of course useful to the chemical microscopist or other microscopists who must examine phase objects.

A part of the optical system of a phase microscope is shown schematically in Fig. 3.46. An opaque diaphragm with an annular opening is placed at the lower focal plane of the condenser (A_2). The illuminating system is arranged to give Köhler illumination. An image of the annular diaphragm is formed in

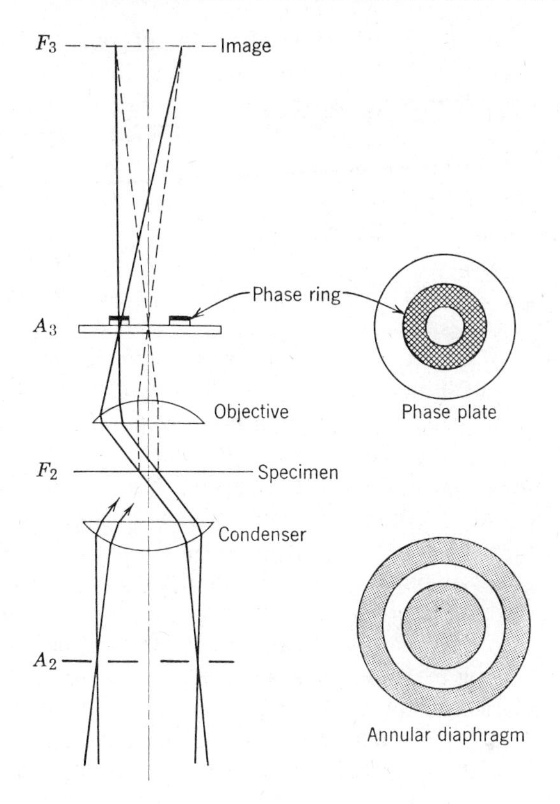

Fig. 3.46 The optics of a phase-contrast microscope.

the back focal plane of the objective (A_3). A phase plate placed at A_3 has a ring that matches the image of the annular diaphragm. The properties of this ring are specified later. The rest of the phase plate is transparent.

Consider an extended phase object of uniform thickness. Rays passing through this object some distance from its edge are not deviated. They strike the phase ring and continue through it to form an image of the object. Rays that pass through the area surrounding the object at some distance from the edge of the object also pass through the phase ring and form the image of the background. However, the rays that pass near the edge of the object are diffracted. Both the background rays and the rays that have passed through the object are diffracted at the edge of the object. Most of the diffracted rays pass through the phase plate outside the phase ring. Of course, some of the diffracted rays do pass through the phase ring. The rays passing through the phase ring can be considered the reference beam while the diffracted rays can be considered to be the sample beam. All of the deviated rays are focused at or very near the image of the edge. Because of this, the sample and reference beams have appreciable intensity only near the image of the edge of a phase object. Outside the edge of the image, there is an undeviated background beam that has passed through the phase ring and a deviated sample beam that has not. Inside the edge of the image of the object, there is an undeviated sample beam that has passed through the phase ring and a deviated reference beam that has not.

Consider the image area at which there is an undeviated background beam and a deviated sample beam that has undergone a phase shift. In the absence of the phase plate, the situation illustrated in Fig. 3.44 applies. The object wave (*1*) can be considered as the sum of the background wave (*2*) plus another wave (*4*) which differs in phase by $\lambda/4$. Now if the background wave can be retarded by $\lambda/4$, the phase difference between it and wave *4*, which is associated with the deviated sample beam, becomes $\lambda/2$. Under these conditions waves *4* and *2* can interfere destructively. Such a retardation of the background wave can be accomplished by means of a phase plate with a phase-retarding layer of the proper thickness on the phase ring. Since waves *2* and *4* have different amplitudes, they do not completely cancel each other. By putting an absorbing layer on the phase ring, the amplitude of wave *2* can be reduced to equal that of wave *4*, and complete cancellation of the two waves can occur giving maximum edge contrast in the image.

In the above discussion the area just outside the image of the object has been considered. This area is darker than the background or the central area of the object. The area on the other side of the edge in the image is lighter than the background. Thus the image of an object is outlined by dark and light halos. For fine structures the inner halos merge so that the structure appears light and is surrounded by a dark halo.

One type of phase plate is described in the above paragraphs, but other types are possible, for example, the phase of the deviated wave rather than the background wave could be retarded by $\lambda/4$. With this kind of phase plate, wave *4* and wave *2* reinforce each other and the type of edge contrast in the image is reversed.

The discussion of phase microscopy given above is very brief and overly simplified. More complete discussions are given by Françon [2], Martin [3], and Bennett et al. [8].

Reflected Light Interference Microscopes

The principles of interference microscopy can be applied to reflected light microscopy as well as to transmitted light microscopy. Of course the instruments are designed differently, but the same types are used. Reflected light interference microscopes are used primarily for examining the surface contours of a specimen. Complete descriptions of these instruments can be found in the references already cited [2, 3, 8, 9].

The Applications of Interference Microscopy

One of the principal uses of the interference microscope is the observation of colorless specimens producing small phase shifts. However, the fact that most interference microscopes are designed so that phase shifts can be introduced into the sample and reference beams at will, makes it possible to use these microscopes to measure optical path differences. Because the optical path length consists of the thickness of the specimen multiplied by its refractive index, it becomes possible to measure either specimen thickness or refractive index if the other is known or can be determined. A number of specific methods for measuring refractive indices using interference microscopes are given by Francon [2]. Fundamentally, these methods depend on measuring the difference in phase retardation between the object and its background. If the flat tint method is used, phase retardations are measured essentially by observing color differences between the specimen and translating these differences into retardation, using charts that relate the colors in Newton's series with retardation. The use of compensators of various types such as the path length correctors in Fig. 3.43 make this task easier.

The other method of determining phase retardation is by measuring the displacement of an interference fringe that passes over both the object and the background. This method is somewhat more accurate than estimating colors, but its accuracy varies with the type of fringe and with the method of measuring the displacement. Some of the possible types of fringe displacements are shown in Fig. 3.42. Care must be taken to be certain of the order of the displaced fringe if it cannot be traced completely over the object.

The refractive index of a solid-phase object can be determined by comparing its retardation with the retardation of an equal thickness of a mounting medium of known refractive index. Since only path differences are measured, the object can be completely surrounded by the mounting medium. If the thickness of the object is not known, it can be determined by measuring its retardations when immersed successively in two different mounting media. By successive immersion in two different mounting media, sufficient data can be obtained for solving simultaneous equations and thus determining both the thickness and the refractive index of the object.

The immersion technique can be reversed, and if the refractive index of the solid is known that of the liquid can be determined. Françon [2] suggests the use of glass spheres of known refractive indices to determine the refractive index of liquids, such as molten eutectics. Whereas Kofler's method (see p. 244) determines the temperature at which the index of the melt matches that of the standard, Françon's method makes it possible to determine the refractive index at any selected temperature using very few reference standards. The diameter of a sphere can be measured, hence its thickness determined; thus by measuring the displacement of fringes caused by the sphere immersed in a melt, the refractive index of the melt can be determined.

The refractive indices of liquids can also be measured using cells of known thickness or a wedge-shaped cell of known wedge angle. Usually these cells are used in conjunction with fringe displacement methods of measuring retardation.

The ability of the interference microscope to measure refractive index and thickness of objects has been applied to the solution of a number of chemical and physical problems. For example, slopes and changes in the slope of objects can be measured. This makes possible the measurement of contact angles between liquids and solids [2]. Biologists have used interferometry to measure the dry mass of living biological materials [233, 234]. The interference microscope can be used to observe concentration gradients in systems in which diffusion is occurring. Interferometry, especially reflected light interferometry, has been widely used to study film thicknesses and the surface contours of various objects [235, 236]. There are also many other recent applications of interference microscopy that have been reviewed by Cocks [55].

6 PHOTOMICROGRAPHY

A discussion of photomicrography is beyond the scope of this chapter, however, the chemical microscopist often finds it necessary to record his observations photographically. The methods he may use vary widely, from simply placing a camera over the eyepiece and snapping a picture to the use of specialized cameras which are completely automated.

The fundamentals of photomicrography are set forth in a number of books [1, 197, 237–243]. There is an extensive bibliography of photomicrography in *The Particle Atlas* [44] and the recent literature has been reviewed by Cocks [55].

References

1. E. M. Chamot and C. W. Mason, *Handbook of Chemical Microscopy*, Vol. I, 3rd ed., Wiley, New York, 1958.
2. M. Françon, *Progress in Microscopy*, Row Peterson, Evanston, Illinois, 1961.
3. L. C. Martin, *The Theory of the Microscope*, American Elsevier, New York, 1966.
4. J. Rheinberg, *Principles of Microscopy*, 1906, p. 19; *J. Roy. Microscop. Soc.*, **1896,** 373; G. L. Royer, C. Maresh, A. M. Harding, *J. Biol. Photog. Assoc.*, **13,** 123 (1944).
5. F. W. Sears, *Optics*, 3rd ed., Wesley, Reading, Massachusetts, 1948.

6. J. R. Benford in *Applied Optics and Optical Engineering*, R. Kingslake Vol. III, Academic, New York, 1965.
7. A. E. Conrady, *Applied Optics and Optical Design*, 2 Vols., Oxford Univ. Press, New York, 1929; Dover, New York, 1957 (Vol. I) and 1960 (Vol. II).
8. A. H. Bennett, H. Jupnik, H. Osterberg, and O. W. Richard, *Phase Microscopy*, Wiley, New York, 1951.
9. W. Krug, J. Rienitz, and G. Schulz, *Contributions to Interference Microscopy*, J. H. Dickson, Transl., Hilger & Watts, London, 1964.
10. A. Köhler, *Z. Wiss. Mikroskopie*, **10,** 433 (1893).
11. P. Gray and G. Wess, *J. Roy. Microscop. Soc.*, **70,** 287 (1950).
12. D. Kay, *Techniques for Electron Microscopy*, 2nd ed., Blackwell Scientific, F. A. Davis, Philadelphia, Pennsylvania, 1965.
13. D. C. Pease, *Histological Techniques for Electron Microscopy*, Academic, New York, 1964.
14. L. Kofler and A. Kofler, *Mikroskopishe Methoden in der Mikrochemie*, Haim, Vienna, 1936.
15. L. Kofler and A. Kofler, *Thermo-Mikro-Methoden zur Kennzeichnung Organischer Stoffe und Stoffgemisch*, Wagner, Innsbruck, 1954.
16. E. E. Jelley, unpublished work.
17. H. W. Emmerton and D. H. Page, *J. Roy. Microscope. Soc.*, **76,** 113 (1957).
18. R. C. Williams and R. W. G. Wyckoff, *J. Appl. Phys.*, **17,** 23 (1946).
19. G. L. Kehl, *Principles of Metallographic Laboratory Practice*, McGraw-Hill, New York, 1949.
20. American Society for Metals, *Metals Handbook*, Am. Soc. for Metals, Cleveland, Ohio.
21. American Society for Testing and Materials, *ASTM Standards, Part 31 Physical and Mechanical Testing of Metals*, Am. Soc. for Testing and Materials, Philadelphia, Pennsylvania, 1968.
22. T. G. Rochow, *J. Roy. Microscop. Soc.*, **87,** 39 (1967); *Matls. Res. Stds.*, **6,** 545 (1965).
23. Am. Soc. for Testing and Materials, *Resinography of Cellular Plastics*, S.T.P. 414, Am. Soc. for Testing and Materials, Philadelphia, Pennsylvania, 1967.
24. Am. Soc. for Testing and Materials, *Symposium on Resinographic Methods*, S.T.P. 347, Am. Soc. for Testing and Materials, Philadelphia, Pennsylvania, 1964.
25. M. N. Short, *Microscopic Determination of the Ore Minerals*, U. S. Geol. Survey Bull. 914, 2nd ed., 1940.
26. E. N. Cameron, *Ore Microscopy*, Wiley, New York, 1961.
27. F. S. Reed and L. L. Mergner, *Am. Mineralogist*, **38,** 1184 (1953).
28. T. Jones and R. M. W. Hawes, *Microscope*, **14,** 200 (1964).
29. G. R. Rigby, *The Thin Section Mineralogy of Ceramic Materials*, British Ceramic Research Assoc., Stoke-on-Trent, England, 1953.
30. H. Insley and V. D. Fréchette, *Microscopy of Ceramics and Cements*, Academic, New York, 1955.
31. G. Münch, *Mikroskopie*, **20,** 196 (1965).

32. O. W. Richards, *The Effective Use and Proper Care of the Microtome*, The American Optical Co., 1949; O. Glasser, *Medical Physics*, Vol. I, 1950, p. 750; *Trans. Am. Microscop. Soc.*, **75,** 136 (1956).
33. A. N. J. Heyn, *Fiber Microscopy*, Interscience, New York, 1954, p. 142.
34. J. H. Graff, *Pulp and Paper Microscopy*, Institute of Paper Chemistry, Appleton, Wisconsin, 1942.
35. J. M. Mathews, *Textile Fibers*, Wiley, New York, 1924.
36. J. Siedemann, *Staerke*, **14,** 348 (1962).
37. A. M, Gaudin, D. W. McGlashan, *Econ. Geol.*, **30,** 552 (1935); **33,** 143 (1938).
38. AIME, *Technical Publ. 912* (March 1938).
39. Maurice Auger, *Econ. Geol.*, **40,** 345 (1945).
40. C. H. Carpenter et al., *A Photomicrographic Atlas of Woody, Non-Woody and Man-Made Fibers Used in Papermaking*, Coll. of Forestry at Syracuse Univ., Syracuse, New York, 1963.
41. W. Hausner, *Faseratlas-Das Erkennen der Textile Faserstoffe*, V. E. B. Fachbuchverlag, Leipzig, 1962.
42. Verein Deutscher Zement Werke, *Mikroskopie des Zementklinkers*, Bilderatlas, Beton-Verlag, Dusseldorf, 1965.
43. P. F. Konoval, B. V. Volkonskiĭ, and A. P. Khashkovaskiĭ, *Atlas of Microstructures of Cement Clinkers, Refractory Materials and Slags*, Gas. Izd. Lit. po Stroit., Arkhitekt, i Materialam, Leningrad, 1962.
44. W. C. McCrone, R. G. Draftz, and J. G. Delly, *The Particle Atlas*, Ann Arbor Science Publishers, Ann Arbor, Michigan, 1967.
45. E. M. Chamot and C. W. Mason, *Handbook of Chemical Microscopy*, Vol. II, 2nd ed., Wiley, New York, 1940.
46. F. Emich, *Monatsh. Chem.*, **22,** 670 (1901); **23,** 76 (1902); *Ann. Chem.*, **351,** 426 (1907).
47. J. Donau, *Monatsh. Chem.*, **25,** 545 (1904); *Ann. Chem.*, **351** (1907).
48. B. K. Seeley, *Anal. Chem.*, **24,** 577 (1952).
49. H. J. Bremneis, *Mikrochemie*, **9** (1931).
50. E. Wiesenberger, *Proc. Intern. Symp. on Microchemical Techniques*, Interscience, New York, 1961.
51. A. A. Benedetti-Pichler and J. R. Rachelle, *Ind. Eng. Chem. Anal. Ed.*, **12,** 233 (1940).
52. B. Tufts and J. P. Lodge, *Anal. Chem.*, **30,** 300 (1958).
53. H. Behrens and P. D. C. Kley, *Orgasische Mikrochemische Analyse*, Voss, Leipzig, 1922. English transl. by R. E. Stevens, Microscope Publications, Chicago, 1969.
54. F. Amelink, *Rapid Microchemical Identification Methods in Pharmacy and Toxicology*, Interscience, New York, 1962.
55. G. G. Cocks, *Anal. Chem.*, **40,** 158R (1968); **38,** 197R (1966); **36,** 163R (1964) and earlier.
56. H. F. Schaeffer, *Microscopy for Chemists*, Dover, New York, 1966.
57. G. Martin, C. E. Blythe, and H. Tongue, *J. Ceram. Soc.*, **23,** 61 (1924).
58. J. M. Dallavelle, *Micromeritics*, Pitman, New York, 1948.

59. R. D. Cadle, *Particle Size Determination*, Interscience, New York, 1955.
60. G. Herdan, *Small Particle Statistics*, Elsevier, New York, 1953.
61. R. P. Loveland, *Methods of Particle Size Analysis*, in *ASTM Special Publication 143*, 1953.
62. *Physics of Particle Size Analysis*, *Brit. J. Appl. Phys., Suppl.*, **3** (1954).
63. R. R. Irani and C. F. Callis, *Particle Size*, Wiley, New York 1963.
64. H. L. Alling and W. G. Valentine, *Am. J. Sci.*, **214,** 50 (1927).
65. F. Chayes, *Petrographic Modal Analysis*, John Wiley, New York, 1956.
66. S. Becher, *Ann. Physik*, **47,** 285 (1915).
67. E. E. Jelley, *Ind. Eng. Chem. Anal. Ed.*, **13,** 196 (1941).
68. R. W. Wood, *Physical Optics*, 3rd ed., Macmillan, New York, 1934.
69. L. Fletcher, *The Optical Indicatrix*, Frowde, London, 1802.
70. N. H. Hartshorne and A. Stuart, *Crystals and the Polarizing Microscope*, 3rd ed., Arnold, London, 1960, p. 337.
71. J. Mitchell, Jr., and W. M. D. Bryant, *J. Am. Chem. Soc.*, **65,** 128 (1943).
72. E. E. Jelley, unpublished work.
73. E. E. Jelley, *J. Roy. Microscop. Soc.*, **56,** 101 (1936).
74. W. M. D. Bryant, *J. Am. Chem. Soc.*, **63,** 511 (1941).
75. E. E. Jelley, *Phot. J.*, **74,** 514 (1934).
76. W. M. D. Bryant, *J. Am. Chem. Soc.*, **65,** 96 (1943).
77. F. E. Wright, *Methods of Petrographic-Microscopic Research*, Carnegie Inst. Wash. Publ. No. 158 (1911).
78. A. Johannsen, *Manual of Petrographic Methods*, McGraw-Hill, New York, 1918.
79. O. Maschke, *Ann. Physik. Chem.*, **145,** 519 (1872).
80. F. Becke, *Sitzber. Akad. Wiss. Wien.*, **102,** 358 (1893).
81. J. L. C. Schroder vander Kolk, *Kurz*, *Anleitung zur Mikroskopischen Kristallbestimmung*, Wiesbaden, 1898; *Tabellen zur Mikroskopuschen Bestimmung der Mineralen nach Ihrem Brechungsindex*, Wiesbaden, 1906.
82. C. P. Saylor, *J. Res. Natl. Bur. Stds.*, **15,** 277 (1935).
83. E. S. Larsen and H. Berman, *Microscopic Determination of the Nonopaque Minerals*, U. S. Geol Survey Bull. No. 848 (1934).
84. P. Gaubert, *Bull. Soc. Franc. Mineral.*, **45,** 89 (1922).
85. R. C. Emmons, *Am. Mineralogist*, **13,** 504 (1928).
86. C. P. Saylor, *J. Res. Natl. Bur. Stds.*, **15,** 97 (1935).
87. E. Posnjak and H. E. Merwin, *J. Am. Chem. Soc.*, **49,** 1970 (1922).
88. R. C. Emmons, *Am. Mineralogist*, **11,** 115 (1926); **13,** 504 (1928); **14,** 414, 441, 482 (1929).
89. E. E. Jelley, *J. Roy. Microscop. Soc.*, **54,** 234 (1934).
90. R. N. Titus, private communication.
91. E. E. Jelley, unpublished work.
92. H. E. Merwin, *J. Wash. Acad. Sci.*, **3,** 35 (1913).
93. W. M. D. Bryant, *J. Am. Chem. Soc.*, **65,** 96 (1943).
94. A. C. Dunningham, *J. Chem. Soc.*, **105,** 368 (1914).
95. F. E. Wright, *J. Wash. Acad. Sci.*, **4,** 269 (1914).
96. E. Clerici, *Atti. Accad. Lincei.*, **16,** 336 (1907); **18,** 351 (1909).

97. C. Viola, *Atti. Accad. Lincei.*, **19,** 192 (1910).
98. L. Nichols, *Natl. Paint Bull.*, **1,** 12 (1937).
99. A. E. Edwards and C. E. Otto, *Ind. Eng. Chem. Anal. Ed.*, **10,** 225 (1938).
100. H. A. Frediani, *Ind. Eng. Chem. Anal. Ed.*, **14,** 439 (1942).
101. F. E. Wright, *J. Geol.*, **10,** 33 (1902).
102. F. E. Wright, *Am. J. Sci.*, **29,** 416 (1910).
103. A. Johannsen, *Am. J. Sci.*, **29,** 436 (1910).
104. M. Berek, *Centr. Mineral. Geol.*, **1913,** 388, 427, 464, 580.
105. W. Nikitin, *Z. Krist.*, **47,** 378 (1910).
106. E. E. Jelley, *J. Roy. Microscop. Soc.*, **56,** 101 (1936).
107. R. C. Emmons, in *Microscopic Characters of Artificial Minerals*, A. N. Winchell, Ed., 2nd ed., Wiley, New York, 1931.
108. R. C. Emmons, *The Universal Stage*, Geol. Soc. of Am. Monograph No. 8, 1943.
109. E. Mallard, *Bull. Soc. Ind. Mineral. St. Etienne*, **5,** 506 (1882).
110. F. E. Wright, *Am. J. Sci.*, **24,** 336 (1907).
111. F. E. Wright, *Am. J. Sci.*, **31,** 87 (1911); *J. Wash. Acad. Sci.*, **1,** 60 (1911).
112. F. E. Wright, *Am. J. Sci.*, **20,** 288 (1905).
113. A. Johannsen and T. C. Phemister, *J. Geol.*, **32,** 81 (1924).
114. R. G. Wood and S. H. Ayliffe, *J. Sci. Instr.*, **12,** 194 (1935).
115. W. M. D. Bryant, *J. Am. Chem. Soc.*, **60,** 394 (1938).
116. W. M. D. Bryant, *J. Am. Chem. Soc.*, **63,** 511 (1941).
117. W. Haidinger, *Ann. Physik. Chem.*, **68,** 302 (1846); **70,** 571 (1847); **71,** 321 (1847); **76,** 99, 204 (1849); **77,** 89 (1849); **81,** 572 (1850).
118. F. Pockels, *Lehrbuch der Kristalloptik*, Tenbner, Berlin, 1906.
119. W. Haidinger, *Ann. Physik. Chem.*, **75,** 1 (1845).
120. C. Leiss, *Die Optischen Instrumente der Firma R. Fuess*, Engelmann, Leipzig, 1899, p. 220.
121. M. Berek, *Optische Messmethoden im Polarisierten Auflicht*, Deut. Mineral. Gesellschaft, Berlin, 1937. Reprinted from *Fortschr. Mineral Krist Petrog.*, **22,** 1 (1937).
122. S. Yamada, *J. Am. Chem. Soc.*, **73,** 1182 (1951); **73,** 1579 (1951).
123. K. Nakamoto, *J. Am. Chem. Soc.*, **74,** 390 (1952); **74,** 392 (1952).
124. Cf. R. Tsuchida and M. Kobayashi, *Chem. Abstracts*, **33,** 2034 (1939).
125. E. E. Jelley, *J. Roy. Microscop. Soc.*, **54,** 13 (1934).
126. F. I. G. Rawlins, *Trans. Faraday Soc.*, **25,** 762 (1929).
127. M. H. Dufet, *Bull. Soc. Franc. Mineral*, **24,** 373 (1901).
128. E. E. Jelley, *Nature*, **136,** 335 (1935).
129. W. R. Brode, *Chemical Spectroscopy*, 2nd ed., Wiley, New York, 1943, p. 267.
130. G. W. Lewis and M. Calvin, *Chem. Rev.*, **25,** 273 (1939).
131. L. G. S. Brooker, *J. Am. Chem. Soc.*, **62,** 1116 (1940); **63,** 3192, 3203, 3214 (1941); **64,** 199 (1942); *Rev. Mod. Phys.*, **14,** 275 (1942).
132. A. Maccoll, *Quart. Rev.*, **1,** 16 (1947).
133. S. E. Sheppard, *Rev. Mod. Phys.*, **14,** 303, 340 (1942).

134. F. H. Spedding, *J. Chem. Phys.*, **5,** 160 (1947); *Phys. Rev.*, **50,** 574 (1936).
135. C. E. K. Mees, *Theory of the Photographic Process*, Macmillan, New York, 1942, p. 1035.
136. C. E. K. Mees, Ref. 135, p. 992.
137. E. E. Jelley, unpublished work.
138. V. C. Vesce, in *Protective and Decorative Coatings*, J. J. Mattiello, Ed., Vol. II, Wiley, New York, 1942.
139. C. P. Saylor, *J. Res. Natl. Bur. Stds.*, **15,** 97 (1935).
140. W. C. McCrone, *Fusion Methods in Chemical Microscopy*, Interscience, New York, 1957.
141. W. C. McCrone, A. Smedal, and V. Gilpin, *Anal. Chem.*, **18,** 578 (1946).
142. H. B. Bradley, *Anal. Chem.*, **29,** 1239 (1957).
143. W. C. McCrone and S. M. O. Bradovic, *Anal. Chem.*, **28,** 1038 (1956).
144. E. F. Fullam, *Microscope*, **10,** 76 (1954).
145. D. G. Grabar and W. C. McCrone, *J. Chem. Educ.*, **27,** 649 (1950).
146. F. W. Matthews, *Anal. Chem.*, **20,** 1112 (1948).
147. F. Ordway, *J. Res. Natl. Bur. Stds.*, **48,** 152 (1952).
148. J. H. Welch, *J. Sci. Instr.*, **31,** 458 (1954).
149. R. L. Causer, *J. Sci. Instr.*, **41,** 393 (1964).
150. W. J. Gutt, *J. Sci. Instr.*, **41,** 393 (1964).
151. R. A. Mercer and R. P. Miller, *J. Sci. Instr.*, **40,** 352 (1963).
152. J. G. Dodel, *Microscope*, **14,** 302 (1965).
153. J. C. W. Crawley, *J. Sci. Instr.*, **40,** 330 (1963).
154. M. G. Lozinsky, *High Temperature Metallography*, L. Herdan, Transl., Pergamon, Oxford, 1961.
155. R. Mitsche and F. Jeglitsch, *Radex Rundschau*, **2,** 408 (1963).
156. M. G. Lozinsky, *J. Roy. Microscop. Soc.*, **82,** 211 (1967).
157. W. F. Cook, *J. Roy. Microscop. Soc.*, **86,** 79 (1966).
158. W. C. McCrone, *Applications of Thermal Microscopy*, Tech. Info. Bull. 3003, Mettler Instrument Corp. Princeton, New Jersey, 1968.
159. H. G. Drickamer, *Science*, **142,** 1429 (1963).
160. A. Van Valkenburg, *High Pressure Microscopy*, Paper 62-WA-269, ASME, 1962.
161. F. Fisher and G. Karesk, *Mikrochemie*, **33,** 310 (1947).
162. F. Fisher and E. Neupauer, *Mikrochemie*, **34,** 319 (1949).
163. H. H. O. Schmid, H. K. Mangold, and N. O. Lundberg, *Mikrochem. J.*, **7,** 287 (1963); **7,** 297 (1963); **9,** 134 (1965).
164. M. Brandstätter and A. Kofler, *Mikrochemie*, **34,** 364 (1948–1949); **36/37,** 291 (1951).
165. W. Ostwald, *Z. Physik. Chem.*, **22,** 306 (1897).
166. T. Zincke, *Ann. Chem.*, **159,** 381 (1871).
167. J. F. J. Dippy and N. H. Hartshorne, *J. Chem. Soc.*, **1930,** 725.
168. N. H. Hartshorne and A. Stuart, *J. Chem. Soc.*, **1931,** 2583.
169. W. D. Bancroft, *J. Phys. Chem.*, **2,** 143, 245 (1898); **3,** 144 (1899).
170. E. L. Skan and B. Saxton, *J. Phys. Chem.*, **37,** 197 (1933).
171. W. M. D. Bryant, *J. Am. Chem. Soc.*, **60,** 2814 (1938).

172. Liquid Crystals and Anisotropic Melts, *Trans. Faraday Soc.*, **29,** No. 148, Part 9, pp. 881–1084 (Sept 1933) (1933).
173. D. Vorländer, G. Kreiss, and C. Kuhrmann, *Z. Angew. Chem.*, **43,** 13 (1930).
174. C. Weygand and R. Gabler, *Naturwissenschaften*, **27,** 28 (1939); *J. Prakt. Chem.*, **151,** 215 (1938); *Z. Physik. Chem.*, **B47,** 148 (1941).
175. G. H. Brown and W. G. Shaw, *Chem. Rev.*, **57,** 1049 (1957).
176. G. H. Brown, G. J. Dienes, and M. M. Labes, *Liquid Crystals*, Gordon and Breach, New York, 1967.
177. G. H. Brown, *Anal. Chem.*, **41,** 26A (1969).
178. G. W. Gray, *Molecular Structure and the Properties of Liquid Crystals*, Academic, New York, 1962.
179. L. Kofler, *Z. Anal. Chem.*, **133,** 27 (1951).
180. E. Lindpaintner, *Arch. Pharm.*, **277,** 398 (1939).
181. F. Reimers, *Dansk. Tidsskr. Farm.*, **14,** 219 (1940); *Z. Anal. Chem.*, **122,** 404 (1941).
182. C. Arceneaux, *Anal. Chem.*, **23,** 906 (1951).
183. M. Brandstätter, *Z. Physik. Chem.*, **191,** 227 (1942).
184. R. Fischer and G. Kocher, *Mikrochem. Mikrochim. Acta*, **33,** 131 (1946).
185. W. M. D. Bryant, *J. Am. Chem. Soc.*, **60,** 1394 (1938).
186. G. C. Crossmon, *Am. Ind. Hygiene Quart.*, **18,** 341 (1957).
187. G. C. Crossmon, *Anal. Chem.*, **20,** 10 (1948).
188. G. C. Crossman, *Microchem. J.*, **10,** 273 (1966).
189. G. C. Crossmon, *Microscope*, **14,** 498 (1965).
190. Yu. A. Cherkasov, I. Mittin, Transl., *Intern. Geol. Rev.*, **2,** 218 (1960).
191. W. G. Kirchgessner, *Microscope*, **15,** 511 (1967).
192. K. G. Schmidt, *Staub*, **41,** 436 (1955), A. L. Forlini, Transl., in *Proc. Intern. Microscopy Symp., Chicago, 1960.*
193. D. G. Grabar, *J. Air Pollution Control Assoc.*, **12,** 500 (1962).
194. L. Forlini, *The Particle Analyst*, **1,** 97 (1968).
195. K. M. Brown, W. C. McCrone, R. Kuhn, and A. L. Forlini, *Microscop. Cryst. Front*, **13,** 311 (1963); **14,** 39 (1963).
196. A. N. Winchell and H. Winchell, *The Microscopical Characters of Artificial Inorganic Solid Substances: Optical Properties of Artificial Minerals*, 3rd ed., Academic, New York, 1964.
197. R. Kingslake, *Applied Optics and Optical Engineering*, Vol. IV, Academic, New York, 1965.
198. W. von Ignatowsky, *Z. Wiss. Mikroskopie*, **26,** 387 (1909).
199. J. Smiles and J. E. Barnard, *J. Roy. Microscop. Soc.*, **269,** 365 (1924).
200. H. Siedentopf, *J. Roy. Microscop. Soc.*, **49,** 349 (1929).
201. H. Seidentopf and R. Zigsmondy, *Ann. Physik*, **10,** 1 (1903).
202. H. Siedentopf, *Kolloid-Z.*, **36,** 1 (1925).
203. U. Gerhardt, *Z. Physik*, **35,** 697 (1926); **44,** 397 (1927).
204. O. von Baeyer and U. Gerhardt, *Z. Physik*, **35,** 718 (1926).
205. S. Amelinckx, *The Direct Observation of Dislocations*, Academic, New York, 1964.

206. G. Heilmann, *Appl. Opt.*, **4,** 1201 (1965); **5,** 1065 (1966).
207. G. K. T. Conn and F. J. Bradshaw, *Polarized Light in Metallography*, Butterworths, London, 1952.
208. B. W. Mott and H. R. Haines, *Research*, **4,** 24, 63 (1951); *J. Inst. Metal.*, **80,** 629 (1952).
209. A. Köhler, *Z. Wiss. Mikroskopie*, **21,** 129, 273 (1904).
210. T. Casperson, *Skand. Arck. Physiol.*, **73** (1936); *Nature*, **143,** 602 (1939).
211. J. Loofbourrow, *Bull. Basic Sci. Res.*, **5,** 46 (1933); *J. Opt. Soc. Am.*, **29,** 535 (1939).
212. T. Casperson, *J. Roy. Microscop. Soc.*, **83,** 67 (1964).
213. C. D. Felton, E. J. Thomas, and J. J. Clark, *Textile Res. J.*, **32,** 57 (1962).
214. L. P. Zaitseva and T. G. Porokhova, *Zavodsk. Lab.*, **29,** 1088 (1963).
215. R. F. Deacon and D. Barella, *J. Sci. Instr.*, **39,** 521 (1962).
216. A. J. Forty, *Phil. Mag.*, **8,** 663 (1963); **9,** 673 (1964).
217. V. J. Nikitenko and V. L. Indenbom, *Zavodsk. Lab.*, **29,** 222 (1963).
218. H. Luther, O. Abel, and F. Monostory, *Brennstof Chem.*, **44,** 246 (1963).
219. W. Sandritter and K. Kiefer, *Methoden und Ergebnisse der Zytophotometrie und Interferenzmikroskopie*, V. E. B. Gustav Fischer, Jena, 1965.
220. F. Gabler, *Microscope*, **15,** 85 (1966).
221. H. W. Zieler, *Microscope*, **13,** 363 (1963).
222. H. G. Kruger, *Leitz-Mitt. Wiss. Technik*, **1,** 247 (1961).
223. E. A. Hauser and C. J. Frosch, *Ind. Eng. Chem. Anal. Ed.*, **8,** 423 (1936); *J. Opt. Soc. Am.*, **27,** 110 (1937).
224. D. R. Morey, *Textile Res.*, **4,** 491 (1934).
225. H. Eder, H. Fritsche, and H. J. Kinskel, *Z. Wiss. Mikroskopie*, **68,** 42 (1967).
226. H. D. Pflug, *Leitz-Mitt. Wiss. Techn.*, **3,** 183 (1966).
227. G. Sansoni, *Jenaer Rundshan*, **8,** 170 (1963).
228. R. O. Marts, *J. Biol. Phot. Assoc.*, **23,** 151 (1955).
229. M. Haitinger, *Die Fluorezenzanalyse in der Mikrochemie*, Haim, Leipzig, 1937; reprinted by Edwards Bros., Ann Arbor, Michigan 1944, *Fluoreszenzmikroskopie. Ihre Anwendung in der Histologie und Chemie*, Akademische Verlagsgesellshaft, Leipzig, 1938.
230. O. W. Richards, *Medical Physics*, Vol. III, Year Book Publ., Chicago, 1960.
231. J. Eisenbrand, *Fluorimetrie*, Wissenschaftliche Verlagsgesellschaft, Stuttgart, 1966.
232. L. P. Johansson, in *Cytochemical Methods with Quantitative Aims*, D. Lindstrom and R. Brown, Eds., Academic, New York, 1957.
233. R. Barer and D. A. T. Dick, in *Cytochemical Methods with Quantative Aims*, D. Lindstrom and R. Brown, Eds., Academic, New York, 1957.
234. H. G. Davies, in *General Cytochemical Methods*, J. F. Danielli, Ed., Academic, New York, 1958.
235. S. Tolansky, *Multiple Beam Interferometry*, Clarendon, Oxford, 1948.
236. S. Tolansky, *Microstructures of Diamond Surfaces*, N.A.G. Press, London, 1958.

237. R. M. Allen, *The Microscope*, Van Nostrand Reinhold, New York, 1940.
238. C. P. Shillaber, *Photomicrography*, Wiley, New York, 1944.
239. K. Michel, *Die Photomicrographie*, 2nd ed., Band X, in *Die Wissenschaftliche und Angewandte Photographie*, K. Michel, Ed., Springer, Vienna, 1962.
240. D. F. Lawson, *The Technique of Photomicrography*, Macmillan, New York, 1961.
241. G. G. Rose, *Cinemicrography in Cell Biology*, Academic, New York, 1963.
242. R. Schenk and G. Kistler, *Photomicrography*, F. Bradley, Transl., Franklin Publ. Co., Palisade, New Jersey, 1963.
243. R. P. Loveland, *Photomicrography*, Wiley, New York, 1970.

Chapter **IV**

ELECTRON MICROSCOPY

R. G. Scott and A. N. McKee

1 Introduction

2 The Transmission Electron Microscope

Evolution 295
The Microscope 296
The Image 302
General Operating Procedures 307
  Bright Field 307
  Dark Field and Diffracted Beam 312
  Electron Diffraction 313
  Reflection Microscopy and Diffraction 316
Related Equipment 318
Sample Preparation 320
  The Specimen Support 320
  Dispersions 321
  Internal Structure 325
    Sectioning 326
    Cleaving 327
    Thinning 327
    Fragmentation 327
    Extraction 328
    Low-Temperature Techniques 328
    Autoradiography 329
  Surfaces 329
    Replicas 329
    Chemical Sections 332
    Etching 332
    Decoration 333
    Profile Technique 333
Applications 334
  Qualitative Observations 334
    Chemical Reactions 334
    Location and Identification of Chemical Phases 337

Arrangement of Molecules 340
Molecules 345
Quantitative Measurements 345
Concentration 345
Lateral Dimensions 346
Thickness 346
Weight 347
Ordered Structures 348

3 Scanning Electron Microscopy

Introduction 349
Historical 356
Emission Microscopy 356
Secondary Electron Microscopy 357
Scanning Electron Microscopy 359
The Modern Scanning Electron Microscope 361
The Electron-Optical System 361
Vacuum 361
Electron Gun 362
Electron Lens System 362
Lens Aberrations 363
Vibration 364
Detectors Used in the Collection of Signals 365
Secondary Electrons 365
Reflected and Backscattered Primary Electrons 367
Electromagnetic Radiation 368
The Conductive Mode 369
Display and Recording 370
The Interaction of Electrons with Matter 373
Energy Transfer from Incident to Specimen Electrons 373
The Emission of Secondary Electrons 374
Backscattered Electrons 376
Effects of an Electron Beam on Polymeric Specimens 377
Contrast of the Scanning Electron Microscope Image 377
The Emissive Mode 377
The Conductive Mode 380
Low-Voltage Operation 380
Resolution of the Scanning Electron Microscope 380
Secondary and Backscattered Electron Images 380
Other Modes 381
Stereomicroscopy 382
Specimen Preparation 383

Applications of Scanning Electron Microscopy 384
Introduction 384
Applications of the Scanning Electron Microscope in Chemistry 385
The Future of Scanning Electron Microscopy 392

4 The Electron Microprobe

Introduction 398
The Instrument 399
Sample Preparation 403
The Production of X-Rays 404
Dimensions of the X-Ray-Producing Volume 405
The Analysis of X-Radiation in the Electron Microprobe 407
X-Ray Crystal Optics 407
X-Ray Detectors and Their Associated Equipment 411
Nondispersive Analyzing Systems 413
Techniques of Analysis 415
General Procedures 415
The Light Atom Problem 416
Wavelength Shifts 416
Analysis by Sample Current Measurement 417
Note on Standard Sample 418
Qualitative Analysis 418
Quantitative Analysis 419
Dead Time and Instrument Drift 420
Background Radiation 420
Emission Concentration Law and Atomic Number Effect 421
The Absorption Correction 422
Fluorescence Correction 425
Applications 426

I INTRODUCTION

Over the past 30 years electron microscopy has grown from an interesting demonstration of physical theories to a standard technique utilized in many research laboratories. Present estimates indicate that there are about 3000 commercial instruments in the United States alone. As recently as 10 years ago, if one referred to an electron microscope it was generally understood that he meant a system consisting of electric and magnetic fields that behaved as source, condenser, objective, projector, and viewing screen arranged similarly to the same elements in a conventional transmission optical microscope.

Within recent years, however, it has become increasingly difficult to define "electron microscope" in such simple terms. So far, three major classes of instruments have evolved. The transmission electron microscope was established first; then in the early 1950s modern x-ray microscopes evolved from techniques developed for the previous instrument. X-ray microscopy has developed into a broad, relatively independent science, which is discussed by R. V. Ely elsewhere in this volume. A more recent and closely related instrument is the scanning electron microscope. Its acceptance as a research tool in many branches of science has been rapid, and articles and discussions on the subject are numerous. It is considered in this chapter only because its role in chemical research is not clearly defined.

Electron microscopy has developed with the same rapidity characteristic of other modern sciences, and technical literature on the subject has been extensive. A list of general references on this work has been placed at the end of the chapter, and in the discussion presented here, when references are omitted on a specific topic they can be found reviewed at length in the general references. It is not the purpose of this chapter to compete with an already adequate library of information concerning applications of this class of instruments but rather to discuss the problems encountered by chemists who wish to apply these tools to their research programs.

In most of the applications of electron microscopy, there is no clearly defined preferred way of proceeding. Although the techniques are sound and well tested, a wide range of choices exists, and for this reason electron microscopy is still thought of as an art as much as a science. To be proficient the microscopist needs constant practice in a number of meticulous operations. This makes it impractical for a research chemist to do his own electron microscopy. However, to use observations from microscopy, he should have a clear understanding of the potentialities and limitations of the instrument.

In preparing this survey of the application of electron microscopy to chemistry, it soon became evident that the science is not as well advanced or as clearly defined as it is for optical microscopy. This is probably the result of restrictions imposed by the instrument. The general areas of work to be considered here include: (1) the supermolecular structure of preparations, including studies relating physical treatment and chemical reactivity, and the results of chemical treatment; (2) the identification of atoms and molecules in solids; (3) the direct observation of molecules; and (4) mention of preliminary attempts at the direct observation of chemical reactions.

2 THE TRANSMISSION ELECTRON MICROSCOPE

Evolution

It is impossible to assign definite periods to the advance of electron microscopy because characteristics of one point in time are found throughout the development. However, there are three eras during which certain features were stressed.

The first era was one of invention and development. If the original purpose of the transmission electron microscope was to break the barrier of invisibility imposed by Abbe's theory of seeing small objects, by reducing the wavelength of the radiation used, then it might be said that investigations were initiated by the work of Busch [1]. In 1927 he published a paper describing magnetic lenses for electrons and other charged particles. However, one might also consider de Broglie's theory of 1924 [2] that assigned a wavelength to moving particles, the discovery of electrons by Thompson [3] in 1897, or earlier studies on wave motion and interference. In 1928, Ruska started a study of magnetic lenses that led to the announcement of the first electron microscope, which was described by Knoll and Ruska [4] in 1931. The microscope was first introduced into the western hemisphere in the late 1930s by Burton [5] and a number of his graduate students.

The second era started with the introduction and development of commercial high-resolution instruments in the early 1940s. A great deal of effort was also devoted to the improvement of the specimen being studied. This period of application included three major developments: (1) the introduction of replica techniques by Mahl [6] to allow the study of "opaque" objects; (2) the demonstration of shadowing effects to increase contrast [7]; and (3) development of sectioning techniques to allow the use of the full resolving power of the instrument of three-dimensional structures.

The third stage started in the late 1950s with an increased realization of the unique features of the instrument. Manufacturers no longer just build a microscope—it is designed for a specific job. This is particularly true for work in the biological sciences or those related to metallurgy.

Another mark of this era has been the introduction of a variety of specimen stages for heating, cooling, stretching, tilting, or irradiating [8]. New methods have also been developed for extracting information from the electron beam. These include modified photographic techniques, direct measurement of beam currents, and the use of image intensifiers with related television techniques.

At the same time instrument makers improved the resolving power of instruments from 20 to about 2 Å in the better instruments today. The future promises not only a clearer view of finer detail but a more detailed analysis of the electron image by closer coupling of the microscope to computer techniques.

The Microscope

The general components of an electron microscope and principal electron paths are shown in Fig. 4.1. These components, plus shields and a vacuum-tight shell, make up the microscope column. Designs have varied, but in high-resolution instruments the column is vertical and has the gun at the top. This has been done to simplify alignment procedures and to produce maximum stability.

Electrons accelerated in conventional microscopes have associated wavelengths ranging from 0.07 Å for a 30-kV accelerating potential to 0.036 Å for 100 kV. With the exception of the electron gun, focusing is accomplished by strong electromagnetic fields. In conventional lenses the magnetic field is parallel to the microscope axis. This produces a rotation of the image as well as magnification. The focal length of such lenses depends on both the accelerating voltage and the current in the lens. Focusing is accomplished by varying the current rather than moving the lens as in optical microscopy, however, there are a few simpler instruments in which permanent magnets are used and adjustments are made by moving pole pieces.

The electron gun assembly consists of a heated filament, biased grid cap, and anode, including supports and insulators. In most commercial microscopes the filament can be operated in steps between −30 and −100 kV with

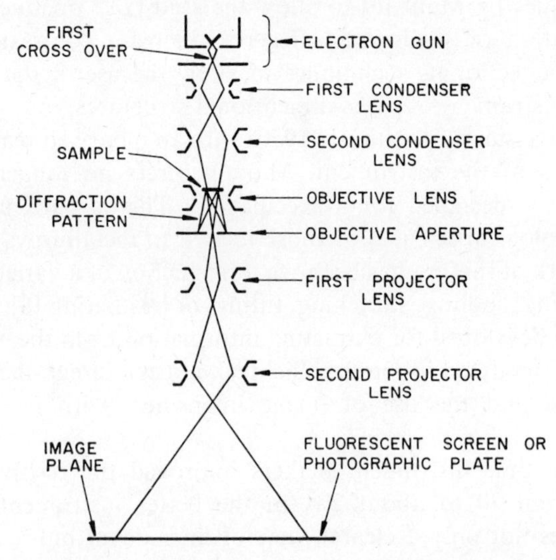

Fig. 4.1 Schematic representation of an electromatic electron microscope.

respect to the anode set at ground potential. Recently, there has been increased interest in both higher and in lower accelerating voltages. Higher voltages (experimental instruments have gone as high as 2 MV) are used for increased penetrating power and reduced damage to the sample, while lower voltages are used to improve the contrast in images of thin samples.

The filament is usually a 5-mil tungsten wire bent into a sharp V and mounted in such a way that it may be changed rapidly with a minimum of adjustments because its average life is only about 50 hr. For high-resolution work it has been found preferable to use a special pointed filament to produce a smaller, more coherent source. Under conventional operating conditions the crossover or point of focus in the gun acts as a 50-μ source. By moving the filament away from the control grid aperture or by increasing the grid bias, the beam current can be reduced to protect radiation-sensitive materials.

As in the optical microscope, the condenser lens controls illumination of the sample. In the out-of-focus condition the angular collimation is determined by the size of the source and the position of the lens. When the condenser is focused on an object, the angular spread of illumination and the area radiated is limited by the condenser aperture. The minimum spot size at the sample with a single-element lens is about 30 μ, while a double lens reduces it to about 2 μ. The arrangement is particularly useful when working with radiation-sensitive materials.

In most microscopes built within the last few years, the condenser lens has two components that allow a wider range of illuminating conditions and restrict the radiated area to the part of the sample under investigation. The first stage produces a demagnified image of the source for the second element to act on. The limiting aperture is in the second stage. Unlike its optical counterpart, the condenser is not required to fill the aperture of the objective lens. This is because the problem is complicated by convergence introduced by the field of the objective lens above the sample and scattering from the object. In fact, it can be shown that although there may be an increase in brightness the image quality deteriorates as the size of the condenser aperture is increased above a limiting value. Agar [9] proposed, from a practical standpoint, that the semiangular aperture be 10^{-3} radians or less. For a condenser object distance of 15 cm, this would require a physical aperture of 300 μ.

One of the major aberrations of the illuminating system is astigmatism caused by unsymmetrical magnetic and electric fields within the microscope. It may result from a built-in asymmetry of a lens, or from as temporary an effect as misalignment or contamination on an aperture. In the illuminating system astigmatism produces a cylindrical component in the lens system, which in turn produces an unsymmetrical image of the source. Adjustable compensating fields are placed between the condenser and objective to produce symmetrical illumination.

Electron beam–tilting devices, required for high-resolution dark-field and reflection studies (see p. 316), may be located between the condenser and the sample. The deflecting is normally accomplished with two carefully designed fields, one that swings the beam away from the column axis and another that brings it back to cross the axis in the sample at a desired angle. To make full use of the technique in diffraction methods, it must be possible to rotate the deflecting system about the microscope axis.

The next part of the column is the specimen chamber. There are two general positions for the sample. The conventional location for high resolution is within the magnetic field of the objective lens. Many microscopes also allow the sample to be studied in a large chamber about 5 cm above the center of the objective lens. Although this does not produce maximum resolution, it does permit the use of larger specimens and more elaborate instrumentation. It also produces longer effective camera lengths for diffraction, and in diffracted-beam studies it allows more critical masking for a given size aperture.

As in the case of most other advances, the introduction of electron microscopy as an aid in seeing very small detail has imposed new and severe restrictions on the nature and form of the specimen. Two general restrictions have been recognized and dealt with from the outset. The sample must withstand high vacuum and it must be thin enough to transmit electrons in a useful form. Other restrictions are more closely related to the interaction of intermediate and high-energy electrons with matter. There are three possibilities:

1. If the sample is thin enough, the electron may pass through with its initial energy and direction.

2. It may interact with an atomic nucleus. Because of the relatively small mass of the incident electron it suffers a change of direction without transferring energy to the sample. This is referred to as an elastic collision, and since there is no energy loss the scattered electron maintains its original wavelength and can form diffraction patterns of crystalline materials or produce other interference effects which are described in other sections. Electrons of 100 kV can penetrate 20 μ of organic or other material of relatively low atomic number. However, in doing so these electrons are scattered through large angles because of multiple collisions and cannot be imaged properly by the microscope. In the preferred sample the average electron that reaches the image plane should not suffer more than one interaction with an atomic potential field within the sample. This limit is usually defined as the mean free path for electrons in the sample [10], and if

this definition is used, it can be shown that the useful thickness range for the single-hit effect can be extended by using small objective apertures. For example, at 60 kV carbon has a limiting thickness of 20 Å but with a 10^{-4} radian angular aperture this becomes 100 Å.

3. It may interact with one of the electrons in the sample. Because of similarity of mass there is usually an energy transfer to the sample. In many cases this transfer within the sample is more important than the chromatic effects produced in the beam. One result of this type of interaction is the production of free radicals. In time these may combine to form a cross-linked network. In some preparations this acts to reduce further damage. For example, organic substrates can be stabilized by prolonged low-level radiation before turning up the beam for work at high magnification. However, a very harmful effect is produced by the rearrangement of bonds in crystalline material. At normal operating conditions the crystallinity in most organic materials is destroyed within seconds. Other compounds are decomposed by radiation. In general, organic preparations are far more sensitive than metals and other inorganic materials.

Three operating conditions affect the extent of radiation damage: flux rate, exposure time, and accelerating potential. Experiments have indicated that total flux through the sample is far more important than flux rate, although the latter determines the temperature of the sample. Since the photographic emulsion is a near-perfect detector (it is noise limited), little can be done about the radiation required to record. The greatest amount of time is usually spent searching or focusing. One can reduce the area radiated at one time by focusing the condenser or reducing the flux rate from the gun. Unfortunately, this reduces visibility and the ease of focusing. This problem has been reduced by the introduction of image intensifiers. Recent experiments have shown that by going to higher voltage radiation damage is reduced. As the voltage is increased, the cross section for inelastic scattering is reduced faster than that for elastic scatter [11].

The same type of interaction takes place between the electron beam and residual organic vapors in the microscope column. Two of the chief sources of this material include diffusion pump oil and grease used on vacuum seals. The gases are decomposed by the beam and condense out on radiated areas to form a cross-linked polymeric contamination. For a given system the rate at which the deposit builds up depends primarily on the nature of the gases present, the intensity of the radiation, and the temperature of the sample. For example, with a double condenser focused for maximum intensity and the sample at ambient temperature a layer forms at the rate of about 1 Å per second. The rate can be reduced appreciably by heating the sample to about 250°C or by surrounding it with a cold chamber at about −65°C.

Another change produced by the inelastically scattered electrons is a temperature rise in the sample. This is a mixed blessing because the increase in temperature reduces the rate of contamination, but it may destroy crystal structure or increase the rate of drift in the image.

The objective lens represents the start of magnification. It determines focus, resolution, and contrast. Modern high-resolution objectives are immersion lenses with focal lengths of about 2.5 mm. Although placing of the sample in the magnetic field of the lens improves performance, it has two undesirable consequences. First, it limits the size of the sample to about 2 mm on a side. Second, for critical work with diffracted-beam techniques (p. 312) the lens introduces a convergence to the incident beam that cannot be controlled by the operator.

In the objective lens the dominant aberrations are spherical and chromatic. In the presence of spherical aberrations, electrons traveling near the edge of the lens are brought to focus sooner than those near the center of the lens. The operator has only partial control of this effect. When the microscope is used without an objective aperture, electrons scattered at large angles ($\sim 10^{-2}$ radians) strike the fluorescent screen so far from focus that they contribute a general background haze. The effect can be reduced by limiting the angular aperture in the focal plane of the objective lens. This should be reduced only to the point where the Airy diffraction effect takes over. For a conventional 2.5-mm focal length lens, this corresponds to a 50-μ aperture. The focal length of an electron lens depends not only on the field strength but also on the energy of the radiation involved. Variations caused by the heterogeneous character of the radiation are referred to as chromatic aberrations. Loss of monochromatic character results from: electrons leaving the emitter with slightly different energies; high-voltage fluctuations; and energy loss in the sample. For a given sample chromatic aberrations can be reduced by lens design. The microscopist can help only by working with very thin samples at high voltage to reduce energy losses.

The projector acts to produce a magnified image of either the image plane (microscopy) or back focal plane of the objective lens (diffraction). Physically, the lens is usually built as a two-element system, both elements being adjustable. Some of the advantages of this type of design include: (1) a wide range of magnifications can be covered without changing pole pieces in the lenses; (2) it shortens the length of the instrument; and (3) it allows normal and small-angle electron diffraction studies with the object in its conventional position.

When used as a microscope the second element is operated near full strength and the current is varied through the first element to determine magnification. In the high range of magnification, the first projector is operated as an adjustable lens to produce a series of magnifications. In the low range the first projector actually reduces the size of the image acted on by the second projector. The first projector knob is usually calibrated to read magnification directly but because of uncertainties of operating conditions this should not be relied on to better than 10%.

Major image imperfections created by the projector lens are classed as distortions. These take two forms. One is a radial variation in magnifying power that produces a "pin cushion" or "barrel" distortion. The second is a rotational distortion attributable to change in image rotation with distance from the center of the lens.

In choosing a final magnification, a number of things must be considered. First, the purpose of magnification is to help the eye to see detail below 0.1 mm in size. If the resolving limit of a device is D (in millimeters), then any magnification above $M = 0.1/D$ is unnecessary. When working within the resolving limit of the electron microscope, it is preferable to choose as low a value as possible. This produces a large image area and allows the flux through the sample to be reduced without impairing visibility on the viewing screen. However, one should be aware of the resolution limit imposed by grain in the screen and photographic materials.

One interesting characteristic of the objective-projector system is the great depth of focus caused by the small angular aperture of the system (about 10^{-2} radians). The strength of the projector can be varied through a very wide range without influencing the quality of the image. This effect is also seen in the fact that the viewing screen and camera are separated by a fairly large distance and no change in focus is required when one switches from one to the other.

Finally, the electrons impinge on a fluorescent screen and are converted to visible light. Compared to the photographic emulsion or other imaging devices, the screen has a rather low contrast and poor resolving power. Grain size and diffusion limit resolution to about 100 μ. Because of this the screen is used only to search the sample for a desired area, find the region of best focus, set illuminating conditions, and compensate for aberrations.

The intensity of an image is limited by the beam current density at the screen, which in turn is limited either by the radiation sensitivity of the sample or the capacity of the source. For a fully dark-adapted eye, a useful lower limit is about 10^{-11} A/cm^2 [12]. In the normal position the eye uses about a 10^{-4} part of the light produced (the screen itself is only 20% efficient), and resolves only a few tenths of a millimeter. Some "image intensification" can be produced by using an optical microscope to view the final screen. Because of the resolution limit in the screen (about 100 μ) a 10× optical system is adequate. Magnification itself reduces the brightness of an image but this is restored by the increase in the aperture of the optical system.

The Image

An object is seen in an electron microscope because it produces variations in intensity on a fluorescent screen. These variations may be produced by one or more of the following effects: amplitude modulation, edge effects, refraction, diffraction, and miscellaneous deflecting fields.

In amplitude modulation part of the intensity passing through the object is subtracted from the final image. The amount is a function of the mass thickness penetration by the beam. The loss takes place when scattered electrons are stopped by the objective aperture or returned to the image so far from focus that they contribute only to a uniform background. This latter case is caused by lens aberrations. Amplitude modulation is a major source of contrast above 10 to 20 Å.

Edge effects are caused by Fresnel diffraction. Consider a partly coherent beam illuminating a thin carbon film that has small holes in it. One part of the beam passes through one of the holes and remains as essentially parallel illumination. Another part of the beam encounters the edge of the film and is scattered. At distances that are very short with respect to the focal length of the objective lens, these two new wave groups overlap to form interference fringes. Owing to the width and incoherence of the source, only one order is usually observed although the number can be increased by working with a pointed filament.

When the objective lens is underfocused, the lens is too weakly energized and the true image plane is beyond the fluorescent screen. At the screen the rays are converging. The opposite conditions exist when the objective lens is overfocused. The edge of an underfocused image is shown in Fig. 4.2*a*. The characteristic feature is a bright fringe, outlining the more electron-dense material. Fig. 4.2*b* shows an overfocused image, and here a dark fringe is the characteristic feature. The distance between the center of the dark and light fringe at a particular objective lens setting x is related to the distance from focus Δf by $x = \lambda \, \Delta f$. The separation at the point where the fringe disappears is used as a measure of resolving power and the performance of the instrument.

Phase effects represent the major source of contrast below 10 Å. Refraction is caused by an inner potential. The potential inside a sample is higher (Δv) than in free space. For carbon Δv is about 10 V. This imposes a phase delay on radiation passing through a carbon film, and it has been found that the effective refractive index can be expressed as $1 + (\Delta v/v)$. For an object thickness T, the resulting phase lag becomes $T \, \Delta v/2v\lambda$ wavelengths compared to the wave in a vacuum. It can be shown that if a region of higher index than the background is viewed in the overfocused condition the region will appear brighter. In the underfocused condition this is reversed.

In the in-focus condition contrast attributable to phase effects has been

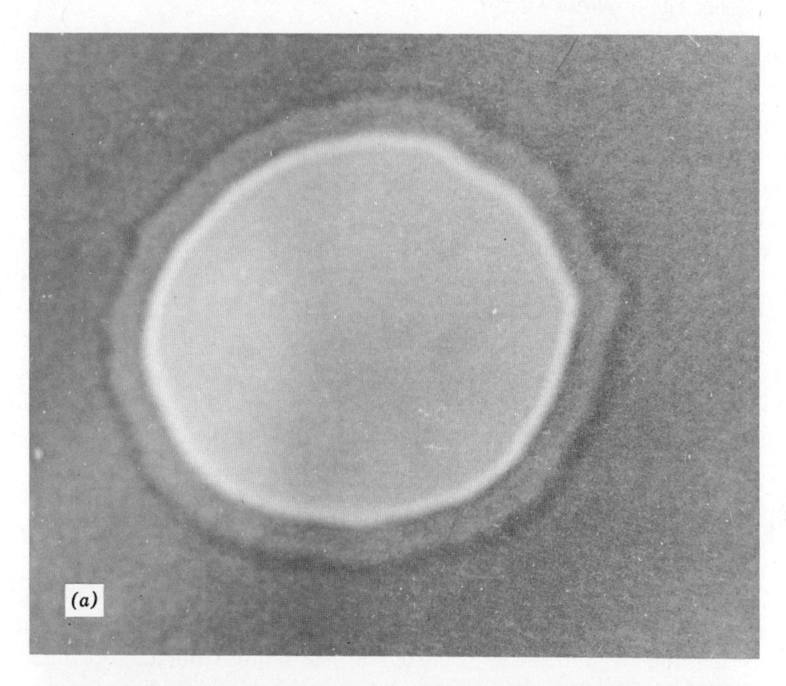

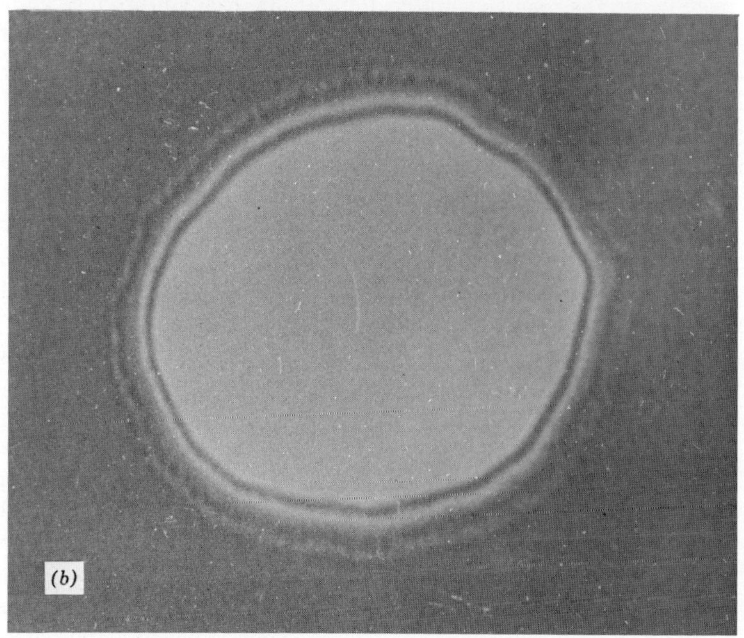

Fig. 4.2 Fresnel fringes. (*a*) Under focus image; (*b*) focus image.

increased by applying the procedures used in optical phase microscopy to the electron microscope. This has been accomplished by recognizing that a coherent incident wave front, after leaving the object, consists of an unscattered wave from the background and a refracted wave that has suffered a small phase change. If a ring-shaped retarder is placed in the back focal plane of the objective lens in such a position that it acts only on the refracted wave, it will increase the phase difference between the two parts of the beam and improved contrast will result in the image plane because of destructive interference. This has been accomplished in the electron microscope by preparing a thin carbon film with a hole in it that transmits only the unscattered beam at the back focal plane [13]. To produce maximum effect at 100 kV it was found that the film should be 250 Å thick. The hole should be about 0.5 μ and the entire unit should be heated to about 200°C during operation to prevent thickness changes from buildup of contamination.

As long as the size of crystalline regions is smaller than the resolving power of the instrument, or is destroyed by the beam, then the image can be interpreted in terms of the previous sections. As the size and perfection of the crystallites increase, dramatic contrast effects can be produced. This is because most Bragg reflections lie outside the objective aperture. The study of this scattered radiation is discussed in the section on dark field. If the objective aperture is removed, the scattered radiation will produce an image because it represents a very small angular spread, but the image will not coincide with the bright field because of spherical aberration.

When the crystallites are small and imperfect, the intensity distribution can be interpreted by simple geometric theory. The fundamental requirement of applicability is that only a small fraction of the incident beam be diffracted. This condition is usually met in poorly crystalline polymers, and as a result of this low intensity, observations are best made using diffracted-beam techniques (p. 312). However, inorganic crystals, particularly metals, are extremely regular in atomic arrangement, and as a result most of the incident radiation is involved in a small number of reflections for a given crystal orientation. Interpretation then requires that the nature of the beam-sample interaction be taken into consideration (General References, A8 and A9).

Large intensity variations may take place across an image of a crystalline preparation such as this because: (1) the angular positions of various parts of the sample may not satisfy the Bragg conditions exactly; (2) certain thicknesses prevent the observation of diffraction at the Bragg angle because of interference effects between the transmitted and scattered waves within the crystal. Thus light and dark contours of equal thickness are produced in the image; and (3) local irregularities of atomic position and lattice displacements may produce isolated changes in the scattered beam. This allows the observation of imperfections in crystals such as dislocations, stacking faults, and strain, even in bright field.

Many crystal lattices have been imaged in an electron microscope. The requirement for this is that the crystal planes be in the diffracting position and the illumination set so that the zero-order beam and one Bragg reflection enter the objective aperture. Contrast is high when the relative scattering is high. This requires a large structure factor; also, the sample should be stable to radiation.

A more indirect method is found in the use of moiré fringes [14]. These may be produced in one of two ways, either from the superposition of crystals of slightly different *d*-spacing or from slightly misaligned crystals having the same spacings. In both cases double diffraction takes place. A beam diffracted by the upper crystal acts as a primary beam, which is further diffracted by the second crystal. The diffraction pattern then consists of primary reflections and satellites. In the image plane these reflections produce a system of fringes that are particularly useful in showing imperfections in the overlap of crystal planes.

The image is also influenced by magnetic and electric fields in the sample. In the case of magnetic fields, any magnetic domain within the sample is not seriously altered by the fields of the incident radiation, and as a result magnetic domain boundaries have been studied extensively. However, the incident electrons can rapidly change charge distribution in semiconductors. Some work has been reported on observations of variations in conductivity. In chemical applications these charge effects are more often detrimental because they cause image distortion in diffraction studies.

When the desired region is in proper focus, the most universal technique for recording is to allow the electron beam to fall directly on a photographic plate. (External cameras are available for use when some image quality can be sacrificed to obtain a print in a few seconds.) The problems of electron micrography have been neglected by most microscopists. In describing the microscope it has been a common practice to say that the effect of medium-energy electrons (50- to 100-kV) on a photographic emulsion is essentially the same as the effect produced by light. This impression has been coupled with the supposition that since the type of radiation remains unchanged the same emulsion and photographic procedures should suffice for all problems. In a number of cases this has placed an unnecessary burden on microscope manufacturers because many of the desired effects can be produced most easily by a careful modification of the recording procedure.

The main difference between the effect of light and electrons is that in the latter case a single electron can make silver grains developable, while in the case of light a number of photons are required. This single-hit process has a number of interesting results. For example, it simplifies exposure-density relationships. Below $D \cong 1.5$ the variation of density with exposure is essentially linear. (Consequences of this are summarized in the following paragraphs. For greater detail see Valentine [15].) It also has the practical effect of eliminating the sensitivity loss on long exposures that is characteristic of the effect of light on emulsions at low-intensity levels.

The principal characteristics of emulsions that concern electron microscopists include speed (sensitivity), resolving power, and contrast. The sensitivity of an emulsion to electrons is mainly influenced by the particle size of the silver halide. Valentine reports sensitivities differing by as much as 10^6, and differences of 10^4 can be produced within a single emulsion by the way it is processed.

Two factors limit the fine detail that can be recorded from an image produced by electrons. The first is called grain, although it is only indirectly related to the size of the reduced silver particles. Grain is usually an order of magnitude larger, and is caused by statistical fluctuations in the radiation, that is, grain is the result of lack of information in an image, not the cause of it. The amount of granularity (noise) has been shown to vary, as $1/\sqrt{N}$, where N is the number of electrons falling in a given area during an exposure. Therefore, a more sensitive (faster) emulsion has more grain. There is a rough balance between the speed of the emulsion and magnification. One has a choice of working at high initial magnification, compensating for the loss of intensity by film speed but losing secondary magnification (photographic) because of increased grain.

Since optical density over its useful range in electron micrographs is essentially a linear function of exposure, the contrast seen by the eye, expressed as γ, increases with density and is relatively free of the emulsion type. It has been shown that this relation can be expressed as $\gamma = 2.3D$. This is not a continually increasing property. If D_s is the maximum developed density of the film, it can be shown $\gamma_{\max} = 0.85D_s$ at $D = 0.65D_s$. One can obtain emulsions that have a wide variety of maximum density values, but for those used most commonly in electron microscopy D_s is usually equal to or greater than 6.

A few techniques are presented here to demonstrate how the amount of information disclosed in a micrograph can be improved by photographic aftertreatment. Consider two general problems in applied electron microscopy.

1. An underexposed negative has resulted in a study in which the work cannot be reported at a corrected exposure. This usually occurs when one tries to reduce radiation damage to the sample by reducing the total electron flux. If one is aware that an underexposure is going to result, developing conditions may be changed. However, if the negative has been developed and

fixed before the underexposure is discovered, a scattered-light technique proposed by Kind and Lau [16] has been found useful. They showed that when a thin negative is bleached and illuminated against a black background by oblique light a direct positive image of improved contrast can be produced. This can then be rerecorded by a variety of means. For example, Fig. 4.3 was prepared using a Polaroid Land camera.

2. Some images have too wide a range of intensities to be photographed by conventional means. Most emulsions used in electron microscopy are designed for a maximum density of 8, while conventional techniques do not use densities much above 2. This problem has been handled in one of two ways. A series of prints can be prepared, each covering a specific density range. This procedure has been improved by Veres and Krug [17], who used a variety of stains. We have found that the holocopy process of Lau [18] is well suited for overexposed negatives. Lau demonstrated that his holocopy process reproduced fine detail over a very wide range of intensities in a single print. In this process black and white negatives are chlorinated. The resulting silver chloride grains are smaller and more transparent than the original silver grains. The usefulness of the technique is shown in Fig. 4; Fig. 4.4*a* shows a conventional part of the diffraction pattern, while Fig. 4.4*b* is a print of the holocopy.

Even with a properly exposed negative, it is possible to simplify the extraction of information by a photographic technique that allows a two-dimensional density analysis of an electron micrograph. This is an application of the broad subject of equidensitometry introduced by Lau and Krug [19]. The procedure can also be used to improve the quality of low-contrast micrographs because it reduces the pictures to a line drawing of a narrow range of densities. An application of the technique is presented in Fig. 4.21.

General Operating Procedures

Many of the operating conditions of an electron microscope are determined by the instrument maker. At present there are six major companies producing microscopes and these built-in conditions are surprisingly standard. However, most modern instruments do not limit the microscopist to taking high-magnification, bright-field pictures. There are three major classes of studies that can be performed: bright field, dark field, and diffraction and reflection. Each of these procedures is further modified depending on the requirements of the study.

Bright Field

The basic operating procedure is the focusing and magnification of the zero-order spot and other electrons scattered through a small angle limited by

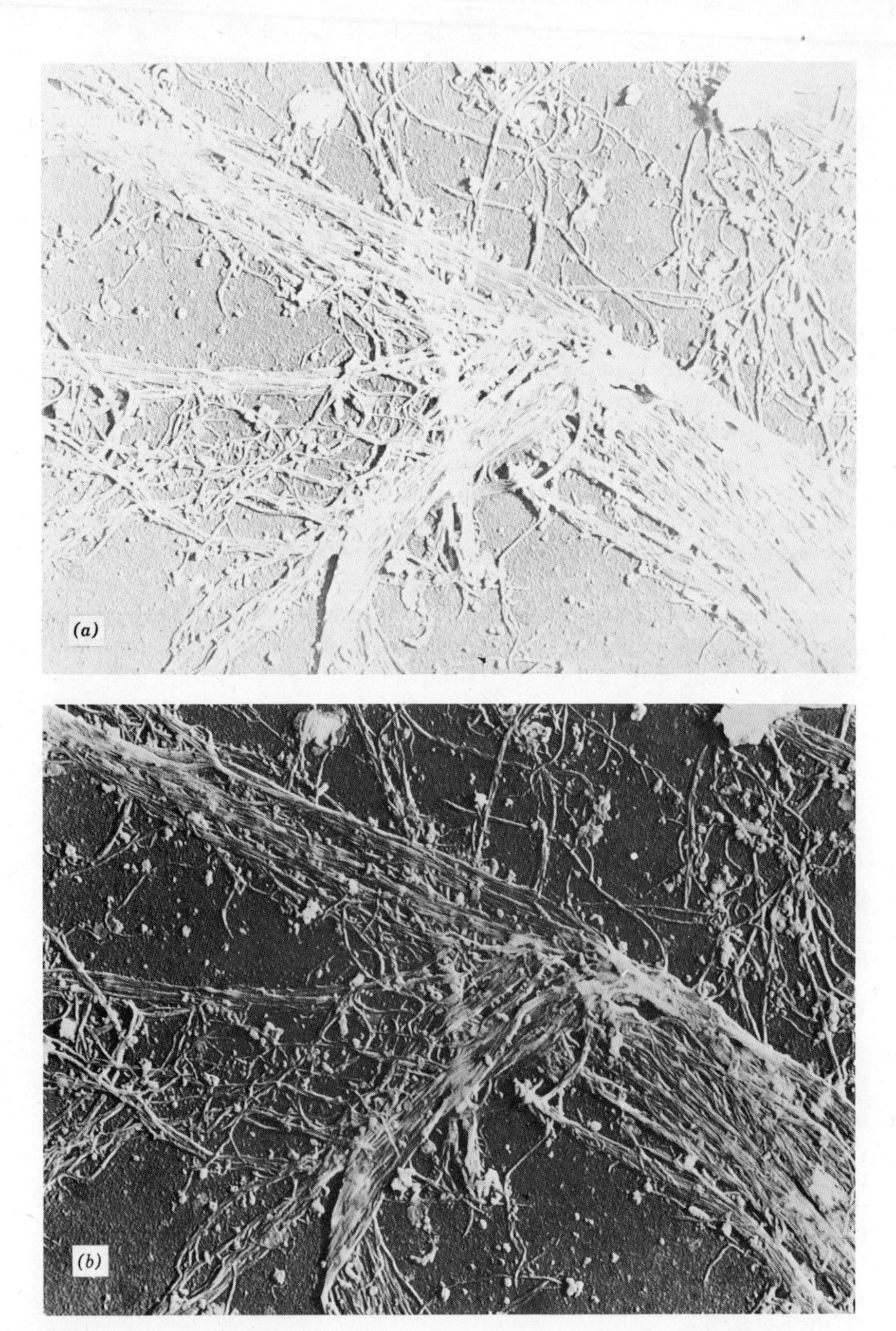

Fig. 4.3 (*a*) Print made from an underexposed negative showing the product of ultrasonic treatment of a synthetic fiber. (*b*) Scattered-light print made from the negative used in (*a*) prepared by the method of Kind and Lau [16].

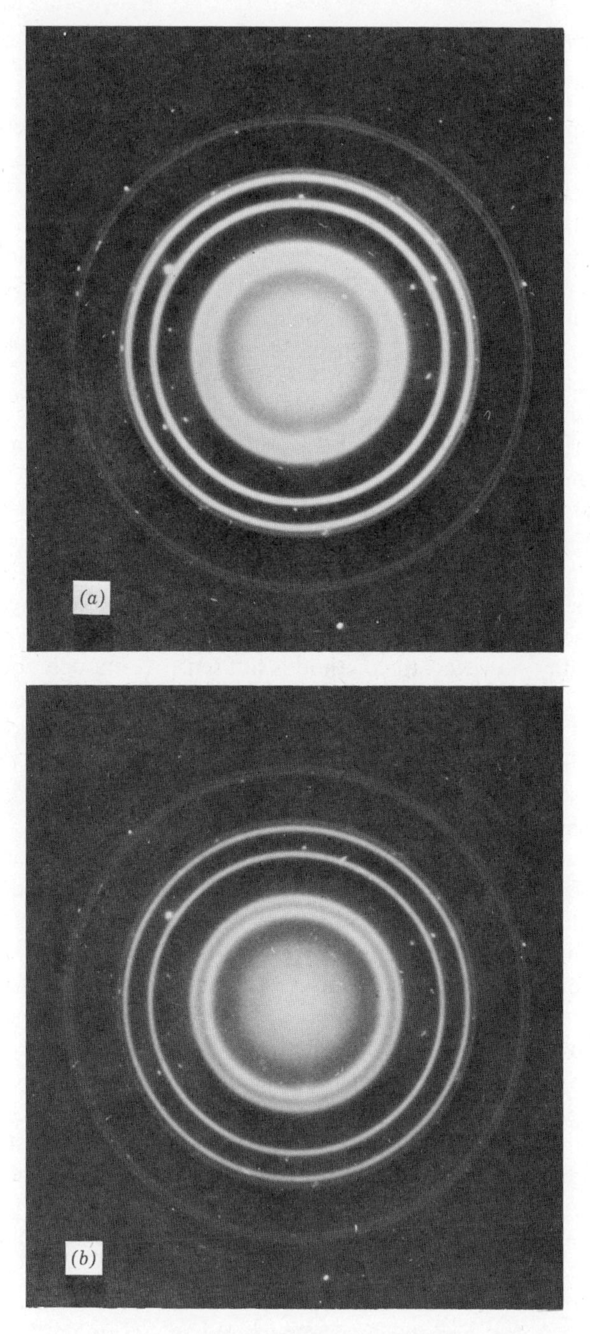

Fig. 4.4 (*a*) An overexposed electron diffraction pattern. (*b*) Holocopy of (*a*). Note that while the outer ring in both prints has the same contrast, the inner rings are better defined in (*b*).

an aperture in or near the back focal plane of the objective lens. This may be a physical aperture or an effective one set by the spherical aberration of the lens. The angular aperture is usually about 10^{-3} radians (corresponding to a physical aperture in the focal plane of about 25 μ). As in optical microscopy, image contrast may be improved by reducing the size of the aperture but eventually a point is reached where it starts to limit field size and resolution.

If the microscope is properly aligned and compensated, and a useful sample has been prepared, the next problem is the preparation of an in-focus micrograph. The method of focusing depends on the magnification and the accuracy desired. Oddly enough, in modern microscopes the position of focus at low magnification is more difficult to locate than at high magnification. Swinging the object lens current back and forth until a sharp image is seen is extremely inaccurate with double-condenser instruments. In the low- and medium-magnification range, some microscopes have devices that produce the effect of multiple sources [20]. These form blurred or double images at all objective lens settings except at Gaussian focus.

One of the simplest and most effective procedures is to remove the objective aperture, focus to the point of minimum contrast, reinsert the aperture, and expose the film: however, the method is not sensitive enough above 10,000×. For high magnification, Fresnel fringes seen at interfaces are used as a focusing aid by most microscopists. However, if the combination of magnification and image intensity does not allow observation of the fringe spacing desired, a series of micrographs should be prepared at small changes in objective lens current through focus.

The method of determining magnification obtained at any particular setting depends on the magnitude and the accuracy required. A standard object should be large enough to be easily measured, yet small enough so that measurements can be made in various parts of the field of view. At low magnification it is convenient to use a 1000-mesh grid. The use of carefully calibrated styrene latex particles as an internal standard is also common practice; however, precision is limited by contamination buildup. This may be about 1 Å/sec. For higher magnification a carbon replica of a ruled diffraction grating ($\sim$50,000 lines/in.) allows an accuracy of 1 or 2%. This accuracy can be used only by employing the grating as a substrate. At high magnification ($\sim$200,000 direct), Heidenreich et al. [21] recommend the use of the 3.4-Å spacing in the lattice of a specially prepared graphite. This material is also used as a check on the performance of the microscope (see p. 348).

The quality of micrographs is dependent on the performance of the microscope, the operator, and the environment. When working at high resolution, it is important to know that at a given time the instrument is capable of a certain level of performance. This is usually about 10 Å. The behavior of the instrument can be adversely affected by misalignment, contamination,

Fig. 4.5 Plate showing the buildup of contamination and image drift.

mechanical or electrical instability, or an unsuitable specimen. It is therefore desirable to make a routine check on the performance of the instrument. The test preferred by many instrument makers is a study of the Fresnel fringes that outline a hole in a thin film at high magnification [22]. The test can be made visually or photographically, depending on the time available and the level of quality desired. If it is to be made visually, a magnification high enough to allow the observation of a 10- to 20-Å fringe spacing on the fluorescent screen is required. If the resolving power of the screen is assumed to be 70 Å, then a magnification of about 70,000× should be used.

The alignment of the instrument should be checked by the method recommended by the instrument manufacturer. The objective aperture to be used in later studies should be inserted and centered. Then at about 70,000× a small hole in a thin carbon film is imaged on the screen. Starting from above focus, decrease the current in the objective lens slowly until either the black fringe starts to disappear or the distance between the center of the black and neighboring white fringe is equal to 10 Å in object space. The fringe pattern is photographed. A second micrograph is prepared by making an exposure and then superimposing a second exposure 2 min later (Fig. 4.5).

In the first micrograph: (1) The fringe spacing at which the black fringe disappears is a measure of the resolving power of instrument at that time. If this loss of the black fringe takes place at a spacing above the expected limit of the instrument, it may be taken as an indication of an instability in the high-voltage or lens currents. (2) If the fringe spacing differs with direction in the micrograph, it is caused by astigmatism and the amount of correction (Δf_A) needed may be determined by noting the fringe spacing at the maximum separation (x_1) and minimum separation (x_2); then from $x = \lambda\,\Delta f$,

$\Delta f_A = \Delta f_1 - \Delta f_2$. (3) If there is a blurring of the fringe in one direction, this indicates an image shift that may be caused by vibration, drift, electric fields caused by contamination, or stray magnetic fields. The second micrograph is recommended to allow one to measure the rate of image drift and contamination buildup.

Dark Field and Diffracted Beam

The technique of operating an electron microscope without using the zero-order beam has two major divisions. The more general technique (dark field), which does not distinguish between electrons that have been scattered elastically or inelastically and is not concerned with any particular scattering direction, is much like dark-field optical microscopy in its type of image and interpretation. In bright field and assuming no Bragg reflections, the image intensity can be expressed as $I_{BF} = I_0 e^{-SW}$, where I_0 is the incident intensity, S the total scattering cross section, and W the mass thickness. The dark-field image intensity can be expressed as $I_{DF} = I_0 e^{-SW}(e^{KSW} - 1)$, where K is the fraction of electrons scattered into the aperture. In the case of bright field, the contrast improves with increased thickness, while in dark field it improves to a point for small thicknesses and then reverses. The conventional method of producing dark field is to use the objective aperture to transmit scattered radiation. For low resolving power the aperture may be displaced laterally. Resolution may be improved by leaving the aperture on the axis of the microscope and tilting the incident beam. Dupouy [23] improved the technique further by inserting a thin wire mask across the center of the objective aperture. With this the zero-order beam is removed and also those electrons that would normally be scattered outside the limiting aperture. The electrons used have suffered only a small energy loss, so chromatic aberration is low.

The second technique involving operation without the zero-order beam uses the elastically scattered radiation that contributes to the Bragg pattern. This type of study has been referred to by Reisner as diffracted-beam microscopy. Simply stated, this technique allows one to observe the regions that are producing a particular reflection that is transmitted by the objective aperture. The technique has been useful in the study of imperfections in crystals, and in chemistry it is used to show the distribution of a particular crystal type or form in a powder sample.

In diffracted-beam microscopy, as its name implies, the microscope images only scattered electrons. The limiting aperture is in the back focal plane of the objective lens, and the effect can be produced in two ways. The simplest method involves a lateral displacement of the objective aperture away from the undeflected beam. Since this uses off-axis radiation, the resolving power is reduced because of spherical and chromatic aberrations. Kobayashi and Sakaoku reduced this effect by using a double aperture situated symmetrically about the central beam. The method is similar to that of Dupouy but uses specific Bragg scattering.

A method that is becoming more popular is that of tilting the illumination about an axis located at the sample. Thus only axial rays are used and there is little deterioration of resolving power.

Electron Diffraction

One of the most significant contributions that the electron microscope makes to the study of chemistry is not usually thought of as microscopy, but it is just as relevant as looking at the interference figure in the back focal plane of the objective lens in a polarizing microscope to characterize a specimen through its optical properties. Davisson and Germer [24] were first to report observations on the diffraction of electrons. They used relatively low-energy radiation (65- to 600-V) and measured the reflected radiation from near-normal incidence. In the same year, Thomson and Reid [25] reported the first studies of transmission diffraction through thin films. Also about the same time, Nishikawa and Kikuchi [26] demonstrated diffraction at grazing incidence. This is the most common form of reflection diffraction used today.

Electron microscopy and diffraction were joined about 1940 when Hillier incorporated diffraction attachments in RCA's first commercial microscope. Since then there have been many refinements, but the two key developments were reported by: (1) Simard and associates [27], who demonstrated the magnification of patterns within the microscope to improve dispersion; and (2) Le Poole [28], who developed a practical system of obtaining patterns from small selected areas. An adjustable aperture is set in the image plane of the objective lens so that all the electrons (scattered and unscattered) from a chosen area of the sample are allowed to pass down the column. The projector lens is then focused on the back focal plane of the objective lens where the first diffraction image is formed.

In this section we consider only the diffraction of electrons by solids because the diffraction of electrons by gases is discussed at greater length by Bartell in another part of this volume. Many of the principles of x-ray diffraction reviewed by Lipscomb and Jacobson in this volume can be applied to electron diffraction with only slight modifications: however, there are four important differences.

First, the wavelength associated with the electrons used in diffraction is much shorter than that encountered in x-ray studies. For example, the wavelength for 100-kV electrons is 0.036 Å, while the copper K_γ-line used in x-ray diffraction is 1.5 Å. One of the most obvious effects of this is to change the Ewald sphere to very near a plane. The effect is illustrated in Fig. 4.6. From a practical standpoint it means that the Bragg angle is very near zero,

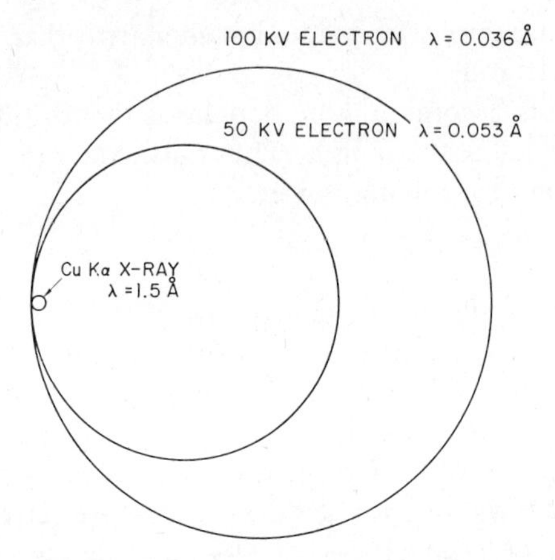

Fig. 4.6 Ewald sphere for x-rays and electrons showing the effect of wavelength. (Reprinted from Ref. 71, p. 81, by courtesy of the American Society for Testing and Materials.)

and when the incident beam is parallel to a family of planes an entire series of reflections may be seen rather than one or two.

Second, the scattering efficiency is much higher for electrons than for x-rays (about 10^6 times). X-rays usually require a sample volume of 15,000 μ^3 and exposures measured in hours, while a sample volume of 0.1 μ^3 requires an exposure of a few seconds for electrons. Another way of expressing this increased scattering efficiency is to say that the electrons have poorer penetrating power. This makes electron diffraction a very useful tool for studying surfaces, particularly at low voltages (~100 V) [29]. Unfortunately, this speed and efficiency is counterbalanced by a greatly increased inelastic cross section that results in increased radiation sensitivity.

Third, the elastic scattering cross section for electrons does not increase as rapidly ($_Z0.7$) as it does for x-rays ($_Z1.2$). Thus in structure analysis, for example, it is easier to identify the position of hydrogen in the presence of heavier atoms [30].

Fourth, and most important, there are a number of advantages in the fact that electrons are relatively easy to focus, for example, short physical camera length with long effective length, and good collimation at high intensity. However, the unique contribution comes from being able to choose a particle position or region to a precision of less than 1 μ and to prepare a pattern in a few seconds. However, problems are imposed: (1) Since the camera length now depends on the position of the sample, a careful voltage and a current calibration has to be made. This is best accomplished by including materials of known spacings in with the sample to be studied. (2) If one wants to relate the reflections in a pattern to a direction in the sample, a careful study of image rotation has to be made. (3) In the preparation of patterns of d-spacing measurements, lens effects such as astigmatism have to be either measured or compensated.

Resolving power in electron diffraction has been specified in two ways. First, in terms of crystal quantities: $RP = d/\Delta d$, where d represents a particular interplanar spacing and Δd the smallest detectable change. A second method proposed by Cowley and Reese [31] is independent of the diffracting angle. In one form, $RP = \lambda L/S$, where $S =$ the smallest measurable difference on the plate, $L =$ camera length, $\lambda =$ electron wavelength. In the selected area position, RP is usually about 700. Figures as high as 10^4 have been obtained.

Electron diffraction patterns can be prepared in a commercial microscope by a number of different lens combinations, each system having a particular advantage.

1. *Below the final projector lens:* No lenses operate below the sample, so the zero spot is focused with the condenser. The advantage is that the camera length is fixed and directly measurable. It has the disadvantage that the selection of an area is not too precise. The size of the area is limited either by the size of the condenser aperture or the openings in the specimen grid. Diffraction in this position can be accomplished either by transmission or grazing incidence reflection. This type of technique can also be carried out at the normal object position by switching off all lenses below the sample, thus increasing the camera length, but it requires the use of wide-bore pole pieces.

2. *At the normal object position:* This is the most conventional position for accomplishing selected area diffraction for d-spacings from about 1 to 100 Å because it also allows a high-resolution view of the desired area and the procedure involved is relatively standard. For heterogeneous samples it is very important to have the first image of the sample focused exactly in the plane of selected area aperture to be sure that the diffraction pattern is created only by material within the selected area.

For small-angle diffraction in this position, the first condenser is at maximum strength to produce a small spot. The second condenser is focused so crossover is formed just below the sample. The objective and second projector lenses are turned off and the intermediate lens images the crossover on the screen.

3. *In the upper part of the specimen chamber:* Many microscopes have a large specimen chamber above the objective lens that allows an increase of objective focal length of about 5×. Here, one can create an effective camera length as high as 5 m. It has the added advantage that the size of the pattern in the back focal plane of the objective lens is increased so that when operating in the diffracted-beam mode a given size of aperture transmits a smaller

area of the pattern (about one-half as much). This produces an improved contrast because a greater percentage of the inelastic component is removed. Unfortunately, the procedure is time-consuming in many instruments because the entire column has to be evacuated after each sample change.

Applications have been found for electron diffraction in many branches of science. These include determinations of perfection, size, and shape of crystals, degree of crystallinity, and crystallite orientation. The two most important applications in chemistry include the study of molecular structure and the identification of crystalline compounds.

The identification of an unknown crystalline substance follows the procedures described in the section on x-ray diffraction by Lipscomb and Jacobson, with particular attention paid to the points discussed here concerning lens effects.

The two most common types of patterns encountered are spots from single crystals or rings from polycrystalline materials. The powder pattern is preferred to the spot pattern in identification because it usually contains all possible reflections, while a crystal pattern only shows one zone at a time.

The first impression one obtains when looking at a diffraction pattern is an idea of crystallite dimensions. Above 100 Å the reflections form sharp rings or discrete spots which expand to diffuse halos in amorphous material. The data required for identification include the distance D of the spot or ring from the central spot and the relative intensity. D is converted to interplanar spacing by $(dD/N) = 2\gamma L$, where γL is the camera constant. These values are then compared to data in the ASTM handbook of x-ray diffraction patterns. For simplicity, the selected area should not contain more than two different crystal forms.

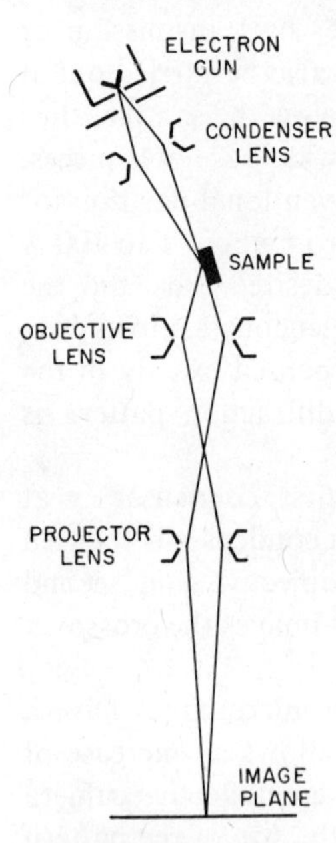

Fig. 4.7 Schematic representation of a reflection electron microscope.

Reflection Microscopy and Diffraction

Reflection electron microscopy was first demonstrated by Ruska [32] in 1933. The angle between the incident and reflected beam was 90° and the resolution was poor. The present system of using grazing incidence was first used by Von Borries and Jansen in 1941 and allows a resolution of a few hundred angstroms. A general outline of the equipment is shown in Fig. 4.7. Since "opaque" samples are used, large energy losses take place and these losses, coupled with

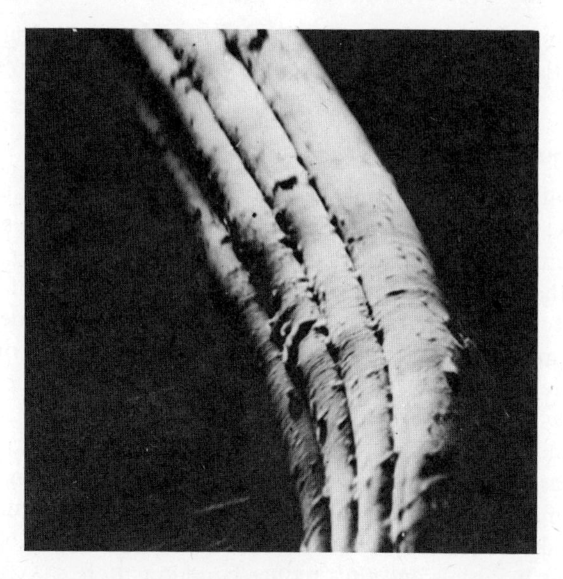

Fig. 4.8 Reflection electron micrograph of a silver replica of a rayon fiber. A thick replica was used to reduce radiation damage. (Reprinted by courtesy of D. E. Bradley.)

chromatic aberration in the objective lens, produce a loss of resolving power. The circle of confusion about a point object is given by: $d = C_c\alpha(\Delta v/v)$, where C_c is the chromatic constant, α the half-aperture angle, and $(\Delta v/v)$ the fractional energy loss.

Experiments in the 1950s demonstrated the following advantages of reflection microscopy over the use of replicas in transmission microscopy: (1) Reflection techniques allow the study of any degree or type of surface topography, while in many cases it is difficult to remove a replica for observation, particularly if the sample is not to be destroyed. (2) The behavior of a surface in response to a variety of treatments can be studied directly. (3) Reflection microscopy makes selected area reflection diffraction possible.

However, this reflection technique has the following disadvantages: (1) Since a large part of the beam is stopped by the sample, thermal effects are severe, particularly in dielectrics. (2) It is a low-intensity technique. (3) At present, resolving power is over 100 Å. (4) Because of the oblique angle of the sample, magnification is not symmetrical and only a limited band of the sample is in focus at one time. Figure 4.8 [33] illustrates the application of the instrument to the study of synthetic fibers.

Although equipment now exists for placing the sample in a more favorable position in the objective lens (to reduce chromatic effects), and effective intensifiers are available, the use of this technique has been almost completely replaced by the scanning electron microscope. Its main use now is in reflection diffraction.

Related Equipment

The activity in finding new ways of extracting information from the electron microscope has created such rapid changes in accessories for the electron microscope that it serves our purpose here to consider only the types of devices that are commercially available. There are few if any standard techniques for such apparatus. Most of the devices are made by all major microscope manufacturers and they represent the best source of information.

First, there is the problem of temperature regulation. In the low-temperature range two devices are available. One cools the sample. This is particularly useful for sharpening diffraction effects by reducing the thermal motion of atoms. It has also been reported that by using liquid helium temperatures radicals formed by the beam can be stabilized to reduce lattice distortions. The second device produces a cool surface near the sample to condense out carbonaceous materials and thus reduce contamination.

Heating devices that produce temperatures up to about 1000°C are in use in annealing studies. They have been refined to allow tilting for diffraction studies. Gas reaction cells have also been used to study chemical reactions and crystal growth [34].

Equipment is also available for deforming samples during observation in the microscope. They have been used on plastics as well as metals, but few applications have been found in chemistry.

A wide variety of tilting devices having differing degrees of sophistication are available. These are designs for: (1) the large chamber between the condenser and objective lens, (2) within the objective lens, and (3) below the projector lens. These stages have been used in diffraction studies, such as the identification of single crystals, and more recently have found application in the interpretation of three-dimensional structures in sections.

The possibility of making wet cells in order to study liquid samples was explored early in the development of the microscope, but work did not progress because of difficulties in making the chamber strong enough to withstand the pressure and thin enough for 60-kV electrons. With 20 years of experience in vacuum techniques and high-voltage instruments, this field is receiving more attention. The preferred material for a substrate or "window" is cleaved graphite [35].

As research workers have become more familiar with the unusual capabilities of the electron microscope, the techniques involved have become more elaborate, the variety of sample types has increased, and in many cases the nature of the information obtained has changed. Development of these attitudes has been seriously restricted by limitations imposed by the method of observing an image in the microscope. The limitation is felt most when one tries to focus properly at low intensity. The variables that determine visibility include: brightness, size, and contrast. In the early instruments the fluorescent screen was observed directly. The intensity was turned up to allow the small structures to be seen well enough to be focused. Then the intensity was decreased by defocusing the condenser to use the full resolving power of the objective. After a short time it was found practical to use a low-power optical microscope to view the final screen. Although the angular aperture of the viewing system was improved, there was still a sacrifice of intensity to gain magnification.

The need therefore has been for a device that produces an image of increased brightness without loss of detail. The most popular method today is the combined use of an intensifier tube and a television display [36]. In such a system the electron beam in the microscope is converted into light at a transparent screen. The light image is coupled to the first stage of an intensifier either by a camera lens or fiber optics. Here the light is converted into electrons, which are accelerated and strike a target, creating more light and creating more electrons. After three or four stages, the light is picked up on a vidicon tube and transferred to a television monitor.

The gain in intensity in the multiplier tube may be as high as 10^6, but this does not represent the advantage to the observer. In most cases the coupling between stages is relatively inefficient, for example, between the electron beam in the microscope and the first stage of the intensifier tube. The efficiency in present systems can be stated in practical terms if one considers a set of operating conditions under which a microscopist can just focus at an intensity on the fluorescent screen required for a 4-sec exposure. The use of an intensifier allows the intensity to be reduced to the equivalent of a 2-min exposure for the same accuracy of focus.

The main advantages of such a system include: (1) The beam current density at the object can be reduced to minimize damage to the specimen without reducing the visibility of the system. (2) Focusing at very high magnification is simplified. (3) Contrast can be increased electronically. (4) If the monitor is being photographed, contrast can be reversed electronically to save additional steps in some photographic techniques. (5) Several people can view the microscope image in a well-illuminated room. (6) High-voltage microscopy is helped by being able to move the viewing screen about independently of the column. This is important in making adjustments or protecting the operator from radiation.

One area that has not been explored very extensively is tape-recording the image. This should be a particularly good method of comparing standard patterns with unknowns or observing transient events.

Sample Preparation

In transmission electron microscopy it is rarely possible just to put an object into the microscope and look at it because of instrument features discussed in the last section. As the performance of the instrument has improved, preparation techniques have become more sophisticated and specialized. Even with 30 years of development, electron microscopy has more art in it than most sciences and most of that is found in specimen preparation. In this section only outlines of procedures are presented to give research workers an idea of the variety of ways a sample can be characterized. For more specific details the books listed in the General References are recommended.

These techniques fall into two general classes: (1) those designed to show a surface; and (2) those used to describe structures beneath a surface. A third class might include a combination of the two, or techniques that show internal structure by working with freshly prepared surfaces.

It is a good practice when considering techniques to be applied to a research program not to confine oneself to those used only in one branch of science. One thing that continues to characterize the practice of electron microscopy is good communication between disciplines, and as a result techniques found effective in one science soon find uses in others. Therefore in preparing this survey of techniques, the list includes methods that have potential usefulness in problems related to chemistry as well as those now in routine use.

The Specimen Support

All samples for the transmission electron microscope are either thin films, less than a few tenths of a micron thick, or are mounted on thin films, neither of which is self-supporting over large areas. In the conventional microscope the area of a sample is usually limited to a few square millimeters at most. Mounting of the films is usually accomplished by the use of multiaperture grids. Although these grids have taken an infinite number of forms, the basic grid is a 200-mesh/in. disk 3 mm in diameter and about 0.8 mm thick. Some of the special forms include: (1) A structure that allows one to identify a particular area for future investigation. This may range from a mark in the center of a disk to a completely coded grid system. (2) Elements that can be used to hold the sample at elevated temperatures. (3) Holders for the study of tensile behavior. (4) Special grids for holding sections. These grids are usually made of copper, but when special mechanical strength or resistance to heat or chemicals is required they can be obtained made of steel, nickel, platinum, gold, rhodium, and so on.

When the sample is too small or too weak to be properly mounted over a grid system, then some sort of substitute film is required. The choice of material and structure should be determined by the nature of the preparation and the information desired. For example, diffraction techniques require a completely amorphous substrate of low scattering power, while staining techniques require uniformity and stability. The substrate always adds to the background noise in the micrograph, and the uniformity of this scattering on a near-molecular level becomes very important at high resolution. The two most popular techniques today are the use of a porous membrane or the preparation of vacuum-evaporated films formed on nearly structureless substrates.

The most widely used plastic materials include collodion and Formvar. The main advantage of this type of support is its ease of formation, flexibility, and strength. Plastic films are useful in preparing carbon grids and perforated films. Carbon grids are used to check microscope performance (see p. 310) and as very fine mesh grids for sections.

Disadvantages of plastic films include: (1) It is difficult to make them thinner than 200 Å. (2) They possess a surface structure about 50 Å in size. (3) Thermal and electrical conductivity are poor. Because of (3), specimens (particularly those containing large particles) drift in the beam. This can be greatly reduced by adding a thin coat of evaporated carbon.

The general method of preparing plastic films is by casting a dilute solution (e.g., 5% collodion in amyl acetate) on water, or by allowing a drop of solution to dry on a glass slide. This film is then stripped off over water. The support can be strengthened and stabilized some by cross-linking the film by exposure to an electron beam of low current density before starting observations.

Although many inorganic materials form films by vacuum evaporation, only carbon has found general acceptance. The evaporation of carbon is usually carried out at a vacuum of better than 10^{-3} torr, using 50 to 80 A. The target is set at about 10 to 20 cm from the source (pointed carbon rods). Carbon films can be stripped over water and picked up on grids for film thicknesses down to about 100 Å. If films below this thickness are to be used, it is best to coat Formvar films and then wash out the plastic. This method has the added advantage that residual plastic between the carbon and grid acts as a cement. It is particularly useful if the sample is to be treated while it is on the grid. Carbon films have the disadvantages that they take longer to prepare than plastic films, are hydrophobic, and tend to be brittle. Other evaporated films that have been used for special problems include beryllium and silica.

Dispersions

On the face of it, the study of dispersions in the electron microscope should be relatively simple. Just spread some of the sample on a grid and observe it.

However, this type of sample has been studied longer than any other, yet the work is farther away from a universal technique than any of the other procedures. The problem can be broken down into three stages: (1) obtaining a representative set of samples; (2) transferring the sample in the desired form to a microscope grid; and (3) maintaining that form during observation and recording.

The problem of obtaining a representative sample has two facets: (1) obtaining representative distribution from a dispersion or powder; and (2) breaking down or dispersing the sample to a desired extent. If the original sample has a uniform particle size below a few tenths of a micron (e.g., colloid or aerosol), the first step presents no problem. However, in the presence of a wide spectrum of sizes, separating techniques using cascade impactors or selective sedimentation along with careful optical or scanning electron microscopy are recommended.

The usual methods of breaking down preparations include mechanical devices, ultrasonic radiation, or chemical means. Mechanical devices have taken a variety of forms, the most popular of which has been the Waring Blendor. In this apparatus the sample in the form of a liquid suspension is beaten by high-speed blades. Another device that has been useful for breaking up small cells is the Mickle disintegrator [37]. This instrument uses two tuned reeds and the grinding is done by glass beads about 0.1 mm in diameter.

Ultrasonic radiation has found greater use in cleaning small microscope parts than in breaking down samples. This is probably attributable to power requirements, although it has been shown to be effective in breaking down fibrous materials and dispersing powders in liquids.

Chemical methods usually involve the solution of one component of a multiphase system, and since the nature of the distribution of phases may be important this procedure is usually made part of another technique (e.g., see extraction replicas).

In preparing particulate samples the problem of dispersion is encountered twice. First, the sample (e.g., a dry powder) must be dispersed in a fluid medium; then the dispersion must be retained when the sample is dried on the substrate. The simplest case is found in a dry powder of particle size less than about 10 μ. A preparation of this kind can be dusted on or dispersed in air and allowed to settle on a plastic-coated grid. Stability in the microscope beam can be improved by vacuum-coating the preparation with a thin layer of carbon. Another conventional method involves allowing a dilute suspension to dry on a coated grid. This suspension can be placed on the grid as a single drop or many small droplets sprayed on with an atomizer. This latter technique reduces the effect of agglomeration on drying.

The powder may also be mixed in a material that wets the particles and exerts shearing forces as it spreads out over a surface. Linseed oil or nitrocellulose in dibutyl phthalate have been used extensively. The Langmuir trough method is an important extension of this technique that has been used for dispersing long-chain molecules. It involves the application of a solution containing the molecules to be studied and a spreading protein to the surface of water. The monomolecular layer that forms is picked up on a coated grid and the protein dissolved away.

Freeze-drying [38] is a special method of reducing the agglomeration that takes place when a suspension is allowed to dry on a coated specimen grid. For example, if the sample is dispersed in water, a very thin layer (less than 0.5 mm) is immersed in isopentane and cooled to −150°C. This layer may be prepared by spraying the suspension into a precooled specimen support. The frozen preparation is then dehydrated in a vacuum at a temperature below −40°C. The most important step is the initial freezing, which has to be as rapid as possible to prevent disruption of fine structures by the formation of ice crystals.

In a related study the critical point technique was developed by Anderson [39] to eliminate the collapse of delicate structures caused by forces found in the passage of the liquid-air interface through the sample at the time of drying. Simply stated, in this procedure the sample, as an aqueous suspension, is dehydrated in the liquid state by exchanging water with ethyl alcohol and then amyl acetate. The acetate is replaced by liquid carbon dioxide by putting the preparation in a bomb and flushing at 25°C with liquid carbon dioxide. With the bomb completely filled, the temperature is raised to 35°C. The carbon dioxide critical temperature is 31°C so the final preparation is immersed in gas under pressure. Once the sample is in the desired form on a coated grid, the remaining two problems involve making sure that the form is retained during electron bombardment and that the image contrast is great enough to disclose the desired information. The problem of stability of the sample under electron bombardment is manyfold and depends on the physical and chemical nature of the preparation. In general, charge and thermal effects can be reduced by evaporating a thin coat of carbon onto the sample surface. Radiation damage (including chemical changes) can be minimized by using low beam current, very high voltages, and extremely low sample temperature.

Improving contrast, however, has more general solutions. At a given set of sample conditions, contrast can be varied by changing the accelerating potential. At high voltages, in bright field, contrast is reduced and the sample becomes more transparent, while at low voltages contrast increases but the sample becomes more opaque. Therefore there is an optimum voltage for a given sample and substrate.

At a given accelerating potential, contrast is developed mainly by variations in thickness or mass density. For this reason one of the most popular ways of enhancing surface detail is by the oblique vacuum evaporation of very thin layers of heavy metals on the sample surface. The geometry of the system is shown in Fig. 4.9. Evaporation of metals such as platinum, palladium, or chromium is usually carried out using an electrically heated tungsten filament, molybdenum foil, or a carbon electrode. This technique is extremely effective and widely applied, but it has a few limitations. The shadowing material has a grain structure that adds to the background noise in a sample viewed at high magnification. Table 4.1 shows some average particle sizes reported for a few shadowing materials. It has been shown that careful cleaning of the vacuum system and the simultaneous evaporation of carbon with the shadowing material reduces the average grain size. A second point to remember is that the shadowing effect is highly directional. The limitation

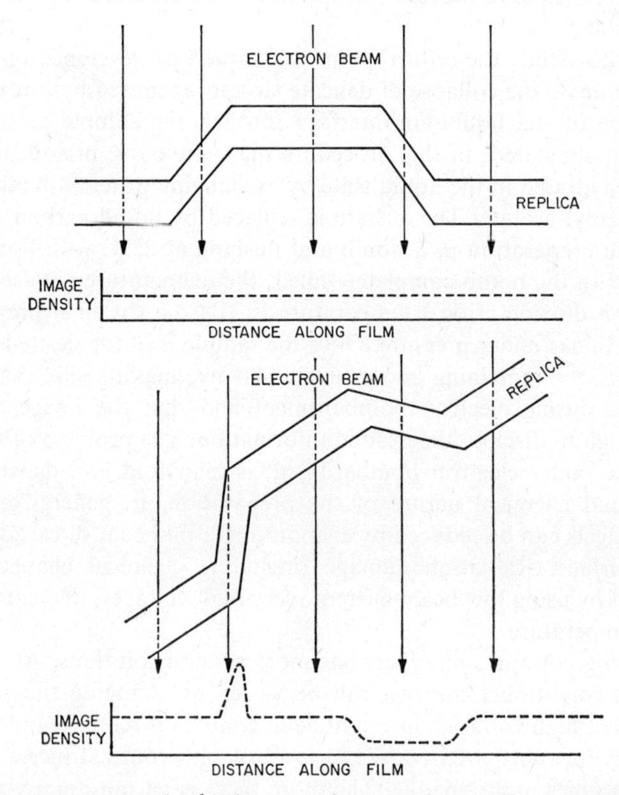

Fig. 4.9 Diagram showing the development of contrast in a replica prepared by vacuum evaporation. (Reprinted from Ref. 43, p. 286, by courtesy of the Textile Research Institute.)

Table 4.1 Metals for Shadowing*

	Granularity (Å)	*Contrast per Unit Thickness*	*Required Thickness* (Å)
Chromium	25–50	6.5	18
Palladium	50–100	6.6	10
Platinum†	<25	6.3	7
Platinum/palladium 80/20	25–50	—	7
Uranium	Tendency to craze	6.0	6

* Data from D. E. Bradley, General References, A10.
† Alloys with tungsten heater.

of good definition in one direction can be removed by one of two methods. First, the sample being shadowed can be rotated about an axis that is parallel to the intended viewing direction. Second, the sample can be shadowed at a direction normal to the surface and then tilted in any desired direction during observation in the microscope. These techniques are illustrated in Figs. 4.11 and 4.24. Staining with heavy metals has also been found useful, particularly for biological systems. The present procedures have been devided roughly into two main classes: positive staining and negative staining.

A positive stain is one designed to increase image contrast by increasing the electron-scattering power of a particular structure through the addition of electron-dense reagents. This addition can be either active, involving chemical groups, or passive, by occupying small spaces in the preparation. Some of the more popular staining groups include double bonds, sulfur, and acids. The techniques are applicable to all forms of matter, and the staining can take place at any time during the preparation of the sample.

In negative staining, as its name implies, the agent is outside the structure being studied. The stain is a highly electron-dense material that causes an increase and reversal in contrast. The main advantages of the technique are: (1) the method is simple enough to allow the study of a great number of preparations; (2) there is good preservation of detail with increased contrast; and (3) only small quantities of sample are required. The technique is applicable to large surfaces as well as small particles. In the latter case a spraying technique is preferred, and best results are obtained if the droplets wet the substrate.

Internal Structure

In the preparation of samples for the study of internal structure, a wide variety of techniques have been developed. These include sectioning, splitting, thinning, casting, fragmentation, and extraction. The choice of the technique depends on the properties of the sample and the form in which the information is required. One also must decide whether he wants to study form or molecular arrangement. The study of one may destroy or reduce the visibility of the other.

SECTIONING

Microtomes for cutting sections thin enough for present commercial electron microscopes (100 Å to 0.1 μ) have taken many forms, ranging from the addition of wedges to reduce the rate of advance in standard microtomes to elaborate systems depending on controlled thermal expansion to advance the sample into the path of the knife. The first knives were made of carefully polished metal and results were poor because of texture in the metal and instability of the cutting edge. The edge of freshly broken glass has become the most widely used cutting tool because it is inexpensive, easily prepared, and has good cutting characteristics. However, the edge deteriorates rather rapidly and cannot be used to cut hard materials. At present the best general purpose cutting edge is a carefully polished diamond a few millimeters long.

In cutting sections there is a wide variation in procedures and the choice is dictated by the sample and the information desired. For example, in biological research one is usually interested in the form (morphology) of groups of molecules stained with heavy metals and dispersed in or perfused by an amorphous polymeric embedding material. The techniques involved have been made fairly standard because chemically and mechanically the systems are fairly uniform.

In the case of semicrystalline polymers, one is usually interested in either distribution of added phases (e.g., pigments, other polymers, or voids) or the nature of molecular arrangement. In most cases in which embedding is used, it is designed to hold the sample only at its surface. There must be a minimum of interaction between sample and embedding material. In isolated cases the structure can be opened without altering fine detail, and then perfusion of the sample with embedding material greatly simplifies sample preparation. Another serious problem is that of plastic deformation from cutting. When noncrystalline polymers are sectioned, they deform but in time recover to a large extent. The recovery can be accelerated by exposing the section to mild plasticizing vapors while it is still on the water bath. Most of the deformation in semicrystalline polymers is not reversible, therefore cutting is best done (1) along natural cleavage planes, (2) at low temperatures, or (3) in a way that the extent of distortion can be measured.

Techniques for preparing thin sections of hard materials such as metals and alloys evolved with the introduction of the diamond knife. In this field, however, the application is not as widespread as in the sectioning of organic materials. Present studies require that all crystalline form be maintained in the production of the film and there are other techniques that are more distortion free (see p. 336). The main advantages of sectioning include: (1) It represents a section of sample unexposed to reagents. This simplifies observations on pore and crack formation. (2) It provides a rapid method for studying phase distributions.

CLEAVING

Several crystalline materials that exist in a layered structure have planes of easy cleavage (for example, mica or graphite). Specimens suitable for the electron microscope can be produced by splitting the sample and then repeatedly stripping the sample between two pieces of an adhesive tape. When the preparation is thin enough, it can be removed from the tape with either petroleum ether or chloroform.

There is a related technique that has been applied to oriented polymer systems such as films and fibers [44]. Consider a synthetic fiber, for example. If its free ends are taped to a glass slide so the fiber is under a slight tension, the fiber can be nicked with a razor blade. Then, by pulling on the nick with a pair of forceps (observed at about 50 to 100 × under a stereomicroscope) a split can be started down the axis of the fiber. The split usually works its way toward the surface of the fiber at a depth of about 0.1 μ below the surface. It propagates for great distances to produce surprisingly distortion-free surface sections. Unfortunately, the final thickness of the "section" is a complex function of the history of the sample, but in many cases it is possible to remove and study repeated layers.

THINNING

This technique is used almost exclusively in the study of inorganic or low-molecular-weight organic preparations. It is usually preferred to sectioning because of obvious distortions in the preparations. The procedure is divided into two steps: (1) the initial thinning, which involves some form of machining to reduce the sample to a few hundredths of a millimeter in thickness; and (2) the final thinning to the point of electron transparency, which is done by electropolishing, chemical polishing, ion bombardment, and so on. The requirements made of the particular agent used in this step are very limiting. Material must be removed at a controlled rate that shows no selective etching over an area of a few square millimeters. This extraction or later handling should not alter the base material. The most popular technique is electropolishing. The literature contains a number of techniques of varying complexity and a useful list of electrolytes (General References, A9, p. 455).

FRAGMENTATION

This technique was used extensively in fiber characterization before the introduction of useful sectioning techniques. Short lengths of sample are dispersed in a liquid (usually water) and exposed to ultrasonic radiation or beaten mechanically. The resulting dispersion is then allowed to dry on a coated grid and either stained or shadowed before observation. Although a lot has been learned about the fibrous character of textiles and other fibrous materials by using this technique, it is not widely used because it can not show exactly where each fibril came from and what its relationship was with the original sample.

EXTRACTION

The characterization of a part of a sample by first removing it from the preparation can be done in many ways. If it is not important to know the relative positions of the structures to be studied, then it is a relatively simple matter to find a solvent for the undesired portion. The residue can then be collected, washed, and transferred as a dispersion to a carbon-coated microscope grid. In choosing a solvent, great care must be taken to be sure of its effect on the phase to be collected. Many reagents listed as nonsolvents for a particular solid can have a marked effect on the same material at the particle size usually dealt with in electron microscopy.

The technique is generally used to study a second component in a multiphase system, such as inclusions in an alloy. The extracted phase is usually identified by electron diffraction. In the general procedure the sample is etched with a reagent that attacks the matrix material. The surface is then vacuum-coated with carbon and the etching continued to free the film.

LOW-TEMPERATURE TECHNIQUES

Probably the most common word in electron microscopy next to resolution is artifact. Sample preparation at reduced temperature is an attempt to reduce the incidence of artifacts and distortions. However, when not used with care, the technique may introduce artifacts of its own.

Until recently, low-temperature microtomy has had a very sketchy history. One of the first people to cool the sample during cutting was Fullam, who used a high-speed microtome [40]. The main purpose was to remove heat generated by the knife. Others have used it for hardening elastomers and reducing irreversible distortion in semicrystalline polymers.

In applying the electron microscope to problems in chemistry, if one is interested in three-dimensional relationships on a small scale (e.g., crystal structure), then the previous paragraph is pertinent. However, if a two-dimensional representation can be used, then the production and study of surfaces may be adequate. If the preparation does not have a volatile phase, then simple fracture at low temperature followed by replication may be sufficient. If not, then the fracture may be followed by chemical or ion etching.

For the case in which one phase of the preparation is volatile (e.g., water),

Steere [41] has developed a special vacuum-etching technique. The preparation is frozen very rapidly to prevent the formation of crystals of a detectable size. While being kept at liquid nitrogen temperature, the preparation is placed in a vacuum system and mechanically cleaned. The temperature is then lowered again and a replica prepared by vacuum evaporation.

AUTORADIOGRAPHY

This technique is used for locating specific atoms or molecules in a bulk preparation. The procedure was developed for optical microscopy and has been successfully extended to the electron microscope to gain approximately a 10-fold increase in resolving power (0.2 μ). In the general procedure relatively thick sections (about 0.1 μ) are prepared of the labeled material and transferred to specimen grids. A thin film of a fine-grain photographic emulsion (Kodak NTE) is placed over the section and allowed to stand in a light-tight box. After this "exposure" has been made, the silver halide is developed with conventional reagents. Overall image quality can be improved by removing the gelatin base, but the silver "image" may be displaced.

Surfaces

REPLICAS

The technique of shadowing to improve contrast was discussed in Section 2, p. 324. If the sample itself does not contribute additional useful information to the image of a shadowed preparation and if the shadowing material can be supported on a grid, then there is no need for the sample to be present in the microscope. This is the basic concept of a broader technique called replication. A replica may be required for a number of reasons. For example: (1) the sample may be too thick to transmit an electron beam properly; (2) one may wish to study only the sample surfaces; and (3) the sample may be altered by conditions in the microscope.

A very large number of replica techniques have evolved over the last 30 years. Each of these seemed to work well for a specific person on a rather specific set of samples. Although these methods appear to be different, they all share a few basic concepts. The first division of techniques is usually determined by the number of steps involved in the process.

In the two-step technique, for example, a relatively thick cast is made of the surface to be studied. The thick casting is used to give stability to the replica while it is stripped from the surface being investigated. The cast is usually made by molding a plastic against the surface under heat and pressure or by coating the surface with a plastic softened with a solvent. After removing the cast from the surface, a thin reproduction is made of the surface of the cast, usually by vacuum evaporation of carbon, silica, chromium, and so on. Finally, the cast is dissolved away and the replica transferred to an uncoated grid.

The procedure outlined above is illustrated by the self-molding technique developed by Sugihara and Yoshida [42]. A small piece of acetylcellulose film about 0.05 mm thick is mounted on a glass slide. Particles or short lengths of fibers are dispersed on the film surface and the preparation exposed to methyl acetate vapor at about 60°C. Exposure time (usually a few minutes) is varied to control the depth of embedding. The sample may then be removed with adhesive tape, and a shadowed, evaporated, carbon replica prepared of the film surface. The film is removed by washing in acetone and the replica is transferred to an uncoated grid.

The single-stage process omits the thick cast. The thin replica is either removed by very careful stripping from a near-structureless surface or by destroying the sample with reagents. We consider the replication of fibers in an illustration of single-step replication because they represent one of the most difficult shapes to reproduce. Not only may they be highly crenulated but they also have very small radii, which makes the use of shadowing somewhat involved [43].

In the most straightforward procedure, a fiber is cemented at its ends to a glass slide in such a way that the fiber is straight without tension. Replicating material is then vacuum-evaporated on to the fiber surface at an oblique angle to the long axis of the fiber. The angle should be about 20° so that the replicating material will produce shadows. Chromium has been used when the replica is to be removed by mechanical stripping [44]. If a suitable solvent can be found to dissolve the fiber, carbon-palladium is better because it has higher resolving power.

The chromium replica can be stripped by placing a section of the coated fiber on a drop of 10% polyacrylic acid in water on a glass slide. When the water has evaporated, the fiber can be removed, leaving the replica in the polyacrylic acid. The replica and polymer are then coated with a 1% solution of polystyrene in carbon tetrachloride to lend strength to the replica and help flatten it on the grid. The polyacrylic acid is washed away with hot water and the replica polystyrene is transferred, metal side down, to a coated grid. The styrene is finally washed away with carbon tetrachloride. The purpose of the step involving polystyrene is to make sure that the replica flattens out properly on the grid.

In this preparation the shadow angle relates directly to irregularities in structure across the fiber. Irregularities running along the fiber are seen in varying degrees of relief because of the curvature of the fiber. The relief is a minimum in a plane that contains the fiber axis and the source of evaporation, and maximum at the edges of the fiber. This effect can be used to disclose any structure by tilting the replica in the microscope. Figure 4.9 illustrates how contrast is produced from an evaporated replica. As an example, Fig. 4.10 shows the replica of an acrylic fiber prepared by evaporating at 90° and viewed

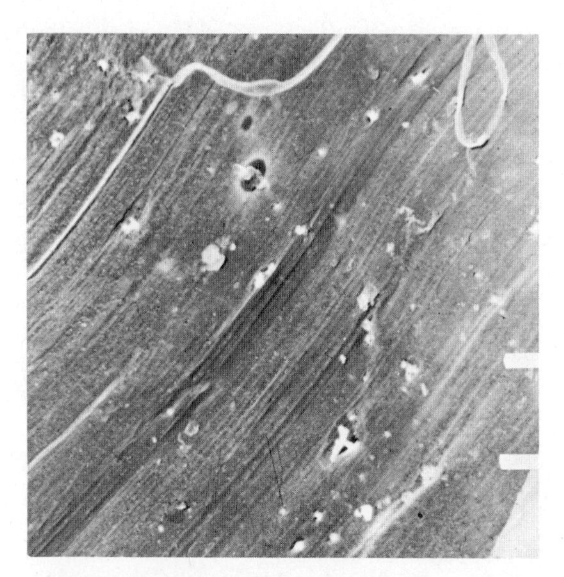

Fig. 4.10 Replica of an Orlon fiber, surface normal to the electron beam. (Reprinted from Ref. 43, p. 287, by courtesy of the Textile Research Institute.)

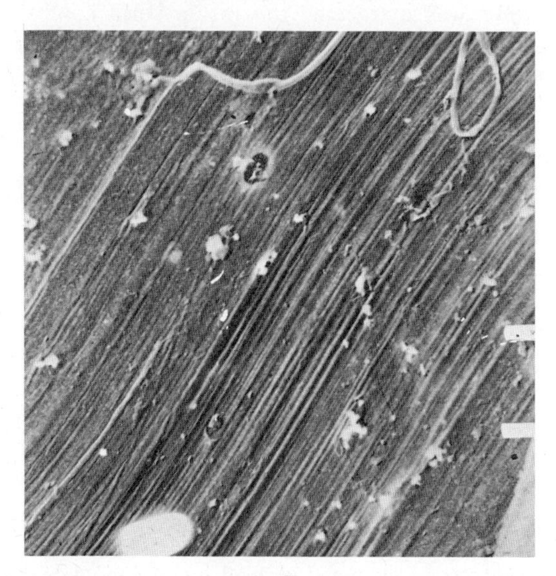

Fig. 4.11 Replica used in Fig. 4.10, tilted 20° to the electron beam. (Reprinted from Ref. 43, p. 287, by courtesy of the Textile Research Institute.)

at a 0° tilt, while Fig. 4.11 shows the same field of view with the replica tilted 20° about the fiber axis. Note that it is not just the angle of evaporation that determines the contrast but rather the relationship between the angle of evaporation and the angle of observation. If one views along the shadowing direction, the contrast should be near zero.

CHEMICAL SECTIONS

In some special cases it is possible to carry the study of surfaces one step further by removing the surface and observing it directly. In the study of metals, this has been done by oxidizing the surface and separating the film by the use of solvents or electropolishing. In the study of polymers, a thin surface layer has been insolubilized either by the use of chemicals or radiation. The soluble portion is then extracted and the resulting film shadowed. This class of techniques is useful only in the characterization of surface morphology, for the crystalline character of the material is either altered or destroyed by the treatment.

One of the unusual features of this type of preparation is shown in Fig. 4.12. To prepare the section an acrylic fiber was treated with hot concentrated sodium hydroxide for a few minutes [43]. This insolubilized a thin layer that the hydroxide came in contact with, including some fine fissures in the fiber. When the polymer was extracted, the inside surface of the section was shadowed. This effectively places the observer inside the fiber looking out. The white bands in the figure are cracks in the fiber surface, most of which are associated with delusterant particles.

ETCHING

The partial removal of one phase while others are left essentially untouched has been the principal technique in the application of optical microscopy to metals and minerals for many years. The extension of this technique to electron microscopy has been obvious and rewarding. It has been shown to be useful in the characterization of pure materials as well as multiphase systems. In the case of single phase systems, detail is seen because of a difference in rate of attack on different crystal faces.

The extent of etching must be very light by standards used in optical microscopy. The obvious requirement is that it be heavy enough to produce detail but light enough not to cause a distortion of the replica when it is removed from the surface. Although etching is usually thought of as being carried out by the partial dissolving of a sample by a liquid or a vapor, a number of different processes have evolved including: (1) the use of electric current in an otherwise inert bath; (2) positive-ion bombardment; and (3) even mechanical polishing, which may remove one phase faster than another because of differences in hardness, thus producing useful information.

The technique has also enjoyed a limited amount of success in the field of

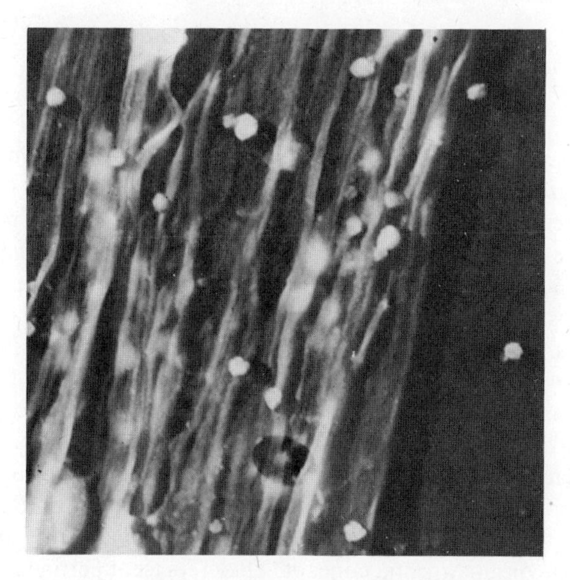

Fig. 4.12 Underside of Orlon chemical surface section.

polymer characterization. Here, however, the requirements of the etching agent are a little different and a little harder to meet. The agent must not have any solvent characteristics, very little or no tendency to act as a plasticizer, and yet be a good degrading agent [44]. Examples of agents used include *n*-propylamine for polyethylene terephthalate, and sulfuric acid for polyethylene.

DECORATION

This method is not considered a true replica technique but rather a method for demonstrating very small steps on the cleavage surface of ionic crystals. These steps are normally as small as a few unit cells in height. Briefly, the procedure is as follows. Vacuum evaporate 5 to 10 Å of gold onto a freshly cleaned surface. Without breaking the vacuum, coat the sample with 50 to 100 Å of carbon by evaporation. The "replica" is then removed from the sample surface and transferred to a specimen grid.

PROFILE TECHNIQUE

Brief mention should be made of a technique that allows relatively thick samples to be placed directly into the microscope. Here one sees a highly magnified shadow image of the edge of a sample. If a dielectric is to be studied, it is preferable to vacuum-coat the sample with a very thin layer of carbon to reduce the buildup of charge from the electron beam. Sikorski has used this technique to study the origin of surface irregularities in fibers [45].

It has also been used in a number of cases to check the accuracy of lateral measurements made in the optical microscope [46].

Applications

The variety of applications of electron microscopy to problems in chemistry has become so great that it is impractical to review all of them. In this section we consider a few examples in each of the major areas of activity. In most studies there are two general approaches, the first of which is qualitative in nature. It is concerned with the identification, location, and appearance of phases or groups. The other approach involves the measurement of quantity and dimensions. In many examples these two divisions are interrelated because a micrograph may contain both types of information.

The discussion of applications starts with the consideration of examples that may be carried out by a well-trained optical microscopist but not with the pictorial clarity that simplifies the job of communicating information. Studies are then reviewed that use the unique properties of the electron microscope. Unfortunately, the choice of examples may not give proper recognition to the originator of a concept. They represent pertinent ideas that are the most familiar to the author.

Qualitative Observations

CHEMICAL REACTIONS

The study of reactions with the electron microscope has followed two distinctly different paths. In the first set of procedures, a reaction is interrupted at measured intervals and the reactants and products are analyzed. One of the classic examples of this approach is the study of the photographic process. The full study of the process has made effective use of every type of technique. The work has concentrated on three stages of the process and we consider them here in the order of their occurrence.

First, there is the unexposed grain. The largest silver halide particle used in photography measures a few microns on a side and so its gross morphology and crystal form has been worked out by optical and related techniques. The important contribution of the electron microscope has been in illustrating the fine structure and how it relates to image formation. To the uninitiated this is a monumental task, for one is trying to study the unaltered structure of a highly radiation-sensitive material with an ionizing radiation. It was recognized at an early stage in the work that although thin crystals could be prepared for transmission microscopy at room temperature they were completely altered by the beam before useful micrographs could be prepared. The first significant information was derived from special carbon replicas which were designed to maintain as much of the total morphology as possible. For example, in work reported by Hamilton and Brady [47], crystals were

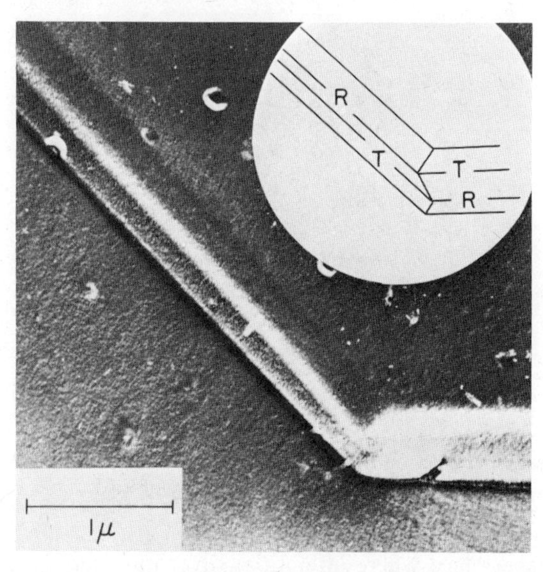

Fig. 4.13 Replica of a silver halide crystal showing the retention of three-dimensional form. (Reprinted from Ref. 47, p. 416, by courtesy of the American Institute of Physics.)

deposited on a continuous plastic support and a carbon coating was applied. Then, the plastic was dissolved away and the fresh surface was carbon-coated. When the crystal was dissolved away, features on either side could be distinguished by the differences in shadow directions. Replication was so complete that tilting studies in the microscope disclosed the nature of twinning. An example is shown in Fig. 4.13.

More recently, Hamilton [48] found that by using a cold stage the radicals formed by the beam could be stabilized well enough to allow the preparation of micrographs showing image contrast resulting from stacking faults. The example shown in Fig. 4.14 was prepared at a sample temperature of about 120°K. Even under these conditions some free silver can be seen.

The second subject is a study of the latent image. This term was used by early photographers to describe the invisible structures that exist in an emulsion between the time of exposure and the developing process. The electron microscope has been instrumental in demonstrating the presence of aggregates of silver formed by exposures to light of an intensity capable of forming a latent image [49]. These sites could contain as few as two or three silver atoms, but larger ones were more efficient. It was demonstrated that these grains exerted a catalytic effect on the development of the entire grain. It has been further shown that this aggregation takes place more readily at some physical imperfection. The imperfection acts to provide traps for

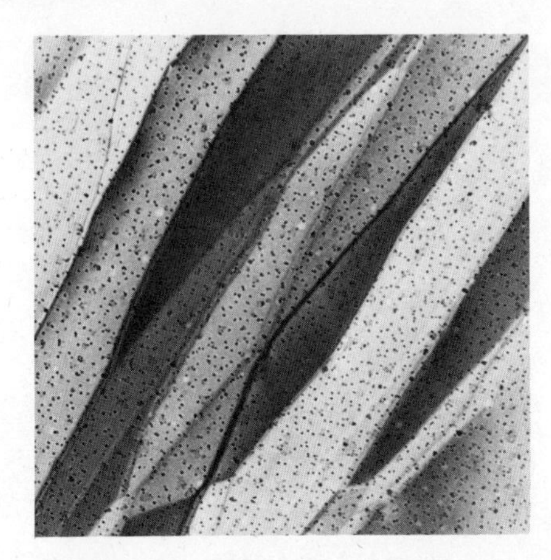

Fig. 4.14 Transmission electron micrograph of a thin silver halide crystal taken at low temperature. (Reprinted from Ref. 48, p. 5, by courtesy of Taylor and Francis, Ltd.)

electrons and as a place where interstitial silver ions may be generated. The crystal surface is the most important source of imperfections because it will be in immediate contact with the developer.

The third area of study has been concerned with the developed grain. The electron microscope has shown that the reduced silver in a photographic image is filamentary in nature (Fig. 4.15). Observation in the optical microscope has indicated that in some cases the developed grain is a sort of pseudomorph of the original halide, but this appearance has been shown to be attributable to the restraining forces in the gelatin. Filament formation from suitable nuclei has been found in many other places such as vapor deposition, and it has been attributed to crystal growth at screw dislocations located on the tips of the fibrils.

As the amount of solution physical development occurring along with direct development increases, the filaments become thicker and shorter. Thus the form of the developed silver depends more on the composition of the developing bath than on the nature of the developer. Koerber [50] demonstrated that for small developed grains (10- to 50-mμ) the size of the grain is more important than the morphology in determining image tone. Morphology is important for large grains. The warmest tones come from loose structures. The tone shifts toward black as the structure becomes more compact.

As another example of this type of study there is the work of Geiss and Lawless [51], who studied the initial stages of oxidation in tantalum. High-purity tantalum that had been thinned for transmission electron microscopy was oxidized in a reactor vessel at pressures of 10 to 100 torr for 30 to 90 min.

Fig. 4.15 A silver halide grain after development showing its fibrillar character along with a partial retention of its original triangular form. (Reprinted by courtesy of J. F. Hamilton.)

The development of the oxide was highly ordered and electron diffraction showed different suboxide phases.

In the second method of studying chemical reactions of this nature with an electron microscope, the observations are made while the reaction is in progress. In the early development of the microscope this method was limited to watching the beam destroy the sample, but with the introduction of higher voltage, special specimen stages, and improved recording equipment, the procedure is becoming increasingly important. For example, Hashimoto [52] and associates developed a specimen chamber for observing reaction processes involving gas and high temperature (Fig. 4.16). The pressure in the chamber can be varied from about 10^{-4} to 300 torr and the temperature varied from room temperature to 900°C.

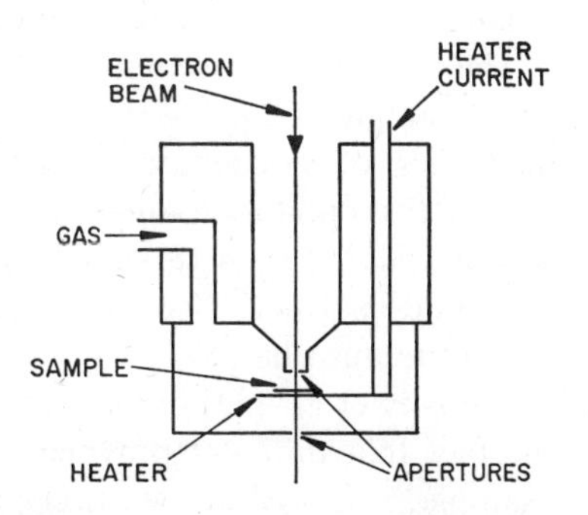

Fig. 4.16 Schematic diagram of a high-temperature stage for observing chemical reactions.

LOCATION AND IDENTIFICATION OF CHEMICAL PHASES

The more-or-less conventional techniques of analytical chemistry are well suited to report the composition of a preparation with great speed and

Fig. 4.17 Longitudinal section of a fiber made from a mixture of polymers. (Reprinted from Ref. 43, p. 291, by courtesy of the Textile Research Institute.)

accuracy. However, many times questions are asked regarding the nature of a dispersed phase, particularly in industrial processes. If: (1) a dispersed phase is small compared to the resolving power of an optical microscope; (2) it is present in a particulate concentration that is large enough to make the job of searching $\frac{1}{8}$-in. grids a practical operation; and (3) the sample is not seriously altered by the requirements of the technique; then the electron microscope can be particularly useful. In this section we consider three general approaches: the use of reagents, tracers, and stains and diffraction. A later part of the chapter reports on the use of electron probes in analysis.

The extraction process itself may be used as a method of localizing phases. For example, Scott [43] identified phases in fibers made from various quantities of polyamides and polyesters by alternately using a solvent specific for a phase on longitudinal sections. Figure 4.17 shows the untreated section, while Fig. 4.18 shows the section treated with hydrochloric acid, a polyamide solvent. The micrographs demonstrate that the polyamide is a discontinuous phase.

Brodsky et al. [53] did essentially the same thing by taking advantage of the fact that different polymers have different responses to radiation. For example, it is well known that polymers such as polyethylene, polystyrene, and nylon are cross-linked by radiation, while polymethyl methacrylate and polytetrafluroethylene decompose in the beam.

Fig. 4.18 Longitudinal section of the sample used in Fig. 4.17 after one phase had been dissolved away. (Reprinted from Ref. 43, p. 291, by courtesy of the Textile Research Institute.)

In the case of polymers there are a number of examples in which different chemical phases were located by chemically substituting groups with high electron-scattering power. Kato [54] made interesting applications of the fact that osmium tetroxide can be added to unsaturated bonds such as those found in polybutadiene. Results of one of these studies are shown in Fig. 4.19. This shows a stained thin section of an ABS (acrylonitrile, butadiene, styrene) plastic. The stained particles are butadiene dispersed in a glassy copolymer of acrylonitrile and styrene. A second glassy polymer was added to make the plastic more transparent. The reason it did not serve its purpose is seen in the phase separation in the glassy portion. The fine structure within the butadiene is occluded copolymer.

Belavtseva [55] found that silver sulfate and mercury nitrate could be used to replace the hydrogen ion in carboxyl groups (e.g., polyacrylic acid) to produce increased contrast.

Senchencov and associates [56] successfully modified the technique of electron microscopic autoradiography used in biology to study metal alloys. The surface containing a radioactive isotope to be studied was coated with carbon by vacuum evaporation. This was coated in turn with a layer of photographic emulsion. After the required exposure the preparation was developed, fixed, and dried. The replica was removed from the metal surface

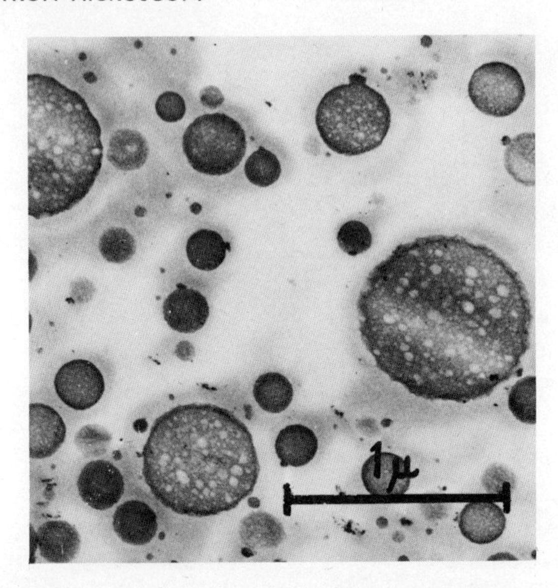

Fig. 4.19 A thin section of an ABS plastic stained with osmium tetroxide. Note the diffusion into the matrix. (Reprinted from Ref. 54, p. 6, by courtesy of the Plastics Age Company, Ltd.)

and the gelatin layer removed with alkali. They pointed out that the carbon that acts as a support for the radiograph can be shadowed at this point so the preparation may also serve as a high-resolution replica. The resolution of the technique was reported to be near the grain size of the M-type emulsion used (about 0.2 μ).

ARRANGEMENT OF MOLECULES

When working with light, polarizing or interference microscopes are useful in studying the arrangement of molecules. This is particularly true for long-chain molecules. These studies are equally effective for amorphous as well as more ordered systems [57]. However, the interpretation of such studies requires a number of assumptions or additional information gained from the use of other instruments.

The majority of studies of this type involving the use of the electron microscope have been most effective in characterization of crystalline materials. The most effective technique is electron diffraction, but since a pattern can be produced by a wide variety of morphologies the studies must be augmented by microscopy.

With the discovery that long-chain polymer molecules are capable of forming complex, well-ordered structures, considerable information regarding molecular arrangement was obtained using the techniques described in

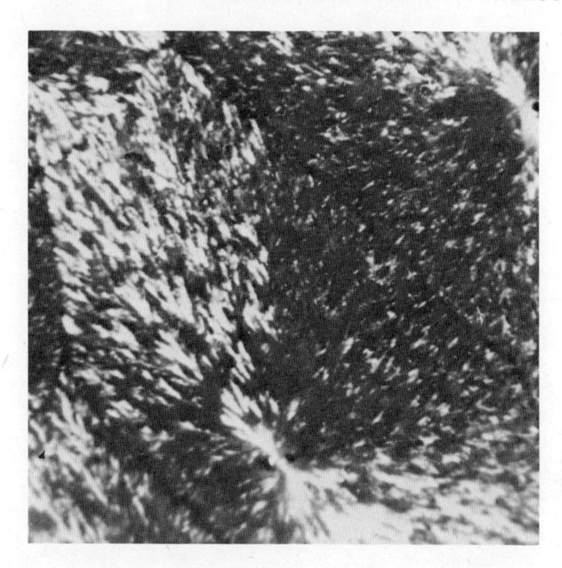

Fig. 4.20 Diffracted beam electron micrograph of a nylon film. (Reprinted from Ref. 60, p. 1090, by courtesy of the American Institute of Physics.)

the last section [58]. Consider, for example, the arrangement of molecules within a spherulite. This is a product of polymer precipitation in which growth starts from a nucleus and expands at an equal rate in all directions. These structures usually have diameters ranging well over 1 μ, so that a thinning procedure is required. Sectioning techniques produce such severe distortions that results obtained from the technique must be checked carefully. Some promising low-temperature sectioning has been reported recently, but one should still make critical tests to measure the degree of distortion. Some success has also been reported from etching, delamination, and fracture techniques applied to bulk polymers. The most important discoveries have been made using spherulites grown in thin films. They are large two-dimensional structures that possess at least a qualitative relationship to the structures seen in bulk polymer. The most well known of these observations is the Maltese cross seen in a spherulite observed between cross polarizers [59].

Working from thin films of nylon 6-6 slowly precipitated from a dilute formic acid solution, Scott [60] showed by use of wide-field diffraction that the molecular chain axis was in the plane of the film-forming layers of hydrogen bond sheets. Then, by using ($hk0$) reflections (planes parallel to the chain axis), it was shown by comparing the direction of objective aperture to the nature of the diffracted beam image that the chain axis was aligned tangential to the growing surface of the "two-dimensional spherulite" (Fig. 4.20). The dark-field picture also gave a measure of the degree of crystallite

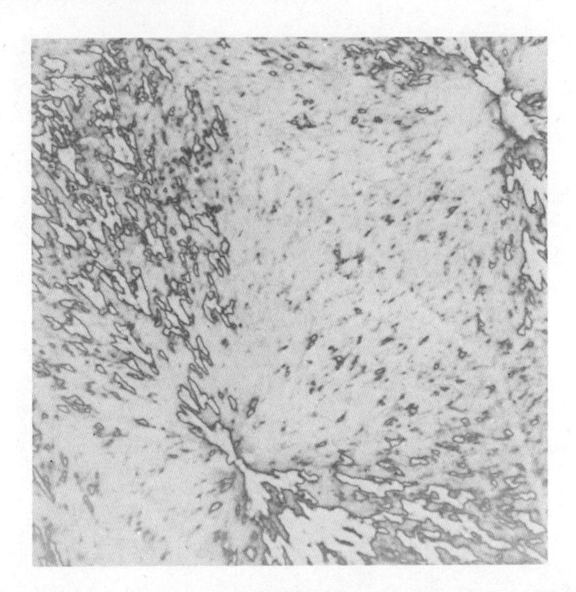

Fig. 4.21 A Sabattier print (equidensity image) of Fig. 4.20 showing the relative intensity of diffraction produced by crystalline regions in various quadrants of spherulites.

alignment because the objective aperture used 10° of arc in the diffraction pattern and 10° within the spherulite diffracted into that part of the pattern. Not until the later work of Keith and Padden [61] was it recognized that these pictures also indicated crystalline cross types between lamellae.

This gross uniplanar order seems inconsistent with the idea of a true spherulite, but investigations such as that of Sakaoku and Peterlin [62] have demonstrated that the basic arrangement in a three-dimensional structure is lamellar. They grew small spherulites from a dilute solution of hot glycerin and observed them on an amorphous substrate on which aluminum crystals had been deposited to serve as a calibration for determining spacings. A composite picture from their study is shown in Fig. 4.22. The brighter area in the center of the picture is the region from which the selected area diffraction pattern was prepared. The micrograph was prepared by superimposing exposures of the general region and the selected area. The diffraction pattern has been tilted slightly to correct for rotation differences between the pattern and the micrograph produced by the magnetic lenses.

The decoration technique originally designed to show active sites on the surface of ionic crystals has recently been applied to the characterization of crystalline polymers. In study of spherulites in thin nylon-6 films, Spit [63] effectively combined decoration with a staining technique. He first demonstrated that phosphotungstic acid selectively stained the (amorphous) space

Fig. 4.22 A bright-field electron micrograph of a nylon spherulite including the electron diffraction pattern of the clear area shown in the micrograph. The outside ring is the (111) of aluminum, used for calibration. (Reprinted from Ref. 62, p. 1035, by courtesy of Interscience Publishers.)

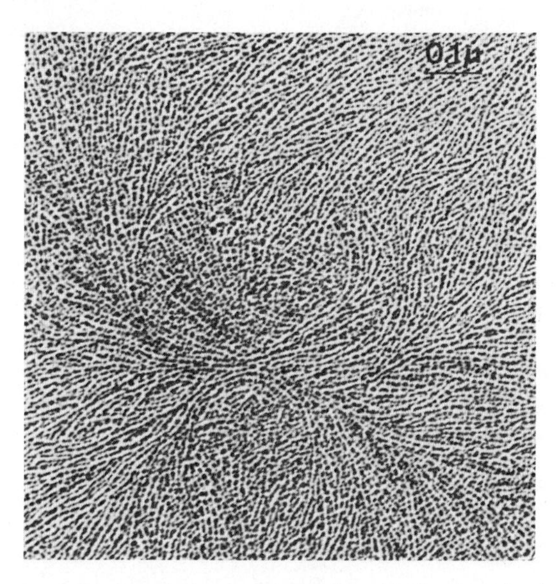

Fig. 4.23 Gold decoration applied to a nylon-6 film. Notice that the decoration shows order in all parts of the spherulite, including the region of the nucleus. (Reprinted from Ref. 63, p. 52, by courtesy of Marcel Dekker, Inc.)

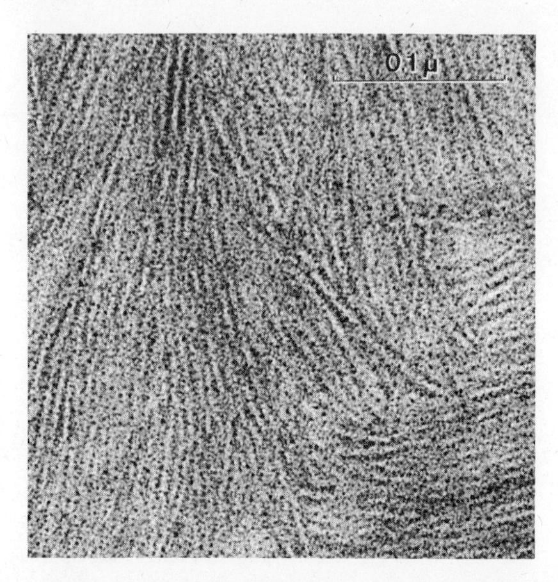

Fig. 4.24 A transmission micrograph of a nylon-6 film that had been stained with phosphototungstic acid and then decorated with gold. Note that the great majority of gold particles lie on the stain bands. (Reprinted from Ref. 63, p. 52, by courtesy of Marcel Dekker, Inc.)

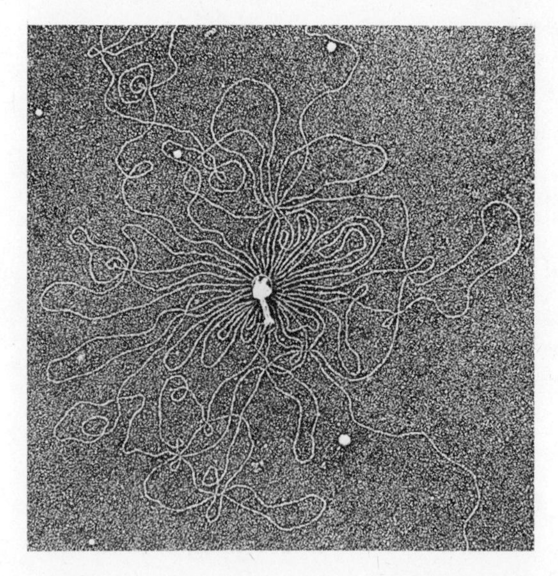

Fig. 4.25 T_2 Bacteriophage, osmotically shocked, showing DNA molecules. Preparation rotated while being shadowed. (Reprinted from Ref. 65, p. 861, by courtesy of Elsevier Publishing Company.)

between crystalline lamellar fibrils. Then, by applying gold decoration first to unstained (Fig. 4.23) and then stained preparations (Fig. 4.24), he was able to show that the stain and decoration marked the same region.

MOLECULES

The goal of high-resolution microscopy is the observation of atoms and molecules. This study has been divided into two projects: first, the design and construction of an instrument capable of the required resolving power and, second, the preparation of a sample having the desired properties.

The electron microscope has been used successfully only in the study of macromolecules. Hall [64] has pointed out that the main problem is not one of resolving power but contrast. He showed, for example, that the mass of a molecular particle could be increased fourfold with stains but that the conditions for such an uptake were drastic. Although shadow casting introduces a fine granularity into the image, it has been the most useful technique for disclosing detail. An example showing a DNA molecule is found in Fig. 4.25 [65]. Another important problem is the use of a substrate that does not add confusing detail. The most structureless substrate in use is freshly cleaved mica, recommended by Hall.

Quantitative Measurements

The separation of qualitative and quantitative observation in an electron micrograph is at times an arbitrary decision because the one thing that distinguishes an electron micrograph is the almost inevitable presence of a mark that indicates the size of a scale (usually in microns) that has been enlarged to the magnification of the image. In this brief review we consider the measurement of number, size, and weight of complete structures. The samples usually encountered in this type of measurement include particles dispersed on a surface, or structures in a matrix that have been revealed either by producing a fresh surface or by sectioning. In the last-mentioned two types, the reconstruction of structures having at least one dimension that is too large and improperly oriented to be completely observed in a single preparation is handled by stereological techniques. Also included in this science is the problem of superposition of particle images. The subject of stereology has developed to a point beyond the scope of this chapter, and the reader is referred to the Proceedings of the Second International Congress for Stereology to obtain information on the present state of the art.

CONCENTRATION

There are two general techniques available to those interested in determining the number of particles per unit volume dispersed in a liquid. In the first procedure Sharp and associates [66] developed a centrifugation technique in which all particles from a known volume of fluid are collected on an agar surface. When the preparation is dry, the particles are transferred to the microscope in much the same way that particles in an extraction replica are transferred. He found a statistically random distribution of particles and demonstrated that with a large enough count as few as 10^5 particles per cubic millimeter could be determined.

Backus and Williams [67] developed a very useful technique from the simple fact that spraying techniques can be refined to the extent that the entire contents of a droplet can be seen in the microscope. By adding a known concentration of recognizable particles (usually a polystyrene latex) to an unknown dispersion and by counting the tracer and unknown particles, the concentration of the unknown group can be found. Use of this technique has been proposed as a method of determining molecular weight of polymer molecules. The general difficulties of the technique are the requirements that only one molecule can be contained in a particle and that the molecules be recognizable. Practical considerations include a grid coat surface that allows the droplet to spread properly, and labeling particles about 100 Å in diameter. This technique is still very much in the development stage.

LATERAL DIMENSIONS

The determination of the length and width of small structures is the most universal and most well-analyzed application of the electron microscope. The study resolves itself into three problems: sample preparation, measurement, and the analysis of data. The degree of difficulty ranges from the measurement of micrographs of electron-opaque spheres with a ruler to the use of a densitometer and computer to study irregular semitransparent structures.

Mounting samples for particle size analysis usually involves the same techniques used in particle counting. Cravath et al. [68] have shown that in spraying techniques one may assume that each droplet contains a full range of particle sizes in the correct proportions.

The study in this area that holds great promise for chemists is the measurement of polymer molecule dimensions. Modern electron microscopes are capable of resolving large molecules. The delay in the useful application of the microscope to this study is the lack of adequate preparation techniques. The problems include those found in the counting technique and, in addition, the difficulty of making precise lateral measurements in small low-contrast objects.

THICKNESS

The two most popular techniques used to measure the dimensions of an object parallel to the electron beam include shadowing and energy absorption.

The principle of the shadowing technique is illustrated in Fig. 4.26. The method operates on the edge of an object, no matter how broad it is, and it

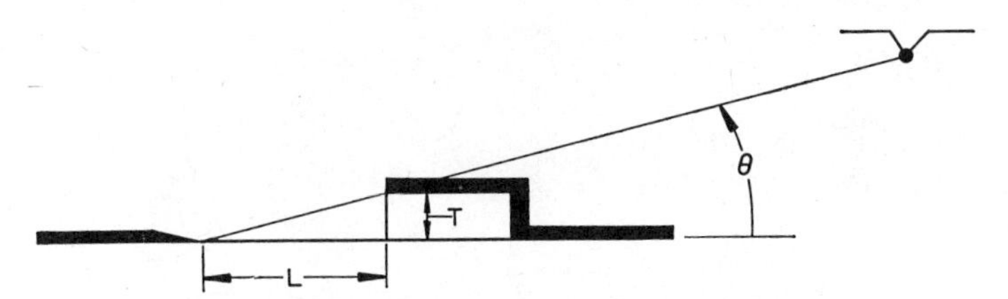

Fig. 4.26 Schematic representation of metal shadowing.

is important to know or to assume an edge profile and how it relates to the rest of the object. The most general case, shown in the figure, is a straight edge. The working equation is: $T = L \tan \theta$, where T is the thickness and L the length of the shadow.

The shadowing procedure has been a very popular method of measuring the thickness of polymer precipitates. The edge is assumed to be straight and the thickness is known to be uniform from observation of electron-beam absorption. Figure 4.27 shows a polyethylene crystal shadowed at an 11° angle. The average shadow length is 500 Å, which was produced by an average step height of 100 Å.

The fundamental equation for image intensity attenuation by the sample is: $I = I_0 e^{-Spd}$, where I_0 represents the intensity of the beam before entering the sample, I the intensity after leaving the sample, S the specific cross section (or cross section per gram), p the sample density, and t the thickness. S is a constant of an instrument under a given set of conditions (particularly, objective aperture diameter), and it is essentially independent of atomic number. Therefore $t = (1/SP) \log (I/I_0)$. S is determined from samples of known thickness (e.g., polystyrene latex particles). The determination of I/I_0 is usually made either directly with a Faraday cage or photographically.

Reisner [69] used the instrument in the selected area diffraction mode with a retractable electrode to read the intensity in the zero-order spot. Thickness measurements apply to areas as small as the illuminated area. He found the technique particularly useful for measuring the rate of contamination buildup on the sample. This is a measurement that is important to all quantitative microscopy and the general performance of the instrument.

WEIGHT

Many people have examined the possible use of the equation discussed in the last section as a method of determining the dry mass of small objects. Bahr and Zeitler [70] have been most active in recent years. They have outlined a photographic technique that allows the study of irregularly shaped objects and have designed all of the necessary equipment to make it a highly routine procedure. Their studies indicate a useful range of dry masses between

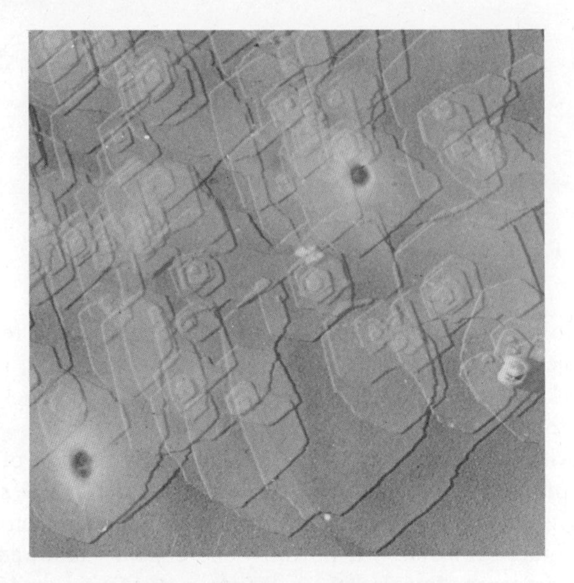

Fig. 4.27 Shadowed preparation of polyethylene crystals.

10^{-11} and 10^{-18} g. The lower area limit is a circle of $D = 200$ Å imposed by the absence of corrections for phase effects.

ORDERED STRUCTURES

When employed in the diffracting mode, it is possible to use the electron microscope to make all the measurements conventionally made by x-ray diffraction. These include degree of crystallinity, crystallite size, and orientation. The techniques complement each other because electron diffraction is best accomplished on very small quantities of sample. Electron diffraction has the advantage that through diffracted-beam microscopy one can observe the diffracting regions. For example, Scott [71] showed that by observing the appearance and disappearance of diffracting crystallites as the sample was rotated a measure of crystallite dimensions parallel to the incident beam was obtained. This has been found to be useful in measuring the degree of distortion produced in sectioning polymers, which in turn allows the usefulness of other studies on the same sections to be judged.

Hess and Heidenreich [21] used the direct observation of lattice spacings seen in graphite as a measure of the resolving power, magnification, and performance of an instrument used at high resolution. Figure 4.28 is a micrograph of graphite that shows a resolving power of 2 Å.

Fischer et al. [72] demonstrated an entirely new approach to the problem of relating information gained from diffraction techniques to physical structure seen in polymers. They found that after appropriate treatment

Fig. 4.28 Phase-contrast image of the layer plane orientation in "graphitized" carbon black. (Reprinted by courtesy of W. M. Hess.)

electron micrographs could be used in light-scattering experiments to produce patterns that agreed quite well with low-angle x-ray studies, indicating that there is a relationship between surface structure and bulk morphology.

3 SCANNING ELECTRON MICROSCOPY

Introduction

Suppose the surface of a specimen is scanned with a finely collimated beam of electrons. The interaction between the beam and the specimen surface may result in the production of secondary electrons, of photons by cathodoluminescence, of x-rays, or of backscattered electrons. In addition, an emf may be set up across the specimen and electrons may be absorbed by the specimen. Information from any of these processes may be analyzed by appropriate methods or collected and converted into a signal from which a permanent record may be obtained, usually by photographing a cathode ray tube (CRT) whose point-by-point brightness is modulated by the signal from the specimen.

Figure 4.29 schematically represents the shape of the volume of the specimen excited by the probe, and the level from which the elements of the various signals can emerge from the specimen and therefore be detected and analyzed. The dependence of the size of the volume, and of the signals

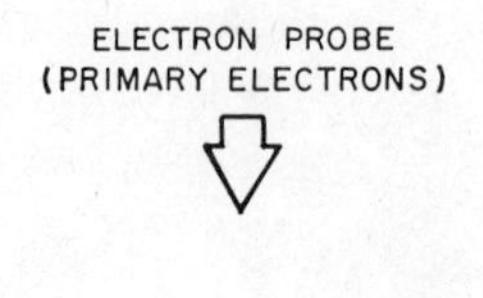

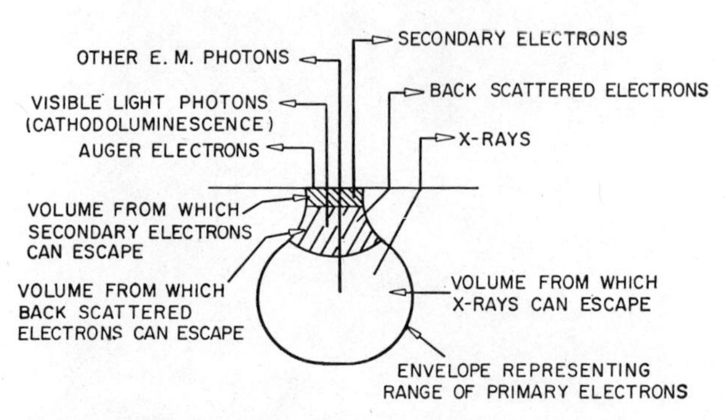

Fig. 4.29 Schematic representation of the penetration of electrons into the surface of a specimen and of the signals that can result from the interaction of the primary electrons with matter.

produced, on instrumental parameters and the physical properties of the material is discussed in the following sections.

The scanning electron microprobe or scanning electron probe microanalyzer utilizes the x-radiation produced by the interaction. Crystal spectrometers and associated counters are commonly employed, and elements between sodium and uranium can easily be detected; specific crystals have been developed to extend this range. The microprobe can also be used to produce a magnified image of the specimen, and a camera system is normally available to produce x-ray, electron, and optical micrographs. A combined high-resolution transmission electron microscope and microprobe analyzer has been constructed by Duncumb (P. Duncumb, *Fifth International Conference on Electron Microscopy*, Paper KK4, Academic Press, New York, 1962). Although this instrument has recently been commercialized (by Associated Electrical Industries, Harlow, England), it is not discussed in this treatment.

The scanning electron microscope utilizes the signals resulting from the other processes mentioned above. An appropriate element is positioned so that the signal is detected. and ancillary equipment converts this signal to, and amplifies, an electrical signal which is applied to the deflection plates of the CRT. In this section we are mainly concerned with the signal resulting from the generation of secondary electrons, in which emissive mode the instrument is employed primarily as a magnifying device. The conductive mode, in which the micrograph records the effects of currents induced in the specimen, is widely used in the study of solid-state processes, especially in microelectronics. The signals from backscattered electrons and from photons arising from cathodoluminescent processes are employed in more specialized fields of research.

In this review we discuss the historical background of the subject, including briefly some nonscanning emission microscopes. We examine the requirements of the instruments and describe some of the practical solutions to the problems by considering the construction of the instruments. Next, there is a short consideration of the interaction between electrons and matter and, finally, we consider some of the work reported in the literature on materials that fall within the limits of this volume.

The great utility of the scanning electron microscope lies in the fact that it combines a great depth of focus with moderate resolutions at magnifications that bridge the gap between the optical and the transmission electron microscopes. Coupled with this combination of attributes is the case of specimen preparation if it is desired to examine the surface of a bulk specimen and, indeed, the size of the specimen that can be readily accommodated. The fact that the micrograph is obtained by photographing a CRT means that rapid-process Polaroid-type film can be used, and the number of specimens that can be examined and micrographs obtained is very large for a given time.

As a magnifying device, the scanning electron microscope occupies a range of applications in that area of magnification, resolution, and depth of field not well covered either by the (light) optical techniques of microscopy and macrophotography or by the transmission electron microscope.

The great depth of focus of electron optics is attributable to the extremely low effective numerical aperture at which the lenses operate. At magnifications in the range of light microscopy, the depth of focus of the scanning electron microscope is at least several hundred times greater than that of the optical microscope. This is illustrated in Fig. 4.30, which also depicts the effect of the size of the final aperture in the objective lens on the depth of focus.

The upper limit of useful magnification of the scanning electron microscope is usually determined by the specimen detail and often by its behavior toward the heating and vacuum conditions. With many specimens, moderate magnifications (of the order of 5,000 ×) are adequate, but for fine structure or the surface of particles, for example, instrumental magnifications of 50,000 × or greater may be desirable. The minimum magnification is of the order of 20 to 100 ×.

A useful feature is that once the instrument has been adjusted for the maximum magnification desired lower magnification micrographs can be

Fig. 4.30 Depth of focus. Micrographs of a steel spring measuring 0.38 in., made at constant azimuth and orientation. (*a*) Photomicrograph using Bausch & Lomb Model L photomicrographic camera fitted with a 32-mm micro Tessar lens (f/16, 30 sec). (*b*) Scanning electron microscope micrograph using a 200-μ final aperture. (*c*) Hollow glass spheres; 2060 ×. (*d*) Scanning electron microscope micrograph using a 50-μ final aperture.

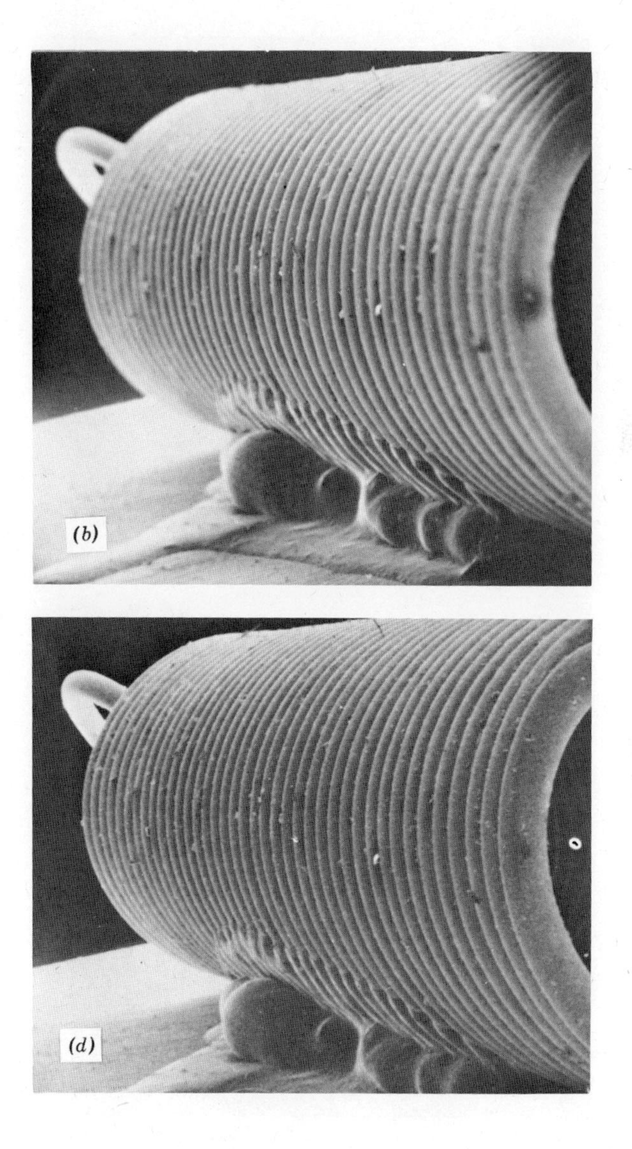

Fig. 4.30 (Continued.)

Fig. 4.31 Various structure levels on a mosquito. (*a*) Structure on individual eyelets; Magnification: 8571× (nominal). (*b*) Eyelets; Magnification: 2571× (nominal). (*c*) The head, showing compound eye, antennae, and proboscis; Magnification: 85.7× (nominal). (*d*) Overall view of the mosquito; Magnification: 21.4× (nominal).

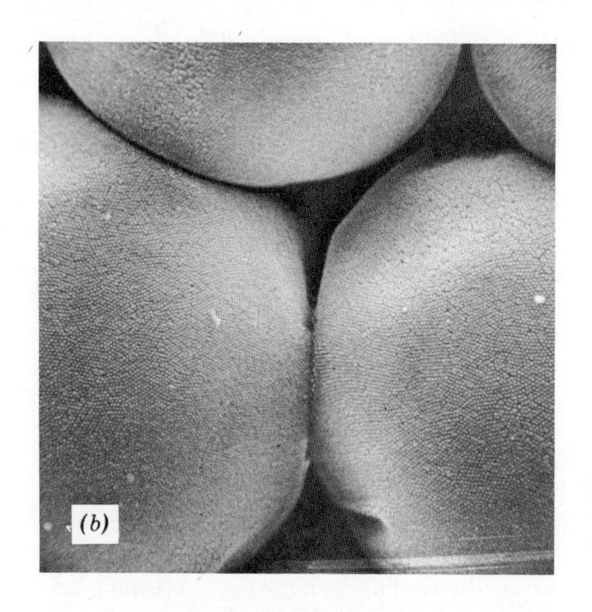

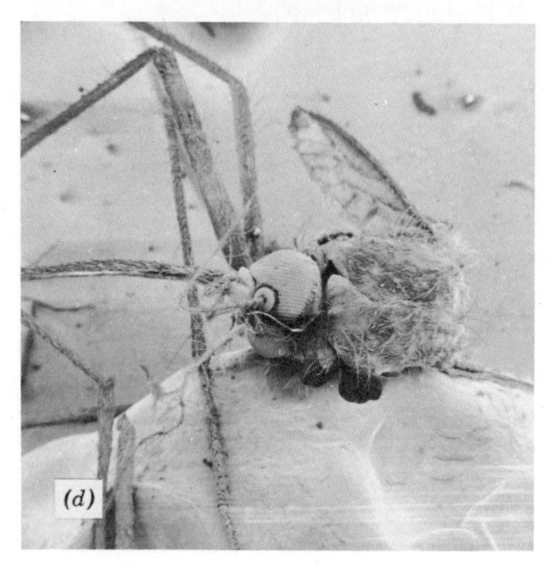

Fig. 4.31 (Continued.)

obtained without refocusing and often without resetting intensity controls. This capability makes the instrument ideal for obtaining sequences, and Fig. 4.31 illustrates three levels of magnification of a mosquito.

Resolution lies in the range between the light and transmission electron microscopes, as is discussed elsewhere.

In some fields, however, the scanning electron microscope is a unique tool. Some very rough materials cannot be satisfactorily replicated, hence cannot be examined with the transmission electron microscope. In the field of microelectronics, biasing currents can easily be externally applied to the device via vacuum feed-throughs; the device is mounted in the microscope, and the effect of these currents made visible by contrast changes.

Historical

Emission Microscopy

In emission microscopy the imaging electrons are emitted by the specimen itself as a result of the effects of some primary radiation or cause. Hence the specimen acts as a cathode, and variations in the current density of the image can be caused by the following factors.

1. Variations in the spatial distribution of the emitted electrons may be attributable to local surface orientation with respect to the direction of the primary radiation or to the optical axis of the imaging system (the collector).
2. The characteristic physical properties of the specimen, such as its secondary electron emission coefficient or its photoelectric work function, may play a part. Furthermore, the crystallographic orientation of grains or other components of the structure may have an effect.
3. Local or overall fields, either magnetic or electric, attributable to variations in the magnetization or the contact potentials of surface elements, affect the paths of electrons near the specimen surface.

Nonscanning electron emission images were first produced when Goldstein [73] performed some experiments on the properties of electron rays. The development of the imaging capabilities of magnetic and electrostatic lenses was carried out by a number of German workers [74], who were interested in the imaging of electrons emitted by ion bombardment and by the thermionic process.

A photoemission microscope was first described by Brüche [75]. Several more modern instruments are discussed by Möllenstedt and Lenz [76], who also discuss some of the factors influencing emission. Some of these, such as oxide layers and contamination by organic polymers, are discussed below with reference to their effect on secondary electron emission.

Secondary Electron Microscopy

Kollath [77] has published data on the energy distribution of secondary electrons emitted from the surface of thick specimens by bombardment with slow primary electrons (Fig. 4.32). In this case the number of secondary electrons of particular energies does not vary greatly with the target material.

Deitrich and Seiler [78] employed an electrostatic high-resolution analyzer to determine the energies of secondary electrons emitted from the exit side

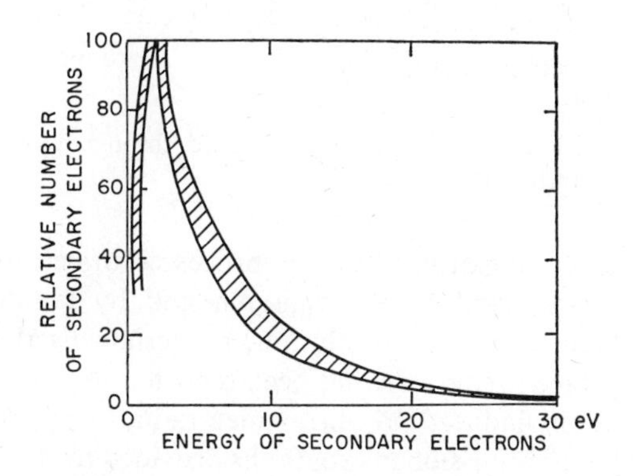

Fig. 4.32 Common representation of the energy distributions of secondary electrons from various targets (Ta, Mo, W, Ag, Zn, Mg, brass, Be ON Mo, NiFe, CuFe, CuBe) measured by Kollath [5]. All experimental points are situated in the hatched region.

of a thin foil bombarded by primary electrons in a moderate energy range. There was no significant difference between the distributions resulting from the 5- and 57-keV primary electrons, and these distributions were as shown in Fig. 4.33.

The dependence of the electron yield, which is defined to be the ratio of the number of secondary electrons to incident primary electrons on the specimen material, has been studied by Kollath [79], and Fig. 4.34 is taken from this source. The secondary yield (η) reached a maximum at a primary energy value (E_p) of the order of hundreds of electron volts for the material studied and decreased gradually as E_p was further increased. At higher values of E_p, η decreases to a value significantly below unity, but there are still appreciable differences between materials [80].

The relation between η and the angle of incidence α of the primary electrons to the normal to the specimen surface was measured by Muller [81]. If α is small,

$$\eta \sim \cos^{-1} \alpha,$$

so contrast is to be expected as a result of variation in η with α across a surface that is not flat.

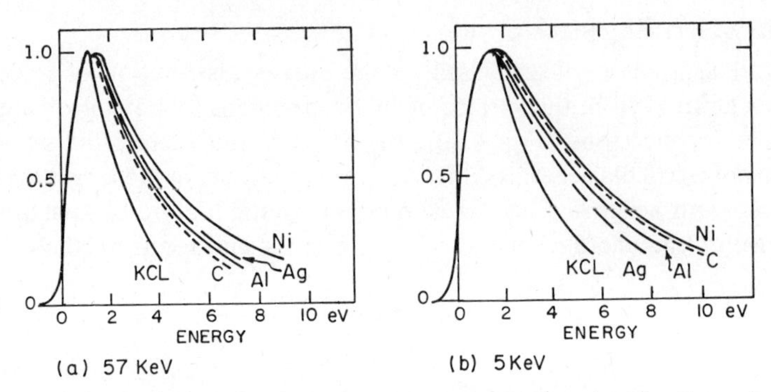

Fig. 4.33 Energy distribution of electrons transmitted through thin films of various materials. (Adapted from Ref. 6.)

Some recent work comparing different processes of release of electrons (thermal, photoelectric, and kinetic emission and by bombardment with neutral particles) in terms of resolution, material discrimination, and reproduction of surface topography has been reported [82]. It was found that photoelectric emission induced by ultraviolet radiation provided the best resolution, while kinetic emission micrographs provided the best reproduction of surface topography. Surface areas of different chemical and crystallographic nature mostly yielded higher contrast with thermal and photoelectric emission than with kinetic emission. A paper describing the use of a commercial emission microscope on polished sections and other metal specimens has appeared [82].

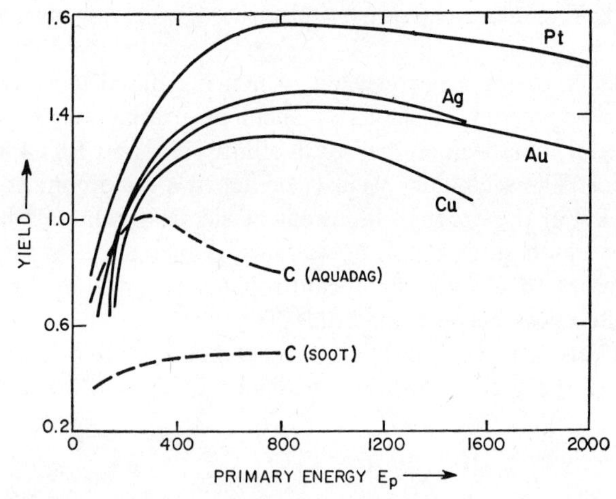

Fig. 4.34 Secondary electron yield versus primary energy for various target materials (Kollath [7]).

Möllenstedt and Lenz [76] describe a number of microscopes with imaging lenses using secondary electron emission as the source of contrast. The final electron-optical magnifications of these instruments is in the range of 1000×, and their theoretical resolution, with suitable limiting apertures, is of the order of 80 Å. Photographic recording is employed.

Scanning Electron Microscopy

The emission microscopes mentioned above depend on the use of lenses to produce a magnified image of the specimen surface. In the case of the scanning electron microscope, the object surface can be considered to be divided into an array of elements, the diameter of which is equal to the diameter of the primary electron beam. This beam is scanned in a square raster covering a specified area of the surface. From each element of the surface is emitted a signal comprising secondary electrons. This signal is detected, amplified, and applied to the grid of a CRT, so that the intensity of each element of the CRT is proportional to the magnitude of the signal of secondary electrons from the specimen element. In this fashion an intensity image of the specimen surface is constructed without benefit of lenses, and the magnification is simply the ratio of the dimensions of the CRT to those of the raster scanned by the beam on the specimen surface.

The first instruments to operate on the principles outlined in the above paragraph were constructed by Knoll [84], and the microscope described in his paper is depicted in Fig. 4.35.

The gun G, provided a beam of primary electrons, which was scanned over the surface of the specimen by the action of the deflection coils D_1, to which sawtooth sweep voltages (magnetic lenses) or currents (electrostatic) were supplied by the sweep generator G. When the specimen S was nonconducting,

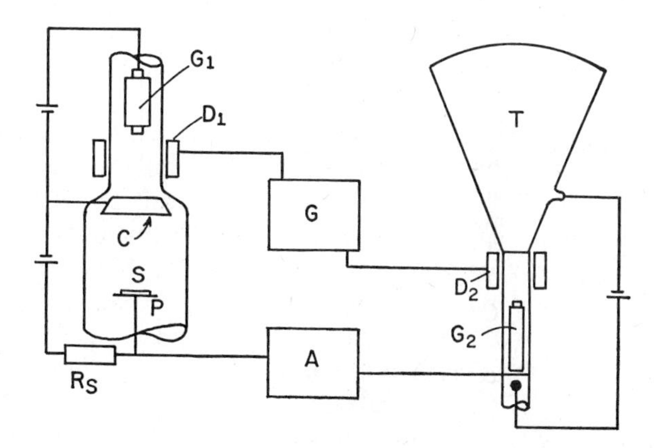

Fig. 4.35 M. Knoll's scanning electron microscope. (Adapted from Ref. 12.)

it was placed on a metallic signal plate P. The secondary electrons emitted by the specimen were detected by the collector electrode C, which was held at about 10 V with respect to P, the secondary emission current flowing through the signal resistor R_s. The voltage drop across R_s was the signal containing the information concerning the secondary electron emission from the surface. This was amplified by a wide-band amplifier A prior to its application to the gun G_2 of the viewing tube T. Hence the signal controlled the emission of the electron gun in the viewing tube, in which the electron beam was scanned over the viewing screen by the deflection coils D_2 in synchrony with the beam of primary electrons over the specimen. The image produced by this brightness modulation could be photographed from the viewing screen.

In Knoll and Theile's microscope, no demagnifying lenses were used, and resolution was limited by the final beam diameter, which was of the order of 100 μ.

Von Ardenne [85] devised an instrument that used fast, scattered electrons in addition to the secondary electrons. The beam was made to pass through the specimen, where it struck photographic paper on a rotating drum which was mechanically actuated in synchrony with the beam. No provision was made for a visual tube, so focusing adjustments were made after each micrograph was recorded; recording time was about 30 min.

Zworykin et al. [86] employed a two-stage electrostatic lens system to demagnify the crossover of an electron gun to a diameter of about 500 Å, and a further long-focus electrostatic lens of approximately unit magnification before the beam impinged on the specimen. The latter lens was necessary to provide space for deflection coils, a fluorescent screen, and a photomultiplier. By suitable selection of the components, such an arrangement could constitute a preamplifier, and the thermal resistor noise could be significantly reduced. The primary beam was modulated by a 3000-kHz square wave, and the photomultiplier output electronically processed before being applied to a facsimile recorder.

French workers constructed a scanning electron microscope in 1946 [87] but, as were earlier instruments, it was limited by noise to a resolution of a few microns. Brachet [88], however, showed theoretically that if the secondary electron current could be collected and amplified without the introduction of further noise the possible resolution was about 100 Å. Davoine [89] constructed a scanning microscope with a resolution of about 2 μ.

The principles involved in these microscopes were developed and improved upon by a group at the Engineering Laboratory, University of Cambridge, under the direction of C. W. Oatley. The work of this group led to the commercialization of the instrument by the Cambridge Instrument Company under the trade name Stereoscan.

The Modern Scanning Electron Microscope

In the following discussion we consider the construction of the scanning electron microscope as it is currently commercially available and touch upon features of some experimental instruments. A block diagram featuring the essential components of the microscope is shown in Fig. 4.36, and reference shall be made to the components in turn.

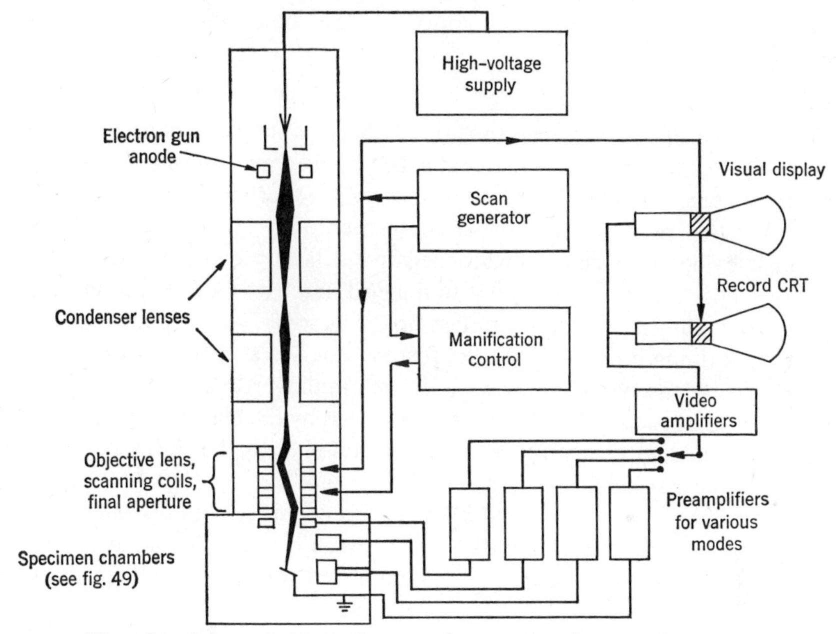

Fig. 4.36 Schematic block diagram of a scanning electron microscope.

The Electron-Optical System

VACUUM

Because of the limited range of electrons in air, the electron-optical system of the scanning electron microscope must be enclosed in a vacuum system. The vacuum is typically maintained by a combination of mechanical and diffusion pumps at a pressure of 10^{-4} to 10^{-5} torr. There is a tendency in commercial and particularly in research microscopes to seek ultrahigh vacuums, down to the region 10^{-10} torr. Such pressure levels require the use of ion pumps, possibly in combination with bakeable metal seals in place of the normal elastomeric seals. It is desirable to have a system of isolation valves between the specimen chamber and the column, and the column and the electron gun, to obviate the need to return the whole column to atmospheric pressure when changing a specimen or a filament. This arrangement reduces pump-down time after these operations and tends to keep the lenses and apertures in a far cleaner condition than would otherwise be possible. An air-lock system could also be used within a larger specimen chamber; this would have the advantage of keeping the detector under a constant vacuum and further would reduce the possibility of residual vapors from the specimen interacting with a detector. The smaller volume being evacuated would shorten pump-down time after a specimen change, but the maximum size of the specimen might be limited by the dimensions of the air-lock chamber.

ELECTRON GUN

The source of primary electrons is normally a tungsten hairpin filament in a triode electron gun. The filament is heated by a current of a few amperes at low voltage, and electrons are emitted by thermionic processes. The filament tip is precisely located at the center of a small hole in a closed Wehmelt cylinder, which is biased negative relative to the filament, hence exercises a control function similar to the grid in a triode valve. The electrons leaving the filament are constrained to "crossover" at a point near the filament, forming a disk of least confusion which is a virtual source for the demagnifying lens system. The anode is an annular plate and is held at earth potential, thus accelerating the electrons emitted by the filament tip. Provision is usually made to center the anode plate relative to the optic axis of the electron-optical system.

The life of a filament is limited by thermal evaporation and by gas attack, and with a vacuum of around 10^{-5} torr average filament life is about 30 to 40 hr. Hence the filament changing and resetting procedure is made as straightforward as possible.

THE ELECTRON LENS SYSTEM

The function of the lens system is to reduce the diameter of the beam at crossover from about 50 to 100 μ to 50 to 200 Å. Such a demagnification is carried out by a lens system comprising one [90] to four [91] lenses. Commercial microscopes have two or three lenses, and the condensing system may be a single lens, a double lens with two pole pieces and one set of windings, or a pair of single lenses. Spray apertures serve to limit the dimensions of the electron beam, and provision may be made to center the aperture and also to select apertures of different sizes.

The final or objective lens completes the demagnification and is used to focus the image. Its design is critical since the aberrations of the lens system are largely attributable to those of the final lens. It operates at a much larger angular aperture than do the condenser lenses, and its focal length must be of the order of a centimeter to provide the desired working distance. The specimen must be outside the magnetic field of the final lens since even a small field interferes with the collection of the low-energy secondary electrons. Metal screens shield this area from external magnetic fields. In order to fulfill these conditions, an asymmetric pole piece is employed, and the lens normally has a smaller final than initial bore. An aperture at the final lens serves to define the angle of convergence of the beam at the specimen plane. Usually, a number of apertures of differing size are provided, and their holder is adjustable to the optic axis of the electron-optical column. It is often convenient to make a first study of the specimen using a small aperture to obtain maximum depth of field.

Magnetic lenses are used almost exclusively, although in principle electrostatic lenses can be used equally well. Magnetic lenses are more easily cleaned, which is vital since it is difficult to avoid contamination from residual gases and their breakdown products. Magnetic lenses are sturdy and retain their alignment well, and the evacuated volume can be reduced compared to electrostatic lenses since the windings can be outside the vacuum. Finally, the preformance of magnetic lenses is slightly better from the point of view of aberrations.

The double-deflection system that scans the probe in the required raster is located between the condenser and objective lenses. Magnetic deflection is customarily used since the coils can be located outside the evacuated bore. A sawtooth generator supplies the current to the pair of coils producing the X- and Y-deflection of the beam and to the spot in the CRT, which also normally uses magnetic deflection. In addition, electrical shifts are normally provided to permit the accurate positioning of the specimen raster at the highest magnifications, and external rotation of the scan coils permits precise alignment of the raster to details in the specimen. A reduced scanning area with increased speed is very useful for fine-focusing.

Finally, it may be desired to modulate the electron beam, or turn it off very rapidly, in order to observe the decay of some excitation. These facilities can be provided by the use of a magnetic or electrostatic field which, suitably excited, deflects the beam in relation to an aperture located near the filament.

LENS ABERRATIONS

1. *Chromatic aberration.* This is a result of a spread of $e\delta V$ about a mean energy eV_0. Its effect is that in place of a point focus the illumination passing through the lens is contained within a circle of least confusion of diameter d_c, which can be written [92]:

$$d_c = C_c \frac{\delta V}{V_0} \alpha,$$

in which C_c is the chromatic aberration constant and α is the semiangular aperture.

2. *Spherical aberration.* Since a lens has a finite numerical aperture, electrons moving in trajectories more inclined to the lens axis experience stronger fields and are more deflected. The diameter of the disk of least confusion d_s is given by [92]:

$$d_s = \tfrac{1}{2}C_s\alpha^3,$$

where C_s, the spherical aberration coefficient, is a function of the focusing field.

3. *Diffraction.* The fundamental limit on the size of the spot is set by diffraction, and the diameter d_d of the Airy disk is dependent on the wavelength, according to:

$$d_d = 1.22d/\alpha = \frac{1.22}{10^8\alpha}\sqrt{150/V}\ \text{cm},$$

where V is the accelerating voltage.

4. *Astigmatism.* This aberration is caused by imperfect rotational symmetry about the axis of a lens, and the diameter of a circle of least confusion d_a has been taken by Smith [93] to be the portion of the image in which about 80% of the total current lies.

These aberrations can occur concurrently, and the procedure generally adopted [94] is to treat the individual estimates as error functions and regard the effective spot size d_{eff} as the root of the sum of the squares of the separate diameters, that is,

$$d_{eff} = [d_0^2 + d_c^2 + d_s^2 + d_d^2 + d_a^2]^{1/2},$$

where d_0 is the diameter of the crossover of the electron gun.

In order to minimize the effects of chromatic aberration, fluctuations of the power supplies must be reduced to less than 1 part in 10^5. The values of C_c and C_s are less for magnetic than for electrostatic lenses. The effects of astigmatism can be minimized by careful attention to column cleanliness, and those effects still occurring can be corrected by the use of a stigmator. This component is usually located in the objective lens and superimposes an asymmetric field, the strength and direction of which can be varied on the electron beam to nullify the noncircularity of the probe. The problem is that the actual field differs from the intended by a component in two perpendicular phases inclined at an unknown angle to some reference axes. The amplitude and angle controls of the stigmator control the currents in the stigmator coils until the field attributable to the stigmator nullifies the astigmatism in the probe.

VIBRATION

Considerable attention is paid to the design of a vibration-free support system for the electron-optical column. Spring shock mounts or pneumatic devices are used, but site specifications still call for an almost vibration-free floor. Special care must be given to the firm attachment of the specimen to the stub or the stage. The bulk of the specimen may be a critical factor, especially in nonrigid samples such as yarns or fabrics.

Detectors Used in the Collection of Signals

SECONDARY ELECTRONS

Scintillator/Photomultiplier Detectors. The problem of noise-free amplification was ameliorated by McMullen's use [95] of an electron multiplier of the then new beryllium-copper dynode type. Further, the specimen was mounted so that the normal to its surface made an angle of about 65° to the incident electron beam.

The following drawbacks indicated that an electron multiplier was not the ideal detector.

1. Only about 20% of the secondary electrons were collected because of the bulky nature of the device, which prevented it from being mounted at desirable proximity to the specimen.
2. The secondary emission coefficient of beryllium-copper is not high, and a proportion of electrons from the specimen produced no secondaries in the multiplier. This resulted in additional noise in the system.

For these reasons, a detector employing a scintillator and photomultiplier was developed by Everhart and Thornley [96], and this detector is used in most commercial instruments. The detector is shown schematically in Fig. 4.37.

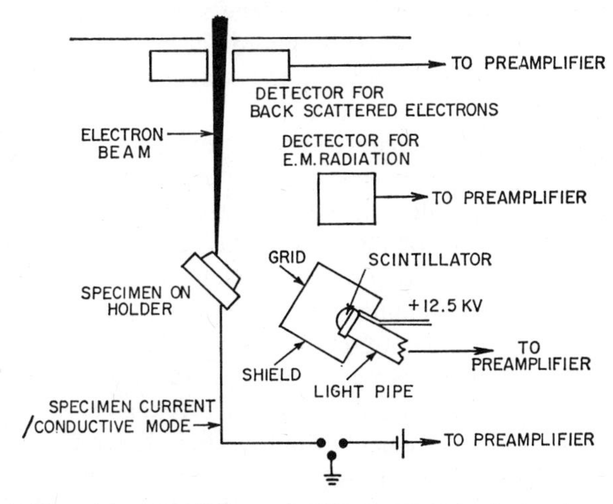

Fig. 4.37 Scintillator/photomultiplier and other detectors in the specimen chamber of a scanning electron microscope.

The cylindrical metal shield is closed at the input end by a copper gauze grid biased (in the commercial Stereoscan) between -30 and $+250$ V, which permits the slow secondary electrons to be repelled or attracted, respectively, by the grid. On entering the collector, the electrons are accelerated toward a hemisphere of plastic scintillator, which is covered by a layer of aluminum about 700 Å thick and maintained at a potential of about 12.5 kV. The electrostatic field causes most of the electrons to strike the scintillator head near its apex, and the photons generated in the scintillator are transmitted along a Lucite acrylic resin light pipe to a photomultiplier, in which the light is converted back into an electrical signal. This arrangement can provide considerable amplification without the introduction of much noise.

The scintillator material employed was that developed for the detection of β-particles, and substantially noise-free amplification of the secondary electron current from the specimen is achieved if the voltage applied to the focusing ring is greater than 10 kV.

Everhart and Thornley [96] found that the light pipe could be bent to any desired configuration without excessive transmission loss but that care had to be taken with the surface finish, or light loss would result. Further, reflection losses could occur, and additional losses could result at points where the light pipe was supported.

The collector functions with a variety of conditions of the aluminum layer, but if it deteriorates seriously the field will be affected and noise will be introduced. The frequency with which the scintillator should be recoated depends largely upon the quality of micrographs required, hence on the desires of the operator.

The secondary electron current leaving the specimen is of the order of 10^{-12} A, while the voltage leaving the photomultiplier is between about 0 and $+4$ V. This voltage is then amplified to a maximum value of the order of 40 V before being applied to the CRTs.

An alternative to the Everhart-Thornley geometry has been described [97]. In this case an additional magnetic field confines the secondary electrons from the specimen and guides them to a detector of the scintillator type located *above* the final lens of the microscope. This geometry permits the specimen to be placed much closer to the final lens, which can therefore be of short focal length, and the diameter of the primary electron beam can be smaller. The authors state that a greater number of secondary electrons can be collected and that voltage contrast is made more sensitive by this method. The arrangement also permits discrimination between secondary electrons of different energies, and the signal attributable to reflected primary electrons can be avoided. The specimen position, however, limits manipulation, and specimen stage size is greatly restricted. Further, detectors for other signals cannot be simultaneously employed.

Solid-State Detectors. Silicon diodes. Gonzales [98] describes the use of silicon diodes in the collection of secondary electrons. The collector is a flat-box diode placed to the side of the specimen and positively biased to attract secondary electrons onto a silicon diode held at a positive 10-kV potential. The secondaries are thus accelerated and electron-hole pairs are created; a current is made to flow in an external circuit and this is applied to the grid of a CRT.

Electron multipliers. The disadvantages of the dynode type of electron multiplier have already been discussed. Recently, channel electron multipliers, in which a coating on the inside of a glass tube provides a continuous distribution of "dynodes," have become available [99]. A grounded screen prevents distortion of the collecting field at the entrance to the detector by the high voltage applied to the tube to provide the amplification of the incident signal of secondary electrons. Ways of using these detectors are described by Thornton [100] and by Ong [101], who reports good signal levels with incident sample currents of 10^{-11} to 10^{-6} A.

REFLECTED AND BACKSCATTERED PRIMARY ELECTRONS

Using reflected electrons, McMullen [102] showed that if a smooth surface is inclined to the beam at 60° it is possible to detect variations of about 0.5° in surface angle. This is about twice as sensitive as for secondary electrons.

Primary electrons can be collected by the Everhart-Thornley arrangement by removing the Faraday cage and holding the scintillator at, or somewhat beneath, the potential of the specimen. In this case the photon output of the secondary electrons is negligibly small, and the scintillator can collect a greater proportion of the reflected primary electrons by being placed close to the specimen, thus increasing the solid angle subtended by the scintillator.

Of greater potential use, however, is the fact that contrast attributable to compositional (i.e., atomic number) variation can be detected by means of backscattered electron images [103]. Two identical solid-state detectors (silicon diodes) are placed as indicated in Fig. 4.38, so that when the beam is in position A they receive the same signal as they do from position C. The signals from A and C will be different if the materials α and β have different atomic numbers, and signal contrast results. At B, however, the slope of the surface of material α causes the reflection of primary electrons to be different on the sides of the projecting material by an amount δ_1 or δ_2 than it would be from A on the horizontal surface of the feature. Hence, if the signals from the diode are added, topographical information attributable to $\delta_1 + \delta_2$ will be recorded, but if the signals are subtracted compositional information will be predominant since the topographical information will be reduced to $\delta_1 - \delta_2$.

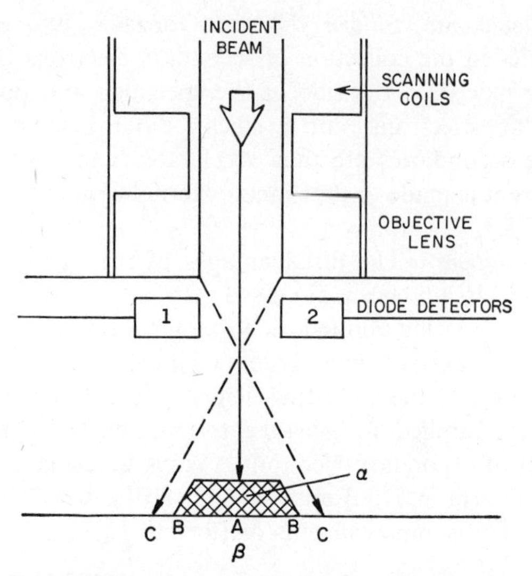

Fig. 4.38 Backscattered electron images attributable to topography and atomic number differences (see text for explanation).

If solid-state detectors are not available, images attributable to backscattered electrons can be obtained by biasing the collector to, for example, −75 V, so that secondary electrons are not collected, and adjusting stage tilt to optimize the signal. The contrast is usually greater than with secondary electrons, and electronic adjustments are normally required to produce an aesthetically acceptable micrograph.

ELECTROMAGNETIC RADIATION

When the scanning electron microscope is used in the luminescent mode, devices for collecting and amplifying radiation emitted by the interaction of the primary electron beam and the specimen must be considered. In the visible and near-infrared regions, photomultipliers can be employed, and their spectral characteristics are of importance. Further, if it is desired to analyze the emitted spectrum as a function of wavelength, a monochromator is needed. For wavelengths greater than about 1.2 μ, semiconductor photovoltaic or photoconductive devices can usefully be substituted for the photomultiplier. Considerable care must be exercised in matching the detectors available for such an analysis with the radiation emitted by the specimen. Such considerations are discussed at some length by Thornton [104], who also provides an examination of the processes of recombination by which radiation is emitted by a luminescent specimen.

Pruden and others [105] noted that instrumental factors influencing resolution in the cathodoluminescent mode include spot size, accelerating potential, beam current, light scattering, and the persistence of luminescence in the materials being studied; very slow scan rates are necessary for materials exhibiting long-persistence luminescence. These investigators constructed a detector comprising individual glass fiber optic light guides disposed in a 320° arc about and just above the surface plane of the specimen.

Another approach was reported by Williams and Yoffe [106], who recorded the cathodoluminescence on one of two channels. One channel is nondispersive—a sample of the total emitted light is transmitted via a quartz light pipe to a photomultiplier and thence to the amplifier and display systems. A quartz lens in the vacuum wall of the microscope focuses an image of the specimen onto the entrance slit of a spectrometer, from which the output is fed via a preamplifier to the videoamplifier of the microscope. In this way an image can be constructed using light of any selected wavelength, and optical microanalysis of materials is possible in a fashion analogous to that of the electron microprobe using x-radiation.

Cathodoluminescence may of course be observed in a crude fashion by simply removing the scintillator tip of the secondary electron collector and employing the light pipe to transmit light from the specimen to the photomultiplier, hence to the display system.

THE CONDUCTIVE MODE

When the scanning electron microscope is used in the conductive mode, the word "detector" is somewhat of a misnomer. In this mode currents induced in the specimen by the incident beam are used to modulate the brightness of the CRT, or can be applied to the *Y*-deflection plates of the CRT to produce a deflection-modulation display [107]. This specimen-absorbed current is the current that flows in the specimen to maintain electrical neutrality, since $I_p > I_s + I_{rp}$, where I_p = primary beam current, I_s = secondary beam current, and I_{rp} = reflected primary beam current. Such a current is small and cannot be detected in specimens containing large currents, such as a *p-n* junction. At a *p-n* junction, charge collects as a result of the field at that junction. Similar charge concentrations occur at any high-field region of a bulk semiconductor, or at a surface barrier.

The great bulk of work utilizing a conductive mode is concerned with semiconductors and similar devices and thus lies outside the scope of this discussion. The absorbed current is related to the average atomic number and may be useful in demonstrating different phases within a specimen. An example of such work is given in the paper by Shirai et al. [108], in which it is stated that there are a number of difficulties, especially with regard to the specimen. It was found that the intensity of backscattered electrons and the magnitude of the specimen current depended monotonically on the atomic number of the specimen material. They also compared the magnitude of the specimen current to values computed by taking into account the mean atomic number of the component parts of some alloys.

Display and Recording

Prior to display and recording, the output of the detector is amplified electronically, and the signal fed to the grid (brightness control) of a CRT, to the Y-deflection plates of a CRT, or to the controls of a facsimile recorder. Oatley et al. [109] consider in some detail the optimum number of lines per picture in terms of image quality and resolution. The human eye cannot readily detect a smaller distance than about 0.02 cm, so a 10×10 cm tube for direct visual examination need not resolve more than about $N = 500$ lines. Photographic film, however, can record detail finer than the eye can see, so if enlargement is to be performed, N may be increased to a value of, for example, around 1000. If the diameter of the Gaussian image of the probe is d_0 and the specimen area scanned is a square of side D, the criterion for satisfactory collection and display of information is that $N = D/d_0$. In this case a considerable time, of the order of minutes, is required for the line-by-line recording of the micrograph, and this influences various stability requirements.

The above considerations led to the provision of two CRTs in some scanning electron microscopes. The almost instantaneous visual display required for the focusing of the image is provided by a lower-resolution CRT with a long afterglow screen and with a frame repetition frequency of about one scan per second. The photographic record is obtained from a high-definition, short-persistence screen. This reduces halation attributable to the spot in a tube with afterglow, and by setting the sweep frequency so that a single frame occupies the whole of the exposure time, any drift in the conditions will not result in blurring, although there may be some minor distortion of the image.

Photographic recording of the image can be on conventional film, either 35-mm roll or various cut sizes, but Polaroid-type film is widely used to produce either positive or negative images. Use of Polaroid film allows the image to be evaluated far more quickly and micrograph deficiencies can be rapidly corrected.

An alternative method of display is deflection modulation of the spot of the CRT [35] (Fig. 4.39). In this technique the signal from the detector is fed to the Y-plates of the CRT and so is superimposed on the frame time-base signal; the intensity of the spot is constant. Let us compare an image recorded as an intensity-modulated micrograph to the same image recorded as a deflection-modulated micrograph. A black region on the former is represented by a horizontal line on the latter, since the detector signal is zero. When the signal is not zero, the spot is deflected and the line thus acquires a

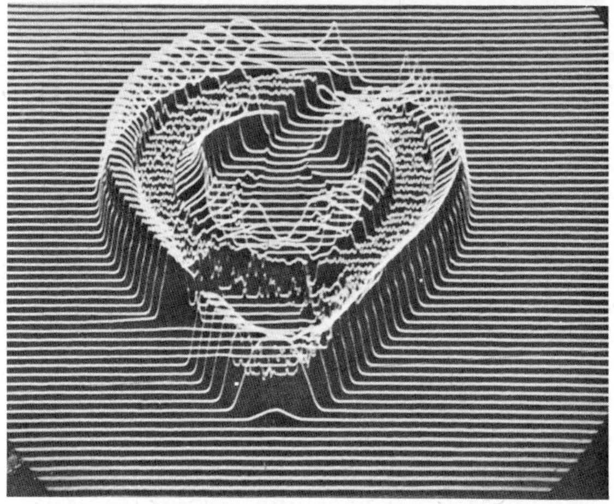

Fig. 4.39 An example of the use of deflection-modulation display to demonstrate the base current induced in an NPN transistor. (*a*) Television display of the base current induced in an NPN transistor by a 15-kV scanning electron beam; 1024-line raster. (*b*) Deflection-modulation display of the same base current used in (*a*); 64-line raster. (*c*) Deflection-modulation display of the base current induced in an NPN transistor by a 21-kV scanning electron beam; 256-line raster. (Reproduced with the permission of T. E. Everhart, Ref. 35.]

(*Continued overleaf*)

Fig. 4.39 (Continued.)

slope, so increased brightness is represented by increased slope. Maximum brightness is represented by a peak in the line seen. The image is constructed of a number of such lines, and the aesthetic quality is usually improved by increasing the number of lines. By reducing the number of lines, however, it is possible to record data on time-dependent events more rapidly than would be the case with intensity modulation.

In intensity-modulated micrographs the number of distinguishable brightness levels or shades of grey is limited, and interpretation is qualitative [110]. The deflection modulation system, however, is quantitative, since the deflection from the base line is proportional to the signal from the detector. Deflection modulation is perhaps most useful when applied to the signal from the conductive mode of the scanning electron microscope, but it can be used with any other signals such as those of the emissive mode or of the cathodoluminescent mode. In the conductive mode the deflection is proportional to the induced base current.

A further type of display has been described by Wittry and Van Couvering [111]. In this method a stereoscopic display is obtained for any signal from a scanning beam instrument by forming two deflection-modulated displays with slightly different perspectives. The perspective is changed electronically to facilitate the display of any function of two variables and to avoid difficulties in image registration.

Instead of applying the signal from the detector system to a CRT, it could be applied to a facsimile recorder and the information recorded in this fashion. Alternatively, the signal can be considered a voltage analog of image brightness, and if this is recorded together with the relevant *X*-*Y* coordinates, the entire micrograph is coded in a fashion that could be presented to a computer. After suitable preconditioning the photograph can be transformed by an analog-to-digital converter with digital form, and this information fed into a digital computer [112]. The computer can also be used to generate the raster on the specimen and on the CRT [113], and the micrographs can be stored, reduced, and analyzed by suitable programming. Examples cited include the problem of correlating changes in secondary electron current with changes in the potential of a semiconductor device [113] and the recognition and measurement of the size, shape, and orientation of particles [112]. The latter approach could permit analysis of texture problems hitherto precluded by the time and tedium involved.

The Interaction of Electrons with Matter

From the viewpoint of scanning electron microscopy, there are a number of basic physical phenomena that should be discussed:

1. The manner in which energy is lost by the incident electron and transferred to the specimen, especially as a function of depth
2. The relation between secondary electron emission and the properties of specimen, beam energy, and other factors
3. Physical processes.

Energy Transfer from Incident to Specimen Electrons

In the energy range of interest in scanning electron microscopy, the incident electrons dissipate their energy mainly in interactions with the bound electrons of the target material. The energy loss associated with emission of radiation by an electron in the Coulomb field of the nucleus can be neglected. The Coulomb field of the bound electrons interacts with that of the incident electron, and energy is transferred from the latter to the former, which is forced into some allowed excited state. An equation can be set up to determine the relative probabilities that each electron will attain a specific state, and Bethe [114] showed that the rate of loss with distance s is:

$$\frac{dE}{ds} = \left(2\pi e^4 \frac{ZN}{E}\right) \ln \left[\frac{E}{E_i}\left(\frac{E}{2}\right)^{1/2}\right],$$

where E is the energy of the electron, Z is the atomic number of the atom, N is the number of atoms per cubic centimeter, and E_i is the average excitation potential of the atom. This equation holds well for gases, when the electron path is long between interactions. In solids, however, experiments must be performed on thin foils, where it is difficult to allow for elastic collisions in which there is little loss of energy but there are important changes of direction.

The Emission of Secondary Electrons

The yield of secondary electrons as a function of the energy of the incident electron was studied by Kollath [115]. His findings for metals are illustrated in Fig. 4.40, in which all the secondary electrons are collected, and σ is the number of emitted electrons per incident primary electron. It is clear that there exists an energy E_{max} at which σ is a maximum, and that σ declines for $E_i > E_{max}$. The maximum may be attributable to the fact that for $E_i > E_{max}$ the secondary electrons are liberated from the electronic structure at a greater distance from the surface, hence are less likely to reach it and be emitted. The yield of secondary electrons is dependent on the smoothness of the specimen surface and on its cleanliness and purity, especially with regard

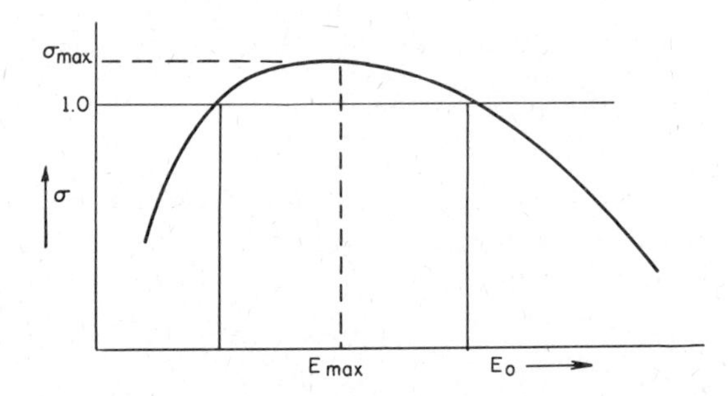

Fig. 4.40 Dependence of the emission yield σ of secondary electrons on varying accelerating voltage E_0. (Adapted from Kollath [40].)

to oxide layers. For a variety of materials, $0.5 < \sigma_{max} < 2.0$ and $100 < E_{max} < 800$ eV, and a normalized plot of E/E_{max} versus σ/σ_{max} results in a universal curve for most materials. In the absence of heating effects, the yield is independent of the primary beam current.

For a beam at normal incidence, the energy distribution of emitted secondary electrons exhibits a broad peak, representing electrons with energies between 0 and 30 eV, and a second sharper peak representing high-energy backscattered electrons; minor satellite peaks flank the high-energy peak. The shape of the energy distribution curve has been the subject of study by Hachenberg and Brauer [116] and by Harrower [117], the latter investigating their shapes at low-beam energies and for samples with extensive heat treatment [118]. In nonconductors the case is rather different. The energy distribution curve is as shown in Fig. 4.41, in which the mean energy of the secondary electrons is reduced to about 5 to 10 eV, while σ_{max} is between 1 and 20, compared to about 1.0 for metals. The increased yield is probably attributable to the reduction in scattering in the specimen of the secondary electrons and their relatively easier path to the specimen surface. The

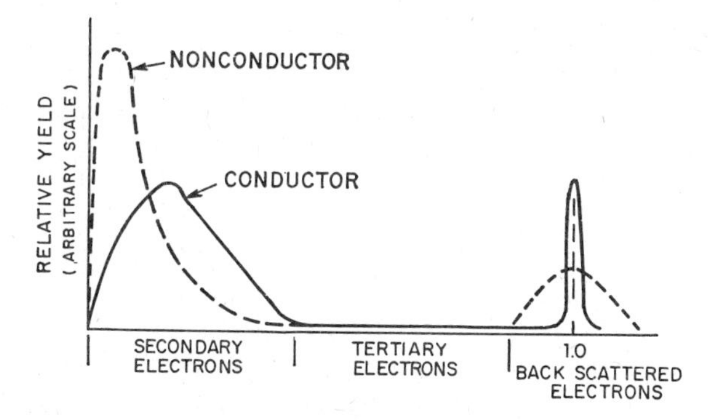

Fig. 4.41 Schematic representation of the energy distribution of electrons emitted from conductors and nonconductors. (Adapted from Thornton [47].)

reduction in the mean energy of the emitted secondary electrons may also be related to the greater depth from which the secondary electrons can escape. This point is more fully discussed by Thornton [119].

The angular distribution of the emitted secondaries is, to an excellent approximation, cosine for energies of 10, 15, and 20 eV [120]. This suggests that the angular distribution of secondary electrons inside the specimen is isotropic, so that they strike the specimen surface in a cosine distribution.

Secondary electron yield varies significantly with the angle of incidence of the primary beam. This is a fact of paramount importance in scanning electron microscopy and results from the fact that as the angle of incidence is altered so is the distance from the point of excitation to the surface for some of the secondary electrons (Fig. 4.42). When the distance is reduced, the number of secondary electrons emitted by the specimen is increased.

Suppose the coefficient of absorption of the secondary electrons is β and that it is exponential, and define the mean escape depth of these electrons as

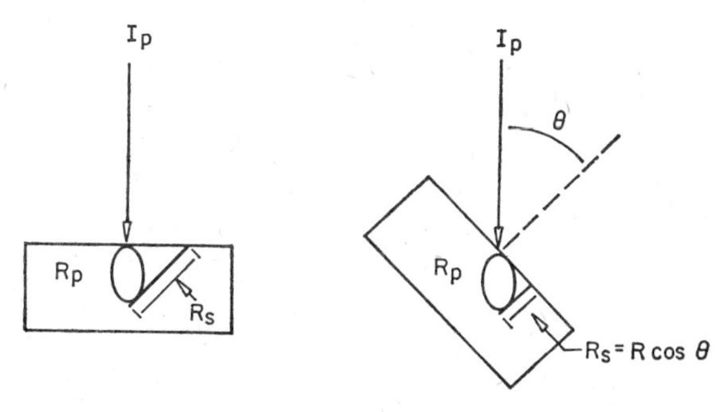

Fig. 4.42 Effect of specimen azimuth on position of excited volume.

d_e. The yield for the beam at normal incidence (Fig. 4.42*a*) is:

$$\sigma(0) \propto \exp(-d_e\beta),$$

while that for the beam at an angle θ (Fig. 4.42*b*) is:

$$\sigma(\theta) \propto \exp(-d_e\beta \cos\theta),$$

and

$$\sigma(\theta) = \sigma(0) \exp[d_e\beta(1 - \cos\theta)].$$

This equation is attributable to Bruining [121] and was confirmed by Lukjanov [122] for $0 < \theta < 80°$, when the beam energy was greater than E_{max} and the surface was not rough.

Secondary electron yield is but little increased by a decrease in the work function of a surface, as was shown by McKay [123], or by increase in temperature, unless this affects a surface layer [124]. However, McKay did draw up an empirical relation between the maximum yield and the work function, and a similar plot relating σ_{max} and density. Sternglass [125] related σ_{max} to location in the periodic table, finding that in each row the yield increases from the alkali metal to the multivalent metal. For nonconductors, however, the yield has a greater dependence on temperature and on surface work function, which is probably attributable to the lower energy of the emitted secondary electrons [126].

Backscattered Electrons

Backscattered electrons are often arbitrarily defined as electrons leaving the surface of a solid specimen with energies greater than 50 eV. Such a definition excludes secondary electrons but includes other electrons, such as those primary electrons that are reflected from the surface or reemerge after losing only a proportion of their initial energy.

Bishop [127] showed that there is a gradual increase in the backscattering coefficient η with the atomic number Z of the target, and further that η increases but slightly with an increase in the energy of the incident electrons in the range of accelerating potentials normally encountered in scanning electron microscopes. Moreover, the distribution of backscattered electron energies varies with the angle of scattering, and the most probable energy and mean energy losses increase with scattering angle. Theoretical considerations of single-event scattering phenomena and diffusion of electrons in matter have been published by Everhart [128], by Archard [129], and by Dashen [130], and fair agreement with the experimental relationship between η and Z can be predicted. Differences in the production and energies between elements of differing atomic number have proved to be of value in micrographs obtained using the emissive mode.

Effects of an Electron Beam on Polymeric Specimens

The radiation dose applied to a polymeric system by the electron beam of a scanning electron microscope at higher magnifications may be sufficient to cause chemical and morphological changes. The effects are largely bond rupture and may result in the formation of new bonds. These effects are dependent on the nature of the bonding of a material, in particular the rigidity of the molecular structure, and upon the presence of impurities or additives which may catalyze changes. Time of examination is also of importance.

The radiation dose depends on the accelerating voltage and current density, the nature of the vacuum and any residual gases in it, the diameter of the electron beam, the temperature and thermal conductivity of the specimen, and upon the area of the scanned raster. The application of a metal coating aids in the examination of radiation-sensitive materials since the incident electrons are reduced in energy before they strike the polymer and the metal coating reduces heat buildup in the scanned area.

The effects and the onset of radiation damage vary with the conditions of examination and with the polymer under consideration. Usual indications are a depressed "raster burn," seen at magnifications lower than the maximum at which the instrument was focused, or a single-line burn produced when the instrument is in the "line-set" condition while the recording intensity is being estimated. If the damage is more severe, cracks may initiate and propagate in the material, and these in turn cause the metal coating to fracture. Lips of such cracks display a characteristic white appearance because of the reflection of electrons from the curled edges. The effects of charging of exposed nonconducting material may cause the crack to appear completely bright. The beam may also result in the production of gases, which may break through the metal-coated specimen surface and escape into the chamber, where the electron trajectory may be disturbed.

Contrast of the Scanning Electron Microscope Image

The Emissive Mode

In the case of the scintillator/photomultiplier collector system, the contrast of the image on the CRT, hence that recorded by the photographic emulsion, depends on the following factors:

1. The signal entering the collector
2. The gain of the amplification system and its signal-to-noise ratio
3. The fractional brightness change which can be detected on a CRT screen and recorded on photographic emulsion

There are three components of the signal collected:

1. *Secondary electrons from the specimen.* The trajectory of these electrons is affected by surface fields on the specimen but is largely controlled by the field of the collector. The collected fraction of the secondary electrons emitted by the specimen depends on the numerical aperture of the collector and on the voltage difference between the collector and the emitting surface element. The energies of these secondary electrons is small, and their paths can be influenced by the electric field, hence they may be collected from the "back side" of a surface projection (as from *A* in Fig. 4.43), while secondary electrons from *B* reenter the specimen, hence do not reach the collector.

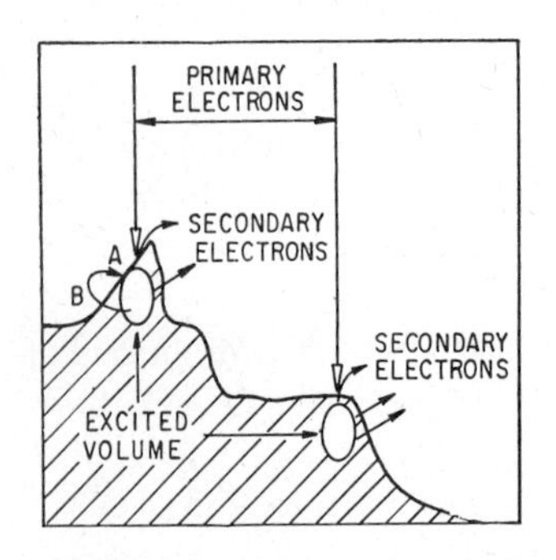

Fig. 4.43 Collection of secondary electrons.

2. *Backscattered electrons.* These electrons are reflected at or near the surface of the specimen, and they possess energies between those of incident electrons and about 50 eV. The paths of these electrons after reflection are not greatly, if at all, affected by the field of the collector, hence they travel a straight-line path. Thus only those that are reflected along a straight line from the image element to the scintillator contribute to image formation.

3. *Tertiary electrons.* These electrons may contribute to the signal from the scintillator. They are produced by the collision of secondary or backscattered electrons from the specimen with the chamber or with the collector. The electrons emitted by such collisions may strike the scintillator, but any contribution attributable to tertiary electrons is usually small.

The signal resulting from electrons striking the scintillator depends on the number of photons produced per electron and upon the number of these photons that strike the photocathode of the photomultiplier. Normally, the signal is primarily attributable to secondary electrons and it is small. Hence noise is a major consideration. Noise may result from a dark current leaving the photocathode of a photomultiplier in the absence of an external signal. Further, there is shot noise resulting from the statistical fluctuations in the probe itself, hence in the signal from the specimen. Oatley et al. [131] state that the basic signal-to-noise ratio of the element being scanned is $n^{1/2}$, where n is the number of electrons in a square probe of side d_0, and:

$$n = jd_0^2 \cdot d_0^2 t / D^2 e,$$

where j = uniform current density inside the probe, D = side of a square being scanned in time t, and e = electronic charge.

We now must consider the minimum difference ΔS between two signals S and $S + \Delta S$ from adjacent elements that can be detected above the noise level of the system. Rose [132] states that the human eye cannot distinguish a region of brightness B from an adjacent region of brightness $B + \Delta B$ unless the signal-to-nose ratio is about $5B/\Delta B$. Hence

$$\sqrt{n} \geq 5B/\Delta B,$$

so the above equation becomes

$$25(B/\Delta B)^2 \leq jd_0^4/D^2e,$$

and further considerations of noise introduced by the system modify this to:

$$100(B/\Delta B)^2 \leq jd_0^4t/D^2e.$$

It can be seen that an increase in beam current is necessary to detect increasingly smaller differences in contrast, but low beam currents are adequate if the time during which a signal is collected from each specimen element is increased to a sufficiently great magnitude.

From the above discussion it is evident that secondary electrons provide more image detail and information than would reflected electrons when a rough surface is being examined, so the collector is normally biased to reject the reflected primary electrons to as great a degree as possible. Further, the total signal attributable to secondaries is usually greater than that attributable to primary electrons, and the signal-to-noise ratio is then better.

The fact that the trajectory of primary electrons is less affected by small electric or magnetic fields can be used to advantage, however, with specimens possessing a tendency to charge, and also on specimens possessing small magnetic fields, when the topography alone is of interest.

With secondary electron images, the importance of contrast resulting from topographical variation cannot be overemphasized. Everhart et al. [133] estimated that an alteration of only 1 or 2° is usually sufficient to result in a significant alteration in the brightness of the final image. Two exaggerated cases are depicted in Fig. 4.43. A thin spike could appear brighter than the rest of the surface since electrons emitted from it in all directions would be drawn into the collector. A step would also have enhanced brightness by virtue of the shorter distance to the surface, and there may be an additional effect attributable to reflection by the face.

If voltages vary across a specimen, contrast will result from changes in the trajectories of secondary electrons between the specimen and the detector [133], although this effect is less than that resulting from the suppression of secondary electron emission in positively biased regions [134]. Kimoto and

Hashimoto state that differences as small as 0.25 V can be detected, and such observations are of great importance in the microcircuit field. Such voltage differences do not affect the trajectories of backscattered electrons.

The Conductive Mode

Contrast observed while employing the scanning electron microscope in the conductive mode can be considered to be attributable to two related physical processes involved in the charge collection mechanism. First, electron-hole pairs are created. Second, these move under the influence of internal fields, and the charge is then collected in specific regions. Furthermore, contrast results from any circumstances that affect these processes, such as backscattering and the production of secondary electrons, and biasing fields as temperature gradients which might speed the diffusion and charge collection of the electron-hole pairs produces contrast effects.

Low-Voltage Operation

It is possible to examine insulating specimens without a conductive coating if the accelerating potential of the electron beam is reduced to about 1 to 2 kV [139]. For such incident electrons the secondary emission coefficient of most insulators is greater than unity, which prevents negative charging. When such a low accelerating voltage is employed, the characteristics of the beam are considerably altered. If the normal electron gun is used, its current density decreases with the voltage, and the size of the crossover is increased. The lens aberrations increase, so the resolving power of the microscope decreases and stray magnetic fields exert greater effects, further reducing the resolution. The penetration of the electron beam is related approximately to the square of the voltage, and low voltages may improve the visibility of surface layers. Diffraction effects may become more prominent, as the effective wavelength of the electron beam is inversely proportional to the square root of its voltage. Thornley [135] has suggested that a resolution of less than 1000 Å should be attainable and has reported a resolution of 2000 Å. Such a loss of resolution implies that unless it is vital to work with uncoated specimens this precaution is desirable for precise work.

In routine operation, when high resolution is not imperative, it is often useful to take advantage of the lower current density obtained at a reduced voltage since this results in reduced charging effects and also in thermal effects which may occur with polymers or biological samples.

Resolution of the Scanning Electron Microscope

Secondary and Backscattered Electron Images

The resolution of the scanning electron microscope is limited first by the size of the beam which, with the conventional tungsten hairpin filament and a typical lens system, is about 100 Å for a usable beam current of 10^{-12} A at 25 kV. Secondary electrons can escape only from a volume within several hundred angstroms of the specimen surface, and the surface area of this volume is not too much larger than that of the probe as is shown in Fig. 4.29. For a probe of 100-Å diameter, an edge resolution of about 150 Å could be expected, although features of these dimensions may not be able to be resolved on a flat surface. Pease and Nixon [136], using a four-lens demagnifying system, obtained a spot size of about 60 Å and showed edge resolution of about 100 Å; another reason for the worsening of resolution attributable to specimen properties is that the secondary electron signal most probably includes electrons generated by backscattered electrons outside the volume depicted as producing secondaries in Fig. 4.29. The energy of backscattered electrons, however, is considerably greater, so they can escape from a considerably larger volume of greater surface diameter. As a result, the resolution of the scanning electron microscope in the backscattered mode is considerably poorer (about 1000 Å). Further, since the size of the excited volume is dependent on the accelerating potential applied to the beam and the composition of the specimen, the resolution of the backscattered electron image depends on these parameters, while that of the secondary electron image is independent of them. In practice, the resolution attainable is often limited by specimen properties other than those already mentioned, such as thermal instability, and inclined planes on the surface result in the spot actually being larger than it would be on a surface orthogonal to the beam by a factor of the reciprocal of the cosine of the angle between the beam and that plane. A final consideration is that in order for optimum resolution to be achieved sufficient contrast is required between adjacent micrograph elements being used to establish the resolving power of the instrument, hence conditions of recording and collection are also of importance. It should be recognized that to achieve optimum resolution critical settings of the stigmator are required. Finally, in many of the applications of the scanning electron microscope, very high resolution is not the critical criterion, although ability to achieve the specified resolution at least implies that the instrument is in good operating condition.

Other Modes

The resolution of the luminescent mode has been discussed by Thornton [137], who points out that it is dependent on the spot size, the depth from which the radiation can escape, the penetration of the incident electrons and the presence or absence of a nonradiative surface "dead layer," in which the excited charge carriers do not emit a signal. The reason that the resolution can be of the order of the size of the spot is attributable to the fact that the image is constructed from the signals from a photomultiplier, hence there are no optical elements in which diffraction limits the resolution to about one-half the wavelength of the illuminating wavelength.

In the conductive mode resolution is replaced in importance by the sensitivity of detection of biasing voltages, and it is normally assumed that potential differences of about 0.5 V can be rendered visible. If the scanning electron microscope is being used as an electron microprobe, the spatial resolution will be determined by the factors discussed in Section 4, p. 405, although in general it will be related to spot size, escape depth of the x-ray photons, fluorescence, and specimen properties.

The questions of resolution and contrast are closely related, and the contrast required for optimum resolution must always be considered.

Stereomicroscopy

Qualitatively, stereopairs are of the most assistance in determining whether a particular doubtful feature is a projection above the specimen surface or a depression beneath the surface. Pairs are also useful in taking full advantage of the three-dimensional appearance of the micrographs obtained with a scanning electron microscope; the construction of an open material, such as a fabric, and the interlacing of fibers therein could be presented with great impact, as well as providing the basis for quantitative work.

Stereopair micrographs are obtained by moving the specimen through a suitable angular displacement between exposures. The angle is small enough that the images are still sufficiently similar to be fused in binocular vision, yet large enough that the resultant stereoscopic parallaxes are large enough to measure. It is important that the axis of tilt of the specimen lie in a plane perpendicular to the visual axis and perpendicular to the interpupillary axis. Garrod and Nankivell [138] showed how to correct for tilt error, which results from a difference in magnification between the two micrographs because of the noncoincidence of the optic axes of the microscope and the axis about which the specimen is tilted. Any foreshortening resulting from the overall tilt of the specimen stage, or of the general topography of the specimen, must be taken into account. Wells [139] discusses the procedures for correction of the tilt error, and in addition for the elimination of perspective errors attributable to foreshortening, which he calculated will be small in practical cases. Such principles were applied in a study of fracture surfaces of metallographic surfaces [140]. The accuracy of these methods has been discussed elsewhere [141], while discussions of photogrammetry appear in standard texts [142].

As an alternative to the tilting method, it is possible to rotate the specimen through a known angle, thus avoiding changes in magnification [143]. An equivalent operation is to rotate the raster on the specimen between exposures.

The pairs can be viewed in a stereoviewer. Quantitative measurements can be made with floating-dot-type instruments and the data computed into heights above particular planes, or into a three-dimensional system of coordinates, as discussed in the references. The computations for precise and small-scale work are likely to be long unless a computer is available.

Specimen Preparation

Specimen preparation for the scanning electron microscope is considerably less complicated than for the transmission electron microscope or, for that matter, for the optical microscope. This is in part attributable to the fact that surfaces are being examined rather than internal structure. Hence specimen preparation comprises providing a surface, if a suitable one does not already exist, and attaching the specimen to a conducting stub with the required surface exposed. If the specimen stage has sufficient freedom of manipulation, a number of surfaces on the same specimen can be readily positioned in an attitude to the beam that permits examination.

In many cases the specimen need only to be reduced to a suitable size and it is ready for metal-coating. In other cases the material may be fractured, which in the case of polymers is often done to advantage at reduced temperatures. Fracture of fibers can cause drawing, heat buildup to softening or melting temperatures, and other artifacts [144]. Ion etching, using an argon source, has been used to reveal structure in dental tissues [145] and in other biological tissues. Cutting produces a rich array of artifacts, such as knife marks, compression, shearing, smearing, cold-drawing, heating effects, and shattering. Part of the material may be removed by chemical or other means, and the resulting structure or residue examined. Alternatively, an rf reactor can be employed to remove some or all of the organic material in a specimen [146].

If the specimen is a nonconductor, its surface is normally metal-coated to allow the charge to be dissipated from the area being examined and to provide a source of secondary electrons. A range of the heavier metals has been used for these purposes, the basic requirements being that their evaporation does not result in structure being introduced into the surface being examined and that they can be conventionally vacuum-evaporated onto the specimen. Standard evaporators are used, although the plate holding stubs is normally rotated during the evaporations and provision is sometimes made for tilting the specimen. The latter two provisions are to ensure that the specimen is coated as uniformly and completely as possible. The requirements are quite different from the metal shadowing employed with or without replication in transmission electron microscopy. Another consideration is that the heat of the evaporation process does not damage the specimen and also that the vacuum be retained when relatively large amounts of metal are evaporated.

The metal and the thickness of the coating are largely at the convenience of the operator. The coating thickness should always be less than the resolution of the instrument, and less than the image detail dimensions sought in the specimen. Metals used include silver, gold, gold/palladium, and aluminum, and thicknesses range from tens to hundreds of angstroms.

A platinum/carbon or a carbon coating also provides a conductive surface and has the advantage [147] of allowing compositional differences to be detected.

Metal coatings allow only topographical detail to produce image detail since the thicknesses are normally greater than the mean free path of electrons of these energies in the metal [76].

An alternative method [148] is to employ a commercial antistatic finish such as is used in textile processing. A short treatment with an aerosol spray permits immediate examination of specimens. Its use appears to be somewhat limited, owing to solids present in the spray and to the fact that "droplets" of antistatic material are often found on the surface being examined. It is useful for fabrics, and individual fibers can be examined to fairly high magnification with fair resolution. The productions of secondaries is not as prolific as with a metal-coated surface, but the noise resulting from higher photomultiplier settings may not be too objectionable. It is also useful when sprayed on the reverse side of a fabric to improve the continuity to the stub. A fabric treated in this way can normally be successfully examined with a single metal coating on the face side, while without the antistatic two coatings may be required to obtain a specimen suitable for examination.

Applications of Scanning Electron Microscopy

Introduction

In recent years there has been a tremendous increase in publications concerning the theory, potential, and applications of the scanning electron microscope. Much of the literature pertaining to the subject has been referred to earlier in this discussion. Several comprehensive bibliographies do, however, exist and these include those edited by Wells [149] and by Johari [150].

Two international conferences were held in 1968, one at Chicago, Illinois [112] and the other at Cambridge, England, sponsored by the Institute of Physics and the Physical Society in association with the Royal Microscopical Society. Papers concerning the scanning electron microscope have appeared at national and international conferences since about 1957, and the relevant literature is cited in the bibliographies listed above.

Two of the major fields of application of scanning electron microscopy fall outside the limits of this article. The first field is biology and medicine, and among the extensive review articles published is that by Hayes and Pease

[151]. The scanning electron microscope has been widely used in the study of dental enamel and skeletal material, cells, microorganisms, chromosomes, whole-mount and sectioned tissue, and even living specimens [152].

The other field covers various aspects of microelectronics, especially examination of devices under operating conditions. Wells's bibliography [149] includes many references to such work, and a review was provided by Matta at the Chicago conference [112]. He discusses secondary electron, junction-induced current and electron-beam-induced conductivity modulation modes and mentions the use of the electron beam in the exposure of photoresist in the fabrication of some devices.

Applications of the Scanning Electron Microscope in Chemistry

Some chemical applications of the scanning electron microscope have been discussed by Drew [153]. They include the study of ice crystals via replication, in which information has been obtained about the growth mechanism of ice crystals and changes in orientation. Other studies considered particles of propellant material that had been quenched by the application of pressure after ignition had commenced. The propellant surface displayed what appeared to be inflated bubbles, which would not be expected if, as was previously thought, ammonium perchlorate decomposed only by a sublimation from solid to gaseous oxidation products. Other work concerned the combustion of aluminum [154], and it was shown that an oxide layer, if it remained intact, could prevent the combustion of aluminum particles passed through a burner flame. Drew also discusses the structure of a thin film (lead sulfide) as depicted directly in the scanning electron microscope and by the transmission electron microscope, making use of an extraction replication method. He also mentions the extensive use to which the scanning electron microscope can be put in the study of the surfaces of solids used in gas chromatography, both before and after coating with a stationary phase. Another study of chromatographic support material was made by Kirkland (Fig. 4.44).

Drew concludes with the observation that the scanning electron microscope cannot examine dynamic processes and that "chemical applications therefore must rely upon inferences gleaned from permanent structures" obtained before, during, or after such a process.

Direct observation of the physical effects of chemical reactions was reported in the work of McAuslan and others on the controlled decomposition of crystals that are explosive under other conditions. In one instance [155] silver azide crystals were deposited on a thin silver plate, and they were then slowly heated to temperatures below which explosive reactions initiated. It was found that small crystals did not need to be metal-coated for examination, but larger crystals did. The heating of the crystals caused a

Fig. 4.44 Zipax (Du Pont) chromatographic support beads. (*a*) Magnification of 15,318 × showing glass microbeads. (*b*) Magnification of 733 × showing a number of beads. (By permission of J. J. Kirkland.)

disintegration of the originally relatively smooth surface into one of a dimpled appearance, presumably as a result of the formation of metallic silver and gaseous nitrogen. The size of the structures suggested that the conversion to metallic silver was not complete. McAuslan and Smith [156] report that dehydration occurs in lead styphnate crystals at 120°C and that this process leads to a breakup of the crystallographic order. Such a change in the surface area of the crystal would be expected to be of importance in an explosive reaction since this is a rate process limited in part by surface area. The low beam current of the scanning electron microscope compared to that of the transmission microscope is of advantage in this type of work, as is the ability to examine surfaces directly.

Other examples of the physical results of chemical reactions include oxidation and corrosion, which may take place under mild or severe conditions, and on metals or on other materials. Pease and Ploc [157] examined the oxidation of steel by cycled heating of the sample in the specimen chamber at atmospheric pressure followed by evacuation and scanning electron microscope examination. The effects of oxidation can be observed at crystal boundaries, and effects attributable to crystal orientation were observed. The use of the scanning microscope permitted the treatment to be performed in situ and aided the examination of the same area after each heat treatment. These investigators also showed that the growth of oxide whiskers could be inhibited by doping the specimen with a bismuth layer. Castle and Masterson [158] studied the corrosion of boiler tubes, in which process the oxidation of mild steel occurs in high-temperature aqueous solutions. The metal oxide samples were fractured at liquid nitrogen temperatures, and crystal size was measured in a search for a porous structure. The earliest work on the growth of oxide films in iron was done by Knoll [159].

The scanning electron microscope is an excellent tool for the examination of the growth habits of crystalline materials at moderate resolution. Thornton et al. [160] examined silicon crystals grown by the vapor-liquid-solid process from gold/silicon alloys. They found that growth varied from fine whiskers at low temperatures (950°C) to epitaxial layers at high temperatures (1100°C). The whiskers had a rounded cross section, but it was found that a short etching process removed an overgrowth deposited in a metastable growth process and revealed the true cross section of the whisker to be of a somewhat irregular hexagonal nature. It was noted that some of the irregularity could be the result of the etching process.

The surfaces of graphite crystals have been examined by Minkoff [161]. The crystals were grown in nickel in the presence of a small amount of lanthanum as an impurity and extracted by dissolving the nickel in dilute acid. Holes, which possessed hexagonal symmetry, were found in the crystal. The inside of the hole was characterized by a stepped appearance, and it was postulated that the steps were related to a blocking of the screw dislocations by the lanthanum, in a fashion similar to the mechanism of whisker growth. In fact, hole formation was likened to the production of a negative whisker. The study was extended to the growth of graphite crystals in iron and nickel alloys [162] when it was confirmed that the growth process could be blocked by the absorption of an element that spheroidizes graphite. Holes may be formed in the crystal surface, but as the impurity content of the melt increased, flake development could be stopped and branched crystals formed. The end result of a number of branches was a spherulitic structure. Gardner and Cahn [163] have also studied whisker growth, using a scanning electron microscope to examine the recrystallization processes in an iron/aluminum alloy. Surface layer fracture of refractory metal carbides (such as TiC, B_4C, and FeC_2) caused by giant laser pulses has been illustrated by scanning electron microscopy [164].

The scanning electron microscope is also useful in the study of the surface structure of films of paint applied to different surfaces. Brooks et al. [165] reported such an examination, in which they observed the effects of TiO_2, $CeCO_3$, acicular talc, calcined clay, and delaminated kaolin on the surface appearance of paints formulated over a range of 35 to 65% PVC with PVAc, acrylic, and linseed oil binders. They found that the resolution was good enough to clearly distinguish the finer pigment and extender particles, as well as to delineate surface structure of individual particles.

Coutts and co-workers [166] have reported an investigation into the wear of phonograph record grooves; the great depth of focus and ease of specimen manipulation made the scanning electron microscope the ideal tool for the job. Although damage resulting from the first play is readily discernible, wear increases regularly with the number of plays (up to 50) and with the weight of the pickup.

The study of man-made fibers has been reported by a number of authors, including Newman and Sikorski [167], who performed high-resolution studies of the surface topography of viscose rayon fibers. They report the construction of a special specimen stage to permit the rotation of the fiber about its axis in the field of view of the microscope and emphasize the compromise that must be made between the optimum probe size and the acceptable noise level in the image. Van Veld et al. [168] reported surface studies over a wide range of man-made fibers, including the use of a peeling technique [169] to reveal interior structural detail (Fig. 4.45). It was noted that care is required in interpreting the results of such an operation since considerable distortion is likely to occur and the effects of peeling vary considerably for different polymers and orientations. A chemical etchant (chromic-sulfuric acid) was employed in a study of polypropylene to reveal the structure of a transcrystalline (columnar) skin [170], random etch patterns being produced in contrast to the radial arrangements observed in spherulitic surfaces which were observed randomly throughout the specimen. Transcrystalline morphology, however, originated at the surface. An investigation of the morphology of polyethylene terephthalate films crystallized under different time and temperature conditions has been reported [171], and attempts were made to obtain a correlation with kinetic data.

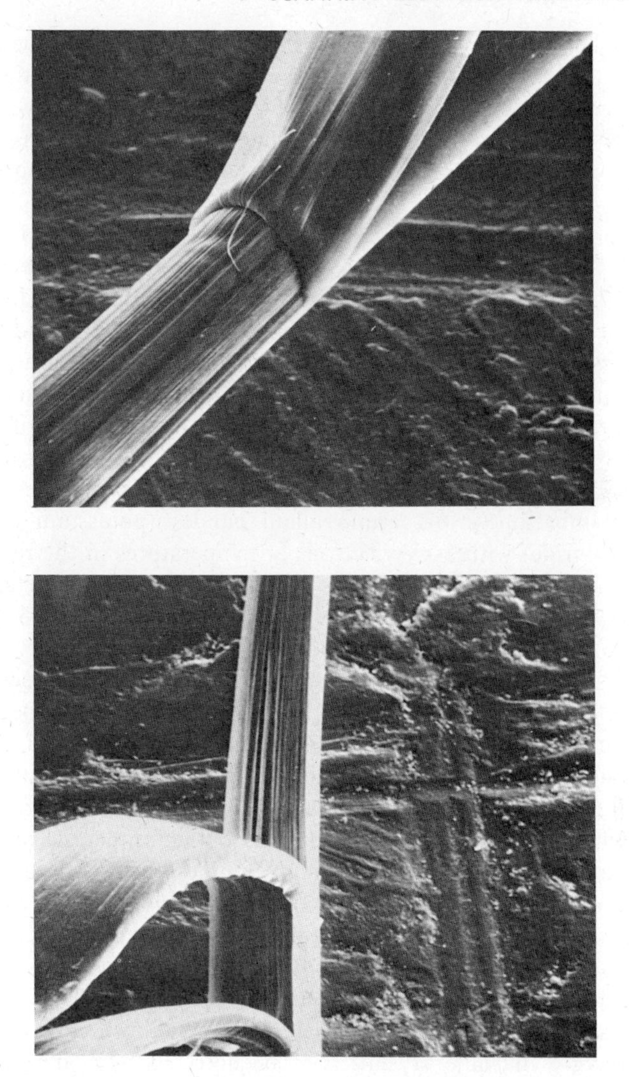

Fig. 4.45 Peeling habit of polyester fibers. (*a*) Skin peeling; 1296×. (*b*) Cord peeling; 1296×. (By permission, from Ref. 96.)

Another field in which the great depth of focus of the scanning electron microscope is of considerable advantage is in the study of fractures, for example, of the fracture surfaces in fiber composite materials [172]. Examination of such micrographs permits assessment of where the composite system failed, be it in the matrix, the reinforcing fiber, or in the adhesion bond between the fibers and the matrix. The topography of the fiber surfaces and the relation between this and adhesion can be studied, as can the alignment and distribution of the whiskers across the fracture face. Use of the scanning electron microscope in fractography should bridge the gap between optical and electron microscopy, where the resolution of the former is not satisfactory and the preparation techniques of the latter are prohibitively time-consuming.

Few reports of cathodoluminescence studies with the scanning electron microscope appear in the literature to date, but progress is being made. Davoine [173] showed that only those parts of a cadmium sulfide crystal with a high concentration of primary defects are luminescent, and that while some of these regions emit monochromatic radiation others are made to emit polychromatic radiation between 3000 and 6000 Å, which was the wavelength range studied. A curved mirror aided the collection of light, and a filter was used as a monochromator in front of the photomultiplier. Davoine [174] also reported the observation of cathodoluminescence produced by bombardment of certain alkali halides (potassium iodide and potassium chloride) with 5-kV electrons at temperatures in the range 90° to 450°K, and radiations of wavelengths 2,000 to 12,000 Å were detected.

Korda et al. [175] examined the microstructure of a p-16 phosphor powder using secondary electrons and light to form the image. A flexible noncoherent fiber optics rod was used, and useful micrographs were obtained at 2400×, the resolution in the cathodoluminescent mode being about 1500 Å. Zinc sulfide phosphor grains were studied by Shaw et al. [176].

Pruden and others [105] report cathodoluminesce mapping of materials with resolution better than 1000 Å in the scanning electron microscope and were able with their arrangement (p. 368) to obtain sequential micrographs using secondary electron emission and luminescence without changing specimen orientation or breaking the vacuum. Williams and Yoffe [106] examined single crystals of hexagonal zinc selenide held at 100°K by a liquid nitrogen mount and have published luminescent micrographs employing both total emitted light and monochromatic wavelength.

Contrast resulting from magnetic fields, such as those corresponding to domain structure in some crystals, can be detected by scanning electron microscopes. Banbury and Nixon [177] devised an electrode structure in which the specimen was located. Secondary electrons emitted from a surface where no magnetic field existed were collected after passage through a limiting aperture, while the effects of even a small magnetic field or the path of an electron were such that it was prevented by the aperture from reaching the collector. The lack of a signal from regions where there was a magnetic field led to contrast between them and zero magnetic field areas. The effects of the magnetic contrast can be switched off by not biasing the electrode and by removing the aperture disk. These investigators also employed deflection modulation display to emphasize the magnetic structure and pointed out the possibility of making the observations quantitative. Among the materials studied were cobalt crystals and magnetic tape, and later work [178] included examination of a magnetic tape erase head.

Other workers have used unmodified instruments to study magnetic domain structures [179]. It was shown that the contrast observed was a result of the deflection of the secondary electrons by Lorentz forces attributable to the demagnetizing fields near the surface of the specimen, and further that the method could detect demagnetizing fields attributable to internal structures as well as to the surface domain structure. The direction of the electron detector was found to be of importance, since the contrast resulting from either surface or internal effects depended strongly on the detector orientation. Detailed investigations were made of large single crystals of cobalt, and it was found that heating the specimen to 240°C caused the magnetic contrast to disappear; the internal domain structure, however, was unaltered, as was shown by the return of the original contrast as the specimen was cooled. Domain patterns revealed by a modified Bitter technique (evaporating iron in an inert atmosphere) [180] were also studied as were crystals of barium ferrite, while barium titanate crystals were examined by other workers [181].

Other studies of ferroelectric crystals include those by LeBihan and Sella [182], who examined sublimed WO_3 crystals. They found best contrast with 5-kV accelerating potential and suggested that the contrast was not solely attributable to crystallographic effects but in part to other effects such as surface charges. They also found that the contrast varied markedly with crystal orientation with respect to the scan direction of the probe and made observations with both backscattered and secondary electrons. Saparin and others [183] made theoretical calculations concerning the contrast to be expected as a result of local magnetic and electric fields.

Mention has been made in the literature of the generation of orientation patterns in the scanning electron microscope in bulk specimens of semiconductor crystals (silicon, germanium, and gallium arsenide) [184] and in crystals containing copper and cobalt [185]. The latter paper discusses the patterns in terms of the anomalous absorption of incident electrons and indicates that the patterns are developed in bulk crystals when the change in the angle of incidence of the electron beam on scanning the specimen is greater than 2θ, θ being the Bragg angle; the effect is best observed at very low magnifications when the regions being scanned, hence producing the patterns, are of the order of thousands of microns across. At higher magnifications this 2θ criterion can be satisfied by deflecting the scanning beam beneath the objective lens, or by rocking the sample, synchronously with the CRT display raster, in two orthogonal directions while keeping the probe

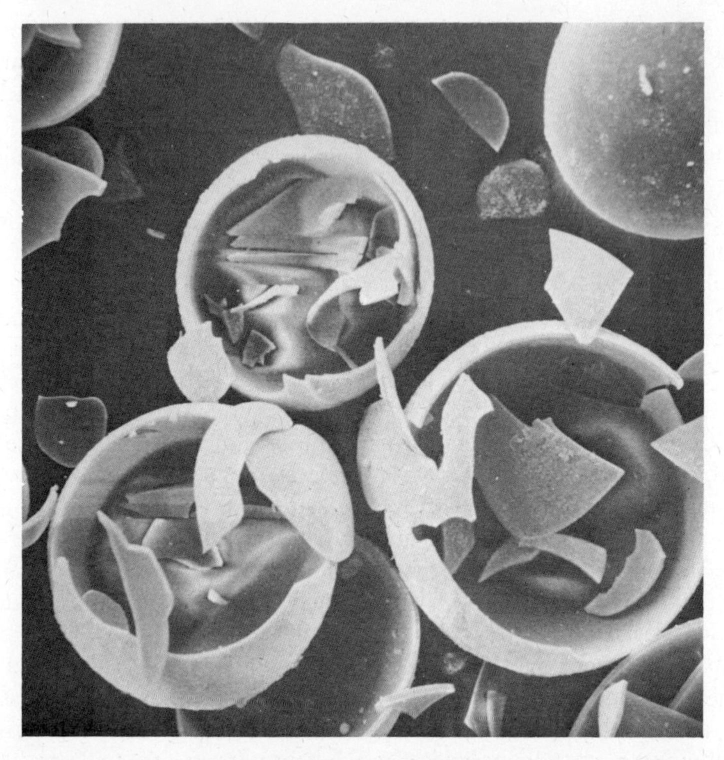

Fig. 4.46 Fracture face produced by a simple bend fracture of a bar comprising Thornel-25 fibers in epoxy matrix; 900×. (Micrographs in Figs. 4.46 and 4.47*a* by the courtesy of J. L. Brown, Georgia Institute of Technology, Engineering Experiment Station, Analytical Instrumentation Labs, and by permission of Engis Equipment Company, Morton Grove, Illinois, in whose publication *Engis "Stereoscan" Colloquium—1969* they appeared.)

static [184]. Coates [184] found that the best results were obtained with a collimated, high-intensity beam attained by turning off the objective lens and defocusing the condenser lenses, respectively. The patterns are not the conventional Kikuchi patterns encountered in transmission electron microscopy, but they do closely resemble these patterns and behave in a similar fashion when the specimen is tilted or rotated. Hence these patterns could prove of considerable importance in determining the crystallographic orientation of small grains in polycrystalline aggregates [185].

The Future of Scanning Electron Microscopy

The question posed by the title of this section cannot be answered in a single and simple fashion. In the future the scanning electron microscope of the type with which we have been primarily concerned here will doubtless

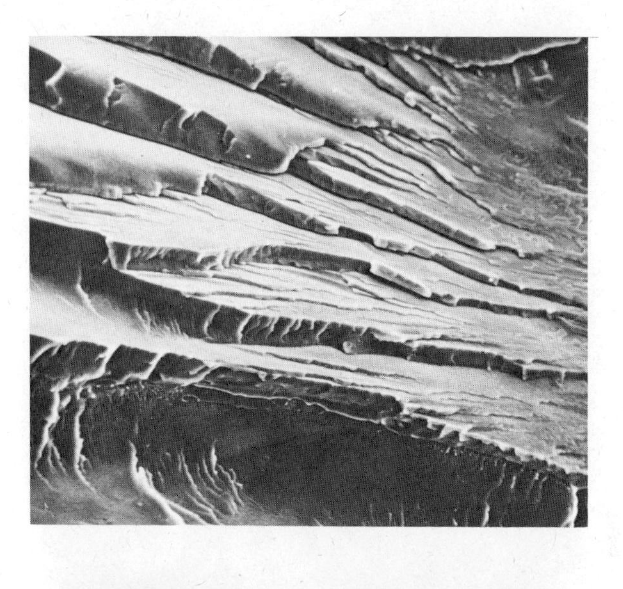

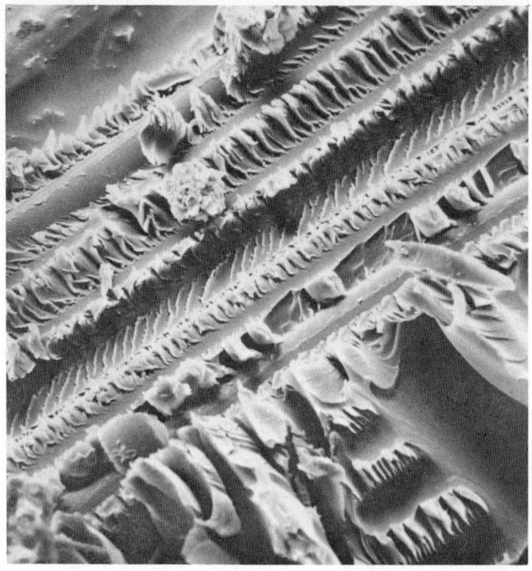

Fig. 4.47 (*a*) Cleavage fracture of polyester resin filled with glass fibers; 1039×. (*b*) Fracture of a block of polyethylene terephthalate showing laminar structure and "ball" structure visible on some of the fracture planes; 693×. (By permission, from Ref. 96.)

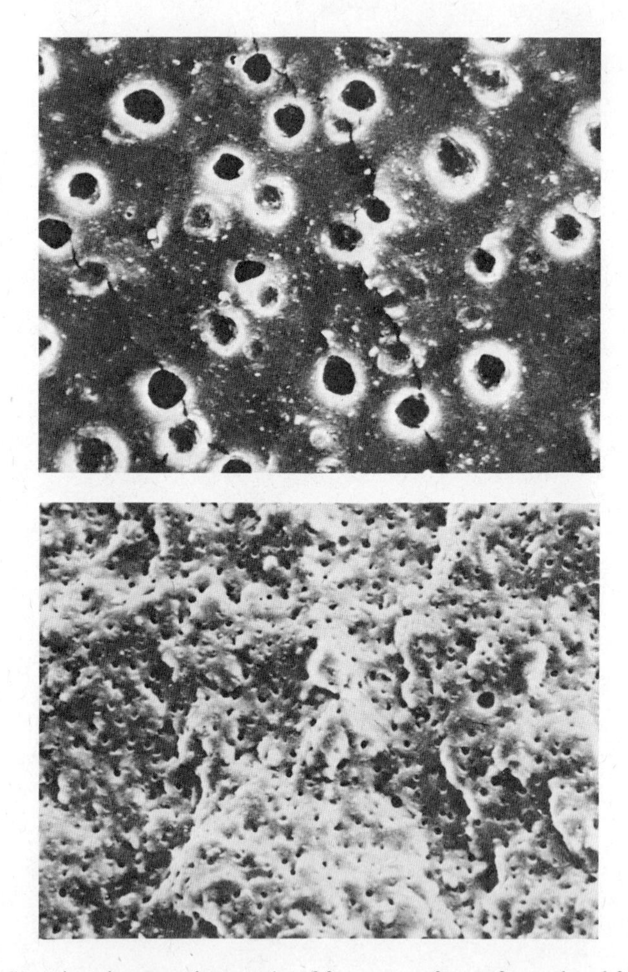

Fig. 4.48 Scanning electron micrographs of fracture surfaces of two closed foam systems. The vulcanized elastomers were fractured at liquid nitrogen temperatures after being pressurized with argon for 48 hr at different pressures. (*a*) and (*b*) SBR styrene-butadiene copolymer; (*c*) and (*d*) Viton (Du Pont) fluoroelastomer; approx. 214×. (Reproduced by permission of C. W. Stewart from a paper to be published in the *Journal of Applied Polymer Science*.)

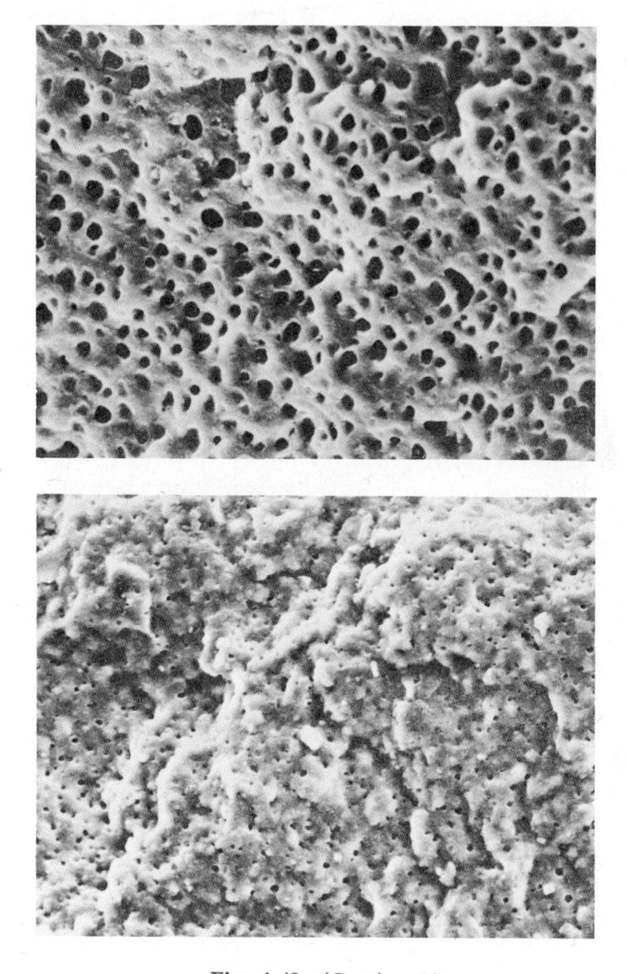

Fig. 4.48 (Continued.)

continue to be developed and to be of great utility in the study of surfaces. The resolution of the secondary electron and other emissive modes will continue to be limited by the size of the excited volume, which will always be larger than the probe itself. Hence the limit of resolution first depends on the size of the probe, and this in turn depends on the size of the virtual source in the electron gun of adequate brightness. Use of a cold cathode, such as the lanthanum hexaboride rod cathode described by Broers [186] permits a brightness of approximately five times that of the tungsten hairpin filament and a minimum probe diameter of 30 ± 7 Å, leading to a point-to-point resolution in scanning reflection of 65 ± 15 Å. Another gun is one with a field emission type of electron source such as that described by Crewe et al. [187], which uses a tungsten tip with a surface oxide layer resulting in increased emission from the (111) planes resulting from field enhancement. This leads to a beam current of about 1 μA at 1 to 3 kV with an apparent source size of about 30 Å. It is extremely important that both these guns be precisely aligned relative to the optic axis of the microscope, and further that the required vacuum be around 10^{-8} to 10^{-9} torr for the field emission source.

When the scanning electron microscope is to be applied to the study of surfaces, large samples are normally encountered and a variable working distance is desirable. The immediate future should see wider applications of special specimen stages to manipulate the specimen—strain it, for instance—and its environment, especially its temperature. More work on its application in modes other than the secondary electron mode will doubtless be reported. More instruments of a combined nature will probably be developed, providing both medium-resolution microscopy and x-xay microanalysis including low-atomic-number-element identification; one such has been described [188] and others can be assembled from accessories. It seems that the nondispersive solid-state detectors will provide an attractive route to x-ray analysis in at least qualitative terms, since they have the advantages of operating under less stringent conditions than do the crystal spectrometers; focusing optics are, however, offered by at least one manufacturer [189].

If the scanning microscope is used in a transmission mode, however, its resolution is of the order of the size of the spot instead of about twice the diameter of the spot as is the case in the emissive mode. The reason for this is that transmission microscopy requires thin specimens, and in this thickness of matter the excited volume does not spread far in a lateral dimension. A field emission electron source of small virtual size requires only a moderate demagnification to provide a beam of suitable dimensions, and Crewe [190] reports the achievement of resolution of less than 4 Å with a microscope consisting of an electron gun and one lens, the system being operated at 20 kV.

An electron gun scanning microscope is reported by Crewe et al. [191]. This simply comprises a field emission gun capable of producing a probe of 100-Å diameter with a beam current of 10^{-10} A, which permits scan times of the order of seconds. The electrons passing through the specimen are detected by a scintillator/photomultiplier combination and used to modulate the brightness of a synchronously scanned CRT, and resolution of better than 200 Å has been obtained using transmitted electrons. A secondary electron detector above the specimen will be incorporated, as well as a magnetic or electrostatic energy analysis system.

Attachment of an energy analysis system to a scanning transmission electron microscope has been discussed by a number of authors, including Crewe [192], who published a series of micrographs of a thin aluminum specimen taken with various energy losses from 0 to 700 V. That obtained with zero loss corresponds to a conventional micrograph, while those taken at different losses contain different information and may be the basis of another contrast mechanism. An advantage of the scanning system is that the information is presented serially, and this is of advantage if endeavors are to be made in the field of aftertreatment of the image by analog computer methods [193]. A high-voltage (600-kV) transmission scanning electron microscope has been described by Cowley and Strojnik [194], and electron diffraction patterns from very small regions of the sample are also discussed.

It can be seen that the field of scanning transmission electron microscopy is in its infancy, but it appears likely that it will have a place in the range of microscopes available to research workers with both dark- and bright-field images possible in addition to selected area diffraction and energy analysis of the transmitted electrons.

Secondary electron spectroscopy by the analysis of Auger electrons is a nondestructive technique capable of identifying extremely small quantities of surface constituents, and it seems likely that it will be extended to quantitative analysis and provide information as to chemical states of the atoms examined. Light elements are more readily detected than by x-ray methods, while the latter is more sensitive for elements of larger Z. The Auger electrons are ejected by low-energy (1- to 5-kV) incident electrons striking an approximately flat specimen at a low angle of incidence. One of the inner shell electrons of the bombarded material is ionized, so that an electron moves into this vacancy; a photon or an Auger electron is emitted to restore the energy balance of the structure. The energy of the x-ray photon or the Auger electron is characteristic of the atom and can be used to detect it, and energy analysis systems for this purpose have been published. Harris [195] used a system similar to that of Lander [196], comprising a 127° sector electrostatic analyzer separated from a secondary emission multiplier by a slit system. In order to screen out the low-energy secondary electrons, no acceleration was imposed between the sample and the first dynode of the multiplier. The distribution of electron energies was observed by slowly varying the analyzer deflection voltage. The effects of noise were reduced by working with the derivative of the distribution taken with respect to energy rather than distribution itself. A vacuum of around 10^{-8} to 10^{-9} torr is required for the analysis and, since the environment must be free of organic materials, ion-pumping with metal seals is necessary.

Auger analysis can profitably be combined with low-energy electron diffraction (LEED) [197], the hemispherical retarding grids used in this apparatus being readily adaptable to the Auger analysis requirements. A commercial instrument that performs both Auger electron and LEED studies has been announced [198].

4 THE ELECTRON MICROPROBE

Introduction

The transmission electron microscope has done much to satisfy man's desire to see and characterize very small structures, but in many cases it has failed to disclose the exact chemical nature of these details. Indirect methods such as selective staining and etching have been of some help, and selected area electron diffraction techniques have added information about crystalline components. However, it was not until spectroscopic techniques were coupled with an electron probe and imaging system that a strong tie was established between high-resolution microscopy and chemical analysis.

The early development of x-ray spectroscopy was established by the classic work of Moseley [199], who investigated the x-radiations from various elements and showed a straight-line relationship between atomic number and the square root of the frequency of certain bright lines in the radiation spectrum. The techniques that developed in this field were combined with electron beam technology in an instrument (the electron microprobe) proposed in a patent by Hillier [200] in 1947 and first applied successfully by Castaing [201] and Guinier [202].

In an electron microprobe, a particular region of the surface of a specimen is bombarded by electrons that have been focused to a relatively fine probe whose diameter is typically between 0.1 and 3 μ. A volume of the specimen of the order of cubic microns is irradiated by the incident electrons, and a complex x-ray spectrum is emitted. This spectrum includes the characteristic radiations of the elements included in this volume of material, and suitable analysis permits these elements to be identified and their concentrations determined.

The most recent step in bringing the technique of spectrum analysis and microscopy together was the addition of scanning attachments to the probe, by Cosslett and Duncumb [203], so that one could prepare a map of the distribution and concentration of various elements within a chosen area of a sample.

One of the more important limitations of the microprobe that has restricted its use in problems of general chemistry has been the inability of past instruments to properly detect and analyze radiation from the lighter elements (see p. 416). However, recent developments in the instrument and refinement of techniques have led to a capability of detecting elements as light as beryllium. This has enabled research workers to apply the microprobe to an increasing number of problems involving the more common elements such as carbon, oxygen, and nitrogen.

An extensive body of work has been built up from the application of the microprobe, particularly in the fields of metallurgy, mineralogy, solid-state electronics, and biology. In addition to conference reports and papers in the literature, a number of books have appeared on the subject [204–210], and a very comprehensive bibliography was prepared in 1964 by Heinrich [211].

The Instrument

The basic requirements of the electron-optical system for an electron microprobe are essentially the same as those for a scanning electron microscope. This can be seen by comparing the block diagram of Fig. 4.49 with that shown in Fig. 4.36.

The electron beam is derived from a heated tungsten hairpin filament, and the accelerating voltage between the filament and the anode plate is normally between a few kilovolts (for example, 2 kV) and 50 kV. Most published work appears to have been performed at between about 30 kV (at which Castaing [202] originally worked) and 10 kV, although there is a trend toward the use of lower accelerating voltages. In this range a vacuum of about 10^{-4} torr is sufficient to minimize scatter of the electrons making up the probe. Greater accelerating voltages lead to greater penetration into the specimen and an increase in the yield of characteristic x-radiation, but also to possible loss of lateral resolving power (p. 405).

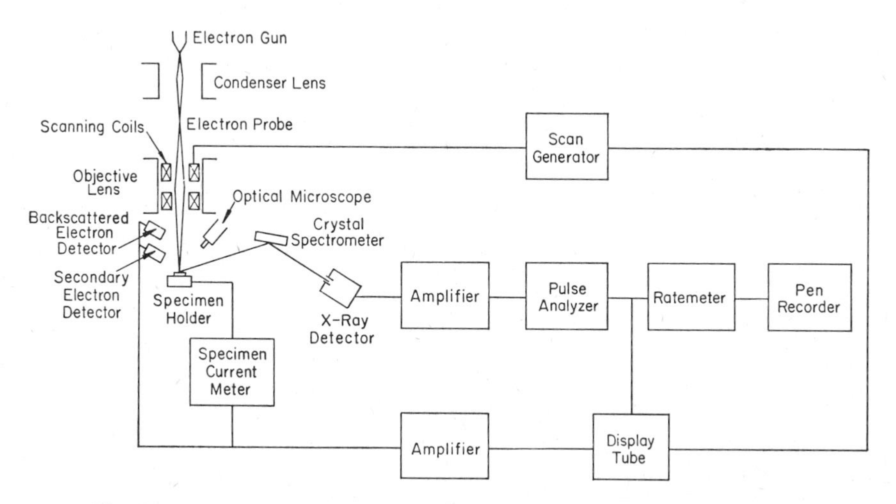

Fig. 4.49 Schematic representation of an electron microprobe analyzer.

Demagnification of the crossover of the electron beam can be attained to achieve a desired beam size of 0.1 to 3 μ by a two-condenser system. Magnetic lenses are normally used for the reasons cited earlier (Section 3, p. 362). The objective (or second) lens is normally of a fixed focal length since the distance between it and the specimen is controlled largely by the space needed to accommodate the x-ray spectrometers and the light optics. The fixed focal length entails a fixed demagnification, and its electron optical properties determine in the main the maximum current density in the minimum probe size. The focal length of the first (condenser) lens can be varied between about 2 mm and infinity and is used to vary the probe size and its current density by controlling the size of the source demagnified by the objective. Various objective lens designs are discussed by Castaing [202], and the condenser lens may be of unitized double-lens design to permit greater flexibility at the lower end of the beam parameters. A stigmator, which controls the shape of the probe spot, is normally provided.

Stability of power supplies over long periods of time (up to $\frac{1}{2}$ hr) is a vital necessity since any change in the lens currents or in the high voltage to the electron gun causes defocusing of the electron beam. Hence feedback control systems are used to maintain the required stabilities (of about 0.01%) to maintain a constant beam size and exciting electron energy to the region being analyzed.

An electron microprobe is normally equipped with a light microscope through which the area of the specimen under bombardment can be examined. The optics may be refracting or reflecting, and polarized illumination may be provided. This microscope fulfills an important function in the precise positioning of the bombarded area on the Rowland circle of the spectrometer. This operation is performed by bringing the probe, the position of which is observed by luminescence or contamination (see p. 416), into exact coincidence with the cross wires of the eyepiece, using the deflection coils; the cross wire is prealigned to the correct position. As an alternative to deflecting the electron beam, the area to be examined can be selected by automatically or manually positioning the stage in the desired manner by the use of mechanical controls.

The resolution of the microscope must necessarily be good for such a procedure ($\sim$0.5 μ), and it can also be used in searching operations. Polarized light is particularly useful in aligning and identifying grain boundaries in geological specimens. In the above discussion we have considered the basic fixed-beam instrument that has been used to make an analysis of a chosen point or a line scan by a synchronous drive on the specimen movement and data recorder.

In many recent microprobes, scanning coils have been added to the electron-optical system, normally in the bore of the objective lens (Fig. 4.49). These serve a twofold purpose: (1) The scanning facility plus suitable detectors can be used during the search and positioning operations, and micrographs prepared in this mode are of assistance in interpreting the x-ray data. (2) A distribution map for a particular element can be obtained by using an x-ray detector and rate meter system set for a particular wavelength. The final signal is seen as a change in intensity on a CRT driven by the scanning coil signal generator.

The specimen stage of an electron microprobe is an important component since its X- and Y-motions are used to locate the area of the specimen to be examined and rough coincidence can also be set. The specimen holder can also be rotated so that a scan can be performed along any desired direction across the specimen. A vertical motion is also required for the positioning of the x-ray source on the Rowland circle since the objective lens is of fixed focal length and the spectrometer height is fixed. The specimen may be at right angles to, or inclined at an angle to, the beam, depending on the geometry of the objective lens, x-ray detection devices, and the optical microscope. Specimen holders may be flat stubs to which thick specimens can be attached, or may be designed to facilitate loading of special shapes or holders. Transmission stages for thin films are available, and provision is normally made for the measurement of specimen current. Some microprobe stages permit x-ray projection microradiography and the preparation of special divergent beam x-ray diffraction patterns (Kossel line patterns) formed by allowing a microcrystal under electron bombardment to serve as an x-ray source as well as a diffraction system [212]. The specimens are normally loaded into the stage after it has been removed from the specimen chamber, and an air lock or specimen chamber/column isolation valve may be provided to facilitate rapid changes. The specimen stage normally has space for a number of samples and also for standard specimens.

X-rays emanating from the specimen are normally made to fall, at an advantageous take-off angle, onto an x-ray spectrometer which disperses the different wavelength components of the characteristic x-spectrum and permits a detector such as a proportional counter to analyze the energy of the radiation at different wavelengths. For soft x-rays the spectrometer must be evacuated, and it is normal for the spectrometer to be a part of the vacuum system of the instrument. Special isolating procedures may be necessary to protect the windows of particular proportional counters from the increase in pressure when the column is brought to atmosphere, so isolation valves may be necessary.

Nondispersive, gas proportional counters and solid-state detectors have recently been developed, and a considerable amount of ancillary electronic equipment is required to analyze the data collected and to correct for the various factors involved in the generation and passage of x-rays from the atom to the counter.

The critical excitation potential of the characteristic x-radiation being used lies typically in the range of 1 to 25 kV, while the accelerating potential applied to the impinging electrons is in the range of 2 to 50 kV with a probe current of about 10^{-8} to 10^{-5} A. The angle of incidence of the electron beam is usually between 45 and 90°, while the take-off angle to the spectrometer, that is, the emergence angle of the x-rays from the specimen, is about 5 to 55°.

The major effect of changing the angle of incidence of the probe from its normal 90° is that the mean primary x-ray generation occurs nearer the surface once penetration is reduced, and allowances must be made as indicated earlier. In addition, backscattering will be enhanced. Variation in take-off angle is of greater importance since absorption of various wavelengths by self or other element absorption will be altered, as will fluorescence. Birks [213] illustrates the effect by studying binary components of chromium in iron, titanium, and aluminum and concludes that take-off angles of less than 6° should be avoided, while any take-off angle of greater than 30° should prove satisfactory.

The above discussion has tacitly assumed that a single spectrometer and analyzing system is employed. In practice, most microanalyzers are fitted with more than one spectrometer and associated rate meter equipment, so simultaneous recording of data is both practicable and desirable. The time required for a complete analysis is considerably reduced, as are the stringent stability requirements. Heinrich [214] considered a number of techniques of multiple rate meter representation. An individual chart recorder could be used for the output of each rate meter, but cost, space, the difficulty of synchronization of the different recorders, and the need to replot the data make this an inefficient approach. Multipen recorders (up to about five) permit simultaneous recording of a number of channels, but a shift of position along the time axis is necessary to permit free travel of all the pens. Oscillograph recording permits perfect registration, but poor contrast of the tube coupled with the necessity of development of the film limit the method. On an oscilloscope any number of channels can be simultaneously registered. Another possibility is to use an *X-Y* recorder by which concentration maps simulating a perspective view of the area scanned can be produced.

An alternative technique is to represent each output by a particular color and to use color filters and film to record multicolored images sequentially [215]. Ficca [216] has developed this idea using a color display computer oscilloscope, which is able to record simultaneously images in each of three primary colors; this obviates the possibility of misregistration and greatly reduces the time involved. The signal from each channel is used to modulate the intensity of one of the guns, and primary colors represent "pure" regions, while additive colors indicate regions where two or more elements are located. The system has been used for multicolor displays of images obtained from specimen current, x-ray scanning, x-ray concentration mapping, and for tricolor display of x-ray intensity profiles. Polaroid color film is used to record the images.

Sample Preparation

The discussion on p. 399 illustrates the similarity of design considerations found between the microprobe and an electron microscope. The requirements that the instruments impose on the specimen are also closely related: (1) A preparation must be able to withstand a vacuum, high temperature, and high-intensity ionizing radiation. (2) It should show sufficient detail so that analytical tests can be related to structural features. (3) The structures to be studied should not be altered by preparation techniques. The few basic concepts of sample preparation for the probe presented here may seem crude compared to those described earlier in the chapter, but the demands on quality and cleanliness are not relaxed. The technique also requires that the exposed surface be free of structural relief. This reduces the possibility of variation in x-ray count caused by topographic changes in the path length of x-rays formed within the sample.

The majority of samples investigated by microprobe techniques fall into two general classes: bulk samples or dispersions. The study of bulk samples requires the preparation of clean, flat surfaces. In the case of hard materials, such as those encountered in metallurgy or mineralogy, grinding and polishing techniques similar to those used in optical microscopy have been generally accepted. Birks [217] has recommended that diamond abrasive be used for the final polish since its cutting characteristics minimize differences between hard and soft regions in the final surface. Etching should be avoided since it may cause localized chemical changes in the specimen.

Softer materials, such as biological preparations, have been studied using standard sectioning procedures, although low-temperature sectioning is preferred if one wishes to eliminate contamination from fixing and embedding procedures. The thickness of sections used ranges from about 50 μ to about 0.05 μ, but the best balance of x-ray intensity and resolving power is usually obtained from 5-μ sections. These preparations are mounted either on stubs or on lightly coated electron microscope grids. These should be made of elements that do not interfere with the analysis, and radiation effects are reduced by vacuum-coating the preparation with a thin conductive layer of carbon or aluminum.

Particulate samples can be studied directly after they have been deposited on a substrate or by treating them as a part of a bulk preparation by embedding them in a plastic and then producing a surface either by polishing or sectioning. Direct observation is preferred because of the ease of preparation and the improvement of contrast with small particles by eliminating the background signal produced by the matrix material. In fact, for the study of fine precipitates in solids (e.g., metals or minerals), it is often found advantageous to reduce the study to one of dispersed particles by the extraction replica technique (see Section 2, p. 328).

The Production of x-Rays

When an element is radiated with a stream of electrons whose energy is in excess of a few kiloelectron volts, the resulting x-ray energy spectrum is characterized by a series of discrete intensity maxima superimposed on a broad continuous band. The origin of this spectrum is most easily understood in terms of the Bohr atom. In fact it was a study of characteristic x-rays that contributed in large part to the acceptance of his model of the atom. In this model, the electrons in the atom are grouped about the nucleus in discrete quantum mechanically prescribed energy levels represented by the *K*-, *L*-, and *M*-shells. The elements differ from each other in the periodic table by the number and disposition of these electrons occupying allowed positions in the energy levels around the nucleus.

If an incident electron penetrates the atom to the *K*-shell and has sufficient energy to eject an electron from it, the vacancy will probably be filled by an electron from the *L*-shell, thus creating a vacancy in that shell which will probably be filled by an electron from the *M*-shell. Hence two characteristic x-ray photons are emitted, and two characteristic lines are observed. In general, excitation of an inner electron shell produces a spectrum of x-rays of all energies lower than that of the initial ionizing event. Moseley [199] showed that the relationship between frequency (ν) of the photon and the atomic number Z of the element could be expressed:

$$\nu = K(Z - 1)^2 = 1/\lambda,$$

where K is a constant depending on the transition involved, and λ is the wavelength of the emitted photon. For a full description of the selection rules governing allowed transitions leading to the so-called diagram lines, the reader is referred to a textbook on modern physics [218].

The magnitude of the energy required to remove an electron from an inner shell of an atom is known as the critical excitation potential, which is a discrete value for each energy shell of each atom, its magnitude increasing with atomic number and decreasing as the shell in question becomes more remote from the nucleus. The energy of the ionizing electron, in the case of the electron microprobe, must be greater than the critical excitation potential if it is to eject an electron from a particular energy state. The efficiency of production of characteristic x-radiation depends in part on the difference between the accelerating potential applied to the incident electron and the critical excitation potential of that radiation.

An electron beam striking a specimen also produces a broad spectral band of radiation as a result of energy losses of the incident electrons in the fields surrounding atomic nuclei. The long wavelength part of this spectrum extends into the infrared region, while the short wavelength portion has a sharp cutoff defined [219] by the equation $\lambda_{min} = 12350/v$, where v is the accelerating potential in volts. The wavelength at maximum intensity for thick samples is located at about $2\lambda_{min}$ but it decreases toward λ_{min} as the thickness of the target is reduced. These radiation characteristics are independent of the composition of the target, although the integrated intensity increases with increasing atomic number as well as accelerating voltage.

If the energy of an x-ray photon resulting from an incident electron is sufficiently great, it can in turn ionize other atoms in the sample. This ionization results in the production of fluorescent characteristic x-rays, which in turn leads to an increment in the total production of x-rays. This enhancement resulting from fluorescence leads to errors in quantitative analysis, and it is discussed on p. 425.

If the energy of an x-ray is slightly greater than the excitation potential of a particular energy level, the photon can be readily absorbed and its energy transferred to an electron in that shell. Such processes produce an abrupt increase in the absorption coefficient, and the wavelength region in which this takes place is called an absorption edge. The result is considerable decrease in the magnitude of a beam of x-rays of that wavelength traversing a volume of that element. The ratio of the intensity of a beam of x-rays after passing through an absorbing medium (I_2) relative to the initial intensity (I_1) is described by a form of Beer's law [220]:

$$I_2/I_1 = \exp{(\mu/\rho)\rho x},$$

in which x is the distance traveled by the beam, and μ/ρ is the mass absorption coefficient for the material. The importance of x-ray absorption is that it limits the detection of a material since it leads to an apparent decrease in the amount or concentration of that constituent.

Dimensions of the x-Ray-Producing Volume

The dimensions of that volume of the specimen which, as a result of electron beam excitation, contributes to the radiation analyzed by the spectrometer are of fundamental importance since they determine the spatial resolution of the technique. In this case minimum spatial resolution may be considered to be the minimum volume (or its corresponding surface area) from which an analyzable emission of x-ray photons is produced. The two major dimensions of the excited volume are its depth and its diameter, and these have been measured experimentally by a number of authors [221, 251].

The diameter of the volume being considered is greater than the diameter of the probe since the electrons on entering the specimen undergo deflection from their initial paths. There is also some spread produced by x-ray fluorescence. Some electrons are ejected from the specimen (backscattered and secondary electrons). The mean path of the electron in the material decreases with increasing atomic number and with decreasing accelerating potential. Hence the size of the excited volume can be decreased by operating at reduced accelerating potential. For optimum line-to-background ratio, the accelerating potential is normally taken to be three times the critical excitation potential, and the edge of the excited volume occurs at that envelope at which the energy of the electrons has been decreased by interaction with the specimen to a value just greater than the excitation potential for a particular line. In order to reduce the accelerating potential below this level, efforts must be made to improve the brightness of the probe and the line-to-background ratio of the spectrometer.

Wittry [222] showed that the volume of the region analyzed can be made arbitrarily small by critical choice of the probe diameter and of the beam-accelerating potential, within limitations set by the need for satisfactory line discrimination in a suitable time. In his discussion the total diameter D of the excited region is equal to:

$$D = d + z_m,$$

where d is the diameter of the electron probe and z_m is the depth of penetration of the electrons. The latter can be written:

$$z_m = 0.033(V^{1.7} - V_k^{1.7})\frac{A}{\rho Z} \text{ microns},$$

where V (in kilovolts) is the accelerating potential, V_k (in kilovolts) is the critical excitation potential of a K-line of the material, Z is the mean atomic number, A is the mean atomic mass of the volume, and ρ is its density. Further, d can be written in terms of V, V_k, and the intensity of the line in question, so it is possible to write an expression for D and determine the value of V that leads to a minimum for D, hence for the minimum analyzed volume. If the ratio V/V_k is between 1.1 (strong concentrations for which the continuous spectrum is negligible) and 1.5 (low concentrations), the volume exhibits a sharply marked minimum; the value is of the order of 0.2 μ^3 for pure copper. The minimum volume decreases as Z decreases since the critical excitation potential decreases. Suggestions have also been made by Duncumb [223] that the soft x-ray lines of the heavier metals should be studied since one can use lower voltages than those required to excite the inner shells, but Castaing [205] states that for precision work the K-lines should be used up to $Z = 35$, and the L-spectrum beyond that. Andersen [224] contends that a "general use of low accelerating potentials is desirable and feasible," and is worthwhile because of the improved spatial resolution.

The Analysis of x-Radiation in the Electron Microprobe

In the discussion on p. 404, it was pointed out that there is a distinctive wavelength and energy for every characteristic line of each element. Thus diffraction of the x-rays by crystals or gratings, followed by energy discrimination by means of a proportional counter and pulse-height (amplitude) analyzers permits the identification of lines excited within, hence elements present in, a specimen.

It is necessary to detect a small peak (the characteristic x-ray line) above a background level of radiation associated with that line. Thus one is confronted by a statistical problem and the standard deviation of n counts per second counted for t seconds is $\overline{nt}$, and the standard deviation of the intensity n is $\overline{n/t}$. It has been suggested that the definition of the limit of detectability is that the peak/background ratio is at least three standard deviations of the background [225]. The intensities of the line and background are integrated over a period of time and depend on factors such as primary beam current, accelerating potential and overvoltage, the matrix surrounding the element to be detected, and the efficiency of the x-ray system. Thus the absolute detection limit depends upon instrumental and specimen characteristics and the spatial resolution since the limit is related to the number of atoms detected within a certain excited volume. The decision as to whether to use K- or L-spectral lines for elements of intermediate atomic numbers depends on the relative detection sensitivity of each line and system. Fergason [226] used a multichannel analyzer to improve the precision of relative x-ray intensity measurements and, with this setup, was able to measure trace elements at 100 to ± 10 ppm for iron, nickel, silicon, and aluminum, in uranium. Sensitivity and detection limits of various elements in particular samples are discussed in detail in standard texts.

We now consider the x-ray crystal optics involved and later the counters and some associated circuitry involved. Nondispersive techniques are then mentioned.

X-Ray Crystal Optics

A crystal diffracts only particular x-ray wavelengths for a given crystal orientation, and this statement is expressed by Bragg's law:

$$n\lambda = 2d \sin \theta,$$

where n is an integer representing the order of diffraction, λ is the wavelength in angstroms, d is the lattice spacing of the crystal, and θ is the angle of diffraction. The principal diffraction is the first order ($n = 1$) and the effects of higher orders can be eliminated by energy discrimination. As different angles (θ) are examined, different wavelengths are observed, hence separated from the spectrum of x-radiation. The shortest wavelength encountered is $\lambda_{min} = 12{,}350/v$, and from a practical standpoint the choice of d-spacing in this region is dictated by the smallest angle the equipment can resolve. The longest wavelength (λ_{max}) that can be diffracted depends only on the lattice spacing d when $\theta = 90°$, that is,

$$\lambda_{max} = 2d.$$

The above equations are written for perfect crystals, and Birks [219] considers the effects of imperfections in structure that occur in real crystals. Such imperfections cause a broadening of the diffracted peak and may alter the height of the diffraction curve. He points out, however, that even with relatively large half-maximum breadths the wavelength resolution is still better than that required for almost any spectrochemical analysis and that imperfect preparation of the diffracting crystal normally broadens the diffracted line more than do crystal lattice imperfections.

The geometry of the spectrometer is of considerable importance since it is related to the wavelength resolution, which is the wavelength difference between two wavelengths that can be just separated. A cylindrically curved crystal of radius R brings x-rays of different wavelength to a focus at points lying on a circle of radius $R/2$, the circle being tangent to the crystal and also passing through the source of the x-rays (Fig. 4.50). This circle is known as the Rowland circle. In the Johann geometry, a crystal is bent to a radius R, the surface of which crystal is tangent to the focal circle at a small region only, and diffraction from other parts of the crystal leads to a reduction in wavelength discrimination as a result of aberrations. In the Johansson arrangement the crystal surface is ground to a radius of $R/2$ so that the whole of its face lies on the focal circle, and x-rays diffracted from all points on the crystal face are brought to a common focus. If it were possible to grind all diffracting crystals, the latter arrangement would be universal, but this is not the case, and the Johann geometry is of considerable importance. It is employed, for example, in spectrometers using surface-ruled diffraction gratings, and those using multiple monomolecular layers of soap, both of which are used in the detection of the very long wavelength radiations associated with low-atomic-number elements.

For optimum sensitivity and resolution, it is necessary to locate the x-ray source (the bombarded area of the specimen), the crystal spectrometer, and the window of the detector on the circumference of the Rowland circle. If

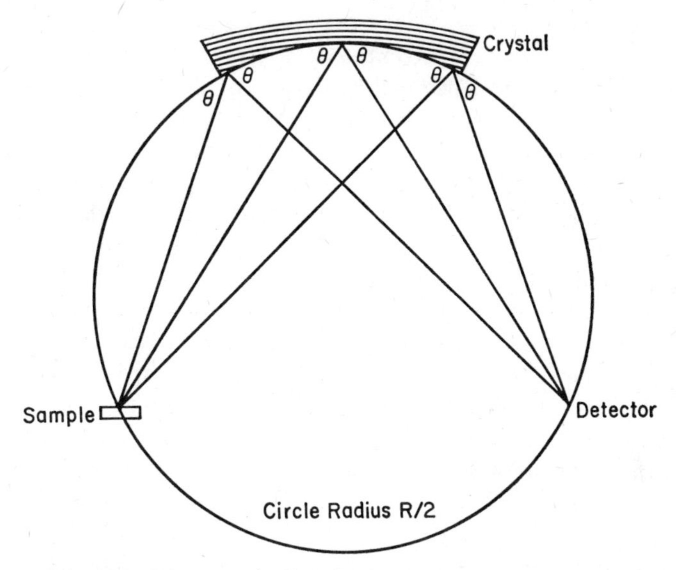

Fig. 4.50 Diagram of a focusing x-ray wavelength analyzer.

the center of the Rowland circle is fixed and the crystal and detector are moved around the circumference of the circle, then the angle of emergence of the x-rays from the specimen surface varies with the Bragg angle used at any time. Such an arrangement [227] implies that the detector moves at twice the angular rate of the crystal, and further that the detector be turned as it is moved so that it always points at the crystal. Another arrangement [228] moves the crystal in a straight line away from the source and rotates about its center so that its distance from the source is expressed by $2r \sin \theta$ when the crystal is at a Bragg angle of θ. The equator of the path of the detector is:

$$\rho = 2r \sin 2\theta,$$

where ρ is the distance of the detector from the source and, since this is not a circular path, the mechanical arrangements are complex. The corrections to be applied to an analysis are considerably simplified if the angle of x-ray emergence from the source is fixed. Another possible arrangement [229] is to flex the crystal as it is rotated so that the radius of the focusing circle changes with the angle of incidence. The positioning of the analyzed region precisely on the Rowland circle of the spectrometer is critical, and the task is aided by the use of an optical microscope with a high numerical aperture objective with a small depth of focus to bring the feature to be studied precisely into coincidence with the cross wires of the eyepiece. An error of a few microns can cause an error of a few percent in comparing the intensities of unknown and standard specimens since malpositioning of the specimen is equivalent to faulty peaking of the spectrometer crystal.

The arrangements discussed in the previous paragraph are all examples of the use of fully focusing optics, but a semifocusing system is also possible. In this system the position of the crystal is fixed relative to the source, and it is rotated such that it makes the desired angle θ with the radiation. The crystal is curved but unground, and a number of crystals are required, each corresponding to a Rowland circle of a different diameter to cover the desired spectral range. The x-ray detector is moved through twice the angle through which the crystal is rotated on the circumference of a circle centered on the axis of rotation of the crystal. The sensitivity and resolution are less because the intensity and the line-to-background ratio are poorer than for focusing crystals. However, the system has the advantage that it is compact, easier to build, and is less affected by the movement of a scanned probe than is a fully focusing system.

An electron microprobe is normally equipped with a number of different crystals, and care is taken by the manufacturer that they can be easily and rapidly exchanged and set with great precision, usually while the column and specimen chamber are held under vacuum. Crystals are selected on the basis of their lattice spacing (wavelength coverage), stability, ease of bending, and diffraction efficiency. For short wavelengths, lithium fluoride ($2d = 4.02$ Å) is suitable, while for longer wavelengths mica ($2d = 19.97$) and gypsum can be used. Mica is stable in the vacuum, can be easily bent after cleaning, and its relatively higher orders of diffraction can be eliminated by pulse-height discrimination. It can also be used [230] as the basis of a single detector for all elements from boron to uranium. Lead stearate decanoate layers are deposited on the surface of a mica crystal, which is then bent; the coating is used for the elements from boron through fluorine, and the underlying mica for sodium through uranium. The vacuum spectrometer must be fitted with a thin detector window and a magnetic shield to repel electrons, and since it operates at a low vacuum must be isolated from the column high vacuum.

The $2d$-spacings of such heavy metal stearates can be around 100 Å, and about 100 monomolecular layers may be involved in the diffraction. Hence the half-width of a spectral line is larger than for a case of diffraction from a typical thick crystal, but unless the elements to be distinguished are very similar in wavelength the width is not a severe problem and intensities are satisfactory. The stearate crystal may be useful for oxygen and nitrogen, but it is likely that diffraction gratings will prove superior for longer-wavelength radiations. In this type of dispersing system, a beam of x-rays is incident upon a solid grating surface at a small angle of incidence and produces a diffracted beam by total external reflection according to the equation:

$$\lambda = d(\cos i - \cos \theta),$$

where θ is the angle of reflection, λ is in angstroms, and i and θ in degrees, and d is the grating spacing [231]. This diffraction occurs with useful intensity only if the angle of incidence does not appreciably exceed the critical angle of total external reflection, so the x-ray collection efficiency is low because the solid angle subtended by the grating at the source is small. Franks has extended his work to curved gratings, which provide a considerable increase in collection efficiency. The wavelength range of such gratings may be between about 20 and 200 Å.

X-ray Detectors and Their Associated Equipment

The most commonly used x-ray detectors are gas ionization devices, which consist of a cylinder along the axis of which is suspended a thin wire insulated from the walls of the cylinder. A window of material transparent to x-rays is situated in the side of the cylinder, and a positive voltage is applied between the wire and the body of the cylinder. When an x-ray photon passes through the window, the collision between it and a molecule of the gas results in the ionization of the molecule and a concomitant loss of energy by the photon. The electron and the ion thus produced are accelerated in opposite directions by the applied potential, and the electron gains sufficient energy to ionize other gas molecules, providing further electrons for subsequent ionizing events until an electron burst is produced. The burst is collected at the wire anode to produce a pulse of current. A pulse is registered for each x-ray photon that enters the window. The gas amplification in the tube depends on detector geometry, gas composition and pressure, and applied potential difference.

It is preferable to operate the proportional counter within a certain voltage range determined by those factors that determine gas amplification. If the voltage exceeds this range, the counter will become a Geiger counter, in which the pulse amplitude is the same for all incident photon energies, and at lower voltages ionization will be insufficient to produce a pulse of adequate amplitude. Not only is the Geiger counter nonproportional, but its dead time is far greater than for a proportional counter. The dead time is that period required by the counter to recover from the previous pulse, and any x-ray photons entering the tube during its inactive period are not counted. The dead time of a Geiger counter is of the order of 200 μsec, which limits the counting rate to 500 cps, while at lower count rates the counter is inactive for a portion of the time, so a percentage of the counts are lost. Corrections for the dead time can be calculated and added to the recorded counting rate. For a proportional counter, dead time is typically of the order of 1 μsec, and the corrections are correspondingly smaller, while the counting rates can be up to about 10,000 per second. At counting rates in excess of about 10,000 per second, the pulse amplitude begins to decrease, and if the decrease is sufficient the amplitude may be moved out of a preselected amplitude range, with an apparent decrease in x-ray intensity. Another factor that must be considered is the dead time of the electronic circuitry, and if this is large enough a correction must also be made for its effects.

The composition of the gas used in the detector is of importance since it plays a part in determining the operation of the detector. A gas proportional counter filled with argon at low pressure is proportional within a voltage range of only a few tens of volts, and short-wavelength radiation is not completely absorbed by the gas, thus reducing the sensitivity of the counter. Further, the thick windows needed to operate the counter under reduced pressure absorb the longer-wavelength, lower-energy photons. These limitations led to the development of higher-pressure, thin-window detectors known as flow counters, through which gas is continuously flowed through a window which is not vacuum tight. The voltage range over which these counters are proportional is much greater (of the order of hundreds of volts) and this permits the counters to be operated at the high gas amplifications required by the longer wavelength radiations. P-10 gas, which consists of 10% methane and 90% argon, is commonly used as the flow gas; the methane acts as a quench gas and permits improvements in detector dead time and reduced pulse shrinkage as a consequence of reduced detector voltage. Incident x-ray photons of different energies produce electron bursts and current pulses of different magnitudes or amplitudes, and the counter can thus discriminate between photons of different energies and identify the characteristic x-ray photons emanating from different elements in a specimen. It is this discrimination that permits a proportional counter to be used as a nondispersive detector of x-radiation (see p. 413).

The magnitude of the pulse from the detector is of the order of millivolts, hence is fed to a preamplifier before being transmitted via cable to a rapid-response video amplifier in which the gain is typically 200 to 1000×, with the voltage range of the output being about 5 to 100 V. In this fashion the magnitude of individual pulses is greatly increased, although the relation between the magnitudes of different pulses is not altered. The energy dispersion characteristics of the proportional counter are normally combined with the discrimination of a pulse-amplitude (or pulse-height) selector. The pulse-amplitude selector consists of an electronic gate circuit in which the lower level is set to eliminate noise emanating from the amplifier's circuits. Thus the lower gate prevents the addition of amplifier noise to the background measured by the detector. The upper limit of the gate is adjusted to accept the pulse amplitude corresponding to the x-ray wavelength being detected. The setting of the upper limit relative to the amplitude of the pulse under consideration is of importance in that this is an effective means of discriminating against higher orders of other wavelengths which might be diffracted by the crystal at the wavelength being studied; the pulse amplitudes of two x-rays of different wavelengths whose different orders are diffracted at the same Bragg angle are considerably different from each other. Pulses having amplitudes greater than the upper limit of the gate are eliminated by an anticoincidence circuit. In this device an inverted pulse is produced of all pulses that are larger than the gate limit. These are fed back into the circuit and cancel the original pulse after it has passed through both gates. Birks [232] cautions against setting too small a range in the pulse-amplitude selector since at very high counting rates the movement of the pulse mentioned above can cause a shift sufficient to introduce appreciable error attributable to losses in the selected pulse.

After passage through the pulse-amplitude discriminator, the pulse is transmitted to other circuits which prepare the electrical information for read out. These electronic components include integrating circuits which produce signals corresponding to the counts per second made by the proportional counter; the rate may be recorded digitally, applied to a meter, or used to modulate the spot of a CRT. An alternative circuit is the scalar type in which the total number of pulses occurring in a desired time unit is recorded in some fashion; the data may be printed on strip charts, or directly on tape in a form suited to computer operation. Scanned x-ray images may also be formed using the output from the pulse-height analysis directly.

Nondispersive Analyzing Systems

In a so-called nondispersive system, the x-radiation from the bombarded specimen enters the detector directly, without first being diffracted by a crystal into different wavelengths. Analysis of the spectrum into component wavelengths is performed by a detector and pulse-amplitude analyzer combination.

Overlapping is a function of the resolving power of the detector, and the principles of measuring neighboring lines have been discussed by Dolby [233] (Fig. 4.51). He assumed that the shape of the pulse-amplitude distributions is Gaussian to a first approximation and that the standard deviation of the curve is equal to the separation between adjacent elements. A set of n simultaneous equations for a system with n unknowns can be set up and solved by determinants for the components of a curve representing the total spectral distribution. Birks [234] discusses the advantages of a multichannel analysis coupled to such a detector and compares the results of this analysis

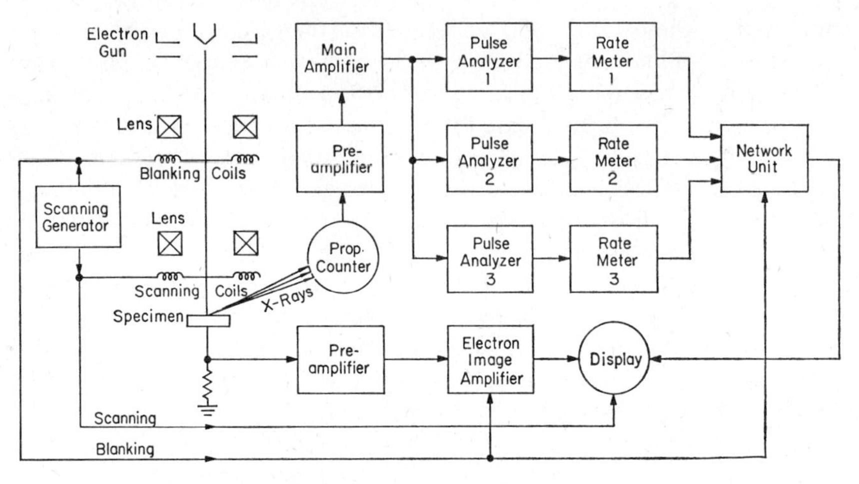

Fig. 4.51 Block diagram of a multichannel pulse analyzer.

with a crystal spectrometer measurement on the same samples. The agreement between the percentage proportions of the constituents of an alloy was excellent.

Commercial nondispersive spectrometer systems are now being offered. The detector is likely to be a semiconductor device in which the absorption of x-ray photons produces free charge carriers which are collected by an applied bias. The detector operates at liquid nitrogen temperatures to improve the signal-to-noise ratio by reducing the thermal noise of the device. A pulse, whose amplitude is proportional to the energy of the photon, is produced in the detector, amplified by a field effect transistor preamplifier, and further amplified before being fed to a multichannel analyzer. The signal is applied to the analyzer for a preselected time period, and the energy spectrum stored in the analyzer is later displayed on a CRT, a chart record, or a paper/tape print-out. Care must be taken to exclude backscattered electrons which add to the x-ray signal; in principle, this can be done by protective fields to stop the electrons, or by protecting the detector by a window that stops the electrons but not the x-rays; the selection of the window material is critical if it is desired to discriminate between the low-energy x-rays from the light elements and backscattered electrons. Energy discrimination of nondispersive spectrometer systems is constantly improving, with results reported to around 200 eV, and a range of detection of between sodium ($Z = 11$) and uranium ($Z = 92$) is claimed.

Techniques of Analysis

In problems of analysis the microprobe is used to identify or show the location of atoms (qualitative) and to specify the quantity in relative or absolute units (quantitative). Before discussing these subjects, it is helpful to consider a few problems common to both methods.

General Procedures

The correct choice of analysis conditions is of great importance since they exert a considerable effect on the accuracy of the analysis. Some of the conditions cannot be optimized since they result from nonadjustable features of the instrument; these sometimes include both the angle of incidence of the beam and the take-off angle of the emergent x-rays. The wavelength of the x-rays used may be determined by the spectrometers available, while the choice of beam currents and accelerating potentials may be the result of a compromise between the time, permissible probable error in the analysis, and the resolution required. In summary, it can be said that the best (constant) analysis conditions to select are those with which the best analysis can be obtained; this implies those conditions for which the necessary corrections can most accurately be made, and this in general occurs when such corrections are small. If the requirements of two necessary corrections are incompatible, then the correction for which the data are less certain should be minimized. An example of this is the effect of emergence angle; the magnitude of the fluorescence correction increases with increased emergence angle, while the magnitude of the absorption correction decreases with increased emergence angle; it has been argued [235] that the highest possible emergence angle should be chosen.

The relative x-ray intensity, that is, the ratio of the characteristic x-ray from an element in a mixed specimen to that of the line from the element in a 100% pure specimen, is not precisely linearly related to the composition of that specimen because of specimen and other effects. Attention paid to corrections applied to ensure an accurate measurement of the intensity of the characteristic lines of the specimen of those of a standard sample can lead to quantitative accuracies of equal to or better than 10% relative. Further attention to the observed intensities to obtain the real, generated intensities can improve this to about 5% or better. In most work, a practical rather than a rigorous approach is employed since there are uncertainties in some fundamental parameters, such as x-ray absorption coefficients, so the most precise mathematics is limited by the uncertainty in these parameters. Further, the number of variables for which rigorous allowance would need to be made is great enough that such an approach would lead to impractically lengthy calculations. Certain approximations can be used to obtain results of acceptable precision.

THE LIGHT ATOM PROBLEM

The problem of analyzing for the lighter atoms mentioned earlier in this report is primarily one of detecting and discriminating between K-lines of relatively long wavelengths. For example, the wavelength for calcium is 3.4 Å, while it is 24 Å for oxygen, 44 Å for carbon, and 113 Å for beryllium. Not only is it difficult to find crystals with *d*-spacings large enough having high x-ray scattering efficiency, but appreciable absorption losses usually take place before the photons reach the sensitive volume of the detector. The most popular technique involves a nondispersive detection system of relatively poor energy resolution but high quantum efficiency followed by a system for pulse-height analysis.

Errors are introduced into the analysis by contamination that builds up on the radiated region during examination. The deposit is caused by the polymerization of adsorbed organic films and pump oils since the level of the vacuum is not high. A contamination spot causes a decrease in x-ray emission, both by absorption of x-rays and by reduction of the energy of the incident electrons before they reach the specimen. Thus the deposit not only increases the amount of carbon in the analysis but also produces a marked reduction in the values for elements such as oxygen and nitrogen through increased absorption. The effect is particularly serious when the specimen and standard contaminate at different rates and when the accelerating potential of the probe is close to the critical excitation potential for the element. The rate depends primarily on the probe diameter (being greater for a smaller probe) and probe current.

WAVELENGTH SHIFTS

The wavelength of the characteristic x-ray line can be affected by the valence state of the element since the screening effects of the electrons in outer shells may be added to those in inner shells, depending on the orbital angular momentum of the electrons. The wavelengths of the characteristic electrons increase with increasing negative ionic charge [236] and a shift from metallic to ionic bonding may cause a shift of up to 1 eV in the energy level [235], which could lead to a reduction of up to 20% in the measured intensity of the peak in going from metallic aluminum to alumina. For the K-lines of elements of $Z \geq 18$ and for the L-lines of elements of $Z \geq 48$, the shift does not usually affect the precision of quantitative analysis since the wavelengths are not long and the transitions do not normally involve electron states close to the valence bond. Andersen [237] notes that the relative distribution of intensities in the L-spectra of metals in the first transition series could provide a powerful tool for the determination of their chemical bonding. For studies concerning the chemical state, however, the shift of the $K\beta_1$ ($M_{II,III} \rightarrow K$) line or the long wavelength satellite $K\beta I$ are recommended by Theissen [247] even for elements of medium Z. To overcome this effect the spectrometer should be precisely set to the peak of the shifted line by profiling the spectrometer over the line; the factor by which the shift changes the relative intensities may also be determined. Backlash in the spectrometer gearing is obviated by the normal procedure of setting the crystal to the peak from the same approach as was used to determine its position.

ANALYSIS BY SAMPLE CURRENT MEASUREMENT

The backscatter coefficient r is defined as the ratio of backscattered electrons to primary electrons, and Heinrich [239] showed (for his instrument, at least) that:

$$r \cong \frac{i_b - i_s}{i_b} = \delta, \tag{4.1}$$

where i_b is the beam current and i_s is the specimen current. Castaing [201] stated that the backscatter coefficient for an alloy r_a was the sum of the products of the backscatter coefficient of each component and its concentration (C), which for a binary system of elements A and B is:

$$r_a = r_A C_A + r_B(1 - C_A), \tag{4.2}$$

and by combining (4.1) and (4.2), one obtains:

$$C_A = \frac{i_B - i_a}{i_B - i_A}, \tag{4.3}$$

where $i_{A,B}$ is the sample current in the pure elements A and B, and i_a is the sample current in the alloy. Heinrich found that the method, although applicable only to binaries or pseudobinaries and requiring a method of suppressing secondary electrons, for example, by biasing the specimen, provided acceptable precision when the atomic number differences between the elements were sufficiently large. The results are obtained instantaneously and continuously, the calculations are very simple, and it is an attractive alternative with the above provisos to x-ray analysis in some cases in which the concentration is greater than 1%. Colby [240] analyzed several binary systems of uranium with much lighter elements by both x-ray techniques and sample current measurements. He mentions that the sample current can be used as the reference charge for a fixed-charge analysis, and he fitted experimental results of sample currents against atomic numbers for a number of pure elements. Those data were used with (4.3) to calculate the weight fraction in a number of uranium alloys, the accelerating potential being best set at 30 or 35 kV. The relative errors compared to the theoretical weight fraction were less than 2%. Poole and Thomas [241] also analyzed binary alloys with large differences in atomic number with like results.

NOTE ON STANDARD SAMPLES

Calibration standards should be similar in composition and distribution to the sample. In some cases this is not possible, especially when dealing with totally unknown specimens, and the calibration standards must be homogeneous on a submicron scale. In addition, the chemical composition must be accurately known on this scale, and the standards must be prepared over a wide compositional range. Both chemical analyses and microprobe analyses should be performed on the standards, and consideration must be given to the problems of aging of the specimen. A technique by which the problems of homogeneity are minimized is a rapid-quench method [242].

An alternative and practical approach is to apply correction procedures that do not involve such specially manufactured standards but which rely upon pure elemental standards. The application of such corrections is discussed in the section on quantitative analysis, but it should be noted that the corrections applied when using standards of pure elements may be larger than when using standards more similar to the specimen.

Qualitative Analysis

The microprobe may be employed in a number of ways to rapidly determine the presence or absence of particular elements in an area of a specimen. If the composition of the specimen is completely in doubt, it is useful to commence by examining the backscattered electron image since this may roughly indicate the disposition of elements of high and low atomic number by either bright or dark regions. To establish that this is an intensity difference and not just a physical difference, one should then examine the x-ray image. In the next more sensitive method, all x-rays from the specimen are allowed to reach the detector and the signal of the rate meter is used to estimate the number of pulses of various heights corresponding to the characteristic wavelengths in the spectrum. This procedure permits discrimination between elements fairly well separated in the periodic table, hence can be a useful guide for subsequent spectrometer settings. Next, x-ray scanning images can be produced by setting the spectrometer to the Bragg angle of one such wavelength and recording the x-rays produced while scanning the probe over the area of interest on the specimen. The image produced will reflect the distribution of the element to whose characteristic line the spectrometer is set in the structure since regions rich in it will produce large numbers of x-rays of that wavelength and thus appear bright in a positive print while regions deficient in the element will be dark. If this procedure is repeated with the spectrometer set to the Bragg angle for each of the elements present, a series of micrographs will be available showing the distribution of these elements in the specimen region studied, and these can be compared with structural information derived from optical or scanning electron micrographs. If an element is present in very small amounts, it may be necessary to set the spectrometer on radiation from a standard of the same element; in this case it is critical that the elevation of the target spot of the specimen and the standard be precisely the same.

In order to improve the contrast of such x-ray scanning images, Melford [243] developed a technique in which the rate meter output, a dc signal, is subtracted from an arbitrary level before modulating the brightness of a CRT. "Concentration mapping" [215] is a procedure by which areas of a particular concentration range of an element are represented by areas of uniform brightness on the image. The amplified output of the x-ray detector is fed into a rate meter, the output of which is applied to the *Y*-amplifier of an oscilloscope whose intensity is kept constant. A mask with a horizontal window in it is placed in front of the screen, so that a photomultiplier behind the mask indicates whether or not the output, hence the concentration, is within the range (or ranges for a number of windows) of interest. In binary systems this technique can be applied to target current readout, leading to a more precise discrimination in a shorter period of time.

It is often desirable to scan the whole x-ray spectrum, and when this is completed, the wavelengths of all the characteristic lines present are identified by comparing them with a published reference. Tables of x-ray wavelength versus the atomic number of the element in general use are those collated by the ASTM [244] and by the AEC [245]. The most intense lines recorded in the scan over all wavelengths usually correspond to the K-spectra, which is also simple in composition since the number of possible transitions is smallest. Two of these lines ($K_{\alpha_{1,2}}$) are unresolved and provide the most intense line for each element. It is the first to be identified and compared to the published values. The K_β-line is then identified, and the procedure repeated for all higher-order lines that can be distinguished from the continuum. The relative intensities of various lines can be a guide to their proper identification, but it must be remembered that such intensity relations can be easily modified (by depopulation of a level, for instance), so their use is merely a guide.

Quantitative Analysis

The basis of the use of the electron microanalyzer in quantitative analysis lies in the exact measurement of the intensities of various x-ray lines produced by the electron bombardment. The x-ray intensities depend both on the composition of the specimen and on the conditions of the analysis.

In practical analyses corrections are first made for the background signal, and corrections are then made for dead time if this is necessary. The specimen corrections are then made, the most important usually being absorption, then fluorescence, and finally atomic number effects. In most laboratories hand calculations have been replaced by the use of computer programs into which are fed the data after correction for the instrumental effects, or by the use of curves which may have been experimentally obtained or completely generated from equations set up on theoretical and/or empirical bases.

DEAD TIME AND INSTRUMENT DRIFT

The dead time of the detecting system and counting circuits represents the shortest period of time in which the system can recognize and record two events. The period is usually of the order of microseconds and produces a reduction in the observed counts compared to ionizing events. This becomes a problem if the counts from a specimen of low concentration are being compared to the counts of a standard of high concentration, and Andersen [246] has proposed a correction method using specimen current to indicate true counts. Corrections for fixed-charge measurements or fixed-time measurements can also be made, and it is further necessary to correct for current-independent counts ("dark current" in the electronics and external radiation) by subtracting them from the observed counts before the true counts are determined. It is difficult to correct for dead time if pulse-amplitude discrimination is employed in the analysis because of variation in the gas amplification of the detector.

Instrumental drift can be a considerable problem, especially with regard to the probe current attributable to the changes in filament position and shape relative to the other components of the electron gun. Such drift is minimized by careful alignment and by allowing the filament to warm up for a sufficient period to equilibrate and by making frequent checks on the standards before and after observing the specimen. Since this drift is usually time dependent, extrapolation with time can usually give precise readings for the beam current, while simultaneous recording of specimen currents allow shorter-term variations to be assessed. The stability of all the power supplies is naturally of great importance, but proper instrument design should take care of this problem.

BACKGROUND RADIATION

Background radiation is comprised of the radiation of the current-dependent x-ray continuum excited in addition to the characteristic line, as well as to the current-independent but time-dependent sources of spurious pulses and cosmic rays. The counts resulting from the latter are likely to be small, while it is the continuum that often sets the lower limit of the detection for a line attributable to a particular element since it is always necessary to discriminate the line from the background. The value of the counts resulting from background continuum is determined on each side of the line in question by scanning the spectrometer, taking care to avoid any other nearby lines so that time background readings are made and the value for the background at the wavelength of the line can be determined by interpolation. The background must be measured on both the specimen and the standard for precise results, and experimental determinations are used instead of theoretical corrections since the latter information is not precisely available.

EMISSION CONCENTRATION LAW AND ATOMIC NUMBER EFFECT

Before considering the corrections in more detail, let us briefly examine the relation between the intensity of characteristic x-rays generated in a sample for an element and the concentration of that element in the sample. It must be assumed that the sample is homogeneous over a relatively extensive volume and, to a first approximation, Castaing [247] showed that the ratio of the intensity of the emission I_A of the sample in the $K\alpha_1$-line of element A to the intensity $I(A)$ of the same $K\alpha_1$-line of a reference target of the pure element is the mass concentration C_A of that element, that is,

$$C_A = I_A/I(A).$$

This approximation did not allow for atomic number (Z) effects, which include changes in energy loss per unit length and backscattering of electrons out of the sample. Castaing then proposed that a coefficient α_i be assigned to each element A_i, α_i being the product of a stopping power coefficient ($S_i = dE/\rho\, dx$) and a backscattering coefficient ($\lambda_i = I/I_B$, where I is the intensity with no backscatter and I_B the actual intensity). The emission-to-concentration ratio then becomes:

$$\frac{I_A}{I(A)} = \frac{\alpha_A C_A}{\sum \alpha_i C_i}.$$

λ increases with increase in Z, while S decreases with increasing Z. These various factors tend to compensate for each other so that the first approximation is often more accurate than may have been thought. There is, however, still some doubt as to the precise values of α, one reason being that it depends on the composition of the matrix as well as on the element under consideration, and Wittry [235] depicted α as a function of the average atomic number of the specimen for a limited number of elements (at 30-kV accelerating potential). Poole and Thomas [241] calculated α-values from experimental data on stopping power and backscattering and applied them to investigations of binary alloys. It is likely that investigations, both experimental and theoretical, will lead to a better understanding of the distribution of emission with depth, hence to better values of α. Heinrich [248] argued that this factor should be taken into account, particularly in the case of the heaviest elements (and especially if the difference in atomic number of different components in a specimen is large).

It should be recognized that the effect of backscattering is also dependent on the accelerating potential (E_0) applied to the electrons. If the energy of the electrons is very much greater than the critical excitation potential (E_C) of the x-ray line in question, much of this energy is lost by backscattering, while if $E_0 \simeq E_C$, the predominant factor is the stopping power or deceleration in the specimen, with little backscattering taking place.

The problem can be overcome in principle by employing standards similar to the specimen in atomic number and atomic weight since the effect is intimately related to the differences in these parameters between specimen and standard.

THE ABSORPTION CORRECTION

The major cause of differences between observed and generated x-rays is absorption of the x-rays as they proceed from the point of excitation to the surface of the specimen and then on to the detector.

In order to establish an absorption correction, the distribution of excitation with depth in the specimen, or its Laplace transform, must be known. The treatment of this problem was considered by Castaing and Descamps [249], and since that work many modifications have been suggested.

For the electron beam incident at 90° to the specimen surface, the absorption correction can be expressed as a function of $X = (\mu/\rho) \csc \theta$, θ being the take-off angle of the x-rays, μ the absorption coefficient (as in Beer's law), and ρ the density. Both the magnitude of the x-ray intensities generated in a layer $d(\rho x)$ and its distance ρx from the surface (Fig. 4.52) depend on the atomic number, so allowances for the effects of atomic number must be introduced into the absorption correction or, more desirably, the effect of

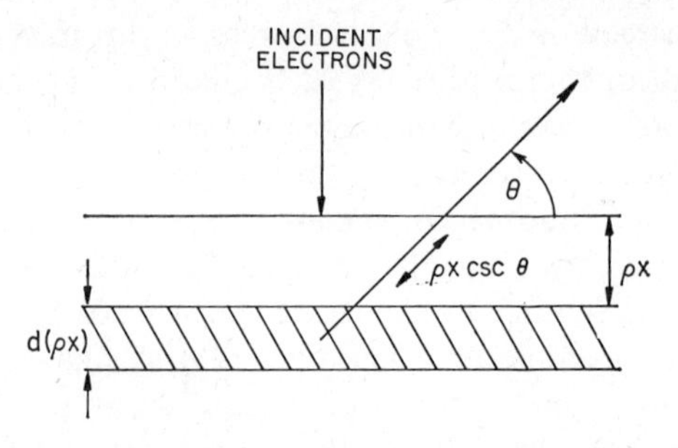

Fig. 4.52 Absorption of electrons in the specimen showing schematically an excited layer $d(\rho x)$ thick at a distance ρx from the surface.

atomic number may be treated separately on the assumption that the absorption correction depends on atomic number only through the depth distribution.

The total intensity within a specimen is given by:

$$I = \int_0^\infty \phi(\rho x)\, d(\rho x),$$

where $\phi(\rho x)$ describes the distribution with depth of the generated radiation, and this corresponds to the Laplace transform $I = F(0)$, that is, the case of no absorption. The measured intensity, which includes absorption effects, is:

$$I = \int_0^\infty \phi(\rho x) \exp\left(-\frac{\mu}{\rho} \cdot \rho x \csc\theta\right) d(\rho x)$$
$$= F(X),$$

and the fraction of the true intensity measured by the detector is

$$\frac{I'}{I} = \frac{F(X)}{F(0)} = f(X),$$

so the concentration of element A is given by:

$$\frac{I_A}{I(A)} = C_A = \frac{I'_A}{f(X)_{\text{spec}}} = \frac{f(X)_{\text{pure}}}{I'(A)}.$$

Wittry [235] lists a number of theoretical and empirical methods by which the relationship between X and $\phi(X)$ can be determined. Castaing [201] used two techniques, first, by measuring the change of x-ray intensity with the take-off angle for a 50% iron/chromium alloy at 38.5 and 27.5 kV. Green [250] improved upon this work and showed that the dependence of X of the $f(X)$ curves of a variety of pure elements is small. Castaing, second, with Deschamps [249] determined $\phi(\rho x)$ in cases in which a thin, isolated layer of a particular material was located at specified distances beneath a layer of second material. The absorption correction curves of Castaing and Deschamps calculated from the experimentally determined $\phi(\rho x)$ curves are shown in Fig. 4.53 and illustrate the correction curves for aluminum, copper, and gold at 25 kV, as well as showing the effects of different accelerating voltages E_0 and different average atomic numbers. The effect of accelerating voltage is attributable naturally to the range effects associated with different energies. Thus the absorption correction depends on Z, E_0 and X, but because of the limited experimental information concerning variations with the former two parameters attempts have been made to calculate the absorption correction with the aid of diffusion theory and theoretical scattering models, with modifications to fit these predictions to the limited experimental data.

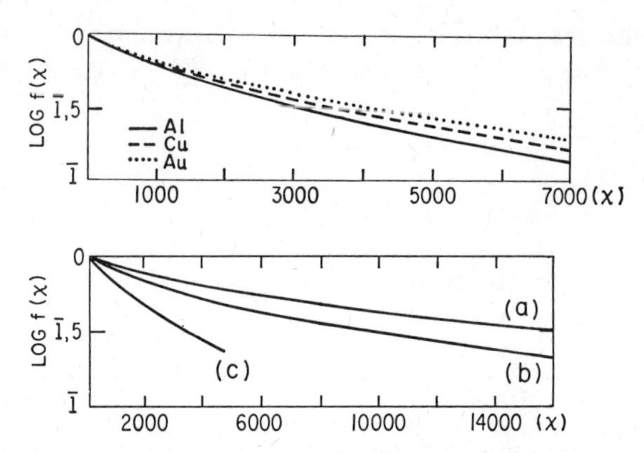

Fig. 4.53(*a*) Absorption correction curves, 29 kV. (Castaing and Descamps.) (*b*) Absorption correction curves for different voltages and average atomic numbers. (*a*) $V = 9.7$ kV, $Z = 13$; (*b*) $V = 15.1$ kV, $Z = 13$; (*c*) $V = 27.5$ kV, $Z = 26$. (Castaing and Descamps.)

Of these attempts, which often produce conflicting results, the most accepted formula appears to be that of Philibert [251] with subsequent modifications. Philibert based a model of electron scattering on Lenard's law of electron scattering, and tabulated values of σ, a modified Lenard coefficient. From the equation developed, $f(X)$ was calculated as a function of Z and E_0 and fitted to Castaing and Deschamps' curves, leading to:

$$F(X) = [(1 + X/\sigma)\{1 + h(1 + X/\sigma)\}]^{-1},$$

and

$$F(0) = 1 + h^{-1},$$

and

$$f(X) = \frac{1 + h}{(1 + X\sigma)1 + h(1 + X/\sigma)},$$

where $h = 1.2\, A/Z^2$ and A the atomic weight. It should be remembered that the approximations involved make this approach unreliable at low atomic numbers. In this range Green's [250] curves may be more precise since he takes into account the dependence of the generation of x-rays on the critical excitation potential E_C, a point also discussed by Duncumb and Shields [252]. The latter authors showed that by replacing Philibert's σ with another coefficient σ_c, the dependence of the depth of x-ray excitation on the penetration of incident electrons and on the critical excitation potential of the x-ray line in question could be taken into account. With this modification, it is predicted that the absorption will fall to zero as E_0 approaches E_C, and further good agreement was obtained with the data of Green [250] and of Castaing and Deschamps [249]. Curves similar to those experimentally obtained by Green have been computed by Archard and Mulvey [253] based upon Archard's theoretical model of electron scattering.

The absorption corrections calculated by using the Duncumb and Shields modification of the Philibert equations, or Green's curves computations based on the former approach, have been published by Colby and Niedermeyer [254]. This procedure is employed by obtaining the mass absorption coefficient of the specimen by an equation of the form:

$$\mu/\rho = \sum_i (\mu/\rho)_i c_i,$$

where c_i is the concentration of element i in the specimen, the c_i being unknown until all the corrections have been made. It is usual to assume, as a first approximation, that c_i is equal to the ratio (intensity from specimen/intensity from standard); the calculated values c' of the concentration using the absorption correction equations are then used to obtain a better value of the coefficient to be used in the absorption correction. Three such calculations usually suffice to cause c' and the second calculated concentration to converge. If this is not so, the measurements may be repeated at a lower accelerating potential or the concentration estimated by some other means.

FLUORESCENCE CORRECTION

The fluorescence correction is of considerably less importance in most cases than is the absorption correction. It must be considered when there is an increase in the observed radiation relative to that generated by the action of the electron beam, and such enhancement can be the result of production of x-rays of the wavelength under consideration either by the continuum or by characteristic radiations of other elements in the specimen. Such a fluorescent enhancement is rarely significant for elements of $Z < 30$ when using the K-lines and $Z < 50$ when using the L-lines [255], and it is likely that enhancement will take place when a suitable element is present in a matrix of high average Z. The quantitative corrections are discussed at some length by Wittry [235] and by Clayton [256]. The effect is minimized in instruments with low take-off angles, because of the shorter path of the x-rays to the surface; in this case absorption corrections are increased.

There are some special cases in which fluorescence corrections are vital to preserve a high degree of accuracy. If an inclusion is being studied in a matrix of material A, the results may indicate the presence of material A in the inclusion when in fact it is not there. Similarly, the concentration of material A in the inclusion may be overestimated because of fluorescence. Another possibility is the result of the fact that x-rays penetrate outside the volume excited by the electron beam. Thus if the size of an inclusion is the same as that of the excited volume, the intensity of x-radiation from this inclusion specimen of pure element A in a matrix with no element A could be less than that from a standard of the same pure element because of fluorescence in the latter which did not occur in the unknown since there was no element A in the matrix. The range of the primary electrons may also be of importance at given boundaries or at an interface between two different chemical systems since it may suggest that diffusion of one into the other has taken place.

Applications

Within a relatively short period of time, the microprobe has become well established in a variety of sciences. The applications of the instrument that have been included in the discussion presented here have been chosen to help show its range of applications and the position that the microprobe occupies among other analytical tools rather than to point out the contribution of the instrument to a particular subject.

For example, an analysis of nuclear fuel melt-down experiments was performed by Stachura and Cooper [257]. The spatial resolution of the probe was of great value in that it permitted the study of the unusual changes in microstructure and the disposition of the materials after melt-down. Considerable interaction between the fuel and the cladding material was found as the molten uranium attached the steel cladding leading to failure, and analyses were made of uranium, copper, iron, chromium, nickel, and carbon.

The measurement of thin films in the micron and submicron range with the electron microprobe has been performed since the early work of Castaing [258], and information as to the thickness of the film as well as its chemical composition can be obtained. This work was extended to determine the thickness of films of elements down below 100 Å, as well as the thickness of superimposed films of different elements. Silver and aluminum films were produced by thermal evaporation under carefully controlled conditions, and the intensities of the aluminum K_α- and the silver L_α lines were measured for fixed accelerating potential and specimen current conditions. A pulse-height analyzer suppressed the silver L_α-second-order line, which occurred close to the aluminum K_α first-order line. Calibration curves for each element were set up for ratio of intensity to film thickness, and a linear curve resulted, permitting the measurement of unknown thin films. An accuracy of about 10% is claimed for 100-Å thick films, and it is suggested that this is an attractive method for the measurement of ultrathin films on either microprobes or, with an accessory, on electron microscopes.

The electron microprobe has been used in the study of glasses and films used in silicon device technology [259]. The object of this study was to measure the thickness of the phosphosilicate glass layer formed during the phosphorous diffusion into the silicon. A variable accelerating voltage technique [260] at constant beam current was employed, and after calibration of the kilovolt intercept against thickness had been obtained on samples characterized by other techniques, the kilovolt intercept at the turnover of a plot of phosphorus K_α intensity against accelerating voltage could be used to measure the phosphosilicate glass thickness since the intensity decreased as the electron beam penetrated through the coating into the substrate. The authors claim an accuracy for this technique of about 150 Å in the range of 700 to 4000 Å. Compositional changes in glasses during analysis have been reported [261], often resulting in topographical changes. The effect is impeded by the presence of multivalent elements such as calcium, magnesium, and aluminum and is most pronounced for monovalent elements such as sodium, potassium, and rubidium. It has been shown [262] that the diffusion of the alkali ions commences after a critical temperature has been reached in the specimen; hence the important parameters were ambient temperature, incident beam current, beam spot diameter, and composition of the glass. The length of time before the diffusion of the ion was detected was indicated by a change in the electron flux. The chemical composition of the phases in borosilicate glasses has also been examined [263].

The microprobe has also been used in conjunction with optical metallographic and x-ray diffraction studies to investigate the compatibility of selected refractory metals (such as Ta, Nb-IZr, and W-Re alloys) with various commercial grade and pure ceramic insulation materials (such as Al_2O_3, MgO, BeO, and ThO_2) at temperatures up to 1540°C for times as long as 1000 hr [264]. Reactions were encountered with Al_2O_3 or with commercial-grade ceramics with SiO_2 as an impurity; the other combinations were compatible. It was suggested that the reactions were influenced by the level of silicon impurity, as well as by zirconium, the alloy which acts as an oxygen sink to drive the reactions forward to form ZrO_2 and the intermetallic compounds Nb_5Si_3, Nb_3Al, and Nb_2Al.

Faulring and Malizie [265] investigated the strontium-silicon system, while Koffman and Moll [266] examined the differences in the K-spectra of silicon in the metallic state and in combination in a wide range of silicon-containing minerals. The silicon spectra were similar in all the compounds but different from the spectrum of the metallic form, leading to the question of what form the standard specimen should take. The effects of chemical combinations on some soft x-ray K- and L-emission spectra from low-atomic-number elements were reported by Fischer and Baun [267]. The K-spectra of boron, carbon, potassium, and nitrogen, and the L-spectra of chlorine and sulfur were examined for compounds including KCl, $KHSO_4$, PbS, $PbSO_4$, KSCN, MoS_2, and $BeSO_4$. Large shifts were found for the sulfur peaks at the low (5-kV) accelerating potentials employed. The same authors [268] investigated silicon, magnesium, and aluminum, while others presented preliminary results for L- and M-spectra in the same wavelength range (~2 to 85 Å), including work on Co and CoO, Ni and NiO, Cu, Cu_2O and CuO, and Zn, ZnO, and $ZnCO_3$.

The application of the instrument in qualitative analysis can be illustrated by the work of Permingreat and Weinryb [269], who found that inclusions in chalcopyrite identified in other studies as linnaeite (Co_3S_4) were in fact carrolite (M_3S_4, where M = Co + Cu + Ni + Fe). In another study Phan and Capitant [270] showed that black particles in an iron ore that were too small to identify by other methods were chromates and that most of the chromium in the ore could not be removed by mechanical means since it was associated with geothite.

In quantitative analysis the information can be presented either in terms of relative composition or mass per unit volume. For example, Temple et al. [271] have shown in their study of variations in the content of Fe and TiO_2 in individual grains of ilmenite ore that the TiO_2 content may range from 53 to almost 100%. This variation was found to relate to structural changes observed with an optical microscope.

The problem of determining mass per unit volume in biological studies has been discussed by Boyde and Schwitsur [272]. They observed that the accuracy of the technique is limited by the problems of specimen preparation. Sections have to be prepared of known thickness smaller than the depth in which characteristic x-rays are produced. Then one has to either make the density uniform over the area being investigated or measure the differences and correct for them in the data.

A consideration of thin sections leads to the possible application of the microprobe to studies in transmission electron microscopy. This combination has been brought about both by using two instruments and by designing one instrument to do both jobs. Using the first method White and co-workers [273] analyzed microcrystalline powders. After making observations in the transmission electron microscope, they prepared an electron micrograph at low magnification so they could locate the desired particle with the optical microscope of the probe. They found that they could work with particles that absorbed only a fraction of the incident beam, interference from the matrix could be reduced, and sample preparation techniques were simple enough to reduce the possibility of contamination.

Others have felt that the two-instrument method suffers from practical difficulties of making the transfer of sample and from the fact that it is not normally possible to plan an investigation well enough in advance to necessitate only one change of sample. As a result, instruments have been developed that include the basic features of the microprobe and transmission electron microscope. A simplified diagram of Duncumb's [274] instrument is shown in Fig. 4.54. Moll and Ogilvie [275] have demonstrated general applications of a

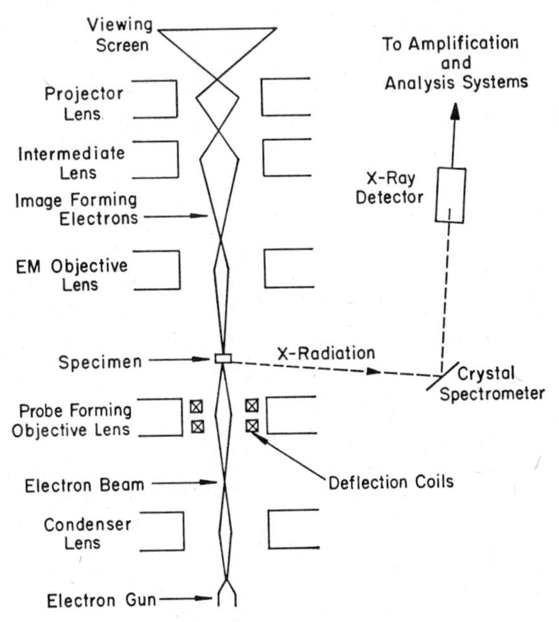

Fig. 4.54 Schematic representation of a combined electron microprobe and transmission electron microscope.

similar instrument to studies of particulate preparations and biological tissues.

Although the examples presented in this section show the limited application of the microprobe in chemical analysis, imposed mainly by the restrictions on the sample, its introduction has greatly extended the usefulness of high-resolution chemical microscopy. However, it should be obvious that the instrument cannot be used independently, and its full usefulness will be realized only when it is integrated into a program with other analytical tools.

References

1. H. Busch, *Arch. Electrotech.*, **18,** 583 (1927).
2. L. deBroglie, *Phil. Mag.*, **47,** 466 (1924).
3. J. J. Thompson, *Phil. Mag.*, **4,** 293 (1897).
4. M. Knoll and E. Ruska, *Z. Physik*, **12,** 389 (1931).
5. E. F. Burton, J. Hillier, and A. Prebus, *Phys. Rev.*, **56,** 1171 (1939).
6. H. Mahl, *Z. Physik*, **21,** 17 (1940).
7. R. G. Williams and R. W. G. Wyckoff, *J. Appl. Phys.*, **15,** 712 (1944).
8. J. H. Reisner, *Sci. Instr.*, **6,** 1 (1961).

9. A. Agar, General References, A10, p. 1.
10. R. E. Burge and G. H. Smith, *Proc. Phys. Soc.* (*London*), **79,** 673 (1962).
11. K. Kobayashi and K. Sakaoku, *Bull. Inst. Chem. Res. Kyoto Univ.*, **42,** 473 (1964).
12. K. Sadashige, *Sci. Instr.*, **10,** 3 (1965).
13. K. Kanaya and H. Kawokatsu, *4th Intern. Congr. Electron Microscopy, Berlin*, 1958, **1,** 3388, Springer-Verlag, Berlin, 1960.
14. G. Oster and Y. Nishijima, *Sci. Am.*, **208,** 54 (May 1963).
15. R. C. Valentine, *Advan. Opt. Electron Microscopy*, **1,** 180 (1967).
16. E. G. Kind and E. Lau, *Optik*, **18,** 646 (1961).
17. I. Veres and W. Krug, *Proc. Conf. Electron Microscopy, Rome, 1968*, **1,** 607, Tipographia Poliglotta Vaticana (1968).
18. E. Lau, *Optik*, **21,** 637 (1964).
19. E. Lau and W. Krug, *Equidensitementry*, Focal Press, London, 1968.
20. J. LePoole, *Proc. Conf. Electron Microscopy, Delft, 1949*, **21,** Nijhoff, The Hague (1949).
21. R. D. Heidenreich, W. M. Hess, and L. L. Ban, *J. Appl. Cryst.*, **1,** 1 (1968).
22. J. H. Reisner, *Sci. Instr.*, **9,** 1 (1964).
23. G. Dupouy, *Advan. Opt. Electron Microscopy*, **2,** 167 (1968).
24. C. Davisson and L. H. Germer, *Phys. Rev.*, **30,** 707 (1927).
25. G. P. Thomson and A. Reid, *Nature*, **119,** 890 (1927).
26. S. Nishikawa and S. K. Kuch, *Proc. Imp. Acad., Japan*, **4,** 475 (1928).
27. G. L. Simard, C. J. Burton, and R. B. Barnes, *J. Appl. Phys.*, **16,** 832 (1945).
28. J. B. Le Poole, *Philips Tech. Rev.*, **9,** 33 (1947).
29. J. W. May, *Ind. Eng. Chem.* (*London*), **57,** 18 (1965).
30. B. K. Vainshtein, in *Structure Analysis by Electron Diffraction*, E. Feigl and J. A. Spink, Eds., Macmillan, New York, 1964.
31. J. M. Cowley and A. L. G. Reese, *Rept. Progr. Phys.*, **21,** 165 (1958).
32. E. Ruska, *Z. Physik*, **83,** 492 (1933).
33. D. E. Bradley, *Brit. J. Appl. Phys.*, **6,** 191 (1955).
34. H. Hashimoto, T. Naiki, T. Eto, K. Fugiwara, M. Watanabe, and Y. Nagahama, *6th Intern. Cong. Electron Microscop., Kyoto*, **1,** 495, Maruzen, Tokyo, (1966).
35. H. Fernandez-Moran, in *The Interpretation of Ultrastructure*, R. J. C. Harris, Ed., Academic, 1962.
36. K. Sadashige, *Appl. Opt.*, **6,** 2179 (1967).
37. I. M. Dawson, *The Nature of the Bacterial Surface*, Blackwell, Oxford, 1949.
38. R. C. Williams, *Exptl. Cell Res.*, **4,** 188 (1953).
39. T. F. Anderson, *Trans. N. Y. Acad. Sci.*, **13,** 130 (1951).
40. A. Claude and E. F. Fullam, *J. Exptl. Med.*, **83,** 499 (1946).
41. R. L. Steere, *J. Biophys. Biochem. Cytol.*, **3,** 45 (1957).
42. K. Sugihara and M. Yoshida, *J. Electronmicroscopy* (*Tokyo*), **17,** 156 (1968).
43. R. G. Scott and A. W. Ferguson, *Textile Res. J.*, **26,** 284 (1956).
44. R. G. Scott, *ASTM Tech. Publ.* No. 257, 121 (1959).
45. J. Sikorski, *Fibre Structure*, J. W. S. Hearle and R. H. Peters, Eds., Butterworths, London, 1963, p. 391.

46. D. S. Skene, *J. Microscopy*, **89,** 63 (1969).
47. J. F. Hamilton and L. E. Brady, *J. Appl. Phys.*, **35,** 414 (1964).
48. J. F. Hamilton, *Phil. Mag.*, **16,** 1 (1967).
49. J. F. Hamilton and F. Urbach, in *The Theory of the Photographic Process*, T. H. James, Ed., 3rd ed., Macmillan, New York, 1966, p. 87.
50. W. Koerber, *Ergeb. Intern. Konf. Wissenschaftliche Photographic, Koln, 1956*, Verlag Dr. O. Helwich, Darmstadt, (1958).
51. R. H. Geiss and K. R. Lawless, *EMSA 26th Ann. Meeting*, *Abstracts* 428, Claitors Book Store, Baton Rouge (1968).
52. H. Hashimoto, T. Naiki, T. Eto, and K. Fugiwara, *6th Intern. Congr. Electron Microscopy*, *Kyoto*, **1,** 495, Maruzen, Tokyo, (1966).
53. P. H. Brodsky, G. G. Cocks, and C. C. Winding, *Proc. EMSA 26th Ann. Meeting*, 416, Claitors Book Store, Baton Rouge (1968).
54. K. Kato, *Japan Plastics* (April 6, 1968).
55. E. M. Belavtseva, K. Z. Gumargalieva, A. I. Kitaigorodsky, and A. V. Vlassov, *3rd Regional Conf. Electron Microscopy*, *Prague*, **A419,** Czechoslovak Acad. Sci., Prague (1964).
56. E. P. Senchencov, S. Z. Bockstein, S. S. Ginsburg, S. G. Kishin, and L. M. Moroz, *4th European Reg. Conf. Electron Microscopy*, *Rome*, **1,** 525, Tipographia Poliglotta Vaticana (1968).
57. M. J. Richardson, *5th Intern. Congr. Electron Microscopy*, *Philadelphia*, **1,** BB-12, Academic Press, New York (1962).
58. P. H. Geil, *Polymer Single Crystals*, Interscience, New York, 1963.
59. H. O. Keith and F. J. Padden, Jr., *J. Polymer Sci.*, **39,** 123 (1959).
60. R. G. Scott, *J. Appl. Phys.*, **28,** 1089 (1957).
61. H. D. Keith, F. J. Padden, Jr., and R. G. Vadimsky, *J. Polymer Sci.*, A-2, **4,** 267 (1966).
62. K. Sakoaku, H. G. Clark, and A. Peterlin, *J. Polymer Sci.*, A-2, **6,** 1035 (1968).
63. B. J. Spit, *J. Macromol. Sci. Physics*, **2,** 45 (1968).
64. C. E. Hall, General References, A11, p. 395.
65. A. K. Kleinschmidt, D. Lang, D. Jacherts, and R. K. Zahn, *Biochim. Biophys. Acta*, **61,** 857 (1962).
66. D. G. Sharp, E. A. Eckert, D. Beard, and J. W. Beard, *J. Bacteriol.*, **63,** 151 (1951).
67. R. C. Williams and R. C. Backus, *J. Am. Chem. Soc.*, **71,** 4052 (1949).
68. A. M. Cravath, A. E. Smith, J. R. Vinograd, and J. N. Wilson, *J. Appl. Phys.*, **17,** 309 (1946).
69. J. H. Reisner, *J. Appl. Phys.*, **31,** 1835 (1960).
70. E. Zeitler and G. F. Bahr, *J. Appl. Phys.*, **33,** 847 (1962).
71. R. G. Scott, *ASTM Tech. Publ.* No. 348, 79 (1963).
72. E. W. Fisher, H. Goddar, and G. F. Schmidt, *Kolloid-Z.*, *Z. Pol.*, **226,** 30 (1968).
73. E. Goldstein, *Ann. Physik*, **11** (3), 832 (1880).
74. H. Busch, *Ann. Physik*, **81** (4), 974 (1926). M. Knoll and E. Ruska, *Ann. Physik*, **12** (5), 607 (1932). M. Knoll et al., *Z. Physik*, **78,** 34 (1932). M. Brüche and H. Johannson, *Naturwissenschaften*, **20,** 353 (1932).

75. M. Brüche, *Z. Physik*, **86,** 448 (1933).
76. G. Möllenstedt and F. Lenz, *Advan. Electron. Electron Phys.*, **18,** 251 (1963).
77. R. Kollath, *Ann. Physik*, **1** (16), 357 (1947).
78. W. Deitrich and H. Seiler, *Z. Physik*, **157,** 576 (1960).
79. R. Kollath, *Lamdolt-Bürnstein Zahlenwerte und Funktionen*, Vol. II, Pt. 6, Springer, Berlin, 1959, p. 1008.
80. J. G. Trump and R. L. Van de Graff, *Phys. Rev.*, **75,** 44 (1949).
81. H. O. Muller, *Z. Physik*, **104,** 475 (1937).
82. W. Engel, *6th Intern. Conf. Electron Microscopy, Kyoto*, Maruzen, Tokyo, 1966.
83. L. Wegmann, *Pract. Metallog.*, **5,** 241 (1968).
84. M. Knoll, *Z. Tech. Physik* **16,** 467 (1935); M. Knoll and R. Theile, *Z. Physik*, **113,** 260 (1939).
85. M. Von Ardenne, *Z. Physik*, **109,** 553 (1938).
86. V. K. Zworykin, J. Hillier, and R. L. Snyder, *ASTM Bull.*, **117,** 15 (1942).
87. C. W. Oatley, W. C. Nixon, and R. F. W. Pease, *Advan. Electron. Electron Phys.*, **21,** 181 (1965).
88. C. Brachet, *Bull. Assoc. Tech. Maritime Aeron.*, **45,** 369 (1946).
89. F. Davoine, Dissertation, University of Lyons, 1957.
90. A. V. Crewe, see Section 3, p. 396.
91. R. F. W. Pease and W. C. Nixon, *J. Sci. Instr.*, **42,** 81 (1965).
92. W. Glaser, *Z. Physik.*, **117,** 285 (1941).
93. K. C. A. Smith, in *Encyclopedia of Microscopy*, L. Clark, Ed., Van Nostrand, Reinhold, New York, 1961, p. 241.
94. V. E. Cosslett, *Practical Electron Microscopy*, Butterworths, London, 1951.
95. D. McMullen, *Proc. Inst. Elec. Engrs.* (*London*), **B100,** 245 (1953).
96. T. E. Everhart, and R. F. M. Thornley, *J. Sci. Instr.*, **37,** 246 (1960).
97. H. Tammura and H. Kimura, *J. Electronmicroscopy* (*Tokyo*), **17,** 106 (1968).
98. A. J. Gonzales, *Record IEEE 9th Ann. Symp. Electron, Ion Laser Beam Technol.*, *Berkeley*, *California*, R. F. W. Pease, Ed., 1967.
99. J. Adams and B. W. Marley, *Philips Tech. Rev.*, **28,** 116 (1967).
100. P. R. Thornton, *Scanning Electron Microscopy*, Chapman and Hall, London, 1968, p. 194.
101. P. S. Ong, *Proc. 3rd Electron Microprobe Conf.*, *Chicago*, *1968*.
102. D. McMullen, Ph.D. Thesis, Cambridge University, 1952.
103. S. Kimoto and H. Hashimoto, in *The Electron Microprobe*, T. D. McKinley, K. F. J. Heinrich, and D. B. Wittry, Eds., Wiley, New York, 1966, p. 480.
104. Ref. 100, pp. 262 et seq.
105. L. H. Pruden, E. J. Korda, D. P. Smith, and J. P. Williams, Ref. 101.
106. P. M. Williams and A. D. Yoffe, *Nature*, **221,** 952 (1969).
107. T. E. Everhart, *Proc. IEEE*, **54,** 1480 (1966).
108. S. Shirai et al., Ref. 82.
109. Ref. 87, pp. 190 et seq.
110. T. H. P. Chang and W. C. Nixon, *J. Roy. Microscop. Soc.*, **88,** 143 (1968).
111. D. B. Wittry and A. Van Couvering, *J. Sci. Instr.*, **44,** 294 (1967).

112. E. W. White, H. A. McKinstry, and G. G. Johnson, Jr., in *Scanning Electron Microscopy—1968*, O. Johari, Ed., Illinois Inst. of Technology, Chicago, 1968.
113. N. C. MacDonald, *Proc. 26th Ann. EMSA Meeting*, C. J. Arceneaux, Ed., Claitors Publ. House, New Orleans, 1968.
114. H. A. Bethe, *Ann. Phys. (Leipzig)*, **5,** 325 (1930).
115. R. Kollath, *Z. Phys.*, **38,** 202 (1937).
116. O. Hachenberg and W. Brauer, *Advan. Electron Electron. Phys.*, **11,** 413 (1959).
117. G. A. Harrower, *Phys. Rev.*, **104,** 52 (1956).
118. G. A. Harrower, *Phys. Rev.*, **102,** 340 (1956).
119. Ref. 100, p. 105.
120. J. L. H. Jonker, *Philips Res. Rept.*, **12,** 249 (1957).
121. H. Bruining, *Physica*, **5,** 901 (1938).
122. S. J. Lukjanov, *Physic Sowjet Univ.*, **13,** 173 (1938).
123. K. G. McKay, *Advan. Electron.*, **1,** 65 (1948).
124. Ref. 100, p. 102.
125. E. J. Sternglass, *Phys. Rev.*, **80,** 925 (1950).
126. Ref. 100 p. 105.
127. H. E. Bishop, Ph.D. Thesis, Cambridge University, 1966, Ref. 100, p. 87.
128. T. E. Everhart, *J. Appl. Phys.*, **31,** 1483 (1960).
129. G. D. Archard, *J. Appl. Phys.*, **32,** 1505 (1961).
130. R. F. Dashen, *Phys. Rev.*, **A134,** 1025 (1964).
131. Ref. 87, p. 187.
132. A. Rose, Ref. 123, p. 131.
133. T. E. Everhart, O. C. Wells, and C. W. Oatley, *J. Electron. Control*, **7,** 97 (1959).
134. S. Kimoto and H. Hashimoto, Ref. 112.
135. R. F. M. Thornley, Ph.D. Thesis, Cambridge University, 1960.
136. R. F. W. Pease and W. C. Nixon, *J. Sci. Instr.*, **42,** 81 (1965).
137. Ref. 100, p. 207.
138. R. I. Garrod and J. F. Nankivell, *Brit. J. Appl. Phys.*, **9,** 2121 (1958).
139. O. C. Wells, *Brit. J. Appl. Phys.*, **11,** 199 (1960).
140. S. F. Tipper, D. I. Dagg, and O. C. Wells, *J. Iron Steel Inst. (London)*, **193,** 133 (1959).
141. S. Kimoto and T. Suganuma, SM 68018, Japan Electron Optics Laboratory Co.
142. For example, *Manual of Photogrammetry*, Am. Soc. Photogrammetry, 3rd ed., Falls Church, Virginia, 1966.
143. Ref. 87, p. 215.
144. I. H. Hall, *J. Appl. Polymer Sci.*, **12,** 739 (1968).
145. A. D. G. Stewart and A. Boyde, *Nature*, **196,** 187 (1962).
146. V. Peck, Ref. 113.
147. See Ref. 76.
148. J. Sikorski et al., *J. Sci. Instr.*, **1** (2), 29 (1968).
149. O. C. Wells, Ref. 98.
150. O. Johari, Ed., *Scanning Electron Microscopy—1969*, Illinois Inst. of Technol., Chicago, 1969.

151. T. L. Hayes and R. F. W. Pease, *Advan. Biol. Med. Phys.*, **12** (1968). See also Ref. 87.
152. W. J. Humphreys, T. L. Hayes, and R. F. W. Pease, Ref. 113.
153. C. M. Drew, Ref. 112.
154. C. M. Drew, R. H. Knipe, and A. S. Gordon, *Pyrodynamics*, **4,** 325 (1966).
155. F. B. Bowden and J. H. L. McAuslan, *Nature*, **178,** 408 (1956).
156. J. H. L. McAuslan and K. C. A. Smith, in *Proc. Stockholm Conf. Electron Microscopy*, F. S. Sjostrand and J. Rhodin, Eds., Academic, New York, 1957, p. 341.
157. R. F. W. Pease and R. A. Ploc, *Trans. Metal. Soc. AIME*, **233,** 1949 (1965).
158. J. E. Castle and H. G. Masterson, *Corrosion Sci.*, **6,** 93 (1966).
159. M. Knoll, *Phys. Z.*, **42,** 120 (1941).
160. P. R. Thornton et al., *Phil. Mag.*, **14,** 165 (1966).
161. I. Minkoff, *Phil. Mag.*, **12,** 1083 (1965).
162. I. Minkoff and W. C. Nixon, *J. Appl. Phys.*, **37,** 4848 (1966).
163. G. A. Gardner and R. W. Cahn, *J. Mater. Sci.*, **1,** 211 (1966).
164. T. J. Bastow et al., *Nature*, **222,** 27 (1969).
165. L. E. Brooks, P. Sennett, and H. H. Morris, *J. Paint Tech.*, **39,** 473 (1967).
166. M. D. Coutts, E. R. Levin, and J. G. Woodward, Ref. 113.
167. P. H. Newman and J. Sikorski, *Proc. 4th European Conf. Electron Microscopy, Rome, 1968*, p. 549.
168. R. D. Van Veld, G. Morris, and H. Billica, *J. Appl. Polymer Sci.*, **12,** 2709 (1968).
169. R. G. Scott, *ASTM Special Tech. Publ. No. 257*, 1959.
170. D. Fitchman and S. Newman, *Am. Phys. Soc. Meeting, Philadelphia, 1969.*
171. K. G. Mayhan, Ref. 150.
172. H. D. Blakelock, reported by *Anon.*, *Nature*, **219,** 326 (1968).
173. F. Davoine, *J. Phys. Radium*, **21,** 121 (1960).
174. F. Davoine, *Proc 3rd European Conf. Electron Microscopy, Prague, 1964.*
175. E. J. Korda, L. H. Pruden, and J. D. Williams, *App. Phys. Letters*, **10,** 205 (1967).
176. D. S. Shaw, R. C. Wayte, and P. R. Thornton, *Appl. Phys. Letters*, **8,** 289 (1966).
177. J. R. Banbury and W. C. Nixon, *J. Sci. Instr.*, **44,** 889 (1967).
178. J. R. Banbury and W. C. Nixon, Ref. 167.
179. D. C. Joy and J. P. Lakubovics, *Phil. Mag.*, **71,** 61 (1968).
180. R. I. Hutchinson, P. A. Lavin, and J. R. Moore, *J. Sci. Instr.*, **42,** 885 (1965).
181. G. Y. Robinson and R. M. White, *Appl. Phys. Letters*, **10,** 320 (1967); **11,** 141 (1967).
182. R. LeBihan, and C. Sella, Ref. 167.
183. G. V. Saparin, G. V. Spivak, N. N. Sedor, and L. F. Gomolva, Ref. 167.
184. D. G. Coates, *Phil. Mag.*, **16,** 1179 (1967); Ref. 150.
185. G. R. Booker, A. M. B. Shaw, M. J. Whelan, and P. B. Hirsch, *Phil. Mag.*, **16,** 1185 (1967).
186. A. N. Broers, *J. Appl. Phys.*, **38,** 1991 (1967); **38,** 3040 (1967).
187. A. V. Crewe et al., *Rev. Sci. Instr.*, **39,** 576 (1968).

188. V. G. Macres et al., Ref. 113, p. 368.
189. D. Kynaston and A. D. G. Stewart, Ref. 150.
190. A. V. Crewe, Ref. 150.
191. A. V. Crewe, D. Johnson, and M. Isaacson, Ref. 113, p. 360.
192. A. V. Crewe, *Science*, **154,** 729 (1966).
193. M. G. R. Thomson and A. V. Crewe, Ref. 113, p. 358.
194. J. M. Cowley and A. Strojnik, Refs. 150, 167.
195. L. A. Harris, *J. Appl. Phys.*, **39,** 1419 (1968).
196. J. J. Lander, *Phys. Rev.*, **91,** 1382 (1913).
197. R. W. Weber and W. T. Paria, *J. Appl. Phys.*, **38,** 4355 (1967).
198. *Industrial Res.*, November 1968, p. 23. Varian Associates have produced a selected bibliography on auger electron spectroscopy.
199. H. G. J. Moseley, *Phil. Mag.*, **26,** 1024 (1913).
200. J. Hillier, U.S. Patent 2,418,029, 1947.
201. R. Castaing, Thesis, University of Paris 1951, Publ. O.N.E.R.A., No. 55.
202. R. Castaing and A. Guinier, *Proc. 1st Intern. Conf. Electron Microscopy, Delft, 1949*, p. 60, 1950.
203. V. E. Cosslett and P. Duncumb, *Nature*, **177,** 1172 (1956).
204. C. A. Andersen, *Methods of Biochemical Analysis*, Vol. XV, D. Glick, Ed., Interscience, New York, 1967, p. 147.
205. R. Castaing, *Advan. Electron. Electron Phys.*, **13,** 317 (1960).
206. V. E. Cosslett and W. C. Nixon, *X-Ray Microscopy*, Cambridge Univ. Press, London, 1960.
207. L. S. Birks, *Electron Microprobe Analysis*, Interscience, New York, 1963; *X-Ray Spectrochemical Analysis*, Interscience, New York, 1969.
208. P. Duncumb, J. V. P. Long, and D. A. Melford, *Electron Probe Microanalysis*, Hilger and Watts, London, 1967.
209. R. Theisen, *Quantitative Electron Microprobe Analysis*, Springer, New York, 1968.
210. J. A. Belk and A. L. Davis, Eds., *Electron Microscopy and Microanalysis of Metals*, Elsevier, Amsterdam, 1969.
211. K. F. J. Heinrich, in *The Electron Microprobe*, T. D. McKinley, K. F. J. Heinrich, and D. B. Wittry, Eds., Wiley, New York, 1966, p. 841.
212. A. Lutts, in *Advances in X-Ray Analysis*, Vol. 11, J. B. Newkirk and H. G. Pfeiffer, Eds., Plenum, New York, 1968, p. 345.
213. L. S. Birks, Ref. 207, p. 122.
214. K. F. J. Heinrich, in *Advances in X-Ray Analysis*, Vol. 7, W. M. Mueller, Ed., Plenum, New York, 1964, p. 382.
216. J. J. Ficca, Jr., *Proc. 3rd Electron Microprobe Conf.*, Electron Probe Analysis Soc of Am., Chicago, 1968.
217. L. S. Birks, Ref. 207, p. 64.
218. For example, F. K. Richtmeyer, E. H. Kennard, and T. Lauritsen, *Introduction to Modern Physics*, 5th ed., McGraw-Hill, New York, 1955, Chapter 8.
219. L. S. Birks, Ref. 207, p. 74.
220. H. A. Liebhafsky and H. G. Pfeiffer, *J. Chem. Educ.*, **30,** 450 (1953).
221. R. Castaing and J. Descamps, *J. Phys. Radium*, **16,** 304 (1955).

222. D. B. Wittry, *J. Appl. Phys.*, **29,** 1563 (1958).
223. P. Duncumb, in *X-Ray Microscopy and X-Ray Microanalysis*, A. Engström, V. Cosslett, and H. Pattee, Eds., Elsevier, Amsterdam, 1960, p. 365.
224. C. A. Andersen, Ref. 211, p. 58.
225. L. S. Birks, *X-Ray Spectrochemical Analysis*, Interscience, New York, 1959.
226. L. A. Fergason, in *Advances in X-Ray Analysis*, Vol. 9, G. R. Mallett, M. Fay, and W. M. Mueller, Eds., Plenum, New York, 1966, p. 265.
227. L. S. Birks, E. J. Brooks, and G. W. Gourlay, *Rev. Sci. Instr.*, **29,** 425 (1958).
228. J. W. Kemp and G. Anderman, *Proc. 5th Conf. Ind. Appl. X-Ray Anal., Denver, 1956*, Plenum, New York (1956).
229. P. Duncumb, *Brit. J. Appl. Phys.*, **10,** 420 (1959).
230. P. S. Ong, Ref. 211, p. 43.
231. A. Franks and K. Lindsey, Ref. 211, p. 83.
232. L. S. Birks, Ref. 207, p. 105.
233. R. M. Dolby, *Proc. Phys. Soc.* (*London*), **73,** 81 (1959).
234. L. S. Birks, Ref. 207, p. 101.
235. D. B. Wittry, Ref. 214, p. 395.
236. A. Faessler, *Proc. Xth Colloq. Spectroscopicum Internationale, Washington, D.C. 1963*, p. 307, Spartan Books, New York (1964).
237. C. A. Andersen, *Natl. Conf. Electron Probe Microanalysis*, Electron Probe Analysis Soc of Am., College Park, Maryland, 1966.
238. R. Theisen, Ref. 209, p. 8.
239. K. F. J. Heinrich, Ref. 214, p. 325.
240. J. W. Colby, Ref. 214, p. 357.
241. D. M. Poole and P. M. Thomas, in *X-Ray Optics and X-Ray Microanalysis*, H. H. Pattee, V. E. Cosslett, and A. Engström, Eds., Academic, New York, 1963, p. 411; *J. Inst. Metals*, **90,** 228 (1962).
242. J. I. Goldstein, F. J. Majeske, and H. Yakowitz, in *Advances in X-Ray Analysis*, Vol. 10, J. B. Newkirk and G. R. Mallett, Eds., Plenum, New York, 1967, p. 447.
243. D. A. Melford, *J. Inst. Metals*, **90,** 217 (1962).
244. ASTM Data Series DS-37, American Society for Testing and Materials, Philadelphia, 1965.
245. A. J. Bearden, NYO-10586, U.S. Atomic Energy Comm., Oak Ridge, Tennessee, 1964.
246. C. A. Andersen, Ref. 204, p. 193.
247. R. Castaing, Ref. 205, p. 319.
248. K. F. J. Heinrich, Ref. 214, p. 326.
249. R. Castaing and J. Descamps, *J. Phys. Radium.*, **16,** 304 (1955).
250. M. Green, Ref. 241, p. 361.
251. J. Philibert, Ref. 241, p. 379.
252. P. Duncumb and P. K. Shields, Ref. 211, p. 284.
253. G. D. Archard and T. Mulvey, *Brit. J. Appl. Phys.*, **14,** 626 (1963).
254. J. W. Colby and J. F. Niedermeyer, NLCO-914, Clearinghouse of Federal Scientific and Technical Information, Springfield, Virginia, 1964, J. W. Colby, NLCO-944, 1966.

255. S. J. B. Reed, *Brit. J. Appl. Phys.*, **16,** 913 (1965); Ref. 253.
256. D. B. Clayton, Ref. 210, p. 191.
257. S. J. Stachura and L. Cooper, Ref. 242, p. 422.
258. R. Castaing, Ref. 205; W. E. Sweeney, R. E. Seebold, and L. S. Birks, *J. Appl. Phys.*, **31,** 1061 (1960).
259. J. J. Gniewek and N. G. Knopman, Ref. 216.
260. B. W. Schumacher and S. S. Miltra, *Proc. 1962 AIEE Electronic Components Conf., Washington, D.C., 1962*, p. 152.
261. A. K. Baird and D. H. Zenger, in *Advances in X-Ray Analysis*, Vol. 9, G. R. Mallet, M. Faney, and W. M. Mueller, Eds., Plenum, New York, 1965, p. 497.
262. L. F. Vassamillet and V. E. Caldwell, Ref. 216.
263. T. Tran and C. Sella, *Compt. Rend.*, **259,** 1325 (1964).
264. D. E. Fornwalt, B. R. Gourley, and A. V. Manzlone, Ref. 211, p. 581.
265. G. M. Faulring and E. S. Malizie, Ref. 242, p. 409.
266. D. M. Koffman and S. H. Moll, Ref. 261, p. 323.
267. D. W. Fischer and W. L. Baun, Ref. 263, p. 329.
268. D. W. Fischer and W. L. Baun, Ref. 263, p. 371.
269. F. Permingeat and E. Weinryb, *Bull. Soc. Franc. Mineral. Crist.*, **83,** 65 (1960); and Ref. 43, p. 282.
270. K. D. Phan and M. Capitant, *Proc. Intern. Symp. Mining Res.*, **1,** 399 (1962).
271. A. K. Temple, K. F. J. Heinrich, and J. F. Ficca, Jr., Ref. 211, p. 784.
272. A. Boyde and V. R. Schwitsur, Ref. 241, p. 499.
273. E. W. White, P. J. Denny, and S. M. Irving, Ref. 211, p. 791.
274. P. Duncumb, Ref. 211, p. 490.
275. S. H. Moll and R. E. Ogilvie, *Anal. Chem.*, **39,** 867 (1967).

General

A. BOOKS

1. G. F. Bahr and E. H. Zeitler, *Quantitative Electron Microscopy*, Williams and Wilkins, Baltimore, 1965.
2. R. Barer and V. E. Cosslett, *Advanc. Opt. Electron Microscopy*, **1** (1967); **2** (1968).
3. E. F. Burton and W. H. Kohl, *The Electron Microscope*, 2nd ed., Van Nostrand, Reinhold, New York, 1946.
4. V. E. Cosslett, *Practical Electron Microscopy*, Butterworths, London, 1951.
5. R. B. Fisher, *Applied Electron Microscopy*, Indiana Univ. Press, Bloomington, Indiana, 1954.
6. C. E. Hall, *Introduction to Electron Microscopy*, 2nd ed., McGraw-Hill, New York, 1966.
7. M. E. Haine, *Electron Microscope*, Spon, London, 1961.
8. R. D. Heidenreich, *Fundamentals of Transmission Electron Microscopy*, Interscience, New York, 1964.
9. P. B. Hirsch, A. Howe, R. B. Nicholson, D. W. Pashley, and M. J. Whelan, *Electron Microscopy of Thin Crystals*, Plenum, New York, 1965.

10. D. H. Kay, *Techniques for Electron Microscopy*, 2nd ed., F. A. Davis, Philadelphia, 1965.
11. B. M. Siegel, *Modern Developments in Electron Microscopy*, Academic, New York, 1964.
12. R. W. G. Wyckoff, *Electron Microscopy Technique, 2nd Applications*, Interscience, New York, 1949.
13. E. E. Zworykin, G. A. Morton, E. G. Ramberg, J. Hillier, and A. W. Vance, *Electron Optics and the Electron Microscope*, Wiley, New York, 1945.

B. PROCEEDINGS (PUBLISHED IN BOOK FORM)

1. Electron Microscopy Society of America, Annual Meeting from 1967. Abstracts of previous meetings were listed annually in *J. Appl. Phys.*
2. European Regional Conference on Electron Microscopy, every four years. The latest one was 1968.
3. International Congress for Electron Microscopy, every four years. The latest one was 1966.

C. ABSTRACTS

1. *Bulletin Signaletique*, 761 Microscope Electronique, Diffraction Electronique, Centre de Documentation Dv C.N.R.S.

Chapter **V**

MICROWAVE SPECTROSCOPY

W. H. Flygare

1 Introduction 439

2 Rotational Spectra 441

Linear: c Is the Internuclear Axis 443

Symmetric Top: c Is The Symmetry Axis, Oblate 446

Symmetric Top: a Is The Symmetry Axis, Prolate 448

Asymmetric Top 448

3 Experimental Apparatus 452

4 Molecular Stark Effect 458

5 Molecular Structures 461

6 Molecular Rotation and the Effects of Vibration 464

Vibrational Satellites 464

Conformational Analysis 466

Internal Rotation 469

7 Molecular Zeeman Effect 473

8 Nuclear Quadrupole Interactions 487

9 Nuclear Spin-Rotation Interactions 494

1 INTRODUCTION

Microwave spectroscopy involves primarily the observation of pure rotational transitions in the frequency range of 8.0 to 40.0 gHz. The purpose of this chapter is to outline the techniques of microwave spectroscopy and show how one can extract from the spectra meaningful parameters relating to the electronic distribution and structure of a molecule.

The chapter starts with an introduction to molecular mechanics leading to the evaluation of the moments of inertia for a variety of molecules. The pure rotational Hamiltonian is given and the energy levels for various systems are explained and listed. The arbitrary but convenient breakdown into linear, symmetric top, and asymmetric top molecules is discussed. Symmetric top molecules have two equal principal moments of inertia, and asymmetric top molecules have three unequal principal moments of inertia. From the initial prediction of the moments of inertia, we outline the methods of assignment and check leading to the experimental moments of inertia and the experimental molecular structure.

Section 3 deals with the apparatus commonly used in microwave spectroscopy. A standard Stark modulation inexpensive spectrometer which is adequate for molecular structure determination is outlined. More sophisticated frequency stabilization and sweep systems are also outlined, which can lead to high-resolution work in microwave spectroscopy. The newer techniques of microwave-microwave double-resonance spectroscopy are also outlined for common configurations of three-level systems.

The molecular Stark effect which leads to the common method of modulation in microwave spectroscopy is outlined. The valuable measurement of molecular electric dipole moments is illustrated for linear and asymmetric top molecules. The Stark effect is also a very useful tool in the identification of rotational transitions.

Molecular structures are obtained, by the substitution method, from the experimental moments of inertia. The criteria for molecular planarity are discussed as well as the determination of the structures of several interesting molecules.

The effects of vibrations on the observed pure rotational spectrum lead to three rather distinct areas of study. The first area is the observation of centrifugal distortion, which can lead to a determination of the vibrational motion of the molecule through the observed vibration-rotation coupling. The second area is the observation of vibrational satellites or the pure rotational spectrum of molecules in excited vibrational states. This method can be used to study molecular structures in excited vibrational states as well as to identify the vibrational energy spacings. The third area is the study of rotational conformers and barriers to internal rotation. The energetics involved between the ground states of different rotational conformers has long been of interest to chemists. Microwave spectroscopy offers a unique and precise method for measuring these energy differences. In addition, the coupling of the vibration or internal rotation from one conformer to another is observed as a perturbation on the pure rotational energy. Thus the coupling leads directly to an observation of the internal motion and the energy barriers separating the different conformers.

Section 7 discusses the relatively new area of research in microwave spectroscopy that involves the measurement of both the linear Zeeman effect (the molecular magnetic moments or molecular *g*-values) and the quadratic

Zeeman effect (the anisotropy of the molecular magnetic susceptibility). Combining the linear and quadratic Zeeman parameters leads to a direct measurement of the quadrupole moments of the molecule. Also available are the diagonal elements in the total, diamagnetic, and paramagnetic susceptibility tensors. The sign of the electric dipole moment can be obtained from a careful measurement of the molecular quadrupole moment in two different center-of-mass coordinate systems.

Sections 8 and 9 deal with two nuclear-electronic interactions: the nuclear quadrupole interaction and the nuclear spin-rotation interaction. The methods of extracting meaningful data about the distribution of electrons in molecules from these data are discussed.

2 ROTATIONAL SPECTRA

The Hamiltonian describing the energy of a rigid rotating molecule is:

$$\mathscr{H} = \frac{P_a^{\,2}}{2I_{aa}} + \frac{P_b^{\,2}}{2I_{bb}} + \frac{P_b^{\,2}}{2I_{cc}}. \tag{5.1}$$

P_a, P_b, and P_c are the angular momentum components along the *principal inertial axes.* The principal inertial axis system is the center-of-mass coordinate system in which the moment of inertia tensor is diagonal. The principal moments of inertia I_{aa}, I_{bb}, and I_{cc} are defined as *the second moments of atomic mass.*

$$\begin{aligned} I_{aa} &= \sum_n M_n(b_n^2 + c_n^{\,2}) \\ I_{ab} &= 0, \end{aligned} \tag{5.2}$$

where b_n is the center-of-mass coordinate (along the principal inertial b-axis) to the atom indicated by n with mass M_n. The center of mass is identified from any arbitrary origin by the *first moment of the atomic mass.*

$$\bar{X} = \frac{1}{M}\sum_n M_n x_n, \tag{5.3}$$

where

$$M = \sum_n M_n. \tag{5.4}$$

The principal moments of inertia are obtained by first computing the center-of-mass moments in any arbitrary but convenient orientation. The Cartesian second-moment matrix has the matrix elements:

$$\begin{aligned} I_{xx} &= \sum_n M_n(y_n^{\,2} + z_n^{\,2}) \\ I_{xy} &= -\frac{1}{2}\sum_n M_n y_n x_n. \end{aligned} \tag{5.5}$$

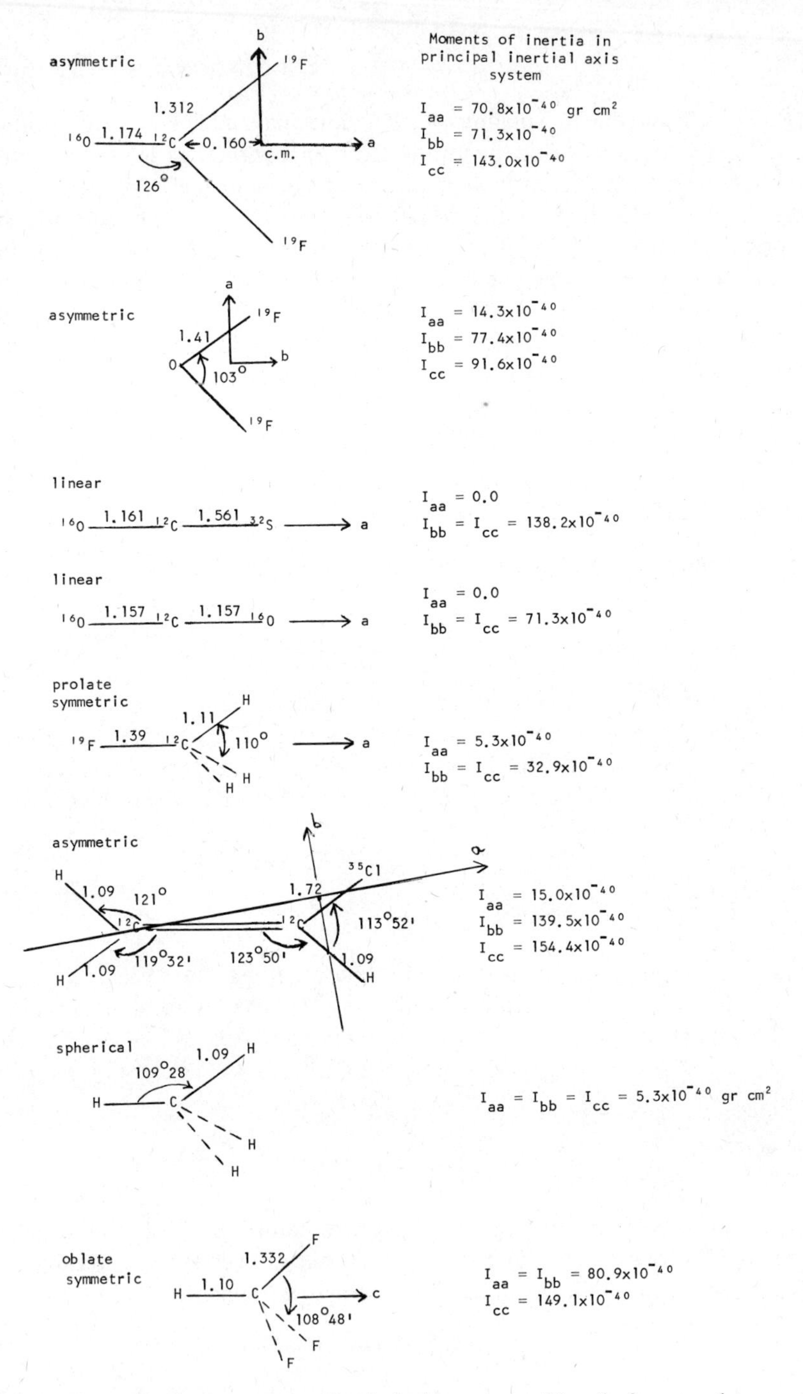

Fig. 5.1 The molecular structures and principal moments of inertia for several types of molecules. c.m., Center of mass.

This 3×3 matrix is then diagonalized to yield the principal moments defined in (5.2). The *a*-, *b*-, and *c*-principal axes are defined to satisfy:

$$I_{aa} \leq I_{bb} \leq I_{cc}. \tag{5.6}$$

The molecular structures and principal moments of inertia of several molecules are shown in Fig. 5.1.

Returning to (5.1) we must obtain the energy levels of the rotating molecule from this equation and an appropriate basis set. An appropriate basis set is obtained by noting that the total molecular rotational angular momentum **P** can be decomposed into the three components along both an axis system fixed in the molecule (a, b, and c) and along an axis system fixed in the laboratory coordinate system (x, y, and z). The Eulerian angles describe the angular dependence between these two axis systems. The matrix elements of the Hamiltonian in (5.1) in this $\langle J, K, M|$ basis are [1]:

Diagonal:

$$\langle J, K, M|\, \mathscr{H}\, |J, K, M\rangle = \frac{\hbar^2}{4}\left\{\left[\frac{1}{I_{aa}} + \frac{1}{I_{bb}}\right][J(J+1) - K^2] + \frac{2K^2}{I_{cc}}\right\}$$

Off diagonal:

$$\langle J, K+2, M|\, \mathscr{H}\, |J, K, M\rangle$$

$$= \frac{\hbar^2}{8}\left\{\left[\frac{1}{I_{bb}} - \frac{1}{I_{aa}}\right][(J-K)(J-K-1)(J+K+1)(J+K+2)]^{1/2}\right\}. \tag{5.7}$$

K represents the projection of J on the molecular fixed-axis system and M is the projection on the space or laboratory fixed-axis system. Equation (5.7) is now described in terms of various types of molecules.

Linear: c Is the Internuclear Axis

In linear molecules, $I_{aa} = I_{bb}$ and the off-diagonal elements in (5.7) vanish. The diagonal elements reduce to:

$$E(J, K = 0, M) = \frac{\hbar^2}{2I_{bb}} J(J+1), \tag{5.8}$$

where $K = 0$, as $I_{cc} \to \infty$.

The frequencies of the rotational transitions of linear molecules are easily obtained by (J is the lower of the two states):

$$\frac{E(J+1, M) - E(J, M)}{h} = \nu_{J \to J+1}$$

$$= \frac{\hbar^2}{hI_{bb}}(J+1)$$

$$= \frac{\hbar}{2\pi I_{bb}}(J+1) = 2B(J+1), \tag{5.9}$$

where the rotational constant B is defined by:

$$B = \frac{\hbar}{4\pi I_{bb}}. \tag{5.10}$$

The units of B are hertz as defined in (5.9). Figure 5.2*a* shows a schematic of a microwave spectrum of a typical linear molecule with a rotational constant of 5 gHz. Molecules with rotational constants near 5 gHz are listed in Table 5.1 [2]. 1 Hz = 1 cps, 1 kHz = 10^3 cps, 1 MHz = 10^6 cps, and 1 gHz = 10^9 cps. Thus each of these molecules exhibits a microwave spectrum similar to Fig. 5.2*a* Each of the above molecules also shows different spectra for the

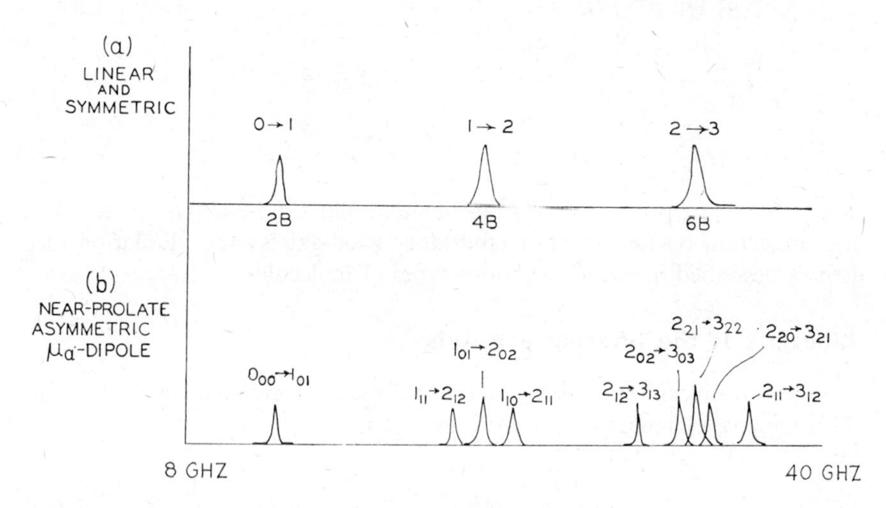

Fig. 5.2 Typical microwave spectra of representa tive molecules. (*a*) Rigid-rotor linear and symmetric top molecule spectrum. (*b*) μ_a-Dipole near-prolate asymmetric top molecule spectrum. See legend for Fig. 5.4 for notation.

Table 5.1 Rotational Constants and Moments of Inertia for Some Typical Linear Molecules

Molecule	B (MHz)	$I_{bb}(10^{-40}\text{ g cm}^2)$
$^{23}Na^{35}Cl$	6536.86	128.363
$^{23}Na^{79}Br$	4534.51	185.046
$^{133}Cs^{19}F$	5527.27	151.809
$^{16}O^{12}C^{32}S$	6081.48	137.975
$^{35}Cl^{12}C^{14}N$	5970.82	140.532
$H{-}^{12}C{\equiv}^{12}C{-}^{35}Cl$	5684.24	147.617
$H{-}^{12}C{\equiv}^{12}C{-}^{12}C{\equiv}^{14}N$	4549.07	184.453

different isotopic species. In OCS we have the rotational constants noted below in addition to the main isotopic species listed in Table 5.1 ($^{16}O^{12}C^{32}S$ is 94% abundant).

Molecule	B (MHz)	*Natural Abundance* (%)
$^{16}O^{13}C^{32}S$	6061.87	1.00
$^{17}O^{12}C^{32}S$	5883.67	0.04
$^{18}O^{12}C^{32}S$	5704.83	0.20
$^{16}O^{12}C^{34}S$	5932.82	4.00
$^{16}O^{12}C^{33}S$	6004.91	0.72
$^{16}O^{13}C^{34}S$	5911.73	0.04

The $J = 0 \rightarrow J = 1$ ground vibrational state rotational spectrum of OCS is shown in Fig. 5.3. The relative intensities of the various isotopic species are illustrated. As noted later, the $J = 0 \rightarrow J = 1$ rotational transitions for molecules in the excited vibrational states are also present and some of these vibrational satellites are as intense as the weaker isotopic transitions. A major problem in microwave spectroscopy is the differentiation of weak isotopic transitions from the rotational spectra of excited vibration states of the main isotopic species. As line widths are normally less than 1 MHz, the different isotopic species are easily resolved if the sensitivity is sufficiently high. Using the above data on the OCS isotopes allows a determination of the moments of inertia for the several isotopes and a determination of the two bond parameters (see Fig. 5.1).

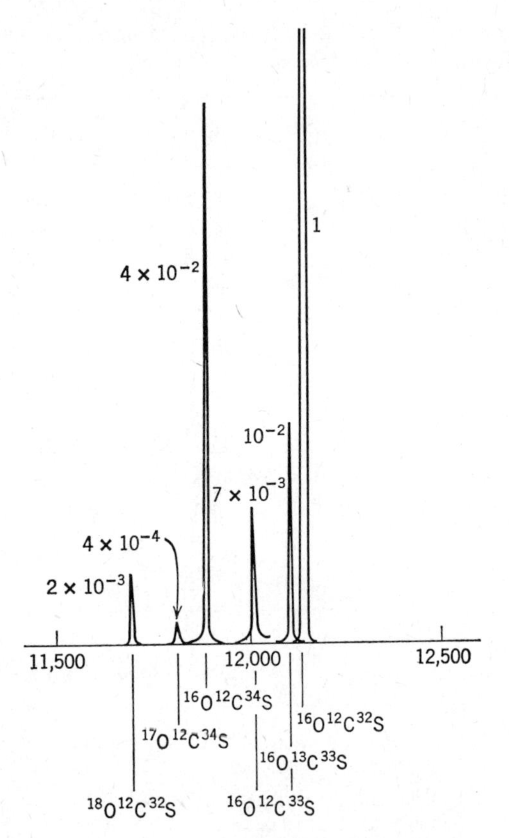

Fig. 5.3 The $J = 0 \rightarrow J = 1$ microwave spectrum of the various isotopes of OCS. The relative intensities of the transitions are according to the natural abundance of the isotopes.

Symmetric Top: c Is the Symmetric Axis, Oblate

$$I_{aa} = I_{bb} < I_{cc}$$

$$A = B > C \tag{5.11}$$

The off-diagonal elements in (5.7) are zero leading to the following energy for a symmetric top:

$$E(J, K, M) = \frac{\hbar^2}{2I_{bb}} J(J+1) + \frac{\hbar^2}{2}\left(\frac{1}{I_{ce}} - \frac{1}{I_{bb}}\right)K^2$$

$$= h\{BJ(J+1) + (C - B)K^2\}. \tag{5.12}$$

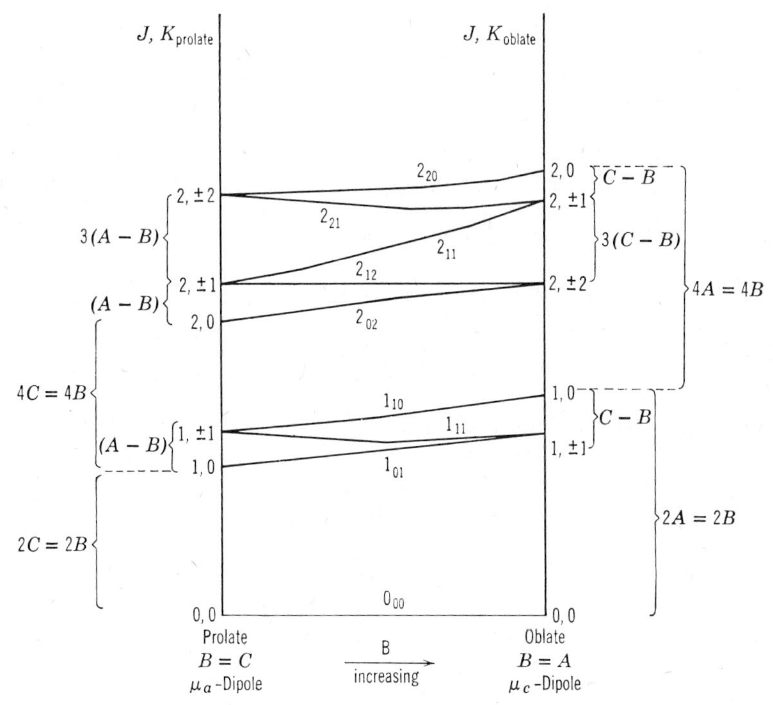

Fig. 5.4 Prolate, oblate, and asymmetric top energy levels. $B = C$ on the left and B increases from left to right until $B = A$ on the right.

The energy levels from (5.12) are shown on the right-hand side of Fig. 5.4. The frequencies between the $\Delta J = +1$, $\Delta K = 0$, $\Delta M = 0$ transitions are identical to the linear molecule as shown in Fig. 5.2*a*. The moments of inertia and rotational constants for several oblate symmetric tops are listed in Table 5.2 (see HCF_3 in Fig. 5.1).

Table 5.2 Rotational Constants and Moments of Inertia for Some Typical Oblate ($I_{aa} = I_{bb} < I_{cc}$) Symmetric Tops

	$I_{aa} = I_{bb}(10^{-40}$ g cm$^2)$	$A = B$ (MHz)	$I_{cc}(10^{-40}$ g cm$^2)$*	C (MHz)
HCF_3	81.08	10,348.74	149.10	5627.70
NF_3	78.56	10,680.96	87.32	9609.37
PF_3	107.30	7,819.90	111.61	7518.06

* Symmetry axis.

Symmetric Top: *a* Is the Symmetry Axis, Prolate

$$I_{aa} < I_{bb} = I_{cc}$$
$$A > B = C \tag{5.13}$$

$$E(J, K, M) = \frac{\hbar^2 J(J+1)}{2I_{bb}} + \frac{\hbar^2}{2}\left(\frac{1}{I_{aa}} - \frac{1}{I_{bb}}\right)K^2$$
$$= h\{BJ(J+1) + (A - B)K^2\} \tag{5.14}$$

The energies for the oblate and prolate symmetric tops are shown in Fig. 5.4. Again, the prolate symmetric top transition frequencies ($\Delta J = +1, \Delta K = 0, \Delta M = 0$) are identical to a linear molecule as shown in Fig. 5.2. The moments of inertia and rotational constants for several prolate symmetric tops are listed in Table 5.3 (see FCH_3 in Fig. 5.1).

Table 5.3 Moments of Inertia and Rotational Constants of Some Typical Prolate ($I_{aa} < I_{bb} = I_{cc}$) Symmetric Tops

	I_{aa} (10^{-40} g cm^2)*	A(MHz)	$I_{bb} = I_{cc}$ (10^{-40} g cm^2)	B(MHz)
FCH_3	5.300	158,319.0	32.859	25,536.12
$ClCH_3$	5.310	158,020.9	63.123	13,292.95
$BrCH_3$	5.320	157,723.8	87.696	9,568.19
ICH_3	5.350	156,839.0	111.859	7,501.31
CH_3—CF_3	154.40	5,434.5	161.830	5,185.00
CH_3—SiH_3	10.610	79,084.9	76.497	10,968.96
CH_3—C≡C—H	5.320	157,723.8	98.180	8,545.84

* Symmetry axis.

It is apparent that most molecules fall in the prolate symmetric or near-prolate symmetric top category.

Asymmetric Top

The asymmetric top energy levels lie between the prolate and oblate limits shown in Fig. 5.4. The energy levels are obtained directly by diagonalizing the matrix described by (5.7) and the designation, $J_{K(\text{prolate}),K(\text{oblate})}$, is demonstrated in Fig. 5.4. The energies of a few of the lower J-energy levels are shown below in Table 5.4 [3].

It is also evident from Fig. 5.4 that a prolate top has an electric dipole moment along the a-axis (μ_a-dipole) and the oblate top has the dipole moment along the c-axis (μ_c-dipole). An asymmetric top can have a dipole moment

Table 5.4 Rigid Rotor Energies in Terms of the Rotational Constants

Level	*Energy/h*
0_{00}	0
1_{01}	$B + C$
1_{11}	$A + C$
1_{10}	$B + A$
2_{02}	$2A + 2B + 2C - 2\sqrt{(B - C)^2 + (A - C)(A - B)}$
2_{12}	$A + B + 4C$
2_{11}	$A + 4B + C$
2_{21}	$4A + B + C$
2_{20}	$2A + 2B + 2C + 2\sqrt{(B - C)^2 + (A - C)(A - B)}$
3_{03}	$2A + 5B + 5C - 2\sqrt{4(B - C)^2 + (A - B)(A - C)}$
3_{13}	$5A + 2B + 5C - 2\sqrt{4(A - C)^2 - (A - B)(B - C)}$
3_{12}	$5A + 5B + 2C - 2\sqrt{4(A - B)^2 + (A - C)(B - C)}$
3_{22}	$4A + 4B + 4C$
3_{21}	$2A + 5B + 5C + 2\sqrt{4(B - C)^2 + (A - B)(A - C)}$
3_{31}	$5A + 2B + 5C + 2\sqrt{4(A - C)^2 - (A - B)(B - C)}$
3_{30}	$5A + 5B + 2C + 2\sqrt{4(A - B)^2 + (A - C)(B - C)}$

along any of the three principal inertial axes. It is easy to establish the rotational selection rules for an asymmetric top by the use of group theory. The selection rules are given in Table 5.5.

As mentioned previously, a large number of asymmetric top molecules fall in the near-prolate symmetric top category and it is convenient to think in terms of the symmetric top limits. A few examples of near-prolate symmetric top molecules are listed in Table 5.6.

Table 5.5 The Selection Rules for an Asymmetric Top*

μ_a-*Dipole*	μ_b *Dipole*	μ_c *Dipole*
ee $\leftrightarrow$ eo	oo $\leftrightarrow$ ee	ee $\leftrightarrow$ oe
oo $\leftrightarrow$ oe	eo $\leftrightarrow$ oe	eo $\leftrightarrow$ oo

* Even (e) and odd (o) refer to the prolate-oblate *K*-quantum numbers (see Fig. 5.4).

Table 5.6 Near-Prolate ($A > B \sim C$) Symmetric Tops That Have μ_a-Electric Dipoles

b / a	A(MHz)	B(MHz)	C(MHz)	*References*
	20,153.8	7,466.50	5,871.87	[4]
	47,353.3	4,659.44	4,242.79	[5]
	77,510.7	12,055.0	10,416.2	[6]
	5,663.5	2,570.64	1,767.94	[7]
	11,854.4	3,524.6	2,759.7	[8]

The near-prolate asymmetric top molecules listed in Table 5.6 all have characteristic $\Delta K_{\text{prolate}} = 0$ spectra (see Fig. 5.4 for transitions) given by the frequencies listed in Table 5.7 and illustrated in Fig. 5.2*b*.

Note the dominant $\Delta K_{\text{prolate}} = 0$ selection rules in these transitions. Thus the symmetric top rigid rotor spectrum that overlaps the linear molecule is split according to Fig. 5.2 in the slightly asymmetric top. Note also that there are several checks and symmetries that appear in this asymmetric top

Table 5.7 Transition Frequencies for the Low-J, $\Delta J = +1$ Series of Transitions in a Near-Prolate μ_a-Dipole Asymmetric Top

$0_{00} \rightarrow 1_{01}$	$B + C$
$1_{11} \rightarrow 2_{12}$	$B + 3C$
$1_{01} \rightarrow 2_{02}$	$2A + B + C - 2\sqrt{(B - C)^2 + (A - C)(A - B)}$
$1_{10} \rightarrow 2_{11}$	$3B + C$
$2_{12} \rightarrow 3_{13}$	$4A + B + C - 2\sqrt{4(A - C)^2 - (A - B)(B - C)}$
$2_{02} \rightarrow 3_{03}$	$3(B + C) - 2\sqrt{4(B - C)^2 + (A - B)(A - C)} + 2\sqrt{(B - C)^2 + (A - B)(A - C)}$
$2_{21} \rightarrow 3_{22}$	$3(B + C)$
$2_{20} \rightarrow 3_{21}$	$3(B + C) + 2\sqrt{4(B - C)^2 + (A - B)(A - C)} - 2\sqrt{(B - C)^2 + (A - B)(A - C)}$
$2_{11} \rightarrow 3_{12}$	$4A + B + C - 2\sqrt{4(A - B)^2 + (A - C)(B - C)}$

spectrum. A few of the obvious symmetries are:

$$\nu(0_{00} \rightarrow 1_{01}) = B + C = \tfrac{1}{3}\nu(2_{21} \rightarrow 3_{22}) = \tfrac{1}{4}[\nu(1_{11} \rightarrow 2_{12}) + \nu(1_{10} \rightarrow 2_{11})].$$

Also, it is evident that the $2_{02} \rightarrow 3_{03}$, $2_{21} \rightarrow 3_{22}$, and $2_{20} \rightarrow 3_{21}$ transitions form a perfect triplet with separations of

$$\Delta\nu = 2[4(B - C)^2 + (A - B)(A - C)]^{1/2} - 2[A(B - C)^2 + (A - B)(A - C)]^{1/2}.$$

As the molecules listed in Table 5.6 are near-prolate tops ($B \sim C$), the $\Delta J = +1$ spectra are grouped with separations approximately $2B$ as shown in Fig. 5.2. Many near-prolate tops also have μ_b-dipole transitions with approximate $\Delta J = +1$ separations equal to $A + C$. Near-oblate tops ($B \sim A$) also show similar patterns which are easily obtained as shown here for near-prolate tops.

In summary, it should be relatively easy to predict the nature of the $\Delta J = +1$ spectra of an asymmetric top from an initial assumed structure. The principal moments of inertia are estimated in order to determine the axes containing the dipole moment and an approximate microwave spectrum is predicted on the basis of the calculated rotational constants, the frequencies listed in Table 5.7, and the patterns shown in Fig. 5.2. The spectrum is searched, assigned, and the experimental moments of inertia are extracted. Of course the analysis is complicated many times by the occurrence of a number of $\Delta J = 0$ (Q-branch) transitions which can occur throughout the entire range for which the $\Delta J = +1$ (R-branch) transitions have been predicted. Many times, a Q-branch assignment is made prior to an R-branch assignment. There are also regular patterns of spectra that occur in the Q-branch series, and these are also of help in analyzing a complicated microwave spectrum. The Stark effect is a great help in assigning a rotational spectrum (see Section 4).

We conclude the discussion at this point and merely point out that the methods of computing the principal moments of inertia and predicting an initial microwave spectrum from these estimates are well developed and in many cases computerized so that the beginner need only have access to a digital computer to initially predict the spectra [9].

3 EXPERIMENTAL APPARATUS

Rotational spectra can be satisfactorily recorded on the simple apparatus shown in Fig. 5.5. This is a basic but adequate spectrometer. The main absorption cell is a standard waveguide vacuum isolated with Mylar windows and connected to the microwave input and output by standard waveguide components. The absorption cell has an electrically isolated Stark electrode in the center of the waveguide which is connected to the square-wave generator. The zero-based square-wave generator, at frequency f, produces a Stark effect when the high voltage is on. Thus if the microwave frequency is at rotational resonance, the molecules will absorb microwave energy only when

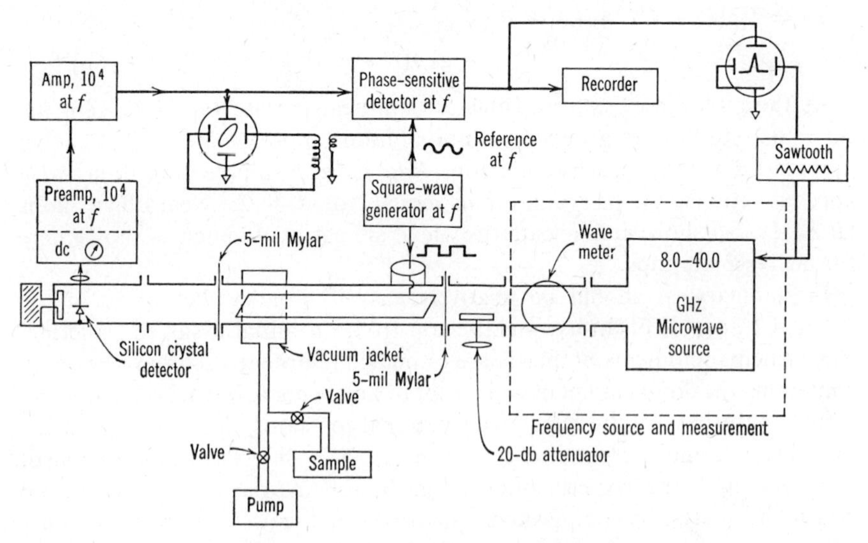

Fig. 5.5 Basic microwave spectrometer.

the field is off. When the high field is on, the rotational energy levels are shifted, giving no rotational absorption. Thus the square-wave voltage modulates the microwave absorption at frequency f. The microwave source in Fig. 5.5 is swept electrically or mechanically. The microwave power is detected at the silicon crystal in a preamplifier with a bandwidth around 10 to 1000 Hz centered at f (the modulation frequency). After further amplification the signal is compared on the first oscilloscope with the reference sine wave at f. The phase of the reference can then be adjusted to give a closed loop. The frequency is then compared in a phase detector to further narrow the bandwidth around the modulation frequency f. Slow sweeps (bandwidth about 1 Hz) are recorded on the recorder. Faster electrical sweeps by a sawtooth applied to the reactive element in the source microwave tube are passed through the same amplifying and phase-detection systems at bandwidths about 10 times the sawtooth voltage. The output is then displayed by the faster-response oscilloscope which is horizontally swept with the sawtooth. The slower sweeps lead to higher signal-to-noise ratios because of the lower bandwidths. However, there are lower limits on the bandwidths that can be applied to reduce the noise. At bandwidths of less than 0.1 Hz or time constants above 10 sec, the amplifier drift and other low-frequency mechanical noise becomes important. Thus multisweep techniques have been developed to increase the signal-to-noise ratios beyond those achieved with integration techniques. These techniques are similar to those developed in nmr. The recorder and oscilloscope are simply replaced with a multichannel analyzer or an analog-to-digital converter with a magnetic storage. The signal from the phase detector is fed into the analog-to-digital converter where a voltage-time trace is digitized along the time axis at a sufficient number of points to assure faithful reproduction of the information. The voltage-time information in digital form is then stored and added to subsequent sweeps over the same frequency range. The random noise averages to zero as the signal grows. The signal-to-noise ratio increases with the square root of the number of sweeps. In this manner there are virtually no limits except patience in the extraction of high sensitivity.

The frequency-measuring apparatus shown in Fig. 5.5 is crude but effective. The wave meter shown in series with the microwave transmission is a commercially available direct-reading resonant cavity device which is typically calibrated to 0.3%. Thus at 10,000 MHz the frequency could be in error up to 30 MHz or more. However, the long-term stability of the wave meter is quite good and it can be calibrated by observing rotational transitions from known molecules. Recently, the National Bureau of Standards compiled a very convenient list of transitions of molecules listed by frequencies [10]. This very complete tabulation lists transitions almost every 1 MHz throughout the entire 8- to 40-GHz region. Thus a wave meter can be calibrated to read

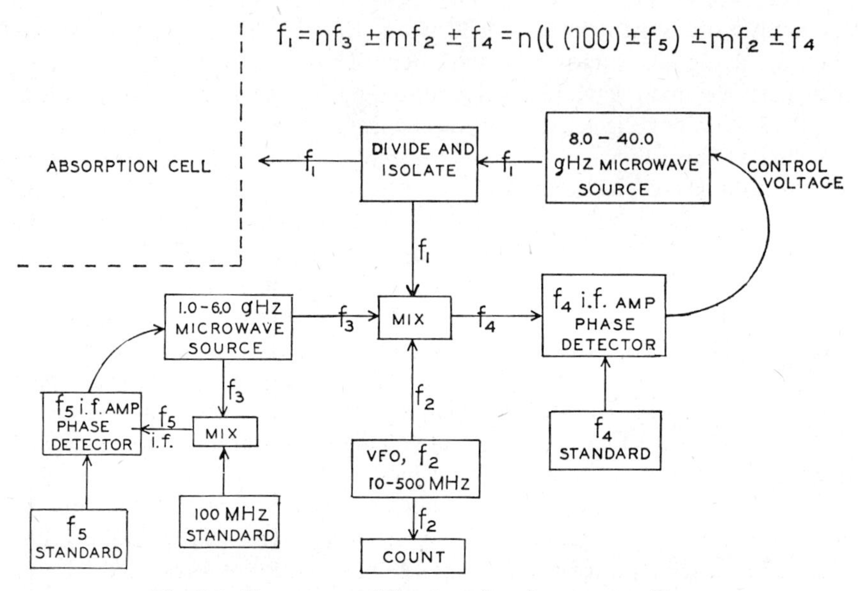

Fig. 5.6 Frequency stabilization of a microwave oscillator.

frequencies to ± 5 MHz or about 0.05%, which is adequate for the determination of molecular structures in most cases. As microwave line widths are typically 1 MHz or less it is desirable to measure the frequencies more accurately. However, the measurement of more accurate frequencies as shown in Fig. 5.6 represents a considerable escalation in price and effort and this discussion emphasizes that information on molecular structures can be obtained with the calibrated wave meter measurements displayed in Fig. 5.5.

Either complete microwave spectrographs or the components in Fig. 5.5 are commercially available. The largest item is the microwave frequency source. Four klystrons can be purchased to cover the 8- to 40-GHz range for approximately $3,000.00. Backward-wave oscillators, which are noisier but much more convenient and reliable over a long period of time, are more expensive, being $1,200.00 each for the 8- to 12-GHz and 12- to 18-GHz bands, $2,500.00 for the 18- to 26-GHz band, and $4,000.00 for the 26- to 40-GHz band. The power supply for either klystrons or backward-wave oscillators is another $2,000.00 to $3,000.00. Wave meters are about $250.00 for each of the four bands, and attenuators, transition sections, detectors, and silicon crystals cost an additional $3,000.00. The square-wave generator and amplifiers can be obtained for about $3,000.00. Oscilloscopes, vacuum apparatus, and recorders are readily available. In summary, the entire 8- to 40-GHz microwave region can be purchased and arranged according to

Fig. 5.5 for a cost of $10,000.00 to $20,000.00. A single more narrow band could be set up for considerably less.

A much more sophisticated frequency control and drive system is shown in Fig. 5.6. The general object is to frequency-stabilize the microwave source to a high harmonic of a very stable crystal oscillator plus a low harmonic of a variable-frequency oscillator (VFO) which is displayed on a counter. Referring to Fig. 5.6, we note that f_3, the 1- to 6-GHz oscillator, is phase-stabilized to a harmonic of a 100-MHz oscillator. Thus f_3 is available throughout the 1- to 6-GHz range at frequencies given by $f_3 = l(100) \pm f_5$ where l is the harmonic and f_5 is the intermediate or control frequency in the phase detector. f_3 is then mixed with the microwave source along with a VFO, f_2, to stabilize the 8- to 40-GHz microwave source. The microwave frequency is given in megahertz by:

$$f_1 = n[l(100) \pm f_5] \pm mf_2 \pm f_4. \tag{5.15}$$

f_5 and f_4 are low-frequency (30.0- to 60.0-MHz) i.f. or control frequencies. m is small, usually 1, to give a direct reading of f_2, and the frequency change of f_1, on the counter.

All of the components or the integrated package shown in Fig. 5.6 are available commercially. More-or-less complicated schemes similar to the plan in Fig. 5.6 are available, but they all revolve about the same principles.

Microwave double-resonance experiments are sometimes used for various reasons in microwave spectroscopy. The general principle used in microwave double-resonance experiments is to employ a high-power microwave oscillator at one frequency to saturate two levels in a three-level system. The second oscillator is used to sample the effects on the remaining two-level transition. A typical experimental arrangement and a set of energy levels for a double-resonance experiment is shown in Fig. 5.7 [11]. The high-power radiation is employed at the lower frequency, ν_1 (X-band), and the higher frequency, ν_2 (Ku-band), low-power radiation is swept over the upper transition. The upper transition can be broadened, and at very high powers at the lower frequency it is completely lost. The key to this microwave-microwave double-resonance experiment is in noting that the high-power radiation in the X-band is not passed through the X- to Ku-band transition or taper section. Thus only the higher frequency, lower-wavelength Ku-band radiation used to observe the upper transition appears at the detector. The high-power radiation perturbs the lower two levels and the effect is indirectly observed by viewing the upper two levels. An extension of this concept is to use double-resonance modulation by pulsing the high-power radiation at frequency f as shown in Fig. 5.8 [12]. In this experiment, which uses the waveguide arrangement of Fig. 5.7, the lower-frequency high-power radiation frequency is modulated at an audio square-wave frequency f. Thus the lower frequency

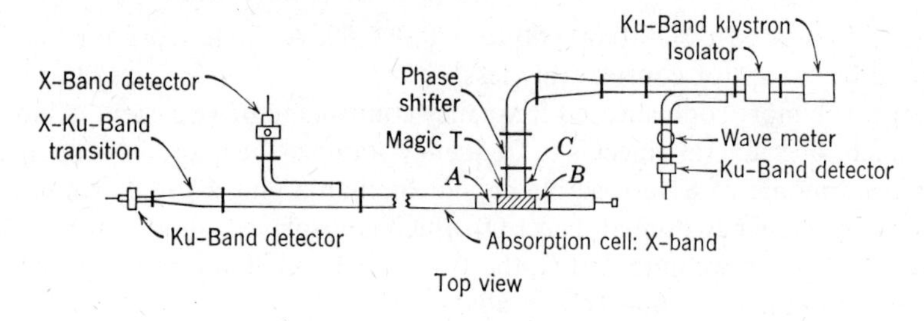

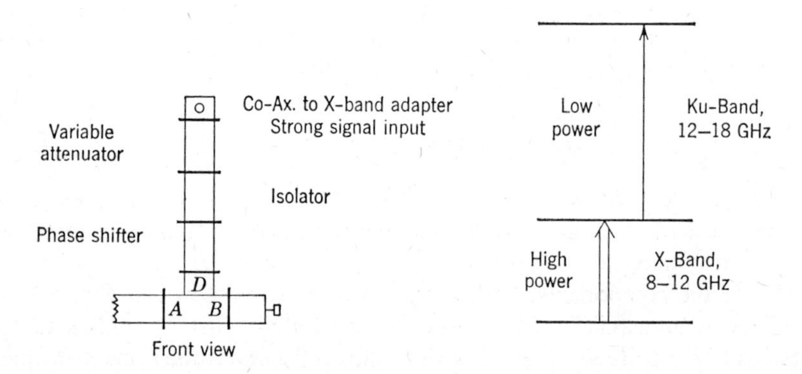

Fig. 5.7 Typical microwave-microwave double-resonance arrangement in which two different frequency bands are employed. The energy level diagram illustrates the double-resonance experiment. The main requirement of this experimental arrangement is that the high-power radiation must be at least one-half the frequency of the low-power radiation. The high-power radiation is attenuated in the smaller wavequide used in the detection section for the higher frequency radiation.

alternates between two values. If one of the two modulated lower-frequency signals is on resonance, the perturbation on the lower two levels will be modulated at f. This modulated perturbation will appear at the higher microwave frequency detector. If the detector at the higher microwave frequency is tuned to f, only the modulated signal will appear which is produced entirely by the lower frequency perturbation. Phase-sensitive detection can be used in this apparatus and the Stark electrode eliminated. This type of modulation is an attractive possibility for special cell arrangements in which a Stark electrode is not feasible. The microwave-microwave modulated double-resonance spectrograph has been used to identify transitions partially obscured because of very dense spectra.

Many times double-resonance experiments are necessary when both microwave frequencies are in the same band [11]. An apparatus for this type of experiment is illustrated in Fig. 5.9 [11]. This apparatus employs

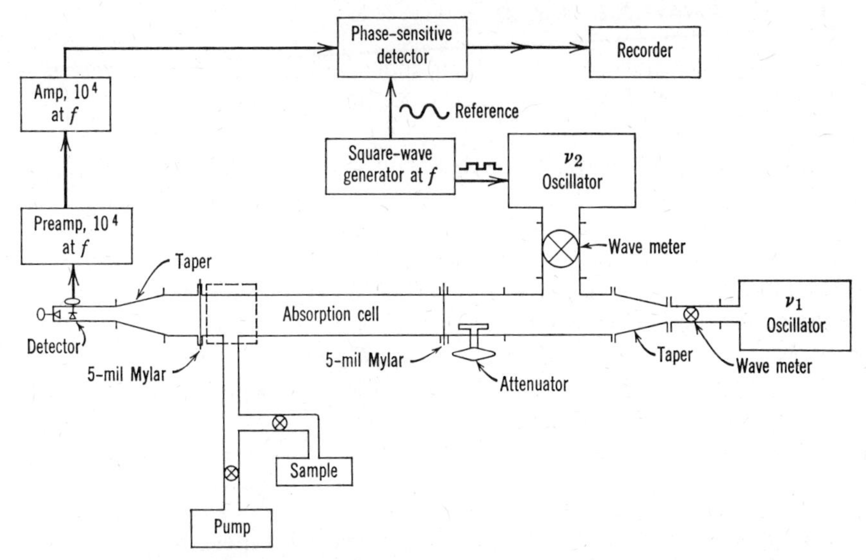

Fig. 5.8 A microwave-microwave double-resonance modulation spectrometer. The waveguide arrangement is similar to the arrangement in Fig. 5.7. ν_2 is frequency modulated on and off resonance and the perturbation on ν_1 is detected at the modulation frequency.

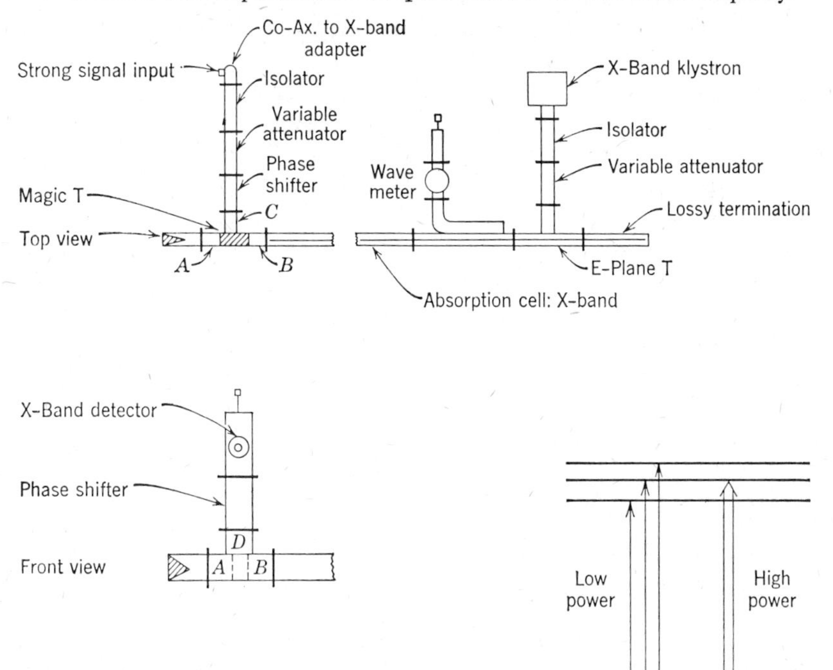

Fig. 5.9 A microwave double-resonance apparatus in which both frequencies are in the same microwave transmission band.

three-way T-junctions and orientated polarization of the microwaves to isolate the two nearly identical microwave frequencies.

4 MOLECULAR STARK EFFECT

The Stark effect is quite important in microwave spectroscopy because of the utilization of Stark modulation (see Section 3). Owing to the phase-sensitive detection at the Stark modulation frequency, the zero-field absorption and the high-field absorption are recorded with opposite polarity of the voltage. Thus the Stark or high-field absorptions can interfere with the absorption lines by partial cancellation of the opposing voltages. The Stark effect is described by the perturbation:

$$\mathscr{H} = -\boldsymbol{\mu} \cdot \mathbf{E} = -(\mu_a \phi_{az} + \mu_b \phi_{bz} + \mu_c \phi_{cz}) E_z, \tag{5.16}$$

where μ_a, μ_b, and μ_c are the molecular fixed electric dipole moments and E_z is the high voltage applied by the Stark electrode as shown in Fig. 5.5. ϕ_{az}, ϕ_{bz}, and ϕ_{cz} are the direction cosines between the molecule fixed and space fixed-axis systems. The matrix elements for the Hamiltonian in (5.16) in the $|J. K, M\rangle$ basis are easily computed and added to the original zero-field matrix elements in (5.7) [1]. The entire matrix can then be diagonalized to yield the energy levels in the presence of the electric field. This analysis is illustrated in Fig. 5.10 in which the energy levels and $\Delta M = 0$ transitions for a linear molecule are shown. The Stark effect for linear and asymmetric tops are normally second order in the electric field. The expected Stark effect for the $0 \rightarrow 1$ and $1 \rightarrow 2$ R-branch spectra for a typical μ_a-dipole near-prolate top is shown in Fig. 5.11. Convenient tables are available for computing the Stark effect for these low-J transitions [13]. However, computer programs are also available for calculation of the Stark effect [9].

In the normal arrangement of the Stark electrode in a microwave transmission cell (see Fig. 5.5), the Stark electrical field is parallel to the electric field of the microwave radiation. Thus $\Delta M = 0$ selection rules are always present. The relative intensities of the various $\Delta M = 0$ transitions are given several other places [1–3]. They are:

$$\begin{aligned} &R\text{-branch, intensity} \propto J^2 - M^2 \\ &Q\text{-branch, intensity} \propto M^2, \end{aligned} \tag{5.17}$$

where J and M in the R-branch transition refer to the higher J-state. The Stark effect in several R-branch transitions for a near-prolate μ_a-dipole asymmetric top is illustrated in Fig. 5.11. The Stark effect in the $3_{13} \rightarrow 3_{12}$ Q-branch transition in formaldehyde is illustrated in Fig. 5.12.

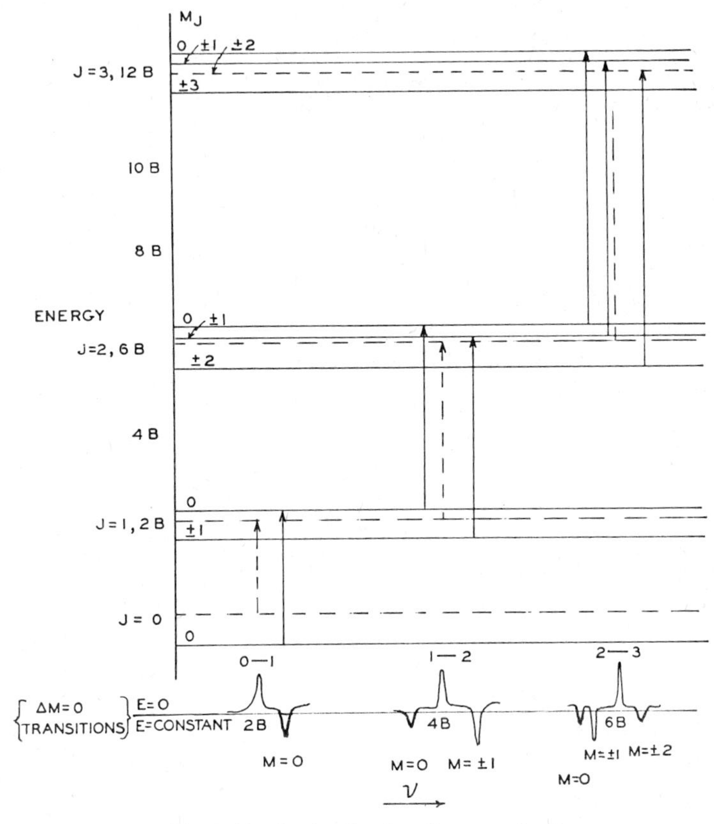

Fig. 5.10 Stark effect in a linear molecule.

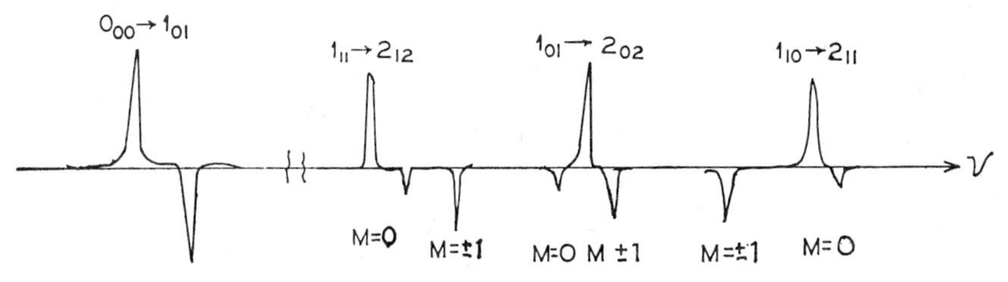

Fig. 5.11 Microwave spectrum showing the Stark effect in a near-prolate μ_a-type molceule (see Table 5.6 and Fig. 5.2*b*).

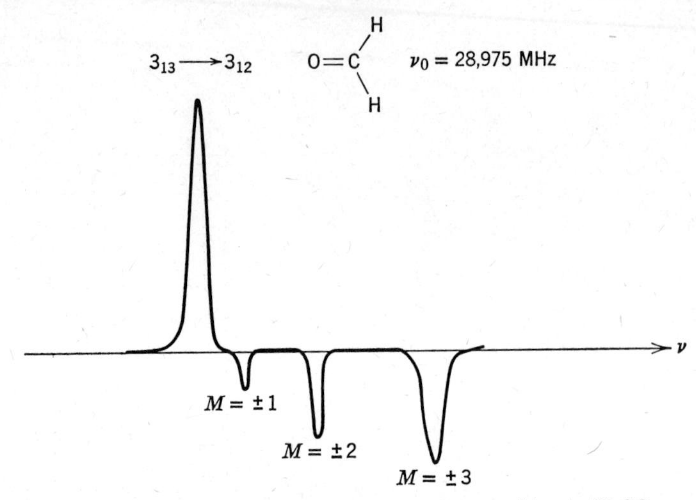

Fig. 5.12 The Stark effect in the $3_{13} \rightarrow 3_{12}$ transition in H_2CO.

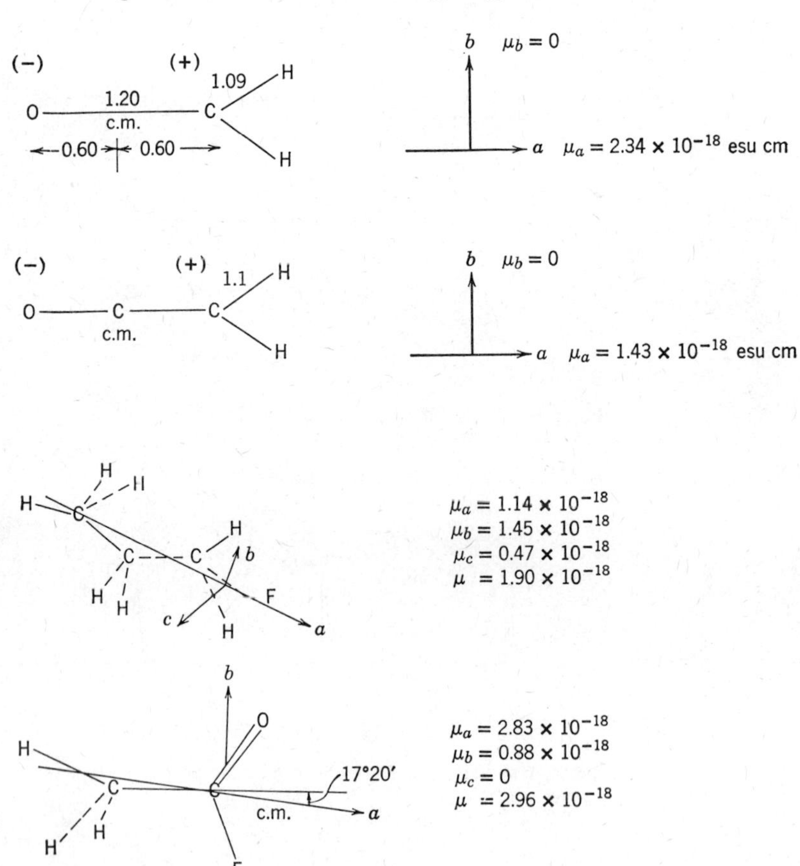

Fig. 5.13 Molecular dipole moments in some typical molecules.

Many molecular dipole moments have been measured by the Stark effect in microwave spectroscopy and many of these data have been tabulated. A few illustrative examples are shown in Fig. 5.13.

The Stark effect for $K \neq 0$ symmetric top transitions shows a first-order Stark effect with the first-order energy given by:

$$E^{(1)} = -\frac{\mu EMK}{J(J+1)}, \tag{5.18}$$

where μ is the dipole moment along the symmetry axis. Thus the Stark effect when $K \neq 0$ is symmetrically distributed about the zero-field transition.

In summary, the Stark effect is an effective tool in identifying the rotational transition, and in many cases an understanding of the Stark effect is essential for the assignment of the microwave spectrum. The Stark effect also leads to very accurately measured molecular dipole moments.

5 MOLECULAR STRUCTURES

We consider here a molecular structure as the definition of the relative positions of the point mass nuclei in the molecule. If the molecule were a truely rigid rotor, we would merely have to collect as many independent moments of inertia as structural unknowns in order to obtain a complete molecular structure. For instance, there are five structural parameters in the *cis*-1,2-difluoroethylene molecule (d_{CF}, d_{CH}, d_{CC}, $<$ CCF, and $<$ HCH). Thus in principle a complete structure can be obtained from the six moments of inertia of two isotopic species. Even though the structure obtained by these methods is quite good, it has a considerable number of errors attributable to vibration-rotation interactions. It has been shown that the best structure or near-equilibrium structure for the ground vibrational state is obtained by a complete substitutional structure [14]. That is, the best structure is obtained by measuring, by isotopic substitution, the coordinates of each atom in the molecule from the main species center of mass. The coordinates to the isotopically substituted atom n from the original isotopic species center of mass are given (along the principal inertial x-axis) by [15]:

$$\begin{aligned}|x| = \Bigg\{ &\frac{1}{2\mu}[(I'_y - I_y) + (I'_z - I_z) - (I'_x - I_x)] \\ &\times \left[1 + \frac{(I'_x - I_x) - (I'_y - I_y) + (I'_z - I_z)}{2(I_x - I_y)}\right] \\ &\times \left[1 + \frac{(I'_x - I_x) + (I'_y - I_y) - (I'_z - I_z)}{2(I_x - I_z)}\right]\Bigg\}^{1/2}. \end{aligned} \tag{5.19}$$

Several complete substitutional structures have been determined which include isotopic substitution of each atom in the molecule. A few of the more complex of these studies include fluorobenzene, in which it was found that the fluorine atom had a negligible effect on the ring structure. Other molecules include pyridine, furan, formic acid, formaldehyde, and many others.

Many times it is impractical or impossible to obtain complete isotopic substitution for each atom. For instance, the fluorine atom has only a single stable isotope. However, good structures can still be obtained from a careful interpretation of the moments of inertia.

One of the more useful initial structural determinations available from a single isotopic species is the determination of molecular planarity. If all of the atoms of a molecule are in the plane containing the principal inertial a- and b-axes, the difference in moments given below should be zero:

$$I_{aa} + I_{bb} - I_{cc} = 2 \sum_n Z_n c_n^2, \tag{5.20}$$

where c_n is the out-of-plane mass coordinate. The values of $I_{aa} + I_{bb} - I_{cc}$ for several molecules are listed in Table 5.8 [4]. Thus an initial assignment shows the general conformation of the molecule (see below).

It is evident from Table 5.8 that the moment differences are incremented in steps of about 3.3 amu Å² for each pair of out-of-plane protons.

Several interesting and reactive molecules have been observed initially by microwave spectroscopy. Also, several interesting structural problems have

Table 5.8 Differences in the Moments of Inertia in Several Planar Molecules and Molecules Containing Out-of-Plane Protons*

Molecule	$I_{aa} + I_{bb} - I_{cc}$ (amu Å²)	*Number of Out-of-Plane Protons*	*Reference*
trans-Acrolein	0.0212	0	[5]
1,2,5-Oxadiazole	0.0707	0	[16]
Furan	−0.0481	0	[17]
Cyclopropene	3.31	2	[18]
Cyclopentadiene	3.11	2	[19]
Ethylene oxide	6.83	4	[20]
β-Propiolactone	6.50	4	[21]
Cyclopropanone	6.70	4	[4]
Cyclobutanone	9.98	6	[22]

* Note that these molecules are all typical near-prolate symmetric tops.

been solved by microwave spectroscopy. We list here a few selected examples in order to illustrate the scope of work. Two isomers of the S_2F_2 molecule were recently identified [23]:

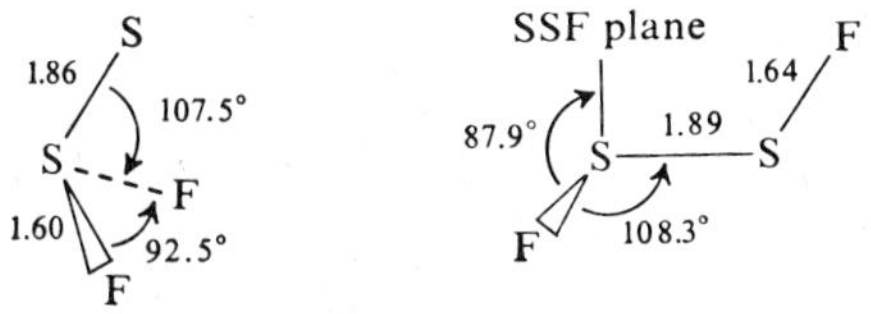

All fluorine atoms in the SF_4 molecule are on one side of the sulfur atom [24]:

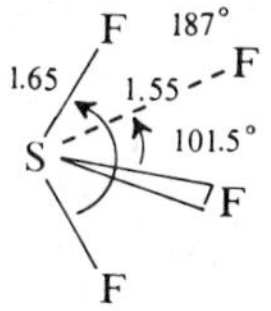

S_2O [25] and NSF [26] have also been identified and have the similar structures given by:

The reactive SiF_2 fragment was also identified [28]:

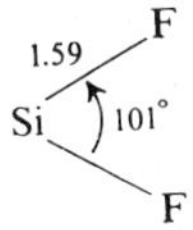

Numerous other examples of inorganic compounds could be sited. However, there are still a large number of molecules that could be studied by microwave spectroscopy. For instance, there are a number of substituted phosphorus compounds that should be examined.

Several organic-type molecular structures have also been identified by microwave spectroscopy. Cyclopropanone was recently identified and found to have a normal ring-closed structure [4]:

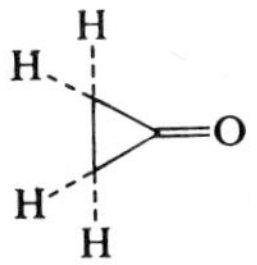

This work on cyclopropanone was interesting, as a reliable molecular orbital calculation indicated that the ring-closed form was unstable. Diazirine was also found to have a ring-closed structure [28]:

Many more structures are discussed in later sections. However, hundreds of structures have been determined by microwave spectroscopy and our aim here is merely to indicate the methods used in extracting the information from the spectra and not to attempt an exhaustive review of the actual structures.

6 MOLECULAR ROTATION AND THE EFFECTS OF VIBRATION

Vibration-rotation interactions are not discussed in detail here. However, it is well known that no molecule is a completely rigid rotor and that significant corrections attributable to centrifugal distortion must be made to the rigid-rotor high-J rotational states. Centrifugal distortion is the change in molecular structural parameters with the increased speed of rotation. The effects of centrifugal distortion are evident in the unequal spacings of R-branch transitions in a linear molecule. The rigid-rotor ideal spacings are shown in Fig. 5.2. Similarly, the different K-states in the $\Delta K = 0$, $\Delta J = 1$ transitions of the symmetric top are separated in frequency because of centrifugal distortion. Thus the $J = 2 \rightarrow J = 3$ symmetric top transition shown in Fig. 5.2 as three overlapping $\Delta K = 0$, $|K| = 0$, 1, and 2 transitions is actually split into three distinct lines because of the centrifugal distortion. The relative intensities go as $(J + 1)^2 - K^2$, where J and K are for the lower J-transition. Centrifugal distortion constants have been measured for several linear, symmetric top, and asymmetric tops, and a complete analysis can lead to the vibrational frequencies of the molecule. A complete study of OF_2, including rotational transitions in states above $J = 30$, has yielded vibrational frequencies in good agreement with those observed directly in the infrared [29]. Thus the vibration-rotation interactions present in molecules can be studied to yield information about the vibrational potential function. However, we now turn to a study of rigid rotors in different vibrational states or different rotational conformers.

Vibrational Satellites

Vibrational satellites are observed in microwave spectra as the rigid-rotor spectra of molecules in excited vibrational states. For instance, the spectra for the idealized isotopic relative intensities in Fig. 5.3 are greatly complicated by the spectra of excited vibrational states. The vibrational energy spacings of OCS are approximately $v_1 = 800\ \text{cm}^{-1}$ (symmetric stretch), $v_2 = 2100\ \text{cm}^{-1}$ (asymmetric stretch), and $v_3 = 500\ \text{cm}^{-1}$ (bend). Thus at 300°K the vibrational satellite for the first excited state of v_1 has a relative intensity:

$$\frac{I(v_1 = 1)}{I(v_1 = 0)} = e^{-4} \sim \tfrac{1}{55},$$

and many of the vibrational satellites are more intense than the lower-abundance isotopic species. As the molecular structures are slightly different in the excited vibrational states, the spectra are scattered near the ground-state transition. Thus extreme care must be exercised in assigning isotopic spectra. The temperature dependence of the spectra usually identifies a vibrational satellite from a ground-state transition. As the temperature is lowered, the vibrational satellite loses intensity and a ground-state transition gains intensity.

One of the more successful applications of the study of vibrational satellites is the study of low-frequency bending motion in small-ring compounds. Molecules such as the four-membered ring compounds trimethylene oxide, trimethylene sulfide, methylenecyclobutane, cyclobutanone, and 1,1-difluorobutane all possess double minimum potential functions describing the out-of-plane vibrational potential function. It appears that these potential functions are best fit with a quartic minus a quadratic potential function:

$$V(Q) = kQ^4 - k'Q^2,$$

where k is the quartic constant and k' is the quadratic constant. For instance, the potential function for 1,1-difluorobutane is shown in Fig. 5.14 [30]. Rotational assignments are available for the first seven vibrational states in this molecule. There are actually four levels below the top of the barrier (barrier height is $241\ \text{cm}^{-1}$) with the lowest levels split by only $0.062\ \text{cm}^{-1} = 1.86$ GHz, and the splitting between the next two levels is $3.87\ \text{cm}^{-1}$. Figure 5.14 indicates that the skeletal structure is not planar in the lowest four levels. This relatively low barrier has also been observed in the other four-membered rings mentioned above. However, cyclobutene, β-propionlactone, and cyclopentadiene were found to be essentially planar.

Five-membered ring systems have additional complications as the ring-bending motion moves around the ring, giving rise to a psuedorotational motion. However, analysis of the microwave spectrum of several five-membered rings does lead to a satisfactory fit of the microwave spectrum.

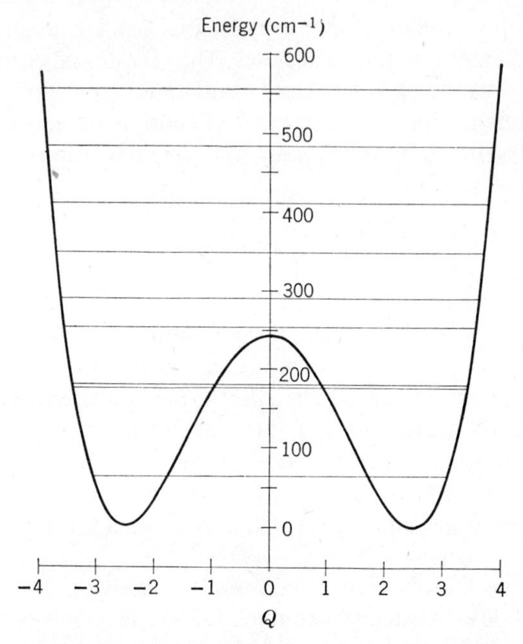

Fig. 5.14 Potential function for the ring-bending mode in 1,1-difluorobutane. The potential function is $V(Q) = 6.184Q^4 - 77.30Q^2$.

Conformational Analysis

Conformational analysis is the identification of various rotamers or conformers of the same molecule. For instance, fluoroacetyl fluoride was found in two forms [31]:

trans *cis*

The rotational constants clearly identify the two forms and relative intensities between transitions for the two forms show that the *cis* form is higher in energy than the *trans* form by 910 ± 100 cal/mole. Both *cis* and *trans* forms are found, which indicates a dominant V_2 potential term. The addition of a positive V_3-term gives a wider potential for the higher-energy *cis* conformer and a lower-energy narrow potential for the *trans* conformer. The double bonds in the isoprene molecule were found to be *trans* by a combination of rotational constants and measurement of the barrier to internal rotation [32, 33].

```
           CH3    H
             \   /
   H          C=C
    \        /   \
     C=C         H
    /   \
   H     H
```

This is similar to the same *trans* conformation of acrolein [5]:

```
          H
           \
   H        C=O
    \      /
     C=C
    /   \
   H     H
```

The *cis* forms of these molecules, butadiene, and 1,1-difluorobutadiene have all avoided detection in spite of the very distinct possibility of a higher energy minimum in the *cis* form. However, acrylyl fluoride was found in two forms separated by only 90 ± 100 cal/mole in energy [34]:

```
          F                     O
           \                     \\
   H        C=O          H        C—F
    \      /              \      /
     C=C                   C=C
    /   \                 /   \
   H     H               H     H
     s-trans                s-cis
```

The *s-trans* is the lower-energy form. Further work on this series of molecules would be most interesting. For instance, the following molecule has four conformers

```
          H                     O
           \                     \\
   F        C=O          F        C—H
    \      /              \      /
     C=C                   C=C
    /   \                 /   \
   H     H               H     H

          H                     O
           \                     \\
   H        C=O          H        C—H
    \      /              \      /
     C=C                   C=C
    /   \                 /   \
   F     H               F     H
```

It would be interesting to study the energy differences in these conformers. As we know so little about the forces involved in determining which of the above forms has the lowest or highest energies, it would be very useful to have data on a controlled system of molecules. Also of interest would be an extension of the above-mentioned work to five-membered chains such as

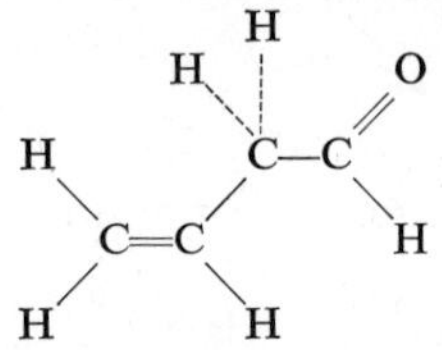

It would be interesting to see whether or not all the heavy atoms prefer to stay in the molecular plane. This would determine whether or not the π-delocalization stabilization in the butadienes passes through the central CH_2 group in the molecule shown. Other work that illustrates the utility of microwave spectroscopy in determining conformations is the work on propionaldehyde [35], in which the *cis* form was observed to be 900 cal/mole lower in energy than the *quache* form:

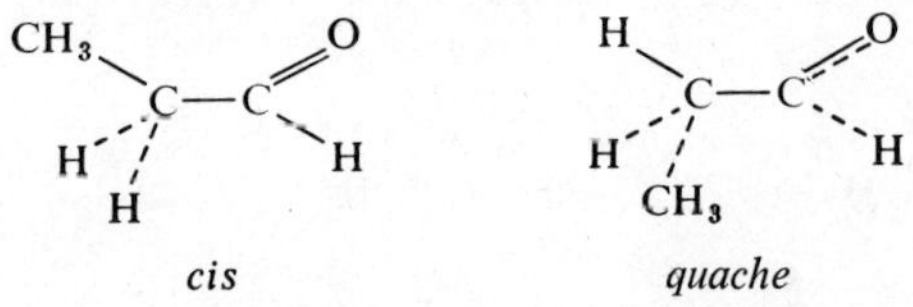

cis *quache*

The *cis* and *qauche* forms of 3-fluoropropene were identified [36]:

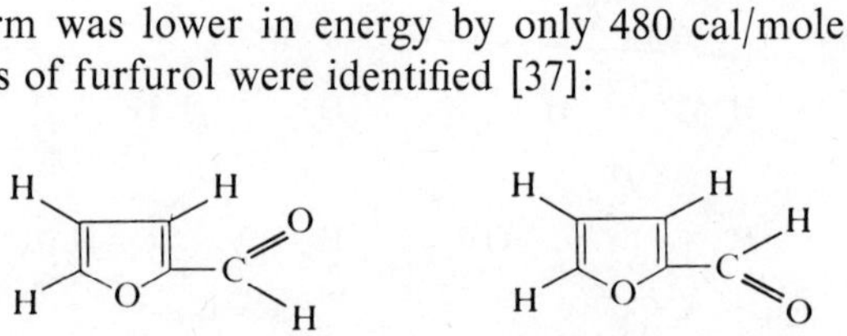

cis *quache*

and the *cis* form was lower in energy by only 480 cal/mole. Both the *cis* and *trans* forms of furfurol were identified [37]:

cis *trans*

Several other examples could be given here to illustrate conformational analysis. However, the work in this field is just starting and there are endless interesting examples that could be studied by microwave spectroscopy.

It also appears possible in the next few years that the kinetics of a unimolecular conformational change will be studied by microwave spectroscopy.

Internal Rotation

Internal rotation is the coupling between the overall molecular rotation with the internal rotation about a single bond. The barriers of methyl groups have been most thoroughly studied. The Hamiltonian describing this coupling is:

$$H = H_r + F(\mathbf{p} - \mathbf{P})^2 + V(\alpha). \tag{5.21}$$

H_r is the rigid-rotor Hamiltonian given in (5.1); **p** is the angular momentum of the internal rotor and **P** is the overall rotational angular momentum. The inverse of the reduced moment of inertia is defined as $F = \hbar^2/(2rI_\alpha)$ and $V(\alpha)$ is the potential, which is a function of the torsional angle α. $V(\alpha)$ is given by:

$$V(\alpha) = \frac{1}{2}\sum_n V_n(1 - \cos n\,\alpha). \tag{5.22}$$

An illustration of the use of this potential function is shown in Figs. 5.15 to 5.17.

The rotational energy levels are affected primarily through the coupling of the total angular momentum of the internal top **p** along its symmetry axis and the angular momentum of the frame, **P**. That is, the effect on the rotational energy levels is attributable to the cross term 2F**pP** in the $F(\mathbf{p} - \mathbf{P})^2$ term in the Hamiltonian. This Hamiltonian gives an energy matrix which is nondiagonal in the torsional quantum number v; these nondiagonal elements arise from the (−2F**Pp**) cross term. Formally, successive Van Vleck transformations can be applied which will reduce to negligible values those matrix elements that are nondiagonal in v. The transformed Hamiltonian matrix can then be factored into smaller rotational matrices $H_{v\sigma}$, one for each torsional state. These may be written [38]:

$$H_{v\sigma} = H_r + f\sum_n W_{v\sigma}^{(n)} P^n, \tag{5.23}$$

where $W_{v\sigma}^{(n)}$ are perturbation coefficients which may be obtained from Herschbach's tabulation of $W_{v\sigma}^{(n)}$ as a function of the reduced barrier parameter $s = 4.66139\ (V_3/F)$ [38]. V_3 is the barrier height expressed in calories per mole and F is expressed in gigahertz. The P^n matrix elements up to $n = 4$ are also tabulated in Herschbach's paper.

Therefore to calculate the effect of internal rotation on the rotational spectrum of a molecule with a high barrier (> 500 cal/mole), one must: (1) determine F from assumed structural parameters, (2) estimate the barrier height V_3 and obtain the necessary $W_{v\sigma}^{(n)}$-coefficients, (3) calculate the P^n-matrix for as high an n as necessary, (4) compute the $H_{v\sigma} = H_r + \sum_n W_{v\sigma}^{(n)} P^n$ matrix elements in a symmetric top basis, (5) diagonalize the result, and (6) calculate the spectra using the $J = 0, \pm 1, \Delta\sigma = 0$ selection rules.

The barriers to internal rotation have been measured for a large number of methyl groups [39]. Analysis of these barriers indicates the existence of an intrinsic barrier for a characteristic linkage as suggested by Wilson [40]. For instance, the intrinsic barrier of the substituted acetaldehydes should be taken as the barrier in acetaldehyde of 1150 cal/mole. The three fold V_3-barrier is illustrated in Fig. 5.15. The intrinsic barrier in the substituted propylenes should be taken as the higher value of 2000 cal/mole as in propylene. The additional effects on the barrier as a result of the substitutions lead to smaller effects. However, there are certain trends. Substitution on the same carbon to which the methyl rotor is attached appears to have the same effect as the corresponding substitution in ethane. In this same light it is interesting

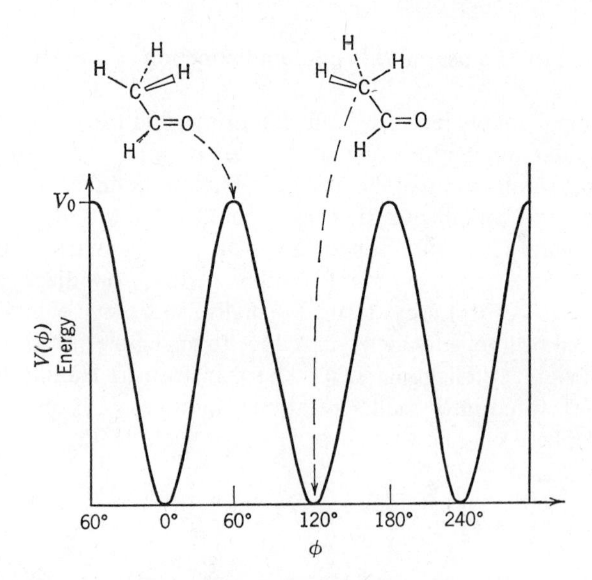

Fig. 5.15 The threefold V_3-potential function in acetaldehyde. The potential minima is where the methyl group proton is eclipsed with the carbonyl double bond. The same lower-energy methyl conformation is observed in propylene.

to note that differences in the barriers between the series of molecules:

CH_3 H
C=C
R H

and

CH_3
C=O
R

where R = F, Cl, Br, CH_3, and $H_2C{=}CH$ are all about 1400 cal/mole with the larger barriers in the substituted propylenes [41, 33]. Furthermore,

Table 5.9 Barrier Changes in the Halosubstituted Propylenes*

Molecule	*Barrier* (cal/mole)	*Change* (cal/mole)
CH_3 H C=C H H	2000	0
CH_3 H C=C F H	2440	+440
CH_3 H C=C H F	2150	+150
CH_3 F C=C H H	1057	−943
CH_3 H C=C Cl H	2671	+671
CH_3 H C=C H Cl	2170	+170
CH_3 Cl C=C H H	620	−1380

* See Ref. 42.

substitution in the *trans*-1-position of propylene:

$$\begin{array}{ccc} CH_3 & & H \\ & C{=}C & \\ H & & R \end{array}$$

has very little effect on the barrier. Substitution in the *cis*-1-position leads to a substantial lowering of the methyl group barrier. Double halosubstitution in the propylenes has been treated by concepts of additivity and the intrinsic barrier [42]. The barrier changes for fluoro- and chlorosubstitution are shown in Table 5.9.

The barriers for the doubly and triply substituted propylenes can be computed by summing the increments given in Table 5.9.

In some molecules the value of V_3 is zero by symmetry leading to a dominant V_6-barrier. This case is illustrated for toluene in Fig. 5.16. The V_6-barriers are normally much smaller than the V_3-barriers. V_6 in toluene and parafluorotoluene are both 14 cal/mole [43, 44].

In some cases both V_3 and V_6 have been measured for the same molecule. The combination of $V_3 = 2000$ cal/mole and $V_6 = -40$ cal/mole is illustrated in Fig. 5.17 [45]. The opposite signs for V_3 and V_6 lead to a sharpening of the top of the potential barrier and a flattening of the minima. The sharper maxima probably arise as a result of repulsion between the protons on the methyl group and the adjacent vinyl carbon proton. The opposite relative signs for V_3 and V_6 have been observed in 1-methylcyclopropene [46].

There are many interesting trends that could and should be pursued experimentally in the measurement of the barrier to internal rotation. Active theoretical work is also being performed in this area [41].

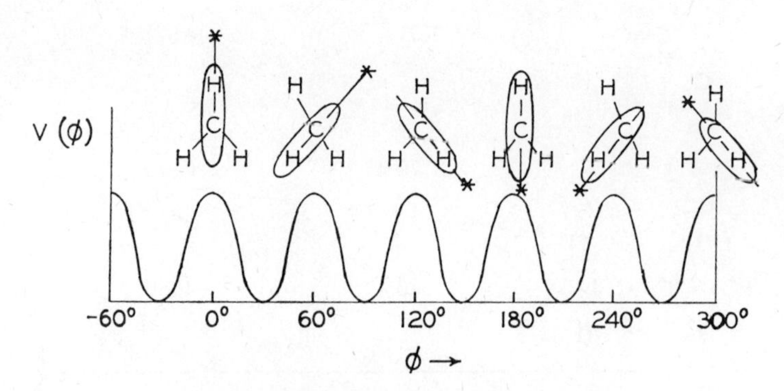

Fig. 5.16 The sixfold V_6-potential in toluene.

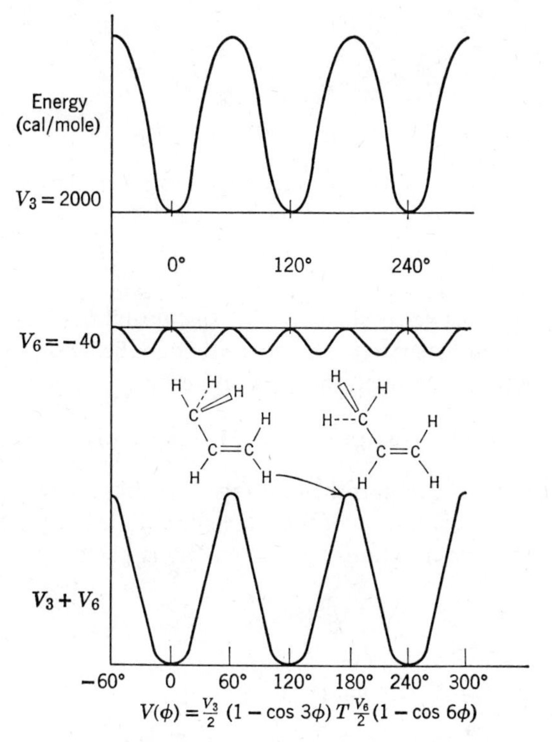

Fig. 5.17 The combination of the $V_3 = 2000$ cal/mole and $V_6 = -40$ cal/mole potential functions in propylene.

7 MOLECULAR ZEEMAN EFFECT

The energy of a molecule in the presence of a magnetic field H can be described by a Taylor expansion about the zero field given by:

$$E = E^0 + \sum_{\alpha=x,y,z} \left(\frac{\partial E}{\partial H_\alpha}\right)_{H_\alpha=0} H_\alpha + \frac{1}{2} \sum_{\alpha,\beta=x,y,z} \left(\frac{\partial^2 E}{\partial H_\alpha \, \partial H_\beta}\right)_{H_\alpha=0, H_\beta=0} H_\alpha H_\beta + \cdots \tag{5.24}$$

E^0 is the zero-field energy. Making the definitions for the *molecular magnetic dipole moment*:

$$\mu_\alpha = -\left(\frac{\partial E}{\partial H_\alpha}\right)_0 \tag{5.25}$$

and the *molecular magnetic susceptibility*:

$$\chi_{\alpha\beta} = -\left(\frac{\partial^2 E}{\partial H_\alpha \, \partial H_\beta}\right)_0 \tag{5.26}$$

gives

$$E = E^0 - \sum_{\alpha} \mu_\alpha H_\alpha - \tfrac{1}{2}\sum_{\alpha,\beta} H_\alpha \chi_{\alpha\beta} H_\beta + \cdots$$
$$= E^0 - \boldsymbol{\mu} \cdot H - \tfrac{1}{2} H \cdot \boldsymbol{\chi} \cdot H + \cdots. \tag{5.27}$$

We now define the magnetic dipole moment in terms of the molecular g-value tensor $\mathbf{g}$ and the rotational angular momentum:

$$\boldsymbol{\mu} = \mathbf{J} \cdot \mathbf{g}. \tag{5.28}$$

Furthermore, we must relate the magnetic susceptibilities and g-values in the molecular fixed-axis system to the magnetic field in the laboratory axis system. This transformation is through the direction cosines:

$$E = E^0 - (\mathbf{J} \cdot \mathbf{g})\tilde{\Phi} \cdot \mathbf{H} - \tfrac{1}{2}\mathbf{H} \cdot \tilde{\Phi}\boldsymbol{\chi}\Phi \cdot \mathbf{H} + \cdots. \tag{5.29}$$

The above energy expression is then averaged, to first order, using the $|J, M\rangle$ basis describing the orientation of the molecule with respect to the laboratory fixed-axis system. The resultant Zeeman energy is [47]:

$$E(J, M_J) = -\tfrac{1}{2}\chi H^2 - \mu_0 \frac{M_J H}{J(J+1)} \sum_g g_{gg} \langle J_g{}^2 \rangle$$
$$- H^2 \left[\frac{3M_J{}^2 - J(J+1)}{(2J-1)(2J+3)}\right] \times \left[\frac{1}{J(J+1)}\right] \sum_g (\chi_{gg} - \chi)\langle J_g{}^2 \rangle. \tag{5.30}$$

$\chi = \frac{1}{3}(\chi_{aa} + \chi_{bb} + \chi_{cc})$ is the average magnetic susceptibility, with χ_{aa}, χ_{bb}, and χ_{cc} being the components along the principal inertial axes in the molecule. H is the external magnetic field, μ_0 the nuclear magneton, J and M_J the rotational quantum numbers, and g_{gg} the molecular g-value along the gth principal inertial axis given by:

$$g_{xx} = \frac{M_p}{I_{xx} m}\left[m \sum_n Z_n (r_n{}^2 - x_n{}^2) + 2 \sum_{k>0} \frac{|\langle 0 | L_x | k \rangle|^2}{E_0 - E_k}\right], \tag{5.31}$$

where M_p and m are the proton and electron masses, I_{xx} is the principal moment of inertia, Z_n is the atomic number of the nth nucleus at distance r_n and projection x_n from the molecular center of mass, L_x is the electronic angular momentum operator, and the sum is over all excited electronic states $|k\rangle$ with energy E_k. The magnetic susceptibility terms χ_{xx} are defined later. $\langle J_g{}^2 \rangle$ is the average value of the squared rotational angular momentum (in units of $\hbar^2$) along the gth principal inertial axis. The $-(\chi H^2)/2$ term cancels out in our observation of an energy difference. Thus we can measure the absolute values of the three g-values and two independent magnetic anisotropy parameters. The magnetic anisotropy components are:

$$\begin{aligned}\chi_{aa} - \chi &= \tfrac{1}{3}(2\chi_{aa} - \chi_{bb} - \chi_{cc})\\ \chi_{bb} - \chi &= \tfrac{1}{3}(-\chi_{aa} + 2\chi_{bb} - \chi_{cc})\\ \chi_{cc} - \chi &= \tfrac{1}{3}(-\chi_{aa} - \chi_{bb} + 2\chi_{cc}).\end{aligned} \tag{5.32}$$

Only two of these equations are independent. The third anisotropy component is the negative sum of the first two as the trace is zero.

The values of $\langle J_a^2\rangle$, $\langle J_b^2\rangle$, $\langle J_c^2\rangle$ are evaluated by standard techniques using the rotational assignment for the molecule.

The values of $\langle J_a^2\rangle$, $\langle J_b^2\rangle$, and $\langle J_c^2\rangle$ are easily evaluated for symmetric tops or linear molecules in the $|J, K, M\rangle$ basis:

$$\begin{aligned}\langle J_a^2\rangle &= \langle J_b^2\rangle = \tfrac{1}{2}[J(J+1) - K^2]\\ \langle J_c^2\rangle &= K^2.\end{aligned} \tag{5.33}$$

$K = 0$ for a linear molecule, which reduces (5.30) to:

$$E(J, M) = 2BJ(J+1) - \mu_0 M g_{aa} H - \frac{H^2}{3}\left[\frac{3M^2 - J(J+1)}{(2J-1)(2J+3)}\right](\chi_{aa} - \chi_{bb}), \tag{5.34}$$

where the b-axis is the internuclear axis. The frequencies for the $\Delta J = +1$ transitions of the linear molecule are:

$\Delta M = 0$ *selection rules:*

$$\nu = \nu^0 + \frac{2H^2}{15h}(\chi_{aa} - \chi_{bb}) \tag{5.35}$$

$\Delta M = \pm 1$ *selection rules:*

$$\nu = \nu^0 \pm \frac{\mu_0 g_{aa} H}{h} - \frac{H^2}{15h}(\chi_{aa} - \chi_{bb}). \tag{5.36}$$

The linear molecule's energy levels involved in the $J = 0 \rightarrow J = 1$ transition are shown in Fig. 5.18. Thus the values of g_{aa} and $(\chi_{aa} - \chi_{bb})$ are easily extracted from the $J = 0 \rightarrow J = 1$ spectra. The linear molecule's energy levels involved in the $J = 1 \rightarrow J = 2$ transition are illustrated in Fig. 5.19, and the $J = 1 \rightarrow J = 2$, $\Delta M = 0$ transition in $^{16}O^{12}C^{32}S$ is shown in Fig. 5.20. The values of g_{aa} and $(\chi_{aa} - \chi_{bb})$ obtained from these types of experiments on OCS [48], NNO [49], and H–C≡C—F [50] are listed in Table 5.10.

The molecular quadrupole moments can be computed directly from the molecular g-values and magnetic susceptibility anisotropies. The appropriate

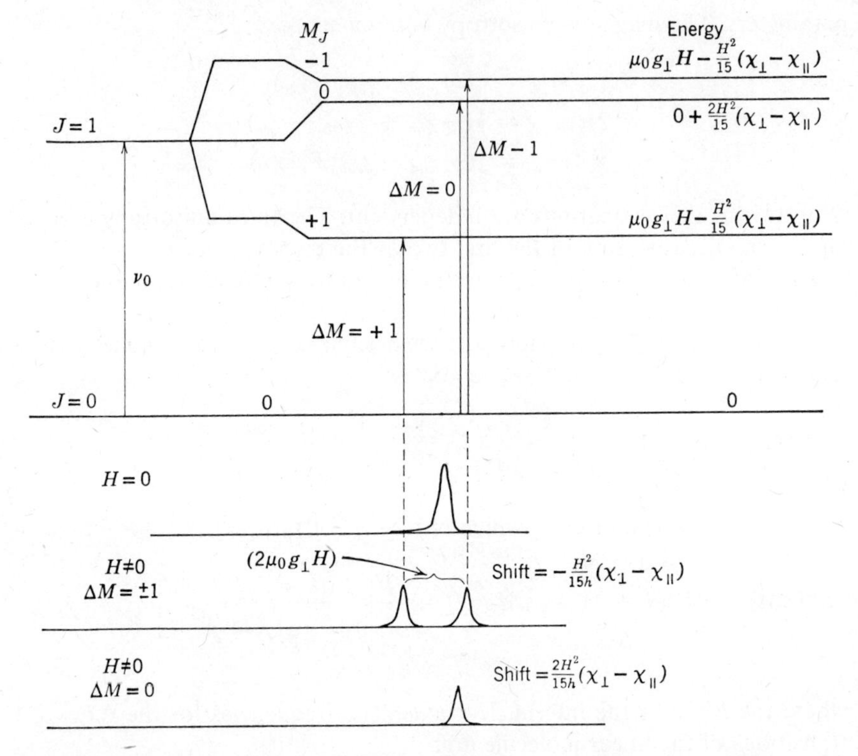

Fig. 5.18 The $J = 0$ and $J = 1$ Zeeman energy levels of a linear molecule in a high magnetic field.

equation is [51]:

$$Q_{zz} = \frac{|e|}{2} \sum_n Z_n(3z_n^2 - r_n^2) - \frac{|e|}{2} \langle 0 | \sum_i (3z_i^2 - r_i^2) | 0 \rangle$$

$$= -\frac{\hbar |e|}{8\pi M}\left[\frac{2g_{zz}}{G_{zz}} - \frac{g_{xx}}{G_{xx}} - \frac{g_{yy}}{G_{yy}}\right] - \frac{2mc^2}{|e|\, N}(2\chi_{zz} - \chi_{xx} - \chi_{yy}). \quad (5.37)$$

$|e|$ is the electronic charge, Z_n the charge on the nth nucleus, and z_n and z_i are the nuclear and electronic center-of-mass coordinates summed over all n nuclei and i electrons. $\langle 0 | \, | 0 \rangle$ indicates the ground electronic state average value. M is the proton mass, $\hbar$ is Planck's constant divided by 2π, G_{zz} is the rotational constant along the zth principal inertial axis, c is the speed of light, m is the electron mass, and N is Avogadro's number. Substituting the parameters for the linear molecules in Table 5.10 gives the molecular quadrupole moments also listed in the table.

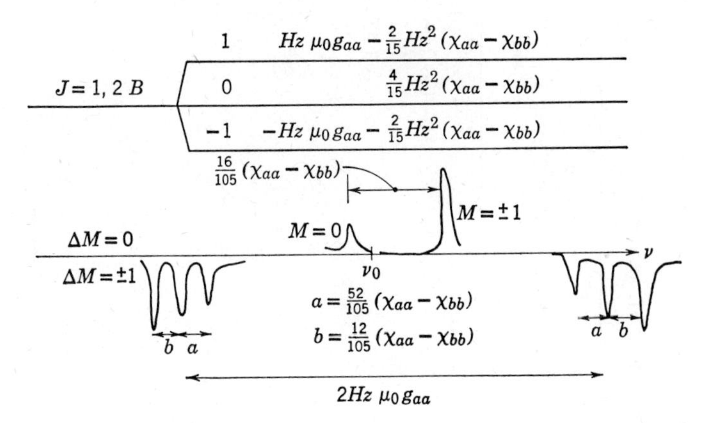

Fig. 5.19 The $J = 1 \rightarrow J = 2$ transition in a linear molecule in a high magnetic field.

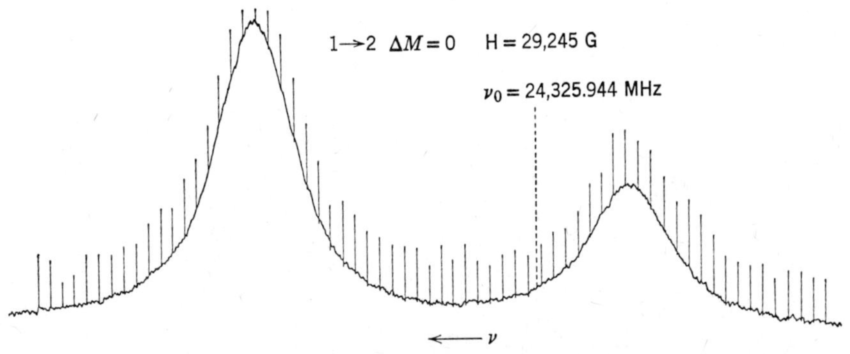

Fig. 5.20 The $J = 1 \rightarrow J = 2$, $\Delta M = 0$ transition in $^{16}O^{12}C^{32}S$ at $H = 29{,}245$ G. The splitting between the high-frequency, more intense $M = \pm 1$ and the weaker $M = 0$ lines is 305 kHz.

Table 5.10 Magnetic Zeeman Data for Some Linear Molecules

a ↑, → b	$^{16}O^{12}C^{32}S$	$^{15}N^{15}N^{16}O$	$H—^{12}C{\equiv}^{12}C—^{19}F$
g_{bb} (10^{-6}	0	0	0
$g_{aa} = g_{cc}$	-0.02871 ± 0.00004	-0.07606 ± 0.0001	-0.0077 ± 0.0002
$\chi_{aa} - \chi_{bb}$ (10^{-6} erg/G^2 mole)	9.27 ± 0.10	10.1 ± 0.15	5.2 ± 0.1
Q_{bb} (10^{-26} esu cm^2)	-0.88 ± 0.15	-3.65 ± 0.25	3.96 ± 0.14
$\sum_i Z_n b_n^2$ (10^{-16} cm^2)	41.495	20.256	28.341
$\langle 0 \vert \sum_i b_i^2 \vert 0 \rangle$ (10^{-16} cm^2)	46.5	23.9	31.0
$\langle 0 \vert \sum_i a_i^2 \vert 0 \rangle$ (10^{-16} cm^2)	4.5	3.0	3.5
χ_{aa} (10^{-6} erg/G^2 mole)	-29.3	-15.5	—
χ_{bb} (10^{-6} erg/G^2 mole)	-38.6	-25.6	—
χ_{aa}^p (10^{-6} erg/G^2 mole)	186.2	98.6	—
χ_{bb}^p (10^{-6} erg/G^2 mole)	0	0	—
χ_{aa}^d (10^{-6} erg/G^2 mole)	-215.5	-114.1	—
χ_{bb}^d (10^{-6} erg/G^2 mole)	-38.6	-25.6	—

The anisotropies in the center-of-mass average values of x^2, y^2, and z^2 for the electronic charge distribution are also directly available from the molecular Zeeman parameters and the known molecular structure.

The total magnetic susceptibility χ_{xx} along any axis is a sum of diamagnetic χ_{xx}^d and paramagnetic χ_{xx}^p components defined by:

$$\begin{aligned}
\chi_{xx} &= \chi_{xx}^p + \chi_{xx}^d \\
\chi_{xx}^d &= -\frac{e^2N}{4mc^2}\langle 0|\sum_i (y_i^2 + z_i^2)|0\rangle \\
\chi_{xx}^p &= -\frac{e^2N}{2m^2c^2}\sum_k \frac{|\langle 0|L_x|k\rangle|^2}{E_0 - E_k} \\
&= -\frac{e^2N}{2mc^2}\left[\frac{\hbar g_{xx}}{8\pi G_{xx}M} - \frac{1}{2}\sum_n Z_n(y_n^2 + z_n^2)\right].
\end{aligned} \tag{5.38}$$

We now define the average values of the second moment of the electronic charge distributions:

$$\begin{aligned}
\langle x^2\rangle &= \langle 0|\sum_i x_i^2|0\rangle \\
\langle y^2\rangle &= \langle 0|\sum_i y_i^2|0\rangle \\
\langle z^2\rangle &= \langle 0|\sum_i z_i^2|0\rangle.
\end{aligned} \tag{5.39}$$

Returning to (5.37), we can relate the anisotropies of the second moments in (5.39) to the observable g-values, molecular structure, and magnetic susceptibility anisotropy. The appropriate equation is:

$$\begin{aligned}
\langle y^2\rangle - \langle x^2\rangle &= \sum_n Z_n(y_n^2 - x_n^2) + \frac{\hbar}{4\pi M}\left(\frac{g_{yy}}{G_{yy}} - \frac{g_{xx}}{G_{xx}}\right) \\
&\quad + \frac{4mc^2}{3e^2N}[(2\chi_{yy} - \chi_{xx} - \chi_{zz}) - (2\chi_{xx} - \chi_{yy} - \chi_{zz})].
\end{aligned} \tag{5.40}$$

Each of the individual elements of the second moment of the electronic charge distribution can be obtained by using the bulk magnetic susceptibility:

$$\chi = \tfrac{1}{3}(\chi_{aa} + \chi_{bb} + \chi_{cc}).$$

The above bulk values can be combined with the experimental anisotropies to yield the individual components.

We can now compute the individual elements in the second moment of the electronic charge distribution given by:

$$\begin{aligned}
\langle 0|\sum_i x_i^2|0\rangle = \langle x^2\rangle &= -\frac{2mc^2}{e^2N}[\chi_{yy}^d + \chi_{zz}^d - \chi_{xx}^d] \\
&= -\frac{2mc^2}{e^2N}[(\chi_{yy} + \chi_{zz} - \chi_{xx}) - (\chi_{yy}^p + \chi_{zz}^p - \chi_{xx}^p)]
\end{aligned} \tag{5.41}$$

Table 5.11 Zeeman Parameters in Some Symmetric Top Molecules

Molecule	$g_{\perp}$	$g_{\parallel}$	$(\chi_{\perp} - \chi_{\parallel})$ (10^{-6} erg/G^2 mole)	$Q_{\parallel}$ (10^{-26} esu cm^2)	*Reference*
$^{+}H_3C—C{\equiv}C—H^{-}$	$+0.00347 \pm 0.0001$	0.310	7.4 ± 0.5	$+4.5 \pm 0.6$	[52]
$^{+}H_3CCN^{-}$	-0.0338 ± 0.0008	0.310	10.5 ± 0.5	-1.8 ± 1.3	[52]
H_3CNC	-0.0546 ± 0.0005	0.310	13.5 ± 0.7	-2.7 ± 2.0	[52]
$H_3C^{127}I$	-0.00655 ± 0.00025	0.310	11.0 ± 0.6	5.5 ± 1.0	[52]
$H_3C^{79}Br$	-0.00544 ± 0.00010	0.310	8.5 ± 0.4	3.5 ± 0.8	[52]
$H_3C^{35}Cl$	-0.0165 ± 0.00010	0.310	7.8 ± 0.4	1.1 ± 0.8	[52]
H_3N	0.563 ± 0.001	0.500 ± 0.001	0 ± 0.9	-1.9 ± 1.0	[53]
PF_3	-0.0659 ± 0.001	0.0810 ± 0.002	1.3 ± 0.3	$+24.1 \pm 3.1$	[52]

Rewriting (5.41) in terms of the molecular g-values and molecular structure [see (5.38)] gives:

$$\langle x^2 \rangle = -\frac{2mc^2}{e^2N}\left[(\chi_{yy}+\chi_{zz}-\chi_{xx})+\sum_n Z_n x_n^{\,2}\right]-\frac{\hbar}{8\pi M}\left[\frac{g_{yy}}{G_{yy}}+\frac{g_{zz}}{G_{zz}}-\frac{g_{xx}}{G_{xx}}\right]. \tag{5.42}$$

The resultant magnetic susceptibilities for the linear molecules are listed in Table 5.10. Also listed are the nuclear and electronic second moments of the charge distributions.

The Zeeman spectra of symmetric tops is somewhat more complicated than a linear molecule as there are two nonzero molecular g-values. A few recent results in symmetric tops are listed in Table 5.11. H_3CCl, H_3CBr, and H_3CI have strongly coupled nuclei (see Section 8), which make the interpretation of the spectra difficult. However, the theory has been developed [47] and the results have been obtained. An intermediate case is the weakly coupled ^{14}N nucleus. The zero and high-field spectra of H_3CCN are shown in Fig. 5.21 [52]. The high-field $\Delta M_J = \pm 1$ trace shows the near complete uncoupling to the M_J- and M_I-states.

The Zeeman spectra of asymmetric tops is more difficult to interpret,

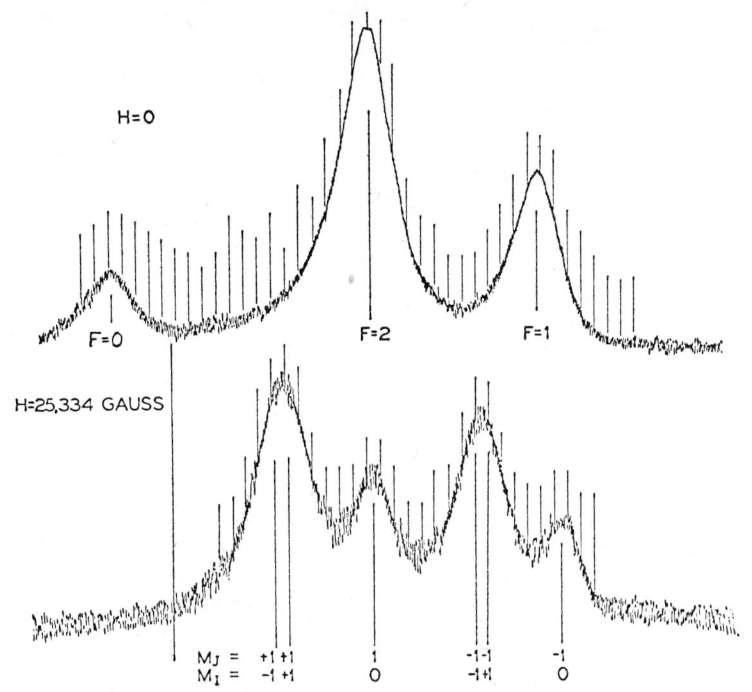

Fig. 5.21 The zero- and high-field $J = 0 \rightarrow J = 1$, $\Delta K = 0$ spectra of H_3CCN.

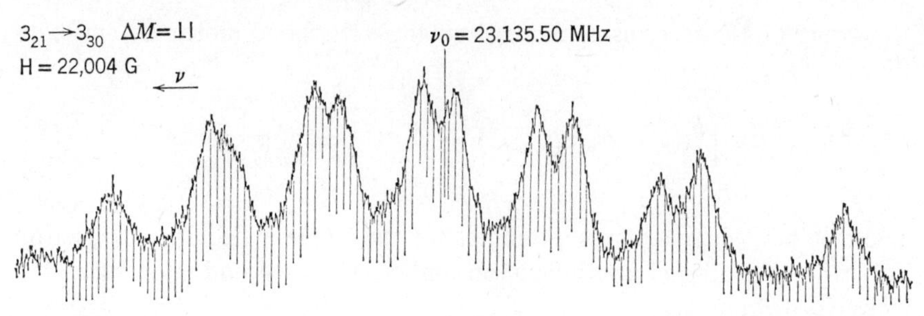

Fig. 5.22 The $3_{21} \rightarrow 3_{30}$ $\Delta M = \pm 1$ ethylene oxide spectra at high field. The markers are every 50 kHz.

however, a few simple patterns are expected from first-order spectra [53]: $\Delta J = 0$, $\Delta M = 0$, $2J$ components; $\Delta J = 0$, $\Delta M = \pm 1$, $4J$ components; $\Delta J = +1$, $\Delta M = 0$, $2J + 1$ components; $\Delta J = +1$, $\Delta M = 0$, $2J + 1$ components; $\Delta J = +1$, $\Delta M = +1$, $2(2J + 1)$ components.

The $3_{21} \rightarrow 3_{30}$ $\Delta M = \pm 1$ spectrum in ethylene oxide shown in Fig. 5.22 [54]. Note that the 12 expected components are visible and considerable asymmetry or second-order character is present. The $3_{12} \rightarrow 3_{13}$ $\Delta M = 0$ transition in formic acid is shown in Fig. 5.23 [55]. Again the expected six components are visible with considerable asymmetry leading to both the first- and second-order Zeeman parameters.

Complete Zeeman data for furan and thiophene are listed in Table 5.12 [56]. Table 5.13 lists the molecular quadrupole moments and the second moment of the out-of-plane electron distribution for several ring compounds. Table 5.14 lists similar data for some other molecules. Table 5.15 lists the

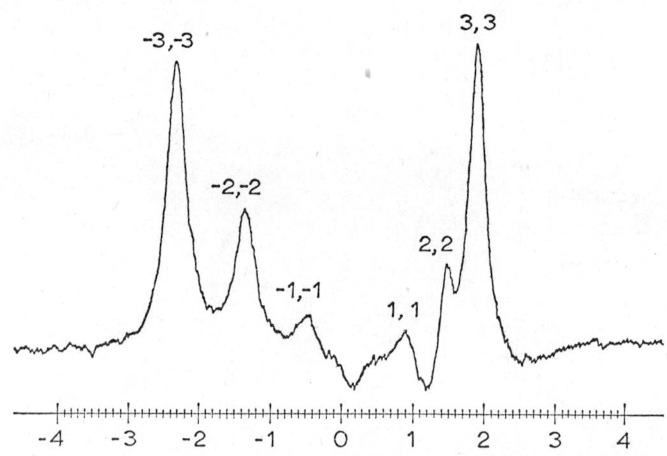

Fig. 5.23 The $3_{12} \rightarrow 3_{13}$, $\Delta M = 0$ formic acid spectra at high field. The lower scale is in megahertz.

Table 5.12 Zeeman Parameters, Molecular Quadrupole Moments, and Second Moment of the Charge Distributions in Furan and Thiophene*

	Furan	*Thiophene*
g_{aa}	-0.0911 ± 0.0007	-0.0862 ± 0.0023
g_{bb}	-0.0913 ± 0.0002	-0.0662 ± 0.0006
g_{cc}	$+0.0511 \pm 0.0001$	$+0.0501 \pm 0.0005$
$2\chi_{aa} - \chi_{bb} - \chi_{cc}$	43.04 ± 0.24	49.6 ± 1.1
$2\chi_{bb} - \chi_{cc} - \chi_{aa}$	34.39 ± 0.20	50.6 ± 1.3
Q_{aa}	0.2 ± 0.4	1.7 ± 1.6
Q_{bb}	5.9 ± 0.3	6.6 ± 1.5
Q_{cc}	-6.1 ± 0.4	-8.2 ± 2.2
χ_{aa}^{p}	159.0 ± 0.4	184.8 ± 2.0
χ_{bb}^{p}	149.2 ± 0.2	244.3 ± 1.5
χ_{cc}^{p}	243.3 ± 0.2	347.3 ± 1.6
$\langle a^2\rangle - \langle b^2\rangle$	-1.6 ± 0.2	13.9 ± 1.0
$\langle b^2\rangle - \langle a^2\rangle$	31.0 ± 0.1	36.1 ± 0.7
$\langle c^2\rangle - \langle a^2\rangle$	-29.3 ± 0.1	-50.0 ± 0.8
$\chi = \frac{1}{3}(\chi_{aa} + \chi_{bb} + \chi_{cc})$	-44.80 ± 1.5	-57.40 ± 0.86
χ_{aa}	-30.5 ± 1.6	-40.9 ± 1.2
χ_{bb}	-33.3 ± 1.6	-40.5 ± 1.3
χ_{cc}	-70.6 ± 1.6	-90.8 ± 1.7
χ_{aa}^{d}	-189.5 ± 1.8	-225.7 ± 3.0
χ_{bb}^{d}	-182.5 ± 1.8	-284.8 ± 3.0
χ_{cc}^{d}	-313.9 ± 1.8	-438.1 ± 3.0
$\sum_n Z_n a_n^2$	30.20 ± 0.04	51.43 ± 0.30
$\langle a^2\rangle$	36.2 ± 0.7	58.6 ± 1.3
$\sum_n Z_n b_n^2$	32.62 ± 0.06	38.16 ± 0.34
$\langle b^2\rangle$	37.8 ± 0.7	44.6 ± 1.2
$\sum_n Z_n c_n^2$	0.0	0.0
$\langle c^2\rangle$	6.8 ± 0.7	8.5 ± 1.2

* The a- and b-axes are in the plane and the a-axis bisects the COC or CSC angle. The magnetic susceptibilities are listed in units of 10^{-6} erg/G^2 mole. The quadrupole moments are in units of 10^{-26} esu cm^2 and the values of $\langle a^2\rangle$ are in units of 10^{-16} cm^2.

Table 5.13 Molecular Quadrupole Moments and the Second Moment of the Electronic Charge Distribution for the Out-of-Plane Coordinate

Molecule (axes: y ↑, x →)	Q_{xx}, Q_{yy}, Q_{zz} (10^{-26} esu cm^2)	$\langle z^2 \rangle$ (10^{-16} cm^2)	*Reference*
	2.8 ± 1.0 2.8 ± 1.0 -5.6 ± 2.0	7.7 ± 1.5	[57]
	-1.9 ± 0.8 5.1 ± 1.0 -3.2 ± 1.0	8.4 ± 0.6	[58]
	1.7 ± 1.6 6.6 ± 1.5 -8.2 ± 2.2	8.5 ± 1.2	[56]
	0.2 ± 0.4 5.9 ± 0.3 -6.1 ± 0.4	6.8 ± 0.7	[56]
	-0.6 ± 0.3 2.4 ± 0.2 1.8 ± 0.3	—	[60]
	-0.5 ± 0.7 1.2 ± 0.8 -0.7 ± 0.7	8.0 (estimated)	[59]
	-4.2 ± 0.5 3.0 ± 0.8 1.2 ± 0.8	6.8 ± 0.4	[54]

Table 5.14 Molecular Quadrupole Moments of Some Representative Molecules

Molecule	Q_{xx}, Q_{yy}, Q_{zz} (10^{-26} esu cm^2)	*Anisotropy of the Electronic Second Moments of the Charge Distribution*	*Reference*
H_2O	0.34 ± 0.6 1.52 ± 0.7 -1.86 ± 0.6	—	[53]
$H_2C{=}O$	-0.1 ± 0.3 0.2 ± 0.3 -0.1 ± 0.3	$\langle y^2\rangle - \langle z^2\rangle = 1.8 \pm 0.1$	[52]
$H_2C{=}C{=}O$	-0.7 ± 0.4 3.8 ± 0.5 -3.1 ± 0.5	$\langle y^2\rangle - \langle z^2\rangle = 0.85 \pm 0.4$	[61]
HC(=O)OH	$+5.1 \pm 0.3$ -5.3 ± 0.3 0.2 ± 0.3	—	[55]
HC(=O)CH$_3$	1.0 ± 0.9 -1.2 ± 0.9 0.2 ± 0.3	—	[62]
$F_2C{=}O$	-3.67 ± 0.7 -0.22 ± 0.5 3.89 ± 1.0	—	[52]

magnetic susceptibility anisotropies for the ring compounds. Note that the value of $\chi_{zz} - (\chi_{xx} - \chi_{yy})/2$ decreases in magnitude down the molecules in Table 5.15. This may be an indication of decreasing mobility or delocalization of the ring compounds.

The final point in this section is to note that only the first nonzero electrostatic multipole moment is independent of the center of mass. The molecular quadrupole moment in the molecule with center of mass at m, Q^m_{xx}, is related to the quadrupole moment at center k, Q^k_{xx}, by:

$$Q^k_{xx} - Q^m_{xx} = 2X\mu_x, \tag{5.43}$$

Table 5.15 Magnetic Susceptibility Anisotropies in Some Ring Compounds

Molecule y ↑ → x	$(\chi_{xx} - \chi_{yy})$ (10^{-6} erg/G^2 mole)	$\chi_{zz} - \frac{1}{2}(\chi_{xx} + \chi_{yy})$ (10^{-6} erg/G^2 mole)	*Reference*
H H H— —H H H	0	−59.7	[57]
H H H— —F H H	−3.6 ± 0.6	−58.3 ± 0.8	[58]
H H S H H	−0.1 ± 0.6	−50.1 ± 1.1	[56]
H H O H H	2.8 ± 0.4	−38.7 ± 0.5	[56]
H H H H	−6.4 ± 0.4	−17.3 ± 0.4	[52]
H H S H H	−2.2 ± 0.4	−15.4 ± 0.4	[59]
H H O H H	−5.7 ± 0.4	−9.5 ± 0.4	[54]

where X is the distance (with sign) from m to k and μ_x is the electric dipole moment along the x-axis. Substituting (5.37) into (5.43) gives:

$$\frac{|e|}{M}[g_{\perp}{}^{k}I_{\perp}{}^{k} - g_{\perp}{}^{m}I_{\perp}{}^{m}] = -2X\mu_x. \tag{5.44}$$

The signs of the electric dipole moment have been determined in several molecules which are listed in Table 5.16.

Table 5.16 Dipole Moment Signs

$Q^k_{xx} - Q^m_{xx} = 2X\mu_x$	*Reference*
$-$ O(H)(H) $+$ (water, O bonded to two H)	[60]
$-$ O=C(H)(H) $+$ (formaldehyde, C bonded to two H)	[51]
$-$ H—C≡C—CH_3 $+$	[52]
$-$ O—C—S $+$	[48]
$-$ N—C—CH_3 $+$	[52]
$+$ H_2C< ring with C=C bearing H, H $-$ (cyclopropene)	[60]

8 NUCLEAR QUADRUPOLE INTERACTIONS

The nuclear quadrupole interaction is the coupling of the nucleus with angular momentum I with the rotational angular momentum J to give the total angular momentum F.

$$I + J = F \tag{5.45}$$

The mechanism of coupling is through the electric quadrupole moment of the nucleus and its orientational energy with respect to the molecule fixed-field gradient produced by all of the charges in the molecule. The first-order correction to the rigid-rotor energy levels is given by the energy expression:

$$E = \frac{2f(IJF)}{J(J+1)}\sum_{g}\chi_{gg}\langle J_g{}^2\rangle. \tag{5.46}$$

$f(IJF)$ is Casimir's function, which has been conveniently tabulated for a large number of I-, J-, and F-values [2].

χ_{aa}^{N} is the nuclear quadrupole coupling constant along the principal inertial axis labeled by a. Typical microwave spectra described by (5.46) are shown in Fig. 5.21. In the case of ^{14}N, $I = 1$ and the possible F-states for $J = 1$ are $F = 0$, 1, and 2. These states are labeled in the spectrum shown in the top trace in Fig. 5.21, in which increasing frequency is from right to left. It is easy to show from (5.46), where $\langle J_g^2 \rangle$ has been given in 5.33 for a symmetric top, that the value of $\chi_{\|}^{N}$ in H_3CCN is negative and equal to -4214 kHz [63]. It is interesting to note that the corresponding isocyanide, H_3CNC, has a positive $\chi_{\|}^{N}$ given by $\chi_{\|} = 483$ kHz [63]. The nuclear quadruple coupling constant is given by:

$$\chi_{aa} = \left(\frac{\partial^2 V}{\partial a^2}\right)_{\text{ave}} Q, \tag{5.47}$$

where V is the electronic potential at the nucleus attributable to all the other changes in the molecule:

$$\left(\frac{\partial^2 V}{\partial z^2}\right)_{\text{ave}} = |e| \sum_{n}{}' Z_n \left(\frac{3z_n^2 - r_n^2}{r_n^5}\right) - |e| \langle 0| \sum_{i} \frac{3z_i^2 - r_i^2}{r_i^5} |0\rangle, \tag{5.48}$$

where the sum over n is over all nuclei except the origin, and the sum over all i is over the electrons. Thus the value of the field gradient $(\partial^2 V/\partial a^2)_{\text{ave}}$ is a sum of positive nuclear and negative electronic contributions. Because of the large inverse power on r in the field gradient operators, the local electrons normally make the dominant contribution to the field gradient. Therefore the second term or the electronic term in (5.48) is expected to be dominant in most atoms, with hydrogen being a notable exception (see below). Thus the sign of $(\partial^2 V/\partial z^2)_{\text{ave}}$ is determined by the asymmetry in the $3z_i^2 - r_i^2$ component in the operator in (5.48). If z is along a C—Cl bond and we are considering the ^{35}Cl nuclear quadrupole coupling, we would expect the average value of $(3z_i^2 - r_i^2)/(r_i^5)$ to be negative as the perpendicular x- and y-directions at the chlorine atom contain lone pairs in the atom and the z-axis contains the bonding electrons that are less dense than the lone pairs. Thus we expect the value of $(\partial^2 V/\partial z^2)_{\text{ave}}$ along the bonds in halogen atoms to be positive.

The value of the nuclear quadrupole moments for several nuclei including ^{37}Cl are given in Table 5.17. They are defined by:

$$Q = |e| \int \rho_n (3z_n^2 - r_n^2)\, d\tau, \tag{5.49}$$

where the density ρ_n is the charge density of the nucleus. This definition is similar to that given in (5.37) for molecules. Multiplying the negative value

Table 5.17 Nuclear Quadrupole Moments and Nuclear Magnetic g-Values for Several Common Nuclei

Nucleus	*Spin*	Q (10^{-34} esu cm^2)	*g-Values*
H	$\frac{1}{2}$	0	2.7927
F	$\frac{1}{2}$	0	2.6273
D	1	0.0130	0.8574
^{14}N	1	0.096	0.4036
^{17}O	$\frac{5}{2}$	−0.126	−0.7572
^{35}Cl	$\frac{3}{2}$	−0.420	0.5473
^{37}Cl	$\frac{3}{2}$	−0.320	0.4555
^{79}Br	$\frac{3}{2}$	1.58	1.3960
Br	$\frac{3}{2}$	1.34	1.5084
^{127}I	$\frac{5}{2}$	−2.94	1.1175
^{23}Na	$\frac{3}{2}$	0.48	1.4774
^{85}Rb	$\frac{5}{2}$	13.40	0.5393

of $Q(^{35}\text{Cl})$ in Table 5.17 by the expected positive value of $(\partial^2 V/\partial z^2)_{\text{ave}}$ gives an expected negative quadrupole coupling constant which is found as shown in Table 5.18 for several molecules.

The results in Table 5.18 indicate a rough correlation with ionic character. As the z-axis p-orbital near the chlorine atom becomes filled with electrons to yield the final lone pair and a spherically symmetric distribution, the field gradient goes to zero.

A detailed study of the chlorine nuclear quadrupole coupling constants in the substituted vinyl chlorides has also been completed [64]. The principal

Table 5.18 ^{35}Cl Nuclear Quadrupole Coupling Constants in Some Typical Molecules Demonstrating Increasing Ionic Character

Systems	$\left(\frac{\partial^2 V}{\partial z^2}\right)_{\text{ave}}$ $Q(^{35}\text{Cl})/h$ (MHz)
Cl (atomic)	−109.6
FCl	−146
ICl	−82.5
H_3CCl	−74.8
ClCN	−83.3
HCl	−15.8
NaCl	0

Table 5.19 Principal Elements in the ^{35}Cl Nuclear Quadrupole Coupling Constant Tensor in the Principal Field Gradient Axis System (C—Cl Bond Axis System) for Several Substituted Ethylenes and Propenes*

Molecule	$\chi_{\alpha\alpha}$	$\chi_{\beta\beta}$	$\chi_{\gamma\gamma}$	*Covalent Character* $>C{=}C{<}^{Cl}$	*Ionic Character* $>C{=}C^{+}{<}\ Cl^{-}$	*Double-Bond Character* $>C^{-}{-}C{<}{=}Cl^{+}$
$H_2C{=}CHCl$	−70.16	+40.07	+30.09	0.717	0.230	0.053
cis-$HFC{=}CHCl$	−73.7	+40.3	+33.4	0.766	0.198	0.036
$F_2C{=}CHCl$	−77.40	+40.64	+36.76	0.817	0.163	0.020
cis-$HClC{=}CHCl$	−72.3	+40.4	+31.9	0.745	0.210	0.045
$H_2C{=}CCl_2$	−78.70	+43.58	+35.12	0.814	0.141	0.045
$H_2C{=}CFCl$	−74.4	+40.9	+33.5	0.772	0.189	0.039
trans-$CH_3HC{=}CHCl$	−71.2	+39.6	+31.6	0.735	0.223	0.042
$CH_3ClC{=}CH_2$	−68.52	+37.48	+31.04	0.712	0.254	0.034
cis-$CH_3FC{=}CHCl$	−73.49	+39.84	+33.64	0.765	0.202	0.033

* α is along the C—Cl bond axis and γ is perpendicular to the skeletal frame. Also listed are the fractional covalent, ionic, and double-bond characters.

quadrupole coupling constants for ^{35}Cl are listed for several molecules in Table 5.19. The coupling constants are actually measured in the principal inertial axis system and are then rotated into the principal field gradient axis system either by measuring directly the angle of rotation by second-order effects or by measuring the principal inertial axis field gradients in two different inertial axis systems. The ^{35}Cl principal quadrupole coupling constants are remarkably similar. However, small changes are apparent and lead to information on the nature of the bonding in these molecules. Also listed in Table 5.19 are the fractional covalent, ionic, and double-bond character as obtained from the Townes and Dailey interpretation of halogen field gradients [2]. The double-bond character or contribution of the C^{-}—C=Cl^{+} structure to the C—Cl bond was calculated from the asymmetry in the field gradient perpendicular to the C—Cl bond axis. The covalent character (C=C—Cl) of the carbon–chlorine bond was calculated assuming 15% *s*-character and no *d*-character in the σ-bonding orbital of the chlorine atom. The ionic contribution (C=$C^{+}Cl^{-}$) completes the C—Cl bond. Although the values in Table 5.19 are arbitrary, the changes in these values on substitution do show trends that are consistent with our present knowledge of the bonding properties of the atoms involved.

In general, successive halogen substitution on chloroethylene decreases the double-bond character of the C—Cl bond and fluorine substitution has a slightly greater effect than chlorine substitution. However, the position of the substituent relative to the original chlorine has essentially no effect on the C—Cl bond. The decrease in the double-bond contribution would be expected with successive halogen substitution as these atoms would also interact with the ethylenic π-bond, delocalizing the π-system over more of the molecule. Thus a smaller double-bond character would be expected in the carbon–chlorine bond. The increase in covalent character of the bond is consistent with the experimental results showing a shortening of the carbon–halogen bond distance as more halogens are substituted. One would also expect the more electronegative fluorine to have a larger effect than chlorine.

Substitution of a methyl group *trans* to the chlorine has essentially no effect on the C—Cl bond. However, substitution of the methyl group on the same carbon atom to which the chlorine atom is bonded decreases the covalent character and increases the ionic character of the bond noticeably. Since the methyl group is an electron-donating group, the field gradient in the neighborhood of the chlorine atom can be expected to be more spherically symmetric and thus the bond more ionic. The changes in the bonding properties of the C—Cl bond on substitution of the fluorine on the 2-position in *trans*-1-chloropropene follow the same pattern as in the substituted ethylenes.

Nuclear quadrupole coupling constants have been reported for several ring compounds, and a detailed study of the nature of the molecular orbitals is available to describe the ^{14}N field gradient in ethylenimine [64] and the ^{33}S field gradient in ethylene sulfide [66].

Several studies have been completed on deuterium nuclear quadrupole coupling. In the case of the deuterium atom, the local electron distribution is expected to be essentially spherical. Thus the field gradient at the deuterium nucleus is dominated by the nearest nucleus, which is partially shielded by the electrons leading to a positive field gradient along the D—X bond. The smaller the field gradient, the more effective the shielding of the adjacent nucleus. This is emphasized in a study of the substituted acetylenes by Weiss and Flygare [67]. They considered first the deuterium quadrupole coupling in D—C≡C—D. From an approximate molecular structure of $d_{CD} = 1.0$ Å and $d_{CC} = 1.21$ Å, the deuteron quadrupole coupling constant attributable to the nuclei can be estimated from:

$$\chi D(\text{nuclear}) = eQ \sum_n{}' Z_n[(3\cos^2\theta_n - 1)/r_n{}^3]$$
$$= 2eQ \sum_n{}' (Z_n/r_n{}^3),$$

where e, Q, and Z_n are the electrostatic charge, deuterium nuclear quadrupole moment as given in Table 5.17, and atomic number of the nth nucleus, respectively. The result summed over the nuclei is:

$$\chi D\ (\text{nuclear}) = 206(6.00 + 0.56 + 0.03)\ \text{kHz},$$

where the first term in the parentheses is attributable to the adjacent carbon nucleus, the second term to the next nearest carbon nucleus, and the last term to the other deuterium nucleus. It is clear from this determination of the nuclear contribution of $\chi D = 6.6 \times 206$ kc/sec and the experimental value of $\chi D \cong 210$ kc/sec that approximately five-sixths of the nuclear contribution has been canceled by the electrons. It is also clear from the nuclear expression that a 10% change in the electronic contribution near the adjacent carbon nucleus will have the same effect on the experimental χD as a 100% change in the electronic shielding of the next nearest carbon nucleus. Thus the deuteron field gradient in D—C≡C—D is primarily attributable to the positive nuclear contribution of the adjacent carbon atom which has been shielded by the electrons in the C—D bond.

The results for several molecules are listed in Table 5.20. A typical spectrum demonstrating the deuterium coupling in D—C≡C—CH_3 is shown in Fig. 5.24.

We note from the results in Table 5.20 that the values of $\chi_{\text{bond}}(D)$ appear to scale with the D—X internuclear distance. We also note that the deuterium field gradients in the substituted acetylenes are relatively insensitive to substitution at the other end of the molecule.

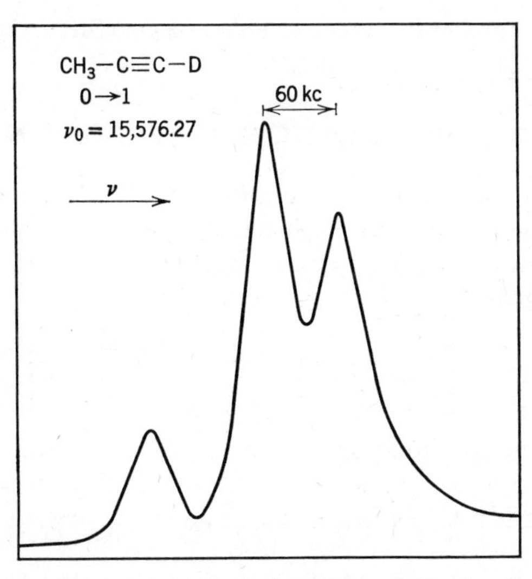

Fig. 5.24 The $J = 0 \rightarrow J = 1$ transition in D—C≡C—CH_3 showing the deuterium nuclear quadrupole coupling. Note the splitting between the stronger lines is only 60 kHz.

Table 5.20 The C—D Bond Quadrupole Coupling Constants in Several Molecules

	χ_{bond}(kHz)	*Reference*
O=C(D_2) (O=C with two D)	171 ± 3	[68]
D—C≡C—D	200 ± 10	[69]
CH_3—C≡C—D	208 ± 8	[67]
CD_3—C≡C—H	176 ± 15	[70]
F—C≡C—D	212 ± 8	[67]
Cl—C≡C—D	228 ± 12	[67]
O=C=C(D_2) (O=C=C with two D)	240 ± 15	[71]
D_2O	308 ± 3	[72]
NH_2D	290 ± 3	[73]
HCOOD	391 ± 4	[74]
DCOOH	249 ± 3	[74]

Studies on a number of other nuclei have led to further information on the electron distribution of molecules.

9 NUCLEAR SPIN-ROTATION INTERACTION

The nuclear spin-rotation interaction is the coupling of the nuclear magnetic moment with the magnetic field at the nucleus caused by the molecular rotation. The effect on the rigid-rotor energy levels of this $I + J = F$ coupling is given by:

$$E(I, J, F) = \left[\frac{J(J+1) + I(I+1) - F(F+1)}{2J(J+1)}\right] \sum_g M_{gg}\langle J_g^{\,2}\rangle, \quad (5.50)$$

where M_{aa} is the spin-rotation constant given, at the kth nucleus [75]:

$$M_{aa}^k = \frac{2e\mu_0 g_k A}{\hbar c}\left[\sum_l{}' Z_l r_{lk}^{-3}[r_{lk}^2 - (r_l - r_k)_a^{\,2}] + \frac{1}{m}\sum_{k>0}\frac{\langle 0|\, L_a\,|k\rangle\langle k|\, L_a/r^3\,|0\rangle + \langle 0|\, L_a/r^3\,|k\rangle\langle k|\, L_a\,|0\rangle}{E_0 - E_k}\right]. \quad (5.51)$$

g_k is the nuclear g-value (see listing in Table 5.17) and all other terms have been defined previously. According to (5.51), if the value of g_k is positive, the spin-rotation constant will be a combination of positive terms depending on the molecular structure and negative terms depending on the projection of excited electronic states. This sum of nuclear and electronic terms is the same as described previously for the molecular g-values in (5.31). Typical spectra representing the spin-rotation interaction are shown in Fig. 5.25 for the ketene molecule. In ketene the two equivalent proton spins add to give total I-states of 0 and 1. The $\Delta F = 0$, $4_{14} \rightarrow 4_{13}$, $F = 3$, 5, and 4 lines in ketene are identified [71]. Other spin-rotation interactions have also been identified. The ^{13}C and ^{17}O spin-rotation interactions in formaldehyde have been observed by high-resolution microwave spectroscopy [76, 77].

The nuclear spin-rotation constants are related to the nuclear paramagnetic shielding in a manner similar to the correspondence between the molecular g-values and the paramagnetic susceptibility as shown in (5.33). This relationship is [78]:

$$\begin{aligned}\sigma_{xx}^k &= \frac{e^2}{2mc^2}\langle 0|\sum r_{ik}^{-3}(r_{ik}^2 - x_{ik}^2)\,|0\rangle \\ &\quad + \frac{e^2}{2mc^2}\left[\frac{\hbar c M_{xx}^k}{2e\mu_n g_k G_x} - \sum_l{}' z_l r_{lk}^{-3}\,(r_{lk}^2 - x_{lk}^2)\right] \\ &= \sigma_{xx}^k(\text{diamagnetic}) + \sigma_{xx}^k(\text{paramagnetic}) \end{aligned} \quad (5.52)$$

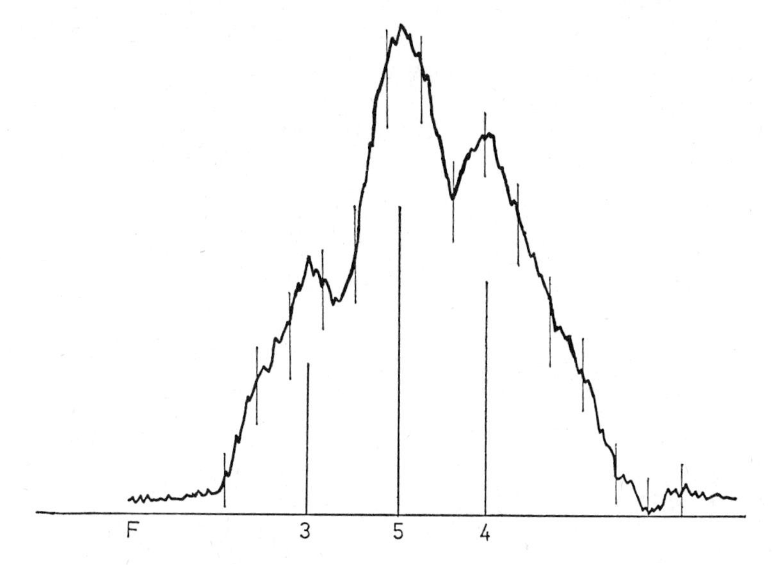

Fig. 5.25 The $4_{14} \rightarrow 4_{13}$ transition in $H_2C{=}C{=}O$. The $I_1 + I_2 = 1$ state shows three $\Delta F = 0$ transitions with $F = 3$, 5, and 4. The markers are every 5 kHz.

σ^k_{xx} is the component of total magnetic shielding at the kth nucleus along the x-axis, which is a sum of a diamagnetic and paramagnetic component. The prime on the sum over the l nuclei excludes the kth nucleus. Several analyses have been completed on extracting the diamagnetic shieldings from the spin-rotation interactions and molecular structure [78].

Acknowledgment

The support of the National Science Foundation is gratefully acknowledged.

References

1. M. W. P. Strandberg, *Microwave Spectroscopy*, Methuen, New York, 1954.
2. C. H. Townes and A. L. Schawlow, *Microwave Spectroscopy*, McGraw-Hill, New York, 1955.
3. W. Gordy, W. V. Smith, and R. F. Trambarulo, *Microwave Spectroscopy*, Dover, New York, 1966.
4. J. M. Pochan, J. E. Baldwin, and W. H. Flygare, *J. Am. Chem. Soc.*, **90,** 1072 (1968).
5. E. A. Cherniak and C. C. Costain, *J. Chem. Phys.*, **45,** 104 (1966).

6. R. Trambarulo, A. Clark, and C. Hearns, *J. Chem. Phys.*, **28,** 736 (1958).
7. B. Bak, D. Christensen, L. Hansen-Nygaard, and E. Tannenbaum, *J. Chem. Phys.*, **26,** 134 (1957).
8. R. A. Beaudet, *J. Chem. Phys.*, **38,** 2548 (1963).
9. Quantum Chemistry Program Exchange, Department of Chemistry, Indiana University, Bloomington, Indiana.
10. M. S. Cord, M. S. Lojko, and J. D. Petersen, *Microwave Spectral Tables*, National Bureau of Standards Monograph 70, Vol. V, 1968.
11. M. L. Unland, V. W. Weiss, and W. H. Flygare, *J. Chem. Phys.*, **42,** 2138 (1965).
12. R. C. Woods, A. M. Ronn, and E. B. Wilson, *Rev. Sci. Instr.*, **37,** 927 (1966); M. L. Unland and W. H. Flygare, *J. Chem. Phys.*, **45,** 2421 (1966).
13. S. Golden and E. B. Wilson, *J. Chem. Phys.*, **16,** 669 (1948).
14. C. C. Costain, *J. Chem. Phys.*, **29,** 864 (1958).
15. J. Kraitchman, *Am. J. Phys.*, **21,** 17 (1953).
16. E. Sagebarth and A. P. Cox, *J. Chem. Phys.*, **43,** 166 (1965).
17. M. H. Sirvetz, *J. Chem. Phys.*, **19,** 1609 (1951).
18. P. H. Kasai, R. J. Myers, D. F. Eggers, Jr., and K. B. Wiberg, *J. Chem. Phys.*, **30,** 512 (1959).
19. V. W. Laurie, *J. Chem. Phys.*, **24,** 635 (1956).
20. G. L. Cunningham, Jr., A. W. Boyd, R. J. Myers, and W. D. Gwinn, *J. Chem. Phys.*, **19,** 676 (1951).
21. N. Kwak, J. H. Goldstein, and J. W. Simmons, *J. Chem. Phys.*, **25,** 1203 (1956).
22. L. H. Scharpen and V. W. Laurie, *J. Chem. Phys.*, **49,** 221 (1968).
23. R. L. Kuczkowski and E. B. Wilson, *J. Am. Chem. Soc.*, **85,** 2028 (1963).
24. W. M. Tolles and W. D. Gwinn, *J. Chem. Phys.*, **36,** 1119 (1962).
25. D. J. Meschi and R. J. Myers, *J. Mol. Spectry.*, **3,** 405 (1959).
26. W. H. Kirchhoff and E. B. Wilson, *J. Am. Chem. Soc.*, **85,** 1726 (1963).
27. V. M. Rao, R. F. Curl, P. L. Timms, and J. L. Margrave, *J. Chem. Phys.*, **45,** 2032 (1966).
28. L. Pierce and V. Dobyns, *J. Am. Chem. Soc.*, **84,** 2651 (1962).
29. L. Pierce, N. DiCianni, and R. H. Jackson, *J. Chem. Phys.*, **38,** 730 (1963).
30. A. Luntz, *J. Chem. Phys.*, **50,** 1109 (1969).
31. E. Saegebarth and E. B. Wilson, *J. Chem Phys.*, **46,** 3088 (1967).
32. D. R. Lide and M. Jen, *J. Chem. Phys.*, **40,** 252 (1964).
33. S. L. Hsu, M. K. Kemp, J. M. Pochan, R. C. Benson, and W. H. Flygare, *J. Chem. Phys.*, **49,** 1482 (1969).
34. J. J. Keirns and R. F. Curl, *J. Chem. Phys.*, **48,** 3773 (1968).
35. S. S. Butcher and E. B. Wilson, *J. Chem. Phys.*, **40,** 1671 (1964).
36. E. Hirota, *J. Chem. Phys.*, **42,** 2071 (1965).
37. F. Moennig, H. Dreizler, and H. D. Rudolph, *Z. Naturforsch.*, **20,** 1323 (1965).
38. D. R. Herschbach, *J. Chem. Phys.*, **31,** 91 (1959).
39. For a review of theories of the barrier to internal rotation and a tabulation of experimental measurements, see J. P. Lowe, in *Progress in Physical Organic Chemistry*, Vol. 6, A. Streitweiser and R. W. Taft, Eds., Interscience, New York, 1968.

40. E. B. Wilson, *Proc. Natl. Acad. Sci. U.S.*, **43,** 811 (1957), and E. B. Wilson, *Advan. Chem. Phys.*, **2,** 367 (1959).
41. W. H. Flygare, *Ann. Rev. Phys. Chem.*, **18,** 325 (1967).
42. R. G. Stone, S. L. Srivastava, and W. H. Flygare, *J. Chem. Phys.*, **48,** 1890 (1968).
43. H. D. Rudolph, H. Dreizler, A. Jaeschke, and P. Wendling, *Z. Naturforsch.*, **22a,** 940 (1967).
44. H. D. Rudolph and H. Seiler, *Z. Naturforsch.*, **20a,** 1682 (1965).
45. E. Hirota, *J. Chem. Phys.*, **45,** 1984 (1966).
46. M. K. Kemp and W. H. Flygare, *J. Am. Chem. Soc.*, **91,** 3163 (1969).
47. W. Hüttner and W. H. Flygare, *J. Chem. Phys.*, **47,** 4137 (1967).
48. W. H. Flygare, W. Hüttner, R. L. Shoemaker, and P. D. Foster, *J. Chem. Phys.*, **50,** 1714 (1969).
49. W. H. Flygare, R. L. Shoemaker, and W. Hüttner, *J. Chem. Phys.*, **50,** 2414 (1969).
50. R. L. Shoemaker and W. H. Flygare, *Chem. Phys. Letters*, **2,** 610 (1968).
51. W. Hüttner, M. K. Lo, and W. H. Flygare, *J. Chem. Phys.*, **48,** 1206 (1968).
52. R. L. Shoemaker, J. M. Pochan, R. G. Stone, R. C. Benson, D. L. VanderHart, R. Blickensderfer, J. Wang, and W. H. Flygare, unpublished results.
53. S. G Kukolich and W. H. Flygare, *Mol. Phys.*, **17,** 127 (1969).
54. D. H. Sutter, W. Hüttner, and W. H. Flygare, *J. Chem. Phys.*, **50,** 2869 (1969).
55. S. G. Kukolich and W. H. Flygare, *J. Am. Chem. Soc.*, **91,** 2433 (1969).
56. D. H. Sutter and W. H. Flygare, *J. Am. Chem. Soc.*, **91,** 4063 (1969).
57. R. L. Shoemaker and W. H. Flygare, *J. Chem. Phys.*, **51,** 2988 (1969).
58. W. Hüttner and W. H. Flygare, *J. Chem. Soc.*, **50,** 2863 (1969).
59. D. H. Sutter and W. H. Flygare, *Mol. Phys.*, **16,** 153 (1969).
60. R. C. Benson and W. H. Flygare, *J. Chem. Phys.*, **51,** 3087 (1969).
61. W. Hüttner, P. D. Foster, and W. H. Flygare, *J. Chem. Phys.*, **50,** 1710 (1969).
62. W. Hüttner and W. H. Flygare, *Trans. Faraday Soc.*, **65,** 1953 (1969).
63. M. K. Kemp, J. M. Pochan, and W. H. Flygare, *J. Phys. Chem.*, **71,** 765 (1967).
64. R. G. Stone and W. H. Flygare, *J. Chem. Phys.*, **49,** 1943 (1968).
65. M. K. Kemp and W. H. Flygare, *J. Am. Chem. Soc.*, **90,** 6267 (1968).
66. R. L. Shoemaker and W. H. Flygare, *J. Am. Chem. Soc.*, **90,** 6263 (1968).
67. V. W. Weiss and W. H. Flygare, *J. Chem. Phys.*, **45,** 8 (1966).
68. W. H. Flygare, *J. Chem Phys.*, **41,** 206 (1964).
69. N. F. Ramsey, *Am. Scientist*, **49,** 509 (1961).
70. W. H. Flygare, unpublished results.
71. V. W. Weiss and W. H. Flygare, *J. Chem. Phys.*, **45,** 3475 (1966).
72. H. Bluyssen, J. Verhoeven, and A. Dymams, *Phys. Letters*, **25A,** 214 (1967).
73. S. G. Kukolich, *J. Chem. Phys.*, **49,** 5523 (1968).
74. S. G. Kukolich, unpublished results.
75. W. H. Flygare, *J. Chem. Phys.*, **41,** 793 (1964).
76. W. H. Flygare and V. W. Weiss, *J. Chem. Phys.*, **45,** 2785 (1966).
77. W. H. Flygare and J. T. Lowe, *J. Chem. Phys.*, **43,** 3645 (1965).
78. W. H. Flygare and J. Goodisman, *J. Chem. Phys.*, **49,** 3122 (1968).

Chapter **VI**

ELECTRON SPIN RESONANCE

Philip H. Rieger

---

1 Introduction 500
The Magnetic Resonance Phenomenon 502
The Bloch Equations 504

2 The Spin Hamiltonian 509
Isotropic Solution Spectra 510
First-Order Solution 510
Higher-Order Solutions 511
Single Nucleus 512
Several Equivalent Nuclei 514
Nonequivalent Nuclei 515
Single Crystal Spectra 519
Polycrystalline Spectra 521

3 Spectrometer Design and Operation 525
Microwaves and Waveguides 525
Microwave Components 528
Klystrons 530
Attenuators and Terminations 531
Directional Couplers and Magic Tees 531
Tuners 532
Phase Shifters 532
Isolators and Circulators 532
Cavities 533
Slow-Wave Structures 533
Esr Spectrometers 534
Homodyne Detection Spectrometer 534
Superheterodyne Spectrometer 536
Homodyne versus Superheterodyne Detection 538
Commercially Available Spectrometers 538
Microwave Frequency and Power 541
Choice of Microwave Frequency 541
Effect of Microwave Power Saturation 542

Cavities 546
Impedance Matching 546
Sample Placement 551
Rectangular Cavities 552
Cylindrical Cavities 555
Optimization of Sample Size 556

4 Sensitivity Enhancement Techniques 557

The Signal-to-Noise Ratio 557
Sensitivity Enhancement by Filtering 558
The Matched Filter 560
RC Filters 561
Time-Averaging Methods 568

5 Resolution 570

Field Modulation Effects 570
Low-Frequency Modulation 570
Small-Amplitude Modulation 571
Large-Amplitude Modulation 572
High-Frequency Field Modulation 576
Small Modulation Index 578
Large Modulation Index 581
Resolution Enhancement Techniques 586
Computer Spectrum Simulation 586
Chemical Resolution Enhancement 587
Resolution Enhancement by Filtering 588
Optimum Resolution Enhancement Filter 588
Derivative Line Sharpening 589
References 590
Bibliography 592

---

1 INTRODUCTION

Electron spin resonance (esr) spectroscopy, also known as electron paramagnetic resonance (epr), paramagnetic resonance (pmr), or electron magnetic resonance (emr), has been an active field of inquiry since the first experiments of Zavoisky [1] in 1945 and has attracted the increasingly active interest of chemists since the early 1950s. The technique has been extensively applied in the study of free radicals, both in liquid solutions and in glasses; indeed, it has been responsible for the opening of several now rather large fields of chemistry to detailed investigations. These investigations have taken several courses, for not only have the products of many radical reactions been identified by means of esr and the mechanisms elucidated, but in many cases startlingly detailed information has been obtained concerning the electronic structure of the radicals and the interaction of the paramagnetic species with its surroundings. A large body of information has been obtained concerning electronic structure and environmental effects in systems containing paramagnetic transition metal ions, the studies taking place in dilute single crystals, glasses, and in liquid solution. Esr has been extensively applied in the study of semiconductors; indeed, in several areas of solid-state physics, esr has proven an indispensable tool.

Through all these developments, an extensive literature has accumulated which has been reviewed, cataloged, discussed, summarized, abstracted, and otherwise dissected and displayed for the curious. There is therefore some doubt about the proper function of an additional article on the subject inasmuch as there are already several excellent introductions to the field, the instrumentation and techniques encountered have been more than adequately described, and space does not permit a comprehensive review of the chemically interesting results.

It is assumed therefore that the reader of this article already has some knowledge of, and more than a passing interest in, esr and that he can best be served by providing enough general information on instrumentation and experimental techniques so that with little or no background in electronics he can understand the operation of his spectrometer and obtain the best possible performance in terms of sensitivity and resolution. Attention is therefore focused on those parameters of esr spectrometers that are susceptible to adjustment by the average chemist.

This article contains a select bibliography of books, reviews, compilations of data, and reports of symposia and conferences. Capsule reviews are provided in some cases for further guidance. Readers interested in a general introduction to the field of esr or in discussions of the various chemical applications are thus referred to one of the several excellent treatises or topical reviews listed among the general references.

The article starts, in Section 1, with a fairly conventional introduction to magnetic resonance so as to have certain of the concepts available for later reference. We proceed in Section 2 to a discussion of the extraction of spin Hamiltonian parameters from spectra, and in Section 3 to a discussion of microwave components and some aspects of esr spectrometers, with consideration of some of the options available to the designer or purchaser of a spectrometer. We then consider in some detail the choice of spectrometer operating parameters—microwave power and frequency, cavity type, and sample size and position. In Sections 4 and 5, we consider filtering, time-averaging, field modulation, and various techniques for resolution enhancement. The choice of topics is somewhat arbitrary, and in most cases more complete discussions can be found elsewhere. It is hoped, however, that the typical reader, interested in obtaining optimum performance for his spectrometer, will find the contents sufficient for most purposes and helpful in directing him to a source of further information when they are not.

The Magnetic Resonance Phenomenon [2]

Application of a magnetic field to a system containing an unpaired electron lifts the degeneracy of the electron spin energy levels, introducing an energy difference between the two states of $\Delta E = g\beta H$, where β is the Bohr magneton ($\beta = eh/4\pi mc = 0.92732 \times 10^{-20}$ erg/G) [3] and g, the gyromagnetic ratio, has the value $g = 2.002319$ for a free electron [4]. At equilibrium and in the absence of electromagnetic radiation, the populations of the two levels are given by the Boltzmann equation:

$$\begin{aligned} N_+^0 &= N \exp(-g\beta H/kT)/[1 + \exp(-g\beta H/kT)] \\ N_-^0 &= N/[1 + \exp(-g\beta H/kT)], \end{aligned} \tag{6.1}$$

where $N = N_+^0 + N_-^0$ is the total number of electron spins in the sample. The equilibrium population difference is then:

$$n_0 = N_-^0 - N_+^0 = N \tanh[(1/2)g\beta H/kT], \tag{6.2}$$

or, for temperatures and magnetic fields normally encountered in the laboratory, the high-temperature approximation is sufficiently accurate:

$$n_0 = (N/2)(g\beta H/kT).$$

At 300°K in a magnetic field of 3000 G for example, $n_0/N = 0.00067$. If the system is perturbed, it may be expected to return to equilibrium following the rate equations:

$$\begin{aligned} \frac{dN_+}{dt} &= N_-W_\uparrow - N_+W_\downarrow \\ \frac{dN_-}{dt} &= N_+W_\downarrow - N_-W_\uparrow, \end{aligned} \tag{6.3}$$

where $W_\uparrow$ and $W_\downarrow$ are the probabilities of upward and downward transitions, respectively. With $n = N_- - N_+$,

$$\frac{dn}{dt} = N(W_\downarrow - W_\uparrow) - n(W_\downarrow + W_\uparrow). \tag{6.4}$$

Since at equilibrium the net rate of change of populations is zero, the relaxation transition probabilities are easily related to the equilibrium population and thus to the Boltzmann factor:

$$n_0 = N(W_\downarrow - W_\uparrow)/(W \;\; + W_\uparrow). \tag{6.5}$$

Defining T_1, the spin-lattice relaxation time:

$$1/T_1 = W_\downarrow + W_\uparrow, \tag{6.6}$$

the rate equation becomes:

$$\frac{dn}{dt} = (n_0 - n)/T_1, \tag{6.7}$$

the solution of which is:

$$n = n_0 + A \exp(-t/T_1), \tag{6.8}$$

where A is a constant of integration depending upon the initial conditions.

In the presence of an electromagnetic radiation field, transitions between the levels are induced with equal transition probabilities for both the absorption and emission of energy. Thus additional terms are added to (6.3) and (6.4), obtaining:

$$\frac{dn}{dt} = -2Wn + (n_0 - n)/T_1, \tag{6.9}$$

where W is the probability per unit time of an induced emission or absorption of a photon. The steady-state solution to (6.9) is:

$$n = n_0/(1 + 2WT_1). \tag{6.10}$$

The power absorbed from the radiation field is given by:

$$\begin{aligned} P_a &= dE/dt = nh\nu W \\ P_a &= n_0 h\nu W/(1 + 2WT_1). \end{aligned} \tag{6.11}$$

As noted below, W is proportional to the radiative power incident on the sample, that is, to the square of the alternating magnetic field. Thus for $2WT_1$ small the power absorbed by the sample is proportional to the incident power. In this region esr is analogous to other forms of spectroscopy at higher energies in that Beer's law is obeyed. At sufficiently high incident power that $2WT_1$ is no longer negligible, however, saturation of the resonance absorption occurs, and the power absorbed actually decreases with increasing incident power. The quantity

$$Z = 1/(1 + 2WT_1) \tag{6.12}$$

is frequently encountered in the literature and is called the saturation factor. $Z = 1$ when the resonance is unsaturated and decreases to zero when the transition is completely saturated.

We now proceed to compute the induced absorption or emission probability W. The term in the Hamiltonian giving rise to the transitions between electron spin energy levels is attributable to an oscillating magnetic field in the x-direction with amplitude H_1 and angular frequency ω:

$$\mathscr{H}_1 = \mu_x H_1 \cos \omega t. \tag{6.13}$$

Since the magnetic moment operator is related to the spin angular momentum operator **S** (6.13) can also be written:

$$\mathscr{H}_1 = h\gamma S_x H_1 \cos \omega t. \tag{6.14}$$

According to time-dependent perturbation theory, the transition probability is

$$W = (1/4)(\gamma H_1)^2 \, |\langle \tfrac{1}{2}| \, S_x \, |-\tfrac{1}{2}\rangle|^2 \, \delta(\omega_0 - \omega), \tag{6.15}$$

where δ is the Dirac δ-function, $\omega_0 = \gamma H$ is the resonant frequency, and $\langle \frac{1}{2}| \, S_x \, |-\frac{1}{2}\rangle$ is the matrix element of the operator S_x between the states $M_S = \frac{1}{2}$ and $M_S = -\frac{1}{2}$. Equation (6.15) is an idealization in the sense that it predicts an infinitely sharp absorption line. In actual practice of course, the line is broadened by a variety of mechanisms. We therefore modify (6.15) to anticipate these line-broadening mechanisms by the introduction of a line shape function $G(\omega)$, normalized such that:

$$\int_0^\infty G(\omega)\, d\omega = 1, \tag{6.16}$$

obtaining, after evaluating the matrix element,

$$W = (1/16)(\gamma H_1)^2 G(\omega). \tag{6.17}$$

The saturation factor, defined in (6.12), is then:

$$Z = 1/[1 + (\pi/4)(\gamma H_1)^2 G(\omega) T_1]. \tag{6.18}$$

The maximum value of $G(\omega)$ is usually denoted by:

$$G(\omega)_{\text{max}} = T_2/\pi, \tag{6.19}$$

where T_2 is called the transverse relaxation time. The significance of T_2 becomes clearer in the discussion of the Bloch equations below. The saturation factor at the absorption maximum may then be written in the more familiar form:

$$Z_{\text{max}} = 1/[1 + (1/4)(\gamma H_1)^2 T_1 T_2]. \tag{6.20}$$

The Bloch Equations [2]

The set of phenomenological equations first derived by Felix Bloch [5] is very useful in the understanding of many aspects of magnetic resonance. These equations deal with the bulk magnetization of the sample rather than energy level populations. Two types of motion are considered in the simplest form of the Bloch equations: (1) the motion of the macroscopic magnetization under the influence of the applied magnetic field, and (2) the relaxation of the magnetization toward its equilibrium value.

According to classical theory, the time rate of change of the angular momentum **L** of a magnet of magnetic moment $\boldsymbol{\mu}$ in a magnetic field **H** is

$$\frac{d\mathbf{L}}{dt} = \boldsymbol{\mu} \times \mathbf{H}. \tag{6.21}$$

The bulk magnetization of a sample **M** is simply the sum of the individual magnetic moments, which in turn are proportional to the individual angular momentums with proportionality constant γ, the magnetogyric ratio. Thus the equation of motion of the magnetization is, in the absence of relaxation effects:

$$\frac{d\mathbf{M}}{dt} = \gamma \mathbf{M} \times \mathbf{H}. \tag{6.22}$$

In the presence of a static magnetic field H_0 in the z-direction the equilibrium bulk magnetization $M_0 = \chi_0 H_0$ is also oriented along the z-axis. If the z-component of the magnetization differs from the equilibrium value, one expects the magnetization to relax toward the equilibrium value, most reasonably by a first-order process:

$$\frac{dM_z}{dt} = (M_0 - M_z)/T_1, \tag{6.23}$$

where T_1 is called the spin-lattice relaxation time, the name referring to the fact that energy is transferred from the spin system to the surroundings (or lattice) during the relaxation process. As noted below, T_1 has the same significance here as in (6.6). The x- and y-components of the magnetization should vanish at equilibrium, so that an additional relaxation time T_2, called the transverse relaxation time, is introduced:

$$\begin{aligned} \frac{dM_x}{dt} &= -M_x/T_2 \\ \frac{dM_y}{dt} &= -M_y/T_2. \end{aligned} \tag{6.24}$$

The relaxation processes represented by T_1 and T_2 are physically different. The spin-lattice relaxation represents a transfer of energy from spin system to the lattice, whereas the transverse relaxation time, for a system of spins precessing about an applied magnetic field, is the characteristic time for loss of phase coherence in the x-y plane. In thermodynamic terms, the free energy driving the T_1-relaxation is primarily enthalpy, whereas the T_2-relaxation involves entropy maximization.

Combining the precession term with the relaxation terms, we have:

$$\frac{d\mathbf{M}}{dt} = \gamma \mathbf{M} \times \mathbf{H} - (\hat{i} M_x + \hat{j} M_y)/T_2 + \hat{k}(M_0 - M_z)/T_1. \tag{6.25}$$

Taking account of the usual experimental situation in which a large applied magnetic field H_0 along the z-axis is combined with a small oscillating field along the x-axis, the total magnetic field is:

$$\mathbf{H} = \hat{i} H_1 \cos \omega t + \hat{k} H_0. \tag{6.26}$$

Evaluating the cross product and separating the equation into its components, we have the Bloch equations:

$$\begin{aligned} \frac{dM_x}{dt} &= \omega_0 M_y - M_x/T_2 \\ \frac{dM_y}{dt} &= -\omega_0 M_x - M_y/T_2 + \gamma H_1 M_z \cos \omega t \\ \frac{dM_z}{dt} &= -\gamma H_1 M_y \cos \omega t + (M_0 - M_z)/T_1, \end{aligned} \tag{6.27}$$

where $\omega_0 = \gamma H_0$.

There is unfortunately no general closed-form solution to the Bloch equations, although they may be solved approximately for a number of limiting conditions. In the case of esr, the most common condition is of relatively low microwave power so that little saturation occurs and M_z is not appreciably different from M_0. Furthermore, the magnetic field is usually changed relatively slowly (compared with the relaxation times) so that dM_z/dt is very nearly zero. Under these steady-state (or slow passage) conditions, it is intuitively apparent that the x-component of the magnetization will undergo periodic motion under the influence of the driving field $H_1 \cos \omega t$. If this is assumed to be the case, it is convenient to write the driving field in complex form $H_1 e^{-i\omega t}$, and to define a complex susceptibility:

$$\chi(\omega) = \chi'(\omega) + i\chi''(\omega), \tag{6.28}$$

such that the x-component of the magnetization is the real part of the expression: $\chi(\omega) H_1 e^{-i\omega t}$, or

$$M_x(t) = H_1(\chi' \cos \omega t + \chi'' \sin \omega t). \tag{6.29}$$

Thus $H_1\chi'$ is the component of M_x in phase with the driving field, and $H_1\chi''$ is the component 90° out of phase, which contributes to the absorption of power from the microwave field.

By making the steady-state assumption, $dM_z/dt = 0$, and taking the above form for $M_x(t)$ and a similar form for $M_y(t)$, straightforward manipulation of the Bloch equations yields:

$$\chi'(\omega) = (1/2)\gamma M_z \left\{ \frac{(\omega_0 - \omega)T_2^2}{1 + (\omega_0 - \omega)^2 T_2^2} + \frac{(\omega_0 + \omega)T_2^2}{1 + (\omega_0 + \omega)^2 T_2^2} \right\} \tag{6.30a}$$

$$\chi''(\omega) = (1/2)\gamma M_z \left\{ \frac{T_2}{1 + (\omega_0 - \omega)^2 T_2^2} - \frac{T_2}{1 + (\omega_0 + \omega)^2 T_2^2} \right\}. \tag{6.30b}$$

Near resonance, $\omega \sim \omega_0$, the first terms of the above expressions dominate, and the second may be neglected. It should be noted that neglect of these terms at this point in the derivation is equivalent to the frequently encountered assumption of a magnetic field rotating in the x-y plane instead of the actual field oscillating along the x-axis. The assumption of a rotating magnetic field has important advantages in that it invites a transformation to a rotating coordinate frame with considerable subsequent simplification both of the algebraic manipulation and of the conceptualization of the motion of the magnetization. We have so far avoided the rotating coordinate system to emphasize the equivalence of the two procedures.

If we now assume a rotating magnetic field:

$$\mathbf{H} = (1/2)H_1(\hat{i}\cos\omega t - \hat{j}\sin\omega t) + \hat{k}H_0, \tag{6.31}$$

the Bloch equations become:

$$\begin{aligned} \frac{dM_x}{dt} &= \omega_0 M_y + (1/2)\gamma H_1 M_z \sin\omega t - M_x/T_2 \\ \frac{dM_y}{dt} &= -\omega_0 M_x + (1/2)\gamma H_1 M_z \cos\omega t - M_y/T_2 \\ \frac{dM_z}{dt} &= -(1/2)\gamma H_1(M_x \sin\omega t + M_y \cos\omega t) + (M_0 - M_z)/T_1. \end{aligned} \tag{6.32}$$

We now write, for the component of M in the x-y plane that rotates in phase with the field:

$$u = M_x \cos\omega t - M_y \sin\omega t \tag{6.33}$$

and, for the component that lags the field by 90°:

$$v = M_x \sin\omega t + M_y \cos\omega t. \tag{6.34}$$

Note that $M_x = u\cos\omega t + v\sin\omega t$, so that $u = H_1\chi'$ and $v = H_1\chi''$. The equations of motion in the rotating coordinate frame are

$$\frac{du}{dt} = (\omega_0 - \omega)v - u/T_2 \tag{6.35a}$$

$$\frac{dv}{dt} = -(\omega_0 - \omega)u - v/T_2 + (1/2)\gamma H_1 M_z \tag{6.35b}$$

$$\frac{dM_z}{dt} = -(1/2)\gamma H_1 v + (M_0 - M_z)/T_1, \tag{6.35c}$$

and the steady-state solutions are easily found to be:

$$u = \frac{(1/2)\gamma H_1 M_0 T_2^2(\omega_0 - \omega)}{1 + T_2^2(\omega_0 - \omega)^2 + (1/4)(\gamma H_1)^2 T_1 T_2} \tag{6.36a}$$

$$v = \frac{(1/2)\gamma H_1 M_0 T_2}{1 + T_2^2(\omega_0 - \omega)^2 + (1/4)(\gamma H_1)^2 T_1 T_2} \tag{6.36b}$$

$$M_z = M_0 \frac{1 + T_2^2(\omega_0 - \omega)^2}{1 + T_2^2(\omega_0 - \omega)^2 + (1/4)(\gamma H_1)^2 T_1 T_2} . \tag{6.36c}$$

According to classical electromagnetic theory, the instantaneous power absorption by a sample of magnetization M in a magnetic field H is:

$$P(t) = \mathbf{H} \cdot \left(\frac{d\mathbf{M}}{dt}\right). \tag{6.37}$$

Since in the present case $\mathbf{H}$ is given by (6.26) and we have assumed that the z-component of the magnetization is nearly constant, the dot product reduces to

$$P(t) = \frac{dM_x}{dt} H_1 \cos \omega t. \tag{6.38}$$

Making use of (6.29) and (6.36), and averaging over one cycle, we obtain

$$P(\omega) = (\omega/2)H_1^2\chi''. \tag{6.39}$$

Thus the power absorbed by the sample from the oscillating magnetic field is proportional to the out-of-phase component of the susceptibility.

For sufficiently low applied power (H_1^2) or sufficiently short spin-lattice relaxation time T_1, the resonance is essentially unsaturated, and the term $(1/4)(\gamma H_1)^2 T_1 T_2$ in the denominator of (6.36*b*) may be neglected, and the absorption line has the familiar Lorentzian line shape:

$$G(\omega) = \frac{1}{\pi} \frac{T_2}{1 + T_2^2\omega^2} . \tag{6.40}$$

The corresponding dispersion line has the shape function:

$$F(\omega) = \frac{1}{\pi} \frac{T_2\omega}{1 + T_2^2\omega^2} . \tag{6.41}$$

The power absorbed by the sample may then be written as:

$$P(\omega) = (\pi/4)H_1^2\chi_0\omega\omega_0 G(\omega)Z, \tag{6.42}$$

where Z is the saturation factor defined by (6.18) above. The absorption line shape $G(\omega)Z$ is plotted for several values of the saturation parameter $(1/4)(\gamma H_1)^2T_1T_2$ in Fig. 6.12.

2 THE SPIN HAMILTONIAN

In order to interpret the results of an esr experiment, it is in principle necessary to know something of the energy states between which the observed transitions occur. This seems a formidable task since one must deal not only with the kinetic energy of the electrons and nuclei and with the electrostatic interactions between the electrons and nuclei but also with the interaction of the external magnetic field with the magnetic moments associated with the electron and nuclear spins and with the orbital motion of the electrons; with the magnetic interaction of the electron spin and orbital magnetic moments; with the magnetic interactions between electrons, between electrons and nuclei, and between nuclei; and with higher-order electrostatic effects such as the electronic interactions with the quadrupole moments of the nuclei. This rather discouraging list of effects is made much more tractable by noting that the absolute value of the energy is of no consequence; we are concerned only with the energy differences between levels, many of which are degenerate at zero magnetic field. Indeed, it can be shown that all of the above effects, to the extent that they are relevant to esr absorption, can be reduced to five terms in an effective Hamiltonian operator called the spin Hamiltonian [6].

The spin Hamiltonian makes use of two operators: **S**, the effective electronic angular momentum operator, which includes both spin and orbital angular momentum; and $\mathbf{I}_i$, the spin angular momentum operator associated with the ith nucleus. The spin Hamiltonian,

$$\mathscr{H}_s = \beta \mathbf{H} \cdot \bar{\bar{g}} \cdot \mathbf{S} + \mathbf{S} \cdot \bar{\bar{D}} \cdot \mathbf{S} - \sum_i g_N \beta_N \mathbf{H} \cdot \mathbf{I}_i + \sum \mathbf{I}_i \cdot \bar{\bar{A}}_i \cdot \mathbf{S} + \sum \mathbf{I}_i \cdot \bar{\bar{Q}}_i \cdot \mathbf{I}_i, \quad (6.43)$$

contains four tensors, $\bar{\bar{g}}$, $\bar{\bar{D}}$, $\bar{\bar{A}}_i$, and $\bar{\bar{Q}}_i$, which describe, respectively, the Zeeman interaction of the electron spin and orbital magnetic moment with the applied field, the electron spin–spin interactions, the electron-nuclear magnetic interactions, and the electron-nuclear quadrupolar interactions. In (6.43) **H** is the magnetic field, β is the Bohr magneton, β_N is the nuclear magneton, and g_N is the nuclear g-factor. The nuclear Zeeman term is assumed isotropic here, an excellent approximation for the purposes of discussions of electron resonance.

It has been found generally possible to completely describe the positions and intensities of experimentally observed resonance lines in terms of the parameters of the spin Hamiltonian. The interpretation of the significance of the parameters once found is beyond the scope of this article [7], but we shall devote the next few pages to a discussion of the extraction of parameters from experimental spectra.

Isotropic Solution Spectra

In liquid solution the anisotropies in the various tensors are effectively averaged, perhaps contributing to line widths, but no longer affecting the line positions. The quadrupole interaction tensor is identically zero for nuclear spins less than one, and the term is usually very small, even for higher spin nuclei; we therefore neglect it in the following discussion. The spin–spin interaction tensor $\bar{\bar{D}}$ is of significance only for systems containing more than one unpaired electron and so can be ignored for most esr work. With these simplifications, the spin Hamiltonian reduces to:

$$\mathscr{H}_s = g\beta \mathbf{H} \cdot \mathbf{S} - \sum_i g_N \beta_N \mathbf{H} \cdot \mathbf{I}_i + \sum_i A \mathbf{I}_i \cdot \mathbf{S}. \tag{6.44}$$

Defining the z-axis by the direction of the applied magnetic field, we have:

$$\mathscr{H}_s = g\beta H S_z + \sum_i [-g_N \beta_N H I_{iz} + A_i I_{iz} S_z + (A_i/2)(S_+ I_{i-} + S_- I_{i+})], \tag{6.45}$$

where we have introduced the "raising and lowering operators" $S_\pm$ and $I_\pm$:

$$S_\perp = S_x \pm iS_y. \tag{6.46}$$

Solutions of the spin Hamiltonian (6.45) are relatively straightforward. Exact solutions are always possible, although usually not in closed form, so that various approximations are usually employed to facilitate interpretation of experimental spectra.

First-Order Solution

To first order in perturbation theory, the energies are simply the diagonal Hamiltonian matrix elements:

$$E(m_S, m_I) = \langle m_S, m_I | \mathscr{H}_s | m_S, m_I \rangle. \tag{6.47}$$

Thus the first-order energies are readily found to be:

$$E(m_S, m_I) = g\beta H m_S + \sum_i (A_i m_S m_{Ii} - g_N \beta_N H m_{Ii}). \tag{6.48}$$

Applying the usual selection rules for esr transitions, $\Delta m_S = \pm 1$, $\Delta m_I = 0$, the energies of the allowed transitions are then:

$$\begin{aligned} h\nu &= E(\tfrac{1}{2}, m_I) - E(-\tfrac{1}{2}, m_I) \\ h\nu &= g\beta H + \sum_i A_i m_{Ii}. \end{aligned} \tag{6.49}$$

Since esr spectrometers normally operate at constant frequency, (6.49) is conveniently rearranged to give the values of the field at which resonance lines occur:

$$H = h\nu/g\beta - \sum_i a_i m_{Ii}, \tag{6.50}$$

where $a_i = A_i/g\beta$ is the nuclear hyperfine splitting in units of gauss. Thus we find that the esr spectrum is predicted to first order to be symmetrical about the center with the spacing between lines simply related to the sums and differences of splitting constants. It should be noted that the nuclear Zeeman term has no effect, at this level of approximation, on the positions of the resonance lines.

The vast majority of esr spectra recorded for liquid solution spectra may be interpreted at least qualitatively using this simple first-order theory. Experimental spectra range in complexity from very simple spectra containing only a few hyperfine lines which may be interpreted immediately by inspection to those of large organic radicals which may contain thousands of lines and which require more elaborate schemes for their interpretation. Frequently, interpretation is hindered by poor resolution, and it may be worthwhile in some cases to attempt to improve the experimental spectrum by altering the sample preparation procedure or by employing a resolution enhancement technique (see Section 5, p. 586).

Generally speaking, in the interpretation of complex spectra from organic radicals in solution, it is possible to identify several of the smaller splitting constants by careful examination of the wings of the spectrum. Since the overall width of the spectrum is $\sum 2I_i a_i$, where I_i is the spin of nucleus i and the sum is over all nuclei contributing a hyperfine splitting, subtracting the small splitting constants determined by examination of the wings from the total width gives a relation among the larger splittings. Working from this starting point, it is usually possible to find a unique interpretation or, failing this, several interpretations which may then be checked by computer simulation of the spectrum and comparison with the experimental spectrum. It is quite possible to be seriously misled in an interpretation by starting with the stronger lines that may be accidental superpositions of several weaker lines. There are numerous cases in the literature in which just such an error has led to interpretations later shown to be erroneous.

For many purposes the first-order theory is sufficiently accurate for both the qualitative and quantitative interpretation of spectra. When the splitting constants are sufficiently large, however, substantial errors on the order of a/H may be encountered in the calculation of g- and a-values. In some cases difficulties may be encountered in the qualitative interpretation as well.

Higher-Order Solutions

In cases in which the first-order solution is inadequate, second-order perturbation theory may be applied. While adequate for most cases, such a treatment has the disadvantage of not readily leading to third- and higher-order corrections, and furthermore is sometimes awkward to apply to cases involving more than one hyperfine splitting constant. It is preferable to write down the exact solution and then to use approximate forms as appropriate to the particular problem. To illustrate this procedure, we derive here general expressions for a single nucleus, for several equivalent nuclei, and indicate the procedure to be followed for several nonequivalent nuclei.

SINGLE NUCLEUS

Writing $\Delta_S = g\beta H$ and $\Delta_N = g_N\beta_N H$ for the electron and nuclear Zeeman splittings, the matrix elements of the Hamiltonian (6.44) for a single nuclear hyperfine term are:

$$\begin{aligned}
\langle m_S, m_I | \mathscr{H}_s | m_S, m_I\rangle &= \Delta_S m_S - \Delta_N m_I + A m_S m_I \\
\langle m_S - 1, m_I + 1 | \mathscr{H}_s | m_S, m_I\rangle &= (A/2)[I(I+1) - m_I(m_I+1)]^{1/2} \\
&\quad \times [S(S+1) - m_S(m_S-1)]^{1/2} \qquad (6.51)\\
\langle m_S + 1, m_I - 1 | \mathscr{H}_s | m_S, m_I\rangle &= (A/2)[I(I+1) - m_I(m_I-1)]^{1/2} \\
&\quad \times [S(S+1) - m_S(m_S+1)]^{1/2}.
\end{aligned}$$

Inspection of these elements indicates that the Hamiltonian matrix will factor into a collection of 2×2 blocks of the form:

$$\begin{array}{c|cc|}
 & |+\tfrac{1}{2}, m_I\rangle & |-\tfrac{1}{2}, m_I+1\rangle \\
\langle +\tfrac{1}{2}, m_I| & \Delta_S/2 - m_I\Delta_N + Am_I/2 & (A/2)[I(I+1) - m_I(m_I+1)]^{1/2} \\
\langle -\tfrac{1}{2}, m_I+1| & (A/2)[I(I+1) - m_I(m_I+1)]^{1/2} & -\Delta_S/2 - (m_I+1)\Delta_N - A(m_I+1)/2
\end{array} \qquad (6.52)$$

Diagonalization of this matrix is straightforward, and results in the energies:

$$E = -A/4 - (1/2)(2m_I + 1)\,\Delta_N \pm (1/2)(\Delta_N + \Delta_S)[1 + (2m_I + 1)x + (2I+1)^2x^2/4]^{1/2}, \qquad (6.53)$$

where $x = A/(\Delta_N + \Delta_S)$. At moderately high magnetic fields in which the two states are not very greatly mixed, it is reasonable to continue to designate the states by the quantum numbers m_S and m_I even though they are no longer strictly "good" quantum numbers. Making the appropriate identification, we obtain the general equations for the energies:

$$\begin{aligned}
E(\tfrac{1}{2}, m_I) &= -A/4 - (1/2)(2m_I + 1)\,\Delta_N + (1/2)(\Delta_N + \Delta_S) \\
&\quad \times [1 + (2m_I + 1)x + (1/2)(2I+1)^2x^2]^{1/2} \\
E(-\tfrac{1}{2}, m_I) &= -A/4 - (1/2)(2m_I - 1)\,\Delta_N - (1/2)(\Delta_N + \Delta_S) \\
&\quad \times [1 + (2m_I - 1)x + (1/2)(2I+1)^2x^2]^{1/2}.
\end{aligned} \qquad (6.54)$$

These expressions are equivalent to the well-known Breit-Rabi equation [8], which gives the energies of the various states in terms of the total angular momentum quantum numbers and which is thus somewhat less convenient for our purposes. The equations, although exact, are not easily used in the analysis of experimental spectra. For this purpose, we expand the square root term in a power series in x. Collecting terms and rearranging, we obtain, through third order:

$$E(+\tfrac{1}{2}, m_I) = \Delta_S/2 - m_I\Delta_N + Am_I/2 + (1/4)[I(I+1) - m_I(m_I+1)]Ax - (1/8)[I(I+1) - m_I(m_I+1)](2m_I+1)Ax^2 + \cdots \tag{6.55a}$$

$$E(-\tfrac{1}{2}, m_I) = -\Delta_S/2 - m_I\Delta_N - Am_I/2 - (1/4)[I(I+1) - m_I(m_I-1)]Ax + (1/8)[I(I+1) - m_I(m_I-1)](2m_I-1)Ax^2 + \cdots. \tag{6.55b}$$

The energies of the "allowed" transitions are then:

$$h\nu = \Delta_S + Am_I + (1/2)[I(I+1) - m_I^2]Ax - (1/2)[I(I+1) - (m_I^2 + \tfrac{1}{2})]m_IAx^2 + \cdots. \tag{6.56}$$

Rearranging, we find resonance lines at magnetic fields given by

$$H = H_0 - am_I - (1/2)[I(I+1) - m_I^2]ax + (1/2) \times [I(I+1) - (m_I^2 + \tfrac{1}{2})]m_Iax^2 + \cdots, \tag{6.57}$$

where $a = A/g\beta$ is again the hyperfine splitting in gauss. The parameter x may be conveniently written

$$x = (a/H)/(1 + g_N\beta_N/g\beta). \tag{6.58}$$

Since the ratio of the nuclear magnetic moment to that of the electron is very small, the denominator of (6.58) may usually be taken as unity with no significant error.

That the correction to the first-order theory is significant is amply demonstrated by considering a common example, the hyperfine spectrum of the vanadyl ion, VO^{2+}. The vanadium nuclear spin is $\frac{7}{2}$, and the hyperfine splitting in aqueous solution is around 116 G [9]. The first-order theory thus predicts eight lines with equal spacings of 116 G. Table 6.1 gives the second-, third- and fourth-order corrections to the line positions, and the resulting spacings assuming $H_0 = 3400$ G and $a = 116$ G. It is particularly noteworthy that

Table 6.1 Corrections to First-Order Line Positions of Vanadyl Hyperfine Spectrum

m_I	C_2*	C_3*	C_4*	H	*Separation*
$-\frac{7}{2}$	−6.20	−0.57	−0.04	3799.19	
					127.96
$-\frac{5}{2}$	−17.41	−1.30	−0.06	3671.23	
					124.01
$-\frac{3}{2}$	−25.60	−1.21	0.03	3547.22	
					120.03
$-\frac{1}{2}$	−30.43	−0.50	0.12	3427.19	
					116.02
$\frac{1}{2}$	−31.50	0.53	0.13	3311.17	
					112.03
$\frac{3}{2}$	−28.39	1.49	0.04	3199.14	
					108.08
$\frac{5}{2}$	−20.68	1.84	−0.10	3091.06	
					104.11
$\frac{7}{2}$	−7.88	0.92	−0.09	2986.96	

* Second-, third-, and fourth-order corrections from (6.57).

since the entire spectrum is shifted to lower field the center of gravity of the experimental lines no longer determines H_0 and thus the g-value.

In practice, it is usually most convenient to compute g- and a-values from spectra by iterative application of (6.57), retaining an appropriate number of terms, to each line of the spectrum generating a first-order spectrum from which the splitting constant and g-value can be determined directly. These are then used in (6.57) in another stage of iteration. An initial estimate of the hyperfine splitting may be obtained by averaging the $2I$ line separations. This average value is usually very close to the true splitting constant.

SEVERAL EQUIVALENT NUCLEI

The procedure for more than one nucleus follows closely that for a single nucleus, except that for several equivalent nuclei it is necessary to find the appropriate basis functions corresponding to particular quantum numbers K and m_K of the total nuclear spin angular momentum $\mathbf{K} = \sum_i \mathbf{I}_i$ [10]. In the following we present the general results for spin $\frac{1}{2}$ and spin 1 nuclei.

With n equivalent spin $\frac{1}{2}$ nuclei, there is one set of basis functions corresponding to $K = n/2$, having $m_K = \pm(n-1)/2 \ldots, \pm\frac{1}{2}$, or 0. There are $n - 1$ sets of basis functions corresponding to $K = (n/2) - 1$, and in general

$$N(K) = \frac{n!}{[(n/2) - K]!\,[(n/2) + K]!} - \frac{n!}{[(n/2) - K - 1]!\,[(n/2) + K + 1]!} \tag{6.59}$$

sets of basis functions corresponding to the quantum number K. Since there are in general several sets of basis functions corresponding to the same value of K, the corresponding energy levels are degenerate, and the degeneracies are not lifted by the application of the spin Hamiltonian (6.45). The degeneracies of the various states are given in Table 6.2 for up to nine equivalent spin $\frac{1}{2}$ nuclei.

With n equivalent spin 1 nuclei, there is one set of basis functions corresponding to $K = n$, having $m_K = \pm n, \pm(n-1), \ldots, \pm 1, 0$. There are $n - 1$ sets of basis functions corresponding to $K = n - 1$, and so on. Again, degeneracies are encountered, and are given in Table 6.3 for up to four equivalent spin 1 nuclei.

The Hamiltonian matrix elements have the same form as (6.51). It can be shown that the off-diagonal elements couple only states within the same set, so that the Hamiltonian matrix factors into 2×2 blocks of the form (6.52) as above, with the resulting energy levels given by (6.54) and (6.55). The selection rules require transitions involving only changes in the electron spin quantum numbers, so that the transition energies are given by (6.56) and the line positions by (6.57).

The resulting spectrum is as if there were several different species, each having a different nuclear spin and concentration, but all having the same hyperfine splitting constant. Thus, for example, four spin $\frac{1}{2}$ nuclei give rise to a spectrum that is the same as would be obtained from three independent species having nuclei with spin 2 and relative concentration 1, spin 1 and relative concentration 3, and spin 0 and relative concentration 2.

In most practical cases it is sufficient to include terms in (6.57) only to second order. The coefficients of a^2/H, computed from (6.57) are given in Table 6.2 and 6.3 below the corresponding quantum numbers and degeneracies.

NONEQUIVALENT NUCLEI

For several nonequivalent nuclei, the line positions are given by (6.57) with additional terms added for each nucleus or, in the case of equivalent nuclei, for each set of basis functions corresponding to a particular K_i. The second-order solution is usually sufficient except for the important cases in which the splitting constants of different nuclei are nearly equal. When the differences in the splitting constants are on the order of the second-order

Table 6.2 Second-Order Structure of the Esr Spectra of Radicals Containing n Equivalent Spin $\frac{1}{2}$ Nuclei*

	m_K:	4	3		2			1				0				
	K:	4	4	3	4	3	2	4	3	2	1	4	3	2	1	0
$n = 0$																1 (1)
$n = 2$											1 (1)				1	1 (2)
$n = 4$							1 (1)			1	3 (4)			1	3	2 (6)
$n = 6$				1(1)		1	5 (6)		1	5	9(15)		1	5	9	5(20)
$n = 8$		1(1)	1	7(8)	1	7	20(28)	1	7	20	28(56)	1	7	20	28	14(70)
	C_2:	2	11/2	3/2	8	4	1	19/2	11/2	5/2	1/2	10	6	3	1	0

	m_K:	9/2	7/2		5/2			3/2				1/2				
	K:	9/2	9/2	7/2	9/2	7/2	5/2	9/2	7/2	5/2	3/2	9/2	7/2	5/2	3/2	1/2
$n = 1$																1 (1)
$n = 3$											1 (1)				1	2 (3)
$n = 5$							1 (1)			1	4 (5)			1	4	5 (10)
$n = 7$				1(1)		1	6 (7)		1	6	14(21)		1	6	14	14 (35)
$n = 9$		1(1)	1	8(9)	1	8	27(36)	1	8	27	48(84)	1	8	27	48	42(126)
	C_2:	9/4	25/4	7/4	37/4	19/4	5/4	45/4	27/4	13/4	3/4	49/4	31/4	17/4	7/4	1/4

* The lines that are degenerate in the first-order treatment are grouped under the value of m_K, the total z-component of the nuclear spin angular momentum. The relative intensities of the lines are given by the table entries; the numbers in parentheses are the intensities in the first-order treatment. The values C_2 are the displacements of the various lines from the first-order positions in the units of a^2/H.

Table 6.3 Second-Order Structure of the Esr Spectra of Radicals Containing ***n*** Equivalent Spin 1 Nuclei*

m_K:	4	3		2			1				0				
K:	4	4	3	4	3	2	4	3	2	1	4	3	2	1	0
$n = 0$															1 (1)
$n = 1$										1 (1)				1	0 (1)
$n = 2$						1 (1)			1	1 (2)			1	1	1 (3)
$n = 3$			1(1)		1	2 (3)		1	2	3 (6)		1	2	3	1 (7)
$n = 4$	1(1)	1	3(4)	1	3	6(10)	1	3	6	6(16)	1	3	6	6	3(19)
C_2:	2	11/2	3/2	8	4	1	19/2	11/2	5/2	1/2	10	6	3	1	0

* The lines that are degenerate in the first-order treatment are grouped under the value of m_K, the total z-component of the nuclear spin angular momentum. The relative intensities of the lines are given by the table entries; the numbers in parentheses are the intensities in the first-order treatment. The values C_2 are the displacements of the various lines from the first-order positions in units of a^2/H.

corrections, some important line shifts not accounted for by the above treatment may be observed [11]. This effect is best illustrated by working through a simple example, the case of two nonequivalent spin $\frac{1}{2}$ nuclei.

Application of the spin Hamiltonian (6.45) generates the following matrix elements:

$$\begin{aligned}
\langle m_S, m_{I1}, m_{I2}| H_s |m_S, m_{I1}, m_{I2}\rangle \\
&= \Delta_S m_S - \Delta_N(m_{I1} + m_{I2}) + A_1 m_S m_{I1} + A_2 m_S m_{I2} \\
\langle -\tfrac{1}{2}, +\tfrac{1}{2}, \pm\tfrac{1}{2}| H_s |+\tfrac{1}{2}, -\tfrac{1}{2}, \pm\tfrac{1}{2}\rangle &= A_1/2 \\
\langle -\tfrac{1}{2}, \pm\tfrac{1}{2}, +\tfrac{1}{2}| H_s |+\tfrac{1}{2}, \pm\tfrac{1}{2}, -\tfrac{1}{2}\rangle &= A_2/2 \\
\langle +\tfrac{1}{2}, -\tfrac{1}{2}, \pm\tfrac{1}{2}| H_s |-\tfrac{1}{2}, +\tfrac{1}{2}, \pm\tfrac{1}{2}\rangle &= A_1/2 \\
\langle +\tfrac{1}{2}, \pm\tfrac{1}{2}, -\tfrac{1}{2}| H_s |-\tfrac{1}{2}, \pm\tfrac{1}{2}, +\tfrac{1}{2}\rangle &= A_2/2.
\end{aligned} \tag{6.60}$$

The diagonal elements

$$\langle +\tfrac{1}{2}, +\tfrac{1}{2}, +\tfrac{1}{2}| H_s |+\tfrac{1}{2}, +\tfrac{1}{2}, +\tfrac{1}{2}\rangle \quad \text{and} \quad \langle -\tfrac{1}{2}, -\tfrac{1}{2}, -\tfrac{1}{2}| H_s |-\tfrac{1}{2}, -\tfrac{1}{2}, -\tfrac{1}{2}\rangle$$

factor in the secular determinant, leaving two 3 × 3 blocks such as:

$$\begin{array}{c|ccc|}
 & |+\tfrac{1}{2}, -\tfrac{1}{2}, -\tfrac{1}{2}\rangle & |-\tfrac{1}{2}, +\tfrac{1}{2}, -\tfrac{1}{2}\rangle & |-\tfrac{1}{2}, -\tfrac{1}{2}, +\tfrac{1}{2}\rangle \\
\langle +\tfrac{1}{2}, -\tfrac{1}{2}, -\tfrac{1}{2}| & \Delta_S/2 + \Delta_N - (A_1 + A_2)/4 & A_1/2 & A_2/2 \\
\langle -\tfrac{1}{2}, +\tfrac{1}{2}, -\tfrac{1}{2}| & A_1/2 & -\Delta_S/2 - (A_1 - A_2)/4 & 0 \\
\langle -\tfrac{1}{2}, -\tfrac{1}{2}, +\tfrac{1}{2}| & A_2/2 & 0 & -\Delta_S + (A_1 - A_2)/4
\end{array}. \tag{6.61}$$

For $A_1 - A_2$ sufficiently small and $(A_1 + A_2)/\Delta_S$ sufficiently large, significant mixing of the states $|-\frac{1}{2}, +\frac{1}{2}, -\frac{1}{2}\rangle$ and $|-\frac{1}{2}, -\frac{1}{2}, +\frac{1}{2}\rangle$ occurs, and the simple second-order theory is no longer accurate. An exact solution is then virtually required, necessitating the use of a computer to diagonalize the secular determinants. This procedure, while tedious in determining the spin Hamiltonian parameters, does provide one important piece of information: the relative signs of the splitting constants. This technique has been used by Fessenden [12] for the determination of the relative signs of splitting constants for a number of simple radicals.

To give some idea of the magnitude of the shifts involved, consider two spin $\frac{1}{2}$ nuclei with an average splitting $a = (\frac{1}{2}) \times (|a_1| + |a_2|) = 50$ G. To first order, there are four lines with the two central lines separated by the difference in the magnitudes of the splitting constants. A simple second-order correction shifts all the lines by approximately 0.37 G, maintaining the same separations as predicted by the first-order treatment. If the two splitting constants have the same sign, however, the exact treatment introduces quite appreciable deviations from the first-order spacings. Table 6.4 gives the separation of the two central lines for a range of splitting constant differences.

Table 6.4 Separation of the Two Central Lines in the Four-Line Spectrum Attributable to Two Spin $\frac{1}{2}$ Nuclei*

First-Order Separation	*Equal Signs*	*Opposite Signs*
2.00	2.13	2.00
1.00	1.25	1.00
0.50	0.89	0.50
0.00	0.74	0.00

* Assuming an average splitting constant of 50 G. The separation to first order is simply the difference in the splitting constants, $|a_1| - |a_2|$.

Note that if the splitting constant signs are opposite, the first-order theory gives the correct spacings, but that with equal signs quite easily measurable deviations may occur.

Single Crystal Spectra

The analysis of esr spectra of single crystals is complicated by the anisotropies in the g- and A-tensors. Although it is possible to find an axis system that diagonalizes the g-tensor, it is not in general the same axis system that diagonalizes the A-tensor. Indeed, in organic radicals, each nuclear hyperfine interaction has in general a different set of principal axes. This problem is simplified to some extent by the fact that in organic radicals the g-tensor is nearly isotropic, so that attention may be focused on the determination of the components of the hyperfine tensors. In transition metal complexes, in which the symmetry of the orbital(s) containing the unpaired electron(s) is largely determined by the crystal field, the principal axes of the g-tensor and the metal nuclear hyperfine tensor are generally coincident. The hyperfine tensors associated with ligand hyperfine interactions, however, usually have different sets of principal axes.

The procedure used in finding the components of the g- and A-tensors involves measurement of the spectrum at different orientations of the crystal with respect to the applied magnetic field. The tensor components may be deduced through careful analysis of the results. Although practically any set of measurements at known angles is in principle capable of generating the required numbers, it is clearly of some importance to choose a systematic approach in advance in order to simplify both the collection and analysis of the experimental data. Several such procedures are described in the literature

[12, 13]; in order to illustrate the method, we discuss one of the methods in detail, following Weil and Anderson [13].

We consider here only the electron Zeeman term of the spin Hamiltonian

$$\mathscr{H}_s = \mathbf{H} \cdot \overline{\overline{g}} \cdot \mathbf{S}. \tag{6.62}$$

The effective electron angular momentum $\mathbf{S}$ is quantized along the vector $\mathbf{H} \cdot \overline{\overline{g}}$, rather than along the applied field to give energy levels

$$E(m_S) = m_S \beta \, |\mathbf{H} \cdot \overline{\overline{g}}| \tag{6.63}$$

and esr transition energies

$$h\nu = \beta \, |\mathbf{H} \cdot \overline{\overline{g}}|. \tag{6.64}$$

Thus given the applied magnetic field and the microwave frequency, the experimentally observed resonance at any angle may be described simply by the magnitude of the vector $\hat{n} \cdot \overline{\overline{g}}$, where $\hat{n}$ is a unit vector in the direction of the field. In practice, it is somewhat more convenient to work with the parameter

$$g^2 = (\hat{n} \cdot \overline{\overline{g}}) \cdot (\hat{n} \cdot \overline{\overline{g}}) = \hat{n} \cdot \overline{\overline{G}} \cdot \hat{n}, \tag{6.65}$$

the square of the experimentally determined "g-value." The matrix $\overline{\overline{G}}$ is the matrix product of $\overline{\overline{g}}$ and its transpose; since the principal values of the g-tensor are positive, there is no loss of information in dealing with the tensor $\overline{\overline{G}}$, and a transformation that diagonalizes $\overline{\overline{G}}$ also diagonalizes $\overline{\overline{g}}$.

To define the direction of the various vectors, we choose a rectangular coordinate system, fixed in the crystal and usually coincident with one or more of the crystal axes. The direction of the magnetic field is then described by the polar and azimuthal angles θ and ϕ in the crystal-fixed coordinate system,

$$\mathbf{H} = H(\hat{\imath} \sin\theta \cos\phi + \hat{\jmath} \sin\theta \sin\phi + \hat{k} \cos\theta). \tag{6.66}$$

Combining (6.65) and (6.66), we easily find that

$$\begin{aligned} g^2 = {} & (G_{11} \cos^2\phi + 2G_{12} \sin\phi \cos\phi + G_{22} \sin^2\phi) \sin^2\theta \\ & + 2(G_{13} \cos\phi + G_{23} \sin\phi) \sin\theta \cos\theta + G_{33} \cos^2\theta. \end{aligned} \tag{6.67}$$

An experimental procedure is quite clearly suggested by the rearranged form of (6.67):

$$\begin{aligned} g^2 = {} & (1/2)[G_{33} + (1/2)(G_{11} + G_{22}) + (1/2)(G_{11} - G_{22}) \cos 2\phi \\ & + G_{12} \sin 2\phi] \\ & + (1/2)[G_{33} - (1/2)(G_{11} + G_{22}) - (1/2)(G_{11} - G_{22}) \cos 2\phi \\ & - G_{12} \sin 2\phi] \cos 2\theta \\ & + (G_{13} \cos\phi + G_{23} \sin\phi) \sin 2\theta. \end{aligned} \tag{6.68}$$

We see that for rotations through the angle ϕ with θ set at 90° or for rotations through the polar angle θ holding ϕ constant, g^2 goes through a variation of the form

$$g^2 = A + B \cos 2\sigma + C \sin 2\sigma,$$

or (6.69)

$$g^2 = A + (B^2 + C^2)^{1/2} \cos 2(\sigma - \sigma_+),$$

where σ represents either θ or ϕ under the above conditions and $2\sigma_+ = \cos^{-1} B/(B^2 + C^2)^{1/2}$. The procedure then is to perform three rotations: (1) rotation through ϕ with θ held at 90°; (2) rotation through θ with ϕ fixed at the value that maximized g^2 on the first rotation; and (3) rotation through θ with ϕ fixed at the value that minimized g^2 on the first rotation. For each of these three rotations, a set of data is obtained that is fitted to (6.69), obtaining nine parameters in all. It can be shown that the components of the G-tensor are related to these fitted parameters by:

$$\begin{aligned}
G_{11} &= A_1 + (A_2 - A_3)D \\
G_{22} &= A_1 - (A_2 - A_3)D \\
G_{33} &= A_2 + A_3 - A_1 \\
G_{12} &= D(A_2 - A_3)C_1/B_1 \\
G_{13} &= (1 + D)^{1/2}(A_3 - A_1)C_2/2^{1/2}B_2 + (1 - D)^{1/2}(A_2 - A_1)C_3/2^{1/2}B_3 \\
G_{23} &= (1 - D)^{1/2}(A_3 - A_1)C_2/2^{1/2}B_2 - (1 + D)^{1/2}(A_2 - A_1)C_3/2^{1/2}B_3,
\end{aligned} \tag{6.70}$$

where the subscripts on A, B, and C refer to the number of the rotation, and $D = B_1/(B_1^2 + C_1^2)^{1/2}$. The components of $\bar{\bar{G}}$ are somewhat overdetermined by the nine experimental parameters, so that internal consistency checks are available, such as

$$A_2 + B_2 = A_3 + B_3 \tag{6.71}$$

Once the components of the G-tensor are known, it is straightforward to diagonalize its matrix to determine the principal values and thus the principal values of the g-tensor.

Determination of the principal values of an electron-nuclear interaction tensor by analysis of the hyperfine structure, or the principal values of an electron-electron interaction tensor by analysis of the fine structure, follows much the same procedure, and can usually be made with the same set of measurements.

Polycrystalline Spectra

The esr spectra of polycrystalline, powdered, or glassy samples are generally difficult to interpret accurately. In order to understand the reasons for the difficulties and to see the directions in which one must go to obtain reliable interpretation of experimental results, we first discuss the line shapes expected from esr powder spectra.

The line shape function is derived by considering the probability of absorption by a crystallite when the magnetic field is between H and $H + dH$, and is oriented at angles θ and ϕ within the solid angle $d\Omega$ with respect to the crystallite axis system, usually chosen as the principal axes of the G-tensor. If it is then assumed that the orientations of the crystallites are entirely random, the normalized absorption probability function is

$$S(H)\,dH = (4\pi)^{-1}\int_{H}^{H+dH} I(\Omega)\,d\Omega, \tag{6.72}$$

where

$$\int_0^\infty S(H)\,dH = 1. \tag{6.73}$$

The function $I(\Omega)$ is the transition probability when the crystallite is oriented at solid angle Ω with respect to the magnetic field.

The integral (6.72) can be solved analytically only for a few simple cases such as an axial G-tensor and Lorentzian crystallite line shapes [14] or an anisotropic G-tensor and δ-function crystallite line shapes [15]. In the present discussion, we limit ourselves to axial symmetry and δ-function line shapes. Furthermore, we neglect variations in the transition probability with orientation. This is tantamount to assuming a small G-tensor anisotropy since Bleaney [16] has shown that for axial symmetry the transition probability is propotional to

$$g_\perp{}^2[(g_\parallel/g)^2 + 1], \tag{6.74}$$

where g is a function of angle

$$g^2 = g_\parallel{}^2 \cos^2\theta + g_\perp{}^2 \sin^2\theta. \tag{6.75}$$

For organic radicals, this is obviously a very good assumption, but for large g-tensor anisotropies, the errors may be large. For example, with $g_\parallel = 2g_\perp$, the transition probability changes by a factor of 1.5 as the orientation is varied.

Writing $d\Omega = \sin\theta\,d\theta\,d\phi$, we note that for axial symmetry H is not a function of ϕ, so that the integration over ϕ in (6.72) can be performed immediately. With

$$\begin{aligned} H_\parallel &= h\nu/g_\parallel\beta \\ H_\perp &= h\nu/g_\perp\beta, \end{aligned} \tag{6.76}$$

a straightforward calculation gives:

$$S(H) = H_\perp{}^2 H_\parallel (H_\perp{}^2 - H_\parallel{}^2)^{-1/2} H^{-2} (H_\perp{}^2 - H^2)^{-1/2}$$

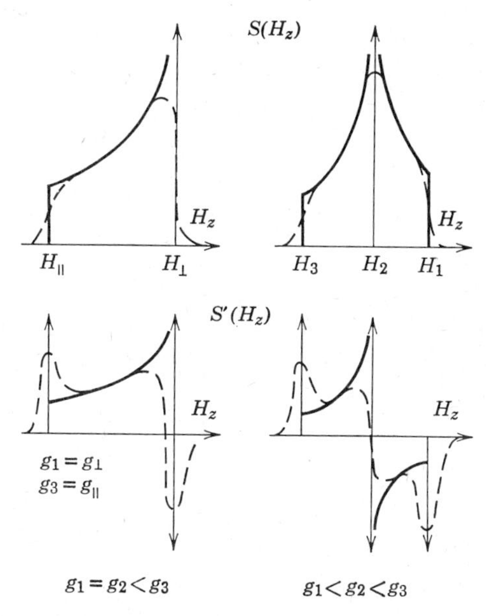

Fig. 6.1 The shape functions $S(H)$ and their first derivatives for (*a*) an axially symmetric *g*-tensor, and (*b*) an anisotropic *g*-tensor. (From Ref. 15.)

for $H_{\perp} \leq H \leq H_{\parallel}$. The general shapes of this function and its first derivative are shown in Fig. 6.1*a*.

Kneubühl [15] has extended the above treatment to the case of a completely anisotropic *g*-tensor. His qualitatively predicted line shapes are shown in Fig. 6.1*b*. Sands [17] and others [18, 19] have included an axial hyperfine tensor in the calculations of powder patterns. In those cases in which the crystallite line widths are sufficiently small, the parameters of the spin Hamiltonian may be obtained at least approximately directly from the experimental spectrum.

Ibers and Swalen obtained an analytical solution for an axial *g*-tensor and Lorentzian crystallite line shapes [14]. The effect of finite line width is shown quite clearly in Fig. 6.2. Clearly, when the crystallite line width is appreciable, substantial errors may be made in reading the spin Hamiltonian parameters from the spectrum, and for very broad lines little or no information is obtainable directly. In these cases it is necessary to simulate the line shape by computer calculation.

Computer line shape simulations have usually taken the form of substitution of the appropriate line shape function, together with the spin Hamiltonian parameters into the integral over orientations (6.72). Taylor and Bray [20] have pointed out, however, that a considerable simplification can be

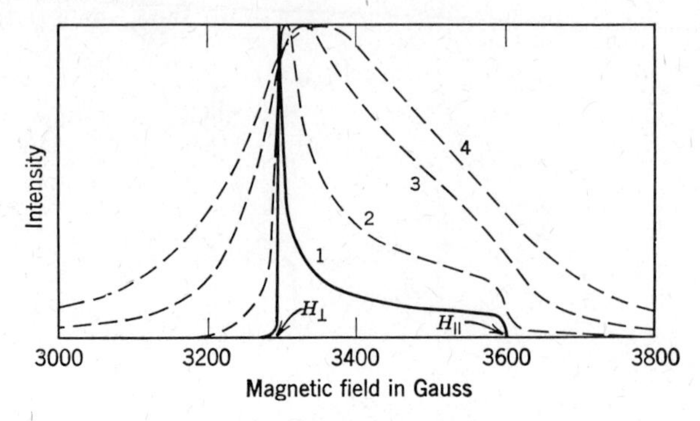

Fig. 6.2 Calculated shape function $S(H)$ for $H_\parallel = 3600$ G, $H_\perp = 3300$ G and crystallite line half-widths of *(1)* 1 G, *(2)* 10 G, *(3)* 50 G, and *(4)* 100 G. The intensities are normalized to unity; the relative values of the true maxima are 963:35:3:1. (From Ref. 14.)

Table 6.5 Summary of Esr Line Shape Calculations for Polycrystalline Samples

g-Tensor	*Hyperfine Tensor*	*Crystallite Line Shape*	*Reference*	*Comments*
Isotropic	None	Delta	[21]	$S = \frac{3}{2}$
Isotropic	None	Gaussian	[22]	$S = 1$
Isotropic	Anisotropic	Delta	[23, 24]	—
Isotropic	Anisotropic	Gaussian	[24, 25]	—
Axial	None	Delta	[26]	—
Axial	None	Gaussian	[27, 28]	Small anisotropy
Axial	None	Lorentzian	[14, 28]	—
Axial	Axial	Delta	[17]	—
Axial	Axial	Gaussian	[18, 19]	—
Axial	Axial	Lorentzian	[19]	—
Anisotropic	None	Delta	[15, 29]	$S > \frac{1}{2}$
Anisotropic	None	Delta	[30]	—
Anisotropic	None	Gaussian	[31]	$S = 1$
Anisotropic	None	Lorentzian	[32]	Small anisotropy
Anisotropic	None	Gaussian and Lorentzian	[33]	Computer programs
Anisotropic	Anisotropic	Gaussian	[34]	Computer programs any number of nuclei up to spin $\frac{3}{2}$
Anisotropic	Anisotropic	Gaussian and Lorentzian	[20]	Computer programs one nucleus only more than one magnetic site

effected if the average is performed for zero line width and the resulting line shape function convoluted with a crystallite line shape function to obtain the broadened powder pattern (see Section 4, p. 559).

The wide variety of approaches found in the literature to calculations of esr powder patterns is summarized in Table 6.5.

3 SPECTROMETER DESIGN AND OPERATION

Microwaves and Waveguides

The microwave region of the electromagnetic spectrum covers the wavelength region roughly from 1 mm to 1 m and is thus characterized by wavelengths of the order of magnitude of commonly encountered laboratory equipment. The wavelength is neither long enough so that microwave circuits can be treated by the ordinary methods of electrical circuits, not short enough that geometrical optics can be applied, although both limits occasionally provide useful insights. Indeed, the only generally correct way of approaching microwave behavior is through Maxwell's equations and the wave equation derived from the Maxwell relations [35, 36]:

$$\nabla^2\mathbf{E} = (4\pi\mu/c^2\rho)\frac{\partial\mathbf{E}}{\partial t} + (\mu\epsilon/c^2)\frac{\partial^2\mathbf{E}}{\partial t^2}\,. \tag{6.77}$$

In (6.77) $\mathbf{E}$ is the electric field associated with the electromagnetic radiation; μ, ϵ, and ρ are, respectively, the permeability, the dielectric constant, and the resistivity of the medium; and c is the velocity of light. If we assume a solution to (6.77) representing a wave of definite frequency $\nu = \omega/2\pi$ traveling along the z-axis,

$$\mathbf{E} = \mathbf{E}_0 \exp i(\omega t - \gamma z), \tag{6.78}$$

we find that the wave equation is satisfied if

$$\gamma^2 = (\mu\epsilon/c^2)\omega^2 - i(4\pi\mu/c^2\rho)\omega. \tag{6.79}$$

If the propagation constant γ is real, as would be the case if the resistivity of the medium were infinite, the wave is propagated along the z-axis without dimunition. If the resistivity of the medium were finite, however, an imaginary term in the propagation constant leads, when substituted into (6.78) to an exponential damping term. If the resistivity is small enough that the first term of (6.79) may be neglected, a strongly damped wave results with E decreasing by $1/e$ in a distance:

$$\delta = (c^2\rho/2\pi\mu\omega)^{1/2}, \tag{6.80}$$

called the skin depth. Typical skin depths of metals for microwave with frequencies around 10 GHz are on the order of 10^{-4} to 10^{-5} cm.

We now consider the wave propagated through a hollow rectangular metal pipe called a waveguide. Figure 6.3 shows the coordinate system. The boundary conditions on the solution are, assuming that the walls are perfect conductors with zero skin depth,

$$\begin{aligned} E_x &= 0 \quad \text{at} \quad y = 0 \quad \text{and at} \quad y = B \\ E_y &= 0 \quad \text{at} \quad x = 0 \quad \text{and at} \quad x = A. \end{aligned} \tag{6.81}$$

If a perfect vacuum with infinite resistivity and unit dielectric constant and permeability inside the waveguide is assumed, the wave equation (6.77) becomes

$$\nabla^2\mathbf{E} = (1/c^2)\frac{\partial^2\mathbf{E}}{\partial t^2}. \tag{6.82}$$

We consider only solutions representing waves propagated along the waveguide with no longitudinal electric field component. Other solutions are obviously possible but these, called the transverse electric field (TE) modes, are by far the most common. With the boundary conditions (6.81), we obtain:

$$\begin{aligned} E_x &= -E_0(nA/mB)\cos(m\pi x/A)\sin(n\pi y/B)\exp i(\omega t - \gamma z) \\ E_y &= E_0 \sin(m\pi x/A)\cos(n\pi y/B)\exp i(\omega t - \gamma z) \\ E_z &= 0, \end{aligned} \tag{6.83}$$

with the restriction that:

$$(m\pi/A)^2 + (n\pi/B)^2 + \gamma^2 = \omega^2/c^2 = (2\pi/\lambda_0)^2, \tag{6.84}$$

where λ_0 is the wavelength of the same wave in free space, and m and n are integers. Rearranging (6.84), we find:

$$\gamma^2 = (2\pi/\lambda_0)^2 - (m\pi/A)^2 - (n\pi/B)^2. \tag{6.85}$$

In order for wave propagation to take place, we have seen that γ must be real, which implies that λ_0 must be less than a critical cutoff wavelength λ_c:

$$(1/\lambda_c)^2 = (m/2A)^2 + (n/2B)^2. \tag{6.86}$$

The length of the wave in the waveguide is seen from (6.83) to be

$$\lambda_g = 2\pi/\gamma = \lambda_0/[1 - (\lambda_0/\lambda_c)^2]^{1/2}. \tag{6.87}$$

Thus the guide wavelength approaches infinity as the free space wavelength approaches cutoff. When the free space wavelength is short compared with the cutoff wavelength, however, the guide wavelength is approximately equal to the free space wavelength.

The integers m and n characterize the mode of propagation; the different allowed waves are designated TE_{mn} modes. By far the most commonly used is the TE_{10} mode, which is the longest wavelength mode propagated, having a cutoff wavelength $\lambda_c = 2A$; throughout the remainder of this chapter, we assume that the TE_{10} mode is the only one of importance. The electric and magnetic fields for this mode are found from (6.83) and Maxwell's relation between E and H [35]:

$$\begin{aligned} E_x &= 0 \\ E_y &= E_0 \sin(\pi x/A) \exp i(\omega t - \gamma z) \\ E_z &= 0 \\ H_x &= -(\lambda_0/\lambda_g)E_0 \sin(\pi x/A) \exp i(\omega t - \gamma z) \\ H_y &= 0 \\ H_z &= i(\lambda_0/\lambda_c)E_0 \cos(\pi x/A) \exp i(\omega t - \gamma z). \end{aligned} \tag{6.88}$$

The electric and magnetic fields for the TE_{10} mode are shown in Fig. 6.4.

Note that the dimension B, in the y-direction of Fig. 6.3 is of no significance for the TE_{10} mode; since it is important to suppress other modes, however, B is usually made on the order of one-half of dimension A. Since the electric field of the radiation is parallel to the y-z plane, this plane is called the E-plane of the waveguide. Similarly, since the magnetic fields lie in the x-z plane, this plane is called the H-plane. A simple mnemonic which makes this notation easy to remember refers to waveguide bends. Two varieties of bends, shown in Fig. 6.5, are possible, and are designated E-bends and H-bends, depending upon which surface remains planar. Clearly, the E-bend is the *E*asy way to bend the waveguide and the H-bend is the *H*ard way.

Waveguide and microwave components are available commercially in certain discrete sizes, commonly given a letter designation, for example, X-band. Unfortunately, there is incomplete agreement between different

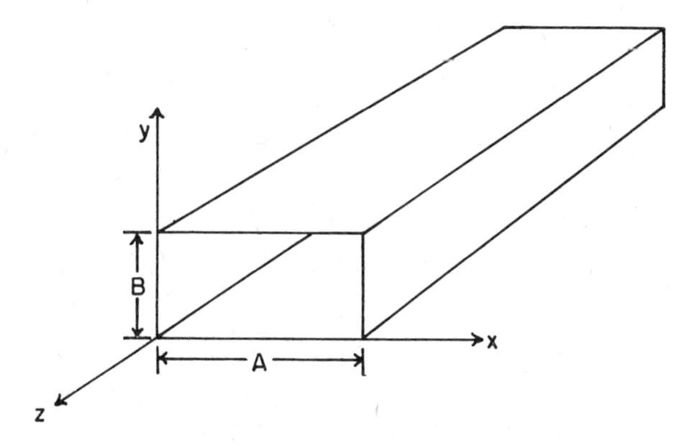

Fig. 6.3 Rectangular hollow metal waveguide.

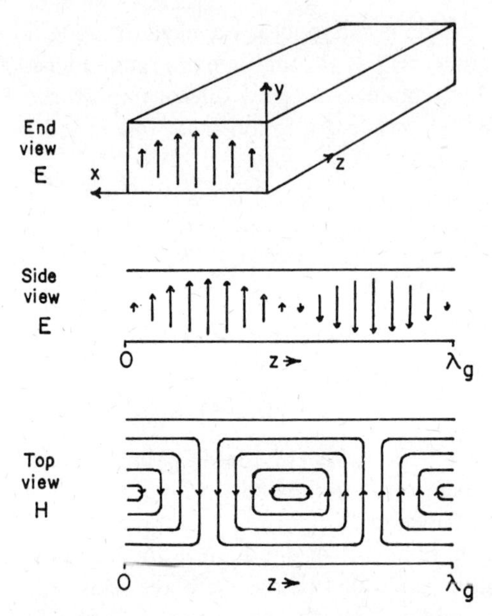

Fig. 6.4 Electric and magnetic fields for TE_{10} mode microwave propagation in a rectangular waveguide.

manufacturers on these designations, so that some caution is needed in designating microwave frequency ranges by lettered bands. Two other designations in common use which provide a more certain identification are the joint Army-Navy (JAN) designations and the Electronics Industry Association (EIA) designations. These various designations, together with some pertinent data on waveguide types, are given in Table 6.6. The majority of all esr spectrometers operate in X-band at around 9 GHz, although spectrometers are commercially available that operate in L-band (1 GHz), S-band (3 GHz), K-band (23 GHz), KA- or Q-band (35 GHz), V-band (70 GHz), as well as in the radiofrequency range of 100 to 800 MHz.

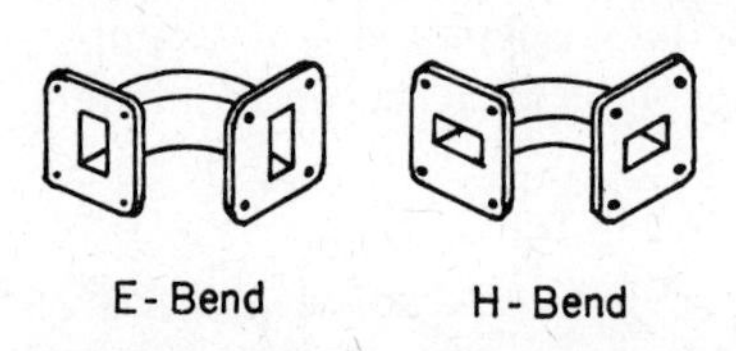

Fig. 6.5 *E*-Plane and *H*-plane waveguide bends.

Microwave Components

In the following discussion we consider briefly some of the more common microwave circuit elements found in esr spectrometers. Further information on these components and others can be found in the books by Poole [36],

Table 6.6 Standard Waveguide Data

Designations			*Rectangular Waveguide Inside Dimensions* (in.)	TE_{10} *Mode*		
*Band**	*EIA*	*JAN*		*Frequency Range* (GHz)	*Frequency Cutoff* (GHz)	*Free Space Wavelength* (cm)
L	WR650	RG-69/U	6.500 × 3.250	1.12–1.70	0.908	26.77–17.63
LS (R)	WR430	RG-104/U	4.300 × 2.150	1.70–2.60	1.375	17.63–11.53
S	WR284	RG-48/U	2.840 × 1.340	2.60–3.95	2.080	11.53–7.59
C (G or H)	WR187	RG-49/U	1.872 × 0.872	3.95–5.85	3.152	7.59–5.12
XN (XB, XC, or J)	WR137	RG-50/U	1.372 × 0.622	5.30–8.20	4.301	5.66–3.66
X (XS)	WR90	RG-52/U	0.900 × 0.400	8.20–12.4	6.557	3.66–2.42
KU (P or Y)	WR62	RG-91/U	0.622 × 0.311	12.4–18.0	9.487	2.42–1.67
K (Q)	WR42	RG-53/U	0.420 × 0.170	18.0–26.5	14.08	1.67–1.13
KA (V, R, or U)	WR28	RG-96/U	0.280 × 0.140	26.5–40.0	21.10	1.13–0.749
Q (V or K)	WR22	RG-97/U	0.224 × 0.112	33.0–50.0	26.35	0.908–0.600
V (M or W)	WR15	RG-98/U	0.148 × 0.074	50.0–75.0	29.90	0.600–0.400
E (Y)	WR12	RG-99/U	0.122 × 0.061	60.0–90.0	48.40	0.500–0.333

* Alternative band designations are given in parentheses.

Alger [37], Townes and Schawlow [35], as well as Volumes 8–11 and 14 of the *MIT Radiation Laboratory Series* [39].

Klystrons

Virtually all esr spectrometers employ a reflex klystron as the microwave power source. Although other microwave sources are occasionally used and have some advantages, we confine our attention to the klystron and the reader is referred to Poole [36] for an introduction to other microwave sources.

A klystron is an electron vacuum tube with conventional electrode elements arranged in an unconventional way [35–37]. A schematic diagram of a reflex

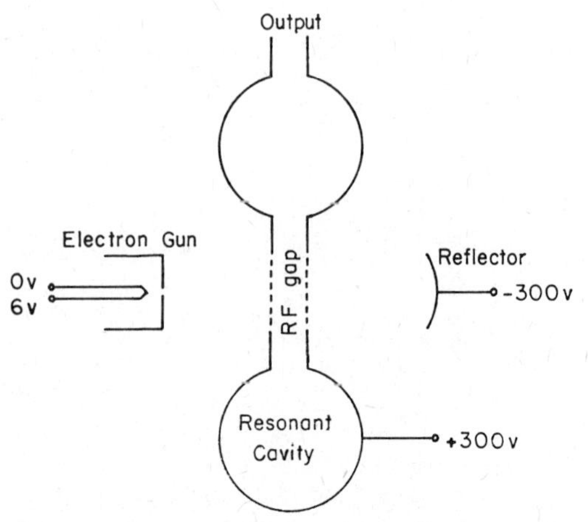

Fig. 6.6 Schematic diagram of a reflex klystron.

klystron is shown in Fig. 6.6. Thermionic electrons emitted from the cathode are attracted into the rf gap by the positive voltage applied to the walls of the resonant cavity (anode). The microwave electric field in the rf gap tends to alternately speed up and slow down the electron beam so that it forms "bunches" at some distance beyond the rf gap. Instead of allowing the beam to continue, however, a negatively polarized reflector (or repeller) electrode turns the electron beam around, and it arrives in the rf gap in the maximum bunched condition. Energy is transferred from the bunched electron beam to the microwave electric field to maintain the oscillations. The dimensions of the tube and the voltages on the reflector electrode and anode are clearly of critical importance in determining the operating frequency of the klystron. In general, several different combinations of reflector and beam voltages result in stable operation, giving rise to the so-called klystron modes. The cavity dimensions of a reflex klystron can be adjusted mechanically so that the klystron is tunable over a frequency range, typically about 10 to 20%.

The electronic tuning range achievable by varying the reflector voltage is smaller, usually on the order of a few tenths of a percent. The Varian V-153/6315 klystron, as a typical example, is tunable mechanically from 8.5 to 10.0 GHz; the electronic tuning range varies from about 90 MHz at 8.5 GHz to about 65 MHz at 10.0 GHz. Klystrons are usually designed to deliver microwave power in the range of tens to hundreds of milliwatts. The V-153/6315 is fairly typical with an output power of about 100 mW when operating with a beam voltage of 250 V.

A great deal of additional information on klystrons is given by Poole [36].

Attenuators and Terminations

Used to control the microwave power level, attenuators consist of a resistive element inserted into the waveguide. Electric currents, induced by the microwave electric field in the resistive element, dissipate power by ohmic losses. Matched loads or terminations are used to absorb microwave power at the end of a waveguide with a minimum of reflection. They usually consist of a tapered section of lossy material centered in the waveguide parallel to the *E*-plane.

Directional Couplers and Magic Tees

A directional coupler is a microwave component that allows one to monitor the microwave power in a waveguide by extracting a small amount of it. A wide variety of types of directional couplers exist, but most consist of two adjacent sections of waveguide with microwave power coupled between them by means of holes, slots, or probes. For example, in the so-called two-hole directional coupler, two waveguides are mounted with adjacent *E*-plane walls, and two coupling holes are placed in the waveguide one-quarter wavelength apart. In this configuration waves traveling in the main guide couple power into the secondary guide, producing a wave traveling in the same direction. If it is desired to monitor only power flowing

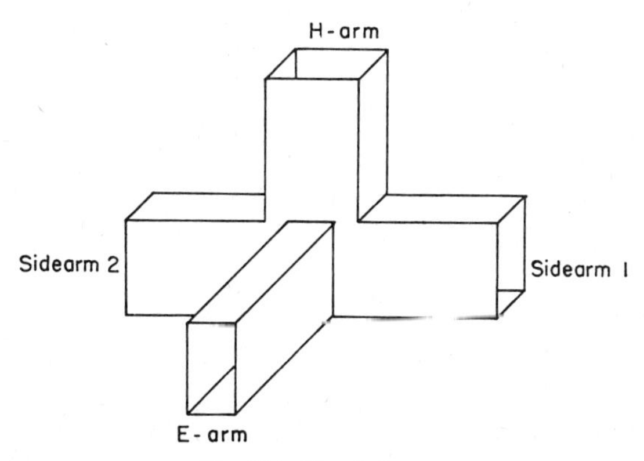

Fig. 6.7 Magic tee.

in one direction in the main guide, one end of the secondary guide is terminated with a matched load (see above). Coupling factors for directional couplers vary with the type from 2 to 30 dB.

A magic tee, shown in Fig. 6.7, is a kind of directional coupler with equal power distribution between two arms. Microwave power entering through the *E*- or *H*-arms couple through their respective *E*- or *H*-fields to the two side arms but not to each other. Posts and/or diaphragms are introduced to match the impedances of each arm.

Tuners

A slide-screw tuner employs a screw obstruction which may be inserted into the waveguide and moved along the waveguide for at least one-half a guide wavelength. The function of the tuner is to match cavities or detectors to the characteristic impedance of the waveguide. This is achieved by reflections from the screw, the magnitude of which are adjustable by the penetration depth and the phase of which is adjustable by the position of the screw along the waveguide.

Tuning of the same kind can also be achieved by two- and three-screw tuners in which fixed screws are located one-eighth or one-fourth of the guide wavelength apart.

Another type of tuning is obtained with an iris which is a metallic partition extending partially across the waveguide. Its function basically is impedance matching, since partitions extending from the *H*-faces of the waveguide are capacitive, and those extending from the *E*-faces are inductive. Partitions with centered holes thus act as resonant circuits. The most common use of an iris in esr spectrometers is as a coupling hole between a waveguide and a cavity. A tuning screw, adjustable into the coupling hole, is frequently provided for fine adjustments of impedance matching.

Phase Shifters

Phase shifters are used to adjust the phase of microwaves arriving at a resonant cavity, detector, and so on. The most direct means of accomplishing a phase shift is to change the waveguide length, and the so-called trombone-type phase shifter does precisely this. An alternative and equally direct method is to squeeze the waveguide so that by reducing the cross section of the guide, the guide wavelength is increased and the effective waveguide length reduced. More sophisticated phase shifters employ a section of dielectric, the position of which may be adjusted in the microwave electric field to retard the phase of the microwaves.

Isolators and Circulators

An isolator is a ferrite device which permits the transmission of microwaves in one direction but not in the other. A ferrite is a magnetic material with extremely high resistivity; the skin depth is then large enough so that microwaves may be propagated through a piece of ferrite material. If a ferrite rod is placed in a circular waveguide, as shown in Fig. 6.8, and the electron spins in the ferrite are aligned by a magnetic field along the direction of microwave propagation, microwaves passing through the ferrite have their plane of polarization rotated (the Faraday effect). If the circular section of waveguide is joined to rectangular sections oriented at 45° with respect to one another, and the length of the ferrite section appropriately adjusted, microwaves transmitted into port 1 have their plane of polarization rotated so that they may be propagated in the rectangular waveguide of port 2, whereas those entering at port 2 have their plane of polarization oriented at 90° to that required for propagation in port 1. Addition of sheets of lossy material perpendicular to the plane of the electric field of the allowed waves in each port help to attenuate the disallowed waves.

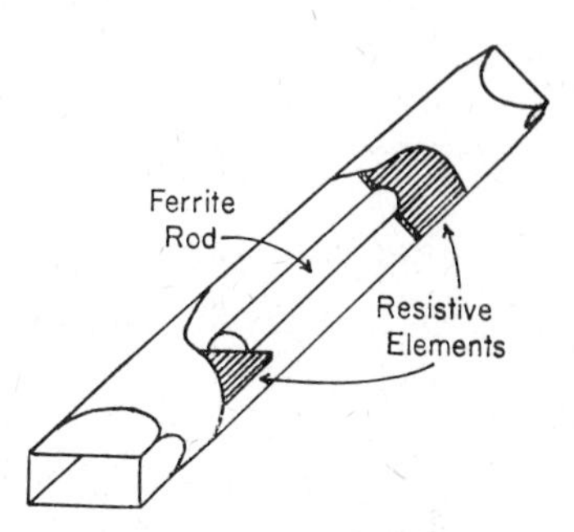

Fig. 6.8 Ferrite isolator.

Circulators are ferrite devices which may most simply be viewed as a collection of isolators arranged in a ring such that microwaves propagated into port 1 emerge only at port 2, those propagated into port 2 emerge only at port 3, and so on.

Cavities

A resonant cavity is essentially a section of rectangular or cylindrical waveguide, closed on both ends, but with some provision for coupling of microwave power. The dimensions of the cavity are such that a standing wave can be maintained within the cavity. Since most esr spectrometers employ a resonant cavity to house the sample, cavity design and properties take on considerable importance. For this reason, more detailed discussion is deferred to p. 546 below.

Slow-Wave Structures [39]

A slow-wave structure in its simplest form is simply a waveguide operating nearly at cutoff. Under these conditions, where, according to (6.87), the guide wavelength is very large, the wave is propagated through the guide very slowly, hence the name slow wave. Since the wave is propagated slowly, the stored energy per unit length is increased. Hence a slow-wave structure can be used in much the same way as a waveguide transmission cell but without the problems of requiring a very long cell to achieve sufficient absorption to be detectable. In practice, it is possible to construct a slow-wave device in the form of an open wire helix, which has enormous advantages with regard to access of the sample, penetration of the field modulation signal, and so on. The difficulty is that both the electric and magnetic fields are maximum along the helix axis where the sample would ordinarily be placed, so that lossy samples cannot be used. Other types of slow-wave structures show promise for esr use, but few data are available in the literature.

Esr Spectrometers

A wide variety of esr spectrometers have been constructed, and many are now commercially available. A complete discussion of the various types, even of those represented by the commercial instruments, is beyond the scope of this chapter, and the reader is referred to other sources for further information [36, 37, 39].

The prospective designer (or purchaser) of an esr spectrometer is faced with what may be a bewildering choice of design features. The following discussion is intended as an introduction to some of these choices, but it is suggested that the reader refer to a more comprehensive source, particularly Chapter III of Alger's book [37], before choosing a spectrometer system.

Most esr spectrometers make use of the same basic microwave circuit in which the klystron oscillator is locked to the resonant cavity, and detection of magnetic resonance absorption takes place by monitoring the microwave power reflected from the cavity with a crystal detector. Crystal noise effects, which usually go as 1/frequency, are suppressed by modulating the magnetic field, usually at 100 kHz, and processing the resultant 100-kHz signal from the crystal detector. Such a spectrometer, which is characterized by its use of homodyne detection, is described below.

The second most common type of spectrometer attacks the crystal noise problem by incorporating a local oscillator klystron so that the signal at the crystal detector is the intermediate frequency resulting from beating the two klystron frequencies. Such a spectrometer is referred to as a superheterodyne spectrometer.

Homodyne Detection Spectrometer

The block diagram of a typical esr spectrometer employing homodyne detection, 100-kHz field modulation, and klystron stabilization by locking to the sample resonant cavity is shown in Fig. 6.9.

Microwave power from the klystron oscillator is routed through an isolator and attenuator used, respectively, to prevent reflected microwaves from interfering with the operation of the klystron and to allow for the control of the radiation power level at the cavity. Power is then directed by the circulator

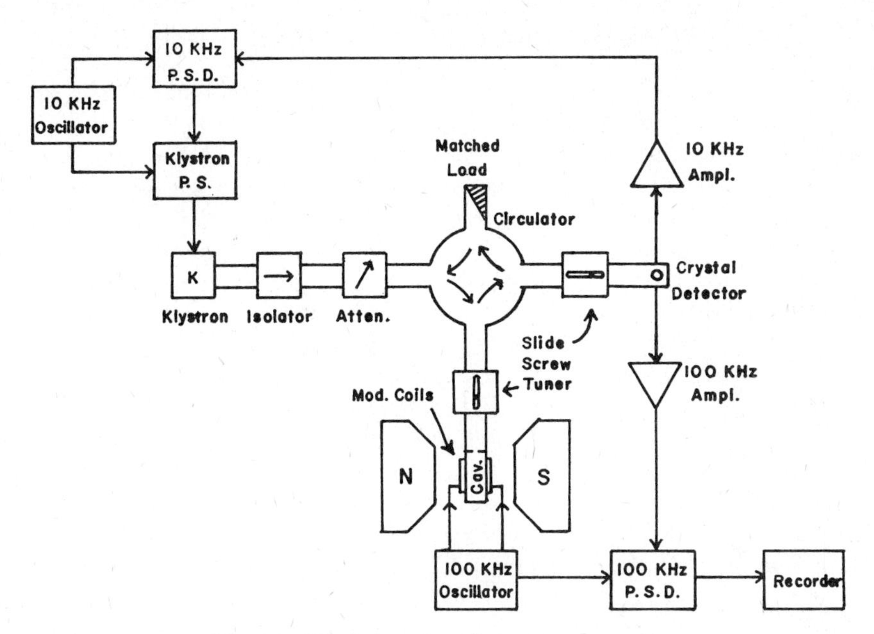

Fig. 6.9 Homodyne detection esr spectrometer.

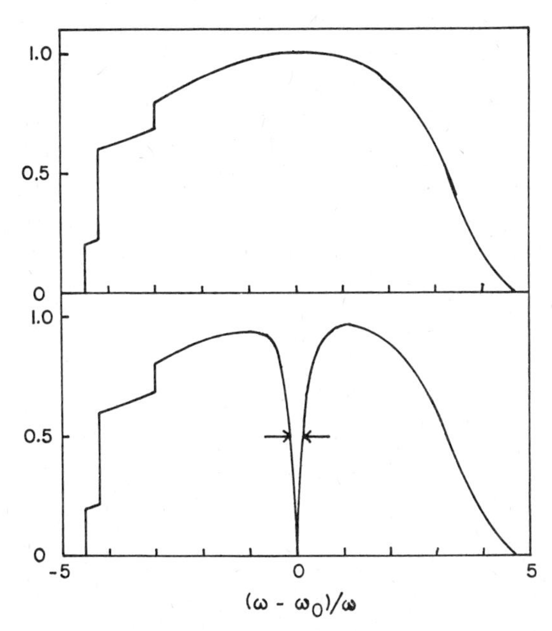

Fig. 6.10 Klystron power to cavity (top). Microwave power reflected from cavity showing cavity resonance (bottom). The width at half height of the cavity dip is $1/\sqrt{3}Q_c$.

to the resonant cavity with impedance matching provided by a slide-screw tuner and a variable coupling iris. Microwave power reflected by the cavity is routed by the circulator to a crystal detector, which is also tuned by a slide-screw tuner. Any power reflected from the tuner arm is routed by the circulator to a matched load where it is absorbed.

Figure 6.10 shows a plot of microwave power arriving at the cavity and microwave power reflected from the cavity as a function of microwave frequency (and thus of klystron reflector voltage). The incision in the reflected power curve, known as the cavity dip, is attributable to the fact that the cavity is resonant at that frequency. If the klystron reflector voltage is modulated by a small-amplitude 10-kHz voltage, the signal at the microwave detector contains a 10-kHz signal, the amplitude and phase of which carry information regarding the position of the klystron center frequency relative to the cavity resonant frequency. The 10-kHz signal is amplified and phase-sensitive-detected so that a dc voltage is obtained which is negative for frequency deviations in one direction and positive for the opposite frequency deviations. This signal is then applied to the klystron reflector power supply to provide a lock of the klystron frequency to the cavity resonant frequency.

Modulation coils, mounted on the cavity, are driven at 100 kHz to modulate the magnetic field. When magnetic resonance absorption takes place within the cavity, the microwaves reflected from the cavity are amplitude-modulated at 100 kHz. The signal from the crystal detector is then routed through a narrow band amplifier and a lock-in detector operating at 100 kHz which provides a dc signal to drive a recorder.

Numerous variations on the above basic spectrometer are possible. For example, the klystron could be stabilized on an external cavity or locked to an external frequency standard. Additional field modulation schemes are possible, some of which are discussed in detail in Section 5. A magic tee or a directional coupler can be used in place of the circulator as the bridge element, but with a loss in sensitivity.

Superheterodyne Detection Spectrometer

A typical superheterodyne esr spectrometer is diagrammed in Fig. 6.11. The principal difference between homodyne detection and superheterodyne detection is the use of a local oscillator or reference klystron locked to the signal klystron with a fixed-frequency difference (typically 30 MHz), and so arranged that the crystal detector supplies an intermediate frequency (i.f.) signal.

Microwave power from the signal klystron is routed, as with the homodyne system, through an isolator and attenuator, and by way of a circulator to the resonant cavity. Power reflected from the resonant cavity is directed by the circulator to the crystal detector, operating in this case as a mixer. Microwaves from the signal klystron are also coupled into the reference circuit

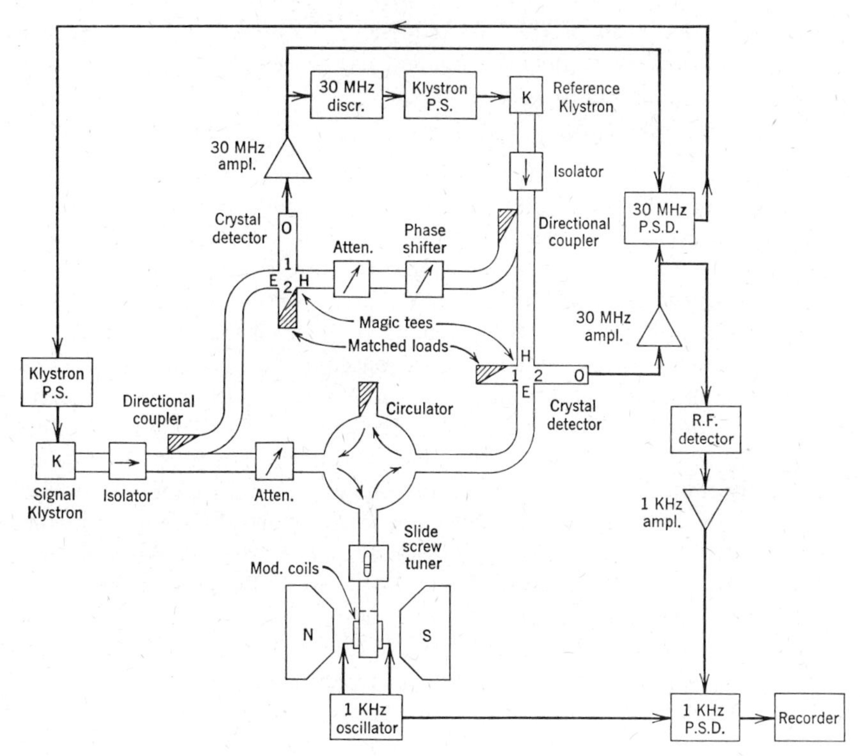

Fig. 6.11 Superheterodyne esr spectrometer.

by means of a directional coupler between the signal klystron attenuator and the circulator.

Microwave power from the reference klystron is directed through an isolator and attenuator to the signal crystal detector where it is mixed with signal klystron power reflected from the cavity to provide the i.f. signal. Reference power is also diverted by a directional coupler through a phase shifter and attenuator to a second crystal mixer where a reference i.f. signal is produced. The i.f. signal from the reference mixer is amplified and applied to a Foster-Seeley discriminator which produces a dc automatic frequency control signal to lock the reference klystron frequency to the signal klystron frequency plus the i.f. frequency.

The i.f. signal from the signal crystal mixer is amplified and, together with an i.f. reference signal, fed into a phase-sensitive detector which provides a frequency lock of the signal klystron to the cavity resonant frequency in much the same way as in the case of the homodyne spectrometer described above. The i.f. signal from the signal crystal is also amplitude-modulated at the field modulation frequency if magnetic resonance absorption is taking place. After detection of the i.f. frequency, the signal, now at the field modulation frequency, is further amplified and processed by a phase-sensitive detector for application to a recorder.

Homodyne versus Superheterodyne Detection

Homodyne detection has the advantages of simplicity of operation, lower initial cost of equipment, and sensitivity and stability at high microwave power levels. At lower power levels, locking of the klystron frequency to the sample cavity becomes more difficult, and recourse is frequently made to an external frequency lock, either to another cavity or to a frequency standard. Unfortunately, the sample cavity resonant frequency is then apt to drift away from the locked klystron frequency with consequent drift in the spectrometer sensitivity. Sensitivity of the homodyne system relies in part on high-frequency field modulation to overcome crystal noise, with the most common modulation frequency being 100 kHz. As is discussed in Section 5, p. 576, this can lead to difficulties if narrow spectral lines are expected since modulation side bands at ± 35 mG occur on the detected resonance lines. Modern commercial spectrometers that make use of so-called back diodes in the microwave detector can use modulation frequencies as low as 5 kHz without appreciable sacrifice of signal-to-noise ratio.

Superheterodyne spectrometers can use any field modulation frequency since the signal-to-noise ratio is not directly dependent upon this parameter; crystal noise is circumvented by use of an i.f. signal of 30 MHz or so. Thus resolution is not hampered by modulation frequency broadening with superheterodyne detection. Considerably less power is required with a superheterodyne system to frequency-lock the signal klystron to the sample cavity, so that at very low power levels the superheterodyne system is more stable than the homodyne system. However, the sensitivity of the superheterodyne bridge becomes a liability at high power levels where the stability decreases. Microphonic problems are generally more acute with a superheterodyne system employing audiofrequency field modulation so that experiments with flow systems or dewars with bubbling refrigerants pose a greater problem with the superheterodyne system.

Generally, then, the superheterodyne detection system is to be preferred over the homodyne system when (1) very narrow lines are to be resolved, or (2) high sensitivity at very low power levels is to be achieved. This means in effect that a superheterodyne spectrometer is best suited for high-resolution esr spectroscopy of organic free radicals in solution; homodyne spectrometers are to be preferred for most other applications.

Commercially Available Spectrometers

In the following discussion we consider some of the esr spectrometers distributed in the United States; information given was current in July,

1969 [40]. Frequency band designations used follow Table 6.6. Sensitivities quoted below refer to a nonlossy, nonsaturable sample giving a line width of 1 G between points of maximum slope on the absorption curve.

Alpha Scientific Laboratories, Inc., Oakland, California. Model AL-340, operating in the rf range of 320 to 380 MHz, has a sensitivity of around 10^{14} spins with 100 mW of rf power. Modulation frequencies of 60 Hz and 100 kHz are supplied. The sample is mounted in a open helical structure. The standard magnet supplied has 6-in.-diameter pole pieces and is capable of fields up to 500 G. Accessories include a low-temperature system, Hall effect crystal-controlled field regulation, a time-averaging computer, and an analog-to-digital converter for direct computer entry of the spectrometer output.

Alpha Scientific also makes accessories for X-band spectrometers including a helical slow-wave structure.

Bruker Scientific, Inc., Elmsford, New York. Two basic models, B-ER 414s and B-ER 418s, are available with the 414s being convertible to the more expensive 418s by the addition of accessories. Operating frequencies available are 200 to 800 MHz, S-band, X-band, K-band, and Q-band. Field modulation frequencies of 713 Hz and 100 kHz are standard with maximum peak-to-peak modulation amplitude of 30 G. Multiple-frequency modulation for resolution enhancement is also possible with an accessory. With the basic X-band instrument, klystrons operating at 30 and 300 mW maximum power are available; the corresponding sensitivities are 10^{11} and 5×10^{10} spins, respectively. Four magnets are available with pole diameters 7, 9, 10, and 12 in. and maximum fields, with a 2-in. air gap, of 8.6, 12.4, 13.8, and 15.9 kG, respectively. The magnets are field-regulated with a Hall effect probe. X-band cavities available include a standard TE_{102} rectangular cavity, a TE_{103} optical transmission cavity, a TE_{104} dual-sample cavity, a TE_{011} cylindrical cavity with rotatable modulation coils, a TE_{011} cavity system for high-temperature work (400 to 1000°C), and a TE_{102} cavity system for work down to liquid helium temperatures. Accessories available include time-averaging computers (1024 and 4096 word units), and integrator, a variable-temperature unit (−170 to 300°C), a field-following proton resonance oscillator for field measurements, a digital microwave frequency-measuring unit, and a conversion unit for wide-line nmr at 10 and 30 MHz.

JEOLCO (U.S.A.) Inc., Medford, Massachusetts. The JEOLCO model JES-3BS esr spectrometers operate at X-, K-, or Q-band. The sensitivity with 100-kHz field modulation (80 Hz also available) is, respectively, 5×10^{10}, 3×10^{10}, and 6×10^{9} spins for the X-, K-, and Q-band instruments. Maximum microwave power available is 300 mW at X-band, 30 mW at K band, and 20 mW at Q-band. The standard 12-in. magnet provides fields up to 13 kG; a complete line of other types of magnets is available as well. The standard cavities are cylindrical and operate in the TE_{011} mode for the X-band spectrometer and in the TE_{012} mode for the K- and Q-band spectrometers. A low-temperature cavity for liquid helium work is available for all three bands, and a TE_{01n}-mode cavity for large nonlossy samples is available for X-band. Other accessories include a variable-temperature unit (−170 to 300°C), a liquid-nitrogen dewar for use in the X-band cavity, a superheterodyne adapter (available for all three frequency bands), an integrator, an nmr gauss meter and a small computer to be used as a spectrum accumulator; conversion equipment for ENDOR is also available.

The JES-P-10S spectrometer is a less expensive instrument operating in X-band with 100-kHz field modulation. Maximum sensitivity is 1×10^{11} spins. The electromagnet is integral with the spectrometer housing. The standard cavity operates in the TE_{102} mode and can accommodate a variable-temperate accessory and a liquid nitrogen dewar. Many of the accessories mentioned above are also applicable to the P-10S system.

Micro-Now Instrument Company, Inc., Chicago, Illinois. Model 801 is an L-band spectrometer which makes use of a transistorized 1-GHz oscillator and a quarter-wave reentrant cavity [38]. Microwave power at the cavity can range from 5 to 300 mW, and with an optional unit, can be extended down to 0.1 mW. Sensitivity is 2×10^{14} spins; lossy samples are usable and the sample can be as large as 10 ml. A Helmholtz coil magnet available from Ventron/Magnion is recommended.

Micro-Now also makes the Model 810 X-band spectrometer, intended primarily as an educational tool. This spectrometer makes use of a transmission cavity and 25-kHz field modulation. No AFC is provided. Magnets manufactured by Eastern Scientific and a lock-in amplifier from Princeton Applied Research are recommended by Micro-Now for use with this instrument.

Micro-Now makes conversion kits for Varian's V-4502 spectrometer: (1) with Model 506, to increase the power levels attainable at the cavity and improve the detector sensitivity; and (2) with Model 504, to convert the homodyne detection system of the 4502 to superheterodyne operation.

Varian Associates, Palo Alto, California. Varian V-4500 series esr spectrometers are available in X-band (4502) and KA-band (4503) versions. Field modulation frequencies of 100 kHz and 20, 40, 80, 200, and 400 Hz are available with a peak-to-peak amplitude of 35 G. Maximum klystron output power is 300 mW at X-band and 50 mW at KA-band, with maximum sensitivities of 5×10^{10} and 5×10^{9} spins, respectively. The klystron frequency is stabilized by locking to the sample cavity for high-power operation, and to an external cavity for low-power operation. Magnets with pole diameters of 6, 9, and 12 in. are available giving maximum fields of 8, 10, and 13 kG with 2.625-in. air gaps. The magnet power supplies are field-regulated with a Hall effect probe. X-Band cavities available include a standard TE_{102}

rectangular cavity, a TE_{103} optical transmission cavity, a TE_{104} dual-sample cavity, a TE_{102} bimodal cavity, a TE_{102} cavity system for liquid helium studies, a TE_{011} cylindrical cavity with rotatable modulation coils, and a TE_{01n} multimode cylindrical cavity for large, nonlossy samples; at KA-band, a TE_{011} cylindrical cavity is standard. Accessories include a conversion unit for superheterodyne operation, a liquid nitrogen dewar, a variable temperature accessory (-185 to 300°C), a 1024-word time-averaging computer, a field-frequency lock unit, a magnetic flux stabilizer, and a conversion unit for wide-line nmr at 2 to 16 MHz.

Varian's E-series spectrometers, which employ more modern electronics, operate in X-band with 200-mW maximum microwave power to the cavity. Modulation frequencies of 35 and 270 Hz, and 1, 10, and 100 kHz are available with up to 40-G peak-to-peak amplitude; maximum sensitivity is 5×10^{10} spins with either 10 or 100 kHz field modulation. Magnets with pole diameters of 4, 6, 9, 12, and 15 in. are available. Cavities and other accessories include all those available for the V-4500 series spectrometers plus facilities for ENDOR and ELDOR. The wide-line nmr conversion kit is for the frequency range 3 to 35 MHz.

Ventron/Magnion, Burlington, Massachusetts. Ventron MVR Series esr spectrometers are available in X-, K-, KA-, and V-band versions. Field modulation frequencies of 6 and 100 kHz are standard, but a low-frequency unit operating at 20, 40, 80, 200, and 400 Hz is also available. Maximum klystron output powers for the four frequency bands are, respectively, 500, 600, 400, and 300 mW, and maximum sensitivities of 5×10^{10}, 1×10^{10}, 5×10^{9}, and 1×10^{9} spins are achievable. Four magnets with pole diameters of 4, 7, 9.5, and 12 in. are available, permitting maximum fields of 6 kG (1.5 in.), 7 kG (2.625 in.), 15 kG (1.75 in.), and 15 kG (1.75 in.), respectively, with the indicated air gaps. The magnet power supplies are field-regulated with a rotating coil magnetometer. A wide variety of cavities for low-temperature work, optical irradiation, and standard work at 100-kHz modulation are available at X-band. Cylindrical cavities operating in the TE_{011} mode for low-temperature operation and/or optical irradiation are available for K- and KA-bands. A TE_{012} cylindrical cavity for operation down to liquid helium temperatures is available for V-band operation. Accessories include liquid nitrogen and liquid helium dewars, a variable-temperature (-175 to 300°C) unit, various kinds of sample cells, and an ENDOR unit operating over the range 2 to 50 MHz.

Microwave Frequency and Power

Choice of Microwave Frequency

The choice of the operating frequency in esr spectroscopy is governed by a number of factors. X-Band microwave components are more readily available and less expensive than components for other frequency bands so that for economic reasons, if no others, this is likely to be the most popular band for esr spectrometers. From the point of view of the chemist, lower-frequency bands are more convenient in the sense that the resonance cavities are larger, providing more room for sample tubes, dewars, and so on. However, for very low frequencies, the required magnet gap becomes very large and magnetic field homogeneity is a problem. As discussed in Section 2, p. 511, higher-frequency operation yields a spectrum for which the high-field approximation is more accurate, and first-order interpretations frequently suffice. However, high frequency spectrometers not only suffer from the requirements of very small cavities (and samples), but also require very large magnets. A Q-band spectrometer operating at 35 GHz requires a magnetic field of approximately 12.5 kG for detection of a resonance at $g = 2$; fields of this magnitude are of course routinely obtained for 60-MHz proton nmr, and so present no problem other than expense. Going up in frequency to V-band (70 GHz), however, enters into the range (around 25 kG) where magnetic fields are barely attainable with ordinary magnets. From these practical considerations, it is apparent that the standard choice of X-band for esr spectroscopy is quite a reasonable one, and operation at another frequency is usually only undertaken for one of two reasons: (1) to change the effect of the spectral density of a relaxation process, or (2) to improve sensitivity.

The dependence of sensitivity on frequency has been discussed by Hyde [41]. For nonlossy, nonsaturable samples, where the sample size can be increased to optimize the filling factor (see Section 3, p. 556), the sensitivity is determined largely by the Boltzmann factor (see Section 1, p. 502), and increases with $\omega^{1/2}$. If the sample is saturable, however, the sensitivity decreases with increasing frequency as $\omega^{-1/4}$, and if the sample is lossy, sensitivity depends on the detailed dielectric properties, but generally speaking lower frequencies are preferred. Sensitivity considerations are therefore not as clear-cut as might be implied from perusal of the specifications of a commercial instrument, since manufacturers always refer to nonlossy, nonsaturable samples in stating their spectrometer sensitivities.

Effect of Microwave Power Saturation

As noted in Section 1, p. 503, saturation of the electron resonance may occur at high microwave power levels. For a homogeneously broadened resonance, the power absorbed by the sample is reduced by the saturation factor given in (6.18). Since the saturation factor is in turn a function of the unsaturated line width, we find that the reduction in absorption is greater near the center of the line than in the wings, resulting in an apparent broadening of the line. This effect is illustrated in Figs. 6.12, 6.13, and 6.14, in which the Lorentzian absorption and first- and second-derivative line shapes are plotted for the saturation parameter $(\gamma H_1/2)^2 T_1 T_2 = 0.3$, 1.0, and 3.0.

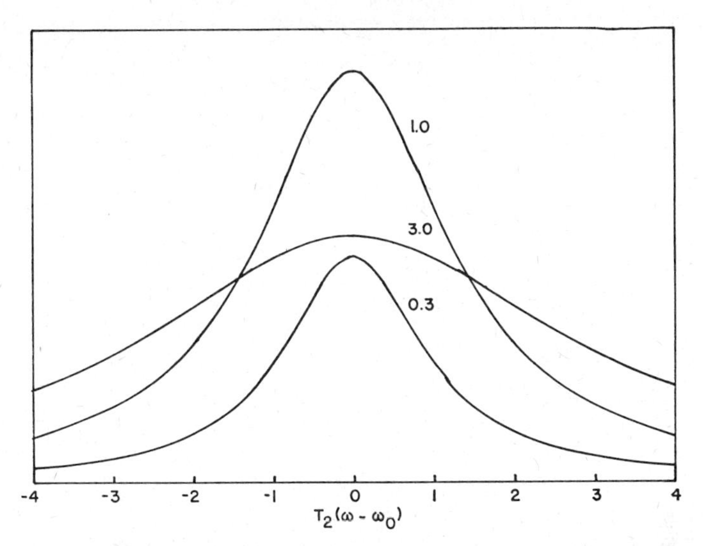

Fig. 6.12 Lorentzian absorption line shapes for saturation parameters of 0.3, 1.0, and 3.0.

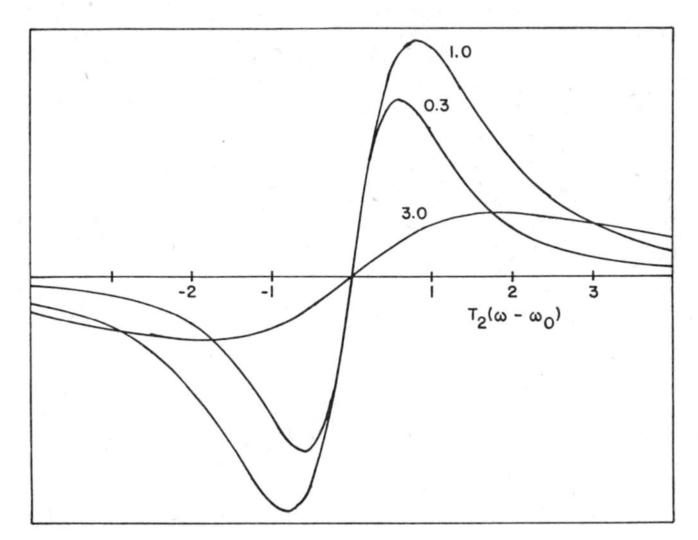

Fig. 6.13 Lorentzian first-derivative line shapes for saturation parameters of 0.3, 1.0, and 3.0.

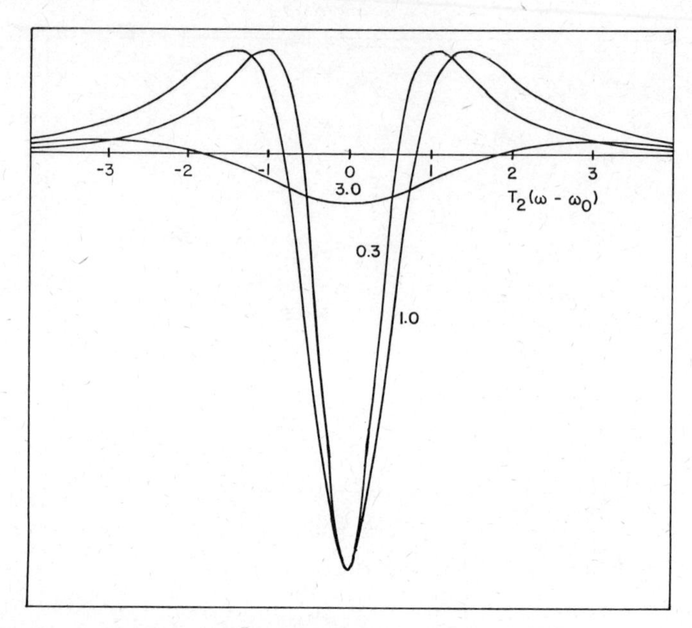

Fig. 6.14 Lorentzian second-derivative line shapes for saturation parameters of 0.3, 1.0, and 3.0.

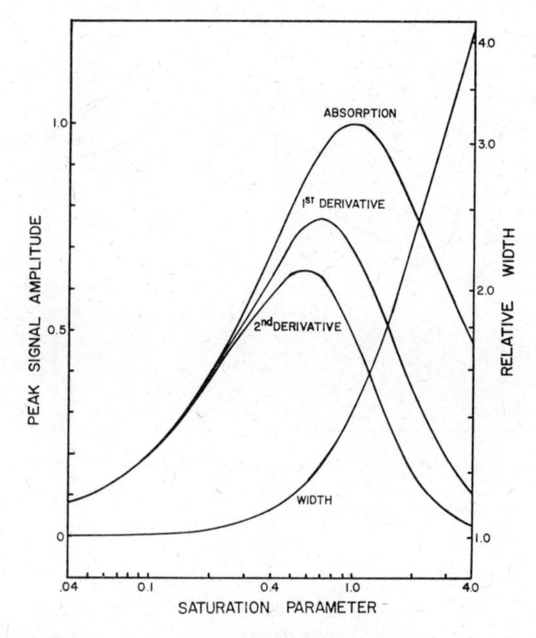

Fig. 6.15 Peak amplitudes and relative widths of Lorentzian absorption, and first- and second-derivative lines as a function of the saturation parameter $(\gamma H_1/2)^2 T_1 T_2$.

To measure accurate line widths or line shapes, it is obviously necessary to work at low microwave power levels. However, at low power, the signal intensity is linear in incident power. Thus a compromise must be reached between the goals of maximizing the signal amplitude and minimizing distortion. This problem is illustrated in Fig. 6.15, in which the peak amplitudes of the absorption and first and second derivatives of absorption are plotted as functions of the saturation parameter. The apparent relative width behaves in the same way for all three functions, and is also plotted in Fig. 6.15. Thus, for example, we see that the signal amplitude of the first derivative may be maximized by adjusting the saturation parameter to $1/\sqrt{2}$, but that the apparent line width is then approximately 20% greater than that of the unsaturated line.

In the discussion thus far, a two-level system has been assumed in which the two relaxation times T_1 and T_2 describe the relaxation behavior of the system completely. In multilevel systems, for example, in free radicals exhibiting hyperfine structure, the situation is more complicated, and it is found that in general the various lines of a spectrum exhibit different saturation behavior. Thus, for example, in the three-line spectrum of 2,5-di-*tert*-butylsemiquinone, in which the theoretical intensity ratios are 1:2:1, the unsaturated spectrum shows amplitude ratios of 1:2.2:1, whereas at high power the ratio changes to 1:1.3:1 [42]. Apparently both T_1 and T_2 are longer for the central line than for the two outer lines. Such behavior is to be expected theoretically since relaxation processes couple various states within the complex of energy levels, usually in such a way that the approach of the population of a particular level to its equilibrium value may be through several distinct mechanisms. Thus there is no single spin-lattice relaxation time for the system as a whole, but rather individual relaxation probabilities for energy transfer between particular levels. This problem has been considered in detail by a number of authors [42, 43] and the reader is referred to the original papers for further information.

Although saturation effects are often troublesome in esr work in which the goal is interpretation of well-resolved spectra, they may be put to practical advantage in the measurement of spin-lattice relaxation times. By noting the microwave power level necessary to maximize the signal intensity, T_1 can be calculated providing H_1 and T_2 are known. The transverse relaxation time is obtainable from the unsaturated line widths, and H_1 may be calculated at least approximately from the cavity geometry and the measured incident microwave power.

Since electron spin-lattice relaxation times are generally very short (on the order of 10^{-6} sec for semiquinones in solution at room temperature, for example), the techniques commonly used in nmr [44] for T_1 measurements are not usually applicable, and the saturation method remains the only practical method for the measurement of electron T_1 values by employing only simple esr techniques. Other methods involving pulse measurements, and other techniques not usually available with basic commercial spectrometers have been reviewed by Alger [37].

In cases in which the resonance line is inhomogeneously broadened by unresolved hyperfine structure, inhomogeneities in the magnetic field or molecular surroundings, and so on, the observed absorption envelope is the superposition of a number of individual components, and frequently has an overall Gaussian line shape. If the envelope width is large compared with the individual component line widths, and the saturation behavior of all components is essentially the same, the envelope shape will not change on saturation but will simply decrease in amplitude. An analysis of this case has been given by Portis [45].

Cavities

Most esr spectrometers employ a reflection cavity coupled to a source and detector by a magic tee or circulator and a length of tunable waveguide. Some form of impedance matching involving the iris coupling to the cavity, the slide-screw tuner in the connecting waveguide, and the size and placement of the sample within the cavity is normally required for operation of the spectrometer. In this section we deal with the details of the matching process and discuss the effects of cavity design and sample geometry on the esr signal.

Impedance Matching [36, 46]

The coupling of a microwave cavity to a length of waveguide is most conveniently discussed in terms of the equivalent electrical circuit, treating the cavity as an *RLC* resonant circuit, and the waveguide as a transmission line.

The equations governing the voltage V and current I of a sinusoidal signal in a transmission line of length d are [36, 47]:

$$\frac{dV}{dx} = -i(\omega/Z_t v)I \tag{6.89a}$$

$$\frac{dI}{dx} = -i(\omega Z_t/v)V, \tag{6.89b}$$

where x is the distance along the transmission line, i is the square root of -1, ω is the angular frequency of the signal, and v and Z_t, the phase velocity and impedance, respectively, are characteristics of the transmission line. The general solutions to these equations are

$$V = A \exp(-i\omega x/v) + B \exp(i\omega x/v) \tag{6.90a}$$

$$I = [A \exp(-i\omega x/v) - B \exp(i\omega x/v)]/Z_t, \tag{6.90b}$$

where A and B are constants to be determined by the boundary conditions. If the transmission line is terminated by an impedance Z_c, in this case representing the cavity, then at $x = d$,

$$V = IZ_c, \tag{6.91}$$

which establishes one of the boundary conditions. The voltage may then be written:

$$V = A \exp(-i\omega x/v) + A \frac{Z_c - Z_t}{Z_c + Z_t} \exp(-2i\omega d/v) \exp(i\omega x/v), \tag{6.92}$$

where the first term represents a signal traveling from the origin to the terminating impedance, and the second term represents a signal reflected back up the transmission line. Clearly, if the impedance of the waveguide and cavity are matched, $Z_c = Z_t$, no signal is reflected. The voltage reflection coefficient, defined as the ratio of the voltage of the reflected signal to the voltage of the incoming signal at the termination (the cavity) is:

$$\Gamma = \frac{Z_c - Z_t}{Z_c + Z_t}. \tag{6.93}$$

In turning attention now to the cavity equivalent circuit, the impedance of an RLC circuit is

$$Z_c = R + i\left(\omega L - \frac{1}{\omega C}\right). \tag{6.94}$$

In the analysis of resonant circuits, an important figure of merit is the quality factor Q, defined as [36, 47]:

$$Q = 2\pi \text{ (energy stored)/(energy dissipated per cycle)}. \tag{6.95}$$

In an RLC circuit,

$$\begin{aligned} Q &= 2\pi(\tfrac{1}{2}LI_m^2)/(RI_m^2/2f) \\ Q &= \omega L/R, \end{aligned} \tag{6.96}$$

where I_m is the amplitude of the (sinusoidal) current and $f = \omega/2\pi$ is the frequency. In terms of Q the impedance of the circuit is

$$Z_c = R[1 + iQ(\omega/\omega_0 - \omega_0/\omega)], \tag{6.97}$$

where $\omega_0 = (LC)^{-1/2}$ is the resonant frequency. For small deviations from the resonant frequency,

$$Z_c = R[1 + 4iQ(\omega - \omega_0)/\omega_0]. \tag{6.98}$$

Since the impedance of the transmission line is to a good approximation purely resistive, one may define a Q-parameter by:

$$Q_r = \frac{\omega_0 L}{Z_t}, \tag{6.99}$$

such that the voltage reflection coefficient can be written as a function of the frequencies and quality factors only:

$$\Gamma = \frac{Q_r[1 + 4iQ_c(\omega - \omega_0)/\omega_0] - Q_c}{Q_r[1 + 4iQ_c(\omega - \omega_0)/\omega_0] + Q_c}. \tag{6.100}$$

Here we see that the condition that the cavity be perfectly matched at resonance, $\Gamma = 0$, is that $Q_c = Q_r$.

Following Poole [36], we now compute the cavity Q, applying the original definition of the quality factor. The energy stored in the electromagnetic field is:

$$U = \frac{1}{2}\int |H_m|^2\, dV, \tag{6.101}$$

where H_m is the maximum magnetic field of the microwaves and the integration is over the volume of the cavity. Energy is dissipated in the cavity through ohmic losses in the walls, by dielectric loss in the sample or other dielectric material in the cavity, and by magnetic resonance absorption in the sample. The power loss attributable to currents induced in the cavity walls is:

$$P_1 = \tfrac{1}{2}R_s\int |H_{tm}|^2\, dS, \tag{6.102}$$

where R_s is the effective skin effect resistance of the cavity walls, H_{tm} is the tangential component of the magnetic field at the walls, and the integration is over the surface of the cavity. Power loss in the dielectric is:

$$P_2 = \frac{\omega}{2}\int \epsilon''\, |E_m|^2\, dV, \tag{6.103}$$

where E_m is the maximum electric field of the microwaves and ϵ'' is the absorptive part of the dielectric constant. Power loss attributable to magnetic resonance absorption is

$$P_3 = \frac{\omega}{2}\int \chi''\, |H_m|^2\, dV, \tag{6.104}$$

where χ'' is the imaginary part of the magnetic susceptibility, given by (6.30*b*). It is customary to define a separate quality factor for each of the three dissipation terms. Thus we define the "unloaded Q":

$$Q_u = \frac{\omega}{R_s}\,\frac{\int |H_m|^2\, dV}{\int |H_{tm}|^2\, dS}, \tag{6.105}$$

the dielectric Q:

$$Q_\epsilon = \frac{\int |H_m|^2 \, dV}{\int \epsilon'' \, |E_m|^2 \, dV}, \tag{6.106}$$

and the magnetic resonance absorption Q:

$$Q_\chi = \frac{\int |H_m|^2 \, dV}{\int \chi'' \, |H_m|^2 \, dV}. \tag{6.107}$$

If the susceptibility χ'' is assumed to be constant over the sample, it may be removed from the integral and the integration restricted to the sample volume. The resulting ratio of integrals depends only on the cavity mode and the sample geometry and is referred to as the filling factor:

$$\eta_\chi = \int_{v_s} |H_m|^2 \, dV \Big/ \int_{v_c} |H_m|^2 \, dV,$$

or

$$\eta_\chi = \frac{V_s \, \langle H_1^2 \rangle_s}{V_c \, \langle H_1^2 \rangle_c}, \tag{6.108}$$

$$Q_\chi = \frac{1}{\eta_\chi \chi''}. \tag{6.109}$$

If the sample is the only important source of dielectric loss, a similar treatment is convenient for the dielectric Q; thus:

$$\eta_\epsilon = \frac{V_s \, \langle E_1^2 \rangle_s}{V_c \, \langle E_1^2 \rangle_c}, \tag{6.110}$$

$$Q_\epsilon = \frac{1}{\eta_\epsilon \epsilon''}. \tag{6.111}$$

The total cavity Q then is:

$$\begin{aligned} \frac{1}{Q_c} &= \frac{1}{Q_u} + \frac{1}{Q_\epsilon} + \frac{1}{Q_\chi} \\ \frac{1}{Q_c} &= \frac{1}{Q_u} + \eta_\epsilon \epsilon'' + \eta_\chi \chi'', \end{aligned} \tag{6.112}$$

or

$$Q_c = Q_c^0(1 - Q_c^0\eta_\chi\chi''), \tag{6.113}$$

where

$$Q_c^0 = \frac{Q_u}{1 + Q_u\eta_\epsilon\epsilon''} \tag{6.114}$$

is the cavity Q off resonance, and it is assumed that power losses attributable to magnetic resonance absorption are small compared with the other losses. If it is assumed that the microwave frequency is locked to the cavity resonance frequency, the voltage reflection coefficient is

$$\Gamma = \frac{Q_c^0 - Q_r}{Q_c^0 + Q_r}. \tag{6.115}$$

Magnetic resonance absorption is detected by a change in the voltage reflection coefficient. The magnitude of this change is:

$$\Delta\Gamma = \frac{\partial\Gamma}{\partial Q_c}\Delta Q_c. \tag{6.116}$$

ΔQ_c, the change in the cavity Q when the resonance is traversed, is

$$\Delta Q_c = (Q_c^0)^2\eta_\chi\chi'', \tag{6.117}$$

so that

$$\Delta\Gamma = \frac{2Q_r(Q_c^0)^2\eta_\chi\chi''}{(Q_c^0 + Q_r)^2}. \tag{6.118}$$

Maximizing the change in the voltage reflection coefficient with respect to the waveguide Q_r, we find that the maximum signal is obtained when the cavity is critically coupled to the waveguide, $Q_r = Q_c$,

$$\Delta\Gamma = \tfrac{1}{2}Q_c^0\eta_\chi\chi'' \tag{6.119a}$$

$$\Delta\Gamma = \frac{1}{2}\frac{Q_u\eta_\chi\chi''}{1 + Q_u\eta_\epsilon\epsilon''}. \tag{6.119b}$$

In general, we note that the signal from magnetic resonance absorption increases with increasing Q_u, with increasing magnetic filling factor and magnetic susceptibility, but decreases with increasing electric filling factor or dielectric constant.

There is a further condition on coupling of the cavity to the waveguide which must be considered, however. Returning to the cavity equivalent circuit, we may incorporate the effects of magnetic resonance absorption by assuming that the inductance is a coil containing a sample of susceptibility χ and filling factor η_χ. The inductance is:

$$L = L_0(1 + 4\pi\eta_\chi\chi). \tag{6.120}$$

Writing $\chi = \chi' - i\chi''$, the impedance of the *RLC* circuit is:

$$Z = R[1 + 4\pi Q\eta_\chi\chi''] + 4iQ[(\omega - \omega_0)/\omega + \pi\eta_\chi\chi'], \tag{6.121}$$

where $\omega_0{}^2 = 1/L_0C$ and $Q = \omega_0 L_0/R$, and we have assumed that $(\omega - \omega_0)/\omega_0$ is small. Note that the imaginary part of the susceptibility appears as a resistive component and thus contributes to the dissipation of power in the cavity, as expected. The real part of the susceptibility, however, is reactive, and results in a shift in the cavity resonant frequency as the resonance is transversed. If the microwave frequency is held constant, locked to an external cavity, for instance, the shift in the cavity resonant frequency as the magnetic resonance is traversed will have the effect of mixing a dispersion component into the signal. In order to avoid this, most spectrometers frequency lock the microwave source to the resonant cavity so that in following the frequency shifts attributable to dispersion the pure absorption mode is detected.

If the cavity is critically coupled to the waveguide, further difficulties may be encountered. The cavity Q is affected by the dispersive part of the susceptibility as well as by the absorptive part and, neglecting absorption, swings both above and below its value off resonance. Thus the cavity, if critically coupled off resonance, swings to both sides of the critical coupling condition as the resonance is traversed. Since the reflected signal changes sign when the cavity swings from overcoupled to undercoupled, serious distortions can result. The solution to this problem is usually to operate the cavity in the overcoupled condition, $Q_c > Q_r$, but sufficiently close to critically coupled that (6.119*b*) is still a reasonable approximation for the reflected signal.

Sample Placement [36, 37, 46]

As noted above, the magnetic resonance signal increases with the magnetic filling factor but decreases with the electric filling factor. Thus it is of considerable importance to position the sample in the cavity at a maximum in the rf magnetic field and at a minimum in the rf electric field. Both rectangular and cylindrical cavities operating in TE modes are commonly employed in esr spectrometers, and these are considered in the following discussion.

RECTANGULAR CAVITIES

The normal mode fields for a rectangular cavity operating in the TE_{lmn} mode are [36, 37, 39]:

$$\begin{aligned}
E_x &= -\frac{m\lambda}{2B}\cos\left(\frac{l\pi x}{A}\right)\sin\left(\frac{m\pi y}{B}\right)\sin\left(\frac{n\pi z}{C}\right)\\
E_y &= \frac{l\lambda}{2A}\sin\left(\frac{l\pi x}{A}\right)\cos\left(\frac{m\pi y}{B}\right)\sin\left(\frac{n\pi z}{C}\right)\\
E_z &= 0\\
H_x &= \frac{ln\lambda^2}{4AC}\sin\left(\frac{l\pi x}{A}\right)\cos\left(\frac{m\pi y}{B}\right)\sin\left(\frac{n\pi z}{C}\right)\\
H_y &= \frac{mn\lambda^2}{4BC}\cos\left(\frac{l\pi x}{A}\right)\sin\left(\frac{m\pi y}{B}\right)\cos\left(\frac{n\pi z}{C}\right)\\
H_z &= -\left(\frac{\lambda}{2}\right)^2\left[\left(\frac{l}{A}\right)^2+\left(\frac{m}{B}\right)^2\right]\cos\left(\frac{l\pi x}{A}\right)\cos\left(\frac{m\pi y}{B}\right)\sin\left(\frac{n\pi z}{C}\right),
\end{aligned} \tag{6.122}$$

where A, B, and C are the x-, y-, and z-dimensions of the cavity, l, m, and n are integers, and λ is the cavity resonant wavelength:

$$\lambda = 2\Big/\left[\left(\frac{l}{A}\right)^2+\left(\frac{m}{B}\right)^2+\left(\frac{n}{C}\right)^2\right]^{1/2}. \tag{6.123}$$

Several considerations are important in choosing the cavity mode. First, the cavity dimensions should be chosen so as to avoid mixed-mode operation, that is, so that only one mode is excited by the microwave source. Second, the dimensions of the cavity should be chosen sufficiently small that it will fit between the magnet poles and that the magnetic field will be homogeneous over the sample. Since only the rf field perpendicular to the static magnetic field is of use in the magnetic resonance experiment, it is convenient, taking the static field in the y-direction, to set $H_y = 0$, by choosing $m = 0$. With this choice, we must have $l > 0$, and the usual choice is $l = 1$. The unloaded Q for a TE_{10n} mode cavity is [36]:

$$Q\frac{\delta}{\lambda} = -\frac{4ABC}{\lambda^3\left[\frac{C}{A}\left(1+\frac{2B}{A}\right)+\frac{n^2A}{C}\left(1+\frac{2B}{C}\right)\right]}, \tag{6.124}$$

where δ is the skin depth. Inspection of the electric and magnetic field equations indicates that the optimum sample location is at $z = pC$ where p is an integer between zero and n. Filling factors for $p = n - 1$ and for a variety of sample shapes are listed in Table 6.7 for cavities operating in the TE_{102} and

Table 6.7 Filling Factors and Optimum Sample Geometries

Cavity Mode	Sample Geometry	η_ϵ	η_χ	$\epsilon'' = 2$		$\epsilon'' = 20$	
				d_{op} (mm)	$\Delta\Gamma/\chi''$	d_{op} (mm)	$\Delta\Gamma/\chi''$
TE_{102}; $Q_u = 5000$	Flat, $a \times b \times d$ at $z = c/2$	$0.82(d/a)^3$	$0.50(d/a)$	0.90	32.8	0.42	15.2
	Cylinder, diameter d, at $z = c/2$, $y = b/2$	$1.09(d/a)^4$	$0.88(d/a)^2$	2.22	10.5	1.25	3.3
	Cylinder, diameter d, at $z = c/2$, $x = a/2$	$0.87(d/a)^4$	$0.78(d/a)^2$	2.39	10.8	1.34	3.4
	Cube, thickness d, at $z = c/2$, $x = a/2$, $y = b/2$	$3.70(d/a)^5$	$2.25(d/a)^3$	3.02	5.2	1.91	1.3
TE_{1010}; $Q_u = 5600$	Flat, $a \times b \times d$ at $z = c/10$	$0.16(d/a)^3$	$0.10(d/a)$	1.48	12.1	0.69	5.6
	Cylinder, diameter d, at $z = c/10$, $y = b/2$	$0.22(d/a)^4$	$0.18(d/a)^2$	3.65	6.3	2.05	2.0
	Cylinder, diameter d, at $z = c/10$, $x = a/2$	$0.19(d/a)^4$	$0.16(d/a)^2$	3.80	6.1	2.14	1.9
	Cube, thickness d, at $z = c/10$, $x = a/2$, $y = b/2$	$0.74(d/a)^5$	$0.45(d/a)^3$	4.08	2.9	2.58	0.8
TE_{011}; $Q_u = 17{,}500$	Cylinder, diameter d, and length L	$11.3(d/a)^4$	$5.29(d/a)^2$	1.69	36.8	0.95	11.6
	Cylinder, diameter d, and length d	$22.6(d/a)^5$	$10.6(d/a)^3$	3.05	13.7	1.93	3.4
TE_{012}; $Q_u = 20{,}900$	Cylinder, diameter d, and length L	$11.3(d/a)^4$	$3.68(d/a)^2$	1.94	28.0	1.09	8.8
	Disk, diameter $2R$ and thickness d	$6.58(d/a)^3$	$0.25(d/a)$	0.62	21.2	0.29	9.9

TE_{1010} modes. In calculating the filling factors, a "square mode" cavity in which $C = nA$ has been assumed. The B/A ratio for most cavities is on the order of one-half; for the purposes of illustration, we have assumed in Table 6.7 and elsewhere in this discussion that $B/A = 4/9$, which is strictly accurate only for a cavity constructed from standard (RG-52/U) X-band waveguide.

It is clear that insertion of the expressions given for the filling factors in Table 6.7 in (6.119*b*) will result in an optimum sample size for maximum esr signal. This is shown graphically in Fig. 6.16, in which the voltage reflection

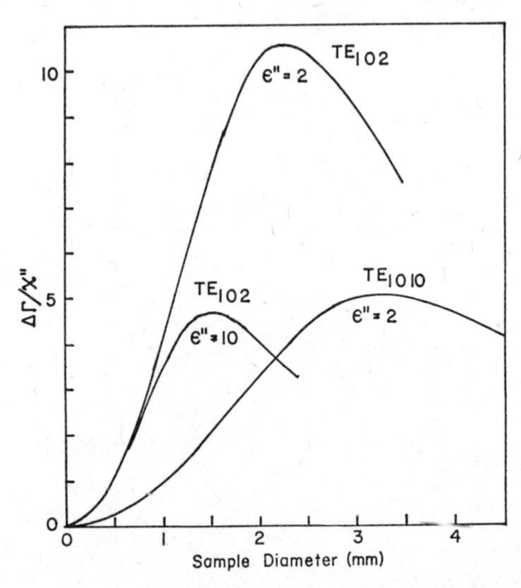

Fig. 6.16 Voltage reflection coefficient (proportional to absorption signal amplitude) as a function of cylindrical sample diameter for TE_{102} and TE_{1010} rectangular cavities.

coefficient is plotted against the diameter of a cylindrical sample. In Fig. 6.16 we have assumed $\lambda = 3.23$ cm and $\delta = 1.22 \times 10^{-4}$ cm, giving an unloaded Q for the TE_{102} cavity of 5000, and for the TE_{1010} cavity of 5600.

The optimum sample size and signal amplitude is easily computed for the sample geometries listed in Table 6.7, given a value of the unloaded Q and the imaginary part of the dielectric constant. These are also given in the table assuming $\epsilon'' = 2$ and 20. It should be noted that by far the best sample geometry is the flat sample extending the width and height of the cavity. Although the esr signal is invariably smaller for the higher-mode cavity, use of a TE_{1010} cavity, for example, allows the use of a larger sample which is on occasion of importance. Furthermore, since there are $n - 1$ potential sample sites available in the TE_{10n} cavity, it is possible to conveniently mount more than one sample; TE_{104} cavities with two sample holes are commercially available from Bruker, Varian, and Ventron/Magnion.

It is of interest to note that there is very little difference in either optimum sample size or maximum signal amplitude between cylindrical samples mounted parallel to the x-axis (the vertical axis when the cavity is mounted in the usual way) and those mounted parallel to the y-axis (horizontal and parallel to the magnetic field). Indeed, mounting the sample along the y-axis has some important advantages [42]. First, the rf magnetic field is constant along the sample rather than varying sinusoidally along the sample length; this is a considerable simplification in the interpretation of saturation measurements. Second, the static magnetic field is likely to be more homogeneous over sample volume when the sample is mounted parallel to the field than when mounted perpendicularly. Unfortunately, however, mounting of liquid samples in horizontal tubes presents some difficulties if there are air bubbles in the tube, and mounting of dewars or other auxiliary equipment is very difficult with a horizontal sample tube.

CYLINDRICAL CAVITIES

The normal mode field equations for a cylindrical cavity operating in the TE_{lmn} mode are [36]:

$$
\begin{aligned}
E_r &= -\frac{lR}{x_{lm}r} J_l(x_{lm}r/R) \sin(l\theta) \sin(n\pi z/L) \\
E_\theta &= -J_l'(x_{lm}r/R) \cos(l\theta) \sin(n\pi z/L) \\
E_z &= 0 \\
H_r &= \frac{n\lambda}{2L} J_l'(x_{lm}r/R) \cos(l\theta) \cos(n\pi z/L) \\
H_\theta &= \frac{ln\lambda R}{2x_{lm}Lr} J_l(x_{lm}r/R) \sin(l\theta) \cos(n\pi z/L) \\
H_z &= \frac{x_{lm}\lambda}{2\pi R} \cos(l\theta) \sin(n\pi z/L),
\end{aligned}
\tag{6.125}
$$

where $J_l(x_{lm}r/R)$ and $J_l'(x_{lm}r/R)$ are Bessel functions, x_{lm} is the mth root of $J_l'(x)$, R and L are the cavity radius and length, respectively, l, m, and n are integers, and λ is the cavity resonant wavelength [36],

$$
\lambda = \frac{2\pi}{[(x_{lm}/R)^2 + (n\pi/L)^2]^{1/2}}. \tag{6.126}
$$

The usual choice is to set $l = 0$, providing an rf electric field node along the axis of the cavity. Values of $m > 1$ serve only to increase the size of the cavity and are not commonly used. The unloaded Q of a TE_{01n} cylindrical cavity is [36]:

$$Q_u\left(\frac{\delta}{\lambda}\right) = \frac{\left[x_{01}^2 + \left(\frac{n\pi R}{L}\right)^2\right]^{3/2}}{2\pi\left[x_{01}^2 + \left(\frac{n\pi R}{L}\right)^2\left(\frac{2R}{L}\right)\right]}. \tag{6.127}$$

The Q is maximized for $L = 2R$, and in the following discussion it is assumed that this condition is met.

For $n = 1$, the only feasible sample geometry is a thin cylinder along the cavity axis. For $n = 2$, a disk-shaped sample mounted at $z = L/2$ is also possible; for higher values of n, $n - 1$ disks could be mounted at the nodes in the rf electric field. Filling factors are given in Table 6.7 for cylindrical samples in the TE_{011} and TE_{012} cavities as well as for a disk-shaped sample in the TE_{012} cavity.

Just as for the rectangular cavities, an optimum sample size is found for nonzero values of ϵ''. Optimum sample sizes and signal amplitudes are also given in Table 6.7 for $\epsilon'' = 2$ and 20, assuming $\lambda = 3.23$ cm and $\delta = 1.22 \times 10^{-4}$ cm. These figures are thus comparable to those for the rectangular cavities. We note that a cylindrical sample in a TE_{011} cavity is comparable to a flat sample in a TE_{102} rectangular cavity, but results in a considerably larger signal than a cylindrical sample in a rectangular cavity. Similarly, a small cylinder (with diameter equal to length) results in a considerably larger signal than a small cube in a rectangular cavity, suggesting that small solid samples (e.g., single crystals) are better studied in a cylindrical cavity.

OPTIMIZATION OF SAMPLE SIZE

In experimentally choosing the optimum sample size, it should be noted that the signal amplitude is maximized when the cavity Q (off resonance) is one-half the unloaded Q of the cavity. This condition may be determined either by estimating the cavity Q with and without the sample from the width of the cavity dip on the klystron mode as indicated in Fig. 6.10 or, better, noting that if the cavity is initially exactly matched to the waveguide without the sample the voltage reflection coefficient should be one-third when the sample is inserted. Thus the depth of the cavity dip should reduce to two-thirds its matched value when the sample is inserted.

It should be emphasized that attempts to use samples larger than the optimum not only make the matching problem difficult, but actually result in a reduction in the esr signal amplitude, and that for lossy samples, such as aqueous solutions, the optimum size may be surprisingly small.

4 SENSITIVITY ENHANCEMENT TECHNIQUES

The various means by which the sensitivity of an esr spectrometer may be improved are conveniently divided into three groups: (1) those techniques that involve optimization of instrumental parameters, for example, cavity geometry and sample placement, microwave power, field modulation frequency and amplitude; (2) those techniques that involve processing of the signal as it appears at the output of the lock-in detector, for example, filtering and time-averaging; and (3) those techniques that involve storage of the signal in either analog or digital form, with further processing taking place later, perhaps in a computer.

Sensitivity enhancement techniques in the first category are discussed elsewhere in this chapter. Spectrometer-type selection is discussed in Section 3, p. 538; the choice of microwave frequency and power is considered in Section 3, p. 541; optimization of sensitivity with respect to cavity type and sample placement is dealt with in Section 3, p. 551; and the choice of modulation frequency and amplitude is discussed in Section 5, p. 570.

In the following discussion we consider the methods of filtering and time-averaging. These techniques are reviewed in considerable detail by Ernst [48], along with a number of methods not dealt with below.

The Signal-to-Noise Ratio

Many essentially arbitrarily defined figures of merit could be employed in discussing sensitivity enhancement. Following Ernst [48], we discuss sensitivity in terms of the signal-to-noise ratio, which is defined as the ratio of the peak signal voltage to the root-mean-square (rms) noise voltage:

$$\text{Signal-to-noise ratio (S/N)} = \frac{\text{Peak signal voltage}}{\text{rms noise voltage}} . \tag{6.128}$$

In the case of lines with two extrema, such as Lorentzian first-derivative lines, the definition of the signal-to-noise ratio must be modified since the signal amplitude is the difference of signal voltages measured at two different times. Thus the appropriate average noise voltage in the signal-to-noise ratio is the rms value of the difference of the noise voltages at the two times of measurement:

$$\text{Signal-to-noise ratio} = \frac{(\text{Maximum signal voltage}) - (\text{minimum signal voltage})}{\text{rms difference of noise voltages at } t_1 \text{ and } t_2} . \tag{6.129}$$

The mean square difference of noise voltages may be expressed by the mean square noise voltage and by its autocorrelation function $R(t)$ (see p. 559):

$$\overline{[n(t_1) - n(t_2)]^2} = 2\overline{n(t)^2} - 2R(t_1 - t_2). \tag{6.130}$$

In the case of white noise, or in cases in which $t_1 - t_2$ is long compared with the noise correlation time, $R(t_1 - t_2)$ is essentially zero, and the two definitions of the signal-to-noise ratio are equivalent. However, in cases in which the line extrema are close together in time, or the noise correlation time is long, the signal-to-noise ratio may be much larger for signals with two extrema than for equivalent signals with only one maximum.

Techniques that optimize the signal-to-noise ratio usually produce severe distortions in the shape of the signal. Thus in those cases in which the presence or absence of a line is the important factor, and its accurate position and shape are of secondary significance, the proper goal may be maximizing the signal-to-noise ratio. When the position or shape of the signal is important, however, any optimization of the signal-to-noise ratio must be done carefully and with the appropriate restraints.

Sensitivity Enhancement by Filtering

The output of the lock-in detector of an esr spectrometer is composed of a signal voltage $s(t)$, which is superimposed on a noise voltage $n(t)$ to give a combined signal. In order to improve the sensitivity of the spectrometer, it is usually desirable to put the signal through some kind of filtering device, the characteristics of which result in a greater reduction in the noise voltage than in the signal voltage. In the following discussion we consider briefly the theory of filters and show the effect on Lorentzian absorption lines and the first and second derivatives of filtering with the so-called matched filter as well as one- and two-section RC filters (Fig. 6.17).

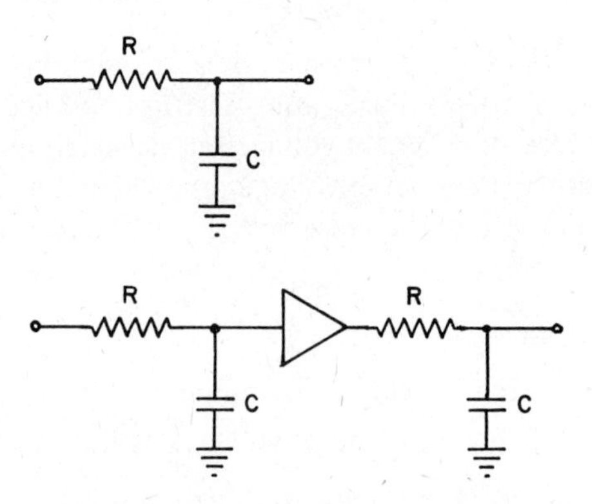

Fig. 6.17 One- and two-section RC filters.

The operation of a filter may be represented mathematically by the convolution integral [48–50],

$$s_{\text{out}}(t) = \int_{-\infty}^{\infty} h(\tau) s_{\text{in}}(t - \tau)\, d\tau, \tag{6.131}$$

that is, the output voltage signal from the filter is the convolution of the weighting factor of the filter $h(\tau)$ with the input voltage signal. The function $h(\tau)$ is normalized,

$$\int_{-\infty}^{\infty} h(\tau)\, d\tau = 1, \tag{6.132}$$

and, for filters operating in real time, must obey the law of causality, $h(\tau) = 0$ for $\tau < 0$; it may conveniently be viewed as the response function of the filter to a unit impulse at $\tau = 0$. Thus, for example, for a single-section RC filter, the weighting factor is an exponential decay function:

$$\begin{aligned} h_1(\tau) &= (1/\tau_c) \exp(-\tau/\tau_c), \quad \tau \geqslant 0 \\ h_1(\tau) &= 0, \quad \tau < 0, \end{aligned} \tag{6.133}$$

where $\tau_c = RC$ is the time constant of the filter. The output voltage from a single-section RC filter is thus a weighted average over the past history of the input voltage, the weighting factor being the exponential decay function (6.133).

In evaluating the convolution integral (6.131), it is frequently useful to deal with the Fourier transforms of the functions. For example,

$$H(\omega) = \int_{-\infty}^{\infty} h(t) \exp(-i\omega t)\, dt \tag{6.134}$$

is called the filter frequency response function. In terms of the Fourier transforms, the convolution integral becomes:

$$S_{\text{out}}(\omega) = H(\omega) S_{\text{in}}(\omega). \tag{6.135}$$

We assume white noise so that the input power spectral density $|N(\omega)|^2 = N_{\text{in}}^2$ is constant for all frequencies. The power spectral density of the noise out of the filter is then, according to (6.135):

$$|N_{\text{out}}(\omega)|^2 = |H(\omega) N_{\text{in}}|^2 \tag{6.136}$$

so that, making use of the power theorem [53], we obtain:

$$\overline{[n_{\text{out}}(t)]^2} = N_{\text{in}}^2 \int_{-\infty}^{\infty} |H(\omega)|^2\, d\omega. \tag{6.137}$$

The autocorrelation function of the noise $R(t)$, which is important in cases in which measurements are made at different timcs, is defined by:

$$R(t) = \overline{n(\tau) n(t + \tau)}, \tag{6.138}$$

and for white noise may be shown to be [48]:

$$R(t) = N_{\text{in}}^2 \int_{-\infty}^{\infty} |H(\omega)|^2 \exp(i\omega t)\, d\omega. \qquad (6.139)$$

The Matched Filter

A detailed investigation of the theory of filters [48, 49, 51] reveals that the maximum signal-to-noise ratio is obtained if the filter frequency response function is proportional to the complex conjugate Fourier transform of the line shape function itself. By the nature of its definition, such a filter makes use of the entire signal function $s_{\text{in}}(t)$ for $-\infty < t < \infty$, and is therefore not physically realizable in real time. Operationally therefore such a filter would be used to process the signal once the complete spectrum was obtained and stored, for example, on magnetic tape. The effect on the line shape by matched filters is dramatic, a Lorentzian absorption line being transformed to a line with the same shape, but twice the width; a first derivative line becomes a second-derivative line with twice the width, and a second-derivative line becomes a fourth-derivative line with twice the width. However, no shift is obtained in the position of the line centers. Although this approach is usually neither convenient nor worth the trouble experimentally, the signal-to-noise ratio obtained is an upper limit approached by physically realizable filters. Furthermore, the functional dependence of the signal-to-noise ratio for the matched filters on scan rate and line width is the same as found for RC filters. In particular, if α is the scan rate and T_2^{-1} is the absorption line half-width at half-height, the signal-to-noise ratios for Lorentzian absorption, first- and second-derivative lines are, respectively:

$$(\text{S/N})_0 = (1/2\pi N_{\text{in}})(\alpha T_2^{-1})^{1/2} \qquad (6.140a)$$

$$(\text{S/N})_1 = (1/2\pi N_{\text{in}})(2\alpha T_2^{-3})^{1/2} \qquad (6.140b)$$

$$(\text{S/N})_2 = (1/2\pi N_{\text{in}})(\tfrac{2}{3}\alpha T_2^{-5})^{1/2}. \qquad (6.140c)$$

Noting that αT_2^{-1} is the reciprocal of the time required to sweep through one-half the line width, we call this time T_t and rewrite (6.140):

$$(\text{S/N})_0 = (1/2\pi N_{\text{in}})T_t^{1/2} \qquad (6.141a)$$

$$(\text{S/N})_1 = (T_2/2\pi N_{\text{in}})(T_t/2)^{1/2} \qquad (6.141b)$$

$$(\text{S/N})_2 = (T_2^2/2\pi N_{\text{in}})(3T_t/2)^{1/2}. \qquad (6.141c)$$

The important features of (6.141) are (1) that the signal-to-noise ratio is in each case proportional to the square root of the time required to traverse the line; and (2) that the signal-to-noise ratio is more sensitive to line width in the cases of the derivatives than in the case of the absorption. The implication of the first point seems to be that employing slower and slower scans could increase the signal-to-noise ratio without limit. In practice, however, this is incorrect since the assumption of white noise is not really correct. Low-frequency noise is usually quite strong so that there is normally a practical upper limit to the signal-to-noise ratio which is independent of the scan rate. It is at this point that time-averaging techniques, discussed in the next section, become advantageous.

Ernst [48] describes several other signal-processing techniques for sensitivity enhancement. These include (1) filters designed to give optimum signal-to-noise ratio subject to the condition of a fixed maximum line distortion error; (2) filters that optimize the signal-to-noise ratio while conserving one or more of the spectral moments; and (3) curve-smoothing techniques involving a least-squares fit of portions of the spectral line to a sum of orthogonal functions. These techniques, which involve processing of a complete spectrum in a computer, are beyond the scope of the present chapter, and the reader is referred to Ernst's article for further details [48].

RC Filters

The frequency response functions of the single- and double-section *RC* filters shown in Fig. 6.17 are, respectively:

$$H_1(\omega) = \frac{1 - i\omega\tau_c}{1 + \omega^2\tau_c^2} \tag{6.142a}$$

$$H_2(\omega) = \left[\frac{1 - i\omega\tau_c}{1 + \omega^2\tau_c^2}\right]^2. \tag{6.142b}$$

Substituting (6.142) into (6.137), we obtain the rms noise outputs from the single- and double-section *RC* filters; assuming white noise inputs:

$$n_{\text{rms}, \, 1\text{-}RC} = N_{\text{in}}(\pi/\tau_c) \tag{6.143a}$$

$$n_{\text{rms}, \, 2\text{-}RC} = N_{\text{in}}(\pi/2\tau_c). \tag{6.143b}$$

Similarly, we find that autocorrelation functions, calculated with the use of (6.139) are:

$$R_1(t) = (\pi/\tau_c)\exp(-t/\tau_c) \tag{6.144a}$$

$$R_2(t) = (\pi/2\tau_c)(1 + t/\tau_c)\exp(-t/\tau_c). \tag{6.144b}$$

Equations (6.143) and (6.144) may then be combined to compute the rms difference of noise voltages, which in turn is needed to compute a signal-to-noise ratio with (6.129).

We now consider the Lorentzian line shape functions:

$$s_0(t) = \frac{1}{\pi} \frac{T_2}{1 + T_2^2\alpha^2(t - t_0)^2} \tag{6.145a}$$

$$s_1(t) = -\frac{2}{\pi} \frac{T_2^3\alpha(t - t_0)}{[1 + T_2^2\alpha^2(t - t_0)^2]^2} \tag{6.145b}$$

$$s_2(t) = \frac{2}{\pi} \frac{T_2^3[3T_2^2\alpha^2(t - t_0)^2 - 1]}{[1 + T_2^2\alpha^2(t - t_0)^2]^3}, \tag{6.145c}$$

where α is again the scan rate in angular frequency units per unit time. It is convenient for the ensuing discussion to make the substitutions:

$$\begin{aligned} x &= \alpha T_2(t - t_0) \\ x' &= \alpha T_2 \tau \\ x_c &= \alpha T_2 \tau_c, \end{aligned} \tag{6.146}$$

where we refer to x_c as the "relative time constant." The output signal resulting from filtering a Lorentzian absorption line through a single section RC filter is then given by

$$s_{\text{out}}(t) = (T_2/\pi x_c) \int_0^\infty \exp(-x'/x_c)[1 + (x - x')^2]^{-1} \, dx'. \tag{6.147}$$

We see that the shape of the output signal is determined solely by the relative time constant x_c. Similar results are obtained for the first- and second-derivative line shape functions for both the single- and double-section RC filters.

The output line shape functions for the two input line shapes (6.145*b*) and (6.145*c*) and the two filter functions have been obtained by numerical integration of (6.147) and similar expressions. The signal amplitudes, normalized to unity for the unfiltered lines, and the signal-to-noise ratios, normalized by the signal-to-noise ratios of the matched filters are plotted in Figs. 6.18 and 6.19. As seen in the figures, the signal-to-noise ratio is maximized for certain values of the relative time constant x_c. The line shapes for these optimum values of x_c are plotted in Figs. 6.20 and 6.21 for the first and second derivatives, respectively. Some data relating to the optimum time constants are given in Table 6.8. As is obvious from the filtered line shapes, the line centers (maxima or zero crossings) are shifted in the direction of the sweep. The first moment of the lines is shifted by the filter time constant τ_c, and in those cases in which the line distortion is not great, and the line center is nearly the same as the first moment, the line center is shifted by τ_c as well. When the lines are more distorted, the shift is less than τ_c, as shown in Figs. 6.22 and 6.23 for the absorption and first and second derivatives,

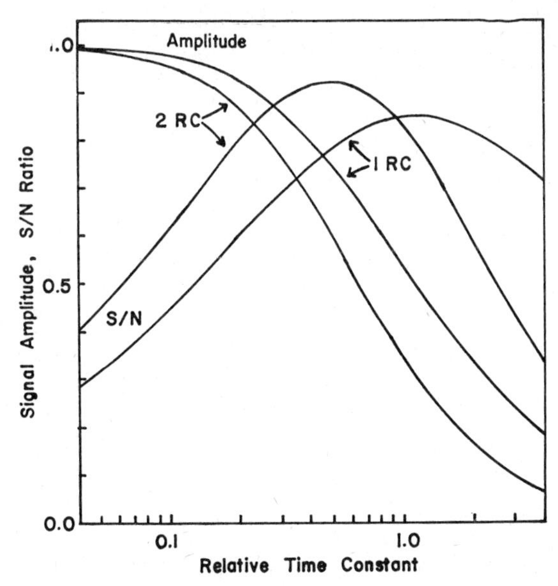

Fig. 6.18 Relative signal amplitude and signal-to-noise ratio for a first-derivative Lorentzian absorption line filtered by one- and two-section *RC* filters.

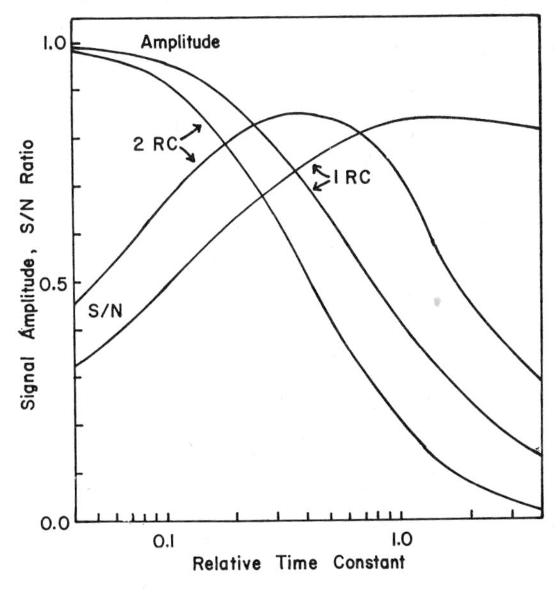

Fig. 6.19 Relative signal amplitude and signal-to-noise ratio for a second-derivative Lorentzian absorption line filtered by one- and two-section *RC* filters.

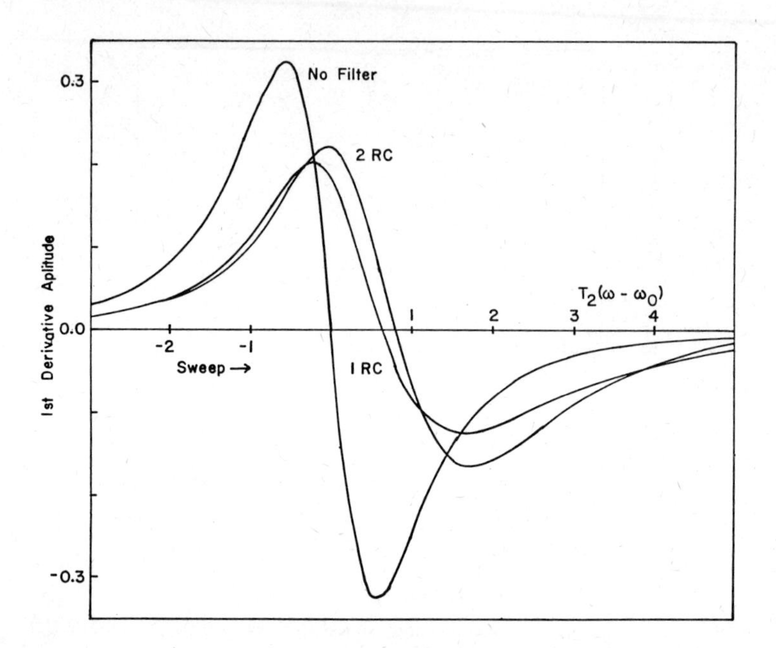

Fig. 6.20 Line shapes of a first-derivative Lorentzian absorption line with no filtering, and with filtering by optimum one- and two-section *RC* filters.

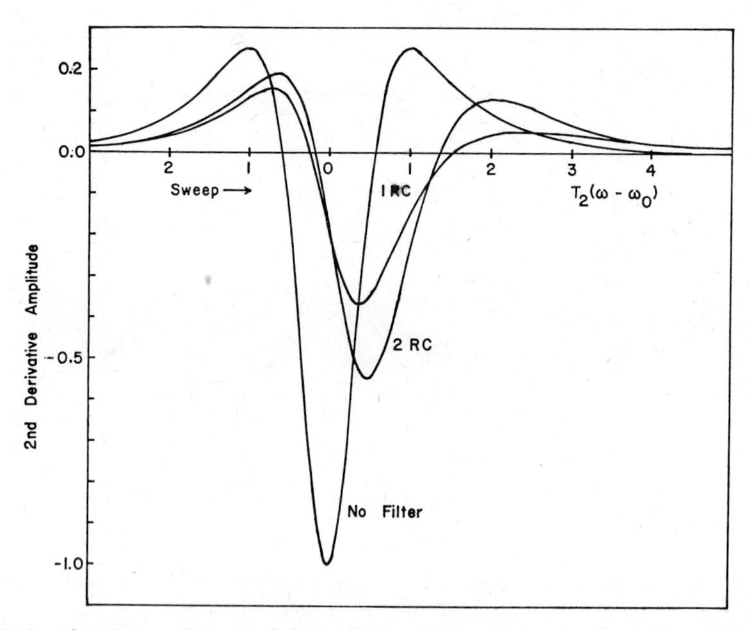

Fig. 6.21 Line shapes of a second-derivative Lorentzian absorption line with no filtering, and with filtering by optimum one- and two-section *RC* filters.

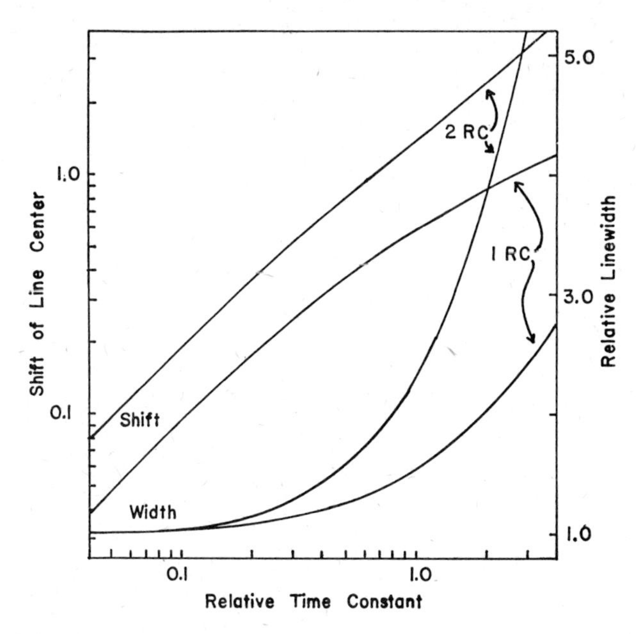

Fig. 6.22 Shift of the line center and relative width of a first-derivative Lorentzian absorption line filtered by one- and two-section *RC* filters.

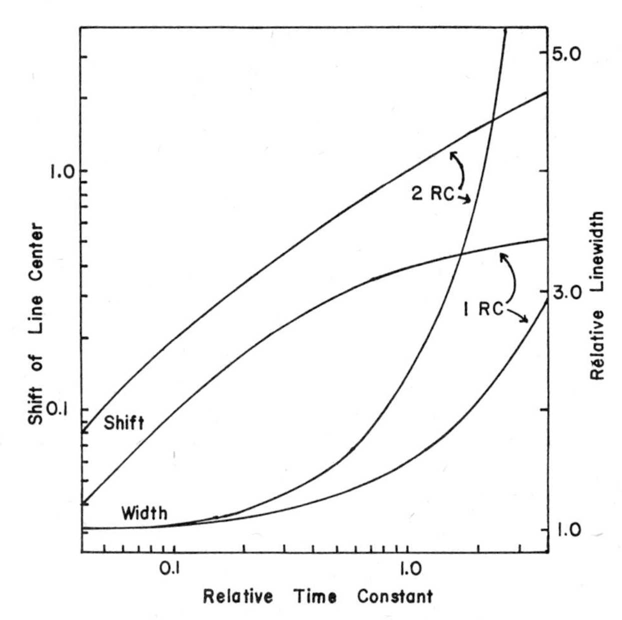

Fig. 6.23 Shift of the line center and relative width of a second-derivative Lorentzian absorption line filtered by one- and two-section *RC* filters.

Table 6.8 Data for Filtering of Lorentzian Lines with Optimum Single- and Double-Section *RC* Filters

Signal Type	*Filter Type*	*Optimum* x_c	*Height of First Peak**	*Height of Second Peak**	*Height of Third Peak**	*Line Shift†*	*Line Width†*	*Signal-to-Noise Ratio‡*
Absorption	None	—	1.00	—	—	0.00	1.00	—
	1-*RC*	2.2	0.54	—	—	0.94	2.15	0.90
	2-*RC*	1.0	0.60	—	—	1.40	1.95	0.96
1st Derivative	None	—	1.00	−1.00	—	0.00	1.00	—
	1-*RC*	1.1	0.62	−0.38	—	0.62	1.60	0.85
	2-*RC*	0.5	0.68	−0.51	—	0.80	1.56	0.93
2nd Derivative	None	—	0.25	−1.00	0.25	0.00	1.00	—
	1-*RC*	1.0	0.15	−0.36	0.05	0.38	1.58	0.84
	2-*RC*	0.33	0.19	−0.55	0.13	0.48	1.33	0.85

* Normalized by maximum peak height of original line.
† Normalized by width (T_2^{-1}) of original line.
‡ Normalized by the signal-to-noise ratio of the matched filter.

Table 6.9 Data for Filtering of Lorentzian Lines with Single- and Double-Section *RC* Filters with a Relative Time Constant of 0.1

Signal Type	*Filter Type*	*Height of First Peak**	*Height of Second Peak**	*Height of Third Peak**	*Line Shift*†	*Line Width*†	*Signal-to-Noise Ratio*‡
Absorption	None	1.000	—	—	0.00	1.00	—
	1-*RC*	0.985	—	—	0.10	1.01	0.35
	2-*RC*	0.976	—	—	0.20	1.03	0.49
1st Derivative	None	1.000	−1.000	—	0.00	1.00	—
	1-*RC*	0.976	−0.971	—	0.10	1.03	0.45
	2-*RC*	0.960	−0.950	—	0.19	1.05	0.62
2nd Derivative	None	0.250	−1.000	0.250	0.00	1.00	—
	1-*RC*	0.244	−0.949	0.241	0.09	1.02	0.49
	2-*RC*	0.238	−0.905	0.233	0.19	1.04	0.66

* Normalized by maximum peak height of original line.
† Normalized by width (T_2^{-1}) of original line.
‡ Normalized by the signal-to-noise ratio of the matched filter.

respectively. The line widths increase with increasing relative time constant as is also shown in Figs. 6.22 and 6.23.

As an example of the significance of these data, consider a first-derivative Lorentzian line filtered through a single-section *RC* filter. If the line width (between extrema) is 0.1 G, the scan rate 1.0 G/min, then the optimum time constant is 5.7 sec. Under these conditions, however, the zero crossing is shifted in the direction of the field scan 0.054 G and the width actually measured between derivative extrema is 0.16 G.

When accurate line width or position measurements are desired, the errors introduced by the optimum filters are too large to be tolerated. A common rule of thumb for choosing a filter time constant is to set the time constant to one-tenth the ratio of the line width to the scan rate, that is, the relative time constant $x_c = 0.1$. The results on line shape parameters, signal-to-noise ratio, and amplitude are given in Table 6.9. For most purposes this compromise is acceptable.

Time-Averaging Methods

As noted above, filtering methods are in most practical cases limited in their ability to enhance the signal-to-noise ratio because of the prevalence of low-frequency noise. This situation has been analyzed in detail by Ernst [48], who showed that if the power spectral density of the random noise has the form:

$$N(\omega)^2 = N_0^{\,2}\omega^{-\lambda}, \qquad 0 < \lambda < 1, \tag{6.148}$$

the signal-to-noise ratio achievable with a single-scan measurement employing a matched filter is given by:

$$\mathrm{S/N} = (\mathrm{S/N})_0 T_t^{(1-\lambda)/2}, \tag{6.149}$$

where T_t is the performance time. Thus if the noise is really white ($\lambda = 0$), the signal-to-noise ratio increases with the square root of the performance time, whereas $1/f$ noise ($\lambda = 1$) results in no increase in signal-to-noise ratio whatever with increasing performance time. The actual case is usually somewhere between these two extremes, but the point is that filtering is limited in its capacity to improve the signal-to-noise ratio.

The method of time-averaging overcomes the above limitation on signal-to-noise ratio enhancement. Instead of a single slow scan through the spectrum, many relatively rapid scans are made through the spectrum and added together such that the desired signals add coherently, whereas the noise adds randomly. The influence of low-frequency noise is strongly reduced in this way, and it can be shown that the signal-to-noise ratio is again proportional to the square root of the performance time, and thus to the square root of the number of scans. As Ernst [48] has pointed out, filtering and time-averaging are complementary methods, not alternatives. Filtering relies on the different spectral properties of the signal and noise, whereas time-averaging is based on the coherence of a repetitive signal and the incoherence of noise.

In practice, the successive scans are summed in the digital storage of a computer which is effectively a multichannel pulse-height analyzer. Since the computer storage locations are discrete, the spectrum is stored as individual points. To achieve this, the field may be swept continuously with the computer storage channels gated consecutively; this method results in effect in integration of the signal over the sampling time for one channel and has the advantage of further reducing high-frequency noise, although the signal is somewhat distorted. Alternatively, a staircase magnetic-field sweep synchronized to the sampling gate may be employed. Again, the signal is averaged over the sampling time, with reduction in high-frequency noise, but since the field is held constant no signal distortion takes place.

The density of sampling points is determined of course by the desired fidelity of reproduction of the signal. As a general rule, Ernst suggests a sampling density of $2f_{10\,\mathrm{dB}}$ samples per second, where $f_{10\,\mathrm{dB}}$ is the 10-dB frequency of the preceding filter. Thus for a single-section *RC* filter the spacing between samples should be approximately equal to the filter time constant *RC*.

In order to successfully apply the time-averaging method to sensitivity enhancement, it is necessary to ensure that the signal is coherent in successive scans. In principle, this can be achieved by starting the scan at precisely the same magnetic field each time. Field scans can be initiated at the same point either by reference to a particular voltage from a Hall crystal probe in the magnet gap, or by reference to a particular proton resonance frequency using a separate nmr probe. This method has the disadvantage, however, that drift of the cavity resonant frequency, particularly when the cavity is operated far from room temperature or when narrow lines are being observed, raises a serious problem of signal incoherence on successive scans, even though the starting magnetic field is the same. This problem may be eliminated by the use of a field-frequency locking device. Such a device works by locking the field to the electron resonance of a sample in a separate nonresonant cavity powered by the microwave source which is locked to the main resonant cavity. Field offset is then achieved by passing a well-regulated current through coils mounted on the external reference cavity. Such a device, capable of field-frequency lock within ± 50 G from $g = 2$ is available commercially from Varian.

An alternative arrangement for ensuring signal coherence on successive scans is to trigger the computer sampling device by a strong esr line from a sample of DPPH or other external standard mounted within the resonant cavity. This method is certainly simpler since it eliminates the need for a field-frequency lock, but in extreme cases of cavity frequency drift and very sharp lines still allows some signal incoherence.

5 RESOLUTION

Field Modulation Effects

Detection of magnetic resonance absorption in principle may be achieved directly by dc amplification of the detector output, or by amplification of the intermediate frequency signal in the case of a superheterodyne spectrometer. Because of the stability problems inherent in dc amplifiers and the requirement of a highly stable microwave source, however, these techniques are rarely used. Recourse is usually made to a modulation scheme so that the signal may be processed by a narrow bandpass amplifier and phase-sensitive detector with elimination of most of the problems of low-frequency noise and drift in the microwave source and associated circuitry.

Since in magnetic resonance spectroscopy two parameters are available—the magnetic field and the microwave radiation—several modulation schemes are possible. The microwave radiation may be either frequency or amplitude modulated, or the magnetic field may be modulated. Although by far the most common choice is magnetic field modulation, the other possibilities have been tried successfully and have the advantage, important in some applications, of not requiring modulation coils [36].

In the sections below, we consider in detail the effects on the observed signal of the magnetic field modulation.

Low-Frequency Modulation

In this section it is assumed that the modulation frequency is sufficiently low that the magnetic field may be assumed essentially constant over a period sufficient for the steady-state solution of the Bloch equations [see (6.36)] to obtain. Thus the field modulation frequency must be much less than the Larmor frequency, or the reciprocal of the relaxation times:

$$\omega_m \ll \omega_0,\ T_2^{-1},\ T_1^{-1}. \tag{6.150}$$

If at any instant the magnetic field is:

$$H = H_0 + H_m \cos \omega_m t, \tag{6.151}$$

the Larmor frequency is:

$$\omega_0 = \gamma H_0 + \gamma H_m \cos \omega_m t. \tag{6.152}$$

Inserting (6.152) into (6.36), again writing $\omega_0 = \gamma H_0$, and neglecting the saturation term in the denominator, we have:

$$\chi''(t) = \tfrac{1}{2}\omega_0 \chi_0 T_2 / [1 + T_2^2(\omega_0 - \omega + \gamma H_m \cos \omega_m t)^2]. \tag{6.153}$$

With $G(x)$ defined by (6.40), we may write more compactly:

$$\chi''(t) = \frac{\pi}{2}\,\omega_0\chi_0 G(x), \tag{6.154}$$

where

$$x = \omega_0 - \omega + \gamma H_m \cos \omega_m t. \tag{6.155}$$

SMALL-AMPLITUDE MODULATION

For γH_m sufficiently small, $\gamma H_m \ll T_2^{-1}$, we can expand (6.154) in a Taylor series about $x_0 = \omega_0 - \omega$:

$$\chi''(t) = \frac{\pi}{2}\,\omega_0\chi_0[G(x_0) + G^i(x_0)(\gamma H_m \cos \omega_m t) + (\tfrac{1}{2})G^{ii}(x_0)(\gamma H_m \cos \omega_m t)^2 + \cdots]. \tag{6.156}$$

Rearranging and collecting terms in $\cos(n\omega_m t)$:

$$\begin{aligned}\chi''(t) = \left(\frac{\pi}{2}\right)\gamma M_0\{&[G(x_0) + \tfrac{1}{4}(\gamma H_m)^2 G^{ii}(x_0) + \tfrac{1}{64}(\gamma H_m)^4 G^{iv}(x_0) + \cdots]\\ &+ (\gamma H_m)[G^i(x_0) + \tfrac{1}{8}(\gamma H_m)^2 G^{iii}(x_0) + \tfrac{1}{192}(\gamma H_m)^4 G^{v}(x_0) + \cdots]\\ &\times \cos(\omega_m t) + (\tfrac{1}{4})(\gamma H_m)^2[G^{ii}(x_0) + \tfrac{1}{12}(\gamma H_m)^2 G^{iv}(x_0) + \cdots]\\ &\times \cos(2\omega_m t) + (\tfrac{1}{24})(\gamma H_m)^3[G^{iii}(x_0) + \tfrac{1}{16}(\gamma H_m)^2 G^{v}(x_0) + \cdots]\\ &\times \cos(3\omega_m t) + \cdots\}.\end{aligned} \tag{6.157}$$

Thus we see that for sufficiently low-amplitude field modulation the component of the microwave absorption signal at the frequency of the field modulation is proportional to the first derivative of the Lorentzian line shape function. Similarly, the second, third, and higher harmonics are proportional to the second, third, and higher derivatives.

Carrying the technique one step further, we see that if the resonance line is also modulated at a second, much lower, frequency, the signal emergent from the phase-sensitive detector operating at the higher modulation frequency will carry a component at the low modulation frequency, as well as a dc component. A second phase-sensitive detector, operating at the low modulation frequency, then converts the signal to a dc component proportional to the second derivative. In principle, additional frequencies could be used, and additional phase-sensitive detectors used to process the signal. The practical limit to this scheme is probably two or three modulation frequencies, although in practice there is little to be gained in using more than two frequencies. Because there are two or more positive and negative extrema in third- and higher-order derivatives, considerable complication is introduced in the interpretation of complex spectra composed of higher-order derivatives. For this reason, either first or second derivatives are recorded or, if resolution enhancement is necessary, recourse is made to some technique whereby a more complex line shape is produced (see p. 589).

In the case of the double-modulation technique, the applied magnetic field is:

$$H = H_0 + H_{m1} \cos \omega_{m1} t + H_{m2} \cos \omega_{m2} t. \qquad (6.158)$$

Inserting the corresponding Larmor frequency into (6.36) and expanding in a Taylor series as above, we find that the term having the desired time dependence is:

$$\frac{\pi}{2} \gamma M_0 (\gamma H_{m1})(\gamma H_{m2})[G^{ii}(x_0) + \tfrac{1}{8}(\gamma^2 H_{m1}^2 + \gamma^2 H_{m2}^2) G^{iv}(x_0) + \cdots] \times \cos \omega_{m1} t \cos \omega_{m2} t. \qquad (6.159)$$

Comparing (6.159) with the second derivative term of (6.157), we see that the double-modulation technique apparently results in a second-derivative signal four times larger than that obtained by second-harmonic detection. The factor of four is somewhat misleading, however, since, as noted below, the modulation amplitudes required to give equal distortions using the two techniques result in equal amplitudes as well. There is, nonetheless, a considerable advantage in the double modulation method since greater noise rejection is possible in two different stages of frequency selective amplification.

LARGE-AMPLITUDE MODULATION

According to (6.157) and (6.159), the signal amplitude increases with the modulation amplitude so that in any practical case it is desirable to use a moderately large field modulation amplitude. Since line shape distortions are expected when the modulation amplitude approaches the line width, it is necessary to know what to expect when the modulation amplitude is increased. In principle, line shapes could be computed from the terms of the Taylor series. This is tedious, however, even for $\gamma H_m T_2$, the ratio of the modulation amplitude to the undistorted line width, as large as one-half. Furthermore, the Taylor series diverges for sufficiently large modulation amplitudes, so that it cannot be used in many situations of experimental interest. Recourse is therefore made to a Fourier analysis of (6.153):

$$\chi''(t) = \frac{1}{2\pi} \chi_0 \omega_0 T_2 \sum_{n=0}^{\infty} a_n(T_2, \omega_0 - \omega, H_m) \cos n\omega_m t, \qquad (6.160)$$

where the coefficients are given by

$$a_n(T_2, \omega_0 - \omega, H_m) = \frac{\omega_m}{\pi} \int_{-\pi/\omega_m}^{\pi/\omega_m} \frac{\cos n\omega_m t \, dt}{1 + T_2^2(\omega_0 - \omega + \gamma H_m \cos \omega_m t)^2}. \qquad (6.161)$$

Integrals of this type have been solved by a number of authors [52, 53].

Wahlquist [52], for example, gives for the coefficients of the dc component, the fundamental component, and the second harmonic component:

$$a_0 = \frac{2}{m^2} \frac{s^{1/2}}{(s-2)^{1/2}(s-r)} \tag{6.162}$$

$$a_1 = \pm \frac{2}{m^2} \frac{(2r-s)^{1/2}}{(s-2)^{1/2}(s-r)} \tag{6.163}$$

$$a_2 = \frac{2}{m^2}\left[2 + \frac{s^{1/2}(1+2r-2s)}{(s-2)^{1/2}(s-r)}\right], \tag{6.164}$$

where

$$r = 1 + \frac{1}{m^2}\left[1 + T_2^2(\omega_0 - \omega)^2\right] \tag{6.165}$$

$$s = r + [r^2 - 4T_2^2(\omega_0 - \omega)^2/m^2]^{1/2} \tag{6.166}$$

$$m = \gamma H_m T_2. \tag{6.167}$$

The functions a_1/m and a_2/m^2 are plotted in Figs 6.24 and 6.25 against $T_2(\omega_0 - \omega)$ for several values of m. As can be seen from the figures, the

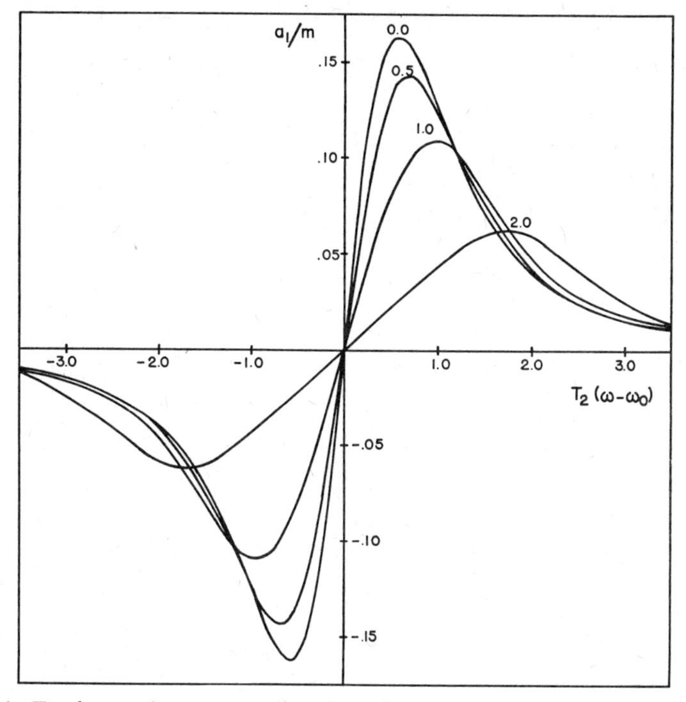

Fig. 6.24 Fundamental component line shape for modulation parameter $m = 0.0, 0.5, 1.0, 2.0$.

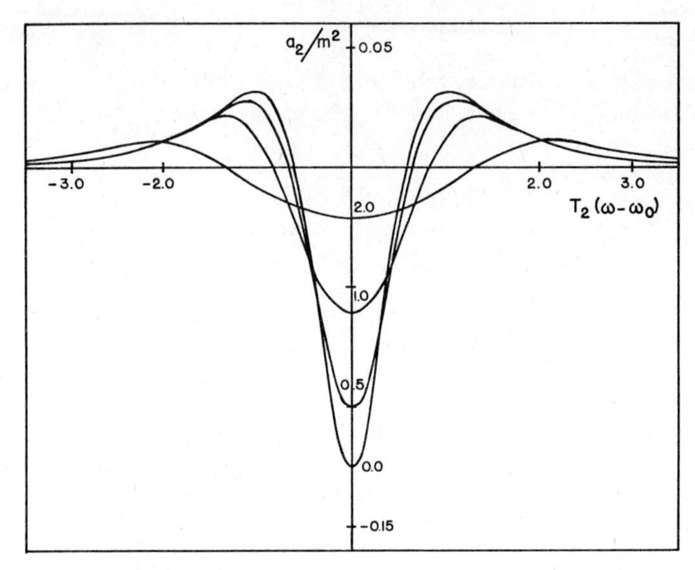

Fig. 6.25 Second-harmonic component line shape for modulation parameter $m = 0.0$, 0.5, 1.0, 2.0.

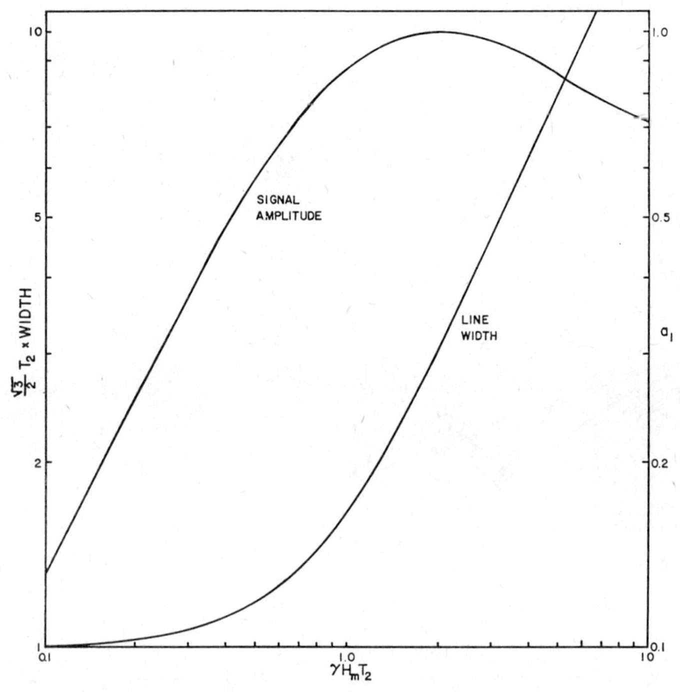

Fig. 6.26 Relative signal amplitude and relative line width of the signal detected at the fundamental field modulation frequency as functions of the field modulation amplitude.

effect of increasing modulation amplitude is to broaden and distort the detected resonance lines from the simple Lorentzian line shapes.

The effect of increasing modulation amplitude is shown in more detail for a_1 and a_2 in Figs. 6.26 and 6.27. In Fig. 6.26 the total signal amplitude of a_1 and a line width parameter are plotted as functions of m. The line width parameter used is $\sqrt{3}T_2/2$ times the full width between positive and negative extrema, which is unity for a first-derivative Lorentzian line. It may be noted that the amplitude goes through a maximum for $m = 2$, but that the line at that point is quite distorted, the width being about three times the normal Lorentzian width. In selecting the modulation amplitude to record an essentially undistorted line, a common rule of thumb is to choose the peak-to-peak modulation equal to one-fourth the width between derivative extrema. This modulation amplitude, corresponding to $m = 0.144$, leads to approximately 1.5% distortion as measured by the increase in the line

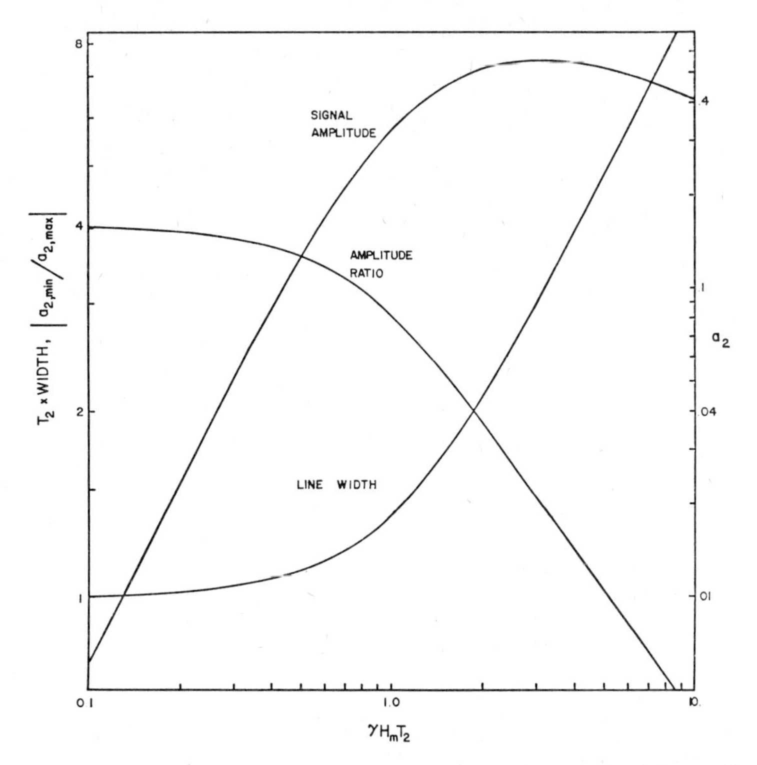

Fig. 6.27 Relative signal amplitude, relative line width, and the ratio of the amplitudes of the negative and positive extrema of the signal detected at the second-harmonic frequency as functions of the field modulation amplitude.

width. The amplitude for this value of m is about one-fifth the maximum. Doubling the modulation amplitude in this case approximately doubles the signal amplitude while increasing the distortion to about 6%.

The total amplitude of a_2 and a line width parameter as well as the ratio of the amplitudes of the negative and positive extrema are plotted in Fig. 6.27. Perhaps the most striking effect of finite modulation amplitude is the rapid decrease of the amplitude ratio parameter. This function decreases rapidly from its value of 4 for the undistorted Lorentzian second-derivative line shape. The amplitude again goes through a maximum, here at approximately $m = 3$, at which point the line width is about three times the width of the undistorted second-derivative Lorentzian line, and the amplitude ratio parameter is approximately 1.45. With the peak-to-peak modulation amplitude equal to $\frac{1}{4}$ the width between positive extrema, $m = 0.25$, the signal amplitude is approximately $\frac{1}{16}$ the maximum value, the width is about 1.027 times the undistorted width, and the amplitude ratio is 3.85. In this case doubling the modulation amplitude again roughly doubles the signal amplitude, but the line width is 10% greater than that undistorted width, and the amplitude ratio decreases to 3.59.

Analysis of the "second-derivative" signal obtained from double-field modulation is more difficult since the integrals required for the Fourier analysis are unavailable. Line shapes have been computed in the range $m_1 =$ 0.1 to 0.5, using the Taylor series expansion, assuming that $m_1 = \gamma H_{m1} T_2 = \gamma H_{m2} T_2$. Terms in the expansion up to the fourteenth order are required at $m_1 = 0.5$, and extension much beyond that point is virtually impossible by this technique. The results are very similar to those obtained for the second harmonic signal from the single-modulation method. Indeed, Fig. 6.27 fairly accurately describes the line shapes obtained from double modulation up to $m = 0.5$, if m is understood to be the ratio of the total modulation amplitude to the undistorted line width, $m = \gamma(H_{m1} + H_{m2})T_2 = 2m_1$. From $m = 0.5$ to 1.0, the double-modulation signal is somewhat better than the second-harmonic signal in every respect. For example, at $m = 1.0$ ($m_1 = 0.5$), both the amplitude and the amplitude ratio of the double-modulation signal are about 15% greater than the values plotted in Fig. 6.27, and the line width is about 10% less than that plotted in Fig. 6.27.

High-Frequency Field Modulation [54]

When the field modulation frequency approaches any of the other characteristic frequencies of the experiment, T_1^{-1}, T_2^{-1}, or ω_0, the above analysis is no longer valid, and the effect of the modulation frequency must be explicitly considered. The most important effect may be easily seen by noting that, according to (6.152), field modulation is equivalent to frequency modulation of the microwave source. The microwave field is written in the generalized form:

$$H_1(t) = H_1 \sin \phi(t), \tag{6.168}$$

where $\phi(t)$ is the total angular displacement at time t. The instantaneous angular frequency is simply $\omega(t) = d\phi/dt$. In the present case:

$$\omega(t) = \omega_0 + \gamma H_m \cos \omega_m t. \tag{6.169}$$

Integrating, we obtain, within a constant-phase angle,

$$H_1(t) = H_1 \sin (\omega_0 t + \beta \sin \omega_m t), \tag{6.170}$$

where $\beta = \gamma H_m/\omega_m$ is called the modulation index. Writing

$$H_1(t) = H_1 \sin \omega_0 t \cos (\beta \sin \omega_m t) + \cos \omega_0 t \sin (\beta \sin \omega_m t), \tag{6.171}$$

and using the relations:

$$\cos (\beta \sin \omega_m t) = J_0(\beta) + 2\sum_{n=1}^{\infty} J_{2n}(\beta) \cos 2n\omega_m t \tag{6.172}$$

$$\sin (\beta \sin \omega_m t) = 2\sum_{n=1}^{\infty} J_{2n+1}(\beta) \sin (2n + 1)\omega_m t, \tag{6.173}$$

we obtain

$$H_1(t) = H_1 \sum_{n=-\infty}^{\infty} J_n(\beta) \sin (\omega_0 + n\omega_m)t, \tag{6.174}$$

where the $J_n(\beta)$ are Bessel functions of the first kind. Thus magnetic resonance absorption could be expected to occur not only at the center frequency ω_0 but also at the side-band frequencies $\omega_0 \pm n\omega_m$. Thus for ω_m comparable to or greater than the line width, line shape distortions, or resolved side bands are to be expected.

A more precise treatment of the problem may be obtained by returning to the Bloch equations, taking explicit account of the time dependence of the magnetic field. Thus the terms $v\gamma H_m \cos \omega_m t$ and $u\gamma H_m \cos \omega_m t$ are added to the right-hand sides of (6.35*a*) and (6.35*b*), respectively. A steady-state solution is no longer possible, and we proceed by (1) assuming H_1 small compared with T_1^{-1}, so that $M_z = M_0$, and (2) writing $w = u + iv$, to obtain:

$$\frac{dw}{dt} = -w[T_2^{-1} + i(\omega_0 - \omega) + i\gamma H_m \cos \omega_m t] + \tfrac{1}{2} i\gamma H_1 M_0. \tag{6.175}$$

Integration of the differential equation is from $-\infty$ to t, so that transient terms may be neglected.

The solution obtained is:

$$w = \left(\frac{i}{2}\right)\gamma H_1 M_0 \frac{\int_{-\infty}^{t} \exp [t'/T_2 + it'(\omega_0 - \omega) + i\beta \sin (\omega_m t')]\, dt'}{\exp [t/T_2 + it(\omega_0 - \omega) + i\beta \sin (\omega_m t)]}, \tag{6.176}$$

where β is the modulation index defined above. Using the expansion

$$\exp [i\beta \sin (\omega_m t)] = \sum_{n=-\infty}^{\infty} J_n(\beta) \exp (in\omega_m t), \tag{6.177}$$

the integral in the numerator becomes:

$$\sum_{n=-\infty}^{\infty} J_n(\beta)[T_2^{-1} + i(\omega_0 - \omega) + in\omega_m] \exp [t/T_2 + it(\omega_0 - \omega) + in\omega_m t]. \tag{6.178}$$

Employing a similar expansion of the denominator and dividing the two series, we obtain:

$$w = \left(\frac{i}{2}\right) \gamma H_1 M_0 \sum_{n=-\infty}^{\infty} \sum_{k=-\infty}^{\infty} J_n(\beta) J_k(\beta) \exp [i(n-k)\omega_m t]$$
$$\times [1 - iT_2(\omega_0 - \omega + n\omega_m)]/[1 + T_2^2(\omega_0 - \omega + n\omega_m)^2]. \tag{6.179}$$

Separating w into its real and imaginary components, we have:

$$\chi'(t) = (\tfrac{1}{2})\omega_0 \chi_0 T_2 \sum_{n=-\infty}^{\infty} \sum_{k=-\infty}^{\infty} J_n(\beta) J_k(\beta)$$
$$\times \frac{T_2(\omega_0 - \omega + n\omega_m) \cos (n-k)\omega_m t - \sin (n-k)\omega_m t}{1 + T_2^2(\omega_0 - \omega + n\omega_m)^2} \tag{6.180}$$

$$\chi''(t) = (\tfrac{1}{2})\omega_0 \chi_0 T_2 \sum_{n=-\infty}^{\infty} \sum_{k=-\infty}^{\infty} J_n(\beta) J_k(\beta)$$
$$\times \frac{\cos (n-k)\omega_m t + T_2(\omega_0 - \omega + n\omega_m) \sin (n-k)\omega_m t}{1 + T_2^2(\omega_0 - \omega + n\omega_m)^2}. \tag{6.181}$$

We now confine our attention to the component of the absorption mode $\chi''(t)$, that oscillates at the modulation frequency; that is, we choose those terms for which $|n - k| = 1$:

$$\chi''(t) = \left(\frac{\pi}{2}\right) \omega_0 \chi_0 \sum_{n=-\infty}^{\infty} \{J_n(\beta[J(_{n-1}(\beta) + J_{n+1}(\beta)] \cos (\omega_m t) G(\omega_0 - \omega + n\omega_m)$$
$$+ [J_{n-1}(\beta) - J_{n+1}(\beta)] \sin (\omega_m t) F(\omega_0 - \omega + n\omega_m)\}, \tag{6.182}$$

where $G(x)$ and $F(x)$ are the Lorentzian absorption and dispersion line shape functions defined by (6.40) and (6.41), respectively. A number of cases may now be discussed.

SMALL-MODULATION INDEX

For β sufficiently small, $J_0 \cong 1$, $J_1 \cong \beta/2$, and all higher-order Bessel functions are very small. The series may then be reduced to the terms with

$n = 0, \pm 1$:

$$\chi''(t) = \left(\frac{\pi}{4}\right) \omega_0 \chi_0 \beta \{[G(\omega_0 - \omega + \omega_m) - G(\omega_0 - \omega - \omega_m)] \cos(\omega_m t)$$
$$+ [2F(\omega_0 - \omega) - F(\omega_0 - \omega + \omega_m) - F(\omega_0 - \omega - \omega_m)] \sin(\omega_m t)\}. \tag{6.183}$$

The component in phase with the modulation signal exhibits two features of Lorentzian absorption line shape, having opposite signs and displaced from the center frequency by $\pm\omega_m$. The out-of-phase component shows three features with Lorentzian dispersion line shape, one at the center frequency, and two side bands of opposite sign and one-half the amplitude at $\pm\omega_m$. These are shown in Figs. 6.28*a* and 6.29*a*, respectively. As the modulation

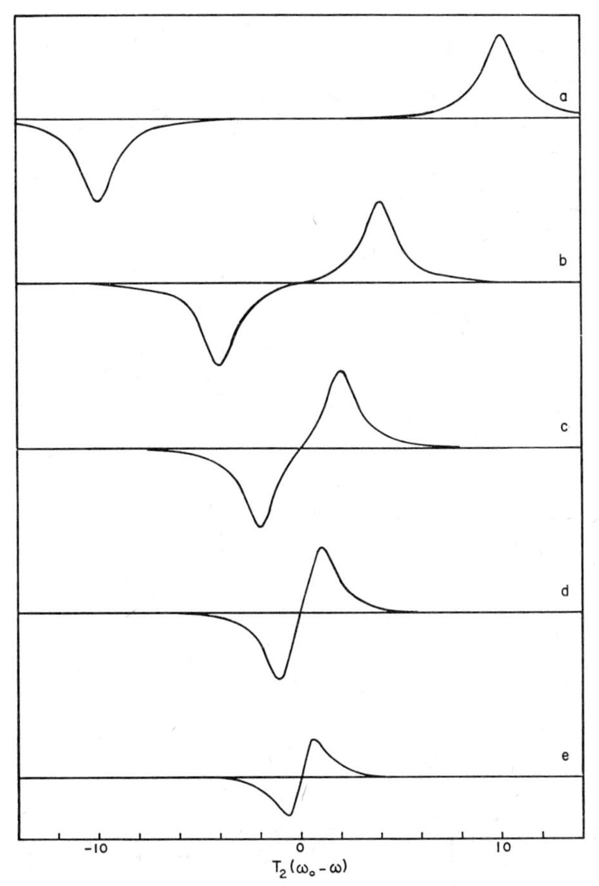

Fig. 6.28 Line shape of the fundamental component in phase with the field modulation signal for various modulation frequencies.

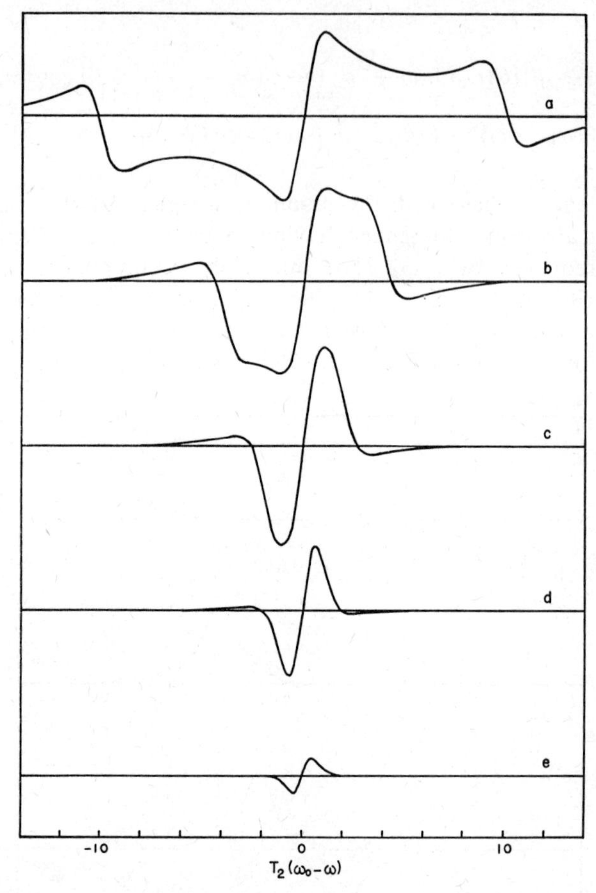

Fig. 6.29 Line shape of the fundamental component out of phase with the field modulation signal for various modulation frequencies.

frequency is decreased relative to the line width, the features at $\pm\omega_m$ move toward the center and coalesce, as shown in Figs. 6.28*b*–*e* and 6.29*b*–*e*. For ω_m sufficiently small compared with T_2^{-1}, (6.183) may be rearranged to give

$$\chi''(t) = \frac{\pi}{2}\,\omega_0\chi_0(\gamma H_m \cos\omega_m t)G^i(\omega_0 - \omega). \tag{6.184}$$

The out-of-phase component vanishes under these conditions, and we note that (6.184) is identical to the corresponding component of (6.156); thus for sufficiently low modulation frequency, the results reduce to the case discussed on p. 571.

LARGE MODULATION INDEX

For larger values of the modulation index, additional side bands appear and, as the modulation index increases, the intensity of the spectrum moves to higher-order side bands, the intensity of the center band and inner side bands decreasing. When the modulation frequency is less than the undistorted line width, the side bands are not resolved, but the line may be substantially distorted from the normal Lorentzian derivative shape.

In esr experiments, the field modulation frequency is commonly 100 kHz or less, and line widths range from around 10 mG up to hundreds of gauss (T_2^{-1} ranges from about 30 kHz up). Thus the experimental conditions under which the present kind of modulation broadening is important are obtained when narrow line spectra are recorded with 100-kHz field modulation. Modulation broadening for field modulation frequencies below 1 kHz is attributable almost entirely to the amplitude of the modulation and is accurately treated in Section 5, p. 570.

Line shape calculations have been performed for $\omega_m T_2$ ranging from 0.05 to 5.0 with $\gamma H_m T_2 = 0.1$ and 0.5 ($\beta = 0.02$ to 10.0). In this range the side

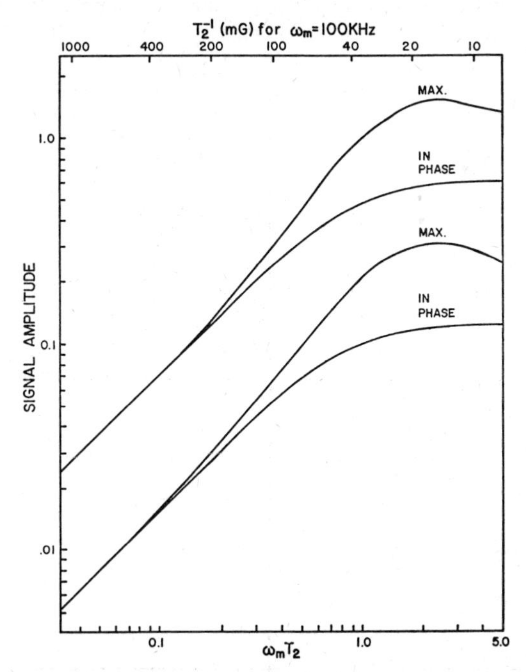

Fig. 6.30 Signal amplitudes (in arbitrary units) of the fundamental component in phase with the field modulation signal, and at the phase angle that maximizes the amplitude, as functions of the field modulation frequency for $\gamma H_m T_2 = 0.1, 0.5$.

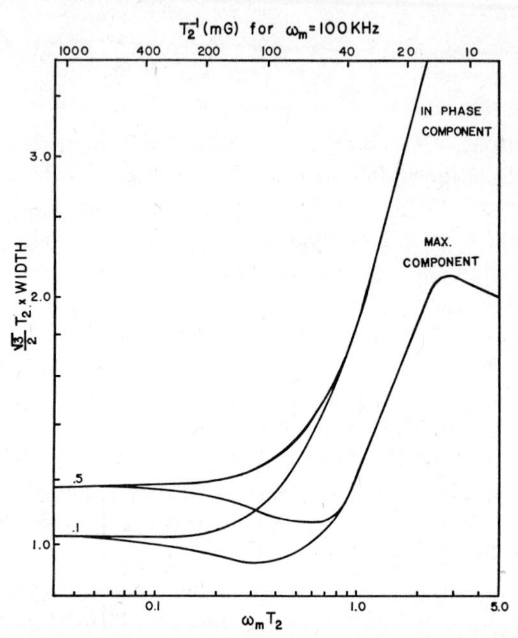

Fig. 6.31 Relative line widths of the fundamental component in phase with the field modulation signal, and at the phase angle that maximizes the amplitude, as functions of the field modulation frequency for $\gamma H_m T_2 = 0.1, 0.5$.

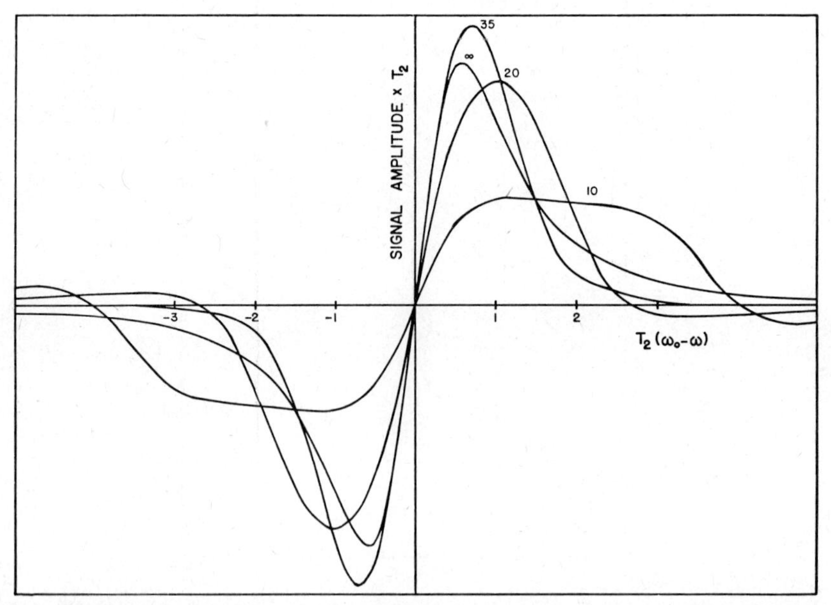

Fig. 6.32 Line shapes of the fundamental component, with the phase angle adjusted to maximize the amplitude, for line widths of 10, 20, and 35 mG and for a very wide line assuming 100-kHz field modulation.

bands are unresolved, although substantial line shape distortions are seen. Figures 6.30 and 6.31 show the variation of the signal amplitude and line width with $\omega_m T_2$ and $\gamma H_m T_2$. Up to $\omega_m T_2 = 0.2$, the line has essentially the expected Lorentzian derivative shape, although slightly broadened by the finite modulation amplitude. Beyond that value, however, the width increases rapidly. The amplitude, which normally increases with T_2^2, levels off beyond $\omega_m T_2 = 1.0$.

It is common experimental practice to adjust the reference phase of the phase-sensitive detector by maximizing the signal amplitude with respect to the phase angle. At low-modulation frequencies, this is equivalent to setting the phase angle to zero since there is no out-of-phase component. When $\omega_m T_2$ becomes significant, however, the out-of-phase component is no longer negligible, and for large $\omega_m T_2$ its amplitude considerably exceeds that of the in-phase component. It is thus somewhat misleading to consider only the in-phase component.

Calculated line shapes, in which the amplitude has been maximized with

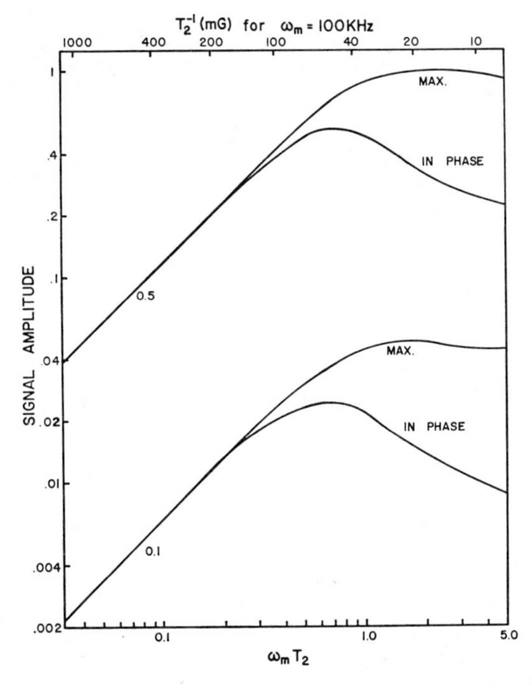

Fig. 6.33 Amplitudes (in arbitrary units) of the component of the double-modulation signal in phase with the higher-frequency field modulation signal, and at the phase angle that maximizes the amplitude, as functions of the higher field modulation frequency for $\gamma H_m T_2 = 0.1, 0.5$.

respect to the phase angle, are plotted in Fig. 6.32 for $\omega_m T_2 = 0.10$, 1.02, 1.78, and 3.57. These values correspond to $(\gamma T_2)^{-1} = \infty$, 35, 20, and 10 mG, respectively, assuming a field modulation frequency of 100 kHz. The amplitude and apparent line width of the maximum component are also given in Figs. 6.30 and 6.31 as functions of $\omega_m T_2$ for $\gamma H_m T_2 = 0.1$ and 0.5. As can be seen, the maximum component is virtually identical with the out-of-phase component for $\omega_m T_2$ greater than about 2.0. It is perhaps somewhat surprising that the apparent width of the maximum component is *less* than the undistorted width for $\omega_m T_2$ in the range 0.1 to 0.9, and beyond that range is still considerably less than the width of the in-phase component. It is of considerable practical significance that unless a correction is made substantial errors can result in line width measurements when 100-kHz field modulation is employed, even for lines several hundred milligauss wide.

If, as is frequently the case, the higher of the two modulation frequencies used in the double-modulation technique to obtain second derivatives is 100 kHz, modulation broadening through unresolved sidebands must be

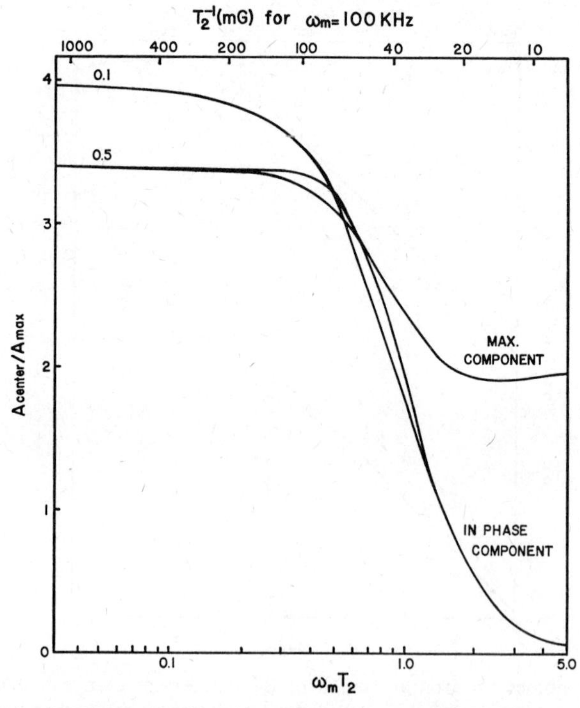

Fig. 6.34 The ratios of the amplitudes of the negative and positive extrema of the component of the double-modulation signal in phase with the higher-frequency field modulation signal, and at the phase angle that maximizes the amplitude, as functions of the higher field modulation frequency for $\gamma H_m T_2 = 0.1$, 0.5.

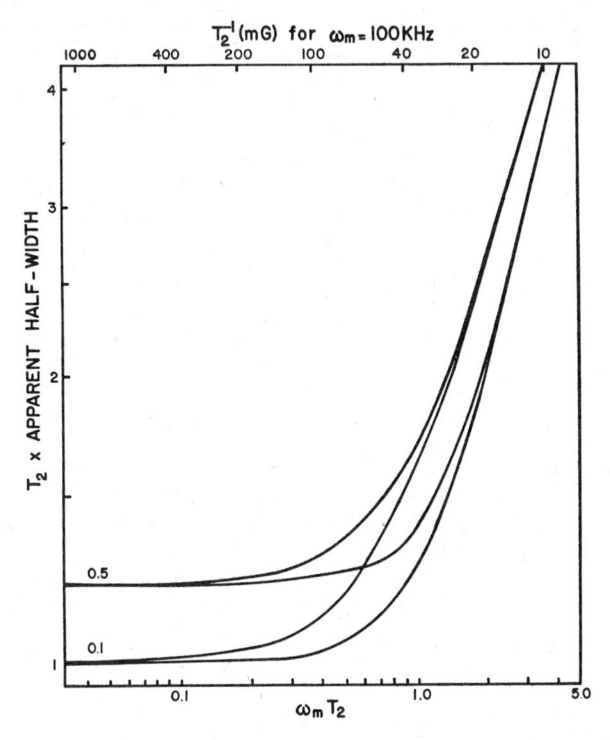

Fig. 6.35 Relative line widths of the component of the double-modulation signal in phase with the higher-frequency field modulation signal, and at the phase angle that maximizes the amplitude, as functions of the higher field modulation frequency for $\gamma H_m T_2 = 0.1, 0.5$.

considered. Assuming that the lower of the two frequencies is sufficiently low that it need not be considered in the solution of the Bloch equations, the line shape may be obtained by either Fourier analysis or a Taylor series expansion of (6.182) in a manner analogous to the treatment of (6.154) on p. 571. Line shapes of the in-phase component have been computed by the latter technique over the range $\omega_m T_2 = 0.05$ to 5.0 for $\gamma H_m T_2 = 0.1$ and 0.5. The results are given in Figs. 6.33, 6.34, and 6.35, in which the signal amplitude, the ratio of the amplitude of the (negative) center peak to that of the (positive) side peaks, and the line width are plotted as functions of $\omega_m T_2$.

Again, the out-of-phase component is appreciable for large $\omega_m T_2$, so that it is necessary to consider the signal with amplitude optimized with respect to the phase angle. Calculated line shapes of the optimum component are shown in Fig. 6.36 for $\omega_m T_2 = 0.0$, 1.02, 1.78, and 3.57. The amplitude, amplitude ratio, and apparent line width, of the maximum component are also given in Figs. 6.33, 6.34, and 6.35. It is worth noting that second-derivative line widths measured by this technique with the higher modulation frequency

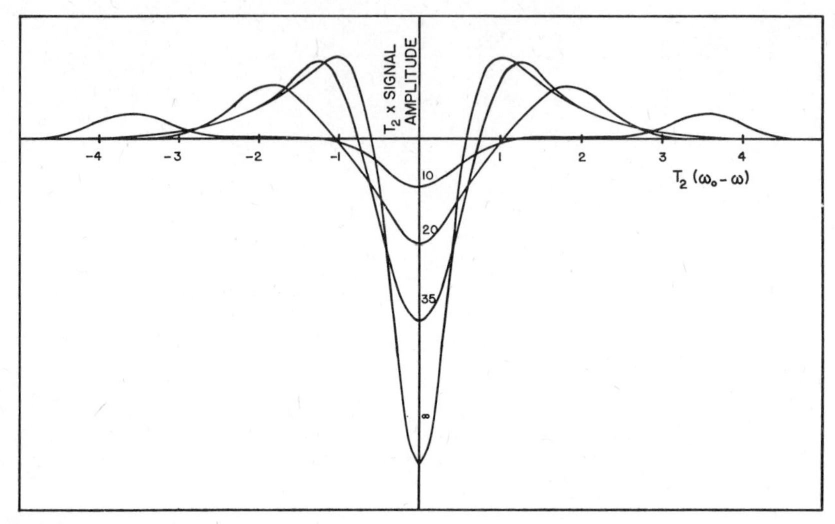

Fig. 6.36 Line shapes of the double-modulation signal, with the phase angle adjusted to maximize the amplitude, for line widths of 10, 20, and 35 mG and for a very wide line assuming 100 kHz for the higher field modulation frequency.

100 kHz are not appreciably in error for line widths in excess of 100 mG, whereas first-derivative line widths differ appreciably from the undistorted widths even for widths as large as 300 mG.

Resolution Enhancement Techniques

In a complex esr spectrum lines frequently overlap, making it difficult and sometimes impossible to interpret the spectrum or to obtain accurate values of the spin Hamiltonian parameters. A number of techniques have been evolved to deal with this problem, and they are considered in the following discussion, where it is assumed throughout that we are dealing with isotropic solution spectra.

Computer Spectrum Simulation

Perhaps the most common approach to the problem of poor resolution is to guess a set of spin Hamiltonian parameters and use them to compute a simulated spectrum, varying the parameters until a fit is obtained that satisfies the experimenter (and perhaps later the referee). A number of programs have been described in the literature for this purpose; perhaps the most widely used is that developed by Stone and Maki [55]. This approach is usually successful, but suffers from the disadvantage of requiring a good deal of computer time, not to mention access to a computer and plotter, but also leaves the skeptical spectroscopist with the nagging feeling that there might be another quite different set of parameters that equally well fits the spectrum. Indeed, there have been a number of cases in which an apparently good fit was obtained by one group, only to have another group obtain better resolution and prove the first results wrong.

An extension of this approach, which systematizes the fitting procedure and increases the chances of being right, is the application of a least-squares procedure [48]. One compares the computed spectrum $g(t)$ with the experimental spectrum $g_i(t)$ and attempts to minimize the error:

$$e^2 = \int_{-\infty}^{\infty} [g_i(t) - g(t)]^2 \, dt \tag{6.185}$$

with respect to each of the parameters used in the simulated spectrum. Iterative methods using a computer are of course necessary. The approach has had somc success in nmr [56] and mass spectrometry [57], and could no doubt be applied to esr as well.

Chemical Resolution Enhancement

There are many essentially chemical parameters that affect resolution which may be adjusted in attempts to improve spectra. This type of approach is not always acceptable of course if the interest is in the system rather than in a particular species, but in cases in which the primary goal is a well-resolved spectrum of a particular chemical species, it is usually worthwhile to try changing such parameters as solvent, concentration of the paramagnetic species, dissolved oxygen concentration, concentration of other species in solution, and temperature. Changes in solvent that affect such things as degree of ion pairing, viscosity, and degree of solvation of the species of interest frequently result in a change in the observed resolution (although not always for the better). Line broadening from spin exchange can frequently be eliminated by working with a lower radical concentration or by removing dissolved oxygen from the solution. Other paramagnetic species in solution may contribute to spin exchange as well, even though their esr spectra may not be observable. Kinetic processes such as electron exchange, proton exchange, exchange of coordinated solvent or other ligands may under some circumstances lead to line broadening if the rates fall in the right range. Resolution can often be substantially improved if the exchanging partner is removed from the system. Solution line widths usually have important contributions from incomplete averaging of anisotropies in the g- and hyperfine tensors which are reduced if the temperature is raised. However, the spin-rotational interaction, which may also be an important contributor to the line width, increases with increasing temperature. In any case, to obtain optimum resolution at the first temperature tried would be an unlikely event.

There are of course many cases in which no combination of chemical parameters can be found that improves the resolution to an acceptable level. In these cases some instrumental resolution enhancement technique may be appropriate.

Resolution Enhancement by Filtering

Ernst [48] has pointed out that a real spectrum composed of lines of finite width may be viewed mathematically as the convolution of a spectrum of infinitely narrow lines (δ-functions) with the real line shape function:

$$g_i(t) = \int_{-\infty}^{\infty} s_i(\tau) g_\delta(t - \tau)\, d\tau. \tag{6.186}$$

This process can of course be reversed mathematically to give:

$$g_\delta(t) = \int_{-\infty}^{\infty} h(\tau) g_i(t - \tau)\, d\tau, \tag{6.187}$$

where $h(\tau)$ is a filter weighting factor that we seek. Taking Fourier transforms of (6.186) and (6.187), we obtain:

$$G_i(\omega) = S_i(\omega) G_\delta(\omega) \tag{6.188}$$

$$G_\delta(\omega) = H(\omega) G_i(\omega), \tag{6.189}$$

so that the filter frequency response function is:

$$H(\omega) = 1/S_i(\omega). \tag{6.190}$$

Unfortunately, however, we saw in Section 4, p. 560, that the optimum filter frequency response function for signal-to-noise optimization was $H(\omega) = S_i(\omega)$ and, as might be guessed, application to the optimum filter for resolution enhancement reduces the signal-to-noise ratio to zero. A useful analogy for this problem, suggested by Ernst [48], is to view resolution and sensitivity as conjugate quantities in an uncertainty relation. Thus for optimum resolution, we have no sensitivity, and for optimum sensitivity, we sacrifice resolution. In practice, of course, some compromise between resolution and sensitivity must be chosen. As we see from the above discussion, however, there is no theoretical limit to resolution enhancement provided we have a sufficiently large signal-to-noise ratio initially.

Several practical approaches to the resolution enhancement problem are suggested or reviewed by Ernst.

OPTIMUM RESOLUTION ENHANCEMENT FILTER

The filter frequency response function for a generalized optimum filter is given by Ernst [48]

$$H(\omega) = \frac{(1 + p) S_i(\omega)}{p S_i^2(\omega) + N_{in}^2}, \tag{6.191}$$

where we have assumed white noise and n_{in} is the noise voltage spectral density (cf. Section 4, p. 558), and p is a parameter that determines the relative emphasis of signal-to-noise ratio and resolution. Thus for $p = 0$, (6.191) reduces to the matched filter of Section 4, p. 560, and for $p \to \infty$, (6.191) reduces to (6.190), the optimum filter for resolution enhancement. The choice of p in practice depends of course upon how much sacrifice in the signal-to-noise ratio can be afforded. Use of this approach in practice involves digitization of the experimental spectrum, numerical computation of the inverse Fourier transform of (6.191), and processing of the spectrum in a computer. This technique suffers from the disadvantage of producing wiggles in the wings of the sharpened lines. Ernst suggests another similar technique that avoids this difficulty [48].

DERIVATIVE LINE SHARPENING

A rather different approach to the development of a resolution enhancement technique is attributable to Allen et al. [58]. We note that in general the resolution of a second-derivative spectrum is better than that of a first-derivative spectrum, which is in turn better than that of the absorption spectrum. The use of still higher derivatives suffers from two drawbacks, however. Third and higher derivatives have subsidiary peaks which may obscure other lines in a spectrum; furthermore, the signal-to-noise ratio deteriorates rapidly with higher derivatives. Allen et al. suggest therefore that a combination of derivatives, for example, first, third, and fifth, added together with appropriate coefficients so that the wiggles in the wings cancel, give a line shape approximating a sharpened first derivative.

Though intuitively clear in these terms, this approach may also be developed in terms of filter theory. The procedure is to expand the frequency response function $H(\omega)$ in a Taylor series and substitute the inverse Fourier transform of the series in (6.187). After some algebra, one obtains the result discussed qualitatively above.

In practicc, there are three methods by which the above sharpening process may be carried out. (1) The spectrum may be digitized, either by hand, by a graph-reading machine connected to a card punch, or by direct digitization of the spectrometer output, and processed by a computer. (2) Analog differentiators may be connected to the output of the lock-in detector with their outputs connected in the way appropriate to the desired result. Electronic circuitry for this approach is described by Allen et al. [58]. (3) The magnetic field may be modulated with a complex waveform containing, for example, frequency components of 20, 33.3, and 100 kHz; if the lock-in amplifier is operated at 100 kHz, the output is automatically a combination of first, third, and fifth derivatives. In order to obtain the desired combination of derivatives, care must be taken to ensure the proper amplitude and phase relationships between the various components of the modulation signal. Glarum [59] has described circuits for this purpose, and such equipment is commercially available from Bruker Scientific.

Acknowledgments

The author is particularly indebted to Professor R. G. Lawler, who compiled the bibliography, and to Dr. R. R. Reeder, who drew some of the figures. Partial support by the National Science Foundation through grants GP-4825 and GP-7811 is gratefully acknowledged.

References

1. E. Zavoisky, *J. Phys. USSR.*, **9,** 245, 447 (1945).
2. See Bibliography, A4, A5, A9, and A11, for more detailed discussions of the topics covered in this section.
3. *NBS Tech. News Bull.*, 175 (October 1963).
4. A. Rich, *Phys. Rev. Letters*, **20,** 967, 1221 (1968).
5. F. Bloch, *Phys. Rev.*, **70,** 460 (1946).
6. M. H. L. Pryce, *Proc. Phys. Soc.* (*London*), **A63,** 25 (1950); A. Abragam and M. H. L. Pryce, *Proc. Roy. Soc.* (*London*), **A205,** 135 (1951); see also Bibliography, A4, A5, A6, A9, A11, and D1, for various discussions of the derivation of the spin Hamiltonian terms.
7. See Bibliography, A6, A11, B11, B13, C8, C13, C25, D6, and D10.
8. G. Breit and I. I. Rabi, *Phys. Rev.*, **38,** 2082 (1931); cf. Bibliography, A11.
9. N. S. Garif'ianov and B. M. Kozyrev, *Dokl. Akad. Nauk SSSR*, **98,** 929 (1954).
10. R. W. Fessenden, *J. Chem. Phys.*, **37,** 747 (1962).
11. R. W. Fessenden and R. H. Schuler, *J. Chem. Phys.*, **43,** 2704 (1965); R. W. Fessenden, *J. Magnetic Res.*, **1,** 277 (1969).
12. See Bibliography, A11 and D8.
13. J. A. Weil and J. H. Anderson, *J. Chem. Phys.*, **28,** 864 (1958).
14. J. A. Ibers and J. D. Swalen, *Phys. Rev.*, **127,** 1914 (1962).
15. F. K. Kneubühl, *J. Chem. Phys.*, **33,** 1074 (1960).
16. B. Bleaney, *Proc. Phys. Soc.* (*London*), **75,** 621 (1960).
17. R. H. Sands, *Phys. Rev.*, **99,** 1222 (1955); D. E. O'Reilly, *J. Chem. Phys.*, **29,** 1188 (1958); R. Neiman and D. Kivelson, **35,** 156 (1961); H. R. Gersmann and J. D. Swalen, **36,** 3221 (1962).
18. M. S. Itzkowitz, *J. Chem. Phys.*, **46,** 3048 (1967).
19. T. Vänngård and R. Aasa, *Paramagnetic Resonance*, Vol. 2, W. Low, Ed., Academic, New York, 1963, p. 509; M. M. Malley, *J. Mol. Spectry.*, **17,** 210 (1965).
20. D. L. Griscom, P. C. Taylor, D. A. Ware, and P. J. Bray, *J. Chem. Phys.*, **48,** 5158 (1968); P. C. Taylor, Ph.D. Thesis, Brown University, Providence, Rhode Island, 1969.

21. L. S. Singer, *J. Chem. Phys.*, **23,** 379 (1955).
22. P. Kottis and R. Lefebvre, *J. Chem. Phys.*, **39,** 393 (1963).
23. H. Sternlicht, *J. Chem. Phys.*, **33,** 1128 (1960).
24. S. M. Blinder, *J. Chem. Phys.*, **33,** 748 (1960).
25. R. Lefebvre, *J. Chem. Phys.*, **33,** 1826 (1960).
26. B. Bleaney, *Proc. Phys. Soc.* (*London*), **A63,** 407 (1950); **A75,** 621 (1960); *Phil. Mag.*, **42,** 441 (1951).
27. D. G. Hughes and T. J. Rowland, *Can. J. Phys.*, **42,** 209 (1964).
28. J. W. Searl, R. C. Smith, and S. J. Wyard, *Proc. Phys. Soc.* (*London*), **A74,** 491 (1959).
29. N. Bloembergen and T. J. Rowland, *Phys. Rev.*, **97,** 1679 (1955).
30. F. K. Kneubühl and B. Natterer, *Helv. Phys. Acta*, **34,** 710 (1961).
31. P. Kottis and R. Lefebvre, *J. Chem. Phys.*, **41,** 379, 3660 (1964).
32. A. K. Chirkhov and A. A. Kokin, *J. Exptl. Theoret. Phys.* (*USSR*), **39,** 1381 (1960); *Soviet Phys.-JETP*, **12,** 964 (1961).
33. T. S. Johnston and H. G. Hecht, *J. Mol. Spectry.*, **17,** 98 (1965)
34. R. Lefebvre and J. Maruani, *J. Chem. Phys.*, **42,** 1480 (1965).
35. C. H. Townes and A. L. Schawlow, *Microwave Spectroscopy*, McGraw-Hill, New York, 1955.
36. Bibliography, F5.
37. Bibliography, F7.
38. *MIT Radiation Laboratory Series:* Vol. 8, *Principles of Microwave Circuits*, C. G. Montgomery, R. H. Dicke, and E. M. Purcell, Eds., 1948; Vol. 9, *Microwave Transmission Circuits*, G. L. Ragan, Ed., 1947; Vol. 10, *Waveguide Handbook*, N. Marcuvitz, Ed., 1951; Vol. 11, *Technique of Microwave Measurements*, C. G. Montgomery, Ed., 1947; Vol. 14, *Microwave Duplexers*, L. D. Smullin and C. G. Montgomery, Eds., 1948; McGraw-Hill, New York.
39. Bibliography, F6.
40. A survey of specifications of commercial esr spectrometers is also found in *Sci. Res.*, **2** (11), 80 (1967) and in Bibliography, F7.
41. J. S. Hyde, in *Tech. Inform. Bull.*, Fall, 1965, Varian Associates, Palo Alto, California; see also Bibliography, F7.
42. J. W. H. Schreurs, G. E. Blomgren, and G. K. Fraenkel, *J. Chem. Phys.*, **32,** 1861 (1960); J. W. H. Schreurs and G. K. Fraenkel, **34,** 756 (1961).
43. J. P. Lloyd and G. E. Pake, *Phys. Rev.*, **94,** 579 (1954); M. J. Stephen and G. K. Fraenkel, *J. Chem. Phys.*, **32,** 1435 (1960); J. H. Freed, **43,** 2312 (1965).
44. J. A. Pople, W. G. Schneider, and H. J. Bernstein, *High-Resolution Nuclear Magnetic Resonance*, McGraw-Hill, New York, 1959.
45. A. M. Portis, *Phys. Rev.*, **91,** 1071 (1953).
46. Bibliography, F1.
47. J. J. Brophy, *Basic Electronics for Scientists*, McGraw-Hill, New York, 1966.
48. Bibliography, F4.
49. S. Goldman, *Information Theory*, Constable, London, 1953, Chapter 8.
50. R. Bracewell, *The Fourier Transform and its Applications*, McGraw-Hill, New York, 1965.

51. J. H. Van Vleck and D. Middleton, *J. Appl. Phys.*, **17,** 940 (1946).
52. H. Wahlquist, *J. Chem. Phys.*, **35,** 1708 (1961).
53. R. Arndt, *J. Appl. Phys.*, **36,** 2522 (1965).
54. O. Haworth and R. E. Richards, in *Progress in Nuclear Magnetic Resonance Spectroscopy*, Vol. 1, J. W. Emsley, J. Feeney, and L. H. Sutcliffe, Eds., Pergamon, Oxford, 1966.
55. E. W. Stone and A. H. Maki, *J. Chem. Phys.*, **38,** 1999 (1963).
56. W. D. Keller, T. R. Lusebrink, and C. H. Sederholm, *J. Chem. Phys.*, **44,** 782 (1966).
57. D. G. Luenberger and U. E. Dennis, *Anal. Chem.*, **38,** 715 (1966).
58. L. C. Allen, H. M. Gladney, and S. H. Glarum, *J. Chem. Phys.*, **40,** 3135 (1964).
59. S. H. Glarum, *Rev. Sci. Instr.*, **36,** 771 (1965).

Bibliography

The general field of esr has experienced such an explosion of information over the past several years that perhaps the most important function to be filled by a chapter on the subject is the direction of the reader toward sources of further information. In the following bibliography an attempt is made to group the references into seven fairly broad categories, though some could be entered under several classifications. Capsule reviews of some of the references are included for further guidance. The present bibliography to some extent emphasizes applications of esr to organic and inorganic chemical studies in liquid solution. A bibliographic selection with a more physical (and solid-state) bias is to be found in C. P. Poole's book (Ref. F5).

A. BOOKS

1. D. J. E. Ingram, *Spectroscopy at Radio and Microwave Frequencies*, Butterworths, London, 1955. Discusses electron spin, nuclear spin, and ferromagnetic resonance as well as microwave spectroscopy, although the examples are now largely out of date.
2. D. J. E. Ingram, *Free Radicals as Studied by Electron Spin Resonance*, Academic, New York, 1958. Still one of the best books for the fundamentals of esr. Experimental results are now out of date, but the sections on techniques are good.
3. Nmr-epr Staff of Varian Associates, *NMR and EPR Spectroscopy*, Pergamon, Oxford, 1959. Deals with many practical details of obtaining and interpreting spectra.
4. G. E. Pake, *Paramagnetic Resonance*, Benjamin, New York, 1962. A good introductory monograph, written for graduate students in physics.
5. C. P. Slichter, *Principles of Magnetic Resonance*, Harper, New York, 1963. An excellent treatment of fundamentals and the dynamic aspects of magnetic resonance. Requires a good physics background for full appreciation, although much of the required physics is developed as needed. Special emphasis is given to the solid state.
6. S. A. Al'tschuler and B. M. Kozyrev, *Electron Paramagnetic Resonance*, Academic, New York, 1964. Useful primarily for discussions of applications of esr to solid-state inorganic systems.

7. L. D. Stepin, *Quantum Radio Frequency Physics*, M.I.T. Press, Cambridge, Massachusetts, 1965. A brief, well-written introduction to nmr, esr, microwave spectroscopy, and masers.
8. M. Bersohn and J. C. Baird, *Introduction to Electron Paramagnetic Resonance*, Benjamin, New York, 1966. The first of the new generation of esr books for chemists that is written at an easily understandable level and could be recommended to the novice as a first introduction to esr.
9. A. Carrington and A. D. McLachlan, *Introduction to Magnetic Resonance*, Harper, New York, 1967. An excellent textbook treating the fundamentals of nmr and esr, with chemical applications; a nice complement to Slichter's book.
10. H. M. Assenheim, *Introduction to Electron Spin Resonance*, Plenum, New York, 1967. Not particularly to be recommended.
11. P. B. Ayscough, *Electron Spin Resonance in Chemistry*, Methuen, London, 1967. The most generally readable and comprehensive book to date, covering all the chemically important aspects of esr.
12. H. G. Hecht, *Magnetic Resonance Spectroscopy*, Wiley, New York, 1967. A good general introduction to magnetic resonance theory, nmr, esr, and double resonance.

B. GENERAL REVIEWS

1. *Ann. Rept. Prog. Chem.*, Chemical Society, London. J. Sheridan, **54,** 9 (1957); M. C. R. Symons, **57,** 68 (1969); M. C. R. Symons, **59,** 45 (1962); A. Carrington, **61,** 27 (1964); D. H. Williams and A. Horsfield, **62,** 213 (1965); N. M. Atherton, A. J. Parker, and H. Steimer, **63,** 62 (1966); A. Horsfield, **63,** 257 (1966); A. Horsfield, **64B,** 29 (1967); C. Thomson, **65B,** 17 (1968); G. R. Luckhurst, **66A,** 37 (1969); C. Thomson, **66B,** 15 (1669).
2. *Ann. Rev. Phys. Chem.*, Annual Reviews, Inc., Palo Alto, California. H. S. Gutowsky, **5,** 333 (1954); J. N. Schoolery and H. E. Weaver, **6,** 433 (1955); C. A. Hutchison, Jr., **7,** 359 (1956); H. M. McConnell, **8,** 105 (1957); J. E. Wertz, **9,** 93 (1958); G. K. Fraenkel and B. Segal, **10,** 435 (1959); R. Bersohn, **11,** 369 (1960); S. I. Weissman, **12,** 151 (1961); R. G. Shulman, **13,** 325 (1962); D. Kivelson and C. Thomson, **15,** 197 (1964); M. T. Jones and W. D. Phillips, **17,** 323 (1966); A. H. Maki, **18,** 9 (1967); A. Carrington and G. R. Luckhurst, **19,** 31 (1968); M. C. R. Symons, **20,** 219 (1969).
3. *Anal. Chem.*, H. Foster, **36,** 266R (1964); D. H. Eargle, **38,** 371R (1966); D. H. Eargle, **40,** 303R (1968).

   The above three series are essentially bibliographic reviews and form excellent entries to the original literature.
4. J. E. Wertz, "Nuclear and Electron Spin Magnetic Resonance," *Chem. Rev.*, **55,** 829 (1955).
5. H. A. Elion and L. Shapiro, "Applications of Electron Spin Resonance Spectroscopy to Organic and Inorganic Chemistry," in *Progress in Nuclear Energy*, Series IX, C. E. Crouthamel, Ed., Pergamon, Oxford, 1961, p. 80.
6. A. Horsfield, "Recent Applications of Electron Spin Resonance in Chemistry," *Chimia* (*Aarau*), **17,** 42 (1963).

7. R. E. Norberg, "Resource Letter NMR-EPR-1," *Am. J. Phys.*, **33,** 1 (1965). Good bibliography.

8. A. G. Redfield, "The Theory of Relaxation Processes," *Advan. Magnetic Resonance*, **1,** 1 (1965). A good review of the abstract theory.

9. C. S. Johnson, Jr., "Chemical Rate Processes and Magnetic Resonance," *Advan. Magnetic Resonance*, **1,** 33 (1965). A good introduction to the theory and a discussion of the possible types of experiments.

10. D. R. Eaton and W. D. Phillips, "Nuclear Magnetic Resonance of Paramagnetic Molecules," *Advan. Magnetic Resonance*, **1,** 103 (1965).

11. S. M. Blinder, "Theory of Atomic Hyperfine Structure," *Advan. Quantum Chem.*, **2,** 47 (1965).

12. T. Kaiser and L. Kevan, Eds., *Radical Ions*, Interscience, New York, 1968. Some good reviews of various aspects of radical ion chemistry. Includes chapters on electron spin densities, metal ketyls, semidiones, radical cations, benzene orbital degeneracy, aromatic anion radicals, Group-IV organometallic radicals, sulfur-containing radicals, ion radicals in low-temperature, γ-irradiated organic solids, small inorganic radicals, irradiated ionic solids, trapped radicals in inorganic glasses, and complex ions of the first-row transition metals.

13. A. J. Freeman and R. B. Frankel, Eds., *Hyperfine Interactions*, Academic, New York, 1967. Contains some good chapters on the theory of electron-nuclear hyperfine interactions.

14. J. C. Verstelle and D. A. Curtis, "Paramagnetic Relaxation," in *Handbuch der Physik*, Vol. XVIII/1, S. Flügge, Ed., Springer, Berlin, 1968, p. 1.

15. D. J. E. Ingram, "Electron Spin Resonance," in *Handbuch der Physik*, Vol. XVIII/1, S. Flügge, Ed., Springer, Berlin, 1968, p. 94. A well written introduction to the field of esr with some discussion on instrumentation.

16. J. E. Wertz, "Structural Information from Paramagnetic Resonance," in *Handbuch der Physik*, Vol. XVIII/1, S. Flügge, Ed., Springer-Verlag, Berlin, 1968, p. 145. A comprehensive discussion of the chemically interesting results of esr.

C. ORGANIC AND BIOLOGICAL APPLICATIONS

1. G. K. Fraenkel, "Paramagnetic Resonance of Free Radicals," *Ann. N. Y. Acad. Sci.*, **67,** 546 (1957).

2. D. H. Whiffen, "Electron Resonance Spectroscopy of Free Radicals," *Quart. Rev. (London)*, **12,** 250 (1958).

3. A. M. Bass and H. P. Broida, Eds., *Formation and Trapping of Free Radicals*, Academic, New York, 1960.

4. R. Bersohn, "Electron Paramagnetic Resonance of Organic Molecules," in *Determination of Organic Structures by Physical Methods*, F. C. Nachod and W. D. Phillips, Eds., Academic, New York, 1962, p. 563.

5. J. W. Boag, "Electron Spin Resonance in Biology," in *Radiation Effects in Physics, Chemistry and Biology*, M. Ebert and A. Howard, Eds., North-Holland, Amsterdam, 1962.
6. M. C. R. Symons, "The Identification of Organic Free Radicals by Electron Spin Resonance," *Advan. Phys. Org. Chem.* **1,** 283 (1963).
7. A. Carrington, "Electron Spin Resonance Spectra of Aromatic Radicals and Radical Ions," *Quart. Rev.* (*London*), **17,** 67 (1963).
8. C. A. McDowell, "Spin Resonance and Hyperfine Interaction," *Rev. Mod. Phys.*, **53,** 528 (1963).
9. B. J. McClelland, "Anionic Free Radicals," *Chem. Rev.*, **64,** 301 (1964).
10. J. R. Morton, "Electron Spin Resonance of Oriented Radicals," *Chem. Rev.*, **64,** 453 (1964).
11. E. deBoer, "Electronic Structure of Alkali Metal Adducts of Aromatic Hydrocarbons," *Advan. Organometal. Chem.*, **2,** 115 (1964).
12. R. N. Adams, "Applications of Electron Paramagnetic Resonance Techniques in Electrochemistry," *J. Electroanal. Chem.*, **8,** 151 (1964).
13. G. G. Hall and A. T. Amos, "Molecular Orbital Theory of the Spin Properties of Conjugated Molecules," *Advan. Atomic Mol. Phys.*, **1,** 1 (1965).
14. F. Schneider, K. Möbius, and M. Plato, "The Use of Electron Paramagnetic Resonance in Organic Chemistry," *Angew. Chem. Intern. Ed. Engl.*, **4,** 856 (1965).
15. A. L. Buchachenko, *Stable Radicals*, Consultants Bureau, New York, 1965.
16. E. Müller, A. Rieker, K. Scheffler, and A. Moosmayer, "Applications and Limitations of Magnetic Methods in Free Radical Chemistry," *Angew. Chem. Intern. Ed. Engl.*, **5,** 6 (1966).
17. G. A. Russell, E. T. Strom, E. R. Talaty, K. Y. Chang, R. D. Stevens, and M. C. Young, "Application of Electron Spin Resonance to Problems of Structure and Conformation. Aliphatic Semidiones," *Rec. Chem. Progr.*, **27,** 3 (1966).
18. F. Gerson, *Hochauflösende ESR Spektroskope*, Verlag-Chemie, Weinheim, Germany, 1967 (English translation in preparation by Wiley-Interscience, New York).
19. S. E. Bresler and E. N. Kazbekov, "Application of Electron Spin Resonance in Polymer Chemistry," *Russ. Chem. Rev.*, **36,** 298 (1967).
20. D. H. Geske, "Conformation and Structure As Studied by Electron Spin Resonance Spectroscopy," in *Progress in Physical Organic Chemistry*, Vol. 4, A. Streitwieser, Jr., and R. W. Taft, Eds., Interscience, New York, 1967, p. 125.
21. R. O. C. Norman and B. C. Gilbert, "Electron Spin Resonance Studies of Short-Lived Organic Free Radicals," *Advan. in Phys. Org. Chem.*, **5,** 53 (1967). The reviews by Geske and by Norman and Gilbert, taken together, summarize most of the information thus far derived from esr studies which is of particular interest to organic chemists.

22. E. deBoer and H. van Willigen, "Nuclear Magnetic Resonance of Paramagnetic Systems," in *Progress in Nuclear Magnetic Resonance Spectroscopy*, Vol. 2, J. W. Emsley, J. Feeney, and L. H. Sutcliffe, Eds., Pergamon, Oxford, 1967, p. 111. A good example of the solution of the equations of motion for nuclear and electron spins in the presence of chemical exchange.

23. G. A. Russell, "Electron Spin Resonance in Organic Chemistry," *Science*, **161,** 423 (1968).

24. M. Szwarc, "Chemistry of Radical Ions," in *Progress in Physical Organic Chemistry*, Vol. 6, A. Streitwieser, Jr., and R. W. Taft, Eds., Interscience, New York, 1968, p. 323. Mostly chemistry, little esr.

25. J. D. Memory, *Quantum Theory of Magnetic Resonance Parameters*, McGraw-Hill, New York, 1968. Hyperfine splitting constants in organic radicals.

26. O. H. Griffith and A. S. Waggoner, "Nitroxide Free Radicals: Spin Labels for Probing Biomolecular Structure," *Accounts Chem. Res.*, **2,** 17 (1969).

D. INORGANIC APPLICATIONS

1. B. Bleaney and K. W. H. Stevens, "Paramagnetic Resonance," *Rept. Prog. Phys.*, **26,** 108 (1953).

2. K. D. Bowers and J. Owen, "Paramagnetic Resonance II," *Rept. Prog. Phys.*, **18,** 304 (1955).

3. J. W. Orton, "Paramagnetic Resonance Data," *Rept. Prog. Phys.*, **22,** 204 (1959).

These three papers provide detailed reviews of early work on esr of transition metal ions in dilute single crystals.

4. W. Low, "Paramagnetic Resonance in Solids," in *Solid State Physics*, F. Seitz and D. Turnbull, Eds., Supplement 2, Academic, New York, 1960.

5. A. Carrington and H. C. Longuet-Higgins, "Electron Resonance in Crystalline Transition-Metal Compounds," *Quart. Rev.* (*London*), **14,** 427 (1960).

6. J. S. Griffith, *The Theory of Transition Metal Ions*, Cambridge Univ. Press, Cambridge, 1961.

7. W. Low and E. L. Offenbacher, "ESR of Magnetic Ions in Complex Oxides," in *Solid State Physics*, Vol. 17, F. Seitz and D. Turnbull, Eds., Academic, New York, 1965.

8. P. W. Atkins and M. C. R. Symons, *The Structure of Inorganic Radicals*, Elsevier, New York, 1967. A very good book on esr and other results pertaining to simple inorganic radicals up to five atoms; contains some very good general material on experimental techniques.

9. G. Lancaster, *Electron Spin Resonance in Semiconductors*, Plenum, New York, 1967.

10. B. R. McGarvey, "ESR of Transition Metal Complexes," in *Transition Metal Chemistry*, Vol. 3, R. L. Carlin, Ed., Marcel Dekker, New York, 1967, p. 90. An excellent review.

11. W. B. Lewis and L. O. Morgan, "Paramagnetic Relaxation in Solutions," in *Transition Metal Chemistry*, Vol. 4, R. L. Carlin, Ed., Marcel Dekker, New York, 1968, p. 33. A very comprehensive review of relaxation processes in solutions of transition metal ions and complexes.

12. J. W. Orton, *Electron Paramagnetic Resonance: An Introduction to Transition Group Ions in Crystals*, Iliffe, London, 1968.

E. SYMPOSIA AND CONFERENCES

1. "Microwave and Radiofrequency Spectroscopy," *Discussions Faraday Soc.*, **19** (1955).

2. "Applications of Electron and Nuclear Resonance in Chemistry," *Chem. Soc. (London), Special Publ.*, **12** (1958).

3. *Free Radicals in Biological Systems*, Proceedings of a symposium held at Stanford University, March 1960, M. S. Blois et al., Eds., Academic, New York, 1961. Papers on most of the areas of biology and biochemistry where esr is now useful.

4. Fifth European Congress on Molecular Spectroscopy, Amsterdam, 1961, *Pure Appl. Chem.*, **4** (1962).

5. *Magnetic and Electric Resonance and Relaxation*, Proceedings of the XIth Colloque Ampere, Eindhoven, July 1962, J. Smidt, Ed., North-Holland, Amsterdam, 1963.

6. *Paramagnetic Resonance*, Proceedings of the First International Conference on Paramagnetic Resonance, Jerusalem, 1962, 2 Vols., W. Low, Ed., Academic, 1967.

7. Symposium on Inorganic Free Radicals and Free Radicals in Inorganic Chemistry, Atlantic City, 1962, *Advances in Chemistry Series*, No. 36.

8. *Electronic Magnetic Resonance and Solid Dielectrics*, Proceedings of the XIIth Colloque Ampere, Bordeaux, September 1963, R. Servant and A. Charru, Eds., North-Holland, Amsterdam, 1964.

9. "Structure de Radicaux Polyatomique," XIVe Reunion de la Societe de Chimie Physique, Bordeaux, May 1964, *J. Chim. Phys.*, **61** (11–12) (1964).

10. *Magnetic Resonance and Relaxation*, Proceedings of the XIVth Colloque Ampere, Ljubljana, September 1966, R. Blinc, Ed., North-Holland, Amsterdam, 1967.

11. Symposium on Esr Spectroscopy, Michigan State University, August 1966, *J. Phys. Chem.*, **71** (1) (1967).

12. *Electron Spin Resonance of Metal Complexes*, Proceedings of the Symposium on Esr of Metal Chelates, Cleveland, March 1968, T. F. Yen, Ed., Plenum, New York, 1969.

13. *Magnetic Resonance and Radiofrequency Spectroscopy*, Proceedings of the XVth Colloque Ampere, Grenoble, September 1968, P. Averbuch, Ed., North-Holland, Amsterdam, 1969.

F. INSTRUMENTATION AND TECHNIQUES

1. G. K. Fraenkel, "Paramagnetic Resonance Absorption," in *Technique of Organic Chemistry: Physical Methods*, Vol. I, Part IV, A. Weissberger, Ed., 3rd ed., Interscience, New York, 1960, p. 2802. The instrumental parts of this review are well written and still useful.
2. R. S. Anderson, "Electron Spin Resonance," in *Methods of Experimental Physics*, Vol. 3, D. Williams, Ed., Academic, New York, 1962, p. 441.
3. T. L. Squires, *An Introduction to Electron Spin Resonance*, Academic, New York, 1964. Electron spin resonance for electrical engineers.
4. R. R. Ernst, "Sensitivity Enhancement in Magnetic Resonance," *Advan. Magnetic Resonance*, **2,** 1 (1966). The best source for the material covered.
5. C. P. Poole, Jr., *Electron Spin Resonance*, Wiley, New York, 1966. By far the best source on instrumentation; includes a great deal of the background material on physics and electronics required to understand or build esr spectrometers. Includes a good bibliography.
6. T. H. Wilmshurst, *Electron Spin Resonance Spectrometers*, Plenum, New York, 1968. Supplements Poole's book in going more deeply into certain design principles of esr spectrometers.
7. R. S. Alger, *Electron Paramagnetic Resonance*, Interscience, New York, 1968. Not as comprehensive as Poole on instrumentation, but an excellent source of information on experimental techniques, treatment of data, and tricks of the trade. Particularly good coverage of esr of solids at low temperatures.

G. COMPILATIONS OF SPECTRA AND DATA

1. Ya. S. Lebedev, N. N. Tikhomikova, and V. V. Voevodskii, *Atlas of Electron Spin Resonance Spectra*, 2 Vols., Consultants Bureau, New York, 1963–1964. Contains simulated first-derivative spectra for various types of multiplets for both Gaussian and Lorentzian line shapes, and is therefore sometimes useful in interpretation of overlapped spectra.
2. K. W. Bowers, "Electron Spin Resonance of Radical Ions," *Advan. Magnetic Resonance*, **1,** 317 (1965). Summary of hyperfine splitting constant data; contains many errors.
3. H. M. Hershenson, *Nmr and Esr Spectra: Index for 1958–1963*, Academic, New York, 1965. Incomplete and difficult to use.
4. H. Fischer, "Magnetic Properties of Free Radicals," *Landolt-Bornstein, New Series*, Gp. II, Vol. 1, Springer, Berlin, 1965. The most complete compilation of *g*-factors and hyperfine splitting constants available.
5. B. H. J. Bielski and J. M. Gebicki, *Atlas of Electron Spin Resonance Spectra*, Academic, New York, 1967. An enormous number of spectra reproduced from the original literature, but arranged rather poorly.

Chapter **VII**

NUCLEAR MAGNETIC RESONANCE

Norbert Muller

---

1 Introduction 601

2 The Resonance Phenomenon 602

Nuclear Energy Levels and the Resonance Condition 602
Longitudinal and Transverse Relaxation 608

3 High-Resolution Nmr Spectrometers 609

4 Chemical Shifts 614

Definitions and Chemical Shift Scales 614
Attempts to Rationalize Proton Shifts 617
  Exact Theory 617
  Effect of Electron Density 617
  Diamagnetic Anisotropy Effect 618
  Ring-Current Effect 621
  Electric Field Effect 622

5 Electron-Coupled Spin-Spin Interactions 623

Interaction Mechanisms 623
Multiplets in First-Order Spectra 625
Deviations from First-Order Behavior 627
  Nearly First-Order Multiplets 627
  Chemical versus Magnetic Equivalence 628
  Deceptively Simple Spectra 628
  Virtual Coupling 629
Proton-Proton Coupling and Molecular Geometry 629
Coupling of Protons with Other Isotopes 630
Signs of Coupling Constants 631

6 Solvent Effects 632

Chemical Shifts 632
  Classification of the Effects 632
  Bulk Diamagnetic Susceptibility 633
  Van der Waals Contribution 634
  Electric Field Effect 635
  Diamagnetic Anisotropy 636

Coupling Constants 638
Solvent Effects on Solute Conformation or Self-Association 638
Solvent Shifts and Spectral Analysis 639

7 Effects of Exchange Processes and Quadrupole Relaxation 639

Chemical Exchange 639
Introduction 639
Rate Measurements 643
Equilibrium Measurements 644
Coupling Constants 646
Quadrupole Relaxation Effects 647

8 Double Resonance 648

Spin Decoupling 648
The Decoupling Process 648
Simplification of Spectra 651
Indirect Determination of Chemical Shifts 651
Signs of Coupling Constants 652
Other Applications 654

9 Relaxation Times and Line Widths in High-Resolution Spectra 655

Systems without Chemical Exchange 655
Effect of Exchange Processes 659
Measurement of T_1 and T_2 660

10 Line Widths and Chemical Shifts in the Presence of Paramagnetic Materials 661

Line Widths 661
Contact Shifts 664

11 Spectra of Elements Other Than Hydrogen 667

General Considerations 667
Fluorine Spectra 669
Carbon-13 Spectra 673

12 High-Resolution Spectra of Macromolecules in Solution 675

Synthetic Polymers 675
Biological Macromolecules 679

13 Spectra of Nonmetallic Diamagnetic Solids 682

Rigid-Lattice Spectra 682
Molecular Motions 686
Adsorbed Species 687

14 Spectra of Oriented Molecules in Nematic Solvents 688
Dipolar Coupling and Molecular Geometries 688
Chemical Shift Anisotropy 693
15 Pulsed Nmr Experiments 694
Identical Nuclei; Relaxation Times 694
Nonidentical Nuclei; Chemical Shifts and Exchange Rates 696
Fourier Transform Spectra 698
Chemical Shifts for Solids 701
16 Computers in Nmr Spectroscopy 701
High-Resolution Spectra 701
Temperature-Dependent Spectra 703
Line Shape Calculations 704
Sensitivity and Resolution Enhancement 705

1 INTRODUCTION

The feasibility of detecting nuclear magnetic resonance (nmr) signals from solid or liquid samples was experimentally demonstrated in 1945, and chemical shifts of nmr absorption frequencies were reported 4 years later. Within an astonishingly short time, it was found that this technique could be used to solve an impressive variety of chemical problems, including the elucidation of the structures of complicated organic molecules, the determination of selected interatomic distances in crystals, the measurement of reaction rates involving time constants of the order of 10^{-2} sec, and the evaluation of equilibrium constants for the formation of hydrogen-bonded complexes, to name only a few. A monograph [1] published in 1959 contained a summary of the earlier history of the field, much of the basic theory, and a survey of the chemical applications then known; it included 474 references. Early in 1968 an informally circulated bibliography of research papers containing results based at least in part on nmr spectra averaged better than 500 new titles each *month*. In addition to the original literature and numerous scattered review articles, a sizable number of texts on nmr are now available [2], subareas of the field are covered in three periodical review publications [2] devoted exclusively to magnetic resonance, and catalogs or data tables covering high-resolution spectra for several thousand organic compounds have been published [2].

The literature of nmr is dominated by descriptions and applications of high-resolution spectra for organic compounds having molecular weights of not more than a few hundred units, examined as liquids or solutions. The principles involved in obtaining and interpreting such spectra form the subject of the first six sections of this chapter. Less common applications become possible when the materials or techniques are changed as, for example, if the solute is a polymer, if the solvent is a liquid crystal, or if two or more irradiating frequencies are used simultaneously, so as to create a situation differing more or less drastically from the conventional high-resolution experiment. The number of such ramifications is increasing steadily, and each one may demand a rather sophisticated mathematical analysis if it is to be fully understood. The objective in the later sections of this chapter is to present as far as possible a nonmathematical account of the main features of each experiment and the type of information that may be derived from it, with references to articles and reviews in which more rigorous presentations can be found.

The instrumentation, particularly for the less common applications, is not described in any great detail. This reflects in part space limitations and the author's preference and in part the fact that a large selection of very sophisticated commercial apparatus is available [3] for various types of nmr spectroscopy and, consequently, very few chemists who use these techniques venture to design or construct their own equipment.

2 THE RESONANCE PHENOMENON

Nuclear Energy Levels and the Resonance Condition

For many purposes, a nucleus is adequately defined by specifying its atomic number Z and its mass number A and may be visualized as an aggregate of Z protons and $(A - Z)$ neutrons in a roughly spherical region about 10^{-12} cm in diameter. Although much remains to be learned about the structure and behavior of such a collection of nucleons, several additional nuclear properties can be quantitatively determined and must be considered in order to account for details of nuclear scattering processes, hyperfine structure of atomic and molecular spectra, and other observed phenomena including nmr [4]. For example, each type of nucleus is normally found with a fixed, characteristic amount of angular momentum, symbolized by the vector $\mathbf{I}$. Its magnitude is given by:

$$|\mathbf{I}| = \sqrt{I(I+1)}\hbar, \tag{7.1}$$

where $\hbar$ is Planck's constant h divided by 2π and I, the *spin quantum number* of the nucleus in question, is some member of the sequence $0, \frac{1}{2}, 1, \frac{3}{2}, 2, \frac{5}{2}, \ldots$. If $I \geq \frac{1}{2}$ the nucleus has a magnetic dipole moment $\boldsymbol{\mu}$, and if $I \geq 1$ it has, in addition, an electric quadrupole moment Q.

Nmr involves the interaction of the nuclear magnetic moment with an applied magnetic field. A finite quadrupole moment may give rise to a competing interaction with the electric field generated by nearby electrons and other nuclei, which may interfere seriously with the nmr experiment. Nuclei with $I = \frac{1}{2}$ thus represent the simplest situation of interest, and it is fortunate that the hydrogen nucleus, almost ubiquitous in organic compounds, belongs to this class. The other two nuclear species most commonly found in organic molecules, $^{12}C_6$ and $^{16}O_8$, are magnetically inert. They illustrate the general rule that for nuclei having even values for both Z and A the spin quantum number must be zero.

When the nucleus is exposed to a magnetic field $\mathbf{H}_0$ parallel to the z axis, the magnetic interaction energy, analogous to that between a magnetized compass needle and an applied field, is given by:

$$E = -\boldsymbol{\mu}\mathbf{H}_0 = -|\boldsymbol{\mu}| \cos\theta \, |\mathbf{H}_0| = -\mu_z H_0, \tag{7.2}$$

where θ is the angle between $\boldsymbol{\mu}$ and the z-axis. The interaction energy is quantized because θ cannot freely take on all values between 0 and 180°. This follows from the space-quantization of angular momentum if one thinks of the nuclear magnetic dipole as arising from the rotation of the positively charged nuclear matter about its own axis, so that $\boldsymbol{\mu}$ remains always parallel to the spin vector $\mathbf{I}$. In quantum mechanics, it is found that the z-component of I can take on only the values:

$$I_z = M_I\hbar; \quad M_I = I, \quad (I-1), \quad (I-2), \ldots, -I. \tag{7.3}$$

The allowed values of $\cos\theta$ are then given by $I_z/|I| = M_I/\sqrt{I(I+1)}$, so that the z-component of $\boldsymbol{\mu}$ can take on the values $|\boldsymbol{\mu}|\, M_I/\sqrt{I(I+1)}$. It is customary to specify the nuclear magnetic moment by the scalar quantity μ which is not the same as $|\boldsymbol{\mu}|$ but rather is the maximum *observable* value of μ_z, that is $\mu = |\boldsymbol{\mu}|\, I/\sqrt{I(I+1)}$. With this definition, the possible values of μ_z are simply $\mu M_I/I$ and the quantized form of (7.2) is:

$$E = -\mu H_0 M_I/I. \tag{7.4}$$

For the special case with $I = \frac{1}{2}$, there are just two allowed energy levels with $M_I = \frac{1}{2}$ or $-\frac{1}{2}$, hence $E = -\mu H_0$ or $+\mu H_0$. Although these levels are sometimes informally described as levels in which $\boldsymbol{\mu}$ is either parallel or antiparallel to the field, it is important later in the discussion to remember that the corresponding values of θ are not 0 and 180°. Indeed, according to the uncertainty principle, the three components of the vector $\mathbf{I}$ cannot simultaneously be exactly known, and to have $\theta = 0$ or 180° would require $I_x = I_y = 0$, $I_z = |I|$. Instead, I_z and $|I|$, and $I_x^2 + I_y^2$ can be specified, but the instantaneous values of I_x and I_y are uncertain.

The basic postulate of spectroscopy is that electromagnetic radiation of frequency ν_0 can induce transitions between two allowed energy levels of an atomic system only if it satisfies Bohr's frequency condition, $h\nu_0 = E_2 - E_1$.

For the two energy states of a nucleus with spin $\frac{1}{2}$, this yields at once the *resonance condition*:

$$h\nu_0 = 2\mu H_0 \qquad \text{or} \qquad \nu_0 = \gamma H_0/2\pi, \tag{7.5}$$

where $\gamma = \mu/I\hbar$ is the nuclear *magnetogyric ratio* (some authors prefer *gyromagnetic ratio*).

Equation (7.5) suggests the basic requirements for an nmr experiment. A collection of nuclei with $\mu \neq 0$ must be placed in a uniform magnetic field and exposed to radiation of the appropriate frequency, preferably monochromatic, in an apparatus that allows absorption of the radiation to be detected. A spectrum may be produced by varying either the frequency or the field through a small range including the resonant value and presented as a plot of absorption intensity as a function of the selected variable. Optimal sensitivity and resolution are obtained when H_0 is of the order of 10^4 G. The μ-value of any known nuclear species then requires that ν_0 be in the radiofrequency range. Specifically, for protons (7.5) is satisfied when the frequency in megahertz is 4.2577 times the field strength in kilogauss.

The most common experimental arrangement used to obtain nmr signals is the crossed-coil apparatus illustrated in Fig. 7.1. Radiofrequency current from a crystal-controlled oscillator is supplied to a transmitting coil placed between the poles of an appropriate magnet with its axis (x) perpendicular to the field direction (z). The second coil is a receiving coil wound around the sample with its axis (y) perpendicular to both x and z. This choice of orientations makes it impossible for the radiation emitted from the transmitter coil

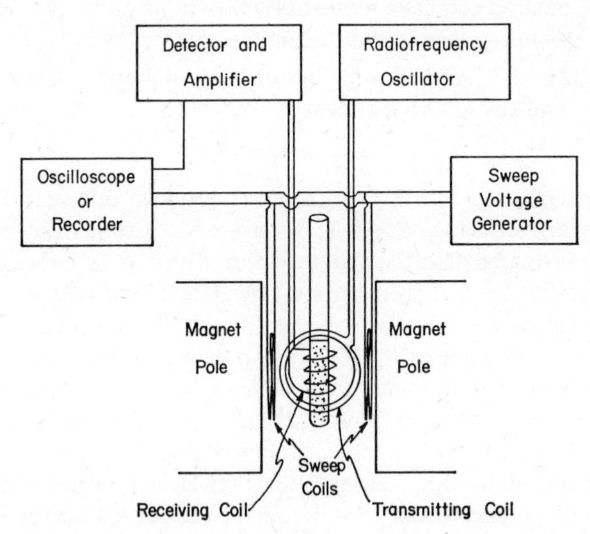

Fig. 7.1 Simplified schematic diagram of a crossed-coil nmr spectrometer.

to induce a voltage in the receiver coil unless the resonance condition is satisfied. The magnetic field is slowly increased by applying a small variable voltage to the sweep coils, a third set of coils wound about the z-axis so that the field generated by a current in the sweep coils adds to the main magnetic field. When the field strength reaches the value required for resonance, a voltage is induced in the receiver coil proportional to the energy absorbed by the nuclei in the sample. This signal passes through a tuned amplifier and detector, and is eventually displayed using a graphic recorder or an oscilloscope.

Some of the applications of nmr spectroscopy can be understood only on the basis of a more detailed description of the process by which the resonating nuclei effect a transfer of energy from the transmitter to the receiver coil. To this end, it is necessary to take a further look at the interaction between the nucleus and the applied field. The approach is to examine the classical motion in a magnetic field of a rigid rotating body possessing a magnetic moment along its axis of rotation [5]. Conclusions based on such an approach may need to be modified to bring them into conformity with the requirements of quantum theory, but the classical description is very helpful by providing ways of visualizing the behavior of the system, and it yields some numerically correct predictions.

The interaction of a magnetic moment $\boldsymbol{\mu}$ and a field $\mathbf{H}_0$ results in a torque $\mathbf{L}$ given by the vector product:

$$\mathbf{L} = \boldsymbol{\mu} \times \mathbf{H}_0. \tag{7.6}$$

If $\mathbf{H}_0$ is along the z-axis and $\boldsymbol{\mu}$ happens to lie in the y-z plane, $\mathbf{L}$ is along the x-axis; its magnitude is $|\boldsymbol{\mu}|\ H_0 \sin\theta$. In classical mechanics the torque is equal to the rate of change of angular momentum, or symbolically,

$$\mathbf{L} = d\mathbf{I}/dt. \tag{7.7}$$

Since the vectors $\boldsymbol{\mu}$ and $\mathbf{I}$ are parallel, $d\mathbf{I}/dt$ is always perpendicular to $\mathbf{I}$, hence $\mathbf{I}$ changes only in direction but not in magnitude. The derivative $d\mathbf{I}/dt$ also remains perpendicular to the z-axis, and this means that I_z cannot change. Thus precisely the quantities $|\mathbf{I}|$ and I_z, which are allowed to have well-defined values when the problem is treated quantum mechanically are constants of the motion in the classical case.

If $\mathbf{I}$ is to change continually in direction with no attendant change in $|\mathbf{I}|$ or I_z, the point of the I-vector must trace out a circle in a plane parallel to the x-y plane. Its tangential velocity is equal to $|d\mathbf{I}/dt|$, and since the circumference of the circle is $2\pi r = 2\pi\ |\mathbf{I}| \sin\theta$, the number of revolutions completed per second is:

$$\nu_0 = |d\mathbf{I}/dt|/(2\pi\ |\mathbf{I}| \sin\theta) = \frac{\gamma H_0}{2\pi}, \tag{7.8}$$

where γ is again the magnetogyric ratio. In other words, the angular momentum *precesses* around the z-axis with a frequency, the *Larmor precession frequency* numerically the same as the resonant frequency of (7.5).

In the crossed-coil spectrometer, the nuclei are exposed not only to the steady magnetic field $\mathbf{H}_0$ but also to an oscillating field generated by the alternating current in the transmitter coil. This may be written:

$$\mathbf{H}_1(t) = 2H_1\mathbf{i}\cos 2\pi\nu t, \tag{7.9}$$

where $\mathbf{i}$ is a unit vector along the x-axis, or it may be thought of as the resultant of two magnetic vectors $\mathbf{H}_R$ and $\mathbf{H}'_R$, each of length $\mathbf{H}_1$, lying in the x-y plane and rotating about the z-axis in opposite directions with the frequency ν. Since $\mathbf{H}_1$ is always much less than $\mathbf{H}_0$, the interaction between $\boldsymbol{\mu}$ and $\mathbf{H}_1$ results in a small additional torque which may be written

$$\mathbf{L}_1 = \boldsymbol{\mu} \times \mathbf{H}_1 = \boldsymbol{\mu} \times \mathbf{H}_R + \boldsymbol{\mu} \times \mathbf{H}'_R. \tag{7.10}$$

If $\nu \neq \nu_0$, the angle between $\boldsymbol{\mu}$ and either $\mathbf{H}_R$ or $\mathbf{H}'_R$ changes very rapidly, and this causes such rapid fluctuations in $\mathbf{L}_1$ that it produces no significant effect on the motion. If $\nu = \nu_0$, this still applies to $\mathbf{H}'_R$, the component of $\mathbf{H}_1$ that is rotating in a direction opposite to that of $\boldsymbol{\mu}$. However, the angle ϕ between $\boldsymbol{\mu}$ and $\mathbf{H}_R$ now remains constant, so that the time derivative:

$$d\mu_z/dt = (L_1)_z = \mu_\perp H_1 \sin\phi \tag{7.11}$$

takes on a constant value, and μ_z can no longer be a constant of the motion. This provides the classical parallel to the quantum-theoretical prediction that the oscillating field will induce transitions between the states with $\mu_z = \mu$ and $\mu_z = -\mu$. The component of $\mathbf{L}_1$ in the x-y plane has a constant magnitude dependent on ϕ. Its direction changes with time because both $\mu_\perp$ and $\mathbf{H}_R$ are rotating, but it maintains a fixed direction relative to $\mu_\perp$. With $\mu_z > 0$, as shown in Fig. 7.2, it slows the nuclear precession somewhat if $|\phi| < 90°$ and accelerates it if $90° < |\phi| < 180°$. As a result, ϕ gradually changes, reaching a limiting value of 90°. When $\mu_z < 0$, the effect is reversed and ϕ approaches $-90°$.

The significance of this becomes clearer when this argument is applied to an assembly of n identical nuclei rather than to just one. The vector sum of the individual magnetic dipoles then yields a resultant macroscopic dipole $\mathbf{M}$. In the absence of any applied field, the nuclear spins are aligned at random, and if n is large the resultant is zero. If the static field $\mathbf{H}_0$ acts alone, each nucleus chooses one of the two allowed states of alignment. Since the state with $\mu_z = +\mu$ is slightly more stable, its population n_+ is slightly larger, and there is a net polarization with:

$$M_z = \mu(n_+ - n_-) = n\mu(1 - e^{-2\mu H_0/kT})/(1 + e^{-2\mu H_0/kT}), \tag{7.12}$$

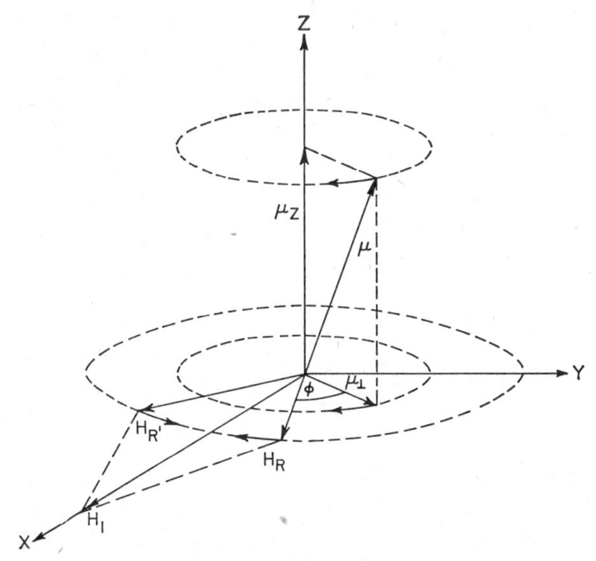

Fig. 7.2 Relation between the nuclear magnetization vector and the components of the oscillating magnetic field. The short, curved arrows indicate the directions of motion of the various vectors.

where $e^{-2\mu H_0/kt}$ is the familiar Boltzmann factor. At ordinary temperatures this is closely approximated by:

$$M_z = M_0 = n\mu^2 H_0/kT. \tag{7.13}$$

Although all the nuclei are precessing at the same frequency, there is no fixed phase relation among them, and the transverse components of **M** remain equal to zero. However, when an oscillating field at the resonant frequency is applied, the nuclei with $\mu_z = +\mu$ tend to come into phase with one another, each $\mu_\perp$ making an angle of $+90°$ with the rotating vector $\mathbf{H}_R$. The nuclei with $\mu_z = -\mu$ also tend to precess in phase but with each $\mu_\perp$ making an angle of $-90°$ with $\mathbf{H}_R$. There is still no net transverse magnetization if the two spin states are equally populated, but because of the excess of nuclei in the more stable state **M** acquires a component $M_\perp$, which rotates in the x-y plane at the Larmor frequency, so that

$$\begin{aligned} M_x &= M_\perp \sin 2\pi\nu_0 t \\ M_y &= M_\perp \cos 2\pi\nu_0 t. \end{aligned} \tag{7.14}$$

The y-component generates a fluctuating magnetic field in the y-direction, which induces the signal in the receiver coil that allows the resonance to be detected.

Longitudinal and Transverse Relaxation

Since the z-component of $\mathbf{M}$ depends primarily on the steady field $\mathbf{H}_0$, while the perpendicular component requires the alternating field, it is not surprising that changes of $\mathbf{M}$ with time are governed by two distinct relaxation times. Thus, if $\mathbf{H}_0$ is abruptly turned on at a time $t = 0$, while $\mathbf{H}_1$ remains zero, M_z grows from zero to M_0 by a first-order process, that is,

$$M_z - M_0 = M_0 e^{-t/T_1}. \tag{7.15}$$

The parameter T_1 controls the rate of change of M_z and is called the *longitudinal relaxation time*. Any change in M_z entails a change in the populations n_+ and n_- and therefore involves a transfer of energy between the spinning nuclei and the thermal motions of neighboring atoms or molecules. Immediately after the field is turned on, the *spin temperature* T_s, defined by:

$$n_-/n_+ = e^{-2\mu H_0/kT_s}, \tag{7.16}$$

is much higher than T, the temperature of the surrounding "lattice." The process in which M_z approaches M_0 can then be regarded as a thermal equilibration in which T_s becomes equal to T. Accordingly, many authors refer to T_1 as the *spin-lattice* relaxation time.

The *transverse* relaxation time T_2 is most simply introduced by considering a system on which $\mathbf{H}_0$ and $\mathbf{H}_1$ together have been acting until M_z and $M_\perp$ have each reached a steady value, and $\mathbf{H}_1$ is then suddenly turned off. The value of $M_\perp$ depends on the degree of phase correlation among the precessing nuclei, and if there is a mechanism that destroys the phase relationships M will decay with:

$$M_\perp(t) = M_\perp(0)e^{-t/T_2} \tag{7.17}$$

This process occurs without any accompanying change in M_z or the spin temperature, so that the time constant T_2 may have quite a different value from T_1.

Two kinds of mechanisms contribute to the transverse relaxation. First, the local magnetic fields acting on particular nuclei within the sample may not all be precisely equal to H_0. Consequently, the precession frequencies of formally identical nuclei may not actually be exactly the same. Two vectors precessing with Larmor frequencies differing by an amount $\Delta\nu$ will be out of phase with one another after a time $t = 1/(2\Delta\nu)$. The local field variations may be the result of inhomogeneity of the applied field, in which case the observed T_2 is more truly a property of the spectrometer than of the sample. In solids the main source of local fields is the magnetism associated with nuclear spins, whence T_2 is also known as the *spin-spin* relaxation time.

The second transverse relaxation process involves the transfer of energy from a nucleus with $M_\mathrm{I} = \frac{1}{2}$ to one initially having $M_\mathrm{I} = -\frac{1}{2}$. This produces no change in the total z-component of magnetization but again destroys the phase coherence of the precessing magnetization vectors and thus reduces $M_{\perp}$.

The operational definition of T_2 is often based on the relationship between relaxation processes and the line width. When the local fields vary over some range, resonance occurs not at a single sharp frequency but over a region of frequencies near ν_0. The nmr signal then appears as a line of finite width, which is large when T_2 is small. It has been shown [6] that when the frequency of H_1 is slowly varied through the resonant value the intensity of absorption is proportional to the Lorentzian function*

$$g(\nu) = 2T_2/[1 + 4\pi^2 T_2^2(\nu - \nu_0)^2]. \tag{7.18}$$

This yields a full width at half-height of $\Delta\nu = 1/\pi T_2$. In most liquids T_2 for protons is of the order of 1 sec so that $\Delta\nu$ may be expected to be less than 1 Hz. In solids T_2 is much smaller, typically 10^{-5} sec, and the signals are correspondingly broader.

An important consequence of the finite relaxation rates is that one cannot indefinitely increase the intensity of an nmr signal by making the observing field H_1 progressively larger. When the frequency exactly satisfies the resonance condition, the rate of energy absorption is proportional to H_1^2 as long as H_1 is small. However, the transitions induced by H_1 tend to raise the spin temperature, that is, reduce the population difference $(n_+ - n_-)$, hence reduce the net magnetization M_z. The relaxation processes, discussed in more detail in Section 9, p. 655, tend to restore n_+ and n_- to their equilibrium values. The relative rates of energy absorption and relaxation determine the steady-state spin temperature eventually reached. This temperature rises as H_1 is increased until n_+ and n_- become essentially equal, the net magnetization approaches zero, and the resonance can no longer be observed. The value of H_1 for which this *saturation* of the signal becomes noticeable depends on the width of the resonance and on T_1. Incomplete saturation reduces the height of the signal by a factor $1/(1 + \gamma^2 H_1^2 T_1 T_2)$. When $H_1 \simeq 1/\gamma(T_1T_2)^{1/2}$, one also finds a pronounced increase in the apparent line width because the rate of energy absorption is reduced more when the frequency is exactly ν_0 than when it is slightly off the center of the resonance.

3 HIGH-RESOLUTION NMR SPECTROMETERS

A number of manufacturers produce and sell equipment for high-resolution, broad-line, and pulsed nmr spectroscopy, and the problem for chemists is usually not one of designing new apparatus but of choosing the most

* The well-known Gaussian function describes a line shape that is not usually found for high-resolution absorption.

suitable instrument from an increasingly large assortment. In this section some of the features and problems of typical high-resolution spectrometers are briefly surveyed. More detailed descriptions of the operation of various spectrometer components may be found [2] in the larger monographs on nmr and in manuals prepared by the companies that make them [3].

The performance of a high-resolution spectrometer is so critically dependent on the quality of the magnetic field that most of the major advances in nmr instrumentation have been contingent on improvements in magnet design. The ability of the spectrometer to resolve signals from similiar nuclei in slightly different chemical environments increases approximately linearly as H_0 increases, while the sensitivity rises as $H_0^{7/4}$ theoretically (actually somewhat more slowly). Hence the trend has been consistently toward the use of higher fields. However, the *homogeneity* of the field over the volume occupied by the sample and its *stability* in time must be carefully optimized to realize the advantage inherent in an increase in field intensity. As mentioned above, the natural width of the resonance signal from a liquid sample is often less than 1 Hz. If H_0 varies over the sample volume by an amount sufficient to make the Larmor frequencies of some nuclei differ by 1 Hz or more from the mean value, the observed line width is determined by this field inhomogeneity rather than by the intrinsic value of T_2. Since for protons ν_0 is of the order of 10^8 Hz, the field must be homogeneous to better than 1 part in 10^8. A similarly rigid requirement applies to the stability, since any variation of H_0 with time entails a variation of ν_0, and if ν_0 changes by more than a fraction of the line width while the signal is being traversed, the line again appears broadened. Slower variations of ν_0 may make it impossible to obtain reproducible measurements of the peak position.

Until quite recently, iron core electromagnets were used in most of the commercial spectrometers. These require very closely regulated dc power supplies and must be provided with a water-cooling system which keeps the magnet temperature as nearly constant as possible. They provide fields of the required quality and up to 23 kG intensity, and since the field strength can be varied over a wide range they make it possible to use spectrometer components tuned to a single fixed frequency for work involving several different nuclear species.

Permanent magnets are generally cheaper and simpler to operate, since warm-up time is eliminated, as is the need for water cooling. With pole gaps large enough to accommodate conventional spectrometer probes, they are limited to fields of about 14 kG or less, and since they are not readily adjustable to provide more than one field intensity, separate radiofrequency components are required for each type of nucleus to be studied. Fields in excess of 30 kG can be obtained with permanent magnets if the pole gap is made small enough, and one instrument has been described [7] in which microcoils and samples enclosed in small capillary tubes are used with such a magnet to obtain proton spectra at 104.5 MHz (24.5 kG).

With either type of magnet, auxiliary devices are commonly used to achieve the optimum in homogeneity and stability. Small, printed-circuit coils called shim coils or Golay coils may be attached to the magnet pole faces so that residual field inhomogeneities can be corrected by adjusting the current flowing through them [8]. Field fluctuations may occur for a variety of reasons including, for example, the effects of moving a steel tool or operator's chair in the neighborhood of the magnet. These are corrected with feedback devices consisting of a sensing unit which detects the onset of a field variation, an amplifier which converts the input from the sensing unit into an appropriate dc signal, and a set of coils in which this signal generates a magnetic field opposing the original fluctuation.

Several other rather sophisticated methods for dealing with field instabilities are based on the following approach. If H_M is the main magnetic field, ΔH is a small unwanted change in H_M, and H_s is the field generated by the sweep voltage, a resonance occurs when:

$$h\nu_0 = 2\mu H_0 = 2\mu(H_M + \Delta H + H_s), \tag{7.19}$$

or

$$H_s = h\nu_0/2\mu - H_M - \Delta H. \tag{7.20}$$

The ultimate goal is to have H_s remain constant so that a given peak occurs at the same point in successive scans over the spectrum. This can be attained not only by making $\Delta H = 0$, but alternatively by applying to ν_0 a correction $\Delta\nu$ which depends on ΔH. Writing:

$$H_s = h(\nu_0 + \Delta\nu)/2\mu - H_M - \Delta H, \tag{7.21}$$

the effect of ΔH is canceled if $\Delta\nu = 2\mu\, \Delta H/h$. The most readily understood way of achieving this field-frequency stabilization in practice is the use of what is commonly called "external locking." In addition to the working sample, a control sample is provided which is exposed to H_M and ΔH but not to H_s. By means of feedback circuitry, the radiofrequency is continually adjusted so that the resonance condition $h(\nu_0 + \Delta\nu) = 2\mu(H_M + \Delta H)$ is at all times satisfied for the control sample. Alternatively, with suitable modification of the apparatus, the desired stabilization can be achieved by using an "internal lock" signal originating from a reference compound placed (in solution or in a capillary) in the same tube with the working sample.

Recently spectrometers have been developed in which the magnetic field, 51.7 kilogauss, is generated in a superconducting solenoid, and it seems probable that modifications of these instruments will make it possible to obtain high resolution spectra at even higher fields in the future [J. K. Becconsall and M. C. McIvor, *Chem. Brit.*, **5**, 147 (1969)]. Such spectrometers are rather expensive and require a ready supply of liquid helium, and it seems improbable that they will be widely adopted for routine nmr studies. However, the dramatic improvement in sensitivity and resolving power which results when these very intense fields are used has allowed the nmr technique to be successfully applied in situations where conventional apparatus is simply not adequate. A number of examples are cited below.

Most of the commercial spectrometers are crossed-coil instruments. The two coils, and a cylindrical well into which the sample may be inserted, are built into the spectrometer *probe* placed between the magnet poles. The degree of coupling between the transmitter and receiver coils can be controlled by adjusting a pair of "paddles." Additional coils to produce the field H_s used to scan the spectrum may also be mounted on the probe body. High-resolution probes usually incorporate a provision for spinning the sample tube about its axis by means of a jet of air driving a small plastic turbine. This reduces the effects of small field inhomogeneities. Variable-temperature spectroscopy between about -100 and $+200°C$ is possible using probes in which a stream of heated or chilled nitrogen flows through an annular space surrounding the sample, thermally insulated from the rest of the assembly by a glass vacuum jacket.

Even as simple an item as an nmr sample tube has undergone some evolutionary development. A typical sample consists of about $\frac{1}{3}$ ml of liquid or solution in a glass tube of 5 mm O.D. Most modern spectrometers allow very little variation in this dimension, so that precision-ground tubing must be used. The best sensitivity is obtained with thin-walled tubing, in which the number of resonant nuclei located within the receiving coil is maximal. Special microcells optimize spectra obtained from very small samples, and a variety of coaxial cells is available for the simultaneous display of spectra from nuclei in two or more separate solutions.

The electronic components that make up the remainder of the spectrometer vary considerably from one instrument to another, and no attempt is made here to describe them in detail. A crystal-controlled oscillator and amplifier are used to generate the radiofrequency field, and suitable circuitry is provided to amplify and detect the voltage induced in the receiver coil and to produce a dc signal which can be displayed on an oscilloscope or graphic recorder. Other components provide for scanning the spectrum, which may be done in several ways. A simple way to vary the field is to use a low-frequency "sawtooth" generator to activate the sweep coils and simultaneously to drive the horizontal displacement of the oscilloscope. Frequency sweeping may be accomplished using a variable-frequency oscillator and mixing its output (frequency ν') with that of the fixed-frequency oscillator (frequency ν). This produces variable sideband frequencies ($\nu \pm \nu'$), one of which is used to excite resonances.

Essentially all modern high-resolution spectrometers now have built-in integrating circuits which convert the ordinary spectrum into a curve with a series of ascending steps, each associated with a particular peak. The height of the step is proportional to the area under the corresponding peak, so that relative signal strengths can be read directly from the integral curve, greatly facilitating spectral interpretation and analytical applications. The alternative procedure of using peak heights as a measure of relative intensities may give misleading results since variations in width from signal to signal are not uncommon. Even with integration, the accuracy of quantitative analyses based on nmr intensity ratios leaves a good deal to be desired, especially if the ratios are much different from unity, because amplification of the resonance signals is never absolutely linear.

As noted above, spectrometer *sensitivity* depends on the magnetic field strength and on the *filling factor*, the fraction of the effective volume within the coils that is actually available to the sample. Other variables that determine the optimum signal-to-noise ratio include the width of the signal to be detected and the sample size. The probe may be designed to accommodate sample tubes ranging up to 15 mm in diameter. The sensitivity gain that attends an increase in sample volume is often smaller than one would hope because the broadening effect of residual field inhomogeneities also increases with the sample volume. It is thus difficult to make accurate general statements about the minimum concentration of nuclei required to produce a measurable signal. An approximate indication is provided by the observation that the sharp proton signal of water dissolved in carbon tetrachloride is detectable in 2×10^{-3} M solutions with careful adjustment of a 60-MHz spectrometer but no special enhancement techniques [9].

The introduction of field-frequency stabilization has made it practicable to improve sensitivity by either using extremely slow sweeps over the spectrum or by using multiple scans. Slow sweeping simply permits the use of more filtering without losing the signal. The multiple-scanning technique requires a small computer, several suitable ones being commercially available. The spectral region of interest is divided into about 1000 subintervals, and the computer accumulates the spectrometer output for successive scans in each subinterval. If there is no resonance peak at a given point, the voltage received by the computer represents random noise and fluctuates in sign, so that the accumulated total increases only slowly with an increasing number of scans. A weak resonance, however, corresponds to a voltage of constant sign, so that the accumulated signal grows relatively quickly with repeated scanning. Theoretically, the signal-to-noise ratio rises in proportion to the square root of the number of scans; in practice, the improvement is somewhat smaller. For identical total amounts of time used to record a spectrum, that is, $T_t = n \times s$ for n sweeps of s seconds or a single very slow sweep of T_t

seconds with optimal filtering, the time-averaged accumulation and the single scan yield essentially identical sensitivity. The multiple-scan procedure is, however, easier to use and it has significantly extended the usefulness of nmr for very dilute solutions and for materials available in very small amounts.

4 CHEMICAL SHIFTS

Definitions and Chemical Shift Scales

The sine qua non of high-resolution nmr spectroscopy is the observation that nuclei of the same element in different chemical environments usually produce separate resonance signals. For example, the 60-MHz proton spectrum of an equimolar mixture of chloroform and cyclohexane consists of two peaks with relative intensities 1:12 separated by about 350 Hz. Since signal strengths are directly proportional to the relative numbers of nuclei contributing to each resonance, the stronger signal, which is at the higher field, can immediately be assigned to the cyclohexane. Nuclei in different groups within the same molecule may also give separate signals. The spectrum of methyl dichloroacetate consists of two peaks 134 Hz apart with relative intensities 1:3. The protons of the $CHCl_2$ group give rise to the smaller signal, which lies downfield from the CH_3 proton peak.

This dependence of peak position upon the intramolecular environment results primarily from the influence of the magnetic field on the electrons near the resonating nucleus. Each electron is subject to the *Lorentz* force $\mathbf{F} = e\mathbf{v} \times \mathbf{H}$ in addition to the electric and magnetic forces acting in the absence of an external field. The additional force perturbs the electronic motion in a way that cannot be quantitatively specified except for the simplest molecules. Qualitatively, the perturbation induces a secondary magnetic field which varies from point to point within the molecule, depends on the orientation of the molecule with respect to the external field, and has a strength directly proportional to that of the external field. The induced field at the site of a nucleus i may be written

$$\mathbf{H}'(i) = -\boldsymbol{\sigma}(i)\mathbf{H}_{\text{ext}}, \tag{7.22}$$

where $\boldsymbol{\sigma}(i)$ is a second-rank tensor, so that $\mathbf{H}'$ and $\mathbf{H}_{\text{ext}}$ are not necessarily parallel. The total field at the ith nucleus, $\mathbf{H}_{\text{ext}} + \mathbf{H}'(i)$, is usually somewhat smaller than $\mathbf{H}_{\text{ext}}$ itself. $\boldsymbol{\sigma}(i)$ is accordingly called the *shielding tensor*. If the molecule is rapidly reorienting as would be expected for a liquid sample, the magnitude of the effective field at the ith nucleus is:

$$H(i) = H_{\text{ext}}(1 - \sigma_i), \tag{7.23}$$

where the *shielding constant* σ_i is one third of the sum of the diagonal elements of $\boldsymbol{\sigma}(i)$. In isotropic liquids the individual components of $\boldsymbol{\sigma}(i)$ are not measurable. Resonance occurs for the ith nucleus when $H(i) = h\nu_0/2\mu$, which requires that the external field have the value:

$$H_{\text{ext}}(i) = h\nu_0/2\mu + \sigma_i H_{\text{ext}}(i). \tag{7.24}$$

In this equation any effects of the medium surrounding the molecule of interest are ignored (see Section 6). To achieve resonance for a nucleus j with shielding constant σ_j then requires:

$$H_{\text{ext}}(j) = h\nu_0/2\mu + \sigma_j H_{\text{ext}}(j). \tag{7.25}$$

Measuring the difference between the two values of H_{ext} at once allows $\sigma_i - \sigma_j$ to be determined since

$$\begin{aligned} H_{\text{ext}}(j) - H_{\text{ext}}(i) &= \sigma_j H_{\text{ext}}(j) - \sigma_i H_{\text{ext}}(i) \\ &\cong (\sigma_j - \sigma_i) H_0. \end{aligned} \tag{7.26}$$

The approximation involved in the second half of (7.26) is equivalent to neglecting terms of the order of σ^2 and introduces no appreciable error.

Although $(\sigma_j - \sigma_i)$ may be readily evaluated empirically, the individual shielding constants are often not obtainable. This does not seriously limit the usefulness of the nmr technique, but it does require that a parameter other than the shielding constant be used in tabulating data. The usual procedure is to adopt some species i as a reference and to define the *chemical shift* $\delta_i(j)$ of species j relative to i:

$$\delta_i(j) = 10^6(\sigma_j - \sigma_i) = 10^6[H_{\text{ext}}(j) - H_{\text{ext}}(i)]/H_0. \tag{7.27}$$

$\delta_i(j)$ is a dimensionless quantity which because of the factor 10^6 in the definition is assigned the units parts per million (ppm). Unfortunately, there is no universal agreement on the choice of a reference compound, even for proton spectra. There is even a difference of opinion as to whether the choice of sign in (7.27) is preferable to the alternative, that is, $\delta_i(j) = 10^6(\sigma_i - \sigma_j)$. Figure 7.3 illustrates two schemes that have been in vogue for presenting proton shifts and shows the shifts of some representative compounds. Each scale is based on tetramethylsilane (TMS) as a standard, and the units are parts per million. The τ-scale is defined by assigning a shift of 10.00 to TMS and smaller values to peaks at lower fields. The δ-scale has TMS as its zero point and positive values for peaks at lower fields.

Occasionally, one finds frequency units instead of parts per million used in reporting shift data. The chemical shift difference is readily obtained by dividing the frequency difference, in hertz, by the operating frequency of the spectrometer, in megahertz. In spite of recent efforts to arrive at a consensus, a reader wishing to compare chemical shifts reported by different authors

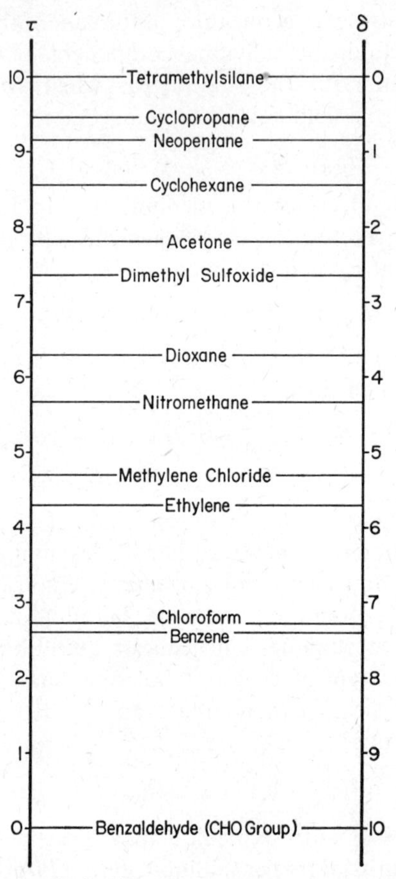

Fig. 7.3 Approximate chemical shifts for protons in representative organic compounds showing two scales in common use.

must take care to ascertain which scheme has been adopted by each. Interconversion between scales based on reference species i and j, but with the same sign convention, is easily made, when medium effects are negligible, using the relation:

$$\delta_i(x) = 10^6(\sigma_x - \sigma_i) = 10^6(\sigma_x - \sigma_j + \sigma_j - \sigma_i) = \delta_j(x) + \delta_i(j). \quad (7.28)$$

The correspondence between peak intensities and numbers of nuclei often suffices to allow an unambiguous identification of each peak in a spectrum. This has made it possible to construct extensive tables listing characteristic chemical shifts for protons in a wide variety of organic functional groups [10, 11]. With the help of such tables and numerous empirical regularities that have been discovered, nmr spectroscopy has been used to solve many analytical problems, especially in the realm of organic structure determination. In view of these successful applications, it is almost paradoxical that there still exists no rigorous predictive theory of the chemical shift.

Attempts to Rationalize Proton Shifts

Exact Theory

A great deal of effort has been expended in efforts to develop a scheme that would allow the chemical shifts in moderately complex molecules to be understood quantitatively on the basis of quantum mechanical principles [12–16]. Ramsey [17] used perturbation theory to derive a formula which, although correct in principle, is often useless in practice because it requires rather detailed knowledge of the electronic wave functions not only for the ground state of the molecule but also for many of the excited states. Other rigorous approaches eventually come up against the same sort of difficulty, which reflects the fact that even in a "simple" organic molecule such as acetic acid the proton shifts depend on the effect of $\mathbf{H}_0$ on the total population of electrons. Adequate molecular wave functions on which successful ab initio calculations could be based are almost never available.

In the absence of a workable rigorous theory, the stage is set for the appearance of diverse semiempirical schemes which allow at least the major features of the data to be rationalized. Over the years four principal factors have emerged which are now widely used in discussing and, to some extent, predicting proton chemical shifts.

Effect of Electron Density

The calculation of the shielding constant for an isolated, spherically symmetrical atom is sufficiently simple that an exact solution has been found [18]. The external magnetic field induces a rotatory motion of the electrons which gives rise to a secondary field opposed to the main field. Accordingly, the shielding constant is positive, and its magnitude is given by:

$$\sigma = \frac{4\pi e^2}{3mc^2} \int_0^\infty r\rho(r)\, dr, \tag{7.29}$$

where $\rho(r)$ is the electron density at a distance r from the nucleus. Although this formula is not applicable to an atom that forms part of a molecule, it suggests that a reduction of the electron density near a given nucleus should entail a decrease in the shielding constant. The proton chemical shift in a compound M—H should then reflect the fractional ionic character of the M—H bond.

This electron-density effect provides a plausible explanation for the fact that the chemical shifts in the methyl halides and other CH_3X compounds decrease linearly with increasing electronegativity of the X atom [19]. (It should be noted that meaningful comparisons of chemical shifts in a series of compounds such as these can be made only if each value is determined using a dilute solution in the same solvent, so that the effect of the solvent medium is constant.) Many other observations illustrate the rule that electron-withdrawing substituents tend to shift the signals of nearby protons to lower fields. A striking case is that of the alkyl carbonium ions [20], in which the positively charged carbon atom has an unusually high effective electronegativity; the CH_3 signal from $(CH_3)_3C^+$ lies about 3.3 ppm downfield from the corresponding signal for $(CH_3)_3CH$.

The charge-density effect makes allowance for changes in the shielding arising within the atom producing the resonance line. However, much evidence has accumulated showing that this effect alone cannot account for all, or even most, of the observed variations in chemical shifts. Thus in ethyl derivatives the methylene proton chemical shift is not simply related to substituent electronegativity although, surprisingly, the difference $\delta(CH_3) - \delta(CH_2)$ is [21]. Again, for ethane, ethylene, and acetylene, the shifts cannot be correlated with reasonable expectations regarding the variation of C—H bond ionic character with carbon orbital hybridization. These and many similar observations call attention to the need for considering contributions to the shielding constant by electrons on neighboring atoms.

Diamagnetic Anisotropy Effect

Neighboring atoms may be treated singly or, for convenience, groups such as C≡N may be considered as a single unit. Qualitatively, the effect of a magnetic field along the z-axis on the electrons involved is similar to the effect for an isolated atom. A circulating motion about the z-axis is superimposed on the previously existing motion, and this disturbance gives rise to a secondary magnetic field. The value of this field at a nearby point may be calculated approximately by replacing the circulating charge distribution by a point magnetic dipole $\boldsymbol{\mu}$. If spherical coordinates are used, with the resonating nucleus at the origin and the dipole representing the neighboring group at the point (r, ϕ, θ), the z-component of the secondary field evaluated at the origin is:

$$H_z' = -(\mu/r^3)(1 - 3\cos^2\theta), \tag{7.30}$$

assuming that the vector $\boldsymbol{\mu}$ is parallel to the z-axis. Depending on the value of θ, H_z' may be either positive or negative. If the magnitude of $\boldsymbol{\mu}$ does not depend on the orientation of the neighboring group with respect to H_0, the group is magnetically isotropic. Then the averaged value of H_z' for a rotating molecule is zero because $(1 - 3\cos^2\theta)_{\text{ave}} = 0$. In other words, the induced dipole of the neighboring group contributes to the shielding only if it is *magnetically anisotropic*.

Intuition correctly suggests that magnetic anisotropy is not at all uncommon. Even if the neighboring group is a single atom, such as the halogen atom of an alkyl halide, there is no reason to expect that the magnetic polarizability will be the same when the field is along the bond axis as when the field is perpendicular to the bond. For this, and other axially symmetric groups, the magnetic properties can be adequately described by two susceptibilities, $\chi_\perp$ and $\chi_\parallel$, defined by:

$$\mu = \mu_\perp = \chi_\perp H_0, \tag{7.31}$$

when H_0 is perpendicular to the symmetry axis, and:

$$\mu = \mu_\parallel = \chi_\parallel H_0, \tag{7.32}$$

when H_0 is along the axis. Since the symmetry axis is not necessarily parallel to the line joining the resonating nucleus and the effective center of the neighboring group, H_z' depends on the angle γ between these two directions as well as on θ. After averaging over all possible orientations of the molecule, it is found that H_z' is related to γ, r, and the magnetic anisotropy parameter $\Delta\chi = \chi_\parallel - \chi_\perp$ by the formula [22]:

$$(H_z')_{\text{ave}} = \left(-\frac{1}{3r^3}\right)\Delta\chi(1 - 3\cos^2\gamma)H_0. \tag{7.33}$$

The implications of (7.33) may be visualized with the help of Fig. 7.4. The arrow at the center represents an axially symmetric anisotropic group. When $\Delta\chi > 0$, a nucleus in the region AA' experiences a positive $(H_z')_{\text{ave}}$, hence its resonance is found at lower H_0 then otherwise expected (decreased shielding). Conversely, the effect of the anisotropic group on a nucleus in the region BB' is to increase the shielding when $\Delta\chi > 0$. The opposite conclusion follows, of course, if $\Delta\chi < 0$. If Fig. 7.4 is rotated about the AA' direction, the curves shown in the figure generate surfaces on which $(H_z')_{\text{ave}}$ is constant.

As an example, the anomalously large proton-shielding constant for acetylene can then be rationalized as follows [23]. The π-electrons of the triple bond can more readily be made to circulate around the bond axis than about a line perpendicular to it so that $\chi_\parallel$ is of larger magnitude than $\chi_\perp$. Since both χ-values are negative, this implies that $\Delta\chi < 0$, and the protons, which lie on the bond axis in region AA', should experience increased shielding, as observed. The estimated magnitude of this effect is [23] 10 ppm. Large anisotropy effects are also anticipated from carbon–carbon double bonds and from carbonyl groups, but these are difficult to deal with because the

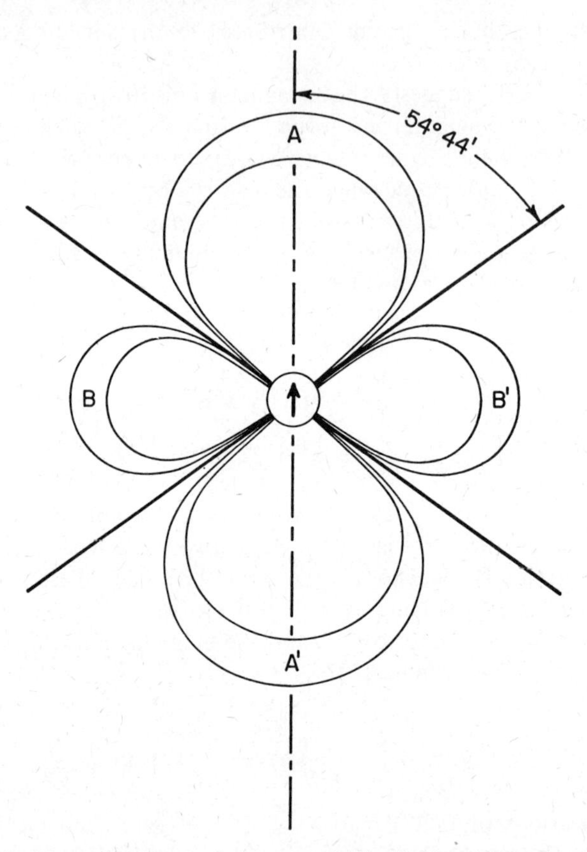

Fig. 7.4 Section through surfaces of constant shielding in the neighborhood of an axially symmetric, magnetically anisotropic group. When $\Delta\chi > 0$, nuclei in the region AA' are deshielded, while those in the region BB' are shielded. The shielding for points on the outer contour is less than that for points on the inner one by a factor of 0.6.

groups lack axial symmetry, and it is not even known positively which component of the susceptibility tensor is largest [24].

Quantitative predictions of chemical shifts using (7.33) involve several difficulties. There is often some doubt as to the best position for the point-dipole representing the anisotropic group, so that both r and γ are somewhat uncertain. Moreover, unless r is fairly large (in practice, several bond lengths) the point-dipole approximation is intrinsically inadequate. A third major problem is that for most groups there is no straightforward, independent method of evaluating $\Delta\chi$. The bulk diamagnetic anisotropy of a single crystal can be used in favorable cases to find $\Delta\chi$-values of molecules or groups within the lattice [24], but this technique has not been extensively applied because detailed knowledge of the crystal structure is required and because most crystals contain several anisotropic subunits, making contributions which cannot be unambiguously separated [25]. Results of theoretical calculations of $\Delta\chi$-values are subject to considerable uncertainty. The most common procedure has been to use the $\Delta\chi$ of a given group as a disposable parameter to be fixed by demanding that (7.33) give answers in agreement with experiment for an appropriate set of compounds. Obviously, this approach must fail unless the anisotropy of the group is independent of its environment. Another serious limitation is that an "experimental" value for the anisotropy contribution to the chemical shift can only be considered reliable if *all* other factors that affect the shift have been *quantitatively* allowed for.

Ring Current Effect

Much attention has been given to a special case of the magnetic anisotropy effect in which the neighboring group is an aromatic ring. When the applied field is normal to the plane of the ring, the properties of the delocalized orbitals of the π-electrons make possible an interatomic circulation of charge around the ring, leading to a sizable induced magnetic dipole. Since no analogous effect can occur when the field is in the plane of the ring, such a group is strongly anisotropic. This ring current effect was first proposed to account for the large bulk diamagnetic anisotropies of crystals of aromatic compounds, and $\Delta\chi$-values for the benzene ring and several related ring systems have been derived from such data [24].

Again Fig. 7.4 may be used to visualize the effect for protons at some distance from the center of the aromatic system. The symmetry axis is now perpendicular to the ring plane, and $\Delta\chi < 0$, so that protons bound to the ring, which are in the region BB', are deshielded. For benzene the contribution of this effect to the chemical shift has been estimated as about 2 ppm, roughly equal to the difference between the measured shifts of olefinic and benzene protons. Nuclei in neighboring molecules or in large substituent groups may of course find themselves in the region AA' and experience enhanced shielding because of the ring current. It should be noted that for protons coplanar with the ring but lying within it, as may be found in macrocyclic compounds, predictions based on the point-dipole approximation are not adequate. Instead, a more accurate analysis shows that the ring current increases the shielding constants of such protons.

The validity of the ring current model has recently been attacked on theoretical grounds [13, 26, 27]. Although the ensuing controversy has not been resolved, the concept appears to provide a better basis than any available alternative for the rationalization of an extensive body of data, and it is unlikely that it will be soon abandoned.

Unusual high field shifts for protons attached to the cyclopropyl group have also been attributed to a ring current effect [28, 29] although cyclopropyl obviously is not an ordinary aromatic system.

Electric Field Effect

The ideas introduced so far still do not suffice to explain all observed features of proton chemical shifts. An outstanding example of the need for an additional factor is the extremely low shielding found for protons in intramolecular hydrogen bonds. Since hydrogen bonds are primarily stabilized by electrostatic forces, several workers were led to explore the possibility that the proton shift is directly influenced by the strong electrostatic field originating at the basic group [29–31].

Again, the simplest model calculation is that for an isolated hydrogen atom exposed simultaneously to an electric field E and a magnetic field. It was shown [30] that the diamagnetic circulation of the electrons is hindered so that the shielding constant changes by an amount

$$\Delta\sigma = -(881/216)a^3E^2/mc^2 = -7.38 \times 10^{-19}E^2. \tag{7.34}$$

This change is negligible for externally applied electric fields, but the local fields near an ionic or strongly dipolar group may reach such large values that the effect becomes important. For example, at a distance of 2 Å from a unit electronic charge, $E = 1.2 \times 10^6$ statvolts/cm, which gives $\Delta\sigma = 1.06$ ppm.

The change in the shielding for a proton of a C—H or O—H bond is more difficult to compute. Approximate calculations yield a formula [31, 32]:

$$\Delta\sigma = -2.9 \times 10^{-12}E_z - 7.38 \times 10^{-19}E^2, \tag{7.35}$$

where E_z is the component along the bond direction. For a field of the order of 10^6 statvolt/cm, the linear term predominates and the total $\Delta\sigma$ is considerably larger than the value for the free hydrogen atom. With this model one can readily explain the shifts of hydrogen-bonded protons and some of the low-field shifts found in charged molecules.

The reality of the electron density effect, magnetic anisotropy effect, and electric field effect seem to be firmly established, but the quantitative application of these effects in predicting or rationalizing spectra still leaves much to be desired. In many molecules all three effects act simultaneously, and it is not feasible to isolate one contribution unambiguously. It is also quite possible that the list of required "effects" is not yet complete. This is suggested particularly by peculiarities in the spectra of the alkyl cyclohexanes and related compounds [33]. Nevertheless, most of the major trends in proton shift values can now be considered understood, and in many cases one can predict with some confidence how the shielding of a proton will be affected by a given structural change in the molecule.

5 ELECTRON-COUPLED SPIN-SPIN INTERACTIONS

Interaction Mechanisms

High-resolution nmr spectra usually contain many more peaks than can be accounted for on the basis of chemical shift differences alone. To cite a very well known example, the aldehyde protons of acetaldehyde, CH_3CHO, produce a multiplet of four lines with intensity ratios 1:3:3:1, 2.8 Hz apart, and the methyl protons give a doublet with components of equal intensity also 2.8 Hz apart. The separation between the centers of the two multiplets in hertz is proportional to the spectrometer frequency as expected for a chemical shift difference, but the spacing within each is independent of the observing frequency. Evidently this fine structure does not have the same physical basis as the chemical shift phenomenon.

To understand these observations qualitatively, one may consider a hypothetical molecule containing two protons, *a* and *b*, which produce well-separated resonance signals, and no other magnetically polar nuclei. Each proton may have either $\mu_z = +\mu$ or $\mu_z = -\mu$. Using the symbols α and β, respectively, for these two states, the four possible nuclear spin states of the molecule are $\alpha(a)\alpha(b)$, $\alpha(a)\beta(b)$, $\beta(a)\alpha(b)$, and $\beta(a)\beta(b)$, or more briefly $\alpha\alpha$, $\alpha\beta$, $\beta\alpha$, and $\beta\beta$. The resonance of nucleus *a* then involves either of the two transitions $\alpha\alpha \rightarrow \beta\alpha$ and $\alpha\beta \rightarrow \beta\beta$, in which the spin of *b* remains fixed. A second pair of transitions, $\alpha\alpha \rightarrow \alpha\beta$ and $\beta\alpha \rightarrow \beta\beta$, corresponds to the resonance of nucleus *b* with no accompanying change in the state of *a*. If each pair of transitions occurs at a single frequency, a two-line spectrum will be observed; however, interactions between the spins of *a* and *b* may cause the frequencies to become unequal so that a four-line spectrum is found.

The most obvious interaction is the direct one, arising from the fact that the local field at *a* includes a contribution from the magnetic dipole of *b*, which obviously depends on the orientation of *b*. This effect plays an important part in determining the shape of the nmr peak from a crystalline sample, but for a liquid it is canceled out by the rapid reorientation of the molecules.

An indirect interaction mechanism suggests itself when one considers the magnetic dipoles arising from the spins of electrons, especially those in the bonds involving *a* and *b*. The simplest case is a molecule with only one pair of electrons in a delocalized molecular orbital extending over both protons.* It may be supposed that at some instant electron 1 is in the neighborhood of nucleus *a* and electron 2 is near *b*. According to the Pauli principle, the spins of the two electrons must be opposed so that, leaving the nuclear spin orientations unspecified for the moment, there are two possible situations,

* This is a hypothetical molecule and not H_2 because the argument requires that the protons have different chemical shifts.

symbolically

I: $\alpha(1)a$; $\beta(2)b$

II: $\beta(1)a$; $\alpha(2)b$

The relative stabilities of arrangements I and II depend on the spin states of the two nuclei because each of them interacts with the magnetic dipoles associated with the electrons.

Several terms may contribute to the magnetic interaction energy between a nucleus and a nearby electron. If the electron is in an s-orbital or a hybrid orbital with some s-character, the major term is the Fermi *contact interaction* [34], which depends on the probability of finding the electron and the nucleus at the same point, that is, on the square of the electronic wave function evaluated at the site of the nucleus. If the electron is not in a pure s-orbital, the normal, through-space, dipole-dipole interaction between the two spins has a finite magnitude depending on the wave function and especially on the distance between the two particles. Also, for an electron not in an s-orbital, the orbital angular momentum gives rise to an additional magnetic dipole which must be included. For present purposes, it suffices to say that each proton interacts primarily with the electron nearest to it and the state of lower energy is that in which the spins of the two particles are opposed. This means that I is more stable than II when proton b is in the α-state, and the reverse is true when b is in the β-state. Since the more stable arrangement makes the larger contribution to the average state of the molecule, there is a small unbalanced electron spin density near nucleus a, with an alignment determined by the spin state of b. This spin density makes a contribution to the local field at a which does not average to zero even if the molecule is rotating. Consequently, the two transitions $\alpha\beta \rightarrow \beta\beta$ and $\alpha\alpha \rightarrow \beta\alpha$ have slightly different frequencies and, by the same token, the two transitions involving reorientation of b do not coincide. The total nmr spectrum then consists of two doublets. Since the numbers of molecules in each of the four spin states $\alpha\alpha$, $\alpha\beta$, $\beta\alpha$, and $\beta\beta$ are very nearly the same, the four transitions are expected to have essentially identical intensities.

Alternatively, the situation can be described by stating that the Hamiltonian representing the total energy of the molecule includes a term of the form:

$$H^{(1)} = hJ_{ab}\mathbf{I}(a) \cdot \mathbf{I}(b). \tag{7.36}$$

Here $\mathbf{I}(a)$ and $\mathbf{I}(b)$ are the nuclear spin vectors and J_{ab} is the *spin-spin coupling constant*, which has the units of frequency and determines how strongly the two spins interact. Depending on the electronic structure of the molecule, J_{ab} may vary not only in magnitude but also in sign. For the case just considered, J_{ab} is positive since the states $\alpha\beta$ and $\beta\alpha$, with $\mathbf{I}(a) \cdot \mathbf{I}(b) < 0$, are stabilized more than $\alpha\alpha$ and $\beta\beta$, which have $\mathbf{I}(a) \cdot \mathbf{I}(b) > 0$. This follows intuitively if one considers that for the $\alpha\beta$ state, for example, the favorable arrangement $\beta(1)\alpha(a)$; $\alpha(2)\beta(b)$ is possible, but for the $\alpha\alpha$ or $\beta\beta$ states at least one nucleus would have an unfavorably aligned electron spin near it.

A negative J_{ab} can easily arise in a fragment such as H^a—C—H^b in which two pairs of electrons provide most of the coupling. This may be shown schematically by considering instantaneous arrangements in which electron 1 is near proton a, electron 2 is in the same bonding orbital as 1 and near the carbon nucleus, electron 3 is also near the carbon nucleus but assigned to the C—H^b bonding orbital, and electron 4 is near b. Since the electron near each of the protons prefers to have its spin antiparallel to the proton spin, the most favorable situation for the nuclear spin state $\alpha\alpha$ is:

$$\text{I: } a(\alpha)\beta(1);\ \alpha(2)\text{C};\ \alpha(3)\text{C};\ \beta(4)\alpha(b)$$

and for the spin state $\alpha\beta$:

$$\text{II: } \alpha(a)\beta(1);\ \alpha(2)\text{C};\ \beta(3)\text{C};\ \alpha(4)\beta(b).$$

For arrangement I the two electrons nearest the carbon atom have parallel spins, while for II the opposite is true. Then one may invoke Hund's rule to predict that I is a lower-energy configuration than II and, accordingly, that the $\alpha\alpha$ state is stabilized relative to $\alpha\beta$. The same argument can be applied to the remaining two spin states with similar results.

The preceding discussion is meant only to make it intuitively apparent that J-values of either sign may be expected in real molecules. Several quantum-theoretical treatments have been developed that allow the coupling constants to be calculated from valence-bond or molecular-orbital wave functions, although not without introducing more-or-less serious simplifying assumptions [16, 35]. The results often agree very nicely with experimental values, some of which are discussed below, and confirm that the J-values may have either sign.

Multiplets in First-Order Spectra

Equation (7.36) provides a starting point for a rigorous, quantum mechanical calculation of the frequencies and intensities of the four nmr transitions for a two-proton system [36]. As long as the protons have chemical shifts sufficiently different to make

$$\nu_0\,|\delta_a - \delta_b| \gg |J_{ab}|, \tag{7.37}$$

the result is the same as that of the approximate discussion just given. The spectrum consists of two doublets each with an intercomponent spacing, in hertz, equal to J_{ab}. The distance between the centers of the doublets is simply the chemical shift difference, and all four components have equal intensities. The somewhat more complicated spectrum that arises when (7.37) is not satisfied is discussed later, except to call attention to the very important finding that in the limiting case, when the two chemical shifts are identical, the spectrum consists of a single line even though J_{ab} is different from zero.

For a molecule that contains more than two protons, the spectrum may still be predicted very simply, providing that any two of the protons either have the same chemical shift or shifts so different that an inequality of the same form as (7.37) applies. Such spectra are called *first-order* spectra, and a reasonably good example is furnished by that of acetaldehyde, described at the beginning of this section.

Because of rapid rotation about the C—C bond, the three methyl protons have identical chemical shifts and the spin-spin couplings among them are not observed. They produce a doublet because they interact with the aldehyde proton, which of course can exist in two spin states. The four-line pattern of the aldehyde group must be explained by considering the spin states of the three methyl protons which are all equally coupled to it. These states can be represented by ordered triplets of Greek letters (e.g., $\alpha\alpha\beta$) and grouped according to the value of M_F, the total z-component of the nuclear angular momentum. Of the eight possible states, one ($\alpha\alpha\alpha$) gives $M_F = \frac{3}{2}$, three ($\alpha\alpha\beta$, $\alpha\beta\alpha$, or $\beta\alpha\alpha$) give $M_F = \frac{1}{2}$, three ($\beta\beta\alpha$, $\beta\alpha\beta$, or $\alpha\beta\beta$) give $M_F = -\frac{1}{2}$, and one ($\beta\beta\beta$) gives $M_F = -\frac{3}{2}$. Since the local field at the aldehyde proton depends on M_F, the appearance of four lines with intensity ratios 1:3:3:1 is thus accounted for.

More generally, if a molecule having n equivalent protons of type A and m of type X, interacting with a coupling parameter J_{AX}, produces a first-order spectrum, then the A-protons produce a multiplet with $m + 1$ components, and the X-signal has $n + 1$ components. For each multiplet the intensities can be predicted by an extension of the argument used for acetaldehyde. The results are summarized in the accompanying diagram, in which the nth row gives the relative intensities of the components of an $(n + 1)$-fold multiplet. Each number in the diagram is the sum of two numbers on each side of the colon immediately above it.

1 : 1
1 : 2 : 1
1 : 3 : 3 : 1
1 : 4 : 6 : 4 : 1
1 : 5 : 10 : 10 : 5 : 1
1 : 6 : 15 : 20 : 15 : 6 : 1
and so on

In the somewhat more complex situation in which the A-protons interact with m type-X protons and s type-Y protons, with coupling constants J_{AX}

and J_{AY}, the splittings are additive, that is, the A-signal is split into $m + 1$ components, J_{AX} hertz apart, and each of these is split again into $s + 1$ lines, J_{AY} hertz apart. The multiplet then has $(m + 1)(s + 1)$ lines, unless some of these happen to coincide. In the special case $J_{AX} = J_{AY}$, such coincidences reduce the A-pattern to a multiplet of $m + s + 1$ components.

Deviations from First-Order Behavior

Nearly First-Order Multiplets

Many problems of structure determination can be solved by applying first-order analysis to the nmr spectrum, but it is obvious that inequality (7.37) is not always obeyed. If not, and there are two sets of equivalent protons, the spectrum is called an A_mB_n pattern and its appearance depends on the ratio $\Delta\nu/|J_{AB}|$, where $\Delta\nu = \nu_0\,|\delta_a - \delta_b|$. The two-proton problem is simple enough to allow the results to be given in closed form [36]. A pattern of four lines is always found unless $\Delta\nu = 0$, but the two lines nearest the center become more intense as $\Delta\nu/|J_{AB}|$ decreases, while the two outer lines become weaker. The intensities for these two pairs of lines are in the ratio $[1 + J_{AB}/(\Delta\nu^2 + J_{AB}^2)^{1/2}]:[1 - J_{AB}/(\Delta\nu^2 + J_{AB}^2)^{1/2}]$. The spacing within each doublet remains equal to J_{AB}, but the distance between their centers is no longer simply $\Delta\nu$ but instead is $(\Delta\nu^2 + J_{AB}^2)^{1/2}$. As already stated, when $\Delta\nu = 0$ the inner two lines coincide and the outer two are reduced to zero intensity, so that the pattern degenerates into a single line.

For larger sets of nuclei, the A_mB_n patterns cannot in general be as simply described, but as long as $\Delta\nu/|J_{AB}|$ is fairly large (i.e., not much less than 10) the observed spectrum resembles a first-order pattern enough to make it possible to arrive at the correct interpretation by inspection. Again, the lines nearest the center of the spectrum appear with enhanced intensities and the outer lines are weakened. Some of the multiplet components are further split so that the number of signals is larger than that predicted from the first-order rules. These discrepancies increase as $\Delta\nu/|J_{AB}|$ becomes smaller, until eventually all resemblance between the observed spectrum and the first-order pattern is lost.

Since $\Delta\nu$ is directly proportional to H_0, while J_{AB} is field-independent, a compound that produces a rather complex spectrum at 10 kG may give a pattern which can be interpreted by first-order analysis if reexamined at 23.5 kG. This has provided a powerful incentive for the development of spectrometers operating at higher and higher field strengths. However, even at the highest fields many spectra remain quite complex, and an extensive literature has grown up dealing with rigorous methods for predicting the appearance of the A_mB_n spectra and of numerous still more complicated types involving more than two set of protons [37–41]. Numerical results are most readily obtained with the help of automatic computers. Details of this work are not given here, but it seems desirable to call attention to several features that may complicate the interpretation of spectra in a somewhat unexpected way.

Chemical versus Magnetic Equivalence

In order to assure the validity of the rule that interactions between equivalent nuclei are not observable in nmr spectra, a rather restrictive definition of "equivalent" must be used. Two nuclei are equivalent if they have the same chemical shift and if, in addition, they are coupled with equal coupling constants to other groups of nuclei that may be present. This usage differs from that found in much of the early literature in which nuclei are defined as equivalent if they merely have the same chemical shift. For example, the protons in vinylidene fluoride

H_1 F_1
C=C
H_2 F_2

are nonequivalent by the strict definition because $J_{H_1F_1} \neq J_{H_1F_2}$ and $J_{H_2F_1} \neq J_{H_2F_2}$, and the proton spectrum [42] does show splittings that depend on $J_{H_1H_2}$. In such cases it is useful to distinguish between *chemical* and *magnetic* equivalence, the two protons being chemically equivalent but magnetically nonequivalent.

Normally equivalent nuclei may become nonequivalent as a result of *isotopic substitution*, thus allowing coupling constants to be determined that are not measurable otherwise. For example, the six protons of ethane form a single chemically equivalent set and $J_{H-C-C-H}$ cannot be observed. If the spectrum is examined with high sensitivity, *satellite* signals attributable to the species $^{12}CH_3{}^{13}CH_3$ (natural abundance 2.2%) can be detected [43]. Spin-spin coupling between the ^{13}C nucleus (which has $I = \frac{1}{2}$) and the directly bonded protons splits the $^{13}CH_3$ signal into a doublet, each component of which has an effective chemical shift differing from that of the $^{12}CH_3$ protons by about 63 Hz. Each of the ^{13}C satellites is then split into a quartet by the proton-proton coupling. The $^{12}CH_3$ protons of the isotopically substituted ethane also produce a multiplet, but this is usually hidden by the much more intense singlet signal of the normal isotopic species.

Deceptively Simple Spectra

When rigorous procedures for predicting complex spectra are systematically applied using a wide range of spectral parameters, a number of special cases can be singled out for which the observed spectrum has fewer lines and yields less information than normally expected. The simplest example is an *ABX*-system, for which the *X*-proton resonance normally consists of two doublets with peak separations which readily allow J_{AX} and J_{BX} to be evaluated. If instead a triplet pattern appears, one is tempted to conclude that $J_{AX} = J_{BX}$ since this is a *sufficient* condition for coalescence of the central pair of lines. It is not obvious that this condition is not *necessary*, but it has been shown that with $J_{AX} \neq J_{BX}$ and $\delta_A = \delta_B$ the X-signal is again a triplet [44], with components separated by $(J_{AX} + J_{BX})/2$. The two coupling constants then cannot be determined from the spectrum, but only their average. A number of such situations have been identified [44], in which the observed pattern resembles a first-order multiplet but the actual interpretation is not straightforward.

Virtual Coupling

Another unexpected result of the exact theory of spin-spin interactions is that a molecule with three (or more) sets of protons, symbolically $A_nB_mX_s$ may produce a spectrum in which the X-proton signal is split by the A-protons even though $J_{AX} = 0$. Numerous examples of such unexpectedly complex multiplets have been observed. Since J_{AX} is most likely to be zero if A and X are several bond lengths apart, the phenomenon has been called virtual long-range coupling, and the conditions under which it may occur have been investigated in detail [45]. A common situation is one in which δ_A and δ_B are not very different and J_{AB} is large. For the ABX_3 case in particular, the X-signal is normally a doublet when $J_{AX} = 0$, with the two components separated by J_{BX} hertz. If there is virtual coupling, two lines J_{BX} hertz apart are still found, but there may be additional fairly strong components between these two so that the expected doublet is not recognizable.

Proton-Proton Coupling and Molecular Geometry

Numerical values of spin-spin coupling constants can often be derived easily from the spectrum, and they furnish important information about the molecular structure beyond what can be inferred from the chemical shifts alone [46]. A very large number of J-values has been tabulated in a recent review [47]. In general, J-values decline as the number of bonds separating the interacting protons increases, but the precise values also depend on other geometrical factors and on such variables as the electronegativities of nearby atoms.

Geminal coupling constants, $J_{H—C—H}$, vary between about -20 and $+5$ Hz and tend to become more positive as the H—C—H angle increases. Substitution of an electronegative atom alpha to the CH_2 group also tends to make $J_{H—C—H}$ more positive, while a similar substitution in the β-position has the reverse effect. These trends, and the effects of nearby π-electrons, were rationalized by Pople and Bothner-By [48] using an extension of a molecular orbital approach developed by Pople and Santry [49].

Vicinal coupling constants, $J_{H-C-C-H}$, range from 0 to about 15 Hz and depend primarily on the dihedral angle ϕ between the two C—C—H planes but also on the nature of the other atoms bound to the two carbons. The equations:

$$J_{H-C-C-H} = 8.5 \cos^2 \phi - 0.28 \text{ Hz}; \quad 0° < \phi < 90° \qquad (7.38)$$

and

$$J_{H-C-C-H} = 9.5° \cos^2 \phi - 0.28 \text{ Hz}; \quad 90° < \phi < 180°, \qquad (7.39)$$

derived by Karplus [50] using valence-bond theory, agree moderately well with observed values. For cyclohexane derivatives of fixed conformation the coupling between adjacent axial hydrogens ($\phi = 180°$) is found to range from 9 to 12 Hz, and the calculated value is 9.2. For adjacent equatorial hydrogens, observed J-values range from 2 to 4 Hz (calculated for $\phi = 60°$, 1.7 Hz). For ethyl derivatives or other compounds in which the conformation is not fixed, the coupling constants represent an appropriate weighted average of the values for the *gauche* and *trans* interactions.

Long-range coupling constants, such as $J_{H-C-C-C-H}$, are usually quite small, but a number of examples are known in which special geometrical relationships or unsaturation in the carbon chain lead to appreciable long-range splittings [51].

Coupling of Protons with Other Isotopes

If in addition to protons a molecule contains one or more nonhydrogen nuclei with $I = \frac{1}{2}$, there will be additional fine structure in the proton spectrum arising from spin-spin interactions with the other nuclei. This is illustrated by the proton spectrum of CH_3F, a doublet with a field-independent splitting given by $J_{H-C-F} = 46.5$ Hz. It has already been mentioned that high-sensitivity spectra of organic compounds usually exhibit extra lines resulting from coupling between protons and naturally occurring ^{13}C nuclei. For hydrocarbons the value of J_{CH} for a proton and the carbon atom to which it is bound depends primarily on the state of hybridization of the carbon atom, in accordance with the empirical equation:

$$J_{CH} = 500\rho_{CH}, \qquad (7.40)$$

where ρ_{CH} is the fractional *s*-character of the carbon atomic orbital involved in the C—H bond [52, 53].

A nucleus with $I > \frac{1}{2}$ may also couple with a nearby proton, splitting its resonance signal into $2I + 1$ components of equal intensity. Deuterium has $I = 1$, and the proton resonance of CH_3D is a 1:1:1 triplet [54], with $J_{H-C-D} = 1.9$ Hz. In general, the coupling constant J_{HD} is equal to the J_{HH} in the corresponding nondeuterated compound multiplied by the quotient of the magnetogyric ratios, γ_D/γ_H, that is,

$$J_{HD} = 0.1535 J_{HH}. \qquad (7.41)$$

This allows J_{HD}-values to be used to evaluate J_{HH}-values not directly observable because the protons in the nondeuterated species are equivalent, for example, the datum just given for CH_3D implies that $J_{HH} = 12.4$ Hz in CH_4. If a proton is coupled to two identical deuterons, it produces a quintet, again with very closely spaced components and with intensity ratios 1:2:3:2:1.

Proton-boron coupling is illustrated by the borohydride ion, BH_4^-, which produces a quartet [55] with peaks separated by 82 Hz superimposed on a relatively weak septet with a spacing of 27 Hz. The quartet represents the species containing the isotope ^{11}B with $I = \frac{3}{2}$ and abundance 81.2%, while the septet is assigned to the ^{10}B species with $I = 3$, abundance 18.8%. Again, the ratio of the two J_{BH}-values is equal to the ratio of the magnetogyric ratios of the two nuclear species.

The expected coupling with nuclei of $I > \frac{1}{2}$ is not always observed, however. CH_3Cl gives a singlet proton signal although both isotopes of chlorine have $I = \frac{3}{2}$. The explanation hinges on the fact that the electric quadrupole moments that characterize nuclei with $I > 1$ interact with the gradient of the local electric field. If the nucleus is in a sufficiently symmetrical environment (e.g., the boron in BH_4^-) the interaction may vanish, and if the quadrupole moment is small it may be relatively weak. Otherwise, it tends to align the nuclear spin vector in a direction fixed within the molecule, and if the molecule is rapidly tumbling the nuclear spin cannot remain long in a state having a well-defined component along the magnetic field direction. The spin-spin interaction then averages to zero. Occasionally, the electric and magnetic interactions are so balanced that the fine structure is observed but each line is severely broadened (see Section 7). This is rather common for compounds with ^{14}N—H bonds, for which the proton resonance may appear as a moderately well-resolved triplet, a broadened triplet, or as a more-or-less severely broadened singlet, depending on the nature of the electron distribution around the nitrogen nucleus.

Signs of Coupling Constants

As stated earlier, theoretically predicted values of J not only vary in magnitude but may be either positive or negative. This raises the question whether or not the calculated signs can be verified experimentally.

For an arbitrary set of nuclei, the predicted high-resolution spectrum is unchanged if the signs of all of the coupling constants are simultaneously reversed. This implies that the sign of J_{AB} cannot normally be inferred from the appearance of an A_nB_m pattern. Moreover, if the spectrum is first order, any one of the J-values can have its sign changed without producing an observable change. To obtain any information about signs, then, one must have at least three sets of nuclei, and at least two of them must be strongly enough coupled to give a pattern appreciably different from a first-order one;

the simplest such case is ABX. The frequencies and intensities of the components of an ABX spectrum can be used to evaluate the magnitude of J_{AB}, J_{AX}, and J_{BX} and to ascertain whether J_{AX} and J_{BX} have like or unlike signs [37–41].

Information about the relative signs of coupling constants can also be obtained from spin-decoupling experiments (see Section 8) and from *double-quantum transitions* [56, 57]. These are transitions in which two nuclei simultaneously undergo spin reorientation as the result of the absorption of two quanta of radiant energy. They are forbidden by the normal nmr selection rules but may be observed when the radiofrequency field is intense enough to saturate the ordinary transitions. If both quanta have the same frequency, and the two nuclei are designated A and B, conservation of energy requires $\Delta E = 2h\nu = \Delta E(A) + \Delta E(B)$ or $\nu = |\nu(A) + \nu(B)|/2$. Therefore the double quantum transition appears at a point on the chemical shift scale halfway between δ_A and δ_B. If the normal signals from A and B are multiplets, the double-quantum transition also shows fine structure and may provide the additional data needed to fix the relative signs of two or more J-values.

Several approaches have been used to obtain absolute signs of coupling constants. Theoretical arguments make it appear extremely unlikely that directly bonded $J_{^{13}C-H}$-values can be negative, hence a proton-proton coupling constant is called positive if it can be shown that it agrees in sign with such a J_{CH}. Other absolute signs can then be fixed by finding the sign of the unknown J relative to one that has been determined in this way. The validity of this procedure has recently been confirmed by means of special techniques which independently allow the signs of certain proton-proton coupling constants to be measured. One of these involved the investigation of the spectrum of a polar molecule partially oriented by means of a strong electrostatic field [58]. Another uses data for molecules partially oriented in a liquid–crystal solvent; such spectra are discussed in Section 14. In some cases the sign of J_{HH} can be inferred from the small changes in its magnitude produced by a change in the dielectric constant of the solvent [59].

6 SOLVENT EFFECTS

Chemical Shifts

Classification of the Effects

For the sake of simplicity, no allowance was made in Sections 4 and 5 for the possibility that chemical shifts and spin-spin interactions might be influenced by changes in the environment of the molecule. However, it has been known for some years that the appearance of a spectrum may be drastically altered by changing the solvent because of attendant changes in the nmr parameters [60]. The interpretation of these solvent effects, like the problem of rationalizing chemical shifts and coupling parameters themselves, is complicated because a number of factors cooperate to produce the observed changes.

The shielding constant for a particular nucleus in a liquid sample may be written $\sigma = \sigma_{\text{gas}} + \sigma_m$, where σ_{gas} is the value appropriate for the isolated molecule and σ_m represents the effect of the medium. The latter term is usually considered as the sum of four contributions:

$$\sigma_m = \sigma_b + \sigma_W + \sigma_a + \sigma_E. \tag{7.42}$$

σ_b reflects the bulk magnetic susceptibility of the medium and σ_W the effect of Van der Waals interactions between neighboring molecules. These two terms must be dealt with even if all molecules involved are inert, nonpolar, and magnetically isotropic. σ_a arises from the magnetic anisotropy of the solvent and is especially important for solutions containing aromatic compounds, and σ_E allows for an intermolecular electric field effect associated with the polarization of the solvent in the neighborhood of a polar solute. Some authors find it desirable to list an additional contribution arising from specific molecular interactions or complex formation, but it seems more logical to discuss the shift in, for example, a hydrogen-bonded complex in terms of the electric field and magnetic anisotropy of the neighboring molecule, that is, as a special case in which σ_E, and possibly σ_a, are abnormally large.

Bulk Diamagnetic Susceptibility

Solvent molecules at some distance from the solute can be treated as if they formed a continuous medium characterized by a volume susceptibility χ_v. The induced magnetic polarization in this medium makes no contribution to the field at the solute molecule provided the sample is in a spherical container. However, convenience normally dictates use of cylindrical nmr samples for which the contribution is [61]

$$(H_b')_{\text{cylinder}} = -\frac{2\pi}{3}\pi\chi_v H_0. \tag{7.43}$$

By recalling the definition of shielding constant [see (7.23)], it follows that

$$(\sigma_b)_{\text{cylinder}} = -(H_b')_{\text{cylinder}}/H_0 = \frac{2\pi}{3}\chi_v. \tag{7.44}$$

For typical organic liquids χ_v lies between -0.4 and -1.2×10^{-6}, so that σ_b is negative and of the order of 0.8 to 2.5 ppm.

The effect of σ_b is especially important when practical considerations make it desirable to use an *external reference*, that is, a reference compound enclosed in a capillary inserted into the nmr tube (as opposed to an internal reference, dissolved in the sample itself). The "true" chemical shift for a peak in the working sample may be defined as the shift that would be found if both sample and reference were in spherical containers, that is,

$$\delta_r(x) = [(\sigma_x)_{\text{sphere}} - (\sigma_r)_{\text{sphere}}] \times 10^6. \tag{7.45}$$

The apparent shift is

$$\begin{aligned}\delta_r(x)_{\text{app}} &= \left[(\sigma_x)_{\text{sphere}} + \frac{2\pi}{3}(\chi_v)_x - (\sigma_r)_{\text{sphere}} - \frac{2\pi}{3}(\chi_v)_r\right] \times 10^6 \\ &= \delta_r(x) + \frac{2\pi}{3}[(\chi_v)_x - (\chi_v)_r] \times 10^6. \end{aligned} \tag{7.46}$$

The difference between the true shift and the apparent shift is the *bulk susceptibility correction* and is often far from negligible, since in proton nmr $\delta_r(x)$ itself seldom exceeds a few parts per million. The popularity of internal referencing arises from the fact that it eliminates the need for such a correction.

With (7.44) or (7.46) as a starting point, several experimental arrangements can be employed to determine unknown susceptibilities from nmr measurements [62, 63]. The method of Frei and Bernstein [62] calls for the use of two reference cells, one spherical and one cylindrical, each filled with the same standard substance and inserted into an nmr tube containing the liquid of unknown susceptibility. The reference compound then produces two signals with a separation proportional to $[(\chi_v)_{\text{ref}} - (\chi_v)_{\text{unkn}}]$. The proportionality constant, empirically evaluated, was 2.058, very nearly the same as the predicted value $2\pi/3 = 2.094$.

Van der Waals Contribution

Experimental values of σ_W are relatively easy to obtain for systems containing only nonpolar, magnetically isotropic molecules, where $\sigma_m = \sigma_b + \sigma_W$, and σ_b can be evaluated from (7.44). Typically, σ_W lies between -0.1 and -0.5 ppm at room temperature. Thus it is about one-sixth as large as σ_b, and the latter must be rather precisely known in order to evaluate the former with good accuracy. An early suggestion [64] (described [65] as "obsolete" in 1960 but still occasionally cited) that (7.44) be replaced by the empirical equation $\sigma_b = 2.6\chi_v$ reflects the fact that σ_b and σ_W have the same sign, and their sum is often approximately $2.6\chi_v$.

An attempt to account theoretically for some observed σ_W values is described in a 1968 publication [66], which briefly reviews earlier efforts. Unlike σ_b, σ_W does not depend on a single property of the solvent, and the theoretical argument not only is not simple but also requires eventually that an empirical, temperature-dependent scale factor be introduced. However, the discussion at least makes clear which factors are important in fixing σ_W.

The results are summarized in an equation:

$$\sigma_W = -(1.99 + 0.0087t) \times 10^6 \frac{\alpha_2 I_2 H_6(y) S_6{}^g}{V_2 r_0{}^3 y^4}, \tag{7.47}$$

which reproduces the trends in the measured values fairly closely. Here t is the centigrade temperature, α_2, I_2, and V_2 are the molecular polarizability, the ionization potential, and the molecular volume for the solvent, respectively, $y = 2(\epsilon/kT)^{1/2}$, ϵ and r_0 are the parameters in a Lennard-Jones (6-12) pair potential function,

$$u = 4\epsilon[(r_0/r)^{12} - (r_0/r)^6], \tag{7.48}$$

for a solvent and solute molecule, $H_6(y)$ is a function defined and tabulated earlier in a discussion of the behavior of imperfect gases [67], and $S_6{}^g$ is a *site factor* which allows σ_W to differ for different protons of the solute molecule. Equation (7.47) is not always a convenient one to apply. For example, to evaluate ϵ and r_0 one must know the corresponding parameters for the pure solute and solvent (derived from the temperature dependence of the second virial coefficient); then $\epsilon = (\epsilon_{\text{solvent}} \cdot \epsilon_{\text{solute}})^{1/2}$ and r_0 is the arithmetic mean of the two r_0-values. The function $H_6(y)/y^4$ is presented as a graph in Ref. 66. The site factor is a function of the geometrical variable $q_0 = r_0/d$, where d is the distance of a given proton from the center of the molecule, and it is also given graphically in Ref. 66. Such a factor is an essential requirement in any successful theory of σ_W, since it is an observed fact that different proton signals from the same compound are shifted unequally when the molecule is transferred from the gaseous state to a liquid medium.

Electric Field Effect

In the discussion of chemical shifts, it was pointed out that an electric field changes the shielding of a proton in a C—H or O—H bond approximately according to the equation:

$$\Delta\sigma = -3 \times 10^{-12} E_z - 7.4 \times 10^{-19} E^2. \tag{7.49}$$

This equation is not restricted to electric fields arising within the molecule. A polar solute tends to polarize or orient neighboring molecules in such a way as to create an electric field, which may give rise to a substantial intermolecular contribution to the chemical shift [31]. The largest effects are found when there are specific solvent-solute (or sometimes solute-solute) interactions, such as the formation of hydrogen-bonded complexes. Alcohols, phenols, carboxylic acids, water, and even much less acidic molecules such as chloroform, all produce nmr spectra which are strikingly dependent on solvent and concentration, and the general finding that protons in hydrogen bonds have resonances at unusually low fields is readily explained by (7.49) if the electron donor is considered as the source of the electric field.

When well-defined solvent-solute complexes do not form, a polar solute still experiences an electric field resulting from the electrical polarization it induces in the surrounding medium. This *reaction field* can be approximately evaluated by a procedure attributable to Onsager and based on an idealized model in which the solute molecules are considered as spheres of radius r and polarizability $\alpha = [(n^2 - 1)/(n^2 + 2)]r^3$, n being the index of refraction of the pure solute. If a point electric dipole $\boldsymbol{\mu}$ is located at the center of the solute, the reaction field in a solvent with dielectric constant ϵ is parallel to $\boldsymbol{\mu}$ and its magnitude is:

$$E = \frac{2(\epsilon - 1)(n^2 - 1)}{3(2\epsilon + n^2)} \frac{\mu}{\alpha}, \tag{7.50}$$

which reduces to

$$E \cong \frac{\epsilon - 1}{2\epsilon + 2.5} \frac{\mu}{\alpha}, \tag{7.51}$$

when $n^2 = 2.5$. For acetonitrile in acetone (7.50) gives $E = 0.25 \times 10^6$ esu. With fields of this magnitude, the term in E_z in (7.49) is considerably larger than the E^2 term. Since $E_z = E \cos \phi$, where ϕ is the angle between $\boldsymbol{\mu}$ and the X—H bond, the reaction-field effect may shift the proton resonance either to higher or to lower fields, depending on the geometry. Moreover, $\cos \phi$ may not be the same for all protons within the solute molecule, so that a change in solvent dielectric constant may induce pronounced changes in the spacings between various peaks in the spectrum.

Various shortcomings of (7.50) have been noted. They reflect the fact that the real molecule is not adequately represented by a polarizable sphere with a point dipole at its center and also that the protons are normally at or near the effective surface of the molecule, rather than buried within it. It is not surprising that (7.50) fails for *p*-dinitrobenzene, in which the dipole vanishes by symmetry, but nevertheless each of the strongly polar nitro groups induces a local reaction field which produces an appreciable σ_E for the nearby protons [31]. A more disturbing observation [68] is that when a given compound is examined in a series of solvent mixtures covering a substantial range of ϵ-values, the empirical values of σ_E do not depend linearly on $(\epsilon - 1)/(2\epsilon + 2.5)$. Instead, σ_E is sometimes found [69] to be approximately a linear function of $\epsilon^{1/2}$. No theoretical basis has so far been proposed for this relationship.

Diamagnetic Anisotropy

Like the electric field contribution, the effect of solvent anisotropy is closely analogous to one of the factors that determines the shielding in an isolated molecule, in this case the neighbor anisotropy effect. If the solvent consists of axially symmetric molecules with a magnetic anisotropy $\Delta\chi = \chi_{\parallel} - \chi_{\perp}$, the effect of each solvent molecule on the local magnetic field acting on a solute proton is given by an equation similar to (7.33), that is,

$$H_z' = (-1/3R^3)\,\Delta\chi(1 - 3\cos^2\theta)H_0, \tag{7.52}$$

where **R** is a vector from the solute proton to the center of the solvent molecule and θ is the angle between **R** and the solvent symmetry axis. The total shift contribution from n molecules of solvent at some instant is:

$$(\sigma_a)_{\text{inst}} = \sum_{i=1}^{n} (1/3R_i^3)\,\Delta\chi(1 - 3\cos^2\theta_i). \tag{7.53}$$

Since the molecules are not usually held fixed with respect to one another, the quantities R_i and θ_i are not constant, and the observed σ_a is a time average of values given by (7.53). If both species of molecules are spherical and all values of θ are equally probable, then R_i is nearly constant and the average value of $1 - 3\cos^2\theta_i$ is zero, so that σ_a vanishes. The appearance of a finite σ_a thus requires not only $\Delta\chi \neq 0$ but also some factor that causes certain mutual orientations of solute and solvent to be preferred over others.

Again, the most straightforward way of assuring a nonvanishing effect is to have a specific interaction between solute and solvent. For example, it appears that chloroform and benzene form a hydrogen-bonded complex with the benzene acting as a π-electron donor and the chloroform proton lying above the ring plane so that $\theta \simeq 0$. Then the average value of $1 - 3\cos^2\theta$ is -2, and since $\Delta\chi < 0$ for benzene, a substantial positive σ_a is found. In dilute solutions in benzene the resonance of chloroform is indeed shifted 1.35 ppm to a higher field from its position in cyclohexane solution [70]. This violation of the usual rule that hydrogen bonding shifts the signal of the participating proton to lower fields serves as an excellent illustration of the fact that the magnitude of a downfield "hydrogen-bond shift" should not be used as a measure of hydrogen bond strength unless allowance is made for possible magnetic anisotropy of the electron-donating species [71].

Formation of a definite solute–solvent complex is not the only mechanism that can lead to a finite σ_a. This is proved by the finding that the screening constant of methane in benzene is 0.33 ppm larger than in hexane, although there is no plausible way for these molecules to form a complex [72]. The situation can be rationalized by considering the many possible relative orientations of a methane and a benzene molecule, two of which are suggested by the following sketch.

For configurations similar to (a) the angle θ is nearly zero, while R is relatively small, so that $(\sigma_a)_{inst}$ is positive and moderately large. For configurations close to (b) θ is near 90° and R is larger, so that $(\sigma_a)_{inst}$ is negative but comparatively small. Averaging over all configurations is thus likely to produce a positive σ_a, in agreement with experiment, although it would be hard to make a quantitative calculation of the effect. The fact that for methane in carbon disulfide, which has rod-shaped molecules, σ_a is negative can be understood in a similar way.

Coupling Constants

Solvent effects on coupling constants were recognized later than the effects on chemical shifts, but numerous examples have now been found [60]. It is thus rather unfortunate that the J-values are still called coupling *constants*, but the usage is so firmly established that there seems little prospect that a less misleading term such as coupling parameters will be generally adopted.

The solvent effects on the couplings are less readily interpretable than solvent-induced chemical shift changes. Mathematical recipes for predicting coupling constants, notably those based on Pople and Santry's molecular orbital approach [49], contain parameters reflecting the electron distribution in the solute molecule, expected to be altered by the reaction field in a polarizable solvent. Solvent effects on J-values should then depend smoothly on solvent dielectric constants. Since the reaction field is approximately parallel to the dipole of the solute, the orientation of the dipole with respect to the bonds connecting the coupled atoms should also have a significant effect. Indeed it has been reported [73] that with increasing solvent dielectric constant the positive geminal J_{HF} increases for vinyl fluoride but decreases for trifluoroethylene and that the difference can be attributed to the different alignments of the dipole moment vectors in the two molecules. However, the *trans*-H—F coupling constant in 1-fluoro-1-chloroethylene [69] is not correlated with any simple function of the dielectric constant of the medium. Instead, this parameter and J_{SiF} in SiF_4, seem to depend on the heat of vaporization of the solvent, suggesting that here the effect is attributable to dispersion forces [74]. Hydrogen bonding also appears to affect coupling constants, presumably because of electron shifts brought about by the electrostatic fields involved.

Solvent Effects on Solute Conformation or Self-Association

In the foregoing discussion it has been implicitly assumed that the solute is a rigid molecule. If this is not true, each observed nmr parameter will represent a weighted average of values characteristic of the conformational species present, assuming that there is rapid equilibrium among them (see Section 7). Such equilibria may be strongly disturbed by the solvent [75, 76].

For example, 1,2-disubstituted ethanes exist as mixtures of *gauche* and *trans* forms, with the more polar *gauche* form usually becoming relatively more stable as the solvent dielectric constant increases. Both the chemical shifts and coupling constants then become solvent dependent [77, 78]. Unfortunately, such data can only be used to obtain reliable information about solvent effects on the relative stabilities of conformers if accurate corrections can be made for all the other nmr solvent effects listed above.

Solute-solute interactions may of course cause spectral changes in much the same way as solute-solvent interactions, and then a change in the medium can affect the spectrum indirectly by modifying solute-solute effects. This is especially noticeable and useful when the solute is self-associated, perhaps by hydrogen bonding, or if two mutually associating solutes are dissolved together in an inert solvent. The spectral parameters are then weighted averages of values characteristic of the monomeric, dimeric, and/or higher polymeric species that may be present. Since the concentrations are related by mass action expressions involving solvent-dependent equilibrium constants, the spectra are strongly solvent, concentration, and temperature dependent. In favorable cases, the equilibrium constants can be evaluated from nmr measurements (Section 7).

Solvent Shifts and Spectral Analysis

As brought out in the discussion of spin-spin interactions, complex spectra such as A_nB_m deviate from the spectra predicted by first-order analysis to an extent that depends on the ratio $\Delta\nu/|J_{AB}|$. Since σ_W, σ_a, and σ_E may all differ substantially for protons of types A and B, it is often found that a change in the solvent has a much more pronounced effect on $\Delta\nu$ than on J_{AB} so that the ratio is strongly solvent dependent. In such cases it may be possible to convert a complex spectrum into an essentially first-order one merely by a judicious change of solvent [79]. Moreover, in $A_nB_mX_p$ systems, virtual coupling will be observed when $\nu_0\,|\delta_A - \delta_B|/|J_{AB}|$ is small, and the occurrence of deceptively simple spectra also hinges on the presence of nonequivalent protons with nearly the same chemical shift. These effects can therefore be brought into play gradually and deliberately by a progressive change in solvent composition [80]. Such experiments may facilitate the evaluation of the relative signs of coupling parameters in a complex spectrum.

7 EFFECTS OF EXCHANGE PROCESSES AND QUADRUPOLE RELAXATION

Chemical Exchange

Introduction

Quite early in the history of high-resolution nmr it was found that the spectra of certain pure liquids or mixtures contained fewer peaks than were anticipated on the basis of the structural formulas and the usual predictive rules. Thus the hydroxyl signal from ethanol, which should be split into a triplet by spin-spin coupling with the adjacent methylene protons, usually appears instead as a singlet. The signal from cyclohexane is also a singlet, although one would expect that the axially bonded protons would not have precisely the same shielding constant as the equatorially bonded ones. Later it was found that in both cases the expected, more complex spectra could be observed under suitable circumstances. For ethanol, rigorous purification of the material was required, and for cyclohexane, a reduction of the temperature to about −95°C.

It is now established that this behavior may arise whenever an exchange process is occurring which transfers a proton (H*) from a site at which it has an effective shielding constant σ_A to a new site where its shielding constant is σ_B. Such a process may be intermolecular, as in

$$\mathrm{ROH + R'OH^* \rightleftarrows ROH^* + R'OH} \tag{7.54}$$

or intramolecular, as in

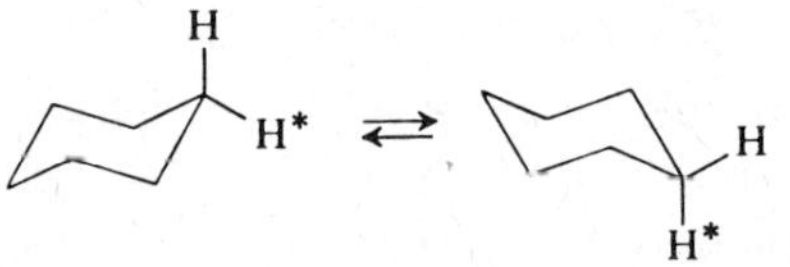

H H* ⇄ H H* (7.55)

Reaction (7.54) involves a transfer of H* from one alcohol molecule to another and is catalyzed by acid or base, while (7.55) requires only a rotation of adjacent methylene groups relative to one another, so that H* is transferred from an equatorial to an axial site without the need for any bonds to be broken. When an exchange mechanism is available, the appearance of the spectrum depends very strongly on the rate of exchange. This can be quantitatively explained on the basis of a rigorous theoretical analysis [81, 82]. A more intuitive approach, which allows the phenomenon to be understood qualitatively without including all the mathematical details, can be developed using the following simple analogy.

Since the transfer of a proton between sites A and B "switches" the resonance frequency between two values, ν_A and ν_B, it is natural to think of a model system consisting of an oscillator that alternately broadcasts one or the other of these frequencies, and a frequency counter with which the output of the oscillator is to be monitored. To obtain a spectrum with the model system, one may count the oscillator output for a period of t seconds. If the number of counts obtained is c, the apparent frequency is $\nu_{\mathrm{exp}} = c/t$. If the process is repeated a large number of times, and $n(\nu)$ is the number of experiments that yields the frequency ν, a graph of $n(\nu)$ versus ν resembles an nmr spectrum and should show peaks in the neighborhood of ν_A and ν_B. For a system in which frequency jumps occur at random times, τ_A and τ_B may be defined as the average lifetimes of periods of oscillations at ν_A and ν_B, respectively, and it is useful to define an average lifetime parameter:

$$\tau = \tau_A \tau_B / (\tau_A + \tau_B). \tag{7.56}$$

The simplest special case, with $\tau_A = \tau_B = 2\tau$, illustrates adequately how the observed spectrum for the model system depends on τ, and shows that it is not possible to resolve the expected two peaks unless exchange is fairly slow.

A useful preliminary step is to examine the response of the model when the oscillator broadcasts only one fixed frequency ν_0. If it is assumed that the frequency count for an interval t is good to the nearest unit, the frequency $\nu_{\text{exp}} = (c \pm 1)/t$ is subject to an error $\Delta\nu$ of the order of $1/t$. The plot of $n(\nu)$ versus ν then shows a single peak centered at ν_0 with a width of about $1/t$. It may be noted in passing that this amounts to a crude derivation of one form of Heisenberg's uncertainty principle, since if the symbol t is replaced by Δt, one has $\Delta\nu\, \Delta t \simeq 1$ or, multiplying through by Planck's constant, $\Delta h\nu\, \Delta t = \Delta E\, \Delta t \simeq h$.

Turning now to the experiment in which the oscillator frequency is changing, it is easy to formulate the conditions under which a spectrum of two sharp lines can be obtained. If the width of each peak is to be much smaller than the separation between them, the counting interval t must be large compared to $1/|\nu_A - \nu_B|$. However, in order to minimize the risk that the frequency will change while counting is in progress, t must be small compared with τ. Thus if τ is very much larger than $1/|\nu_A - \nu_B|$, $n(\nu)$ will have peaks at ν_A and ν_B, and each of these can be made as sharp as one wishes by using long counting periods.

The first effect of reducing τ is to broaden the signals by making it necessary to reduce t. Eventually, when τ reaches a value comparable with $1/|\nu_A - \nu_B|$, it becomes impossible to observe a two-line spectrum. If the counting time is kept small compared to τ, the width of each peak will become larger than their frequency difference. If longer counting intervals are used, the frequency is likely to change one or more times during the run, and a value intermediate between ν_A and ν_B will be observed. With the best compromise value of t, the spectrum is severely broadened, and whether it shows two distinct maxima or only one depends on the precise value of τ.

If τ is still further reduced, and t is kept fixed at some convenient value, the frequency will change more and more often during each counting experiment, and in the limit of very short lifetimes ν_{exp} will always be very nearly

$\frac{1}{2}(\nu_A + \nu_B)$, deviating from this value by much less than $|\nu_A - \nu_B|$. The result is once more a sharp-line spectrum with a single peak at the mean frequency.

The changes in the nmr spectrum of a proton transferred between sites A and B at a variable rate closely parallel the behavior just described. When $\tau \gg 1/|\nu_A - \nu_B|$, the spectrum consists of two sharp lines and exchange is said to be "slow." As τ is gradually diminished, perhaps by raising the temperature or adding a trace of a catalyst, each line broadens somewhat, and the two maxima begin to draw closer together. For $2\tau \cong 1|/\nu_A - \nu_B|$ the rate of exchange is said to be "intermediate" and the spectrum is drastically broadened. Further reducing τ leads to "fast" exchange and a spectrum consisting of a single line at $(\nu_A + \nu_B)/2$, which becomes progressively sharper until it reaches a minimal width depending on other factors such as the field homogeneity.

In observing these effects experimentally, several special cases may be identified differing from one another with respect to the physical origin of the difference in effective chemical shift between A and B sites. In (7.55) an axial proton is in an environment sufficiently different from that of an equatorial proton to make one expect unequal shielding constants. Fast exchange then reduces the number of distinct chemically shifted peaks observed in the spectrum. In (7.54), however, the true shielding constant of the proton is not affected when it is transferred from ROH to R′OH. However, the OH resonance is normally a triplet (for R = ethyl) because at a given instant the molecules in a large population can be divided into three sets according to the spin-states of the methylene protons, which may be $\alpha\alpha$, $\alpha\beta \pm \beta\alpha$, or $\beta\beta$. Each of these species produces one component of the OH triplet and may be considered to have its own effective chemical shift. Whenever R and R′ belong to different sets, the transfer ROH → R′OH does involve a change in proton frequency. Fast exchange in this case causes collapse of the OH multiplet into a single line without affecting the average hydroxyl chemical shift.

The improvement of spectrometer resolution by spinning the samples exemplifies still another kind of exchange phenomenon. A typical molecule in a spinning sample is carried around a circular path, and if the magnetic field is not perfectly homogeneous its resonance frequency changes as it moves. If the maximum spread among the local resonance frequencies is $\Delta\nu$ (usually no larger than a few hertz) the line width without spinning is also about $\Delta\nu$. When the spinning frequency is larger than $\Delta\nu$, fast exchange among all sites along the trajectory is realized and the width of the signal is greatly reduced.

The exact theoretical treatment of the effect of exchange on nmr spectra yields results very similar to those obtained with the nonrigorous approach [81]. If it is assumed that the line widths in the absence of exchange are much less than $|\nu_A - \nu_B|$, the absorption intensity for a given value of τ is found to be proportional to the function:

$$g(\nu) = \frac{\tau(\nu_A - \nu_B)^2}{[\frac{1}{2}(\nu_A + \nu_B) - \nu]^2 + 4\pi^2\tau^2(\nu_A - \nu)^2(\nu_B - \nu)^2}. \tag{7.57}$$

This describes a single peak when $\tau\,|\nu_A - \nu_B| \ll 1$ and a pair of peaks when $\tau\,|\nu_A - \nu_B| \gg 1$. The value of 2τ at which the minimum separating the two peaks just begins to appear is $\sqrt{2}/\pi\,|\nu_A - \nu_B|$.

Rate Measurements

A number of implications of (7.57) have been used as quantitative indications of exchange lifetimes. When exchange is slow enough to allow two peaks to be observed, each one is broadened (see Section 9) to an extent that depends on τ, and the apparent separation is smaller than it would be in the absence of exchange. At slightly higher rates the ratio of the signal amplitude at the maxima to that at the intervening minimum can be used to evaluate τ. When the rate is barely fast enough to coalesce the two peaks, τ has the value given in the preceding paragraph. For still faster exchange, the width of the coalesced signal reflects changes in the rate. Probably the best procedure is to prepare a series of graphs of $g(\nu)$ for various assumed values of τ and find the actual value by finding the member of this series that most closely resembles the observed spectrum.

More general mathematical treatments have been developed [82] that allow this method to be extended to cases in which $\tau_A \neq \tau_B$, in which there are more than two exchanging species, or in which the individual peaks are split by spin-spin interactions. The range of rates in which accurate results can be obtained depends on the resolving power of the spectrometer, on the value of $|\nu_A - \nu_B|$, and possibly on complicating factors such as contributions to the line shapes from unresolved fine structure. A rough rule of thumb is that the rate must be neither very fast nor very slow, that is, $0.1 \lessdot \tau\,|\nu_A - \nu_B| \lessdot 2$. The rate can often be brought into this range by an appropriate adjustment of the temperature. Moreover, the τ-values are simply related to the rate constants, for example, for

$$A \underset{k_B}{\overset{k_A}{\rightleftarrows}} B, \qquad \tau_A = 1/k_A \qquad \text{and} \qquad \tau_B = 1/k_B. \tag{7.58}$$

Thus a series of measurements at different temperatures makes it possible to construct a plot of $\ln k$ against $1/T$ from which the activation energy for exchange can be determined. The procedure was first applied to the study of

restricted rotation of amides [83, 84]

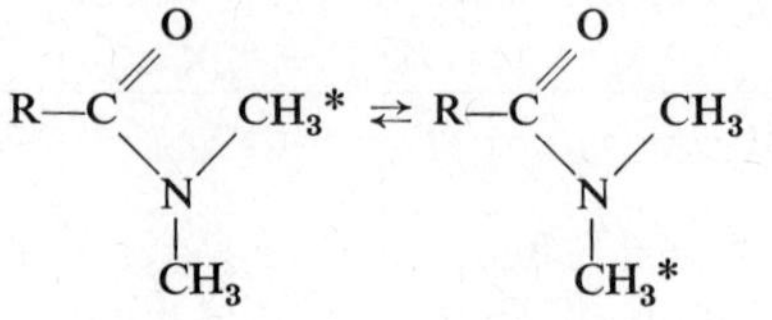

$$\mathrm{R{-}C({=}O){-}N(CH_3)(CH_3^*)} \rightleftarrows \mathrm{R{-}C({=}O){-}N(CH_3^*)(CH_3)} \tag{7.59}$$

When the spectrum is complicated, a modification of the experimental technique is often preferable to a direct theoretical attack. The inversion of cyclohexane [see (7.55)] has attracted a great deal of interest but is difficult to study because of extensive unresolved spin-spin couplings which appear when the conformation is fixed by lowering the temperature. Accordingly, the best measurements of the inversion rate by the high-resolution method are those made with the deuterated species $C_6D_{11}H$ and using double irradiation (Section 8) at the deuterium resonance frequency to eliminate spin coupling between the H and D nuclei [85, 86].

In some cases pulsed nmr experiments (Section 15) can be used to obtain data over a larger range of temperatures than that which can be covered with the high-resolution technique, and the activation parameters can then be evaluated more accurately.

Equilibrium Measurements

For a reaction of type (7.58) involving materials differing in stability, the lifetimes are not equal. In the limit of fast exchange, the spectrum again consists of a single signal at a frequency:

$$\nu = p_A \nu_A + p_B \nu_B, \tag{7.60}$$

where p_A and p_B are the fractional populations of protons at sites A and B, which obey the relations:

$$p_A = \tau_A/(\tau_A + \tau_B), \quad p_B = \tau_B/(\tau_A + \tau_B). \tag{7.61}$$

When there are more than two exchanging sites, (7.60) for fast exchange is replaced by:

$$\nu = p_A \nu_A + p_B \nu_B + p_C \nu_C + \cdots. \tag{7.62}$$

Again the signal position depends directly on the concentrations of the species present at equilibrium.

Equations (7.60) and (7.62) have been used to measure equilibrium constants in a wide variety of reactions. A prominent class of examples consists of conformational equilibria [87, 88]

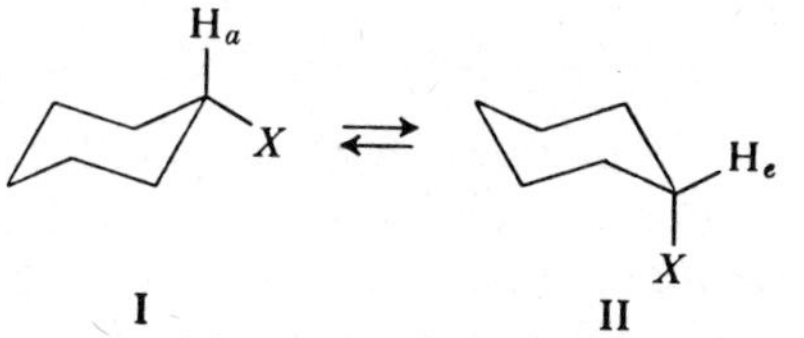

$$\text{I} \rightleftarrows \text{II} \tag{7.63}$$

When X is an electronegative substitutent, the resonance of the CHX proton can readily be singled out. The observed chemical shift is then intermediate between δ_a and δ_e, the shifts for forms I and II, respectively, and equal to:

$$\delta = f_{\text{I}}\delta_a + f_{\text{II}}\delta_e = (\delta_a + K\delta_e)/(1 + K). \tag{7.64}$$

Here f_i stands for the fraction of the material present as form i, and $K = f_{\text{II}}/f_{\text{I}}$. In practice there is often some uncertainty associated with the evaluation of δ_a and δ_e. When values are taken from model compounds of fixed conformation such as

for δ_a, it must be assumed that the *tert*-butyl group does not change the value of δ_a. If the shifts are evaluated by cooling the equilibrium mixture until the exchange becomes slow, one must assume that δ_a and δ_e are each independent of the temperature. Neither of these assumptions is exactly valid, although both appear to be good enough to yield results in approximate agreement with those obtained by other methods.

Equations (7.60) and (7.62) have also been very widely used in studying tautomeric equilibria, protonation equilibria, and intermolecular association, especially the formation of hydrogen-bonded complexes [89]. This is illustrated by the dimerization of carboxylic acids in inert solvents,

$$2\text{RCOOH} \rightleftarrows (\text{RCOOH})_2, \qquad K = x_D/x_M^2, \tag{7.65}$$

where x_D and x_M represent the mole fractions of acid dimers and acid monomers in solution, respectively, and the shift of the carboxyl proton is

$$\delta = (x_M\delta_M + 2x_D\delta_D)/(x_M + 2x_D). \tag{7.66}$$

The reason that x_D is multiplied by 2 where it occurs in (7.66) is of course that each molecule of dimer contains two carboxyl protons. By measuring δ at several concentrations, one obtains enough data to evaluate the key unknowns δ_M, δ_D, and K. Extreme caution must be used to eliminate all traces of certain impurities, such as water, which participate in the proton-exchange equilibrium. For a moist sample,

$$\delta = (x_M\delta_M + 2x_D\delta_D + 2x_{H_2O}\delta_{H_2O})/(x_M + 2x_D + 2x_{H_2O}) \quad (7.67)$$

and misleading results will be obtained, especially for solutions very dilute in acid, unless a correction is made for the water contribution [90, 91].

In any of these exchanging systems, it is likely that δ will be markedly temperature dependent since the populations or concentrations depend on an equilibrium constant which varies with temperature according to the familiar equation:

$$\partial \ln K/\partial T = \Delta H^0/RT^2. \quad (7.68)$$

It thus becomes possible to use the nmr data to calculate enthalpies of reaction. This is simple when the individual shifts of the participating species are independent of temperature. However, small temperature coefficients, $\Delta\delta/\Delta T$, are often found even for nonexchanging species and, in some cases, notably for hydrogen-bonded complexes, $\Delta\delta/\Delta T$ may be as large as 10^{-2} ppm/°C even after correction for the effects of dissociation of the complex [92]. With some reinterpretation (7.62) may provide the basis for an explanation of these temperature effects. If the molecule or complex has several low-lying vibrational levels which are appreciably populated, and if the chemical shift depends on the degree of vibrational excitation, then the sample may be regarded as a mixture of vibrational species with:

$$\delta = \sum_n \delta_n f_n, \quad (7.69)$$

where f_n is the fractional population of the nth level, with shift δ_n, and the temperature dependence is brought in through the relation:

$$f_n = \exp(-nh\nu/kT)/[1 - \exp(-h\nu/kT)] \quad (7.70)$$

if the vibrations are harmonic, or an appropriate analogous relation if they are anharmonic [92].

Coupling Constants

In many equilibrating systems the spin-spin coupling constants obey equations analogous to (7.60) or (7.62). For example, compounds such as 1,1,2,2-tetrachlorofluoroethane, $CHCl_2$—$CFCl_2$, exist as mixtures of two conformers with the hydrogen and fluorine atom either in a *gauche* or *trans* relationship. When these are rapidly interconverting, the average coupling constant is given in an obvious notation, by:

$$J_{HF} = J_{HF}(g)f_g + J_{HF}(t)f_t. \quad (7.71)$$

Thus the coupling constants and their temperature dependences can furnish information similar to that obtained from chemical shift data as just described [78].

Quadrupole Relaxation Effects

An understanding of exchange effects also makes it possible to account for the fact, briefly noted above, that spin-spin coupling between a proton and a nucleus with $I \geq 1$ may split the proton signal into a multiplet, broaden it, or leave it unaltered, depending on the particular example chosen. The fragment ^{14}N—H serves as an illustration [93]. If the interaction of the nitrogen nuclear magnetic moment with the external field were the only factor to be considered, the proton resonance would be expected to appeaı as a triplet because the effective field at the proton should depend on the state of alignment of the nitrogen nucleus, which has $I = 1$ and $M_I = 1, 0,$ or -1. This is indeed observed for NH_4^+, in which the electronic environment of the nitrogen nucleus is so symmetrical that other interactions are relatively unimportant.

If the electric field at the nitrogen nucleus is nonuniform, it interacts with the nuclear electric quadrupole moment, giving rise to a torque tending to align the nuclear spin axis with respect to a direction fixed in the molecule and, since the molecule is tumbling, not fixed relative to the external magnetic field. This force then causes transitions between states of different M_I. Since the proton frequency changes abruptly when such a transition occurs, the effect is very similar to a proton-exchange process even though the N—H bond is not broken. The apparent rate of exchange depends on the lifetime of a state with a given M_I, which is governed by the magnitude of the nuclear quadrupole moment, the gradient of the electric field at the nitrogen nucleus, and the rate of tumbling of the molecule. The latter in turn depends on the sample temperature, viscosity, and the nature of the intermolecular forces. A broadened multiplet, such as is found in the proton spectra of amides, occurs when the lifetime is in the intermediate range, in this case $\tau \cong 1/J_{NH}$. Absence of observable multiplet splitting means that the lifetimes are much shorter. Somewhat surprisingly, the effectiveness of a fluctuating torque in inducing changes in M_I depends on its rate of fluctuation in such a way that increases in temperature cause an increase rather than a decrease in τ [93a].

The same kind of reasoning can be used to explain thc observations, mentioned in Section 5, that protons that interact with deuterons produce multiplets with sharp components, protons that interact with boron nuclei produce multiplets with relatively broad components, and protons that interact with chlorine, bromine, or iodine nuclei produce sharp singlets. Evidently, the rather small quadrupole moment of the deuteron permits fairly long lifetimes for states of fixed M_I. The quadrupole moments of the boron nuclei are considerably larger than that of the deuteron, but for many compounds the field gradient at the boron nucleus is too small to produce fast exchange. For the halogens the field gradients appear to be much larger, so that the proton multiplet is completely coalesced.

Finally, it may be noted that chemical exchange and quadrupole effects may occur simultaneously. Thus quadrupole relaxation in pure liquid ammonia proceeds at such a rate that the proton resonance is a broadened triplet [94], but a trace of water suffices to catalyze fast chemical exchange and causes the signal to collapse to a sharp singlet. Progressive drying of a water-contaminated sample leads to peaks with intermediate shapes depending in a complex way on the rates of both of the time-dependent processes.

8 DOUBLE RESONANCE

Spin Decoupling

The Decoupling Process

A number of workers have explored the effect of subjecting materials simultaneously to oscillating magnetic fields at two different radiofrequencies [95–97]. These *double-irradiation* experiments have proved to be so instructive that most modern high-resolution spectrometers are equipped for this kind of work. Experiments involving more than two frequencies have also been reported [96] but are not discussed here.

To be suitable for double resonance a compound must contain at least two sets of nuclei with different resonant frequencies and with a nonzero spin-spin coupling constant. *Homonuclear* double resonance involves nuclei of the same isotopic species with their resonant frequencies separated because of a difference in chemical shifts. *Heteronuclear* double resonance involves nuclei of different isotopic species, so that the frequency difference is very much larger, and somewhat different apparatus is required. The theory of the effect is, however, essentially the same for both cases.

For the simplest double-irradiation experiment, one might choose a molecule with an AX spectrum and record the signal from the A-species using a weak *observing* field $\mathbf{H}_1$ at a frequency $\nu_A = \gamma_A H_0/2\pi$, at the same time exposing the sample to a much stronger *perturbing* field $\mathbf{H}_2$ at $\nu_X = \gamma_X H_0/2\pi$. The perturbing field causes transitions between the spin states of the X nuclei, and the average lifetime of these states becomes shorter as $\mathbf{H}_2$ increases in amplitude. The use of a strong perturbing field thus creates a situation superficially similar to one in which X is involved in fast chemical exchange or quadrupole relaxation, that is, one expects the A-doublet to collapse to give a single line. This disappearance of multiplet structure is in fact observed, and the phenomenon is aptly described as *decoupling* of the nuclear spins.

The apparent analogy between spin decoupling and exchange phenomena can be seriously misleading if one tries to use it to predict the shape of the spectrum when $\mathbf{H}_2$ is not strong enough to produce complete decoupling. The analogy fails because of a fundamental dissimilarity between double resonance and the other processes that cause collapse of spin multiplets. The latter involve forces that vary with time and have no systematic relationship with the external magnetic field or the individual spin magnetic moments. In contrast with this, the perturbing field in the double-resonance experiment

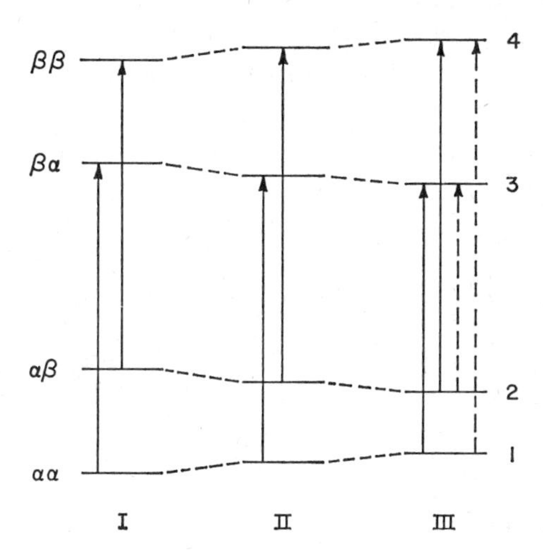

Fig. 7.5 Schematic representation of the energy levels of a two-spin system with $I_A = I_X = \frac{1}{2}$, and transitions involving nucleus A. *I*, unperturbed levels; *II*, effect of spin-spin coupling; *III*, effect of second rf field.

may be decomposed into two vectors rotating in opposite senses in a plane perpendicular to the external field, and its frequency is so chosen that one can find a rotating coordinate system in which $\mathbf{H}_0$, one component of $\mathbf{H}_2$, and the magnetic vector representing the X nuclei are all stationary. Under these circumstances it can be shown that the nuclear spin energy levels are somewhat perturbed by $\mathbf{H}_2$ and that the selection rules are changed, as shown schematically in Fig. 7.5.

On the left, the four spin states of an AX system appear with no perturbing field. The two allowed transitions involving the A-species are indicated by arrows; both occur at the same frequency ν_A. In the center the effect of introducing spin-spin coupling is shown, with $J_{AX} > 0$. The A-resonance is now a doublet with frequencies $\nu_A \pm J_{AX}/2$. The right part of the figure shows that when $\mathbf{H}_2$ is applied the $\alpha\alpha$ and $\beta\beta$ levels are farther raised and the $\alpha\beta$ and $\beta\alpha$ levels lowered so that the previously observed transitions now occur at $\nu_A \pm J'/2$, with:

$$J' = (J_{AX}^2 + 4\gamma_X{}^2H_2{}^2)^{1/2}. \tag{7.72}$$

The selection rules are changed as follows. The two lines just described grow weaker as $\mathbf{H}_2$ increases, and their intensities remain proportional to the ratio J_{AB}^2/J'^2. The normally forbidden transitions $\alpha\alpha \rightarrow \beta\beta$ and $\alpha\beta \rightarrow \beta\alpha$, shown as dashed arrows, become allowed. Each of these occurs by a two-quantum process. For $\alpha\alpha \rightarrow \beta\beta$, a quantum is absorbed simultaneously at two frequencies, one of which is ν_X. By referring to the figure it is easily seen that the other frequency ν', which is equal to the observing frequency when the process occurs, must satisfy:

$$\Delta E = h(\nu' + \nu_X) = h(\nu_A + \nu_X), \tag{7.73}$$

so that $\nu' = \nu_A$. For the transition $\alpha\beta \rightarrow \beta\alpha$, a quantum at ν_X is emitted while one at ν' is absorbed. It follows that

$$\Delta E = h(\nu' - \nu_X) = h(\nu_A - \nu_X), \tag{7.74}$$

so that again $\nu' = \nu_A$. Thus just one new line is observed at the frequency ν_A where the resonance of the A-species would occur if there were no spin-spin coupling.

To summarize, as $\mathbf{H}_2$ increases, the original doublet components move farther apart and lose intensity while a new signal grows in the center of the doublet [98]. When $\mathbf{H}_2$ has become large enough to make $4\gamma_X{}^2H_2{}^2 \gg J_{AX}^2$, only the central component is visible, and the spins are completely decoupled. The progressive changes in signal width characteristic of an exchange process are not observed.

A slight modification of the experiment just described is to keep the magnitude of $\mathbf{H}_2$ fixed at a fairly high level and allow ν_2, the frequency of $\mathbf{H}_2$, to vary over a range of values near ν_X. When ν_2 and ν_X differ only slightly, the preceding discussion remains essentially valid, but the two transitions near the center of the A-doublet no longer coincide exactly. For intermediate values of $\mathbf{H}_2$, four lines are then observed, two of which remain when $\mathbf{H}_2$ is made very large [98].

The theory of double resonance has also been developed for more elaborate spectra, including AX_n, ABX_n, and other cases. If the perturbing field is weak (spin tickling) and its frequency is not exactly equal to ν_X, there are more signals than in an ordinary spectrum, but if $\nu_2 = \nu_X$ and $\mathbf{H}_2$ is large the spectrum of the A-species (or of A and B) will become identical with the much simpler pattern that would be expected if J_{AX} (and J_{BX}) were reduced to zero. The following examples illustrate the practical advantages that ensue.

Simplification of Spectra

The most obvious benefit of using heteronuclear spin decoupling is that merely reducing the number of peaks in a spectrum may make it much easier to interpret. The method is not restricted to molecules in which both nuclear species have $I = \frac{1}{2}$, and it can be particularly useful when the second nucleus is one of those quadrupolar species that ordinarily causes the proton resonance to appear as a very broad, incompletely resolved multiplet. One of the best known systems of this kind is formamide, for which it is not possible to measure proton chemical shifts or coupling constants unless the ^{14}N spins are decoupled [99].

Collapsing a multiplet into a single peak always makes the resonance easier to detect and offers an advantage when working with very small samples, very dilute solutions, or intrinsically weak spectra such as those from ^{13}C in its low natural abundance. In the latter case an additional gain in signal strength arises from a nuclear *Overhauser* effect [99a]. This is, in essence, the result of changes in the relative populations of the nuclear energy levels, brought about by the second radiofrequency field, which enhance the intensity of the ^{13}C transitions [96].

Homonuclear double resonance can also be used to simplify a complex spectrum and make it easier to discover how many different groups of protons are actually present. If it is found that a particular multiplet collapses when some other multiplet is irradiated, this is unambiguous proof that the protons responsible for the two multiplets are spin-coupled to each other. Such information may make it possible to decide which of several suggested structures is the correct one for an unknown material.

Indirect Determination of Chemical Shifts

A most interesting application of spin decoupling is the determination of the chemical shift of the X-species using observations of the A-multiplet. Such a possibility becomes important when the X-resonance cannot be observed directly, perhaps because (when X is not a hydrogen nucleus) an appropriate radiofrequency unit is not at hand or (when X is a proton) because overlapping signals obscure the X-multiplet. A simple hypothetical example is the indirect determination of the difference between the fluorine chemical shifts of CH_3F and CH_2F_2 using the proton resonance spectrum with heteronuclear double irradiation.

First, with ν_1 set at the center of the methyl doublet of CH_3F, one adjusts ν_2 for optimum decoupling by maximizing the signal obtained at ν_1. Then, with ν_1 set at the center of the methylene triplet of CH_2F_2, the procedure is repeated. For each experiment the difference $|\nu_2 - \nu_1|$ can be measured with an electronic counter. The desired fluorine shift difference, in frequency units, is then calculated by means of the following scheme:

$$(\nu_F)_{CH_3F} = (\nu_1)_{CH_3F} + (\nu_2 - \nu_1)_{CH_3F} \tag{7.75}$$

$$(\nu_F)_{CH_2F_2} = (\nu_1)_{CH_2F_2} + (\nu_2 - \nu_1)_{CH_2F_2}. \tag{7.76}$$

Subtracting (7.76) from (7.75) shows that the difference between the fluorine frequencies is equal to the difference between the proton frequencies, readily obtained from the proton spectra, plus the difference in the two values of $(\nu_2 - \nu_1)$. The fact that A and X in this example are different isotopes assures that $|\nu_A - \nu_X| \gg \gamma_X H_2$. When an analogous procedure is used to find a "hidden" proton shift in a proton spectrum, this inequality may not hold. Then the shift obtained is subject to a correction [100] amounting to $(\gamma_X H_2)^2/4\pi\, |\nu_A - \nu_X|$.

Signs of Coupling Constants

When a molecule contains at least three nonequivalent nuclei (APX-case), double irradiation can be used to determine whether J_{PX} and J_{AP} have like or unlike signs. This is most readily explained with reference to a specific example, such as the molecule

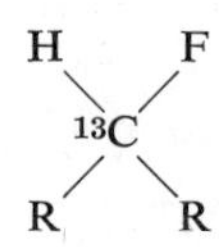

where the R represents a magnetically inactive substituent. The eight spin states of this three-spin system are shown schematically in Fig. 7.6. The states are labeled by giving the M-quantum numbers of the H, ^{13}C, and F nuclei, each of which can take on the values $\pm\frac{1}{2}$. For each level the interaction energy attributable to the spin-spin coupling is indicated. In accordance with known properties of molecules of this type, it is assumed that $|J_{CF}| > |J_{CH}| > |J_{HF}|$ and that $J_{CH} > 0$. The sign of J_{CF} can then be determined by observing the fluorine nmr spectrum using a perturbing field in the proton resonance region, as follows.

Since J_{CH} is positive, the higher-frequency lines in the proton spectrum arise from the transitions labeled ν_2 and ν_4. Let the perturbing frequency be adjusted until it is $(\nu_2 + \nu_4)/2$, so that these two transitions tend to become saturated while ν_1 and ν_3 are approximately unaffected. It is apparent from the figure that the molecules that are perturbed are those that find themselves in levels number 2, 5, 6, and 8 since these are the levels involved in the irradiated transitions. The fluorine signals that arise from these particular molecules correspond to transitions ν_6 and ν_8. These two transitions in the fluorine spectrum therefore show a decoupling effect, while ν_5 and ν_7 remain unchanged, hence the peaks from ν_6 and ν_8 can be identified. If J_{CF} is positive,

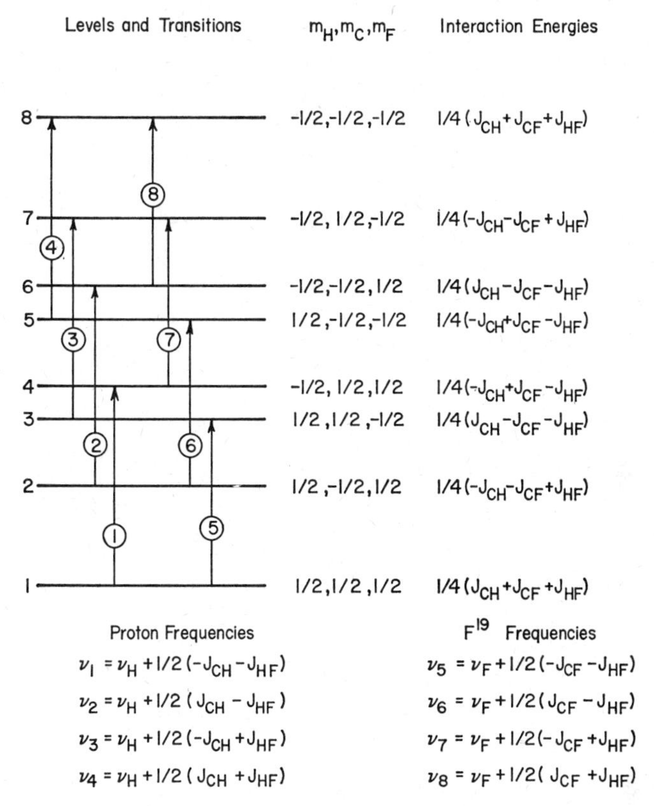

Fig. 7.6 Schematic representation of energy levels and transition for a three-spin system H—C—F. For simplicity, only the transitions involving the H or F nuclei are shown.

they will be the high-frequency members of the fluorine quartet; otherwise they will be the low-frequency components. A similar experiment involving irradiation of some of the ^{13}C transitions, omitted from the diagram for the sake of clarity, would give the relative signs of J_{CH} and J_{HF}.

Analogous homonuclear double-resonance experiments may be used to reveal relative signs of proton-proton coupling constants. However, in a complex spectrum not all the chemical shifts may satisfy inequalities of the type $\nu_0 |\delta_A - \delta_X| \gg |J_{AX}|$, and the details of the analysis may be less straightforward. Freeman and Anderson [101] have developed a procedure in which "weak second r.f. fields" are used at frequencies coinciding precisely with one of the lines in the ordinary proton spectrum. Only those transitions that have a level in common with the transition that is being perturbed are affected, and these are split into doublets. For example, a three-proton *APX* system would have a set of energy levels formally identical with those shown in Fig. 7.6, all of the transitions now being components of the proton spectrum. Irradiation at ν_1 with a weak second field perturbs those molecules that are in levels 1 and 4 and affects the other transitions in which these molecules can participate, that is, ν_5 and ν_7 and the two transitions $1 \leftrightarrow 2$ and $4 \leftrightarrow 6$ which are not shown in the diagram. Thus again the double-resonance experiment yields the relative signs of the J-values by making it possible to determine which transition corresponds to each of the observed lines in the spectrum.

Other Applications

It was mentioned in Section 8, p. 651, that the second radiofrequency field may alter the steady-state populations of some of the nuclear energy levels and consequently change the relative intensities of some of the lines. This *generalized Overhauser effect* [96] is most likely to be important if the second field is strong and the molecules contain some nuclei with long relaxation times. To understand the operation of this effect in detail requires a rather intricate discussion, but for the simplest case, that of a two-spin system, a simplified presentation is possible [101a].

The energy levels for two nuclei with $I = \frac{1}{2}$ are shown in Fig. 7.5. To observe the Overhauser effect, one would use a strong radiofrequency field to saturate the transitions $1 \rightarrow 3$ and $2 \rightarrow 4$ and a weak field to examine the transitions $1 \rightarrow 2$ and $3 \rightarrow 4$. The saturating radiation causes the populations p_1 and p_3 to become equal, as do p_2 and p_4. Let it be supposed that a relaxation mechanism exists that assures that the populations of levels 1 and 4 are kept in thermal equilibrium with one another, that is, $p_4/p_1 = e^{-(E_4-E_1)/kT}$. The populations p_1 and p_2 would normally be governed by a similar relation, $p_2/p_1 = e^{-(E_2-E_1)/kT}$, but this is now incompatible with the enforced conditions $p_1 = p_3$ and $p_2 = p_4$. Instead, the ratio p_2/p_1 becomes approximately equal to $e^{-(E_4-E_1)/kT}$, that is, the population difference is abnormally large, and the transition $1 \rightarrow 2$ has an enhanced intensity. The transition $3 \rightarrow 4$ is similarly enhanced.

For systems with more than two spins, a very elaborate analysis is required. The effect yields information about relaxation mechanisms and may be useful in the interpretation of complex spectra [101b, 101c].

An ingenious double-resonance technique based on a *localized saturation effect* makes it possible to measure the frequency separation between appropriately selected nmr signals with remarkably high precision (± 0.001 Hz) [101d]. In a slightly inhomogeneous magnetic field, not all the molecules have precisely identical transition frequencies. A weak second radiofrequency field, which would normally cause an intensity change in some observed transition, therefore affects only those molecules for which the local field satisfies the resonance condition at frequency ν_2. Accordingly, the loss of intensity in the observed transition (at ν_1) is confined to the contribution of these particular molecules, and the signal suffers a change in appearance which has been vividly described as "burning a hole" in the line. The observed shape is very sensitive to change in the frequency ν_2. The experiment requires extremely precise field-frequency regulation.

9 RELAXATION TIMES AND LINE WIDTHS IN HIGH-RESOLUTION SPECTRA

Systems without Chemical Exchange

In the absence of special effects, such as intermediate-rate exchange processes, the peaks in most high-resolution spectra are so narrow that it is customary to describe spectra by listing only chemical shift values, coupling constants, and peak intensities. Such a description is not in principle complete, however, because it ignores the longitudinal and transverse relaxation times (T_1 and T_2, see Section 2) associated with each resonance. It is found that these relaxation times may vary not only from sample to sample but sometimes even for different sets of protons in the same molecule, suggesting that they might yield chemically interesting information not otherwise accessible. This is especially true when the spectrum is that of a paramagnetic species or when a paramagnetic cosolute is present, as is shown later. This section deals with the measurement and interpretation of relaxation times in diamagnetic samples.

Both T_1 and T_2 are related to the width of the resonance signal, but in somewhat different ways. T_1 is a measure of the time required to establish thermal equilibrium among the nuclear spin states after the main field is turned on, hence it is decreased by any factor that increases the probability of transitions among the spin states. A short T_1 is thus associated with a short average lifetime for a given spin state, and the corresponding change in line width can be regarded as a direct consequence of the Heisenberg uncertainty principle, $\Delta E\,\Delta t \gtrsim h$, if Δt is set equal to T_1 and ΔE is replaced by $h\,\Delta\nu$ so that:

$$\Delta\nu \simeq 1/T_1. \tag{7.77}$$

When other line-broadening effects are negligible, the line width is indeed governed by the rate of spin-lattice relaxation (T_1). However, as shown below, several possible mechanisms may cause the line width to exceed $1/T_1$. As already mentioned, T_2 is defined in terms of the observed line width by:

$$T_2 = 1/\pi\Delta\nu, \tag{7.78}$$

implying that as long as spin-lattice relaxation alone determines the width, $T_1 \simeq T_2$, while the existence of other causes of line broadening may lead to an inequality $T_1 > T_2$.

The interrelation between the factors that determine T_1 and T_2 may be illustrated by considering the effect of magnetic dipole-dipole interactions between nuclei in neighboring molecules in a hypothetical sample for which the viscosity can be adjusted to any desired value. When the sample is very viscous, the unordered arrangement of the molecules causes the field acting on each nucleus to be different, depending on the local environment, but the local field values change only very slowly with time. The local variations of the field may be increased by raising the concentration of magnetic nuclei or by adding a species with a larger magnetic dipole. Such changes increase the line width, that is, decrease T_2, without any major concomitant change in T_1 because changing the static magnetic field at any one nucleus does not affect the average lifetime of its spin states. It is to be expected then that for a very viscous sample T_2 may be very substantially less than T_1.

If the viscosity is now gradually reduced, the rotational and translational Brownian motion of the molecules will cause the local fields to change with time at an increasingly rapid rate. The effective resonant frequency of each nucleus consequently fluctuates about some average value, and as the fluctuations become more and more rapid the line broadening resulting from the differences in the local fields tends to be averaged out. Decreasing the viscosity therefore tends to sharpen the resonance signal, that is, to increase T_2.

The dependence of T_1 on the rate of molecular motions is not quite as simple [102]. Random fluctuations of the local field may be treated as a superposition of periodic changes with frequencies spread out over quite a wide band. As long as the viscosity is high, this band contains almost exclusively frequencies that are much lower than the nuclear resonance frequency. As the speed of random motions is increased, the range of frequencies representing the local field fluctuations is shifted toward higher frequencies, and it eventually overlaps the resonant frequency. When this occurs, the fluctuations are able to induce the transitions $\alpha \leftrightarrow \beta$ in just the same way as an external electromagnetic field of appropriate frequency, so that the lifetime of each nuclear spin state is shortened and T_1 is reduced. If the molecular motions are still further accelerated, the fluctuation frequencies increase until eventually they are several orders of magnitude higher than the Larmor precession frequency. Then they no longer provide an efficient mechanism for nuclear relaxation, so that T_1 once more increases.

It is useful to define a correlation time τ_c, the approximate time between significant changes in the local field strength. For example, if the changes are mainly attributed to molecular rotations, τ_c will be a rotational correlation time, the time required for a molecule to turn through an angle of 1 radian. The minimal value of T_1 occurs when the local field fluctuations are most effective in causing transitions between the nuclear spin states, that is [102, 103], when τ_c is equal to $1/(2\sqrt{2}\pi\nu_0)$.

To summarize, both T_1 and T_2 depend on the magnitude of the local field variations and also on the rate of change of the local fields. As long as the viscosity is high, T_1 is much larger than T_2, and reduction in the viscosity causes T_1 to decrease and T_2 to increase. When the correlation time is less than $1/2\pi\nu_0$, the two relaxation times are equal, and any further reduction in viscosity causes them to increase together, so that the resonance signals become progressively narrower. This behavior is shown schematically in Fig. 7.7, which is based on a detailed mathematical analysis of the processes just described [102].

The preceding discussion can be generalized to allow for a number of line-broadening mechanisms besides intermolecular dipole-dipole interactions. For molecules of some complexity, *intramolecular* dipole-dipole forces obviously give rise to local fields which depend on the orientation of the molecule with respect to the external field and which fluctuate if the molecule is allowed to rotate. The relaxation times then depend on the viscosity as already described, even at high dilution in a magnetically inert solvent. Since molecular motions become more rapid as the temperature rises, the effects of heating on relaxation times parallel those of reducing the viscosity.

Another source of local field variations is overall inhomogeneity of the applied magnetic field. Such variations are not averaged out by the molecular motions, and they may lead to an effective T_2 which is approximately independent of the viscosity, smaller than the intrinsic T_2, hence smaller than T_1, even when it is expected that the two relaxation times should be equal.

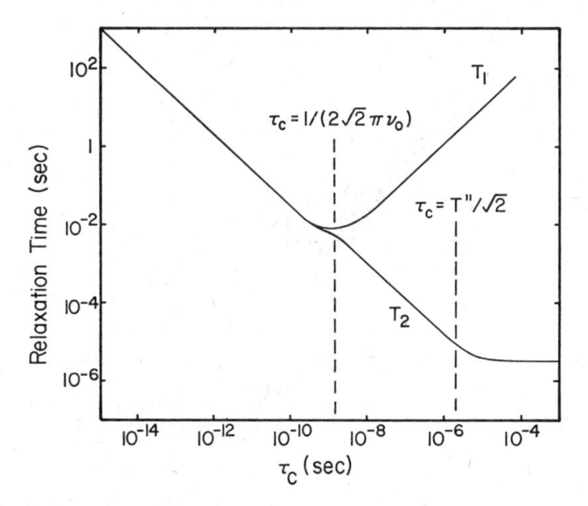

Fig. 7.7 Variation of nuclear relaxation times T_1 and T_2 with correlation time τ_c [102], when $\nu_0 = 30$ MHz. The limiting value of T_2 at high τ_c is $T_2'' = 3 \times 10^{-6}$ sec, corresponding to a typical rigid-lattice line width of 10^5 Hz.

Care must be taken to adjust the magnetic field to minimize inhomogeneities so that a true T_2 can be determined.

The large magnetic moments characteristic of paramagnetic species may lead to pronounced line-broadening effects even if the paramagnetic substance is present at very low concentration. These are more fully described in the next section.

Anisotropy of the shielding tensor provides still another relaxation mechanism. $\boldsymbol{\sigma}$ is anisotropic whenever the magnetic moment induced in a molecule by the applied field depends on the angle between the field direction and some direction fixed in the molecule. For such a molecule the effective field at the site of the nucleus of interest again depends on the state of orientation and varies as the molecule tumbles. This can be observed most readily by minimizing other sources of broadening, for example, by using an anisotropic molecule containing only one magnetic nucleus at low concentration in a magnetically nonpolar solvent, for example [104], $^{13}CS_2$ in natural abundance in CS_2. However, T_1 and T_2 in CS_2 may also be influenced by yet another relaxation process, called spin-rotation interaction [104a]. This occurs because the magnetic moment associated with the overall rotation of each molecule exerts a field at the resonating nucleus, and this field fluctuates as the angular velocity of the molecule is changed by collisions with its neighbors [104b].

The effect of any of these line-broadening mechanisms depends not only on the overall viscosity of the solution, as discussed, but also on the size and rigidity of the molecule containing the resonating nuclei. Thus for small molecules rotational correlation times are ordinarily of the order of 10^{-11} sec, so much less than the period of the nuclear precession (typically 10^{-6} to 10^{-8} sec) that magnetic field fluctuations associated with molecular rotation constitute a rather inefficient relaxation mechanism. Consequently, T_1 for a small molecule is normally large enough to permit the characteristically narrow lines of a high-resolution spectrum to be observed. Larger molecules tend to rotate more slowly, and as τ_c increases T_1 decreases, accounting for the observation that dissolved high polymers, such as globular proteins, often give signals with widths of 100 Hz or more even at concentrations where the viscosity of the solution is little greater than that of the pure solvent. The widths of signals from a protein are strikingly diminished if the material is denatured and changed from a more-or-less rigid form to something approaching a flexible random coil, because in the flexible form the relative motions of various sections of the molecule give rise to local field fluctuations over and above those associated with tumbling of the entire molecule.

For a rigid, spherical molecule of radius a, containing two nuclei with $I = \frac{1}{2}$ separated by a distance b, and rotating in a medium of viscosity η, the intramolecular contribution to the relaxation time is given by the relatively simple formula [105]:

$$(1/T_1)_{\text{intra}} = 2\pi\eta a^3 h^2 \gamma^4 / b^6 kT. \tag{7.79}$$

A somewhat similar formula gives the intermolecular contribution. For molecules that are nonspherical, or flexible, or which contain more than one magnetic nucleus, calculation of T_1 may become quite difficult, but (7.79) correctly suggests that T_1 for a particular nucleus depends very strongly on the geometric arrangement of the nearest magnetically polar neighbors. It is for this reason that chemically nonequivalent nuclei in the same molecule do not necessarily have equal values for T_1. Quantitative analytical applications of nmr therefore require that relative signal intensities be evaluated by measuring integrated peak areas rather than peak heights.

Because of their dependence on molecular size, relaxation time measurements can be very useful in studying the binding between small molecules and polymers [106]. The binding interaction may cause a very noticeable change in the widths of the signals from the small species even when there is no detectable effect on their chemical shifts. If the small molecule is rigid and becomes tightly attached to the macromolecule, its rotational correlation time becomes essentially that of the macromolecule, and all of its resonances will be broadened to approximately the same extent. If the small species is flexible and only one portion of it is tightly held by the macromolecule, the signals from the bound moiety will be broader than those from the remainder of the molecule, for which the freedom of motion is only partially restricted. An example is the binding of penicillin G,

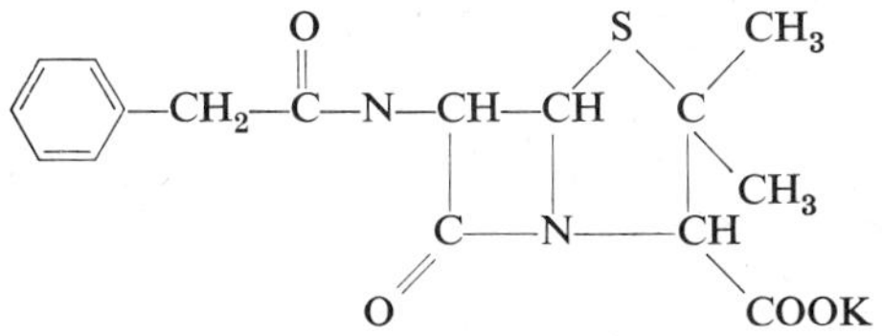

```
                 O             S     CH3
                 ||          /   \   /
  (C6H5)—CH2—C—N—CH—CH     C
                    |    |       |  \
                    |    |       |   CH3
                    C————N———————CH
                   //             \
                  O                COOK
```

by serum albumin, in which it was found [107] that the resonance of the aromatic protons is broadened considerably more than any of the other signals, and it was inferred that the phenyl ring is the part of the molecule directly attached to the protein.

Effect of Exchange Processes

In systems such as the one just cited it often occurs that the small molecules are distributed over two or more sites; in the simplest case, some of them are free while others are bound at identical binding sites. The line width may then depend not only on the relaxation times characteristic of each site but also, as already noted, on the rate of exchange. In the limit of fast exchange between two sites A and B, T_1 and T_2 are expected to be equal and

$$1/T_2 = p_A/(T_2)_A + p_B/(T_2)_B. \tag{7.80}$$

Measurement of the line width over a range of concentrations may make it possible to evaluate $(T_1)_A$, $(T_1)_B$, and the binding constant.

For slower exchange, but still under conditions in which only one signal is observed [108],

$$1/T_2 = p_A/(T_2)_A + p_B/(T_2)_B + 4\pi^2(\nu_A - \nu_B)^2 p_A{}^2 p_B{}^2(\tau_A + \tau_B). \tag{7.81}$$

In such cases T_1 and T_2 are no longer equal. A relatively simple and extensively studied situation [108] is that in which the concentration of one species, for example, B, is kept very small, so that $p_A \gg p_B$. Then,

$$1/T_2 - 1/T_1 = \frac{4\pi^2\tau p_B(\nu_B - \nu_A)^2}{1 + 4\pi^2\tau^2(\nu_B - \nu_A)^2}. \tag{7.82}$$

The last term in (7.81), which represents the contribution of exchange to the line width, depends on the second power of the chemical shift difference in frequency units (not parts per million). Accordingly, in favorable cases $\tau_A + \tau_B$ may be determined by measuring the line width at two different radiofrequencies, for example, 60 and 100 MHz.

Measurement of T_1 and T_2

The actual determination of relaxation times can be carried out by a variety of high-resolution techniques, some of which are briefly described at this time, or by pulsed nmr methods (Section 15).

T_2 may be obtained directly from the observed line width using (7.78), provided that the line is not so sharp or the instrument so poorly adjusted that the width is limited by residual field inhomogeneities.

If T_1 is about 1 sec or larger, it may be determined by the "direct method" of first completely saturating the resonance with a very strong radiofrequency field H_1 and then reducing H_1 to a value far below the saturation level and periodically observing the signal [102]. For complete saturation the population of the two spin states is nearly equal, that is, the spin temperature is very high, and there is no net energy absorption (see Section 2, p. 609). After H_1 is reduced the spin temperature drops at a rate determined by T_1, and resonance absorption again becomes possible. The signal strength S consequently rises from zero to a steady value S_0, according to the equation:

$$S = S_0(1 - e^{-t/T_1}), \tag{7.83}$$

so that T_1 can be evaluated from a record of S as a function of time.

Several other procedures for finding T_1, particularly useful when T_1 is relatively small, depend on observations of the resonance under conditions of incomplete saturation and evaluation of the saturation factor $1/(1 + \gamma^2 H_1{}^2 T_1 T_2)$ introduced in Section 2, p. 609. Perhaps the simplest is to observe the absorption peak at a series of known values of H_1 with constant amplification; the maximal peak height should occur when $H_1 = 1/\gamma(T_1T_2)^{1/2}$. Alternatively, one may observe the derivative of the absorption signal [109] and take advantage of the fact that at a given value of H_1 the peak-to-peak separation for the derivative signal is:

$$\Delta\nu = (1 + \gamma^2 H_1^2 T_1 T_2)^{1/2} / \pi T_2 \sqrt{3}. \tag{7.84}$$

Either result can be used to find T_1 if T_2 is already known, or they can be used together to find both relaxation times.

10 LINE WIDTHS AND CHEMICAL SHIFTS IN THE PRESENCE OF PARAMAGNETIC MATERIALS

Line Widths

A molecule or ion is paramagnetic if it contains a number n of unpaired electrons. Paramagnetic organic species are usually called free radicals or radical ions and have $n = 1$. Inorganic species, commonly free or complexed ions of transition metals, may have several unpaired electrons. In the absence of contributions arising from orbital angular momentum, such a species has a magnetic moment of $[n(n + 2)]^{1/2} \beta_e$. This is about three orders of magnitude larger than the magnetic moment of a proton or other nucleus, so that exceptionally large local magnetic fields exist in the neighborhood of the paramagnetic entity. As a result, even small amounts of paramagnetic substances have pronounced affects on the relaxation times of diamagnetic materials in solution. In some cases so-called contact interactions may also produce very large changes in the chemical shifts [110, 111].

As mentioned in the preceding section, the relaxation times depend on the magnitudes of the local fields within the sample and on the rate of fluctuation of these local fields. The discussion can be extended to include effects of paramagnetic substances, but it must be recognized that the observed relaxation times may reflect an interplay of several physically distinct time-dependent processes including exchange between different sites, rotational or translational motions of molecules, and reorientation, or relaxation, of the electronic magnetic moment. The latter is space-quantized in much the same way as the nuclear magnetic moment, giving rise to electronic spin states with a mean lifetime usually designated τ_s. Perhaps the best illustrative example is the proton signal from water containing dissolved paramagnetic ions at low concentration [112], which has been very widely studied. Very similar results are obtained when the ^{17}O signals from such solutions are examined [113] or when methanol replaces water as the solvent [114].

It is convenient to divide the water molecules into two populations, using the subscripts M and A to designate, respectively, molecules in the first hydration sphere of the ion, hence interacting strongly with it, and molecules elsewhere in the solution. The line width then depends on the populations p_M and p_A, on the relaxation times characteristic of each type of molecule, and sometimes also on the rate of exchange between them.

The energy of interaction between a water proton and a nearby paramagnetic ion may be considered as the sum of two contributions, the dipole-dipole interaction and the contact or spin-exchange interaction. It has been found [115] that the dipole-dipole forces alone determine the proton relaxation times when τ_s for the paramagnetic ion is very small ($\tau_s \leq 10^{-10}$ sec). No simple rules seem to exist that allow one to predict which ions fall into this category, but it is known [115] that Cu^{2+}, Co^{2+}, and Fe^{3+} are such ions, while Mn^{2+} and Gd^{3+} have relatively long τ_s-values. Ions with short τ_s-values present the easier problem and are dealt with first.

The interaction energy between the nuclear magnetic dipole $\boldsymbol{\mu}_N$ and the electronic magnetic dipole $\boldsymbol{\mu}_e$ is given by

$$E_{\text{dipole}} = \boldsymbol{\mu}_N \cdot \boldsymbol{\mu}_e / r^3 - 3(\boldsymbol{\mu}_N \cdot \mathbf{r})(\boldsymbol{\mu}_e \cdot \mathbf{r})/r^5, \tag{7.85}$$

where $\mathbf{r}$ is a vector drawn from the nucleus to the effective position of the electronic spin vector, assumed to coincide with the center of the paramagnetic ion. For a molecule in a given spin state, $\boldsymbol{\mu}_N$ and $\boldsymbol{\mu}_e$ precess around the direction of the magnetic field, each with a constant z-component, but the angle between the $\mathbf{r}$-vector and the field direction takes on all possible values as the molecule rotates. Consequently, even though the spin state of the nucleus remains the same, the value of E_{dipole} may change, either because of the molecular rotation or because of changes in the state of alignment of the electron spin. The relaxation time associated with the dipolar interaction is then equal to the smaller of the two quantities τ_s and τ_c. Usually the rotational correlation time τ_c is in fact the smaller and is used as the dipolar correlation time. Since it is generally much smaller than the nuclear precession period $1/\nu_0$, the arguments given in the preceding section then suggest that T_{1M} and T_{2M} are equal and that the line width, $1/T_{2M}$, should increase if τ_c is increased. The rather complicated expressions for $1/T_{1M}$ and $1/T_{2M}$ that are available in the literature [112] show that indeed the two relaxation times are approximately equal and that $1/T_{2M}$ is directly proportional to τ_c and inversely to r^6.

Because of the rapid dropoff of the dipolar interaction at larger distances, T_{1A} and T_{2A} should be much less different from the relaxation times for pure water than T_{1M} or T_{2M}. Values of T_{2A} may be estimated from data for solutions of complex ions in which the first coordination sphere is filled by ligands other than water so that only the longer-range interactions contribute to the broadening. The overall line width is then governed by the expression [114]:

$$\frac{1}{T_2} = \frac{p_A}{T_{2A}} + \frac{p_M}{\tau_M}\left[\frac{1/T_{2M}^2 + 1/T_{2M}\tau_M + 4\pi^2\,\Delta\nu_M{}^2}{(1/T_{2M} + 1/\tau_M)^2 + 4\pi^2\,\Delta\nu_M{}^2}\right], \tag{7.86}$$

in which τ_M is the mean residence time of a given proton in the hydration sphere of a particular ion and $\Delta\nu_M$ is the chemical shift difference (in frequency units) between coordinated and bulk water molecules. For dilute solutions, p_A is essentially equal to unity, and $p_M = p[M]/55.5$, where $[M]$ is the molarity of the metal ion and p is its hydration number.

Equation (7.86) is greatly simplified if the rate of exchange is so high that the largest terms inside the square brackets are those involving $1/\tau_M$. In the limit of fast exchange, it reduces to:

$$1/T_2 = 1/T_{2A} + p_M/T_{2M}. \tag{7.87}$$

When exchange is much slower, so that $1/\tau_M$ is small compared to $1/T_{2M}$ or $\Delta\nu_M$, the quantity in the square brackets of (7.86) becomes approximately unity and the equation reduces to:

$$1/T_2 = 1/T_{2A} + p_M/\tau_M. \tag{7.88}$$

Experimentally, it may be possible to determine whether (7.87) or (7.88) applies in a given system by following the line width as a function of temperature [112]. Rotational correlation times decrease with rising temperature according to a relation of the form:

$$\tau_c = \tau_c{}^0 \exp{(V_c/RT)}, \tag{7.89}$$

and since $1/T_{2M}$ is proportional to τ_c it likewise decreases as the temperature rises. However, $1/\tau_M$ increases as the solution is heated. Increasing the temperature therefore decreases the observed line width if (7.87) is valid and increases it if (7.88) is valid. In the latter case, the line width should also be strongly dependent on chemical additives or pH changes, which are much more likely to effect major changes in τ_M than in T_{2M}.

The interaction energy of a pair of point dipoles cannot be evaluated using (7.85) if they are in contact, that is, if $r = 0$. Such a possibility cannot, however, be neglected, especially in systems in which the unpaired electron is assigned to a delocalized molecular orbital that encompasses the group containing the resonating nucleus. In such cases the additional interaction [34] may be written:

$$E_{\text{contact}} = hA\,\mathbf{I}\cdot\mathbf{s}, \tag{7.90}$$

where **I** and **s** are the nuclear and electronic angular momentum vectors and the proportionality constant A is called the scalar coupling constant. This is in turn proportional to the probability of finding the unpaired electron in contact with the resonating nucleus, that is, to the square of the magnitude of the electronic orbital wave function evaluated at the site of the nucleus. Only nuclei within molecules in the first coordination sphere of the paramagnetic ion (or within the paramagnetic species itself) are subject to this "contact" or "spin-exchange" interaction.

The line-broadening contributions arising from the contact interaction are proportional to a correlation time τ_e and may be designated symbolically as $(1/T_{1M})_{\text{exch}}$ and $(1/T_{2M})_{\text{exch}}$. The effective relaxation times, allowing also for the dipolar interactions, are:

$$1/T_{1M} = (1/T_{1M})_{\text{dipole}} + (1/T_{1M})_{\text{exch}} \tag{7.91}$$

and

$$1/T_{2M} = (1/T_{2M})_{\text{dipole}} + (1/T_{2M})_{\text{exch}}. \tag{7.92}$$

The exchange contribution is noticeable only if τ_e is larger than or about equal to 10^{-9} sec, and even then the contribution to $1/T_{1M}$ remains much smaller than that to $1/T_{2M}$. As a result, solutions of Mn^{2+} have T_1/T_2 appreciably larger than unity, whereas in the absence of contact contributions the two relaxation times are about equal.

The contact contribution to the line width may depend on the temperature in a somewhat complex way that can be understood by thinking in more detail about the correlation time τ_e. Since the quantity $\mathbf{I} \cdot \mathbf{s}$ in (7.90) remains constant when the molecule rotates, the value of E_{contact} is not affected by molecular rotation and τ_e cannot be identified with the rotational correlation time. τ_e cannot be longer than the mean lifetime of the electronic spin states τ_s, but it may be shorter if the chemical exchange lifetime τ_M is less than τ_s. In general,

$$1/\tau_e = 1/\tau_s + 1/\tau_M. \tag{7.93}$$

For solutions of Mn^{2+} it has been found that at low temperatures chemical exchange is so slow that $1/\tau_M$ is negligible, so that τ_e is essentially equal to τ_s, but at higher temperature τ_M is so greatly reduced that the second term becomes dominant and $\tau_e \simeq \tau_M$.

Because so many factors contribute to the observed line width, it is obviously necessary to interpret such data with great caution. However, when it can be shown that one type of process predominates in bringing about a change in line width, such measurements provide an extremely useful tool for the determination of the rate constant, and they have been used to study the kinetics of atom, ligand, and electron-exchange reactions.

Contact Shifts

In addition to the effect on line widths just discussed, the contact interaction may produce very pronounced changes in the chemical shifts of nuclei in paramagnetic molecules or complexes. The origin of these changes may be understood by considering first an imaginary compound in which the relaxation times are so long that the molecules can be classified according to the value of the quantum number M_S which characterizes the orientation of the total electronic spin vector; in the simplest case in which there is one unpaired electron there are two sets of molecules, with $M_S = +\frac{1}{2}$ or $-\frac{1}{2}$. In such a system the contact interaction splits the nuclear resonance into a doublet because the contribution of the electron spin to the local field at the nucleus changes sign according to the sign of M_S. The doublet splitting is proportional to the scalar coupling constant, and the two components have slightly different intensities because the electronic states in question are unequally populated at thermal equilibrium.

If the correlation time τ_e is reduced, the doublet (or multiplet) components will begin to coalesce. As described in Section 7, the spectrum will consist of very broad signals when the rate of exchange is intermediate. When τ_e is near 10^{-6} sec the broadening is so severe that no nmr signals can be observed. For τ_e-values in the range 10^{-8} to 10^{-9} sec the signals are detectable and have large widths, varying as described in the preceding section. Finally, when τ_e is smaller than about 10^{-10} sec, the collapse of the multiplet structure is so nearly complete that fairly narrow nmr signals are produced. However, because of the difference in populations between the states of different M_S, these exchange-averaged resonances are displaced from the positions in which they would appear if there were no contact interaction. This *contact shift* ΔH is given by [116, 117]:

$$\Delta H = (-Ah/3kT)(\gamma_e/\gamma_N)S(S+1)H_0 \tag{7.94}$$

or, alternatively,

$$\Delta H = (-a/3kT)(\gamma_e/\gamma_N)g_e\beta_e S(S+1)H_0, \tag{7.95}$$

where $a = Ah/g_e\beta_e$ and is the coupling constant, in gauss, that would be observed in the electron spin resonance (esr) spectrum of the species under investigation. γ_e and γ_N are the magnetogyric ratios of the electron and of the nucleus, respectively, and S is the total electron spin quantum number of the paramagnetic molecule.

Contact shifts as large as several hundred parts per million may be found in proton resonance spectra at room temperature. The dependence of ΔH on $1/T$ reflects the fact that raising the temperature tends to reduce the differences between populations of states with different M_S. A more elaborate temperature dependence is found for certain complexes of Ni^{2+}, where a diamagnetic electronic ground state is in rapid thermal equilibrium with a low-lying triplet excited state, so that the number of molecules in the paramagnetic state increases as the temperature rises.

As noted above, nmr signals from paramagnetic materials are sufficiently narrow to allow accurate shift measurements only when τ_e is of the order of 10^{-10} sec or less. Interestingly enough, the short electron-spin relaxation times that allow this condition to be easily satisfied cause extensive broadening of the esr signals for the species in question, so that the conditions for observability of nmr and esr signals are to some extent mutually exclusive. For chelate complexes of paramagnetic transition metal ions, τ_e depends primarily on the nature of the metallic species. Most of the useful results have been obtained for complexes of Co^{2+} and Ni^{2+}, although these are not the only species for which the method is practicable [110]. For solutions of organic radicals or radical ions, τ_e is generally too large to permit detection of nmr signals unless special conditions are met. Adequate reduction of τ_e may be achieved, however, by providing a mechanism for rapid spin-exchange among paramagnetic species. This is favored by using high concentrations of radicals in a solvent of low viscosity and raising the temperature as much as possible [118, 118a], or by dissolving the radicals in a solvent which is itself paramagnetic (e.g., di-*tert*-butylnitroxide) and taking advantage of solvent-solute spin exchange [118b]. Another technique for the observation of contact shifts for organic radicals is to study the materials in the solid state at temperatures a few degrees above absolute zero, where the shifts are so greatly increased that signals from different protons can be resolved even though the line widths are very large [119].

The main objective of these studies has been to determine the scalar coupling constants, hence the electron spin densities at various positions in the radical or ligand, in order to test theoretical methods for predicting these quantities. In particular, it is noteworthy that the contact shift is sometimes to higher and sometimes to lower fields implying that the constant a in (7.95) may be either positive or negative, and that its sign may readily be inferred from the observed shift. The possibility of negative, as well as positive, spin densities arises from the interactions between spins of electrons in different molecular orbitals, which may lead to an excess of electron density with spin "up" near some nuclei and density with spin "down" near others. Although this concept has been extensively applied in rationalizing esr spectra [120], it is impossible to obtain the sign of a directly from an esr spectrum, the predicted spectrum being unaffected if a is replaced by $-a$. Measurement of nmr contact shifts therefore provides the most clear-cut experimental evidence for the occurrence of negative spin densities.

A simple but ingenious application of the contact shift phenomenon makes it possible to study the spectra of solutes in water (or D_2O with traces of HDO) solutions without troublesome interference from the solvent proton signals. Doping the solution with Co^{2+} shifts the solvent peak to lower fields and has essentially no effect on the solute spectrum unless the solute is capable of binding this ion [121].

11 SPECTRA OF ELEMENTS OTHER THAN HYDROGEN

General Considerations

Since any nucleus having either an odd atomic number or an odd mass number possesses a finite magnetic dipole moment, nearly all elements have at least one isotope capable of being observed in an nmr experiment. In many ways such experiments are closely similar to proton magnetic resonance spectroscopy, but a number of characteristic differences and difficulties severely limit the useful information obtainable from all but a few of these species.

Perhaps the most striking difference between nmr spectra from hydrogen and from other elements is the very much larger range of chemical shifts found for the latter. In toluene, for example, the chemical shift difference between the CH_3 protons and any of the ring protons is about 5 ppm, depending somewhat on the solvent; in the ^{13}C spectrum of the same compound [122], the shift difference between the methyl carbon and the five hydrogen-carrying ring carbon atoms is about 107 ppm. Again, for conformationally fixed cyclohexane, the axial and equatorial ring hydrogens differ in shift by just under 0.5 ppm, but the analogous difference between the ^{19}F shifts in 1,1-difluorocyclohexane [123] is nearly 16 ppm. A vast number of other examples could be cited. Moreover, similar differences in magnitude are found for solvent and temperature effects on the chemical shifts. A methyl proton signal from a typical organic solute changes position by perhaps 0.1 ppm or less if the solvent is changed from cyclohexane to carbon tetrachloride, but the fluorine resonance of a compound of the type CF_3R is shifted by about 1.2 ppm by such a solvent change [124].

From a practical point of view, the large spread of chemical shifts for a nonhydrogen nucleus is often advantageous. It permits signals from non-equivalent nuclei to be resolved even when the peaks are relatively broad. It allows subtle changes in molecular environment, such as those accompanying the binding of a fluorine-labeled substrate or inhibitor by an enzyme, to be detected [125, 126]. Taken together with the fact that spin-spin coupling parameters involving nonhydrogen nuclei tend to be only slightly larger than corresponding H—H coupling constants, it reduces the number of instances in which multiplet patterns deviate substantially from the simple first-order rules, since the condition (7.37) is more likely to be satisfied.

However, the factors responsible for these large shift variations also make it extremely difficult to interpret or predict chemical shifts theoretically unless attention is confined to a series of very closely related molecules. The basic difference between the shielding constants for hydrogen and the heavier elements is that the latter contain a much larger number of electrons, capable of a greater variety of behavior. A hydrogen atom is almost always attached to a single nearest neighbor by a single bond, which can be described, in the absence of more elaborate information about the wave functions, by specifying its polarity and the hybridization of the orbital contributed to the bonding wave function by the neighboring atom. At the same unsophisticated level, much more information is needed to define the chemical environment of a carbon atom in a covalent compound. Since the carbon atom may be bonded to two, three, or four neighbors by bonds involving different hybrid orbitals of the carbon atom and/or the attached atom and differing also in polarity and in total bond order, it is hardly to be expected that the shielding constant will be simply correlated with one or two readily identifiable bonding parameters. Even for a fluorine atom, which is usually found bonded only to a single neighbor, the description of the bond is complicated because both atomic orbitals involved in it will be hybrids of *s*- and *p*-orbitals. Moreover, the formally nonbonding electrons of the fluorine atom may become involved in π-bonding as illustrated by resonance of the type

$$\mathrm{X{-}C_6H_4{-}F} \leftrightarrow \mathrm{X{-}\overset{-}{C_6H_4}{=}F^+}$$

which was invoked [127] in a discussion of ^{19}F shifts in *para*-substituted fluorobenzenes.

A second major difference between hydrogen and nonhydrogen magnetic resonance spectra is in the signal strength. Except for the unstable hydrogen isotope ^{3}H, no nuclear species is known that has a larger magnetogyric ratio than the proton, and many have γ-values that are substantially smaller. Assuming that all spectra are determined at the highest field that can be conveniently reached, and therefore making comparisons at constant field strength, the peak intensities are approximately proportional [102, 128] to $\gamma^3 I(I + 1)$ and almost invariably much smaller for heavier nuclei than for protons. The only important exception to this rule is fluorine, since ^{19}F nuclei have a magnetogyric ratio only 4.3% less than that of the proton. In many instances the intensity problem is aggravated by the fact that the magnetically polar species occurs as only one component, often a minor one, of an isotopic mixture. The principal isotopic species of each of the two almost ubiquitous elements, carbon and oxygen, is magnetically nonpolar. The species ^{13}C and ^{17}O do yield usable nmr spectra in their natural abundance of 1.1% and 0.037%, respectively, but neither one is easy to work with, and for ^{17}O the signals are so weak that little work has been done with nonenriched samples [129].

Another consequence of the reduced transition probabilities for species with small magnetogyric ratios is a tendency for relaxation processes involving these species to be relatively slow. This is especially noticeable when the resonating nucleus is buried in the interior of a molecule and separated by several bonds from the nearest neighboring magnetic nucleus. The longer relaxation times prohibit the use of high radiofrequency power levels in attempting to obtain stronger signals because, as mentioned in Section 9, the larger the value of T_1 the smaller the value of H_1 at which the signal begins to be saturated. Among the methods that have been used to circumvent these difficulties are the use of large samples, use of flowing samples in which nuclei that have been irradiated to the point of saturation are constantly replaced by fresh material, introduction of small amounts of paramagnetic substances to hasten relaxation, and use of rapid-passage conditions with display of the in-phase component of the induced magnetization [130]. This produces a signal, similar to the derivative of the usual absorption curve, which does not fall off in intensity as H_1 is increased.

A third source of difficulty is that a majority of nuclear species with nonzero angular momentum have electric quadrupole moments. Examination of a table of nuclear properties reveals that aside from normal hydrogen and its unstable isotope tritium, only ^{3}He, ^{13}C, ^{15}N, ^{19}F, ^{29}Si, ^{31}P, and isotopes of some 15 heavier elements have spin quantum numbers $I = \frac{1}{2}$. For all other magnetically polar nuclei $I \geq 1$, the condition required for existence of a quadrupole moment. A quadrupolar nucleus produces nmr signals which are broadened (Section 7) to an extent that depends on the magnitude of the quadrupole moment and the degree of symmetry of the local electronic environment. Fortunately, the large variations in chemical shift already mentioned often make it possible to resolve peaks from chemically non-equivalent nuclei even when the lines are much broader than ordinary high-resolution signals, but some interesting spectral features, such as long-range coupling constants may be lost, and the problem of low signal strengths may be aggravated by large line widths.

Several reviews concerned with high-resolution nmr spectra of elements other than hydrogen are available [130–133]. The species that offer the greatest promise for extensive applications are ^{19}F and ^{13}C, and only work involving these nuclei is described here in greater detail.

Fluorine Spectra

The high magnetogyric ratio, absence of quadrupole moment, and 100% natural abundance of ^{19}F make it a very attractive species with which to work, and in view of the very large variety of fluorine compounds that have been prepared it is not surprising that many nmr investigations have been reported [133]. For binary compounds of fluorine, there is a rough correlation between the ^{19}F chemical shift and the electronegativity of the second element. This is in accord with results of an attempt to apply Ramsey's theory of chemical shifts to these molecules in a semiquantitative way [134]. Three contributions to the chemical shift are identified, a local diamagnetic term originating from electrons within the fluorine atom, a local second-order paramagnetic term, and a term including all effects of electrons farther away. For ^{19}F (but not for protons) changes in the local paramagnetic term are found to be dominant. This term is zero for a fluoride ion, which has the closed-shell electronic configuration $1s^2 2s^2 2p^6$, and it increases in magnitude as charge is withdrawn from one of the $2p$-orbitals by covalent bond formation. Since the second-order paramagnetism reduces shielding, the nucleus in F^- should be the most highly shielded, with signals from covalent fluorides falling at lower fields as the electronegativity of the neighboring atom rises and the bond polarity decreases, the limit being represented by the nonpolar species F_2.

Although the experimental results support this interpretation to some extent, they also indicate that the ionic character of the bond to fluorine cannot alone determine the chemical shift. In ClF_3, SF_4, certain derivatives of SF_6, and a few other compounds, sets of fluorine atoms are attached by structurally nonequivalent bonds to the same central atom [135]. The shift difference between the two types of fluorine atom in ClF_3 is greater than 100 ppm and must reflect factors other than the electronegativity of the chlorine atom. Spectra of this type provide unequivocal evidence ruling out more symmetrical structures for these molecules (e.g., SF_4 cannot be square planar or tetrahedral). They are usually temperature dependent and thus yield information about rates of exchange of nonequivalent fluorine atoms.

Fluorine atoms in organofluorine compounds also show chemical shifts which range over several hundred parts per million (see Fig. 7.8). These changes cannot be ascribed to concomitant changes in the effective electronegativity of the neighboring carbon atom. Thus CF_3H and CF_3Cl are each related to CF_4 in that one of the fluorine atoms of CF_4 has been replaced by a less electronegative atom, but the shift for CF_3H is more positive than that of CF_4 while that of CF_3Cl is more negative. Several attempts to rationalize these and related data have been offered. It has been proposed that contributions from resonance structures such as

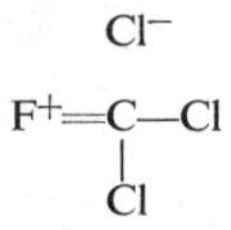

```
     Cl⁻
F⁺═C─Cl
    │
    Cl
```

reduce the shielding [136]. "Repulsive unshielding" has been found in a number of instances in which a heavy atom is close to the fluorine nucleus [137]. This may be related to the fact that the second-order paramagnetic term in the theory depends on the energy gap between the ground state of the molecule and particular excited electronic states, so that substituents (or solvents) may exert their effect by changing these excitation energies [138]. In general, the predictive value of theories in this area is so limited that assignment of peaks in complex spectra is best made with the help of model compounds. Only certain series of compounds, such as the *para*-substituted fluorobenzenes, are exceptional, inasmuch as it has been possible to demonstrate and rationalize a good correlation between the ^{19}F shifts and substituent constants evaluated from independent chemical data [127]. For CF_3—C derivatives, and certain other series of molecules, the shifts could be correlated with the electric field set up by bond dipoles within the substituent groups [139].

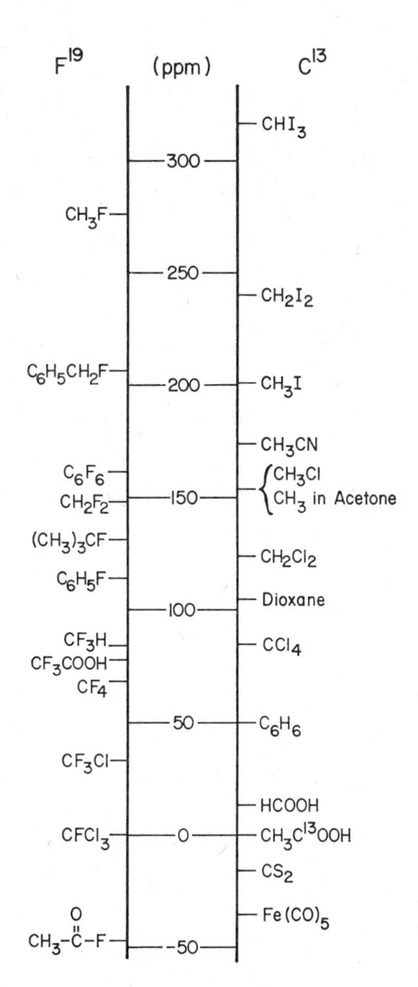

Fig. 7.8 Representative chemical shifts of ^{19}F and ^{13}C nuclei in organic compounds.

In spite of the shortcomings of the theoretical treatments, ^{19}F spectra of perfluoroorganic compounds are useful in structure determinations in much the same way as the proton spectra of the related hydrogen compounds. One of the most startling differences is that in groupings of the type

$$\underset{}{X}—\underset{a}{CF_2}—\underset{b}{CF_2}—\underset{c}{CF_2}—Y$$

the fluorine-fluorine coupling constant J_{ac} is usually in the neighborhood of 10 Hz, while at the same time J_{ab} and J_{bc} may be too small to give resolvable fine structure. Longer-range fluorine-fluorine coupling constants are generally somewhat larger than analogous H—H couplings, hence the observed multiplets, attributable to short-range couplings, are apt to have components that are considerably broadened by unresolved long-range interactions. To an observer accustomed to high-resolution proton spectra, this gives the ^{19}F spectra of larger molecules a characteristically untidy appearance. A similar effect is noted in partially fluorinated organic compounds, such as benzotrifluoride, reflecting the fact that long-range H—F interactions are also relatively large.

Values of ^{19}F chemical shifts reported from different laboratories often agree much less closely than proton shift values. This is partly attributable to the fact that many different fluorine signals have been used as secondary standards, but a more fundamental reason is that even with an internal reference the shift is often strongly dependent on the nature of the solvent, the concentration, and the temperature. Measurements made with a given sample on different days at "ambient probe temperature" may easily disagree by perhaps 0.1 ppm because of an unsuspected change in the operating conditions. Several interesting experiments further illustrate the extreme sensitivity of ^{19}F shifts to subtle environmental changes. The compound $CFCl_3$, often recommended as a chemical shift standard, actually produces three resolvable signals [140] approximately 0.5 Hz apart, corresponding to its three major isotopic components, $CF^{35}Cl_3$, $CF^{35}Cl_2{}^{37}Cl$, and $CF^{35}Cl^{37}Cl_2$. No analogous splittings have been detected for chloroform. The racemic compound 1-phenyl-2,2,2-trifluoroethanol produces the expected doublet (attributable to H—F coupling) in racemic or optically inactive solvents, but a pair of doublets when dissolved in *d*- or *l*-1-phenylethylamine, which interacts preferentially with one of the enantiomers [141]. Similarly, racemic compounds produce two sets of resonances in the presence of enzymes that appear to bind one optical isomer more strongly than the other [125, 126].

The large solvent effects on ^{19}F shifts have been discussed, as have the proton shifts, in terms of the contributions of bulk susceptibility, anisotropy, reaction field, van der Waals, and complex-formation effects. [138, 142]. The two situations differ in that for ^{19}F the van der Waals interactions produce by far the largest single contribution. This gives rise to a rough correlation between the solvent dependence of the shifts and the polarizability of the solvent used [138], but there are so many irregularities that it is clear that other factors are also involved. These must include the shapes of the solvent and solute molecules and the mean intermolecular distances, variables that cannot now be evaluated with a satisfactory accuracy. Recently, it was found in our laboratory that for $CF_3(CH_2)_5OH$ the ^{19}F solvent shifts are better correlated with the solvent surface tension than with any other simple bulk property, probably because the surface tension itself depends both on the polarizability and on molecular shapes and packing properties.

New applications of fluorine nmr are constantly being reported which take advantage of the dependence of the shifts on environment. Isomeric alcohols that differ with respect to the location of a hydroxyl group may produce almost identical proton spectra, but the ^{19}F signals of the corresponding trifluoroacetate esters can often readily be distinguished and assigned [143]. Conformational equilibria in cycloparaffin derivatives can be more readily studied if one of the methylene groups is replaced by —CF_2— because of the large shift differences between fluorines in axial and equatorial positions [123]. For surface-active species such as $CF_3(CH_2)_nX$, where X is a hydrophilic group, the ^{19}F shifts, but not the proton shifts, are strongly altered by micelle formation, thus allowing nmr to be used to study the hydrophobic interactions that cause these molecules or ions to associate [124, 144]. Peaks from such fluorinated detergent ions also change position markedly when the ions are bound by dissolved protein [145].

Carbon-13 Spectra

The feasibility of obtaining ^{13}C nmr spectra from nonenriched samples in spite of the low abundance and long relaxation times was demonstrated [146, 147] in 1956. The potential usefulness of this technique is obvious, but when the area was reviewed [132] in 1965 progress had been rather slow because of the difficulties in working with these signals with the older spectrometers. Three major engineering improvements now greatly facilitate such measurements. First, the development of more powerful electromagnets made it possible to boost the observation frequency for ^{13}C signals from about 10 MHz first to 15.1 and then to 25.1 MHz, which results in a substantial increase in the sensitivity. Second, the development of several methods of achieving improved magnet stability, together with the introduction of signal-averaging computers, made it possible to enhance sensitivity by using multiple scans over the spectrum [148]. The third factor is an innovation in the technique of heteronuclear spin decoupling, called *noise decoupling* [149].*

* A further development which promises to enhance dramatically the usefulness of ^{13}C nmr is the Fourier transform technique, described in Section 15, p. 698.

The ^{13}C spectrum of a typical organic compound shows a large amount of fine structure because of ^{13}C—H spin-spin interactions. Because of the low abundance of ^{13}C, molecules containing more than one ^{13}C nucleus are very rare, and ^{13}C—^{13}C coupling is normally not observed for unenriched samples, although it is a tribute to the present state of the art that these couplings can be detected without isotopic enrichment if a special effort is made to do so [150]. Values of $J_{^{13}C-H}$ range from about 120 to 260 Hz. $J_{^{13}C-C-H}$ may be as large as 50 Hz, although values around 5 Hz are more common [151], and reported values of $J_{^{13}C-C-C-H}$ range from 3 to 7 Hz.

Although these splittings help in assigning the signals, they constitute a very mixed blessing because they further reduce the already small signal-to-noise ratios, and therefore it is often advantageous to decouple the ^{13}C and proton spins by means of an appropriate second radiofrequency [152]. The difficulty here is that to eliminate all the fine structure in a ^{13}C spectrum one must irradiate protons having various chemical shifts, so that a single decoupling frequency is not sufficient. To use a set of frequencies, each adjusted to decouple a particular set of protons, would require very elaborate instrumentation. This problem has been solved by using a radiofrequency signal at a frequency within the range required for proton decoupling, and frequency-modulating this with the output of a noise generator [148]. In effect, this replaces the single frequency by a continuous spectrum, broad enough to comprise appreciable contributions from all the frequencies appropriate to individual protons. It is indeed possible by this means to decouple all the protons simultaneously, although the overall radiofrequency power level is so high that the sample is appreciably heated.

A number of representative ^{13}C chemical shifts are given in Fig. 7.8. They resemble ^{19}F shifts in that they are controlled largely by changes in the local second-order paramagnetism. Again, progress toward a readily and generally applicable, predictive theory has been slow, but a number of empirical rules have been discovered and at least partially rationalized. It appears, for example, that in many aromatic and heterocyclic molecules the ^{13}C shifts are correlated with calculated charge densities [152a, 152b]. The shifts for the halogenated methanes can be correlated using additive substituent parameters, but only if pairwise interaction parameters are also included [152c]. In the polymethylbenzenes, carbon atoms of methyl groups in highly crowded positions produce resonances at fields higher than those in less sterically perturbed groups [152d], and somewhat analogous effects were found in methylcyclohexane spectra [152e]. A very practical application has been the use of ^{13}C nmr to characterize petroleum fractions by evaluating the relative numbers of carbon atoms in saturated groups and in aromatic rings [152f]. The latter were further characterized by determining what fraction of them were in substituted versus unsubstituted ring positions. When ^{13}C spectra were used in conjunction with proton spectra, a much more detailed analysis of carbon types was possible than with either spectrum alone.

It is to be expected that the improved instruments now available will stimulate a great deal of new work with this isotope. Moreover, an increasing variety of ^{13}C-enriched material is being offered commercially, which opens the possibility of performing many tracer experiments, following the chemical transformations involving the ^{13}C nuclei. Such studies will be particularly interesting to biochemists, and indeed techniques have already been developed for growing entire organisms in which a large fraction of the total carbon present consists of the heavy isotope [153].

12 HIGH-RESOLUTION SPECTRA OF MACROMOLECULES IN SOLUTION

Synthetic Polymers

Although the majority of nmr applications has involved compounds of fairly low molecular weight, work on natural and synthetic polymers began to appear in the late 1950s. Again, early progress was impeded by a number of characteristic difficulties which have been greatly alleviated by recent improvements in the sensitivity and resolution of the instruments, and the number of reported applications has increased dramatically [154, 155].

Any polymer may be considered as a chain, possibly branched or cross-linked, of subunits which may or may not all be identical. The nmr spectrum of such a chain differs from that of a solution containing appropriate amounts of the separate subunits both with regard to the chemical shifts and the line widths. The chemical shift of a nucleus in one subunit reflects the local environment and is often influenced by the nature of the directly adjacent subunits. Alternatively, two subunits that are far apart when the polymer is fully extended may be brought into close proximity if the chain assumes some particular coiled configuration, giving rise to an interaction which affects the chemical shifts, and the shifts then become clues from which conclusions about conformational changes may be drawn. The width of each signal is also affected by the structure and composition of the polymer and by its flexibility. For a moderately rigid macromolecule, nuclei in formally identical subunits may find themselves in nonidentical environments so that the observed signal actually consists of many lines, probably too closely spaced to be resolved. By raising the temperature or changing the solvent, the chain can often be made more flexible, whereupon motion of the subunits begins to average out some but not all of these local differences, and narrower lines are observed. Even then, the spectra seldom appear quite as sharp as ordinary high-resolution spectra.

In studying synthetic polymers, it is desirable to optimize the resolution by working with heated samples, which also reduces problems attendant on limited solubility, and by using solvents that tend to disrupt subunit-subunit interactions, such as hydrogen bonds, which restrict the flexing of the chain. Typical objectives of such studies include determinations of polymer composition, of the extent of branching or crosslinking of the chains, and of the type and degree of stereoregularity.

The question of composition naturally arises in connection with copolymers, consisting perhaps of two kinds of repeating unit, A and B. If each of these contains one or more distinctive proton resonances, it is obvious that the overall monomer ratio $A:B$ can be very readily determined. A more subtle question is whether the arrangement of the subunits is random or consists of blocks, as in the sequence $AAABBBBBAABBB\cdots$. Since the chemical shift of a proton in an A-subunit may be subject to influences of the neighboring units, it is sometimes possible to identify separate signals from such a proton in the three possible triads, AAA, AAB, and BAB, and the relative intensities of these must depend in a calculable way on the degree of randomness of the sequence. If the influence of second neighbors is appreciable, analysis of the spectra in terms of longer sequences, such as pentads, may be justified.

In principle the simplest system in which to study chain branching is polyethylene. A perfect, infinitely long chain of $—CH_2—$ groups would yield a single peak, and indeed actual polyethylene samples produce a one-peak spectrum unless the spectrometer sensitivity is very good. Weak additional signals arise from methyl and vinyl groups at the ends of the chains. Branching raises the ratio of methyl to methylene groups and introduces additional signals assigned to such groups as

$$CH_2{=}C\langle\ .$$

Methine resonances, if present, would be expected to fall so close to those of the methylene groups that it would be almost impossible to find them. Characterization of the polymer then requires that the various weak signals, comprising perhaps 1% or less of the total spectral intensity, be separately detected and integrated. This can be done quantitatively using transient-averaging computers with several hundred scans over the spectrum to achieve the necessary signal-to-noise ratios, especially if a spectrometer operating at 100 or 220 MHz is available [156].

The classic examples of the use of nmr to study stereoregularity is polymethylmethacrylate [157], having the repeating unit

$$-CH_2-\overset{\displaystyle CH_3}{\underset{\displaystyle X}{\overset{|}{\underset{|}{C}}}}- \qquad \text{with} \qquad X = -C\!\begin{array}{l} {}^{\displaystyle /\!/ O} \\ {}_{\displaystyle \backslash O-CH_3} \end{array}$$

The spectrum of this material is comparatively easy to analyze because the three main proton signals, from methyl, methylene, and methoxy groups are well separated, and there is no fine structure from spin-spin coupling unless the two protons in a methylene group become magnetically non-equivalent. The key to all questions of stereoregularity [158] is the recognition that in chains of subunits of this type every second carbon atom is asymmetrically substituted, so that the local configuration may be represented by the usual symbols* *d* and *l*. If the sequence of *d*'s and *l*'s is random, the polymer is said to be *atactic*. Other possibilities, arising in response to specific changes in the conditions of polymerization, include the following. The chain may consist exclusively of *d*'s or *l*'s, which makes the polymer *isotactic*. There may be a regular alternation, *dldldldl*, in which case the material is *syndiotactic*. There may be isotactic regions of varying length, such as *dddllllldd* · · · ; such a polymer may be considered as a block copolymer of *d*- and *l*-subunits.

The form of the methylene resonance is determined primarily by the configurations of the subunits containing the methylene group and of the immediately adjacent one. In an isotactic polymer the arrangement must be *dd* or *ll* and the pair of subunits is said to constitute an isotactic dyad. In a syndiotactic chain only syndiotactic dyads, *dl* or *ld*, occur, while atactic or block polymers contain dyads of both types in varying proportions. The drawings of these dyads in the extended configuration correctly suggest that protons *a* and *b* should have different chemical shifts, because of the difference in environment, if the dyad is isotactic, but essentially the same shift if the dyad is syndiotactic. Including further the effect of the coupling constant J_{ab}, the CH_2 signal from an isotactic dyad is predicted, and found, to consist of

CH_3 CH_3 / H_a H / X H_b X H

Isotactic dyad

CH_3 X / H_a H / X H_b CH_3 H

Syndiotactic dyad

* However, with this notation the asymmetric carbon atoms in a subunit of a polymer chain are labeled in a way that may be inconsistent with the labeling of corresponding asymmetric centers in the analogous small molecule. An alternative notation based on the letters *m* and *r*, for *meso* and *racemic* dyads, has been proposed [158a].

four lines, while that from a syndiotactic dyad is a singlet. At 220 MHz, additional structure is resolved which reveals that longer-range interactions slightly modify the chemical shifts. It is then useful to classify the dyads further according to the configurations of the tetrads in which they occur. A *dl* dyad, for example, may find itself in three environments suggested by the symbols *d*(*dl*)*d*, *d*(*dl*)*l*, and *l*(*dl*)*d*. In the *d*(*dl*)*l* and *l*(*dl*)*d* tetrads, the methylene protons within the central dyad are again equivalent, but in the *d*(*dl*)*d* tetrad they are nonequivalent and produce an *AB*-pattern. Similarly, *dd* dyads may have the environments *l*(*dd*)*l*, *l*(*dd*)*d*, or *d*(*dd*)*d*, so that they give rise to three noncoinciding *AB* multiplets.

The chemical shift of a methyl group in this polymer depends primarily on the configurations of the subunit in which the signal originates and of its two nearest neighbors. Three types of triad are possible, the isotactic (*ddd* or *lll*), the heterotactic (*ddl* or *dll*), and the syndiotactic (*dld* or *ldl*). Again drawings that represent the triads in the extended form suggest why there should be differences between the chemical shifts. In the isotactic triad, the central

CH_3 CH_3 CH_3 / H H H / X H X H X H — Isotactic triad

CH_3 CH_3 X / H H H / X H X H CH_3 H — Heterotactic triad

X CH_3 X / H H H / CH_3 H X H CH_3 H — Syndiotactic triad

methyl group is sandwiched between two other methyls, in the heterotactic triad between a methyl and an X, and in the syndiotactic triad between two X's. In polymethylmethacrylate, the result is that the isotactic methyl signal is shifted about 0.25 ppm downfield from that of the syndiotactic methyls, with the heterotactic methyl peak about midway between them.

Polymers of monosubstituted ethylenes are formally analogous to polymethylmethacrylate but more difficult to work with because of the additional

H_a H_c H_a H_c

—C—C—C—C—

H_b X H_b X

spin-spin interactions J_{ac} and J_{bc}. If, as in polyvinylchloride, the chemical shift of H_c is different enough from those of H_a and H_b, the spectrum can be simplified by spin decoupling [159]. In other cases, such as polypropylene [160], the shifts are too close together to allow this, but the necessary simplification of the spectrum can be achieved by selective deuteration which removes some of the signals and reduces the multiplicity of those that remain. The relatively large line widths characteristic of polymers do not permit very closely spaced multiplets to be resolved, and therefore the proton-deuteron spin interaction, which can be observed in small molecules, is unresolved.

From what has been said it should be clear that the overall success of the nmr approach varies a great deal from one polymer to another. Often the key factor is the discovery of a solvent that favorably affects the relative chemical shifts of particular atoms. Comparisons of the polymer spectra with those of model compounds may be helpful, especially in attempts to determine the chain configuration [161]. For vinyl polymers, for example, the corresponding *meso* or *d,l*-2,4-disubstituted pentanes or even the 2,4,6-trisubstituted heptanes may serve as models.

Biological Macromolecules

Biological macromolecules generally differ from synthetic polymers in that they comprise a much larger variety of repeating units in a single molecule, that the molecules in a single sample often have a sharply defined molecular weight, and the chains have more-or-less rigidly fixed configurations as long as the material is in the native form. For proteins, hydrogen signals originate mainly in the 20 or so kinds of sidechain R of the commonly occurring amino acids. Heating, adding denaturants, or using such nonaqueous solvents as trifluoroacetic acid, converts native proteins to a random-coil state in which each subunit

```
      R  O
      |  ‖
 —N—C—C—
   |  |
   H  H
```

produces a spectrum rather similar to that of the corresponding monomeric amino acid, NH_2—CHR—COOH, except for the expected slight broadening of the lines [162]. The overall spectrum then consists of a very large number of signals. Even at 220 MHz not all of these can be resolved, but a number of peaks can be identified and assigned to protons of certain types [163]. Thus the methyl groups of valine (R = isopropyl), leucine [R = $CH_2CH(CH_3)_2$], and isoleucine [R = $CH(CH_3)CH_2CH_3$] together produce a fairly sharp peak at the high-field edge of the spectrum of a typical protein, and a peak at much lower fields corresponds to aromatic protons in phenylalanine

$$(R = CH_2C_6H_5),$$

tyrosine ($R = CH_2C_6H_4OH$), tryptophan

CH_2–(indol-3-yl, N–H)

and histidine

$R = CH_2$–(imidazol-4-yl, N–H)

Such spectra furnish a limited amount of information about the amino acid composition.

If the amino acid composition of a protein is already known, its spectrum in the random-coil form can be simulated with reasonable accuracy by superimposing suitably broadened spectra of the various residues with appropriate relative intensities. One suggested application of this procedure is to test whether or not under specified experimental conditions the protein indeed consists entirely of random coil, since residual secondary structure is apt to restrict the rotational freedom of some of the subunits enough to broaden the corresponding signals so severely that they appear to be missing [163a]. Similarly, specific binding of an added paramagnetic ion by one type of residue may cause a large shift or broadening of the corresponding peaks (Section 10) so that a "missing" signal provides evidence of such interactions [121].

Of even greater interest are the pronounced differences between the spectrum of denatured or randomly coiled protein and of the same material in the native form. As expected, the latter produces considerably broader peaks, and the 60-MHz spectra, although far from featureless, generally show too little detail to be very useful. At 220 MHz, however, even native protein spectra can be resolved into a large number of peaks with chemical shifts which may be greatly different from those found for the denatured species [163]. These differences reveal specific interactions between side chains, usually not on adjacent subunits, which are held together by the folding of the protein chain. An especially striking example is furnished by some of the methyl groups from leucine, isoleucine, and valine in lysozyme, which contribute to a peak in the standard region for methyl groups when the enzyme is denatured, while in the native form they produce a series of peaks some of which are shifted upfield by as much as 1.5 ppm. This has been rationalized by postulating that some of these groups find themselves locked in position near the aromatic rings of other amino acid side chains with the ring anisotropy inducing the observed upfield shifts. Even larger shifts of this sort were found for cytochrome c, in which they were ascribed to the unusually large ring current effect of a prophyrin moiety.

Synthetic polypeptides have interested protein chemists because although they do not show the tertiary structure of globular proteins they are capable of existing either in randomly coiled or in helical form, depending on the solvent and temperature. Transition between these configurations are somewhat analogous to the denaturation or renaturation of proteins and they induce changes in various spectroscopic properties, including the chemical shifts and line widths [164].

The nucleic acids constitute another class of biopolymers with structures that depend strongly on the nature and temperature of the medium. In the native double-helical form, molecular motion is so restricted that the proton signals are very broad. At a "melting" temperature near 75°C the strands separate and the resulting material gives relatively sharp peaks as expected from a randomly coiled polymer [165]. Again the amount of information that can be derived from the spectra increases dramatically when an instrument operating at 220 MHz is used [166]. With such a spectrometer it was found that single-stranded DNA in neutral D_2O solution at about 90°C gave a fairly well-resolved spectrum in which a particularly interesting feature is a pair of peaks 0.12 ppm apart attributed to methyl protons of a thymine group. The corresponding monomer gives only a single peak.

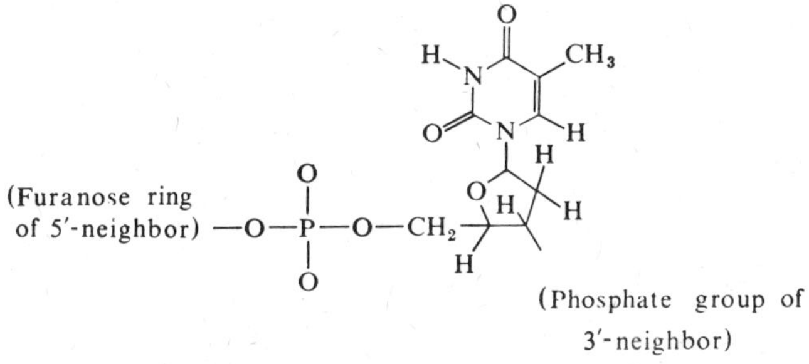

(Furanose ring of 5'-neighbor) —O—P—O—CH_2—

(Phosphate group of 3'-neighbor)

Thymine-containing subunit of DNA

The relative intensity of these two peaks depends on the species of organism from which the DNA is obtained. It appears that thymine methyl groups contribute to the lower-field resonance if they experience the ring current effect of a neighboring purine base, which is readily possible if the purine is in the neighboring 5′-position and the local configuration resembles that in helical DNA. If the corresponding position is occupied instead by a pyrimidine base, the thymine methyl resonance falls at the higher field.

13 SPECTRA OF NONMETALLIC, DIAMAGNETIC SOLIDS

Rigid-Lattice Spectra

The nmr technique has been used to study a great variety of solid materials including metals, alloys, semiconductors, superconductors, ionic crystals, and molecular crystals [167]. The boundary between solid-state chemistry and solid-state physics is so indefinite that it is difficult to justify any selection of topics for inclusion in this survey. Since such a choice was inescapable, it was decided to discuss those applications involving nonmetallic, diamagnetic crystals that provide information either about molecular structures or about the type and extent of molecular reorientation in the lattice.

The most important characteristic of most, although not all, nuclear resonances from solid samples is their large width, often amounting to 1000 ppm or more. When the lines are so broad, satisfactory signals can be obtained with spectrometers in which the magnetic field is much less homogeneous than that required for high-resolution work. For this reason solid-state spectra began to be actively exploited in the earliest days of nmr spectroscopy, and the recent advances in magnet technology have had relatively little impact on this area.

The large line widths usually do not allow one to observe the chemical shift and spin-spin coupling parameters on which most applications of high-resolution nmr depend. Instead, the raw data consist of line shapes, widths, and relaxation times. In discussing these it is convenient to treat separately the nuclear species with $I = \frac{1}{2}$, in which there is no interaction between a nuclear electric quadrupole moment and the electric field within the crystal lattice, and the species with $I \geq 1$, in which this interaction is usually the dominant one. For nuclei with $I = \frac{1}{2}$, the resonant frequency differs from the calculated value, $\nu_0 = 2\mu H_0/h$, primarily because of magnetic dipole-dipole interactions between each nucleus and its near neighbors, which are absent for liquid samples because of the time-averaging effects of molecular motions. A secondary contribution arising from the electron-coupled spin-spin interactions is often negligible and is not included in this discussion.

The simplest situation consists of a lattice in which the magnetic nuclei occur as isolated pairs, so that each spin interacts only with a single neighbor. Because two orientations are possible for the neighboring spin, the resonance is split into a doublet with a separation dependant on the internuclear distance r and the angle θ between the external field direction and the line connecting the two nuclei. For a pair of identical nuclei the splitting is [168]:

$$\Delta H = (3\mu/r^3)(3\cos^2\theta - 1). \tag{7.96}$$

Such splittings were first found for single crystals of gypsum, $CaSO_4{\cdot}2H_2O$, although the observed spectrum consisted of two doublets because pairs of protons, each corresponding to a single water molecule, occur at two crystallographically different sites [169]. In the real crystal the interactions of each spin with more distant nuclei contribute no additional resolvable find structure, but they cause appreciable broadening of each line. Changing the orientation of the crystal with respect to the field causes the factor $3\cos^2\theta - 1$ to take on all values between $+2$ and -1. When the alignment that gives the largest ΔH has been found, one can readily determine r and also calculate the direction of the internuclear vector with respect to the crystal axes.

If it is necessary to use a powdered specimen instead of a single crystal, the spectrum is found to consist of superimposed doublets corresponding to all possible values of θ. It still consists of a double-peaked curve, and the distance between the maxima together with the overall shape allows the internuclear distance to be evaluated.

When the nuclei occur as isolated, equivalent triplets, for example, in compounds containing methyl groups, the resonance is split into a symmetrical triplet [168, 170]. In this situation, and for a few other very simple arrangements, it is still possible to evaluate internuclear distances using only the nmr data, although the problem may be further complicated by effects of molecular motions, such as rotation of a methyl group about its symmetry axis.

For more complex arrangements of the nuclei, the signal no longer exhibits separate maxima but consists of a symmetrical curve which, following Van Vleck, may be characterized by a sequence of "moments." If $\bar{H}$ is the distance in magnetic field units from the center of the resonance, and the absorption intensity is given by the function $g(\bar{H})$, the nth moment is defined:

$$S_n = \int_{-\infty}^{\infty} (\bar{H})^n g(\bar{H})\, d\bar{H} \Big/ \int_{-\infty}^{\infty} g(\bar{H})\, d\bar{H}. \tag{7.97}$$

Because of the symmetry of $g(\bar{H})$, S_n is zero whenever n is an odd number. The most important moment is the second, which is equivalent to the mean-square line width. For a single crystal containing only one species of magnetic nuclei, the second moment is given by:

$$S_2 = (3/2N)I(I + 1)\gamma^2 h^2 \sum_{i>j} r_{ij}^{-6}(3\cos^2\theta_{ij} - 1)^2. \tag{7.98}$$

The sum contains terms representing the interaction of each of the N nuclei in a unit cell with all others in the same or adjacent cells. Because of the factor r_{ij}^{-6}, the total contribution from distant nuclei is small and can be evaluated by approximation methods, so that S_2 depends primarily on the values of θ_{ij} and r_{ij} for near neighbors. An analogous formula involving terms of the same kind has been presented for crystals containing more than one magnetic nuclear species. For polycrystalline specimens S_2 depends only on the r_{ij}-values; the appropriate equation is also available [168].

Evidently, the number of parameters required to fix S_2 is usually too large to allow structural data to be deduced in the absence of other information, but the second moment is very useful when diffraction studies provide all but one or two of the structural parameters. For example, the only unknown in the structure of NH_4Cl as determined by x-ray diffraction is the N—H bond distance. All the needed r_{ij}-values can be expressed in terms of this distance which can then be evaluated from the second moment of the polycrystalline material [171] with an accuracy of about ± 0.004 Å. When single crystals are used, the variation of S_2 with changes in the crystal orientation may provide enough data to make it possible to evaluate more than one unknown.

In special cases the second moment may be small enough to allow chemical shift differences to be observed, especially for nuclei of elements other than hydrogen in which the shifts are much larger [172]. Some crystals contain such low concentrations of magnetic nuclei that all or most r_{ij}- values in the second-moment formula are large, and then no special techniques are required to sharpen the signals. An example is trichloroacetic acid [173], with a unit cell containing two dimer units. If the sample is a single crystal, the protons in each dimer give rise to a doublet with components whose width depends on the orientation and may be as low as 1.2 kHz. The individual peak positions change by more than 100 ppm (10 kHz at 100 MHz) as the crystal is rotated because of changes in the $3 \cos^2 \theta_{ij} - 1$ terms and in the chemical shift. Such data make it possible to determine not only the averaged value of the chemical shift, which determines the resonance position in the liquid phase, but also the shift anisotropy, that is, the dependence of the shift on the alignment of the molecules with respect to the magnetic field.

A somewhat more generally applicable method of obtaining chemical shifts for solids, but one that yields only the usual average value, is to eliminate the dipolar broadening by spinning the sample [174]. For a rapidly rotating specimen, the time-averaged dipole interaction is found to be proportional to $3 \cos^2 \alpha - 1$, where α is the angle between the magnetic field and the axis of rotation. This quantity vanishes when $\alpha = 54°44'$, requiring a rather awkward geometrical arrangement for the magnet and probe. To reduce the line width by this means, the spinning rate in revolutions per second must be at least as large as the line width in hertz for the stationary sample. An air turbine spinner has been described [175] which gives speeds up to 10^4 sec^{-1}, so that the technique appears to be limited to materials in which the resonance is no broader than a few kilohertz initially. Either single crystals or powdered samples can be used.

More recently, a method based on pulsed nmr as opposed to steady-state absorption spectra has been developed which yields chemical shifts and shift anisotropies for solids [176]. This is briefly discussed in Section 15.

Turning to nuclei with $I > \frac{1}{2}$, the simplest case is a crystal containing a single such species all at crystallographically identical sites. The resonance is then split into $2I$ components by electric quadrupole interactions. If the electric field near the nucleus is axially symmetric and the magnitude of the interaction is small compared to the magnetic energy, the quadrupole energy is [177]:

$$E_q = \frac{eQ}{4I(2I-1)} \frac{\partial^2 V}{\partial z^2} [3m_I^2 - I(I+1)]. \tag{7.99}$$

The z-axis is, as usual, parallel to the magnetic field, and Q is the nuclear quadrupole moment. The electric field gradient $\partial^2 V/\partial z^2$ depends on the charge distribution within the lattice and on the now familiar quantity $3 \cos^2 \theta - 1$, where θ is now the angle between the z-axis and the local symmetry axis. The usual transitions are found, with $\Delta m_I = \pm 1$. When appropriate numerical values are substituted into (7.99), it is found that the transitions should be displaced by an amount large enough to make the resulting multiplet readily resolvable, if a single crystal is used, even though each component is broadened by the dipolar interactions. Providing I is half integral, there will be one transition, $m_I = \frac{1}{2} \leftrightarrow -\frac{1}{2}$, which is unaffected by the quadrupole interaction since both levels involved are shifted equally.

A number of factors can complicate the quadrupolar splitting. If more than one type of lattice-site is occupied by a quadrupolar nucleus, the spectrum consists of superimposed multiplets, one from each species. If the sample is polycrystalline, each crystallite again has its own θ-value, and the multiplet becomes so smeared out that only the central $m_I = \frac{1}{2} \leftrightarrow -\frac{1}{2}$ transition (assuming I to be half integral) can be detected. If E_q is not much smaller than the magnetic interaction energy $h\nu_0$ (7.99) no longer applies, and even the $m_I = \frac{1}{2} \leftrightarrow -\frac{1}{2}$ transition is displaced by an amount that depends on θ. It is of course well known that when E_q is very large radiofrequency transitions between states of different m_I can be found even without any magnetic field. The direction of quantization is then fixed in the crystal lattice, and the spectrum is a *pure nuclear quadrupole spectrum*. If the local electric field is not axially symmetric, the field gradient no longer depends on a single angle θ and the $2I$ multiplet components change position in a more complicated way as the crystal is rotated. However, for atoms in sites of cubic symmetry, the field gradient vanishes and the quadrupole effect is eliminated.

Molecular Motions

So far it has been assumed that each atom remains fixed in the lattice to that quantities such as r_{ij} and θ_{ij} are constant. Since the signals are then very broad, the transverse relaxation time, which is defined in terms of the line width as $T_2 = 1/\pi\,\Delta\nu$, is quite short. The longitudinal relaxation time, which is about equal to the lifetime of a given spin state, is several orders of magnitude larger. In that $T_1 \gg T_2$, the solid resembles the limiting state approached by liquids as the viscosity is increased indefinitely, as described in Section 9. In real crystals, however, some of the atomic coordinates quite commonly become time dependent even at temperatures far below the melting point. The process responsible for this may be intramolecular, as is the hindered rotation of a methyl group around its symmetry axis, it may consist of reorientation of an entire molecule with no change in the location of its center of gravity, or it may involve motion of molecules or ions from one lattice site to another, that is, self-diffusion. If the molecular motion is rapid enough, some of the broadening interactions are averaged out. Just as it happens in a viscous liquid when the viscosity is reduced, this leads to an increase in T_2 as the lines become narrower, accompanied by a reduction of T_1.

Molecular reorientation in a solid is always opposed by some potential energy barrier so that unless quantum-mechanical tunneling is invoked an activation energy E_a is required. The fraction of the molecules whose thermal energy exceeds E_a is proportional to $\exp(-E_a/RT)$. Then the effective frequency of reorientation should obey:

$$f = f_c \exp(-E_a/RT), \tag{7.100}$$

where f_0 is a constant. At sufficiently low temperatures f is so small that the lattice is essentially rigid. As the material is warmed, a range of temperatures is reached in which reached f is approximately equal to the line width, in frequency units, of the broadened signal, and the width gradually diminishes as this range is traversed. At higher temperatures the signal again resembles that for a rigid crystal, but one in which the relevant coordinates are replaced by time-averaged values. Several types of reorientation with different activation energies may become effective at different temperatures so that a stepwise reduction of the line width with heating is observed. For substances that exist in more than one solid modification, f may be quite small below the transition temperature and very much larger above it, so that the width of the resonance changes abruptly at the transition point. T_1 is also temperature dependent, passing through a minimum when f reaches the Larmor precession frequency of the nuclei.

In detail, the changes in line shape associated with the onset of molecular reorientations vary considerably from crystal to crystal. If the lattice contains isolated pairs of identical nuclei with $I = \frac{1}{2}$, and these rotate about an axis which makes an angle γ with the internuclear line, the spectrum in the limit of fast rotation is a doublet, as for a rigid crystal, but the separation is given not by (7.96) but by [168]:

$$\Delta H = (3\mu/2r^3)(3 \cos^2 \theta' - 1)(3 \cos^2 \gamma - 1), \tag{7.101}$$

where θ' is the angle between the magnetic field and the axis of rotation. Again the spacing depends on the crystal orientation and the same material in polycrystalline form produces a broadened doublet. In this case the molecular motion does not change r, and the peak-to-peak separation can be used to find r if γ is known. The latter can often be deduced from an approximate knowledge of the molecular structure, and it quite commonly is 90°. In more complicated cases the individual terms contributing to the second-moment formula [see (7.98)] become modified in a somewhat analogous way, but allowance must be made for the fact that not only the angles θ_{ij} but also some of the distances r_{ij} need to be averaged, and considerable computational effort may be required.

For highly symmetrical entities such as NH_4^+ ion, rotation about several different axes is equally probable, and intramolecular dipolar broadening may be completely averaged out. The intermolecular contributions remain, but they in turn are eliminated if self-diffusion rates become sufficiently high. A number of nearly spherical organic molecules form crystals that yield lines almost as sharp as those from the melt until they are chilled far below their freezing points [178]. The best known example is cyclohexane [179], freezing at 6.5°C to form crystals which produce a very sharp line at temperatures to −35°C.

Adsorbed Species

When molecules such as water or benzene are adsorbed on solid surfaces, their motion is restricted and their resonances take on some of the characteristics of signals from solids. The degree of motional freedom depends on the temperature, the extent of surface coverage, and the nature of the bonding between the adsorbate and the surface. In most instances pulsed nmr techniques have been used in such studies to find the temperature variation of the relaxation times [180], but broad-line or even high-resolution spectra may be obtained, especially when the amount adsorbed is considerably more than one monolayer [181]. It has been reported that certain organic liquids produce narrow resonance lines even at low coverages on pyrogenic silicas [182]. The obvious difficulty with this work is that the small amount of adsorbed material present in a typical sample causes the signals to be very weak.

14 SPECTRA OF ORIENTED MOLECULES IN NEMATIC SOLVENTS

Dipolar Coupling and Molecular Geometries

It has been recognized for some years that the usual distinction between solids and liquids, based on the high degree of order in the former and complete absence of long-range order in the latter, is not strictly applicable to all substances. On the one hand, as noted above, molecular motion and a consequent partial breakdown of the order causes characteristic effects in the nmr spectra of many solids well below the melting point. On the other hand, some materials do not at once attain a state of complete disorder above their melting points. For compounds with rather elongated molecules, so-called nematic, smectic, or cholesteric *meso*phases are stable over a temperature range extending from the melting point to a further transition point, above which the substance becomes an isotropic liquid. The nematic mesophase is unique in that it can be used as a medium that allows one to observe a totally new type of high-resolution nmr spectrum which is a sense intermediate between those from crystals and from ordinary isotropic solutions [183–187].

A typical nematogenic compound is *p*,*p*′-di-*n*-hexyloxyazoxybenzene,

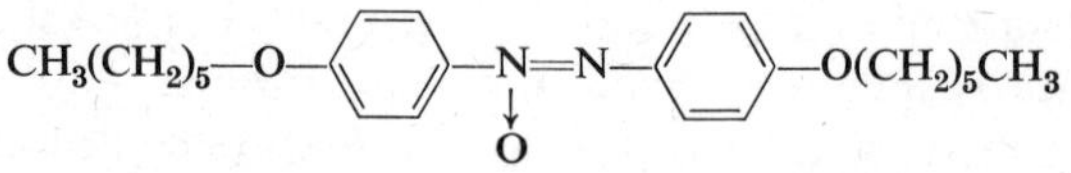

$$CH_3(CH_2)_5\text{—O—}C_6H_4\text{—}\underset{\overset{\downarrow}{O}}{N}\text{=N—}C_6H_4\text{—}O(CH_2)_5CH_3$$

which melts at 80°C, forming a nematic mesophase which persists to 125°C. The mesophase is only slightly more viscous than the isotropic liquid, has a turbid appearance, and is optically anisotropic. These properties apparently reflect the existence of extended regions or domains within which the molecules lie with their long axes approximately parallel although their centers of gravity are randomly distributed and diffusional motion is rapid. When the nematic liquid is placed in a magnetic field, each domain is subject to an orienting force, and if the field strength is a few thousand gauss or more the molecules throughout the liquid become aligned with their long axes parallel to the applied field. The nmr spectrum of such a substance differs from that of a single solid crystal in that the intermolecular dipolar interactions are averaged out by self-diffusion. Because of the persistent orientation of the molecules, the intramolecular dipolar interactions remain and cause the spectrum to consist of a very large number of lines. Any one line is then likely to be too weak to detect, hence any high-resolution signals actually observed are attributable not to the nematic liquid itself but to smaller molecules that have been dissolved in it.

A solute molecule in a nematic solvent finds itself in an anisotropic environment, and its rotational motion is usually restricted to some extent so that it too tends to be oriented with respect to the magnetic field. Again, the situation is somewhat similar to that in a solid except that intermolecular dipolar broadening is absent, but if the number of magnetic nuclei is not too large the *intra*molecular interactions give rise to a limited amount of new fine structure instead of causing the spectrum to be obliterated. In favorable cases these splittings can be used for the quantitative determination of internuclear distances and bond angles with an accuracy that compares favorably with that attainable by rotational spectroscopy or diffraction methods.

The simplest spectrum is that of a molecule containing only two equivalent protons. The intramolecular dipolar interaction is then almost the same as in a single crystal in which the magnetic nuclei occur as isolated pairs, and the spectrum consists of two peaks. The line separation for the crystal is given by (7.96) of the preceding section or, in frequency units, by:

$$\Delta\nu = \frac{3}{4\pi}\gamma^2\hbar r^{-3}(3\cos^2\theta - 1), \tag{7.102}$$

where r is the internuclear distance and θ is the angle between the internuclear axis and the magnetic field. In the liquid the factor $3\cos^2\theta - 1$ must be replaced by its time average, while r remains constant if the solute is considered a rigid molecule. In an isotropic liquid of course, $(3\cos^2\theta - 1)_{\text{ave}}$ is zero, but in a nematic solvent this term has some finite value and the resonance is a doublet. The splitting may be as large as several thousand hertz and can be used to evaluate $(3\cos^2\theta - 1)_{\text{ave}}$ if r is known.

For molecules containing a pair of protons with different chemical shifts, the spectrum consists of two doublets and the splittings depend both on the right-hand side of (7.102) and on J_{ab}. If instead there are two magnetic nuclei of different elements, each with $I = \frac{1}{2}$, each species gives a doublet resonance with:

$$\Delta\nu = J_{ab} + (\gamma_a\gamma_b\hbar/2\pi r^3)(3\cos^2\theta - 1)_{\text{ave}}. \tag{7.103}$$

As the number of magnetic nuclei increases, the spectrum rapidly becomes more complicated. For benzene, for example, some 50 individual peaks have been observed [188], and for molecules much larger than this the analysis of the spectrum becomes extremely difficult as does the task of obtaining adequate signal strength.

Most of the molecules for which structural information has been obtained by this method either contain only chemically equivalent nuclei (protons or fluorine atoms) or are of the A_mX_n type, for example, butyne or propargyl chloride. Even benzene and cyclopropane, with only chemically equivalent protons, produce highly complex spectra because the existence of several distinct spin-spin coupling constants makes the nuclei magnetically non-equivalent. The line positions and intensities are determined by the chemical shift differences $\Delta\nu_{ij}$ (if any), the indirect spin-spin coupling constants J_{ij}, and the direct interaction parameters D_{ij} defined:

$$D_{ij} = -(\gamma_i\gamma_j h/4\pi^2 r_{ij}^3)(3\cos^2\theta_{ij} - 1)_{\text{ave}}. \tag{7.104}$$

Unfortunately, it is seldom possible to evaluate the parameters by inspection from the observed peak separations. For the simplest cases a theoretical analysis first developed for molecules oriented by an electric field [188a] yields equations for frequencies and intensities that allow the spectra to be interpreted without machine computations [188b, 188c]. Otherwise, one must obtain the parameters by an iterative procedure with the help of a digital computer [189]. Trial D_{ij}-values can be calculated from an approximate knowledge of the structure and reasonable estimates for the $3\cos^2\theta_{ij} - 1$ terms. The $\Delta\nu_{ij}$-values and the magnitudes of the J_{ij}-values can be evaluated from high-resolution spectra of the material in question, substituted if necessary with ^{13}C or deuterium (see Section 5, p. 628). If the trial parameters are carefully chosen the computed pattern will be in moderately good agreement with the observed one. The discrepancies can be reduced by successive adjustment of the trial parameters until essentially complete agreement is obtained, and thus the D-values can be accurately determined.

The calculated spectra are unaffected if the signs of all J-values and D-values are simultaneously reversed, but they depend on the relative signs taken for any two of these quantities. Therefore it is only necessary to fix the sign of any one D_{ij}-value by some other method if one wishes to determine the correct sign for all the remaining parameters. This can often be done with the help of certain empirical rules which are becoming increasingly well established as more and more systems are investigated. For example, elongated molecules tend to line up with their long axes parallel to the magnetic field so that $(3\cos^2\theta_{ij} - 1)_{\text{ave}}$ should be positive when the ij line is parallel to the molecular axis. Planar molecules tend to be oriented so that the vector normal to the plane is perpendicular to the field. Unless the molecule is highly asymmetric, rotation about an axis perpendicular to the plane remains essentially free, and then $(3\cos^2\theta_{ij} - 1)_{\text{ave}}$ is positive for any ij line in the molecular plane. In a few cases such an assignment has been verified by comparing the derived sign of one of the J-values with a value obtained from the spectrum of a polar molecule oriented by a strong electric field [183]. The possibility of determining the signs of indirect spin-spin coupling constants with a high degree of confidence has been a major factor stimulating work on liquid-crystal nmr spectra.

It follows at once from (7.104) that information about the molecular structure can be obtained when the D-values are known. The chief obstacle is that this requires a knowledge of the $(3\cos^2\theta_{ij} - 1)_{\text{ave}}$ terms. This problem is greatly simplified if the molecule has a threefold or higher axis of symmetry, which may be called the c-axis. Then the liquid crystal environment does not interfere with rotation about the c-axis, and if the angle between this axis and the field direction is γ, all orientations of the molecule for which γ is the same are equally probably. If the angle between the internuclear line ij and the c-axis is $\bar{\theta}_{ij}$, it can be shown that:

$$(3\cos^2\theta_{ij} - 1)_{\text{ave}} = \tfrac{1}{2}(3\cos^2\bar{\theta}_{ij} - 1)(3\cos^2\gamma - 1)_{\text{ave}}. \qquad (7.105)$$

The quantity $\frac{1}{2}(3\cos^2\gamma - 1)_{\text{ave}}$ is often abbreviated as S_{cc}. The angles $\bar{\theta}_{ij}$ are determined by the molecular geometry, often in a very simple way. In benzene, for example, the sixfold symmetry axis is perpendicular to the plane that contains all the ij lines, and therefore every $\bar{\theta}_{ij}$ is 90° and all the $(3\cos^2\theta_{ij} - 1)_{\text{ave}}$ terms are simply equal to $-S_{cc}$.

Continuing with this example, there are three distinct D-values for the proton-proton interactions, which in an obvious notation may be called D_{12}, D_{13}, and D_{14}. If the carbon–carbon and carbon–hydrogen bond distances are d_{CC} and d_{CH}, it follows from the hexagonal symmetry that $r_{12} = d_{\text{CC}} + d_{\text{CH}}$, $r_{13} = \sqrt{3}(d_{\text{CC}} + d_{\text{CH}})$, and $r_{14} = 2(d_{\text{CC}} + d_{\text{CH}})$. Then,

$$\begin{aligned} D_{12} &= K_{\text{HH}}(d_{\text{CC}} + d_{\text{CH}})^{-3}S_{cc} \\ D_{13} &= (3\sqrt{3})^{-1}K_{\text{HH}}(d_{\text{CC}} + d_{\text{CH}})^{-3}S_{cc} \\ D_{14} &= (\tfrac{1}{8})K_{\text{HH}}(d_{\text{CC}} + d_{\text{CH}})^{-3}S_{cc}. \end{aligned} \qquad (7.106)$$

Any one of these D-values therefore fixes the value of the ratio $S_{cc}/(d_{\text{CC}} + d_{\text{CH}})^3$, but the three quantities cannot be evaluated separately. In principle, the situation could be improved by observing lines from the ^{13}C containing species and determining the D_{CH} for directly bonded carbon and hydrogen, which would be equal to $K_{\text{CH}}S_{cc}/d^3{}_{\text{CH}}$. Even then, one other piece of information, such as a value of d_{CC} from a diffraction study, would be needed in order to find S_{cc} and d_{CH}.

When the spectra of other symmetrical molecules are analyzed, it is again found that at least one structural parameter must be fixed by an independent method and only then can the remaining ones and S_{cc} be determined. Fortunately, this limitation is not as serious as it appears at first glance, because for many of these molecules diffraction studies provide accurate values of all internuclear distances except those involving hydrogen. The nmr method can then be used to locate the hydrogen atoms, and its attractiveness is largely a result of the fact that hydrogen coordinates are the most difficult structural parameters to evaluate by other methods.

It has not been possible to specify quantitatively the reliability of structural data from liquid crystal spectra, but two sorts of evidence are available that bear on this point. First, since the number of D_{ij}-values usually exceeds the number of parameters to be evaluated, two or more different combinations of D_{ij}-values can be used and the results examined for self-consistency. For cyclopropane [185] three calculations of the C—H bond distance gave results which deviated from their average, 1.117 Å, by at most 0.010 Å. The agreement is about as good as that obtained when structures are calculated from rotational constants obtained from various combinations of isotopic species for a typical molecule [190]. The second approach is direct comparison of the nmr results with those of microwave or diffraction studies. The agreement is often very encouraging [187, 191–193], but when larger or somewhat flexible molecules are taken [185, 187], discrepancies up to several hundredths of an angstrom are found.

The fact that the various structural methods do not always give concordant results is not suprising when one considers the fundamental physical differences between them. Microwave spectra yield structures for gaseous molecules in the lowest vibrational state and fairly low-energy rotational states. Internuclear distances are derived from the rotational constants which in the simplest case, that is, for diatomic molecules, measure $(1/r^2)_{\mathrm{ave}}$. Because of the nonrigidity of the molecule, the value of r obtained by taking the square root of the reciprocal of $(1/r^2)_{\mathrm{ave}}$ is not simply equal to the average internuclear distance. Electron diffraction results differ in that the patterns are determined by simple averages of distances and in that the averaging is not only over the zero-point vibrational motion but over the entire assembly of molecules, including some in higher rotational and vibrational states.

Three explanations now suggest themselves for differences between nmr results and those of the other two methods. First, the molecules examined by nmr are in a condensed phase and might be structurally distorted by interactions with the asymmetric medium. Second, the dipolar couplings depend on quantities of the type $(1/r_{ij}^3)_{\mathrm{ave}}$, and the cube root of the reciprocal of such a quantity is not expected to be the same as $(r_{ij})_{\mathrm{ave}}$. Finally, in the analysis of the data, certain geometrical relations are used that hold strictly for a rigid molecule but not for a vibrating one. For example, in a rigid linear molecule the distance between the two end atoms is equal to the sum of the bond distances, but if bending motions are allowed, the bond distances are unaffected while the end-to-end distance is reduced, so that this equality breaks down. Analysis of the observed discrepancies indicates [185] that effects of the third type are the most important. The required correction evidently varies considerably from compound to compound, and no general method for determining it even semiquantitatively has been proposed.

The extent to which solute molecules are oriented in the liquid crystal environment depends on their shape and on the temperature. For symmetric molecules, it is measured by the quantity S_{cc}, which in principle could take on any value between $+1$ and $-\frac{1}{2}$. The reported values actually range between about $+0.3$ and -0.3. The temperature dependence of S_{cc} has been a major source of experimental difficulty because for most nematic solvents the spectra must be determined at temperatures considerably above room temperature. Conventional methods of heating nmr samples often produce small temperature gradients within the tube, so that molecules in different regions show slightly different values of $(3\cos^2\gamma - 1)_{\text{ave}}$. The result is a broadening of the signals, which becomes more pronounced as the distance from the center of the spectrum increases [188]. This has led to a search for solvent systems that exist in the nematic state at the ambient temperature of the nmr probe, and several such mixtures have been reported [194, 195], including the eutectic formed by $C_2H_5O—C_6H_4—N{=}N—C_6H_4—COOC_4H_9$ and $CH_3O—C_6H_4—CH{=}N—C_6H_4—CH{=}C(CH_3)COOC_3H_7$, which melts at 34°C. Such mixtures can be used even below their freezing points because they are readily supercooled. A further advantage of these solvents is that relatively volatile solutes can be introduced at higher concentrations without having the vapor pressure exceed 1 atm.

For molecules lacking a threefold or higher axis of symmetry, it is sometimes still possible to pick out a c-axis and relate the $(3\cos^2\theta_{ij} - 1)_{\text{ave}}$ terms to a single S_{cc}. This applies, for example, to certain substituted benzenes in which the axis perpendicular to the ring is no longer a sixfold axis but the overall molecular shape is so symmetrical that rotation about this axis is unrestricted. The techniques for dealing with these quantities when there is little or no symmetry has been described, and it has been shown that at most five independent parameters are required to describe the state of orientation of any molecule [189]. Of course as the number of motional parameters increases, the utility of the method as a source of structural information is correspondingly reduced.

Chemical Shift Anisotropy

It has been observed several times that the chemical shift of a solute changes when the solvent undergoes the isotropic-nematic transition. In the absence of other factors, such a change reflects the anisotropy of the solute chemical shift. If the molecule has a threefold or higher symmetry axis, the predicted [186] change in the shift is:

$$\delta_{\text{nem}} - \delta_{\text{iso}} = (\tfrac{1}{3})(3\cos^2\gamma - 1)_{\text{ave}}(\delta_{cc} - \delta_{aa}), \tag{7.107}$$

where γ is again the angle between the symmetry axis and the field, δ_{cc} is the chemical shift for a molecule with $\gamma = 0$, and δ_{aa} is the shift with $\gamma = 90°$. Unfortunately, other factors probably do contribute significantly to the shift difference, so that this equation requires correction. The bulk susceptibility and other contributions to the solvent effect on the chemical shift, discussed in Section 6, may vary substantially when the solvent passes from the isotropic to the nematic state [196].

15 PULSED NMR EXPERIMENTS

Identical Nuclei; Relaxation Times

The overwhelming majority of chemical applications of nmr reported during the last 2 decades has been based on the measurement of steady-state absorption spectra. An alternative technique is to expose the sample to one or more short, intense bursts of radiation at the resonant frequency and to observe the transient behavior of the nuclear magnetization after, or between, these pulses. The utility of such pulsed, or "spin-echo," methods for determinations of relaxation times has been recognized for a long time. It has also been shown that this technique can be used to study chemical exchange processes occurring at rates too large or too small to be evaluated by following the changes in the high-resolution spectra described in Section 7. Recently, renewed attention has been drawn to the fact that pulsed experiments can reveal the same information regarding chemical shifts and coupling constants normally obtained from high-resolution spectra and that in certain cases the pulsed technique is capable of higher sensitivity than the conventional one. Use of pulsed experiments to determine chemical shifts with solid samples has also been described. These applications are briefly reviewed in this section.

The effect of a radiofrequency pulse on a collection of nuclei may be understood with the help of the discussion of the resonance phenomenon in Section 2. In the absence of an oscillating field, the total nuclear magnetization is represented by a vector $\mathbf{M}$ parallel to the z-axis. The effect of the alternating field $\mathbf{H}_1$ is associated with a vector component which rotates around the z-axis at the resonant frequency, and it is convenient to visualize this using a rotating set of axes, x' and y' defined so that the effective component of $\mathbf{H}_1$ remains always in the x'-direction. When $\mathbf{H}_1$ is first turned on, the z-component of $\mathbf{M}$ begins to diminish because of transitions among the nuclear spin levels, and simultaneously a component of magnetization parallel to y' appears. The effect is to leave $\mathbf{M}$ unchanged in magnitude but reoriented so that it lies somewhere in the y'-z plane but not along the z-axis. If H_1 is fairly large and is allowed to act during an interval given by

$$\Delta t = \pi/(2\gamma H_1), \tag{7.108}$$

the $\mathbf{M}$ vector will have been rotated through an angle of 90° so that it becomes parallel to the y'-axis. A pulse that conforms with (7.108) is therefore called a 90° pulse. By the same token, a 180° pulse is one for which $\Delta t = \pi/(\gamma H_1)$, and its effect is to reverse the direction of **M**. Experiments involving 90 and 180° pulses are usually the simplest ones to interpret, but more generally a pulse may be described either by specifying the amplitude of H_1 and the pulse duration Δt or by specifying the angle α through which the pulse rotates the magnetization, given in radians by

$$\alpha = \gamma H_1 \, \Delta t. \tag{7.109}$$

Following the application of a single pulse, the perpendicular component of the magnetization decays according to (7.17). Since the signal in the receiver coil is proportional to $M_{y'}$, it also declines exponentially, and observation of this *free induction decay* provides a straightforward method for the determination of T_2. The value so obtained may depend both on the intrinsic relaxation time for the sample and on the degree of homogeneity of the main magnetic field, that is,

$$(1/T_2)_{\text{eff}} = (1/T_2)_{\text{intr}} + (1/T_2{}^*), \tag{7.110}$$

where the last term represents the line broadening attributable to field inhomogeneity [197].

The effect of a series of two or more pulses depends on the duration of each pulse and on the pulse spacing. An especially useful example is the combination of a 90° pulse followed after a time t_c by a 180° pulse [198]. After the first pulse free induction decay occurs, and if the intrinsic line width is small the relaxation time $(T_2)_{\text{eff}}$ may be determined almost entirely by the magnetic field inhomogeneities. As mentioned in Section 2, this happens because the individual magnetization vectors associated with nuclei exposed to unequal fields precess around the z-axis at unequal rates and gradually become out of phase with one another so that the resultant $M_{y'}$ approaches zero. When the 180° pulse reverses the direction of all the magnetization vectors, those that had fallen behind because of their relatively slow precession rates suddenly find themselves ahead of the rest. The same differences in precession frequencies that originally caused dephasing now tend to bring the spin vectors back into coherence. At a time $2t_c$ after the first 90° pulse, the decay of the transverse magnetization is almost completely undone, and a "spin-echo" signal is observed, followed by another decay.

If a further 180° pulse is applied at $t = 3t_c$, a second echo is made to appear at $t = 4t_c$, and the process can be continued at will. A so-called Carr-Purcell train consists of a 90° pulse at $t = 0$ followed by a series of 180° pulses at $t = t_c, 3t_c, 5t_c$, and so on, and produces a signal comprising a free induction decay following the first pulse and a sequence of echoes with maxima at $t = 2t_c, 4t_c, 6t_c$, and so on. It is assumed here that the duration of each pulse is small compared to either t_c or T_2. The echo maxima grow progressively weaker because not all factors leading to transverse relaxation are reversed in the way just described by a 180° pulse. The echo envelope in fact defines an exponential decay curve $y = y_0 \exp [(-t/T_2)_{\text{intr}}]$ and can be used to determine the true T_2 even when $1/T_2{}^*$ in (7.110) is much larger than $(1/T_2)_{\text{intr}}$. However, in an inhomogeneous magnetic field a correction may be required for a contribution of diffusion to the echo decay, since diffusional motion of a molecule causes the local field it experiences to vary irregularly with time. The method can be adapted to allow determination of self-diffusion coefficients [198a].

Several modifications of these echo experiments have been reported which make it possible also to evaluate T_1. One method [198] is to use a 180° pulse at $t = 0$ followed by a 90° pulse at $t = \tau$, where the interval τ can be systematically varied. After the 180° pulse, the magnetization is parallel to the z-axis, but its direction is opposite to that which corresponds to equilibrium. The system then relaxes, and M_z returns to its equilibrium value according to (7.15). In so doing it passes through zero when $t = T_1 \ln 2$. If the time at which the 90° pulse is applied has any value *other* than $\tau = T_1 \ln 2$, the second pulse operates on a magnetization vector of finite size and is followed by a free induction decay signal in the receiver. However, if $\tau = T_1 \ln 2$, there is at that instant no magnetization to be rotated away from the z-axis, and no free induction decay signal can be produced.

Nonidentical Nuclei; Chemical Shifts and Exchange Rates

If the sample contains two sets of nuclei with *slightly* different Larmor frequencies, it is readily possible for the same radiofrequency pulse to rotate the magnetization vectors corresponding to both sets away from the z-axis. The free induction decay observed after the pulse still consists of an exponentially declining signal, but a sinusoidal oscillation is superimposed on this. The perpendicular magnetization is the sum of contributions from the two sets of nuclei, each of which may be represented by a vector which precesses around the z-axis with the appropriate frequency, for example, ν_A or ν_B. These vectors are parallel at $t = 0$, and the angle between them at some subsequent time t is $2\pi t(\nu_A - \nu_B)$. They are completely out of phase whenever this angle is $(2n + 1)\pi$ and in phase whenever it is $2n\pi$, for integer values of n. The receiver signal consequently oscillates with a period $1/(\nu_A - \nu_B)$, and it thus becomes possible to determine the chemical shift difference, which is proportional to $(\nu_A - \nu_B)$. Since the entire time span during which the decay signal is detectable is of the order of $(T_2)_{\text{eff}}$, one can observe several maxima, reflecting the oscillating component, only if $(\nu_A - \nu_B) > (1/T_2)_{\text{eff}}$. This is the same limitation found in high-resolution spectra, since the line width is given by $\Delta\nu = (1/\pi T_2)_{\text{eff}}$, and discrete signals cannot be resolved if the peak separation is very much less than the line width.

Spin echoes can be obtained for a sample of this type using a sequence of pulses such as a Carr-Purcell train. Each echo now shows an oscillating component with the characteristic period $1/(\nu_A - \nu_B)$. If the intrinsic values of T_2 are the same for both set of nuclei, the maximum intensities of successive echoes are, as before, proportional to $\exp(-t/T_2)$.

If the situation is altered by allowing exchange between the two sets of nuclei characterized by a mean lifetime τ (see Section 7), allowance must be made for the loss of phase coherence among the individual magnetization vectors that results from the abrupt changes of the precession frequencies between the values ν_A and ν_B. The spin echo intensities then depend not only on $T_2{}^0$, the relaxation time in the absence of exchange, assumed equal for both sets of nuclei, but also on τ, $\nu_A - \nu_B$, and on the spacing between echoes, $t_{cp} = 2t_c$. Detailed numerical calculations [199] show that over most practically accessible values of these parameters the echo amplitudes still fall exponentially with time, with an effective relaxation time given with reasonable accuracy by an approximate formula first derived by Luz and Meiboom [200],

$$(1/T_2)_{\text{eff}} = (1/T_2{}^0) + 4\pi^2\tau p_A p_B(\nu_A - \nu_B)^2[1 - (\tau/t_c)\tanh(t_c/\tau)]. \quad (7.111)$$

The Luz-Meiboom equation can be further simplified in the limiting cases in which t_c/τ is either very large or very small. When the pulse spacing is large compared to the exchange lifetime, each nucleus changes frequency many times between pulses and the sample acts as though it contained only a single nuclear species precessing at the appropriate average frequency. The relaxation time should then reflect the dependence of the line width on the exchange rate near the limit of fast exchange, discussed in Section 9. In fact, since the limit of $(1/x)\tanh x$ as $x \to \infty$ is zero, the limiting form of (7.111) is:

$$(1/T_2)_{\text{eff}} = (1/T_2{}^0) + 4\pi^2\tau p_A p_B(\nu_A - \nu_B)^2, \quad (7.112)$$

which can be shown, using (7.56) and (7.61), to be equivalent to (7.81). At the other extreme, when the exchange lifetime is much larger than the pulse separation, the whole second term on the right-hand side of (7.111) approaches zero. The exchange then occurs too seldom to have a significant effect on the echo intensities and $(T_2)_{\text{eff}}$ becomes identical with $T_2{}^0$.

It is readily possible to make a series of measurements of $(T_2)_{\text{eff}}$ in which t_c is varied over a wide range, and the results may be used to evaluate the individual quantities $T_2{}^0$, τ, and $\Delta\nu$. The range of exchange lifetimes for which the method is applicable depends on $\Delta\nu$ and $T_2{}^0$ and on the accuracy with which changes in $(T_2)_{\text{eff}}$ can be determined. Since it is necessary to observe a series of echoes that become too weak to detect after a time not much longer than $T_2{}^0$, the exchange process cannot be studied effectively unless τ is smaller than $T_2{}^0$, typically about 1 sec for a liquid. At the other extreme, when exchange is fast so that (7.112) is obeyed, an accurate measurement of τ is possible if the two terms on the right-hand side are similar in magnitude, that is, approximately, if $\tau > 1/(4\pi^2 p_A p_B \Delta\nu^2 T_2^0)$. With $T_2^0 = 1$ sec, $\Delta\nu = 10$ Hz, and $p_A = p_B = \frac{1}{2}$, τ can then be evaluated if it falls roughly in the range $1 \text{ sec} \leq \tau \leq 10^{-3}$ sec. Compounds such as cyclohexane-d_{11} for which the proton T_2^0 is larger than 10 sec are especially suited for this method, and exchange processes involving such materials can be studied over a range of temperatures in which the rate varies by five orders of magnitude [201].

The attractiveness of the pulse method is largely attributable to the wide range of rates that can be covered. In contrast, the major rate-dependent changes in a typical high-resolution spectrum occur when $10^{-1} \text{ sec} \leq \tau \leq 10^{-2}$ sec. The pulse technique is thus more likely to yield reliable values of activation energies. Moreover, in applying the technique over a range of temperatures, it is not necessary to assume that $\Delta\nu$ is itself temperature independent. This not necessarily accurate assumption is usually made in interpreting rate-dependent high-resolution spectra. The pulse method has been extended to cases in which there is coupling between the *a*- and *b*-type nuclei [202]. So far, it appears to be applicable only for compounds that contain no resonating nuclei other than those involved in the exchange process, that is, it has been used [199] to measure the rate of rotation about the C—N bond in

$$Cl_3C{-}C(=O){-}N(CH_3)_2$$

but not in the closely related molecule

$$CH_3{-}C(=O){-}N(CH_3)_2$$

in which the protons of the acetyl group contribute to the spin echo.

Fourier Transform Spectra

For a sample containing nuclei with unequal chemical shifts, the radio-frequency pulses obviously cannot simultaneously be exactly on resonance for all species. It becomes interesting to investigate what happens if a pulse at frequency ν_0 is used to irradiate a nucleus with resonant frequency ν_i. If $\nu_i - \nu_0$ is not too large, the magnetization vector will be tipped away from the z-axis and at $t = 0$ can be assumed to lie in the y'-direction. However, if the y'-axis is defined as one that rotates around the z-axis at the frequency ν_0 while the magnetization precesses at the frequency ν_i, then $M_{y'}$ does not remain constant but undergoes periodic variations at a frequency $\nu_i - \nu_0$. In a crossed-coil experiment, the detector signal is proportional to $M_{y'}$ and consists of a train of sinusoidal oscillations superimposed on the normal free induction decay. If there are additional nuclei with Larmor frequencies ν_j, and so on, oscillating components at frequencies $\nu_i - \nu_0$, $\nu_j - \nu_0$, and so on are simultaneously present. The free induction decay signal then contains the same information normally derived from the high-resolution spectrum. In fact, the high-resolution absorption curve can be generated [203, 204] by taking the Fourier transform of the free precession signal obtained during the interval T, that is,

$$C(n) = \sqrt{2/T}\int_0^T \cos\,(2\pi nt/T)M_{y'}(t)\,dt. \tag{7.113}$$

Ordinarily, it is much easier to obtain the high-resolution data in the conventional way, but the Fourier transform method becomes attractive when the spectrum is complex and so weak that sensitivity enhancement by means of signal accumulation is required.

The superiority of the Fourier method from the viewpoint of sensitivity enhancement may be appreciated in two ways. First, the pulse experiment can be repeated by using successive pulses with a spacing of the order of T_1 or, for a typical liquid, a few seconds. In a period of 2 hr, several thousand decay signals can then be accumulated, whereas a typical scan over a high-resolution spectrum takes perhaps 500 sec, allowing less than 20 repetitions in the same period. An alternative interpretation can be developed by considering the frequency spectrum of a series of pulses at frequency ν_0 with a repetition frequency f_r. Such a set of pulses is equivalent to a continuous signal at the frequency ν_0 with very drastic amplitude modulation, since the amplitude is fixed at some finite value when the pulse is "on" and zero between pulses. This modulated signal is in turn equivalent to a large set of superimposed continuous oscillations with properly selected frequencies and amplitudes. In the terminology of radio communications theory, the pulse train consists of a carrier at ν_0 and a set of side bands with frequencies $\nu_0 \pm nf_r$, where n is any integer. The amplitude of the nth side band diminishes as n rises, and the total spread of frequencies *increases* if the duration of the individual pulses is *reduced.* In a sense, then, the pulses are able to irradiate nuclei in different regions of the nmr spectrum simultaneously, providing that there is a side band at each required frequency. Thus the pulse experiment involves all of the nuclei essentially all the time, whereas in the high-resolution experiment any one nucleus is inactive during the long intervals required to scan relatively distant regions of the spectrum.

Several requirements and limitations of the method emerge if the preceding discussion is somewhat extended. For example, nuclei with chemical shifts, in frequency units, differing by less than f_r respond to the same side band and act as though they were identical, suggesting that the resolution will be improved by making f_r small. Since signals separated by much less than the line width of $\Delta\nu = 1/\pi T_2$ cannot be resolved even in high-resolution spectra, it is not expected that reducing f_r much below this value will yield any further improvement. Moreover, reducing f_r moves all of the side bands closer together, suggesting that it becomes more and more difficult to cover a relatively broad range of chemical shifts as f_r is made smaller. That this is indeed true can be seen also by considering how the Fourier transform is actually evaluated in practice.

The observing period between successive pulses, $T = 1/f_r$, is divided into c equal intervals of duration T/c. The signal strength at the beginning of each interval is read into a computer which, after accumulating such data for the desired number of repetitions, approximates the integral by the sum*

$$C(n) = \sqrt{2T/c}\sum_{k=0}^{c-1} \cos (2\pi kn/c) M_{y'}(Tk/c). \tag{7.114}$$

When the signal is digitized in this way, fluctuations in $M_{y'}$ (i.e., frequencies) having a period smaller than $2T/c$ are overlooked, and therefore the spectrum should contain no lines for which $\nu_i - \nu_0$ exceeds $c/2T$ or $cf_r/2$. The value of c is limited by the capacity of the computer used and is often of the order of 10^3. The method is thus at its best for proton or deuterium spectra, in which the total spectral width is usually 500 Hz or less.

A recurrent theme in the history of nmr spectroscopy is the need for ever-increasing sensitivity, since any major sensitivity gain makes it possible to extend the usefulness of the technique into new territories. The field of biochemistry alone offers several examples, including investigation of enzyme-substrate interactions and examination of diverse materials obtainable, in purified form, only in very small amounts. In this context the Fourier transform technique offers tremendous promise, and consequently it is in a state of rapid development as this chapter is being prepared. It appears certain that the technique will be applied very extensively in the area of ^{13}C spectroscopy (see Section 11) because the low natural abundance of ^{13}C nearly always

* Usually two computers are used, a signal-averaging computer for the original accumulation of data and a high-speed digital computer to calculate the Fourier transform. The latter calculation is not as straightforward as it might appear at first glance. For very weak signals, adjustment of the spectrometer to produce a pure absorption mode signal ("phase adjustment") presents difficulties, and the computer program must include a routine designed to correct the phase [203].

makes it necessary to use some means of sensitivity enhancement. The requirement that the spacing between pulses must be larger than the relaxation time T_1 seemed for a time to limit the utility of the Fourier transform method for ^{13}C because in such very small molecules as CS_2 the T_1-values are of the order of 10^2 sec and it was believed that carbon nuclei in other molecules would have similar T_1-values. However, it was eventually found that special pulse sequences can be used to eliminate this requirement [204a] and that in somewhat larger molecules such as sucrose most of the carbon nuclei actually have T_1-values less than 1 sec [204b], so that even the unmodified Fourier technique can be used to accumulate a great many repetitions in a fairly short time.

Chemical Shifts for Solids*

If one were to attempt to apply the Fourier transform technique to solids, he would encounter the same limitation that prevents detection of high-resolution signals with solids, although in appropriately modified form. To measure a single chemical shift, it is necessary to pick up a periodic component in the induction decay with a period $1/(\nu_i - \nu_0)$. Since the signal becomes too weak to see after a time not much longer than T_2, this requires approximately that $(\nu_i - \nu_0) > 1/T_2$. For most solids, T_2 is expected to be of the order of 10^{-4} to 10^{-5} sec, so that the decay is very short-lived, and if $(\nu_i - \nu_0) \gg 10^4$ Hz the pulse frequency is too far off resonance to be effective. A way to circumvent this difficulty was discovered recently when it was reported that the effects of dipolar interactions, which produce the short T_2-values in solids, are largely canceled in a special type of pulsed experiment [176]. The pulse sequence resembles a Carr-Purcell train except that all of the pulses are 90° pulses and the carrier phase is systematically shifted. For the pulses at $t = (4n - 3)t_c$ the phase shift is $+\pi/2$ and for those at $t = (4n - 1)t_c$ it is $-\pi/2$. The echoes occurring as usual when $t = 2nt_c$ are observed with a phase detector sensitive to the carrier phase. The echo decay has a time constant substantially larger than T_2, and the echo envelope shows periodicity reflecting the chemical shift $\nu_i - \nu_0$.

16 COMPUTERS IN NMR SPECTROSCOPY

High-Resolution Spectra

It has become widely recognized that the scope and power of many instrumental techniques can be greatly enhanced if the conventional apparatus is used in conjunction with an appropriately programmed automatic computer, and nmr spectroscopy furnishes a number of examples [205]. Several of these

* See also Section 8.

have been mentioned in earlier sections of this chapter, and it seems worthwhile to review them briefly and to call attention to one or two additional applications.

It is not surprising that the largest amount of effort has been devoted to the problem of extracting chemical shifts and coupling constants from high-resolution spectra so complex as to prevent immediate evaluation of these quantities from measured peak positions and separations. Mathematically, this problem presents a characteristic difficulty met also in extracting parameters from other types of spectroscopic data, for example, microwave or infrared, or from diffraction curves. The calculation of experimental variables from a given set of parameters is fairly straightforward, but the reverse calculation is far more difficult, or even impossible. This suggests the procedure of calculating a spectrum from a trial set of parameters and then varying them in some systematic way so that successive calculated spectra agree more and more closely with the experimental version. The success of such a method often depends on whether or not one can find a sufficiently good set of initial parameters to assure fairly rapid convergence to the correct solution.

In dealing with a complex multiplet arising from spin-spin interactions between two or more sets of nonequivalent nuclei, one has available the recipe provided by the quantum mechanical analysis [37], in which the chemical shifts determine the elements of a diagonal matrix $\mathbf{H}^0$ and the coupling constants the elements of a matrix $\mathbf{H}'$ having both diagonal and off-diagonal terms. If n is the total number of nuclei involved, each matrix has 2^n rows and columns corresponding to the 2^n possible spin states. The calculations consist in essence of finding the eigenvalues and eigenvectors of the matrix $(\mathbf{H}^0 + \mathbf{H}')$, from which in turn the positions and intensities of all multiplet components can be evaluated. Since many of the off-diagonal elements are zero, the matrix, if it is large, is often readily split into smaller submatrices. At any rate, diagonalization of a matrix is easily handled by digital computers and appropriate programs are widely available.

If the original parameters, symbolically p_i with $i = 1, 2, \ldots, m$, were reasonably close to the best values, the remaining task is to find a set of corrections dp_i such that the new set $p_i + dp_i$ yields a substantially better approximation of the observed spectrum. Of the several techniques that have been used to accomplish this, the conceptually simplest one is first to make small arbitrary changes in the parameters and find, for each observed frequency ν_j the value of $\partial \nu_j / \partial p_i$. The dp_i-values should then be chosen to make

$$\sum_{i=1}^{m} (\partial \nu_j / \partial p_i) \, dp_i = (\nu_j)_{\text{obs}} - (\nu_j)_{\text{calc}} \tag{7.115}$$

for each ν_j, where $(\nu_j)_{\text{calc}}$ is the calculated frequency with the original parameters. The number of frequencies is usually larger than the number of

dp_i-values, so that a least-squares technique is useful in obtaining a best compromise solution. The corrected parameters may then be further improved in the same way until the best possible fit is attained.

Three of the most widely used programs for the analysis of high-resolution spectra of systems containing up to seven interacting nuclei with $I = \frac{1}{2}$, designated LAOCN3, NMRIT, and NMREN, are available in a book [205a] which includes an extended discussion of the rationale, capability, and accuracy of each program. An interesting test of the accuracy of such calculations is presented in Ref. 101d. Numerous other specialized nmr programs are available through the generosity of their originators [40, 205].

A major difficulty with any such method is the lack of assurance that the observed spectrum could not have been fit equally well by some choice of parameters quite different from the one finally arrived at. Relatively trivial examples of this sort of uncertainty are quite common, inasmuch as many spectra contain first-order spin-spin multiplets which can be fit with either a positive or a negative coupling constant. This illustrates the well-known but occasionally forgotten principle that *any results produced by a computer should be treated with a modicum of skepticism.* In practice, such ambiguities can usually be resolved either by supplementary experiments designed to determine the relative signs of key parameters or by appeal to theoretical predictions or to data for model compounds.

The multiplets arising from the combination of spin-spin coupling and partially averaged dipolar interactions for molecules in nematic liquid solvents present a very similar problem, and the same sorts of methods have been used in these cases with considerable success [189].

Temperature-Dependent Spectra

As outlined in Section 7, the spectrum of a mixture of rapidly equilibrating species is characterized by chemical shifts and coupling constants, which are the weighted averages of the parameters for the individual species and which change as the temperature is varied because of changes in the respective opulations. Even for the simplest such systems, it is usually necessary to reduce the number of unknowns by introducing assumptions or, alternatively, to use a computer to find a whole family of solutions, each consisting of a set of parameters which fits the data, and then use auxiliary information to select the most acceptable of these [206, 207]. This may be illustrated by considering a conformational equilibrium between the two species a and b (they might be the *gauche* and *trans* form of a disubstituted ethane) in which the populations are p_a and $p_b = 1 - p_a$, and the observed quantity q is given by the average:

$$q = q_a p_a + q_b(1 - p_a). \tag{7.116}$$

With the helpful, but not necessarily always accurate, assumption that q_a

and q_b are temperature independent, the temperature dependence of q arises from the thermodynamic equation:

$$K_{eq} = (1 - p_a)/p_a = \exp(-\Delta H/RT) \exp(\Delta S/R), \qquad (7.117)$$

and there are thus four parameters to be evaluated, q_a, q_b, ΔH, and ΔS. If $q(T)$ is measured over a fairly small temperature range, it will appear to be a linear function, and it will be possible to determine no more than two of the parameters. Even if data are obtained over the widest possible range of temperatures, it is not generally possible to find a unique solution unless other considerations are brought into play. It may be possible to evaluate ΔS a priori on the basis of symmetry changes, or to fix one of the q-values with the help of a model compound. When such information is lacking, one can program the computer to find values of q_a, q_b, and ΔH with a series of arbitrarily selected ΔS-values and eliminate at least some of them with the help of empirical rules.

Spectroscopic parameters may also be temperature dependent because the molecule has modes of vibration with such low frequencies that one or more excited states are appreciably populated. Such low-energy modes may be torsional oscillations or the stretching of a very weak bond, such as a hydrogen bond [92]. If the quantity q has the value q_i in the ith state, characterized by energy E_i and statistical weight g_i, then,

$$q(T) = \sum_{i=0}^{\infty} q_i g_i \exp[(E_0 - E_i)/kT] / \sum_{i=0}^{\infty} g_i \exp[(E_0 - E_i)/kT]. \qquad (7.118)$$

Again, additional assumptions are required to reduce the number of independent unknowns. For the shift of a proton in a hydrogen bond, the E_i-values and the mean hydrogen bond lengths r_i can be calculated for each i if a potential energy function for the hydrogen bond stretch is assumed, and the shifts δ_i can then be evaluated by assuming a reasonable relation between δ_i and r_i.

Line Shape Calculations

In the spectra of solids or of liquids undergoing intermediate-rate exchange processes, the crucial parameters are obtained not from the positions and intensities of individual maxima but from the shape of a signal, which may or may not show partially resolved fine structure. With a solid sample, the spectroscopic data may be reduced to a single number, the second moment, and used to determine an unknown structural parameter as suggested in Section 8, employing the computer primarily to evaluate the required lattice sums. More information can sometimes be obtained by making use of the intensity at every point in the spectrum rather than merely the overall width. In practice, the intensity $I(\nu_i)$ is read at a finite number of frequencies ν_i

within the band, and a least-squares difference function of the type:

$$F = \sum_i [I(\nu_i)_{\text{calc}} - I(\nu_i)_{\text{obs}}]^2 \tag{7.119}$$

is defined. If the sample is a solid, one or more structural parameters can be systematically varied until F is minimized. For liquid samples with chemical exchange, the relevant parameters are the populations p_j, the exchange lifetimes τ_j, the unperturbed relaxation times T_{2j}^0, the chemical shifts, and possibly the spin-spin coupling constants involving the exchanging nuclei, and again the computer can be programmed to calculate the line shape from a given set of parameters and eventually to select the best set. The problem of determining parameters from spin-echo envelopes, discussed in Section 15 is essentially analogous.

Sensitivity and Resolution Enhancement

These techniques also require conversion of the data to digital form, that is, tabulation of intensity $I(\nu_i)$ at a large number of points on the frequency scale, covering the entire spectrum. The improvement of sensitivity by multiple scanning and use of the computer to accumulate the signal was described in Section 3. The alternative procedure involving repeated observation of free induction decay following a series of pulses and Fourier inversion of the accumulated data, also requires a signal-averaging computer and of course a digital computer for calculating the Fourier transform. As mentioned in Section 15, it may provide a much greater gain in sensitivity in a limited time than the conventional approach.

The objective of resolution enhancement [204, 208] is to find the individual frequencies and intensities of two or more closely spaced lines which collectively give rise to a signal with only partially resolved fine structure. Since each line is characterized not only by its position and intensity but also by one or more numbers which define its shape, there is little hope of successfully decomposing a complex signal without once more resorting to assumptions that reduce the number of unknowns. The usual approach is to assume that all lines are represented either by Gaussian or Lorentzian functions of constant width or by a shape function known only numerically and obtained empirically from an isolated peak observ[illegible] identical instrumental conditions. Preferably, this peak [illegible] be decomposed should arise from different nuclei in [illegible] The use of an empirical shape function is obviously prefera[illegible] isolated signal looks appreciably skewed, since the Gaussian or Lorentzian functions are symmetrical.

The procedure involves estimating from the observed spectrum the positions and intensities of a relatively small number of components, computing a set of $I(\nu_i)$-values based on this choice, and varying the frequencies and intensities until a least-squares difference function is minimized. The resulting spectrum is compared with the experimental one, and additional lines are introduced where the fit is poor. The difference function is again minimized, and the procedure is repeated until a satisfactory fit is obtained.

The most obvious weakness of such a method is that it is not known in advance how many lines are in fact present, and one might suppose that the effort to obtain a nearly perfect fit would result in the introduction of superfluous signals. It appears that this can be avoided in practice by rejecting any line with an optimal intensity that is either negative or positive but no larger than the noise amplitude, or which falls virtually on top of an already included line so that its only effect is to enhance the intensity of the latter. Sample results for a complex multiplet with 12 overlapping components suggest that peak positions calculated by this method are reasonably reliable but that relative intensities are subject to rather large uncertainties [208].

References

1. J. A. Pople, W. G. Schneider, and H. J. Bernstein, *High Resolution Nuclear Magnetic Resonance*, McGraw-Hill, New York, 1959.
2. See the General References on p. 712.
3. *ACS Laboratory Guide*, American Chemical Society, Washington, D.C., 1969, p. 250LG.
4. N. F. Ramsey, *Nuclear Moments*, Wiley, New York, 1953.
5. The mathematical formulation underlying this approach, which has served as the point of departure for several useful theoretical developments, was first presented by F. Bloch, *Phys. Rev.*, **70**, 460 (1946).
6. Ref. 1, p. 37.
7. E. Odeblad, *Acta Obstetr. Gynecol. Scand.*, **45**, Suppl. 2, 1966.
8. M. J. E. Golay, *Rev. Sci. Instr.*, **29**, 313 (1958).
9. N. Muller and P. Simon, *J. Phys. Chem.*, **71**, 568 (1967).
10. J. W. Emsley, J. Feeney, and L. H. Sutcliffe, *High Resolution Nuclear Magnetic Resonance Spectroscopy*, Vol. 2, Pergamon, Oxford, 1965, Chapter 10.
11. J. R. Dyer, *Applications of Absorption Spectroscopy of Organic Compounds*, Prentice-Hall, Englewood Cliffs, New Jersey, 1965, pp. 84–85.
12. W. N. Lipscomb, *Advan. Magnetic Resonance*, **2**, 137 (1966).
13. J. I. Musher, *Advan. Magnetic Resonance*, **2**, 177 (1966).
14. D. E. O'Reilly, *Progr. NMR Spectry.*, **2**, 1 (1967).
15. R. F. Zurcher, *Progr. NMR Spectry.*, **2**, 205 (1967).
16. J. D. Memory, *Quantum Theory of Magnetic Resonance Parameters*, McGraw-Hill, New York, 1968.
17. N. F. Ramsey, *Phys. Rev.*, **78**, 699 (1950).
18. W. Lamb, *Phys. Rev.*, **60**, 817 (1941).
19. H. Spiesecke and W. G. Schneider, *J. Chem. Phys.*, **35**, 722 (1961).

20. G. A. Olah, E. B. Baker, J. C. Evans, W. S. Tolgyesi, J. S. McIntyre, and I. J. Bastien, *J. Am. Chem. Soc.*, **86,** 1360 (1964).
21. B. P. Dailey and J. N. Shoolery, *J. Am. Chem. Soc.*, **77,** 3977 (1955).
22. H. M. McConnell, *J. Chem. Phys.*, **27,** 226 (1957).
23. J. A. Pople, *Proc. Roy. Soc. (London)*, **A239,** 541 (1957).
24. A. A. Bothner-By and J. A. Pople, *Ann. Rev. Phys. Chem.*, **16,** 43 (1965).
25. H. A. Allen and N. Muller, *J. Chem. Phys.* **48,** 1626 (1968).
26. J. M. Gaidis and R. West *J. Chem. Phys.* **46,** 1218 (1967).
27. J. I. Musher, *J. Chem. Phys.*, **46,** 1220 (1967).
28. K. B. Wiberg and B. J. Nist, *J. Am. Chem. Soc.*, **83,** 1226 (1961).
29. D. J. Patel, M. E. H. Howden, and J. D. Roberts, *J. Am. Chem. Soc.*, **85,** 3218 (1963).
30. T. W. Marshall and J. A. Pople, *Mol. Phys.*, **1,** 199 (1958).
31. A. D. Buckingham, *Can. J. Chem.*, **38,** 300 (1960).
32. J. I. Musher, *J. Chem. Phys.*, **37,** 34 (1962).
33. A. Segre and J. I. Musher, *J. Am. Chem. Soc.*, **89,** 706 (1967).
34. E. Fermi, *Z. Physik*, **60,** 320 (1930).
35. M. Barfield and D. M. Grant, *Advan. Magnetic Resonance*, **1,** 149 (1965).
36. Ref. 1, Chapter 6.
37. J. D. Roberts, *An Introduction to Spin-Spin Splitting in High Resolution Nuclear Magnetic Resonance Spectra*, Benjamin, New York, 1961.
38. P. L. Corio, *Structure of High Resolution NMR Spectra*, Academic, New York, 1966.
39. P. Diehl, R. K. Harris, and R. G. Jones, *Progr. NMR Spectry.*, **3,** 1 (1967).
40. E. O. Bishop, *Ann. Rev. NMR Spectry.*, **1,** 91 (1968).
41. E. W. Garbish, Jr., *J. Chem. Educ.*, **45,** 311, 402, 480 (1968).
42. H. M. McConnell, A. D. McLean, and C. A. Reilly, *J. Chem. Phys.*, **23,** 1152 (1955).
43. A. D. Cohen, N. Sheppard, and J. J. Turner, *Proc. Chem. Soc.*, **1958,** 118.
44. R. J. Abraham and H. J. Bernstein, *Can. J. Chem.*, **39,** 216 (1961).
45. J. I. Musher and E. J. Corey, *Tetrahedron*, **18,** 791 (1962).
46. S. Sternhell, *Quart. Rev.*, **23,** 236 (1969).
47. A. A. Bothner-By, *Advan. Magnetic Resonance*, **1,** 195 (1965).
48. J. A. Pople and A. A. Bothner-By, *J. Chem. Phys.*, **42,** 1339 (1965).
49. J. A. Pople and D. P. Santry, *Mol. Phys.*, **8,** 1 (1964); **9,** 311 (1965).
50. M. Karplus, *J. Chem. Phys.*, **30,** 11 (1959).
51. S. Sternhell, *Rev. Pure Appl. Chem.*, **14,** 15 (1964).
52. N. Muller and D. E. Pritchard, *J. Chem. Phys.*, **31,** 768 (1959).
53. C. Juan and H. S. Gutowsky, *J. Chem. Phys.*, **37,** 2198 (1962).
54. M. Karplus, D. H. Anderson, T. C. Farrar, and H. S. Gutowsky, *J. Chem. Phys.*, **27,** 597 (1957).
55. R. A. Ogg, Jr., *J. Chem. Phys.*, **22,** 1933 (1954).
56. K. A. McLauchlan and D. H. Whiffen, *Proc. Chem. Soc.*, **1962,** 144.
57. J. I. Musher, *J. Chem. Phys.*, **40,** 983 (1964).
58. A. D. Buckingham and K. A. McLauchlan, *Proc. Chem. Soc.*, **1963,** 144.
59. S. L. Smith and R. H. Cox, *J. Chem. Phys.*, **45,** 2848 (1966).

60. P. Laszlo, *Progr. NMR Spectry.*, **3**, 231 (1967).
61. N. Bloembergen and W. C. Dickenson, *Phys. Rev.*, **79**, 179 (1950).
62. K. Frei and H. J. Bernstein, *J. Chem. Phys.*, **37**, 1891 (1962).
63. R. F. Spanier, T. Vladimiroff, and E. R. Malinowski, *J. Chem. Phys.*, **45**, 4355 (1966).
64. A. A. Bothner-By and R. E. Glick, *J. Chem. Phys.*, **26**, 1647 (1957).
65. A. A. Bothner-By, *J. Mol. Spectry.*, **5**, 52 (1960).
66. F. H. A. Rummens, W. T. Raynes, and H. J. Bernstein, *J. Phys. Chem.*, **72**, 2111 (1968).
67. A. D. Buckingham and J. A. Pople, *Trans. Faraday Soc.*, **51**, 1173 (1955).
68. P. Laszlo and J. I. Musher, *J. Chem. Phys.*, **41**, 3906 (1964).
69. H. M. Hutton and T. Schaefer, *Can. J. Chem.*, **45**, 1111 (1967).
70. L. W. Reeves and W. G. Schneider, *Can. J. Chem.*, **35**, 251 (1957).
71. P. J. Berkeley, Jr., and M. W. Hanna, *J. Am. Chem. Soc.*, **86**, 2990 (1964).
72. A. D. Buckingham, T. Schaefer, and W. G. Schneider, *J. Chem. Phys.*, **32**, 1227 (1960).
73. S. L. Smith and A. M. Ihrig, *J. Chem. Phys.*, **46**, 1181 (1967).
74. H. M. Hutton, E. Bock, and T. Schaefer, *Can. J. Chem.*, **44**, 2772 (1966).
75. N. Sheppard, *Advan. Spectry.*, **1**, 288 (1959).
76. R. J. Abraham and M. A. Cooper, *Chem. Commun.*, **1966**, 588.
77. A. A. Bothner-By and C. Naar-Colin, *J. Am. Chem. Soc.*, **84**, 743 (1962).
78. H. S. Gutowsky, *Pure Appl. Chem.*, **7**, 93 (1963).
79. R. K. Harris and N. Sheppard, *Mol. Phys.*, **7**, 595 (1964).
80. R. Freeman and N. S. Bhacca, *J. Chem. Phys.*, **45**, 3795 (1966).
81. Ref. 1, pp. 218 ff.
82. C. J. Johnson, Jr., *Advan. Magnetic Resonance*, **1**, 33 (1965).
83. H. S. Gutowsky and C. H. Holm, *J. Chem. Phys.*, **25**, 1228 (1956).
84. M. T. Rogers and J. C. Woodbrey, *J. Phys. Chem.*, **66**, 540 (1962).
85. F. A. L. Anet, M. Ahmad, and L. D. Hall, *Proc. Chem. Soc.*, **1964**, 145.
86. F. A. Bovey, F. P. Hood, III, E. W. Anderson, and R. L. Kornegay, *Proc. Chem. Soc.*, **1964**, 146.
87. N. C. Franklin and H. Feltkamp, *Angew. Chem. Intern. Ed.*, **4**, 774 (1965).
88. W. A. Thomas, *Progr. NMR Spectry.*, **1**, 44 (1968).
89. E. Lippert, *Ber. Bunsenges. Physik. Chem.*, **67**, 267 (1963).
90. N. Muller and P. I. Rose, *J. Phys. Chem.*, **69**, 2564 (1965).
91. N. Muller and O. R. Hughes, *J. Phys. Chem.*, **70**, 3975 (1966).
92. N. Muller and R. C. Reiter, *J. Chem. Phys.*, **42**, 3265 (1965).
93. J. A. Pople, *Mol. Phys.*, **1**, 168 (1958).
93a. J. D. Roberts, *J. Am. Chem. Soc.*, **78**, 4495 (1956).
94. R. A. Ogg, Jr., and J. D. Ray, *J. Chem. Phys.*, **26**, 1515 (1957).
95. J. D. Baldeschwieler and E. W. Randall, *Chem. Rev.*, **63**, 81 (1963).
96. R. A. Hoffman and S. Forsen, *Progr. NMR Spectry.*, **1**, 15 (1966).
97. W. McFarlane, *Ann. Rev. NMR Spectry.*, **1**, 135 (1968).
98. A. L. Bloom and J. N. Shoolery, *Phys. Rev.*, **97**, 1261 (1955).
99. L. H. Piette, J. D. Ray, and R. A. Ogg, Jr., *J. Mol. Spectry.*, **2**, 66 (1958).

99a. K. F. Kuhlmann and D. M. Grant, *J. Am. Chem. Soc.*, **90,** 7355 (1968).
100. W. A. Anderson and R. Freeman, *J. Chem. Phys.*, **37,** 85 (1962).
101. R. Freeman and W. A. Anderson, *J. Chem. Phys.*, **37,** 2053 (1962).
101a. Ref. 1, p. 212.
101b. K. F. Kuhlmann and J. D. Baldeschwieler, *J. Am. Chem. Soc.*, **85,** 1010 (1963); *J. Chem. Phys.*, **43,** 572 (1965).
101c. A. J. Jones, D. M. Grant, and K. F. Kuhlmann, *J. Am. Chem. Soc.*, **91,** 5013 (1969).
101d. R. Freeman and B. Gestblom, *J. Chem. Phys.*, **48,** 5008 (1968).
102. N. Bloembergen, E. M. Purcell, and R. V. Pound, *Phys. Rev.*, **73,** 679 (1948).
103. Some sources, including Ref. 1, p. 205, locate the minimum at $\tau_c = 1/(2\pi\nu_0)$. The disagreement probably arises from differences in the simplifying assumptions introduced in the course of the derivations.
104. H. M. McConnell and C. H. Holm, *J. Chem. Phys.*, **25,** 1289 (1956).
104a. A. Olivson, E. Lippmaa, and J. Past, *Eesti. NSV. Tead. Akad. Toim. Fuus. Mat.*, **16,** 390 (1967).
104b. P. S. Hubbard, *Phys. Rev.*, **131,** 1155 (1963).
105. Ref. 1, p. 202.
106. O. Jardetzky, *Advan. Chem. Phys.*, **7,** 499 (1964).
107. J. J. Fischer and O. Jardetzky, *J. Am. Chem. Soc.*, **87,** 3237 (1965).
108. S. Meiboom, Z. Luz, and D. Gill, *J. Chem. Phys.*, **27,** 1411 (1957).
109. S. Meiboom, *J. Chem. Phys.*, **34,** 375 (1961).
110. D. R. Eaton and W. D. Phillips, *Advan. Magnetic Resonance*, **1,** 103 (1965).
111. E. De Boer and H. van Willigen, *Progr. NMR Spectry.*, **2,** 111 (1967).
112. R. A. Bernheim, T. H. Brown, H. S. Gutowsky, and D. E. Woessner, *J. Chem. Phys.*, **30,** 950 (1959).
113. T. J. Swift and R. E. Connick, *J. Chem. Phys.*, **37,** 307 (1962).
114. Z. Luz and S. Meiboom, *J. Chem. Phys.*, **40,** 2686 (1964).
115. N. Bloembergen, *J. Chem. Phys.*, **27,** 572 (1957).
116. N. Bloembergen, *J. Chem. Phys.*, **27,** 595 (1957).
117. H. M. McConnell and D. B. Chesnut, *J. Chem. Phys.*, **28,** 107 (1958).
118. K. H. Hausser, H. Brunner, and J. C. Jochims, *Mol. Phys.*, **10,** 253 (1966).
118a. G. W. Canters and E. de Boer, *Mol. Phys.*, **13,** 395 (1967).
118b. R. W. Kreilick, *Mol. Phys.*, **14,** 495 (1968); *J. Am. Chem. Soc.*, **90,** 5991 (1968).
119. M. E. Anderson, P. J. Zandstra, and T. R. Tuttle, Jr., *J. Chem. Phys.*, **33,** 1591 (1960).
120. A. Carrington, *Quart. Rev.*, **17,** 67 (1963).
121. P. I. Rose, paper presented at the IUPAC International Symposium on Macromolecular Chemistry, Toronto, 1968.
122. P. C. Lauterbur, *J. Am. Chem. Soc.*, **83,** 1838 (1961).
123. J. D. Roberts, *Chem. Brit.*, **2,** 529 (1966).
124. N. Muller and R. H. Birkhahn, *J. Phys. Chem.*, **71,** 957 (1967).
125. T. McL. Spotswood, J. M. Evans, and J. H. Richards, *J. Am. Chem. Soc.*, **89,** 5052 (1967).
126. E. Zeffren and R. E. Reavill, *Biochem. Biophys. Res. Commun.*, **32,** 73 (1968).

127. R. W. Taft, F. Prosser, L. Goodman, and G. T. Davis, *J. Chem. Phys.*, **38,** 380 (1963).
128. Ref. 1, p. 39.
129. B. N. Figgis, R. G. Kidd, and R. S. Nyholm, *Proc. Roy. Soc.* (*London*), **A269,** 469 (1962).
130. P. C. Lauterbur, in *Determination of Organic Structures by Physical Methods*, Vol. 2, F. C. Nachod and W. D. Phillips, Eds., Academic, New York, 1962, Chapter 8.
131. Ref. 10, Chapters 11 and 12.
132. J. B. Stothers, *Quart. Rev.*, **19,** 144 (1965).
133. E. F. Mooney and P. H. Winson, *Ann. Rev. NMR Spectry.*, **1,** 244 (1968).
134. A. Saika and C. P. Slichter, *J. Chem. Phys.*, **22,** 26 (1954).
135. Ref. 10, pp. 930 ff.
136. L. H. Meyer and H. S. Gutowsky, *J. Phys. Chem.*, **57,** 481 (1953).
137. G. V. D. Tiers, *J. Am. Chem. Soc.*, **78,** 2914 (1956).
138. D. F. Evans, *J. Chem. Soc.*, 877 (1960).
139. J. Feeney, L. H. Sutcliffe, and S. M. Walker, *Mol. Phys.*, **11,** 117, 129, 137, 145 (1966).
140. R. R. Carey, H. W. Kroto, and M. A. Turpin, *Chem. Commun.*, **188** (1969).
141. W. H. Pirkle, *J. Am. Chem. Soc.*, **88,** 1837 (1966).
142. J. W. Emsley and L. Phillips, *Mol. Phys.*, **11,** 437 (1966).
143. S. L. Manatt, *J. Am. Chem. Soc.*, **88,** 1323 (1966).
144. N. Muller and R. H. Birkhahn, *J. Phys. Chem.*, **72,** 583 (1968).
145. T. W. Johnson and N. Muller, *Biochemistry*, **9,** 1943 (1970).
146. C. H. Holm, *J. Chem. Phys.*, **26,** 707 (1957).
147. P. C. Lauterbur, *J. Chem. Phys.*, **26,** 217 (1957).
148. F. J. Weigert and J. D. Roberts, *J. Am. Chem. Soc.*, **89,** 2967 (1967).
149. R. R. Ernst, *J. Chem. Phys.*, **45,** 3845 (1966).
150. F. J. Weigert and J. D. Roberts, *J. Am. Chem. Soc.*, **89,** 5962 (1967).
151. G. Miyazima, Y. Utsumi, and K. Takahashi, *J. Phys. Chem.*, **73,** 1370 (1969).
152. E. G. Paul and D. M. Grant, *J. Am. Chem. Soc.*, **86,** 2977 (1964).
152a. R. J. Pugmire and D. M. Grant, *J. Am. Chem. Soc.*, **90,** 697, 4232 (1968).
152b. T. Tokuhiro and G. Fraenkel, *J. Am. Chem. Soc.*, **91,** 5005 (1969).
152c. W. M. Litchman and D. M. Grant, *J. Am. Chem. Soc.*, **90.** 1400 (1968).
152d. D. M. Grant and B. V. Cheney, *J. Am. Chem. Soc.*, **89,** 5315, 5319 (1967).
152e. D. K. Dalling and D. M. Grant, *J. Am. Chem. Soc.*, **89,** 6612 (1967).
152f. S. A. Knight, *Chem. Ind.* (*London*), 1921 (1967).
153. R. A. Uphaus, E. Flaumenhaft, and J. J. Katz, *Biochim. Biophys. Acta*, **141,** 625 (1967).
154. P. R. Sewell, *Ann. Rev. NMR Spectry.*, **1,** 165 (1968).
155. H. A. Willis and M. E. A. Cudby, *Appl. Spectry. Rev.*, **1,** 237 (1968).
156. R. C. Ferguson and W. D. Phillips, *Science*, **157,** 257 (1967).
157. F. A. Bovey and G. V. D. Tiers, *J. Polymer Sci.*, **44,** 173 (1960).
158. C. E. H. Bawn and A. Ledwith, *Quart. Rev.*, **16,** 361 (1962).
158a. F. A. Bovey, *Accounts Chem. Res.*, **1,** 175 (1968).

159. F. A. Bovey, E. W. Anderson, D. C. Douglass, and J. A. Manson, *J. Chem. Phys.*, **39,** 1199 (1963).
160. F. C. Stehling, *J. Polymer Sci.*, **A2,** 1815 (1964).
161. F. A. Bovey, F. P. Hood, III, E. W. Anderson, and L. C. Snyder, *J. Chem. Phys.*, **42,** 3900 (1965).
162. A. Kowalsky, *J. Biol. Chem.*, **237,** 1807 (1962).
163. C. C. McDonald and W. D. Phillips, *J. Am. Chem. Soc.*, **89,** 6332 (1967).
163a. C. C. McDonald and W. D. Phillips, *J. Am. Chem. Soc.*, **91,** 1513 (1969).
164. W. E. Stewart, L. Mandelkern, and R. E. Glick, *Biochemistry*, **6,** 143, 150 (1967).
165. C. C. McDonald, W. D. Phillips, and S. Penman, *Science*, **144,** 1234 (1964).
166. C. C. McDonald, W. D. Phillips, and J. Lazar, *J. Am. Chem. Soc.*, **89,** 4166 (1967).
167. L. van Gerven, Ed., *Nuclear Magnetic Resonance and Relaxation in Solids*, North-Holland, Amsterdam, 1965.
168. E. R. Andrew, *Nuclear Magnetic Resonance*, Cambridge Univ. Press, New York, 1955, Chapter 6.
169. G. E. Pake, *J. Chem. Phys.*, **16,** 327 (1948).
170. R. E. Richards, *Advan. Spectry.*, **2,** 101 (1961).
171. H. S. Gutowsky, G. E. Pake, and R. Bersohn, *J. Chem. Phys.*, **22,** 643 (1954).
172. P. C. Lauterbur and J. J. Burke, *J. Chem. Phys.*, **42,** 439 (1965).
173. D. C. Haddix and P. C. Lauterbur, in *Molecular Dynamics and Structure of Solids*, R. S. Carter and J. J. Rush, Eds., Natl. Bur. Stds. Special Publ. 301, Washington, D.C., 1969, p. 403.
174. E. R. Andrew and V. T. Wynn, *Proc. Roy. Soc. (London)*, **A291,** 257 (1966).
175. E. R. Andrew, L. F. Farnell, M. Firth, T. D. Gladhill, and I. Roberts, *J. Magnetic Res.*, **1,** 27 (1969).
176. J. S. Waugh and L. M. Huber, *J. Chem. Phys.*, **47,** 1862 (1967).
177. Ref. 168, Chapter 8.
178. J. G. Aston, in *Physics and Chemistry of the Organic Solid State*, Vol. 1, D. Fox, M. J. Labes, and A. Weissberger, Eds., Interscience, New York, 1963, Chapter 9.
179. E. R. Andrew and R. G. Eades, *Proc. Roy. Soc. (London)*, **A216,** 398 (1953).
180. K. J. Packer, *Progr. NMR Spectry.*, **3,** 87 (1967).
181. G. M. Muha and D. J. C. Yates, *J. Chem. Phys.*, **49,** 5073 (1968).
182. J. H. Pickett and L. B. Rogers, *Anal. Chem.*, **39,** 1872 (1967).
183. A. D. Buckingham and K. A. McLauchlan, *Progr. NMR Spectry.*, **2,** 62 (1967).
184. G. R. Luckhurst, *Quart. Rev.*, **22,** 179 (1968).
185. S. Meiboom and L. C. Snyder, *Science*, **162,** 1337 (1968).
186. A. Saupe, *Angew. Chem. Intern. Ed. Engl.*, **7,** 97 (1968).
187. G. Englert, A. Saupe, and J. P. Weber, *Z. Naturforsch.*, **23a,** 152 (1968).
188. A. Saupe, *Z. Naturforsch.*, **20a,** 572 (1965).
188a. A. D. Buckingham and J. A. Pople, *Trans. Faraday Soc.*, **59,** 2421 (1963).
188b. G. Englert and A. Saupe, *Z. Naturforsch.*, **19a,** 172 (1964).
188c. G. Englert and A. Saupe, *Z. Naturforsch.*, **20a,** 1401 (1965).

189. L. C. Snyder, *J. Chem. Phys.*, **43,** 4041 (1965).
190. C. C. Costain, *J. Chem. Phys.*, **29,** 864 (1958).
191. A. D. Buckingham, E. E. Burnell, and C. A. De Lange, *Mol. Phys.*, **15,** 285 (1968).
192. P. Diehl, C. L. Khetrapal, and H. P. Kellerhals, *Mol. Phys.*, **15,** 333 (1968).
193. P. Diehl and C. L. Khetrapal, *Mol. Phys.*, **15,** 201, 633 (1968).
194. D. Demus, *Z. Naturforsch.*, **22a,** 285 (1967).
195. H. Spiesecke and J. Bellion-Jourdan, *Angew. Chem. Intern. Ed., Engl.*, **6,** 450 (1967).
196. A. D. Buckingham and E. E. Burnell, *J. Am. Chem. Soc.*, **89,** 3341 (1967).
197. Ref. 1, pp. 44 ff.
198. H. Y. Carr and E. M. Purcell, *Phys. Rev.*, **94,** 630 (1954).
198a. J. E. Tanner and E. O. Stejskal, *J. Chem. Phys.*, **49,** 1768 (1968).
199. A. Allerhand and H. S. Gutowsky, *J. Chem. Phys.*, **41,** 2115 (1964); **42,** 1587 (1965).
200. Z. Luz and S. Meiboom, *J. Chem. Phys.*, **39,** 366 (1963).
201. A. Allerhand, F. Chen, and H. S. Gutowsky, *J. Chem. Phys.*, **42,** 3040 (1965).
202. A. Allerhand and H. S. Gutowsky, *J. Chem. Phys.*, **42,** 4203 (1965).
203. R. R. Ernst and W. A. Anderson, *Rev. Sci. Instr.*, **37,** 93 (1966).
204. R. R. Ernst. *Advan. Magnetic Resonance*, **2,** 1 (1966).
204a. E. D. Becker, J. A. Ferretti, and T. C. Farrar, *J. Am. Chem. Soc.*, **91,** 7784 (1969).
204b. A. Allerhand, Personal communication.
205. J. D. Swalen, *Progr. NMR Spectry.*, **1,** 205 (1966).
205a. D. F. DeTar, Ed., *Computer Programs in Chemistry*, Vol. 1, Benjamin, New York, 1968.
206. D. F. Koster, *J. Am. Chem. Soc.*, **88,** 5076 (1966).
207. A. A. Bothner-By and D. F. Koster, *J. Am. Chem. Soc.*, **90,** 2351 (1968).
208. W. D. Keller, T. R. Lusebrink, and C. H. Sederholm, *J. Chem. Phys.*, **44,** 782 (1966).

General

BOOKS

Abragam, A., *The Principles of Nuclear Magnetism*, Clarendon, Oxford, 1961.

Aleksandrov, I. V., *The Theory of Nuclear Magnetic Resonance*, Academic, New York, 1966.

Andrew, E. R., *Nuclear Magnetic Resonance*, Cambridge University Press, New York, 1955.

Becker, E. D., *High Resolution Nuclear Magnetic Resonance*, Academic, New York, 1969.

Bible, R. H., *Interpretation of NMR Spectra*, Consultants Bureau, Plenum, New York, 1965.

Bovey, F. A., *Nuclear Magnetic Resonance Spectroscopy*, Academic, New York, 1969.

Carrington, A., and A. D. McLachlan, *Introduction to Magnetic Resonance*, Harper, New York, 1967.

Chapman, D., and P. D. Magnus, *Introduction to Practical High Resolution Nuclear Magnetic Resonance Spectroscopy*, Academic, New York, 1966.

Emsley, J. W., J. Feeney, and L. H. Sutcliffe, *High Resolution Nuclear Magnetic Resonance Spectroscopy*, 2 Vols., Pergamon, Oxford, 1965.

Fluck, E., *Die Kernmagnetische Resonanz und ihre Anwendung in der Anorganischen Chemie*, Springer, Berlin, 1963.

Hecht, H. G., *Magnetic Resonance Spectroscopy*, Interscience, New York, 1967.

Jackman, L. M., and S. Sternhell, *Applications of Nuclear Magnetic Resonance in Organic Chemistry*, 2nd ed., Pergamon, New York, 1969.

Mathieson, D. W., Ed., *Nuclear Magnetic Resonance for Organic Chemists*, Academic, New York, 1967.

Mavel, G., *Theories moléculaires de la Résonance magnétique nucléaire*, Dunod, Paris, 1965.

Pople, J. A., W. G. Schneider, and H. J. Bernstein, *High Resolution Nuclear Magnetic Resonance*, McGraw-Hill, New York, 1959.

Roberts, J. D., *Nuclear Magnetic Resonance*, McGraw-Hill, New York, 1959.

Slichter, C. P., *Principles of Magnetic Resonance*, Harper, New York, 1963.

Strehlow, H., *Magnetische Kernresonanz und Chemische Struktur*, 2nd ed., Steinkopff, Darmstadt, 1968.

Suht, H., *Anwendung der Kernmagnetischen Resonanz in der Organischen Chemie*, Springer, New York, 1965.

SERIAL REVIEW PUBLICATIONS

Emsley, J. W., J. Feeney, and L. H. Sutcliffe, Eds., *Progress in Nuclear Magnetic Resonance Spectroscopy*, Pergamon, Oxford, 1966–.

Mooney, E. F., Ed., *Annual Review of NMR Spectroscopy*, Academic, New York, 1968–.

Waugh, J. S., Ed., *Advances in Magnetic Resonance*, Academic, New York, 1965–.

CATALOGS OF SPECTRA AND DATA COMPILATIONS

Bovey, F. A., *NMR Data Tables for Organic Compounds*, Academic, New York, 1967.

Brügel, W., *Nuclear Magnetic Resonance Spectra and Chemical Structure*, Academic, New York, 1967–.

Howell, M. G., A. S. Kende, and J. S. Webb, Eds. *Formula Index to NMR Literature Data*, 2 Vols., Plenum, New York, 1965–1966.

Nuclear Magnetic Resonance Spectra, Sadtler Research Laboratories, Philadelphia, Pennsylvania, 1967.

NMR Spectra Catalog, 2 Vols., Varian Associates, Palo Alto, California, 1962–1963.

Chapter **VIII**

x-RAY MICROSCOPY

Raymond V. Ely

---

1 INTRODUCTION 716

The Value of x-rays in Microscopy 716
Outline of Methods 717

2 CONTACT MICRORADIOGRAPHY 718

Photographic Techniques 731

3 PROJECTION MICRORADIOGRAPHY 735

Photographic Techniques 743
Specimen Preparation 744
Stereo Techniques 744

4 APPLICATIONS AND RESULTS 746

Medical 746
- Enamel 748
- Dentine 748
- Dental Pulp 749
- Cementum 749
- Jaw Morphology 749
- Bone 749
- Bone Marrow 753
- Ear 753
- Cartilage 753
- Skin and Connective Tissue 754
- Lung 754
- Eye 755
- Liver 755
- Calculi 755
- Kidney 755
- Nervous Tissue 756

Metallurgy and Mineralogy 757
Botanical 766
General 768
- Fabrics and Fibers 768

Paints and Paintings 768
Paleontology and Fossils 769
Paper 769
Plastics 770
Rubber 770
5 FUTURE PROSPECTS 770

---

1 INTRODUCTION

The Value of x-Rays in Microscopy

Following the emergence of the x-ray microscope from the experimental stage to a commercially available instrument, the extension of normal medical and industrial radiographic techniques to the examination of microstructures has been greatly enhanced in many directions. Microradiography embraces all the advantages attributable to x-rays and combines great depth of focus with elemental differentiation at approximately the same range of magnifications as that obtainable with the optical microscope.

In the modern research laboratory microradiography forms a very useful link with optical and electron microscopy techniques and can provide information unobtainable by these or any other known methods, for example, in the nondestructive examination, in detail and depth, of specimens opaque to light or electrons. This offers many attractions, especially if the point focus x-ray projection method is used, since all strata in a comparatively thick specimen are in simultaneous focus.

High-magnification stereoscopy by projection is particularly effective in the delineation of overlying structures, and the relative positions of component elements within a specimen can be accurately and quickly determined by modern stereometry.

In general, the conditions under which a specimen can be examined are more favorable than those needed for electron microscopy because it is not necessary to make exposures in vacuum. The exception arises with very thin specimens of low atomic number, for example, biological tissue, when the x-ray wavelength needed for adequate contrast is sufficiently long to be largely absorbed by air. Normally all work is carried out with the specimen freely disposed in the atmosphere and hot or cold stages can be employed.

x-Ray microscopy cannot, of course, compete in resolution and magnification with electron microscopy, but its utility as a complementary technique is established.

Outline of Methods

Since matter is normally penetrated by x-rays it is not possible to construct focusing reflective or refractive systems similar to those used in light microscopy. Total reflection does, in fact, take place at very small angles of incidence to a flat surface, and systems employing a series of curved mirror surfaces, reflection microradiography, have successfully produced point sources of x-rays. Unfortunately the short wavelengths involved, and the restricted angular aperture resulting from the small glancing angle, introduce serious limitations of a fundamental nature. The small glancing angle also produces considerable geometrical aberrations and the resultant beam intensity is so restricted that long exposures are inevitable.

Jentsch [1] made the first experimental investigation into the possibilities of effectively focusing x-rays, but it was nearly 20 years before Kirkpatrick and Baez [2] overcame the initial difficulties. The work of Ehrenberg [3, 4], Dyson [5], Kirkpatrick and Pattee [6], Pattee [7], Montel [8, 9], and McGee [10, 11] accomplished much toward solving the very considerable problems involved.

Mirror systems used at grazing incidence are subject to severe aberrations and, although a resolution of about 1 μ has been obtained with a two-mirror system, unequal magnifications result from the mirrors not being in the same plane. Increasing the number of mirrors has not resulted in any significant improvement and a practical solution has still to appear.

In the most usual of the two techniques now generally employed, contact microradiography, the specimen is placed directly in contact with a photographic emulsion, resolution being dependent on the intimacy of the contact, the distance of specimen and emulsion from the x-ray source, the size and characteristics of the source, the thickness and constitution of the specimen, and the graininess of the emulsion. The focal spot size of the commercial tube frequently employed in this method is about 1 mm^2 and introduces considerable penumbral blurring unless the distance between the tube and specimen is large, that is, 450 to 700 mm (Fig. 8.1). Equipment of this type is easy to use and a large specimen, or a number of smaller ones, can be covered by a single exposure. There are, however, difficulties when very low kilovoltages are needed for the production of adequate contrast in some specimens, and this is discussed in Section 2.

The alternative method, projection microradiography, requires a more sophisticated type of equipment with a much smaller x-ray source. Direct image enlargement is obtained by positioning the specimen between the source and photographic emulsion, magnification being the ratio of the distance of the source to the emulsion and to the specimen. The resolution obtainable is a function of the source diameter, which can be made extremely small by special electromagnetic, or electrostatic, focusing

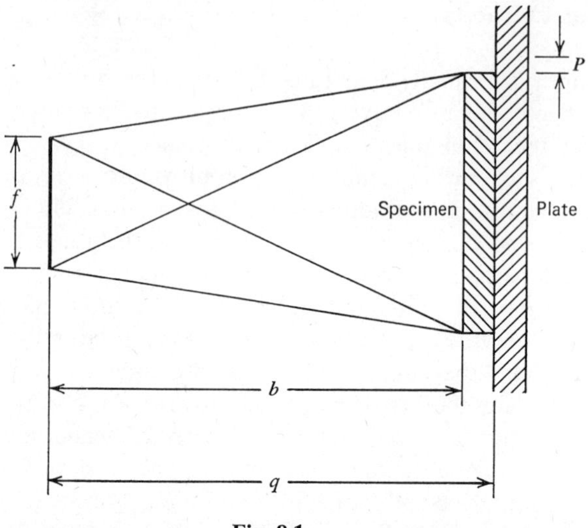

Fig. 8.1

of the electron beam within the tube. A description of these focusing systems is given in Section 3. The graininess of the emulsion does not impose a limitation, as in the contact method, because the initial primary x-ray magnification makes it unnecessary to use high secondary optical enlargement. Other advantages of the method are that (*a*) an area of special interest in a specimen can be selected before the exposure is made by viewing the magnified image on a fluorescent screen and (*b*) a hot, or irradiated, specimen can be placed at a distance from the emulsion to avoid contamination. The great depth of field, coupled with a progressive increase in magnification of the specimen strata, is particularly valuable in the production of negatives for stereoscopic viewing.

Which of the two methods should be used will depend on the characteristics of the specimen and the degree of magnification required to yield the desired information. As will be seen from Sections 2 and 3, a combination of the two techniques may be preferred if a particular area of a specimen needs to be examined over a wide range of magnification, and with minimum expenditure of time.

2 CONTACT MICRORADIOGRAPHY

The essentials of the method were first demonstrated by Goby [12] in radiographs of specimens ranging from diatoms to legs of Lucerta muralis. The negatives could be enlarged up to about ×17 and he termed them microradiographs.

Specimens were placed in direct contact with a fine-grain emulsion and enclosed in a cylinder attached to the x-ray tube so as to exclude all external light. The x-ray beam entered the cylinder through a small hole in a lead disk. Apart from collimating the beam this hole apparently reduced the intensity of the light from the filament to a degree which, together with the opacity of the specimen, prevented fogging of the emulsion. Further progress appears to have been seriously hindered by the slow development of tubes capable of operation at low voltages, and the nonavailability of photographic emulsions of sufficiently fine grain.

Dauvillier [13, 14] designed a tube for continuous evacuation by an oil diffusion pump. It was fitted with an aluminum x-ray window and operated at between 3000 and 7000 V to provide what was then considered to be ultrasoft radiation. The long-wavelength rays were so greatly absorbed in the air column between the tube and specimen that it was necessary to enclose the latter in a cylinder, the open end of which was attached to an airtight coupling flange surrounding the aluminum window. The cylinder was then filled with hydrogen. The principal disadvantage lies in the long exposure times required for the tube to emulsion distances normally employed, for example, 20 to 30 in.

Lippman-type emulsions have become available and contact microradiographs of thin pathological specimens can be enlarged several hundred times. Fricke [15] and Sherwood [16] contributed to these developments and important papers by Lamarque [17], and Lamarque and Turchini [18] followed rapidly. They, like Dauvillier, appreciated the importance of long wavelengths to the production of contrast in histological specimens a few microns thick, and Lamarque termed the technique "historadiography." Injection methods, using salts of heavy elements, were also employed to improve contrast.

In order to extend the voltage range downward they constructed a tube with an x-ray window of lithium, an element known to be exceptionally transparent to x-rays. This tube is said to have operated at as low as 500 V but the radiation passing through the window was so largely absorbed by air that the hydrogen-filled specimen/emulsion chamber technique was again employed. Negatives of high resolution were produced but at the expense of long exposure. This was because of the long distance between source and emulsion dictated by the geometrical blurring produced by the large size of the source.

A tube of the general type described by Lamarque was constructed in 1949 by Ely (Fig. 8.2). An infrared gelatin filter prevented the light from the filament fogging a high resolution plate located in a metal frame, the specimens being placed carefully on the plate to avoid abrasion of the delicate emulsion. The filter, mounted in a slightly larger frame, was then

Fig. 8.2 Early tube for microradiography.

placed over that containing the plate and specimen until a slight bulging of the filter indicated that the specimen had been lightly gripped. The two frames were then clipped together and mounted vertically in the chamber which was then evacuated, together with the attached tube, and the exposure made. A microradiograph (Fig. 8.3) of the complete and unprepared head of a wasp is an example of the work done with this equipment.

Combée [19] introduced a small sealed-off tube, designated C.M.R.-5., for operation between 1.5 and 5 kV. Beryllium, about 50 μ thick, is used for the x-ray window but atmospheric pressure severely limits its diameter and, at 6 cm from the window, the area of the specimen covered is only about 1 cm^2. One of these tubes, attached to a vacuum cassette and mounted on a cabinet enclosing the high-tension generator and control gear, is shown in Fig. 8.4. The tube is mounted flexibly on a block that can be raised and lowered above the cassette. When raised, the cassette can be drawn forward

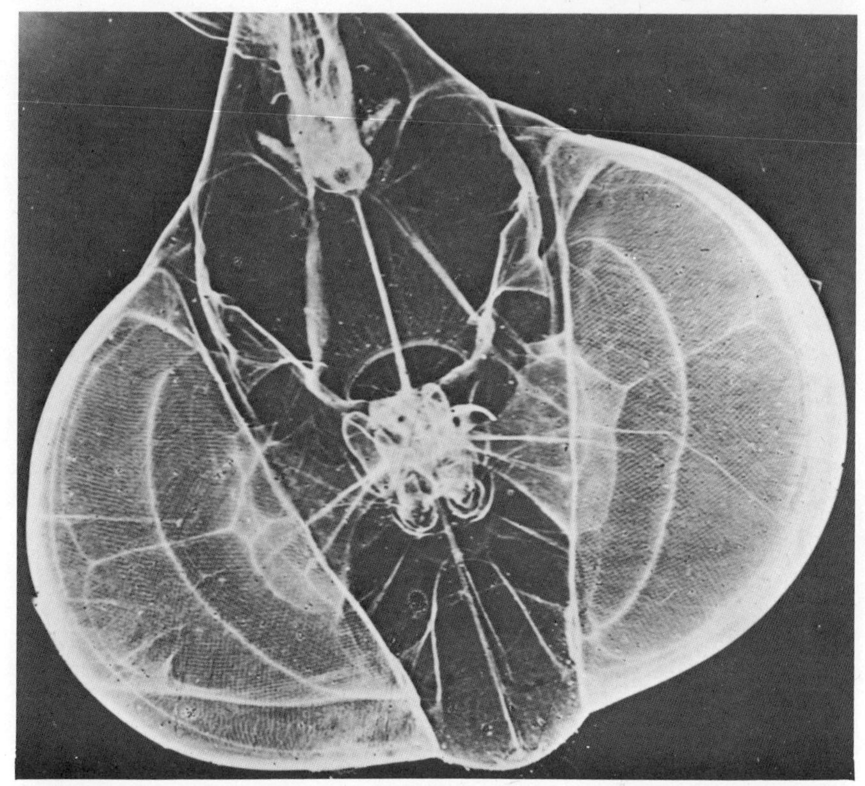

Fig. 8.3 Contact microradiograph of truncated head of wasp (no preparation). Taken with equipment as Fig. 8.2. Chromium radiation at 16 kV. Magnification ×23.

on a slide attached to the top of the cabinet and lifted off its base, to which it is vacuum-sealed by an O-ring joint when in operation. To the base is attached a vertical rod supporting a plate holder, the height of which can be adjusted vertically so that the distance of the plate from the target can be set as needed. Darkroom lighting, of the comparatively bright intensity permissible with very slow emulsions, must be used for the positioning of the plate in the holder and the specimen on the plate, the replacement of the cassette onto its base, and the repositioning of the assembly under the tube. The tube is then lowered to connect with the top of the cassette into which another O-ring seal has been fitted. With normal room lighting the cylinder can now be evacuated and an exposure made by setting the high-voltage meter to the required kV and adjusting the filament control for the appropriate tube current. Having made the exposure the air inlet valve on the cassette is opened; it is important that this be done carefully to avoid displacement of

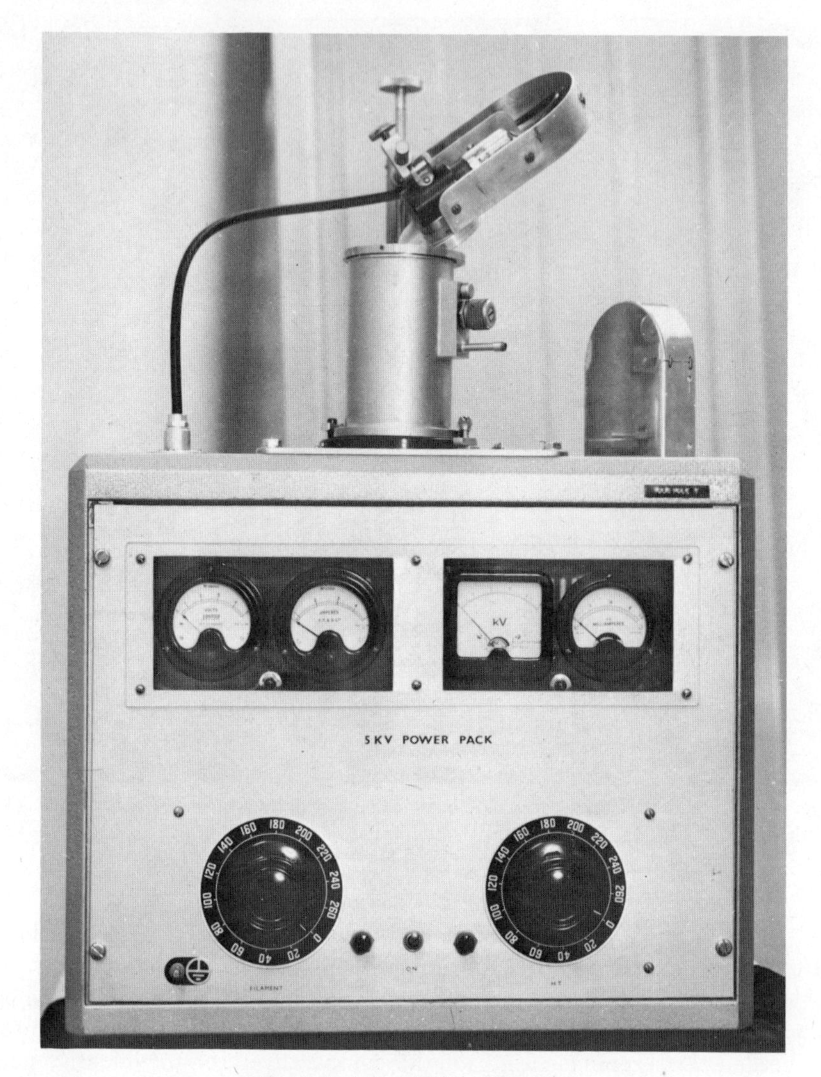

Fig. 8.4 1.5 to 5 kV equipment.

the specimen. In darkroom lighting the plate is then removed for development. Figure 8.5 is a microradiograph taken with this equipment.

Modern sealed-off and demountable tubes, designed primarily for crystallography, can be obtained fitted with beryllium windows and targets of tungsten, chrome, iron, copper, and so forth. The factors involved in the choice of target material are referred to later in this section. The focal spot size of these tubes is generally about 1 mm² and the operating range 5 to 50

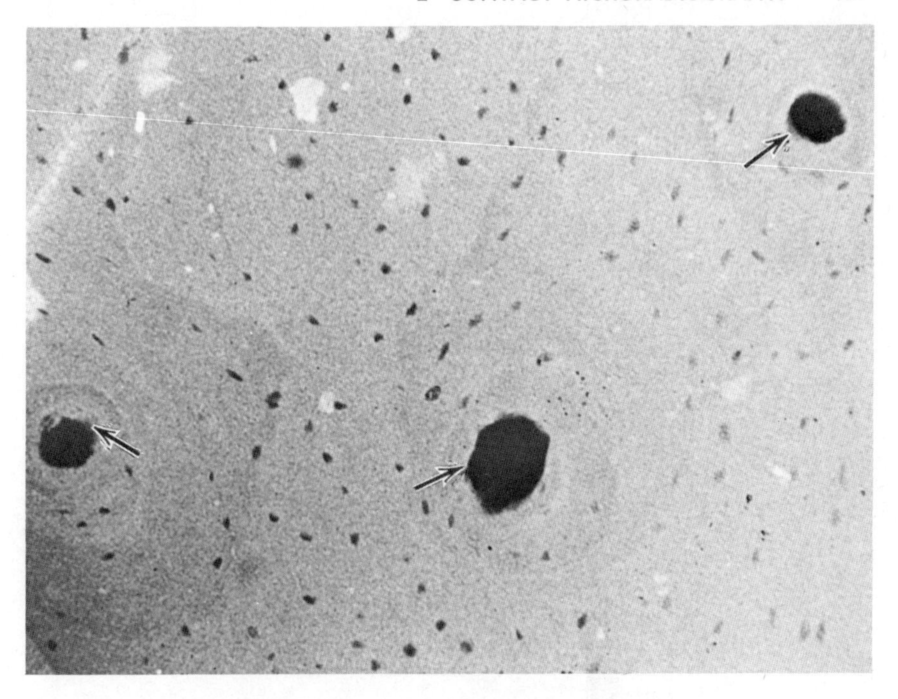

Fig. 8.5 Contact microradiograph of human mandibular cortical bone. Section thickness approximately 40 μ. Magnification $\times 190$. Three transversely cut secondary osteones are indicated by arrows. The shape of the osteocyte lacunae can be seen to vary greatly. Some spots of opaque polishing powder remain adherent to the section. Taken with equipment as Fig. 8.4 at 5 kV.

kV. A typical water-cooled sealed-off tube is shown in Fig. 8.6 (manufactured by Machlett Ltd.). The comparatively large focus requires that the specimen and plate be placed at a considerable distance from the tube to minimize the penumbra effect on resolution, and, at the lower end of the kilovoltage scale, absorption and scatter by the intervening air space can increase exposure times so greatly that the vacuum cassette technique described above may be essential.

A very useful demountable tube design is attributable to Ehrenberg and Spear [20]. This employs an electrostatic focusing system consisting of a biased grid cylinder mounted coaxially with a target, the surface of which is at right angles to the axis of the cylinder. The electrons emitted from a hairpin-shaped filament, the point of which just projects through a small hole in a diaphragm at the end of the cylinder remote from the target, are focused to a spot 40 μ in diameter at the target face. The gap between the tip of the filament and the target is between 2 and 3 cm. The bias voltage applied to the

Fig. 8.6 Typical water-cooled sealed-off tube.

grid is produced by passing the beam current through a variable resistor and effects a semiautomatic control of the beam current by negative feedback. Variants of these focusing and biasing methods have also been incorporated in the design of two special low-kilovoltage tubes for microradiography by Ely [21, 22].

The method Ehrenberg and Spear employed for the measurement of focal spot size is interesting. In the usual pinhole camera technique the size of the hole should be considerably smaller than the electron spot to be measured, but any material in which true holes smaller than 40 μ can be made will not be sufficiently absorbent to x-rays generated at 50 kV. Silver steel rods, 4 mm in diameter and lightly etched, were clamped together in a small frame and separated by 7-μ nylon spacers. It was found that the focal spot size was substantially constant for beam currents between 100 and 500 μA. In comparison with the output obtainable from the diffraction type of sealed-off tube, beam currents of this order result in much longer exposures if all other factors remain the same. In practice the greatly reduced source size permits a much smaller source to film distance without detriment to resolution.

The apparatus constructed and described by Greulich [23] in a planned approach to qualitative microradiography is typical of modern demountable self-contained equipment. The voltage range, 100 to 3000 V, was selected relative to the type of work to be carried out, but the broad specification can be equally well applied to higher voltages. The particular tube employed was based on a design by Henke [24] and is fitted with a movable vacuum seal separating the specimen chamber from the filament and target area so that the specimen can be changed without opening the tube to atmosphere. The high-tension generator, rotary forepump, and oil diffusion pump, together with appropriate vacuum valves and gauges, are mounted in a console which also contains a water-cooling circulating pump and reservoir for the target and diffusion pump. A fan-cooled radiator is attached to the cooling system. A relay automatically shuts down the diffusion pump heater if the water temperature rises above 125°F. The complete equipment weighs 360 lb and is mounted on casters. Maximum output at 3 kV is 2 mA and the tube can be run continuously at 2 kV without overheating.

A demountable tube, constructed recently by Ely for microradiography in the 30 to 100 kV range, is shown in Fig. 8.7. The target/gun chamber is fitted with large ports on all sides so that the target, gun, window, and diffusion pump can be variably positioned. As shown in the figure, the gun is at right angles to the target, which can be moved laterally when working so that a new area can be selected as necessary. This provision is important since continuous bombardment of the target surface by the finely focused electron beam can form small pits on the surface. By this means a large number of exposures can be made before it becomes necessary to recondition the target, an operation normally confined to cleaning up the surface with emery cloth. The flexible experimental construction of this tube involves many more O-ring joints than would normally be required, but the efficiency of this type of connection, when properly made, is so good that no difficulty is experienced in maintaining a vacuum better than 10^{-5} torr, with a 2-in. diffusion pump. A thermoelectric chevron-type baffle is mounted immediately above the diffusion pump and a high vacuum valve is fitted between the baffle and the tube body for bypass pumping so that, should the filament fail, it is not necessary to wait for the diffusion pump to cool and reheat.

The gun and target are both at high voltage and energized by 50 kV negative and 50 kV positive high-frequency generators, respectively. The target is cooled by insulating oil from a small self-contained system through flow-and-return polyvinyl chloride tubes positioned inside the insulator supporting the tungsten element. The focusing cylinder surrounding the filament is

Fig. 8.7 High-voltage tube for microradiography.

biased negatively in steps up to 800 V and the filament is of 0.004-in. tungsten wire in the form of a sharply bent "v."

The gap between the focusing cylinder and the target element, and also the precise positioning of the filament relative to the diaphragm within the focusing cylinder, are both adjustable under vacuum. The gap is set according to the kilovoltage to be used and the filament position is adjusted to focus the image of a test specimen by projection onto a fluorescent screen placed in the film holder. This specimen can conveniently consist of two 10-μ tungsten wires set at right angles and placed as close as possible to the tube window.

In the particular arrangement of the tube elements, as shown in the illustration, the x-ray beam is horizontal. The specimen and film stages are moved independently toward, or away from, the window by small motor drives mounted on the stages. Other motors raise, lower, or move the stages at right angles to the runway; the specimen carrier has two additional motors for vertical and horizontal axial rotation of the specimen, which can then be precisely positioned prior to exposure. The controls for the motors are mounted behind the tube and any backscatter can be prevented from reaching the operator by a lead screen fitted with a lead glass viewing window. The tube, which is completely shockproof, was designed for thick-specimen microradiography, stereoscopic techniques, and work with live or contaminated specimens. These applications require that the x-ray source be small

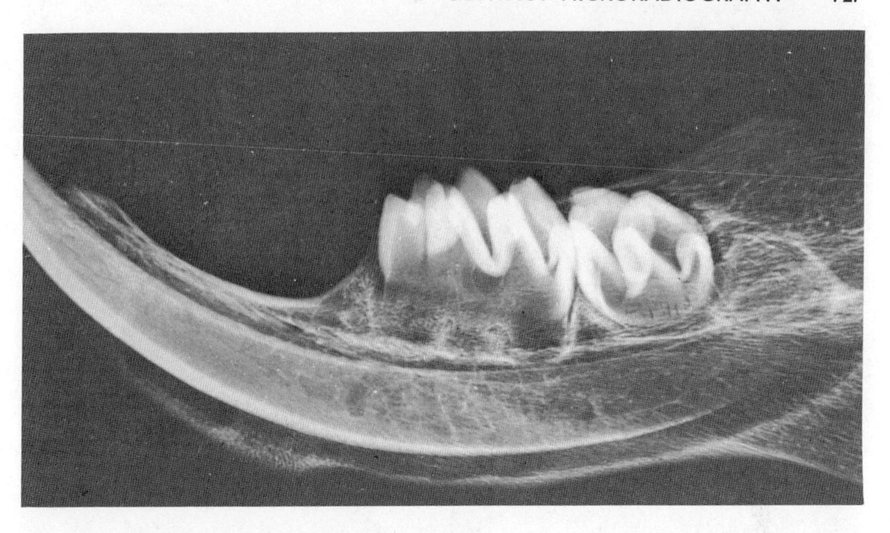

Fig. 8.8 Jaw of rat. Direct x-ray magnification ×8. Taken with equipment as Fig. 8.7. Industrial x-ray film. Tungsten radiation at 40 kV.

enough to permit some displacement of the specimen from the emulsion while retaining acceptable resolution. The source has proved to be sufficiently small to enable the tube to be used under conditions in which larger x-ray magnifications are needed. Reference to these applications is made in the section on projection microradiography. Illustrations of the range of work that can be done with a tube of this type are shown in Figs. 8.8 to 8.10.

The importance of focal spot size is clearly indicated by the geometry of image formation and, as already stated, the ratio of source size to the distance between it and the specimen must be below the limit at which the formation of penumbra, that is, geometrical blurring, is equivalent to the resolution required. The possibility of external vibrations causing relative movements of source, specimen, and emulsion must also be considered. This is termed "kinetic blurring" and points to the avoidance of unnecessarily long exposures. Resolution is also dependent on the thickness of the specimen, its distance from the emulsion, and the thickness of the latter.

Assuming kinetic blurring to be absent,

$$R = \frac{F(S + G + T)}{D}$$

where F is the size of the x-ray source; S, the specimen thickness; G the gap between specimen and emulsion; T, the emulsion thickness; and D, the gap between the source and the specimen. A suitable specimen thickness is partly dependent on the constitution of the specimen but, as a general guide for a test exposure, a thickness of about 25 times the required resolution should give a satisfactory indication.

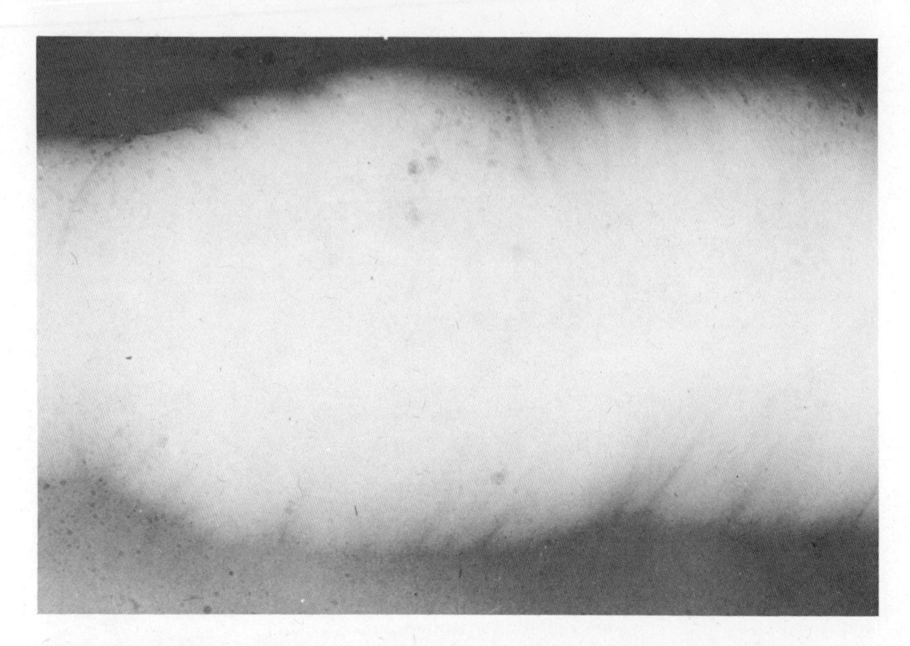

Fig. 8.9 Welded joint. 1.5 mm Al alloy sheet. Direct x-ray magnification ×10. Taken with equipment as Fig. 8.7. Industrial x-ray film. Tungsten radiation at 35 kV.

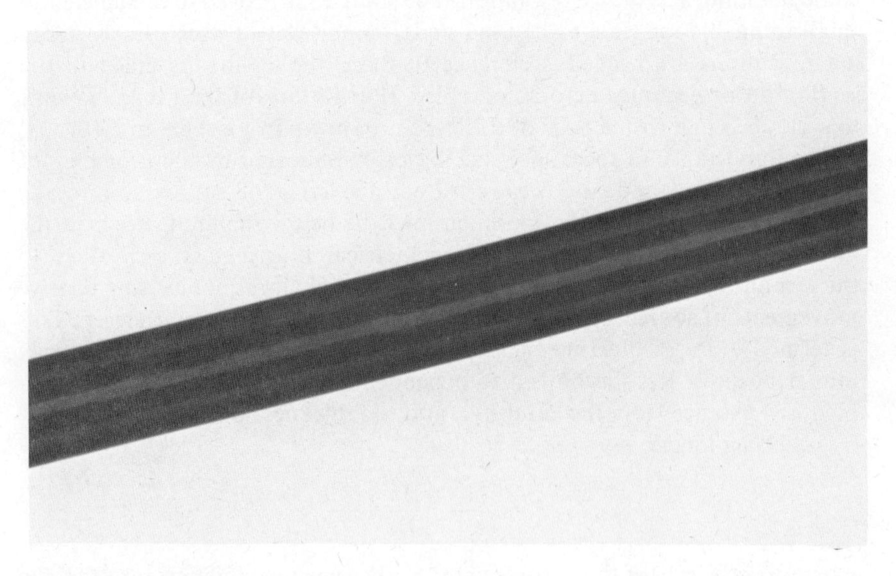

Fig. 8.10 Thermocouple. Direct x-ray magnification ×5. Taken with equipment as Fig. 8.7. Industrial x-ray film. Tungsten radiation at 80 kV.

In this assessment the resolution of the photographic emulsion has been assumed not to be a limiting factor, and it is true that microradiographs on maximum resolution emulsions have been so good that the resolving power of the enlarging microscope has been strained. Nevertheless, it is important to select the most suitable emulsion since the extent of the intended subsequent enlargement must be taken into account. A thick emulsion exposed to x-rays is affected to varying degrees at different internal levels and cannot therefore be enlarged satisfactorily to high degrees of optical magnification. Thin fine-grain emulsions are fundamentally slow and there is no advantage in selecting an emulsion finer than is needed to reach the ultimate magnification required, with the proviso that emulsion response to wavelength charts are consulted.

The size of the individual grains in the several excellent emulsions now available for microradiography is well below that required for the best possible resolution, but unfortunately the ultimate "graininess" of the developed and fixed emulsion must be accepted, together with some thickness of emulsion, as a fundamental limitation pending further developments. Asunmaa [25, 26] investigated the possibilities of irradiated and polymerized plastics in the search for substantially grainless, and extremely thin, recording materials. Ehrenberg and White [27, 28] demonstrated that the thickness of the best commercially available emulsion was too great for maximum optical magnification. With ultrafine experimental coatings they obtained a resolution of about 0.1 μ. It has also been found that an x-ray beam can be scattered in an emulsion by ejection of photoelectrons. In the first of the two papers by Ehrenberg and White the effect was noted when using copper radiation, but it was not apparent with an aluminum target.

A further possible limitation, particularly in the case of thick specimens, is attributable to Fresnel fringes produced by a diffraction effect at the edges of a specimen. The width of the fringe is given by the expression,

$$f = \frac{(d\lambda)}{2}$$

where f is the fringe width; d, the distance of the edge of the specimen from the emulsion; and λ, the x-radiation wavelength. For a thin specimen in contact with an emulsion this effect can be ignored but can become significant if the specimen is thick, or not closely in contact with the emulsion.

Dependent on the elemental constituents of the specimen it is generally possible to improve contrast in qualitative microradiography by substituting characteristic for continuous radiation. Some results obtained by this substitution are referred to in Section 4.

If targets of low atomic number such as Al, Ti, Cr, Fe, Cu, Mo, or Rh are used, strong characteristic radiations are superimposed on the continuous spectra by the $K\alpha$, $K\beta$, and so on, series of lines. In view of its predominant intensity and long wavelength the $K\alpha$ line is normally selected when considering which would be most suitable for a particular specimen. The relationship between wavelengths and the minimum kV excitation levels is obtained from the equation,

$$\text{Vm (in V)} = \frac{12.398.}{\lambda \text{ (in Å)}}$$

Unfortunately the intensity of the characteristic radiations at the minimum kilovoltage levels is also minimum and in practice it is necessary to increase the kilovoltage to between two and four times the minimum for adequate emission.

The continuous spectrum is, of course, extended by this increase in potential but the proportion of characteristic to continuous intensity is sufficiently high to give an appreciable advantage. The precise ratio of $K\alpha$ intensity to that of the integrated continuous spectrum has been the subject of considerable calculation and experimental procedures, and with widely varying results.

Bendit [29] surveyed earlier work and applied further corrections for air-path transmission and Geiger counter efficiency as a function of wavelength to an assumed spectral distribution for the continuous radiation from a Cu target. This assumption was based on a survey of previous literature.

The resulting experimental ratio was 2.2 at 35 kV, that is, less than half the Arndt and Riley figure [30]. Cosslett [31] and Dyson [32] further examined the efficiency of production of characteristic radiation in carbon as well as in copper and estimated that, at a given ratio of applied voltages to that needed to excite the K line, the output from carbon was about 5% of that from copper.

The selection of the most suitable radiation for discrimination between elements in a specimen is dependent on their linear absorption coefficients; maximum contrast is obtained by the use of a wavelength falling between the K absorption edges of two elements. A contrast difference of 0.1 to 0.2 is often adequate and the use of a longer wavelength than is necessary should generally be avoided because it increases exposure time.

If it is necessary to differentiate further between two elements close together in the periodic table, a further refinement in technique makes use of the association of characteristic emission lines with absorption edges. The latter are represented by interruptions in the otherwise regular increase in absorption with wavelength and occur at wavelengths slightly higher than the corresponding emission lines. Maximum contrast is obtained by using a wavelength falling between the K absorption edges of the two elements. An important consideration arising from the fact that a characteristic radiation lies on the long wavelength side of the absorption edge is that an element absorbs little of its own characteristic radiation.

A still greater degree of monochromatization can be obtained by inserting between the source and the specimen a thin foil of an element having an absorption edge that will not seriously impede the $K\alpha$ line radiation, but is capable of largely absorbing the $K\beta$ and other shorter wavelengths, for example, a nickel foil 8 to 9 μ thick with a copper target. With this combination the $K\alpha$ radiation reduces to about 65% of its original intensity whereas the $K\beta$ and all shorter wavelengths reduce to 10% effectively.

Two methods of obtaining quantitative information from microradiographs produced by these techniques are in general use and employ the standard x-ray absorption formula,

$$I = I_{0^e}{}^{-(\mu/\rho)\rho t}$$

where I is the intensity transmitted by the specimen; I_{0^e}, the original intensity; μ, the mass absorption coefficient of the specimen; ρ, its density; and t, its thickness.

One of these methods employs a reference step wedge, for example, of thin layers of pure aluminum or collodion, which is exposed with the specimen. The other method dispenses with the step wedge and is dependent on the accurate assessment of the incident and transmitted radiation as measured directly on negatives produced by a number of consecutive exposures [33–36].

Photographic Techniques

The production of an artefact-free negative by the contact method demands considerable attention to detail. Thin delicate emulsions are very easily scratched and dust or chemical particles can be very troublesome. All solutions should be made up with distilled water and filtered before use. The washing water tap should be fitted with a microfilter and drying carried out in a heated, pressurized clean-air type of cabinet which is never left open to atmosphere when the fan is not working.

Enlargement, depending on the ultimate magnification required, can be made directly on paper with a standard photographic enlarger, preferably fitted with a point source illuminator. The construction and mounting of the enlarger should be such that there is no risk of vibration. Alternatively an intermediate negative can be made using a high-resolution photomicrographic apparatus. Contrast can be varied to some extent by choosing an appropriate type of emulsion for the intermediate negative and a further advantage is that the final print is not reversed, as in the single-stage process.

The range of contrast in an x-ray negative is often too high for the production of a satisfactory print. This difficulty has received considerable attention and several solutions have been suggested. In one of the first, by Thorpe and Davison [256], a sharp positive was made by contact printing from the negative and these were then bound together, in accurate register, with a thin glass plate as a separator. In Frantzell's method [257] a defocused

Fig. 8.11 Contrast control equipment.

and slightly magnified image of the negative was projected onto a ground glass screen and the negative then placed in position on the ground glass. A diapositive was made on another plate placed emulsion downward on the negative. In practice the method proved complicated and Spiegler and Giles [258] proposed an alternative in which the diapositive was accurately registered and printed through for part of the exposure, full density being built up after removal of the diapositive. This method also had disadvantages and, with the benefit of these experiments, an instrument of greater precision and comparatively simple use was constructed for the contrast-controlled printing out of enlargements from microradiographs. As shown in Fig. 8.11 the instrument is set up for low-magnification work but, by swinging away the bottom condenser/mirror stage, a microscope can be substituted. The method of operation is given in a paper by Ely and Hampshire [259] and comparisons between a normal and a contrast-controlled print are shown in Fig. 8.12.

Stereo Techniques. For convenience in reading, applications to contact and projection microradiography are discussed as one in Section 3.

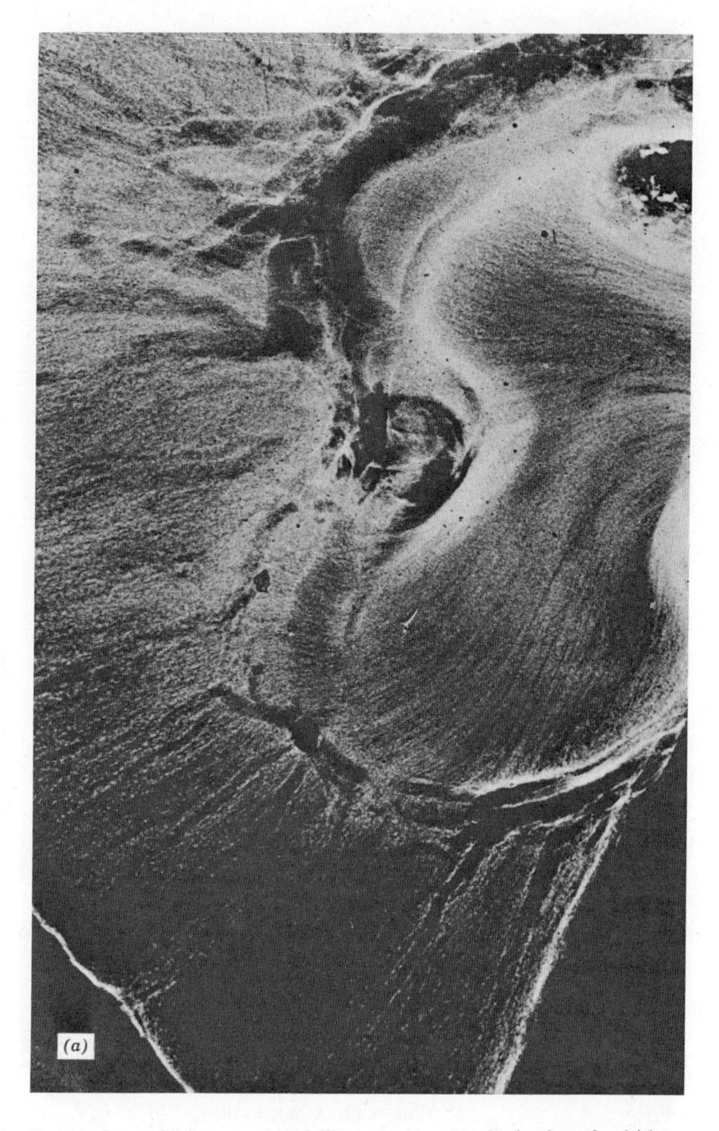

Fig. 8.12 Comparison of (*a*) normal and (*b*) contrast-controlled print of a kidney section.

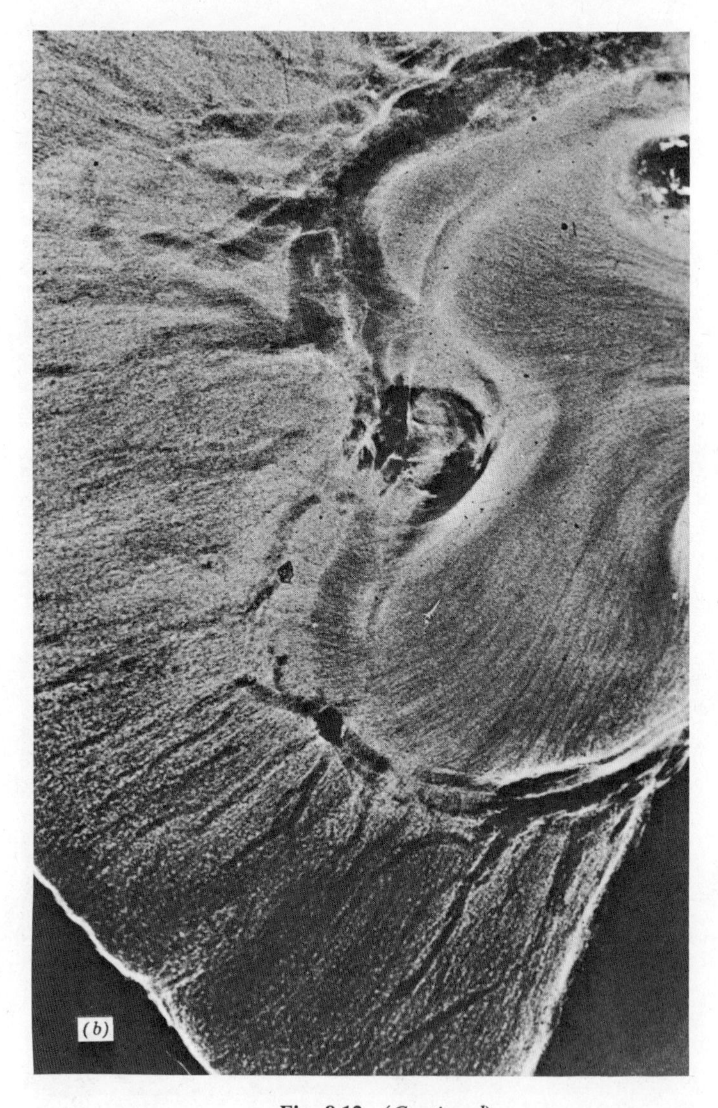

Fig. 8.12 (*Continued*)

3 PROJECTION MICRORADIOGRAPHY

Very early in the history of x-rays the possibilities of pinhole images attracted attention, but it was not until 1936 that the first efforts with any promise were made by Sievert who used a pinhole a few microns in diameter to produce images at a direct magnification of about ×5. Further optical enlargement indicated that reasonable resolution would require still smaller pinholes and better emulsions. Three years later, when electron lenses were being developed, von Ardenne put forward proposals for a tube using electrostatic lenses with a foil transmission target and the alternative possibilities of electromagnetic lenses were investigated by Cosslett. The first practical demonstration of this type of instrument was by Cosslett and Nixon [37] using electron beam generation and focusing methods similar to those employed in electron microscopy. The geometrical relationship between the source, the specimen, and the projected image differs from that in the contact method only with respect to the position of the specimen (Fig. 8.13). The basic details of the microscope are shown in Fig. 8.14, and a complete instrument in Fig. 8.15.

Apart from vacuum and electrical services the Cosslett and Nixon type of x-ray microscope consists of an electron gun, an anode, two electromagnetic lenses, a pole-piece, and a transmission target assembly. The original form of an electron gun has been replaced by one of a shockproof design by Ely. In this gun the filament support takes the form of a molded piston sliding in a based cylinder, and an O-ring let into a groove in the piston forms a vacuum seal between it and the internal wall of the cylinder. All metallic components

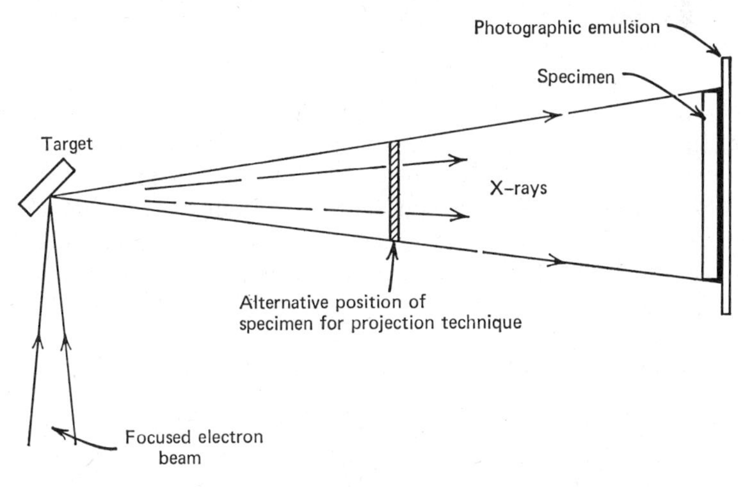

Fig. 8.13

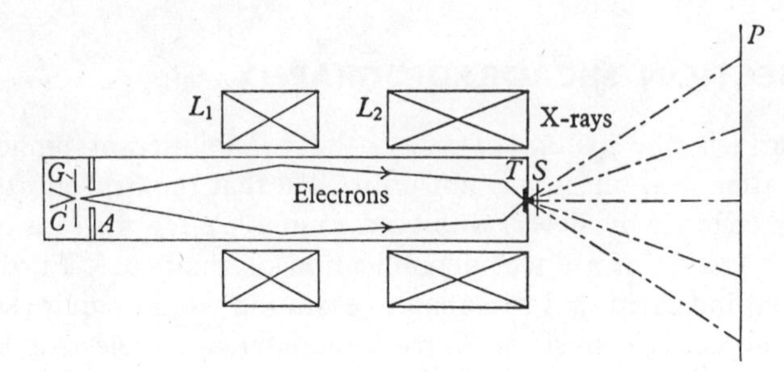

Fig. 8.14 Projection x-ray microscopy. The electron lenses L_1L_2 form a reduced image at T of the cathode C; the x-rays emitted from T project an image specimen S on to the screen (or plate) P.

are of nonmagnetic stainless steel and one end of the cylinder is sealed onto a tubular insulator several inches long and the other carries a cup-shaped diaphragm in which is a small central orifice. The bore of the insulator is at atmospheric pressure. A hairpin-shaped filament is "plugged in" to two metallic studs molded into the piston and which project at the other end for connection to two of the three cores of a high-tension cable passed up through the tubular insulator. The third core is connected to the cylinder for biasing.

Longitudinal movement of the cable is controlled by a clamp and threaded ring from outside the tube. This movement can be made while the tube is in operation and, since the filament mounting is attached to the end of the cable, the extent to which the filament penetrates the orifice in the diaphragm can be varied at will. Positioning of the filament is important in controlling the intensity of electron emission from the gun and so is the spacing of the gun from the anode. The latter is attached to the top of the gun chamber and consists of a nonmagnetic stainless steel disk about 1 mm thick with a central hole about 8 mm in diameter. By virtue of its attachment to the gun chamber, which is in turn bolted to the earthed framework of the equipment, the anode is at earth potential. It can be moved during operation in a plane at right angles to the electron beam passing through the hole by external adjusting screws and this is an advantage when "lining up" the column. No other adjustment is provided. The gap between the gun and the anode, which is normally about 3 mm for 20 kV operation, can be varied either way by a control ring on the gun.

Lens and pole-piece design have followed the work of Liebmann and Grad [38], Liebmann [39], Mulvey [40], van Dorsten and Le Poole [41], and Haine [42]. The first of the two lenses (condenser) is attached to the gun chamber and controls the demagnification of the beam. The second lens (objective)

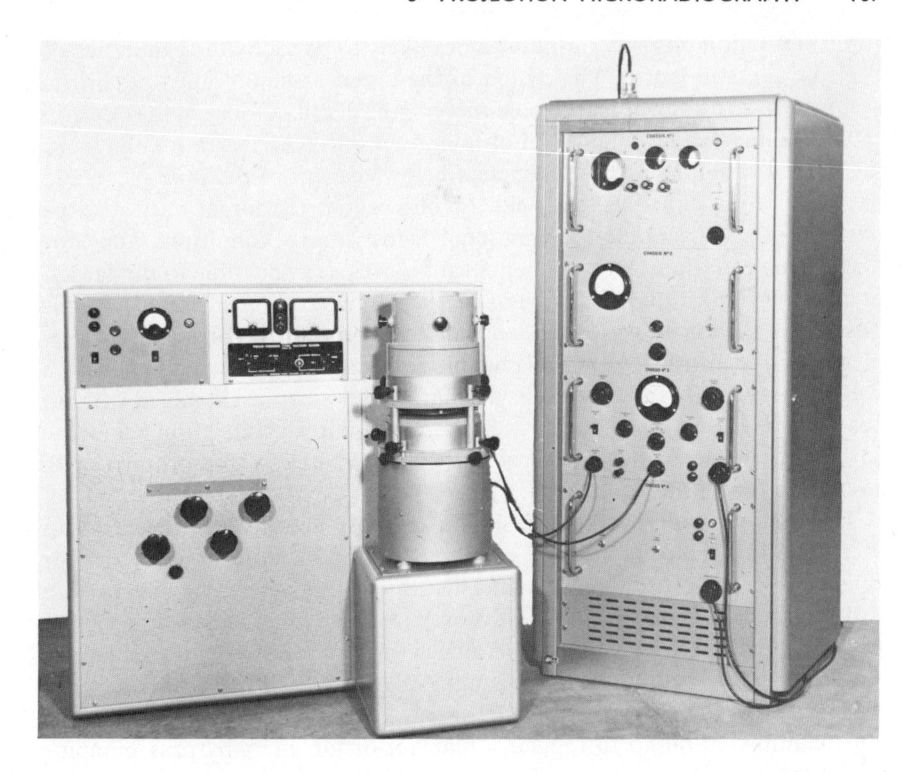

Fig. 8.15 Cosslett and Nixon type x-ray microscope.

focuses the beam onto the target. In spite of its relative weakness the condensing action of the first lens is important as it controls imaging errors which would otherwise reduce the intensity of the beam at the target.

The range of metals suitable for use as transmission targets is more limited than for the solid type since it may not be possible to obtain sufficiently thin vacuum-tight foils. Gold, either beaten out or vacuum deposited on a suitable substrate, has been widely used, but Al and Cu foils, 3 to 5 μ thick, are excellent alternatives, and Ti has also been used with satisfactory results. High-radiation intensity, relative to Z, is obtained with these materials and electron penetration is so small that electron scatter is minimal. On the other hand, absorption of x-rays generated within the target is high and a satisfactory output intensity is obtained only by making the target extremely thin.

Thermal dissipation in a thin foil target does not normally limit loading. For very small focal spots it is almost entirely radial and can be greater than the maximum potential rate of supply.

The limitation on x-ray output is more likely to be set by the properties of electron gun and lenses. The design of the target assembly must permit its accurate disposition within the pole-piece as the depth of focus of the electron spot is limited to a few microns. For minimum spherical aberration the focus must be at a point several millimeters below the top of the pole-piece and within a strong objective lens field. It follows that the target must also be positioned at this point for minimum x-ray source conditions and, for maximum resolution, the specimen must be as close as possible to the target, preferably within 1 mm. The area of a specimen that can be irradiated in this position is obviously very small and may be too limited for some requirements, for example, if it is not possible to remove a small section from a specimen or if individual portions of a large area must be examined without destruction. In such cases the target is raised until it is about 1 mm below the surface of the lens, but this procedure does involve a compromise with resolution for reasons quoted above.

The vacuum requirements are similar to those described in the previous section, and should be capable of maintaining a vacuum of 10^{-5} torr for reliable operation and long filament life. Small leaks can occur in the thin foil targets consequently an oil diffusion pump is recommended, not smaller than 2 in. and preferably with a thermoelectric baffle. Between the baffle and the electron beam column a high vacuum valve should be fitted so that, when this is closed, the column can be let down to atmosphere without shutting off the diffusion pump to replace a filament or target. A by-pass pumping line, with auxiliary valving, allows the column to be evacuated separately and reconnected to the diffusion pumping system by operation of the high vacuum valve. These provisions, and those for high voltage, filament, grid bias, and lens current supplies, are very similar to those required by electron microscopes except that the range and output of the high-voltage generator must be capable of providing lower kilovoltages and a higher output. Stability of supply is also particularly important since exposure times may be much greater. Examples of work done with this instrument are shown in Figs. 8.16 to 8.18.

A microscope of this type can be run for long periods if necessary. In one particular application batches of nuclear fuel coated particles have been examined daily over a long period. An alternative form of assembly employing interchangeable "solid" targets is shown in Fig. 8.19. A number of rings of different elements are mounted on a hollow spindle inserted through the side of a special target chamber sealed by an O-ring joint to the top of the objective lens. A sliding O-ring seal between the spindle and the chamber permits rotation and selection of any target element while the equipment is in operation and characteristic radiation, of a wavelength most suited to the work in hand, can be employed. The interior of the spindle is water-cooled by series

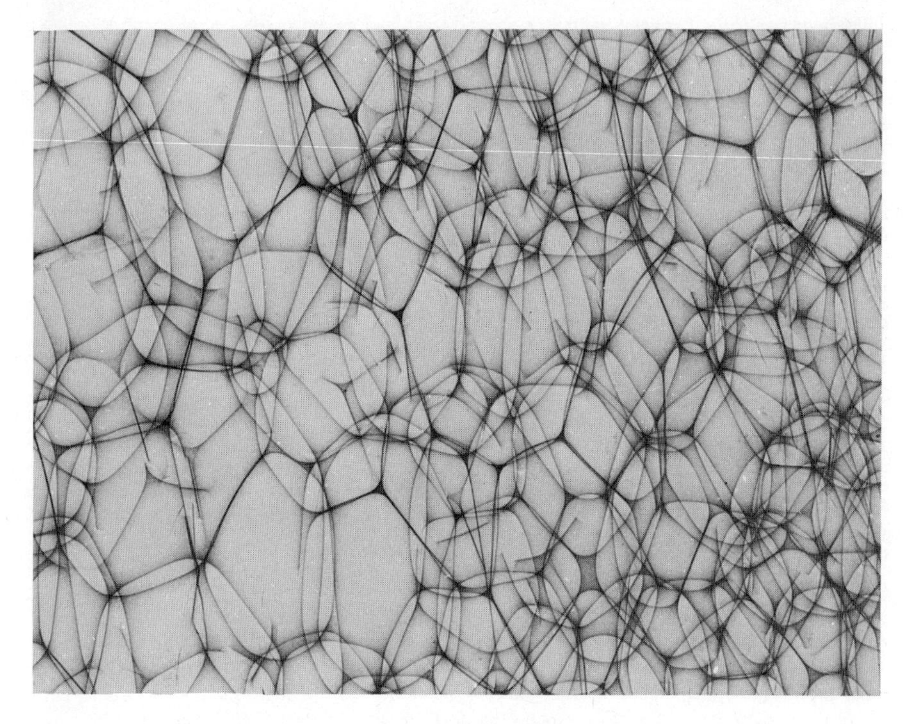

Fig. 8.16 Polyurethene foam. Taken with equipment as Fig. 8.15 at 15 kV. Al target. Direct x-ray magnification ×50. Courtesy of S. B. Newman, U.S. Bureau of Standards, Washington, D.C.

connection with the diffusion pump and thermobaffle cooling system, but contamination of the target surface because of the intensity of the electron beam occurs after the tube has been running for some time. This is indicated by a falling off in x-ray intensity but a slight rotation of the spindle presents a new surface to the electron beam.

A recent development replaces the grounded water-cooled target with an insulated type through which oil is circulated by a small closed-circuit pumping system. The midpoint grounded system, normally employed with sealed-off high-voltage tubes, is thus reproduced in demountable form.

The combination of projection radiography and image intensification is rapidly gaining favor as a solution to a number of problems, for example, the high dose rates inherent in the use of slow, fine-grain emulsions which must be used if fine detail is to be obtained by optical enlargement of contact radiographs taken with conventional equipment, and also the radiography of specimens which cannot be placed in close proximity to a photographic emulsion. Several types of intensifiers are available, some being directly

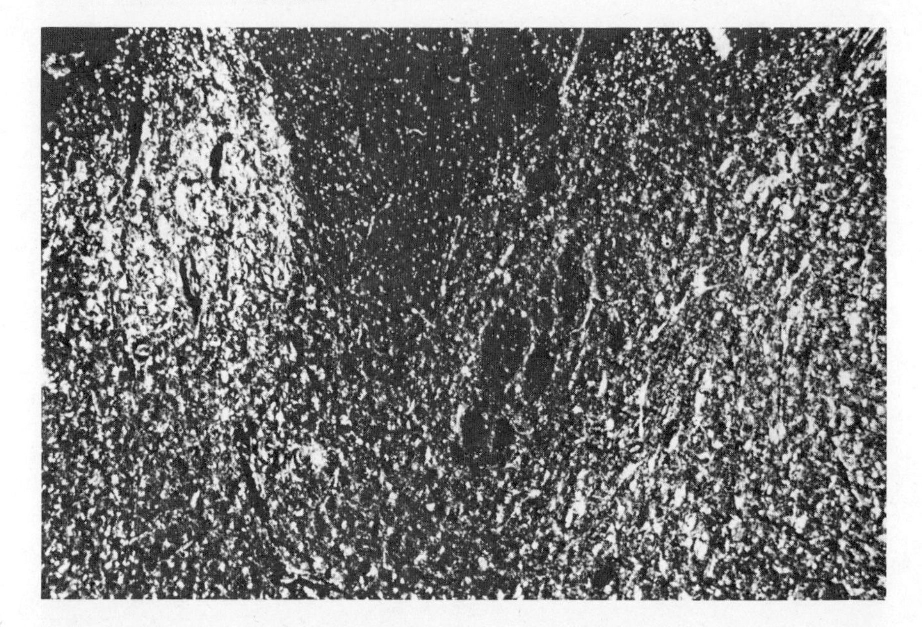

Fig. 8.17 Contact microradiograph of human brain, hypothalamus. Cell bodies of neurones made radiopaque with lead sulfide. Gomori's technique for acid phosphatase. Cells of the supraoptic nucleus appear on either side of the radiolucent fibers of the optic tract. The technique seems useful for emphasizing the position and density of such nerve cell groups. Courtesy of M. A. Bell, Dalhousie University, Nova Scotia.

sensitive to x-rays and others to the light output of an external fluorescent screen excited by an x-ray beam, the simplest form of a directly sensitive instrument consisting of a tubular glass envelope, generally between 5 and 11 in. in diameter.

A thin aluminum former, approximately equal in diameter to that of the tube, is located just inside one end and acts as a support for a fluorescent deposit on which is superimposed a photoelectric layer. X-rays passing through the end of the tube and the aluminum former excite the fluorescent material and cause electrons to be emitted by the photoelectric layer. Traveling down the tube, under the influence of an electrostatic focusing device, the electrons are picked up by a viewing screen, about 1 in. in diameter, which is sealed into the other end of the tube. The very thin, fine-grain, phosphor layer of this screen can be magnified optically for direct viewing, photographic recording, or closed-circuit television presentation. The instrument is robust, simple to operate, and is widely used in medicine and industry with conventional x-ray generators. For the intensification of images produced by fine focus direct x-ray enlargement techniques at kilovoltages below 30 its use is very limited but, above that level, it has many applications.

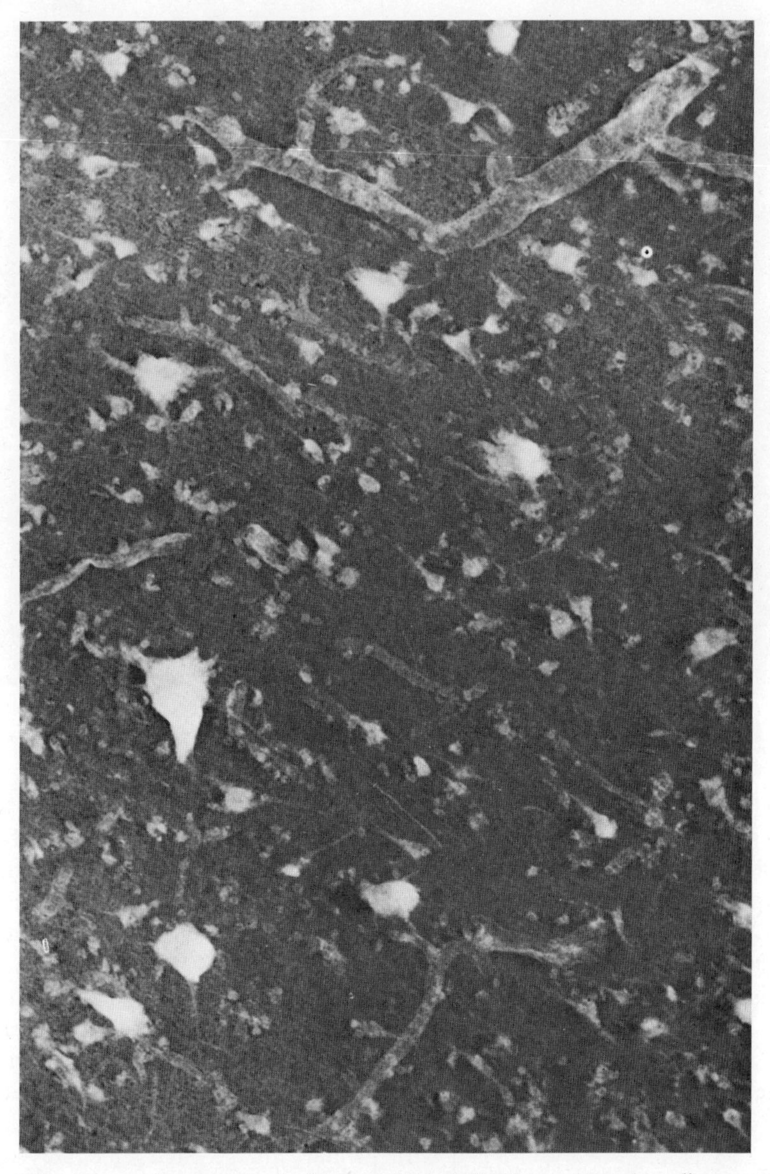

Fig. 8.18 Cortex of human brain. Nerve cells and blood vessels made radiopaque with lead sulfide. Two modifications of Gomori techniques were used on the specimen, one after the other. The first, at pH 5, showed the acid phosphatase activity of the nerve cells, and the second, at pH 9, added a picture of the alkaline phosphatase activity of the vessels. Total magnification ×200 using an Al target. (Direct x-ray magnification ×20 followed by optical ×10.) Courtesy of M. A. Bell, Dalhousie University, Nova Scotia.

Fig. 8.19 Interchangeable solid target chamber for x-ray microscope.

An alternative method is offered by the Dynamicon, a special type of 1-in.-diameter vidicon sensitive directly to x-rays instead of light, and very similar in appearance to the standard vidicon. The faceplate is metallic, the inner side being coated with a thin photoconductive layer which is scanned by an electron beam over an area limited to $\frac{3}{8} \times \frac{1}{2}$ in. On a 14-in. monitor this area is displayed at a magnification of about $\times 22$, and much higher magnifications can be obtained by initially magnifying the x-ray image of the specimen by the projection method before it is sensed by the Dynamicon. A specimen much smaller than the sensitive area of the faceplate can thus be viewed on the monitor, or a large specimen can be scanned. The inherent magnification of the system, when combined with primary image enlargement by x-ray projection, is a very attractive proposition, but the high energy needed for satisfactory operation of the Dynamicon has proved to be a serious handicap.

Indirect systems, incorporating light intensification, at present offer the best solution to the problem. The initial requirement is a very fine-grain fluorescent coating, on a transparent base, to receive the directly magnified x-ray image. This is viewed, through the base, by an intensifier and coupling optics relative to the size of the fluorescent image and that of the light-sensitive window of the intensifier, for example, an image-orthicon. This is a scanning type of instrument having, as its input stage, a 3 or 4-in.-diameter "light to electron" converter cell, the output of which is scanned by a focused beam from an electron gun within the tubular envelope. This beam is modulated and reflected into a multistage electron multiplier consisting of a series of very thin screens or dynodes. At each stage fast electrons, under the influence of an electromagnetic field, are focused to produce a multiplicity of slow electrons on the opposite side of the dynode and these are again accelerated to repeat the process several times. The output at the final stage is then coupled electronically to a viewing monitor. The sensitivity and resolution of the system are very high and these characteristics have been improved still further in the latest form of this instrument, the image-isocon.

Image intensification has also been used by Anderton and Smith [43, 44] in the examination of thin specimens of low contrast. The intensity of the beam, and consequently the brightness of the image on the fluorescent focusing screen, are generally so low that accurate focusing of the microscope by optical viewing is most difficult. To overcome this serious limitation to the development of x-ray microscopy in this sphere a high gain, multidynode, light intensifier is mounted vertically over the axis of the microscope, with suitable optical coupling to the focusing screen. The intensified image is conveniently viewed by an inclined mirror attached to the upper end of the intensifier and, after focusing the microscope, an exposure is made by replacing the screen with a photographic plate. Another advantage of the system is that any particular part of the specimen can be selected with much greater precision than is possible without the intensifier.

The performance of image intensifiers, and associated television systems, is examined in detail by Halmshaw [45].

Photographic Techniques

Primary x-ray magnification of the image makes possible the use of faster emulsions than those needed in the contact method, since optical enlargement is not normally so high and grain not such a major problem. Precautions taken to avoid artefacts because of dust and particles in processing solutions are also diminished. A contrast lantern plate is generally used with the Cosslett and Nixon type of transmission target microscope and the faster industrial or medical x-ray emulsions, single or double coated, for higher energy work with solid target apparatus.

In deciding whether to use a fine-grain emulsion placed near the x-ray source, or one of larger grain further away, it is necessary to take into account the size of the source, the kilovoltage needed for adequate penetration of the specimen, the area of the specimen which must be covered and its thickness.

As stated in Section 2 these factors are related to the detrimental effect of penumbra on resolution if the emulsion to target distance is too small, the absorption of low-energy radiation by air and Fresnel diffraction variation with specimen thickness.

Specimen Preparation

Methods that can be employed for the wide range of work to which microradiography can be employed are many and varied. Reference to the literature is therefore suggested: [34, 78, 153, 198, 260–262].

Stereo Techniques

Since the depth of focus obtainable is virtually unlimited the importance of three-dimensional radiography, as a method of ascertaining nondestructively the relative positions of the structural elements of an opaque object, has long been appreciated. Stereo pairs, for viewing in a stereoscope, can be readily produced by translation or tilting of the specimen or tube between exposures. As early as 1938 Sherwood [263] examined microstructures by mounting a tube on an arc-shaped carrier so that the two exposures could be made at a few degrees on each side of the vertical. He used a special carrier for the specimen and plate so that the former was not moved out of position when a fresh plate was inserted for the second exposure.

Berthold [264] made a detailed study of the possible applications of the technique in the examination of thin metal sections. Engström [265] applied it to the study of bone tissue and Bellman [266] to microangiography. The latter also described a microstereoscope. Saunders, Lawrence, and Maciver [267] found that in stereoscopic arteriograms at tridimensional complex system of vessels resolved easily into one simple basic pattern, and Oden, Bellman, and Fries [268] presented a paper on stereomicrolymphangiography.

Nelson [269] contributed a paper on the stereomicroradiography of carbons. He also described a special attachment for a seal-off tube and a high-power stereoscopic viewer/enlarger. Williams [270] examined metallurgical microstructures. He illustrated the morphology of a single grain from a polycrystalline aluminum alloy and employed an alternative method of angular disposition by keeping the tube stationary and rotating the specimen/plate holder. Other workers experimented with two separate tubes inclined at an angle to the specimen/plate holder, and tubes with two targets and two windows were also constructed. A special stereo bench by Kozlowski is described at the end of Section 4, and Fig. 8.20 illustrates an alternative form of tube mounting as used by the author. One form of stereoscope suitable for low-power viewing of large negatives, and fitted with a stereometer for structural depth measurement, is shown in Fig. 8.21 (manufactured by Hilger and Watts Ltd).

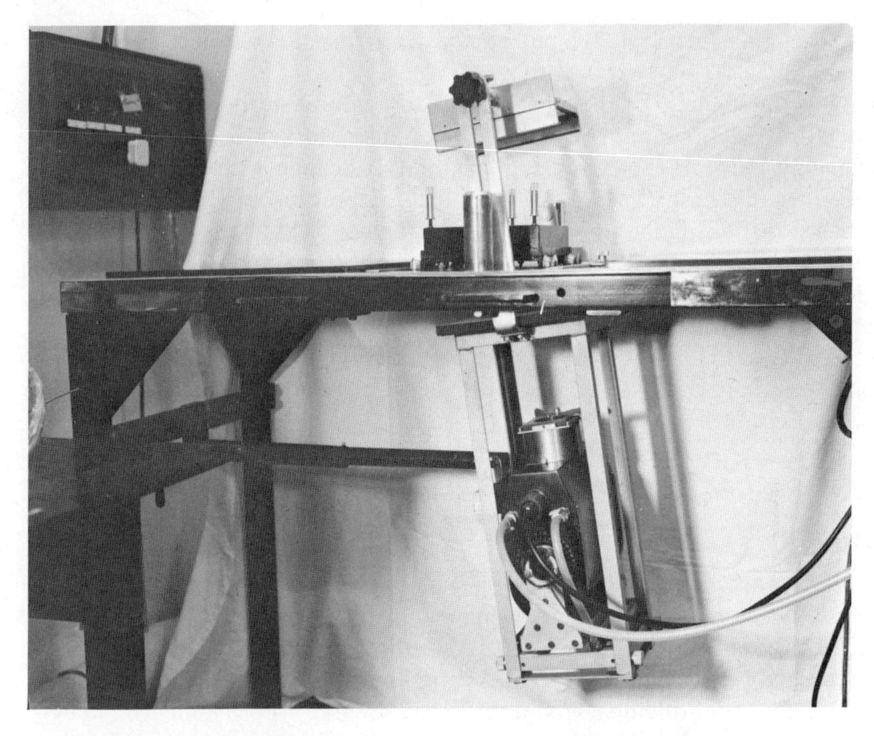

Fig. 8.20 Swinging tube equipment for stereomicroradiography and tomography.

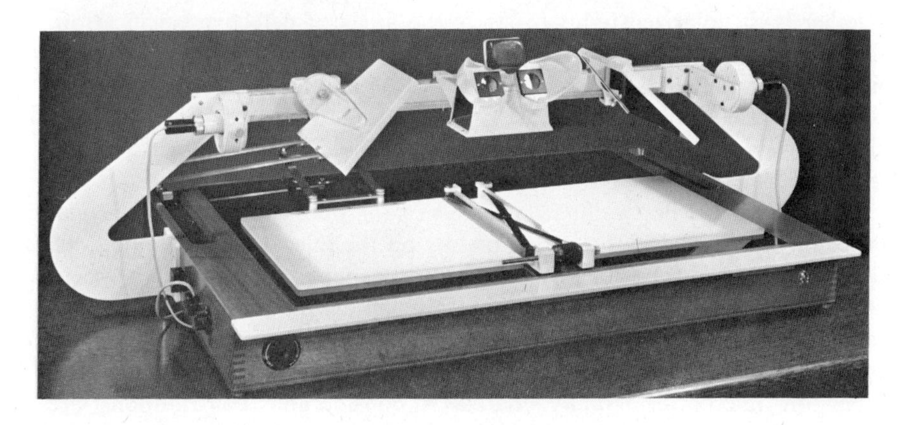

Fig. 8.21 Stereoscope.

Witte and Jütting [271] put forward an ingenious system for stereotomography.

In all these methods the most serious source of distortion lies in possible lateral or rotational movement of the specimen during the process of plate changing between exposures. One method of dealing with this difficulty is described by Latham [272, 273] in his comprehensive reviews of three-dimensional techniques in micro x-radiography.

The increase of geometrical blurring with separation of the specimen planes from the emulsion emphasizes the importance of structural displacements and variations in magnification which occur if a thick specimen is placed close to a point x-ray source. Nixon [34] concluded that they can, in fact, be an aid to stereoscopic viewing since they produce perspective as in normal vision.

4 APPLICATIONS AND RESULTS

Medical

Early work was concentrated largely on the study of dental enamel and the mineralized connective tissues. Applebaum [46] was attracted to the possibilities of judging the degree of mineralization and demineralization of tooth enamel relative to incipient dental caries, and the reception of his results encouraged him to embark on a major survey (see Applebaum, Hollander, and Bödecker [47]). In a later paper [48] he reviewed the literature on fluorine in water as a cause of mottled enamel and employed the same methods, together with polarized light, to study the areas of defective calcification. Sherwood [49] used stereoscopic techniques in the radiography of small biological specimens, and to improve resolution he designed a vacuum cassette. In this a thin membrane, opaque to light from the filament, was sucked down onto the specimen and gently pressed it into contact with the emulsion.

This period was particularly enlivened by the publications of Lamarque, Turchini, and Castel [50] working mainly on the injection of metallic salts. G. L. Clark [51] published the first of several papers on the medical applications of the contact method and the use of characteristic radiations in particular. Endicott and Yagoda [52] studied techniques for locating and measuring radioactive elements in tissues and Barclay published his first paper on microarteriography. Barclay's considerable contributions are summarized in his book, published posthumously [53]. Concurrently Engström and co-workers [54–57] actively encouraged quantitative and qualitative applications of the contact method. In particular, their work covered the determination of the dry mass of minute biological structures, involving microdensitometry, the production of accurate reference absorbers, the response of photographic emulsions, specimen preparation and mounting, the development of special tubes for soft radiation, and the correlation of microradiographs with autoradiographs of bone in following the uptake of radioactive isotopes. Mitchell [58] surveyed applications to biology.

Jowsey, Owen, and Vaughan [59] investigated the injection, retention, and excretion of Sr 90; Röckert [60] studied obliterated dentinal tubules and the calcification of cementum and dentine; and Saunders [61] studied human dental pulp vessels.

A comprehensive study of mineralized dental tissues covering osteopetrosis, and experimentally produced caries and attrition in white rats, was made by Bergman and Engfeldt in several papers [62–64]. Other important contributions were by Bohatirchuk [65–67] on blood vessels, experimental tumors, impregnated cancer tissue, and bone ageing. Lung structure was studied by Cunningham [68], shock wave effects on spiral ganglion cells by Hammar [69], brain autopsy material by Meschan and co-workers [70], osteomyelitis in the extremities by Orlandini and Bo [71], changes in the middle ear ossicles in the course of chronic suppuration by Orlandini and Vidoni [72], and measurement of the variations in calcification in normal rabbit bone by Owen, Jowsey, and Vaughan [73].

Bohatirchuk [74] continued to demonstrate the value of historadiography in the study of undecalcified bone with adjacent cartilage and soft tissue to give complete information about the amount and distribution of calcium salts.

In an effort to determine whether or not is is possible to determine quickly the extent to which osmium tetroxide fixation penetrates muscle tissue, Recourt [75] made contact studies at 2 kV and with magnifications up to ×900.

Fitzgerald [76] employed the contact method to show marked changes in cytoplasmic and nucleolar mass concentration in pancreatic acinar cells following the administration of ethionine. In the same year Fitzgerald, Li, and Yermakov [77] compared mass concentrations of normal and cancer cells of the human uterine cervix and indicated the potential of x-ray microscopy in cancer research.

Freeze drying in the microradiography of renal stones was developed by Carr [78] while investigating the controversial problem of how stones are formed. The belief that the comparison between normal radiographic images and pathological examination of decalcified bone was inadequate prompted Juster, Fischgold, and Ecoiffier [79] to investigate a clavicle from a patient who had been treated for several years for multiple metastases of breast cancer.

An application of the Engström and Lundberg midget tube [80] to chromaffin reaction was made by Hale [81] in following an observation that chromium is deposited in the adrenal medulla and that it remains there even when the color of the chromaffin reaction is removed by exposure to bromine water.

Since 1958 applications of the contact and projection methods have greatly expanded, and the remaining part of this section is dissected under broad headings.

Enamel

Preliminary to an investigation of the relationship between caries susceptibility and variations in enamel structure Gustafson [82–84] studied several hundred ground sections with polarized light and with darkfield illumination. Microradiographic comparisons showed that, although details other than mineralization were lacking, the method was very reliable.

In an attempt to determine the amount and distribution of mineral in all stages of enamel mineralization, Avery, Visser, and Knapp [85] used a combination of four techniques: microhardness tests, microradiography, microdissection, and microsubstitution. The contact x-ray exposures were made at 10 kV and were enlarged to illustrate the calcified rod substance in young calcifying enamel matrix in contrast to the noncalcified interrod substance, older calcifying enamel matrix, and developing human and monkey molars.

Another interesting comparison between optical and x-ray microscopy is reported by Hutchinson, Rowland, and Fosdick (86) when investigating the belief that decrease in caries activity, and permeability to small particles with age, could be explained if all the small changes were primarily linked to the interrod substance of the enamel.

A microradiographic interpretation of abnormal enamel formation induced by subcutaneous sodium fluoride is contributed by Weber and Yaeger [87].

Dentine

Van Huysen [88, 89] showed semiquantitative microscopic differences between the calcification of a normal or unerupted tooth and clinically demonstrable sclerosed dentine. Atkinson and Harcourt [90] studied sections of noncarious bicuspids from young people, and Röckert and Ohman [91] studied the changes in mineralization of replanted human teeth. Minura and his associates [92] analyzed rabbit dentine after injection of sodium fluoride.

Weber [93] demonstrated that, when viewed in the light microscope, the bright rings seen around dentinal tubules could not be related to the peritubular zone as shown by microradiography and concluded that the presence of peritubular matrix could not be determined by transmitted light. He illustrated these findings by a microradiograph and a photomicrograph.

Dental Pulp

Saunders [94] employed both techniques in studying periodental and dental pulp vessels in the monkey and man. Many microangiograms illustrate the wide range of this work.

Cementum

A paper by Glimsher, Friberg, and Levine [95] describes the application of contact microradiography, using nickel-filtered copper $K\alpha$ radiation at 19 kV, to the identification and characterization of a calcified layer of coronal cementum in erupted bovine teeth.

Jaw Morphology

The contact method was used by Park [96, 97] in a radiographic study forming part of an investigation into the development of rat mandibular teeth and bone and the structural changes involved during eruption. The aim was to establish the pattern and degree of calcification during growth. The mandibles used in the survey were all intact and as size increased with age parts of the bone could not be placed in contact with the emulsion so that detail was lost. This difficulty can be largely overcome by employing the projection technique.

Bone

By serial sectioning of undecalcified bone Bohatirchuk [98] found that stain historadiography partly confirmed and partly contradicted data obtained by other methods concerning the different levels of demineralization in the bony lamellae, and concluded that the results obtained demonstrated the true value of historadiography. A preliminary study of the structural appearance of normal bone made by Jowsey [99] indicated the variations noted in microradiographs and their relations to the age of the individual. Ultrasoft radiations were used by Greulich [100] to observe the distribution of organic mass at sites of bone resorption. Sissons, Jowsey, and Stewart [101] used the aluminum reference system to determine the distribution of hydroxyapatite in bone.

Rowland [102] also employed the same form of step wedge but with characteristic radiations emitted from targets placed in the beam from a standard x-ray tube. This method suffers from the disadvantage that the secondary radiations are of very low intensity.

Comparative assessments of the variations in the structure and activity of the cortical bone of the mandible were made by Manson and Lucas [103] using undecalcified sections, 60 to 80 μ thick, from 28 individuals aged from 8 weeks to 76 years.

Hall [104] examined sections of the upper femora of rats after hypophysectomy and found a preponderance of periosteal and endosteal bone compared with the endochondeal bone that was the dominant tissue in the control rat.

The mode of action of vitamin D in relation to alterations in bone development in rats is discussed by Bhussry and Parikh [105] who demonstrated irregular radiolucent patches in microradiographs taken with the Philips CMR tube at 5 kV. This equipment was also used by Mjôr [106] in investigating the microradiographic changes in sections from human lower extremities. He noted that initial decalcification apparently occurs in the more highly mineralized parts of the tissue and that the type of decalcification solution used is important in resolving structural details.

In a paper on dichromatic microradiography Lindström and Omnel [107] showed that two phases in a given specimen can be quantitatively determined *in situ* by absorption measurements. The sensitivity of the technique was dependent on the ratios between the mass absorption coefficients, at the two wavelengths, being as great as possible. Prentice [108] compared the examination of ground bone sections by uv and x-rays.

An extensive study of bone turnover in normal and osteoporotic bone and other metabolic bone diseases was made by Jowsey and co-workers [109] who described their use of quantitative microradiography as micromorphological in approach but focused on the mineral of the skeleton. Specimens were taken from the femur, lumbar vertebra, iliac crest, and rib. The method was found to be reproducible and provided an accurate, direct measure of bone formation and resorption. The results were correlated with other techniques and indicated that bone from the majority of osteoporotic patients differs from normal bone in the amount of resorption. Rowland, Marshall, and Jowsey [110] described the application of the method to a study of radium-burdened human beings who had carried their radioactivity for periods of 20 to 40 years. Stone and Ferguson [111] approached the problem of obtaining quantitative data on bone mineralization by a variation of the combined microradiographic and photographic technique described by Jowsey [112].

The search for further information on the mineral phase in the healing process of fractures was continued by Nilsonne [113] who states in the preface of a lengthy paper that "X-ray physical methods, notably microradiography and microdiffraction, have proved useful in giving reliable information about the mineralizing processes in bone tissue. This is particularly true of microradiography which Engström put on a quantitative basis."

Juster, Fischgold, and Metzger [114] also used microradiography to supplement normal radiographic interpretation in a meningioma of the cranial dome. In studying skeletal lesions in rats, following ingestion of fluoridated water, Röckert and Sunzel [115] found that microradiographs disclosed distinct differences in sections from the spine and indicated that osteosclerosis, visible by x-ray microscopy, appeared to have set in. Bone atrophy in apparently healthy aged persons was studied by Bohatirchuk [116] and microradiographs of ribs from typical cases of rickets and osteomalacia were compared by Bohr and Dollerup [117] with others of iliac crests from patients with osteomalacia or ordinary osteoporosis.

Lloyd [118] used chromium radiation at 20 kV to follow up earlier work on radium poisoning by examination of sections from long bones of patients who had ingested 226_{Ra}.

The purpose of an investigation by Vose [119] was to determine by quantitative microradiography whether the observed increase in ash content of osteoporotic cortical bone results from resorption or from differences in density between extra-haversian sites in the outer cortical layers of osteoporotic and normal bone. A densitometric chart showed that mineral concentration of low-density haversian bone is about 15% less than that of extra-haversian bone at the selected site. Six cases of osteoporosis were made the subject of an intensive individual research by Urist and co-workers [120] and they illustrate the thickness and structure of cortex in patients with and without osteoporosis by microradiographs. Severe osteoporosis, with cirrhosis of the alcoholic and with pathologic fracture, was reported by Summerskill and Kelly [121].

Light, x-ray, and fluorescence microscopy were correlated by Hammar, Radberg, and Röckert [122] in examining the opaque structure, seen by normal radiography, in the posterior part of the sphenoidal sinus some months after transsphenoidal hypophysectomy. Results suggested that the reparative activity in the tissue is greatest immediately after operation, gradually decreasing to comparative stability after 12 months.

Kornblum and Kelly [123] compared cortical bone from the tibia of ischemic limbs with that from limbs of similar age but in which ischemia had not been clinically demonstrated. These workers, using chromium radiation at 15 kV, found a positive correlation and presented the concept of uninhibited centripetal growth of osteons to form bone-filled canals.

Bone biopsies from human autogenous grafts in primary cleft lip and palate cases were subjected to x-ray and tetracycline fluorescence analysis by Breine, Johansson, and Röckert [124].

Smith and Hobdell [125] endeavored to correlate the rate of bone turnover in the iliac crest and femur with the bone mineral as determined by a technique using two parallel scintillation counters. The radiation source was a high-voltage projection instrument fitted with a solid tungsten target, as described earlier.

The volume of bone matrix associated with individual cells was calculated by Hobdell [126]. Great differences were shown between the volume associated with the osteocytes of primary membrane and those of adult bone. The

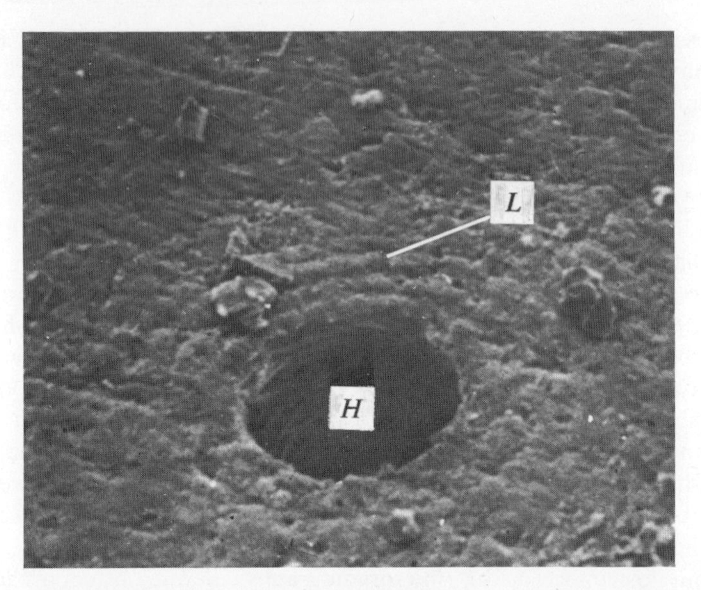

Fig. 8.22 Micrograph of the ground surface of a bone section taken at 10 kV with a scanning electron microscope. Carbon-gold conducting coating after removal of organic matrix with hot 1,2,ethane diamine. A Haversian canal (*H*) is surrounded by circumferential lamellae (*L*). Magnification ×500.

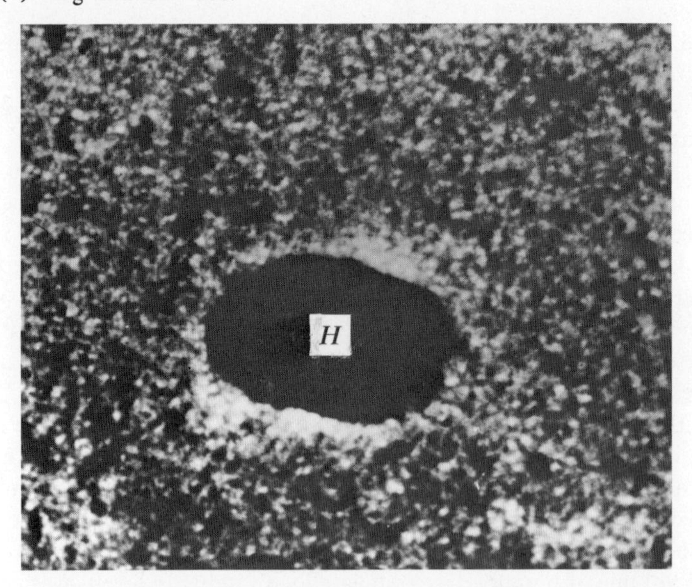

Fig. 8.23 Contact microradiograph of the same bone section as in Fig. 8.22. The image here is produced by the absorption of x-rays by the whole thickness of the bone section. Magnification ×500. Courtesy of M. H. Hobdell.

microradiographic appearance of natural edges within bone sections and different categories of scanning electron microscope images of the mineral front, looked at face-on at those edges, were correlated by Hobdell and Boyde [127]. They found that a microradiograph can record all the information in one negative but considered that the scanning electron microscope can provide more reliable identification of the functional state of the surface if the section is tilted and rotated for the examination of all edges in turn. Both methods were considered to have disadvantages and a full-scale comparison of the two methods is being planned (see Figs. 8.22 and 8.23).

Bone Marrow

A further application of 1.6 kV radiation was made by Harris, Haig, and Watson [128] in the study of cells in unstained smears of bone marrow and lymphoid tissue mounted on a formvar polyvinyl-formaldehyde membrane about 1 μ thick. A clinical and experimental study, following removal of bone marrow by curettage, was made by Branemark and co-workers [129] who used histological microradiographical and visual microscopical techniques. The findings prompted the assumption of a transformation of osteoblasts into reticulum cells in the process of marrow regeneration.

Ear

The projection method, using a projection x-ray microscope, was chosen by Clarke [130] for an examination of undecalcified sections of normal and diseased stapes leading to a comparison of mineral content. With the mineral distribution as a criterion, the otosclerotic stapes were divided into three groups, which were interpreted as representing definite stages in the progress of the lesion.

Röckert and co-workers [131] collaborated in a study on the mineralizing process in otosclerosis, using both contact and projection microradiography and uv fluorescence microscopy after administration of tetracyclines.

Cartilage

Metaphyseal growth in the tibia of newborn rats was studied by Vincent and Dhem [132] who traced the deposition of calcium in hypertrophic cartilage by correlations of microradiography and stain histology. Using the projection technique Saunders [133] examined histological sections by direct x-ray magnification. One illustration, among many others covering a wide field, showed the epiphyseal growth zones at the proximal end of decalcified rabbit tibia.

Apparent neglect of microphormology in studies of calcium-containing tissues lead Bohatirchuk [134] to study cartilage calcification by stain historadiography in order to give specificity in the detection of calcium and possible morphological visualization of calcium in biological tissues at micro levels.

The Lindström absorption spectrophotometric procedure, with some instrumental and measuring technique variation, was used by Howell and Carlson [135] to map *S*-containing compounds in cartilage and other histolic regions.

Skin and Connective Tissue

A capillary haemangioma of the skin, which normal angiography had failed to demonstrate, was described by Veiga Pires [136] who suggested that this case demonstrated that neoplastic lesions of vascular origin may fail to show angiographically if they are limited to the vessels of the skin and are of small dimensions. This was confirmed by Saunders and James [137] who used an injection technique to show the smallest vessels beneath the papillary ridges by the projection method. In a further study Saunders [138] used the same method in a much wider field, including *in vivo* microangiography of rabbit ears.

Following observations on the vascularization of malignant tumors of human limbs, Soila [139] made a survey of available emulsions and concluded that *in vivo* microangiography in humans might be possible at magnifications up to about $\times 30$. In a general review Peterson and Kelly [140] stressed the importance of stereo presentation in the microangiography of thick sections in making possible an understanding of the functional aspects of the circulation by examination of the spatial arrangements of the vessels. The value of the stereo technique was amplified by Saunders [141] in a continuation of his earlier work. In the discussion in this paper Bellman expressed the hope that the unique advantage of microangiography over other techniques would never be overlooked, that is, as the specimen is not largely destroyed by sectioning, it is available for other studies with which comparisons can be made.

The effect of cadmium injection in the testicular vasculation of the rat was studied by Niemi and Kormano [142].

Lung

Since comparatively large air spaces separate important structural detail, sections can be examined by microradiography without objectionable structure overlap. Oderr [143] and Oderr, Dauzat, and Montamat [144] found that sections up to 2 mm thick could be exposed at 2.5 to 5.0 kV, and sections 5 to 20 mm thick required 10 to 30 kV. For the thinner range, evacuation of the space between tube and specimen or replacement of air by hydrogen was normally necessary, and especially so in the study of emphysematous lung. A method of simultaneous air inflation and latex injection was used to delineate the vascular systems. The authors remarked on their inability to show the complete lymphatic pattern but anticipated that this, and also that of the bronchial arterial system, could be profitably demonstrated with suitable equipment.

Eye

A physioanatomical study of the opaqueness of the lamina cribrosa sclerac by perfusion and injection, followed by microradiography, is described by Leroux and co-workers [145]. A second paper by the same authors [146] showed that by microradiography, after perfusion of the anterior chamber of the human iris, the presence of a canalicular system from the vascular network was revealed.

Liver

In two papers Blickman, Klopper, and Recourt [147, 148] detailed the preparation technique used to study the vascular pattern in the liver vessels in rats, using 25-μ sections.

Salmon and Ceccaldi [149] investigated the possible application of the contact method to follow up deposits in the liver and spleen as indicated by autoradiography.

Calculi

The researches of Carr, referred to earlier in this section, were continued by Andrus and co-workers [150] using polarized light, microradiography, and diffraction techniques. Fournier [151], having noted the presence of birefringent bodies during the histological examination of certain intraperitoneal tumors, employed the contact method at 1.3 kV.

Three types of human prostatic concretions were studied by Bengmark, Romanus, and Röckert [152] who used both methods of x-ray microscopy. Thin, plane-parallel sections 75 μ thick were prepared by the method described by Hallén and Röckert [153].

"The composition and structure of human gall-stones" is the title of a very detailed study by Bogren [154], the extent of which can be judged by the fact that 176 were examined by x-ray diffraction and 94 by microradiography.

Kidney

Historadiography was used by Hale [155] to identify the reaction products of eytochemical methods for acid phosphatase, carbonic anhydrase, succinate dehydrogenase, and thiolacetic acid esterase. The contact method at 1.5 kV produced satisfactory microradiographs of rat kidney sections. Comparative photomicrographs are included in his illustrations.

Parallel studies of microangiograms and serial histologic sections of the same specimens are discussed in a paper by Ljungqvist [157] on the intrarenal arterial pattern in the normal and diseased human kidney. Mono- and stereo-microradiographs were produced by the contact method. The importance of the stereo technique to microangiography is emphasized in a later paper by Lagergren and Ljungqvist [158] who quote results on studies of 200 kidneys.

Nervous Tissue

The relationship between glia and nerve cells was investigated by Hydén [159] who devised a method of dissecting out nerve cells from their surrounding glia so that a particular type of nerve cell could be analyzed with its specific type of glial cell.

Grampp, Hallén, and Rosengren [160] measured the mass per unit area of cytoplasm of cells in sections of fixed rabbit spinal ganglion by interference microscopy and two independent microradiographic techniques.

The contact and projection methods, for survey and fine-detail presentation, respectively, were used by Saunders [161] to study the deeply placed and microscopic vessels in the human brain and spinal cord. Stereomicroradiography, by the projection technique, was found to be of particular value in recording the volume pattern on intracerebral vessels as well as showing interesting morphological features of the cortical, transcerebral, and central vessels of the cerebral microcirculation.

An analysis of the angiographic alterations in the brains of rats with cerebral edema was made by Perez, Hodges, and Margulis [162]. Microradiographic studies at high magnifications showed distinct changes characterized by marked tortuosity, decrease in number of branches with narrowing, and dilation of the vessels in white and gray matter.

The microvascular bed of the human cortex and white matter, which appears to be extremely complex in a microangiogram, was found to be capable of anatomical analysis by selective local injection into isolated arterial and venous segments by Saunders, Feindel, and Carvalho [163]. A routine technique was followed, whereby the contact and projection methods were employed to produce survey and detailed microangiograms. The report discusses the arterial components of the human cerebral microcirculation and the venous components of the microvascular system, together with certain neurosurgical considerations of the cerebral microcirculation in man.

The need for further study of the complex cerebral capillary beds was followed up by Saunders and Ely [164] using a high-voltage x-ray microscope and x-ray sensitive vidicon as described in Section 3. The results recorded include sections of freeze-dried human brain showing the radiation of the short cortical vessels within the gray matter, the microvascular pattern of blood vessels around a sulcus of the brain, and detailed trabeculae of the terminal phalanx of a finger *in vivo*.

Histochemical reactions were demonstrated simultaneously by Saunders, Bell, and Carvalho [165] using the contact and projection methods, in conjunction with photomicrography, to detect or identify a small but naturally occurring metal component within a tissue. A new metal was also added to a soft tissue component of low atomic weight which would otherwise be difficult to study.

Microradiographs and control photomicrographs showed features of the human cerebral microcirculation, distribution of the neurons, and neurono-vascular relationships.

Metallurgy and Mineralogy

References to the possibilities of micro x-ray techniques in these fields date back to Röntgen, but interest did not crystallize until Fournier [166] studied alloys, mainly of aluminum and copper, with monochromatic radiations. Trillat [167] included special types of steel in his similar work and Kirchberg and Moller [168] investigated the fine structure of iron ores. Smith and Keller [169] correlated metallographic and radiographic methods in the establishment of standards for the inspection of spot welds and improvements in spot welding techniques, particularly in light alloys. Gross and Clark [170] show magnifications up to ×200 of similar specimens, and Ball [171, 172] surveyed the possible uses of microradiography as a routine tool. He showed examples, at magnifications up to ×50, of gas and shrink porosity.

Segregations and inclusions, particularly in steel, were studied by Betteridge and Sharpe [173] who used specimens about 0.001 in. thick, with cobalt, iron, copper, and chromium as alternative target materials in the x-ray generator. The relative merits of these sources are indicated in a number of illustrations.

Lavender and Jones [174] and Taylor [175] in an examination of banded steels, and the removal of banding by high temperature, used the techniques developed by Betteridge and Sharpe and concluded that microradiography was a useful complementary technique to those normally employed in metallography.

A study of grain boundaries of pure aluminum, and aluminum zinc alloys, was contributed by Berghézan, Lacombe, and Chaudron [176], and Cohen [177] followed up the work of Kirschberg and Moller, referred to above, on the distribution of iron in low-grade ore with the aim of establishing a simple routine for the day-to-day checking of sinters. They compared photomicrographs and micrographs taken at ×125 magnification.

Detection of cracks under tensile strain in individual manganese sulfide inclusions was shown by Homès and Gouzou [178] to be a field in which the method could make an important contribution to the understanding of fracture formation and of deformation.

The complementary techniques of microradiography and optical micrography were used by Honeycombe [179] in a systematic investigation into homogeneities in the plastic deformation of metal crystals. He was concerned especially with their role in the deformation of aluminum as accounting for many of the differences between this metal and cadmium, such as in strain hardening and the extent to which recovery can occur on annealing. Goldschmidt [180] found cobalt K-alpha radiation particularly useful in the study of permanent magnet steels for segregation and microcracks. A foundry development investigation on the effect of superheating and casting temperature on the physical properties and solidification characteristics was made by Goldspiel and Bernstein [181].

In his book Clark [182] devoted a chapter to metallurgical and other applications of microtechnique, and compared the advantages and disadvantages with metallographic methods as shown in the table below.

Microradiography	Photomicrography
Specimen need not be highly polished	High polish required
Requires preparation of two surfaces	One surface required
Shows entire thickness, three-dimensional	Only surface effects observable
Limited by grain size of emulsion	No direct limitation by the emulsion
Differential absorption necessary	Etching reagents, and so on, required
Detects interior cracks: shows variation in density	Requires heterogenous phase and boundary structures
Superimposed images	Single image
Interpretation simple—dependent only on absorption	Interpretation complicated by large variety of reactions required

Nixon and Cosslett [183] and Nixon [184] showed that in a microradiograph of zinc–5% aluminum alloy the dark areas are zinc-rich and the striped regions correspond to a eutectoid of the two constituents, the phase being seen without surface etching. They also studied annealed brass to which a small amount of lead had been added to improve machining properties and demonstrated that the distribution of the lead particles, about 1μ in diameter, was not as uniform as it should be for good machinability.

Neff [185] also found that the x-ray technique had advantages in examining the manganese and sulfur distribution in special steel, and found Cu radiation preferable to Co or Cr in producing contrast. Urlocker, Rutter, and Winegard [186] show comparisons at $\times 200$ of incipient fusion in duralumin. Bermudez de Castro, Fuster Casas, and Boned Sopena [187] made an intensive study of iron ores directed toward a more complete understanding of the fundamental reactions taking place in the process of sintering and determining the nature of the resultant products. Votava, Berghézan, and Gillette [188], following the work of Smoluchowski et al. [189], applied the Bragg effect to a number of metallurgical problems, particularly where a change in orientation and some plastic deformation patterns are to be observed. Their paper is well illustrated with microradiographs, at magnifications up to 625, of precipitates in Al/Ag, crystal orientation and grain boundaries in pure and recrystallized zinc foils, and the diffraction image of deformation produced by rolling and stretching. The need to relate information gained with the results of other methods of examination is particularly stressed by Andrews and Johnson [190] in their report on a wide range of metallic specimens examined with Co, Cr, Cu, and Fe radiations and illustrated at high magnification.

The advantages and limitations of the contact method are discussed in detail by Sharpe [191]. The greatest difficulty normally encountered is that of producing a thin uniform sample, suitable for microradiography, and its subsequent preparation for metallographic examination. The advisability of restricting radiations to those which do not produce fluorescent secondary x-rays in a specimen matrix is pointed out since absorption differences can then be largely masked. Advantages mentioned include the possibility of seeing projected shapes of inclusions and phases in true perspective; moreover, cavities, shrinkage, and porosity defects are not smoothed over, as may happen in polishing for surface examination.

The point projection method was used by Jackson [192] to complement the optical examination of a thucolite ore complex carried out by Bowie [193]. A 25-μ section was examined at 20 kV with a 3 μ copper foil transmission target and the illustrations show the replacement of dolomite by specular iron at $\times 200$ and what are thought to be cellular bodies of pitchblende in hisingerite at $\times 650$. A further example in his paper illustrating the possibilities of this new technique is a stereo pair at $\times 10$ direct x-ray magnification, enlarged optically to $\times 150$, of a thin section of Al–3% Mn alloy in which the Mn is clearly visible as hollow randomly oriented crystals.

The orientation in the rolling direction and the distribution of Al_3/Fe and Al_3/Ni inclusions in aluminum were reported by Erdmann-Jesnitzer and Gunther [194]. Orientation, elongation, and destruction of the enclosed needle-shaped intermetallic compounds were observed as a function of initial position and the degree of cold and hot rolling. Thewlis and Derbyshire [195] employed direct x-ray magnification for the routine examination of reactor fuel element plates and Yamagouchi [196] studied thorium and uranium crystals at $\times 15$, followed by optical magnification to $\times 100$.

The structure and composition of the substrate material in influencing the stress corrosion cracking of alloys, corrosion at grain boundaries, and the decarburization of steel were studied by Foley and Newberry [197] with the General Electric (United States) electrostatic type of x-ray microscope. Pitting corrosion of aluminum alloys was also investigated in stereo and allowed observations to be made relative to the second phase and the pits. Johnson and Andrews [198] showed MnS inclusions and lead particles in free-cutting steel using Co, and alternatively Cr, radiation at $\times 200$. These same radiations, at $\times 75$ magnification, showed MnS and Fe inclusions in Cr–12% steel.

Alloys, in particular those of aluminum-copper, were considered by Ruff and Kushner [199] to represent an ideal subject for investigation by the projection method.

Maximum advantage was taken of the capacity of point projection by using high initial x-ray magnification, any subsequent enlargement by optical means being carried out at low power. A combination of microradiography and other techniques normally used in metallography, as practiced by Spear and Gardner [200], originated with their interest in the manner in which several aluminum alloys solidified, rather than in the identification of constituents. Small molten specimens were cooled at relatively slow rates to various temperatures before quenching in water.

In presenting a paper on the potential applications of the projection method to fuel element materials Sharpe [201] outlined possible interrelations with complementary methods and considered that the physical metallurgist should find it a welcome addition to existing structure-finding techniques. Practical applications in the nuclear energy field included (*a*) the detection of several forms of void, such as Kirkendall holes and grain boundary cavitation during slow deformation of heated metals; (*b*) the concentration variations in the distribution of fissionable particles across the thickness section of a fuel plate; and (*c*) swelling by gas bubble nucleation in an irradiated metal sample and He liberation in beryllium. The possibilities of dynamic studies were thought to have a promising future since a specimen could be placed at a considerable distance from the recording medium. In addition to intergranular and stress corrosion, the initiation and progression of cavitation failure during strain, and bubble growth during irradiation damage in beryllium specimens, were suggested as worthy of investigation.

Shinoda et al. [202] used a modified projection instrument to study the bead of a welded zirconium plate and compared the results with photomicrographs. Etched specimens 0.02 mm thick were studied optically, and the bead in the welded portion showed that several small grains formed a crystalline aggregate which appeared as one grain in the x-ray study.

The improvement in contrast resulting from the use of long-wavelength radiations is emphasized by Williams [203] in comparing microradiographs of a thin section of cast leaded bronze using 47 kV, and alternatively 17 kV. In the latter case the distribution between copper and tin-rich areas and variations in thickness of the specimen are shown to much greater advantage. The alternative method of contrast improvement, that is, by absorption-edge selection, is illustrated in this paper by microradiographs of cast copper–5% iron showing dendritic structure and undissolved particles. The considerable advantage of three-dimensional presentation is particularly stressed and a number of examples show the spatial arrangement of particles, grain corners, and gas pockets. Williams prepared specimens to illustrate the relationship between dihedral angle and microstructure and used these in studies of plastic deformation, recrystallization, grain growth, and transformations as a nondestructive technique.

Microradiography was considered to be particularly useful in the study of complex alloy steels by Taylor [204] who found that it was frequently possible to identify small segregations of highly dispersed complex carbides, the presence of which had not been observed by diffraction and standard photomicrographic examination. Typical microradiographs illustrating his paper are compared with photomicrographs. In one case the former show, with exceptional clarity, the actual dendritic growth of the metal crystals in cast leaded-bronze, whereas the latter show two phases of the copper-tin matrix plus segmented lumps of lead.

Johnson [205] made a comprehensive study of thin transparent geological and mineralized specimens using the same comparative methods. In a section of Jurassic ironstone, ×27, the x-ray technique showed more clearly the variation of iron content in the concentric layers of the ooliths, and the angular shape of the quartz grains was much more precise. An iron ore section comparison, ×45, showed that the presence of mineral film, with strong absorption or of staining within grains, gave too high an estimate of the iron content when examined optically. In the microradiograph the iron content of the ooliths is low and dispersed. The efficiency of mineral dressing operations involving the examination of tailings is shown to be a particularly profitable use of the x-ray method when applied to industrially important ores. Assaying by particle counting was not found to be possible using optical methods but was reasonably easy radiographically at ×113. This was also found to be the case when endeavoring to distinguish gold from accompanying pyrite at ×75.

Johnson also investigated sintered iron ores, the heterogenous structure of granular oxides, holes and slag pools containing dendrites, and Ni radiation was employed at ×18 and ×180.

Microradiographs, ×93, taken with Cu radiation, were found to be an effective aid to operational variations in finding the optimum conditions for the sintering of iron ore with 3% addition of coke and 20% flue dust. An understanding of the structures formed in open-hearth furnaces during service is shown to be greatly assisted by microradiographic examination of chrome/magnesite refractories. In typical illustrations at ×18 to ×38, using Ni radiation, the intimate bonding between the chrome grains and magnesia is clearly shown. The elongated iron-rich spinel crystals are also differentiated. Porosity resulting from loss of material by diffusion is shown in an illustration of chromite grains embedded in columnar iron-rich spinel. Another illustration demonstrates the nucleating effect of the chrome spinel, and orientation of iron-bearing precipitates, in iron-rich magnesia sintered with chrome ore. Further examples of the application of characteristic radiations and comparisons between optical and x-ray techniques are given by Andrews and Johnson [206] with Fe radiation, white streaks in the negative attributable to Cr segregation with dark specks indicating MnS, whereas with cobalt radiation the MnS showed up as white at ×75.

In another microradiograph of the same area, taken with a chromium target, the MnS, and associated cavities or silicates, were seen as dark with no Cr segregation. Some white specks were assumed to be the result of traces of lead or other heavy elements. A photomicrograph of alloy steel, ×230, showed sigma phase in austenite matrix after heating, whereas the corresponding x-ray negative indicated segregation between the phases.

Sharpe [207], using the projection method, subjected coated uranium carbide particles, agglomerates of which form fuel-element cores for high temperature reactors, to microradiographic inspection. The particles, generally 50 to 5000 μ in diameter, were individually coated with carbon or silicon carbide by a pyrolytic decomposition process and it was necessary to check the coating for uniformity, microcracks, and possible migration of uranium into the coating. Information on the uniformity, homogeneity, and presence of surface contamination of the cores was also required. The microsource, about 1 μ in diameter, permitted primary x-ray magnifications of ×5 to ×100 and, using a comparatively fast emulsion (Ilford contrast lantern plate), secondary magnifications of ×5 to ×20 produced results comparable in resolution to that of the optical microscope. Exposure times were much lower than required by the alternative contact method and the obtrusive background blackening of the emulsion, normally produced by a radioactive particle in contact with a photographic emulsion, was not present even though the particles had been subjected to long periods of irradiation.

In the qualitative measurement of coating thickness and core diameter from an enlargement print the inherent geometrical features of the projection technique required corrections, which are set out in the paper. Included also are illustrations demonstrating uniformity and nonuniformity of coatings, lamellar cracks at interfaces between layers, radial cracks, and discrete contamination surface spots. Other examples show a "break-away" process at various stages of development in graphite coated specimens held for 3 hr at 2200°C, and also a carbon coated fuel particle after irradiation. The latter emphasizes the importance of the projection technique in eliminating the autoradiographic background, and there was evidence of a denser band apparently corresponding to the penetration range of fission fragments. Further improvements in the efficiency of the technique are expected from sequential inspection and stereographic presentation.

Trillat [208] followed up his earlier work, in which he applied the contact method to the study of light alloys. For this new research he used the point projection technique with direct x-ray magnification and illustrated the results obtained with Al/Sn, Al/Be, and Al/Mn alloy specimens. Procedures for heat treatment examination are also discussed and results shown.

Another paper by Sharpe [209] enlarges the field of the projection method by use of the higher power rotatable target source, described in Section 3.

Attention is directed to the high image magnification required in the inspection of semiconductor devices, microminiaturized circuits, metrology and microporosity in aluminum sheet, and graphite spheres.

Pyrocarbon-coated PuO_2 microspheres were examined by Morris and Strong [210] using a technique designed to prevent contamination of the emulsion while making a microradiograph by the contact method. Their paper shows the design of a special fixture for transferring and radiographing particles without exposure to atmosphere.

McClung [211] reported the solution of a similar problem concerning nuclear fuel particles at 4 to 10 kV by the contact method using a special helium-filled chamber between the tube and specimen to reduce the air absorption. This chamber was fitted with a 12-μ-thick polythene diaphragm placed over each end to retain the helium; these thin membranes were found to absorb very little of the longest x-ray wavelength energy. He obtained a resolution of between 2 and 3 μ.

The difficulties experienced in studying the effect of heat treatment of powder metal alloys by photomicrography, for example, in preparing specimens in which the constituents differ in hardness and ductility so that the surface will be free from distortion, led Blue [212] to apply microradiographic methods. The results of varying the times of heat treatment on the extent of diffusion and physical properties of 90% Cu/10% Sn material were recorded by serial microradiographs. It was seen that the physical properties increased with sintering time up to a maximum of 90 min at 850°C. Porosity was also shown to increase, the distribution becoming more uniform with sintering time.

Following a comprehensive review of earlier work Sharpe [213] described a technique employing a point focus projection microscope to measure lattice constants, and to observe lattice modifications in single crystal specimens (Fig. 8.24). The accuracy of the lattice constant determinations using a suitable crystal, was reported as likely to attain parity with other x-ray techniques, with the advantage that precision ancillary equipment and the need for corrections or extrapolations are not normally required. Since the technique is nondestructive sequential, thermal or stress measurements could be made. As a further advantage, the ability to position the specimen at a distance from the emulsion eliminated the danger of blackening if the specimen had been irradiated (Fig. 8.25). Specimens could also be irradiated to a higher degree of activity than before. It was pointed out, however, that accuracies of the high order anticipated depended largely on the excellence of temperature control, since variation of this parameter during exposure would result in a corresponding degree of line-blurring.

A similar projection technique was used by Ichinokawa and Shirai [214] who observed localized crystal defects in single crystals. They described those results as pseudo-Kossel lines superimposed on a magnified projection image. Serial microradiographs taken by moving the x-ray source in a plane parallel

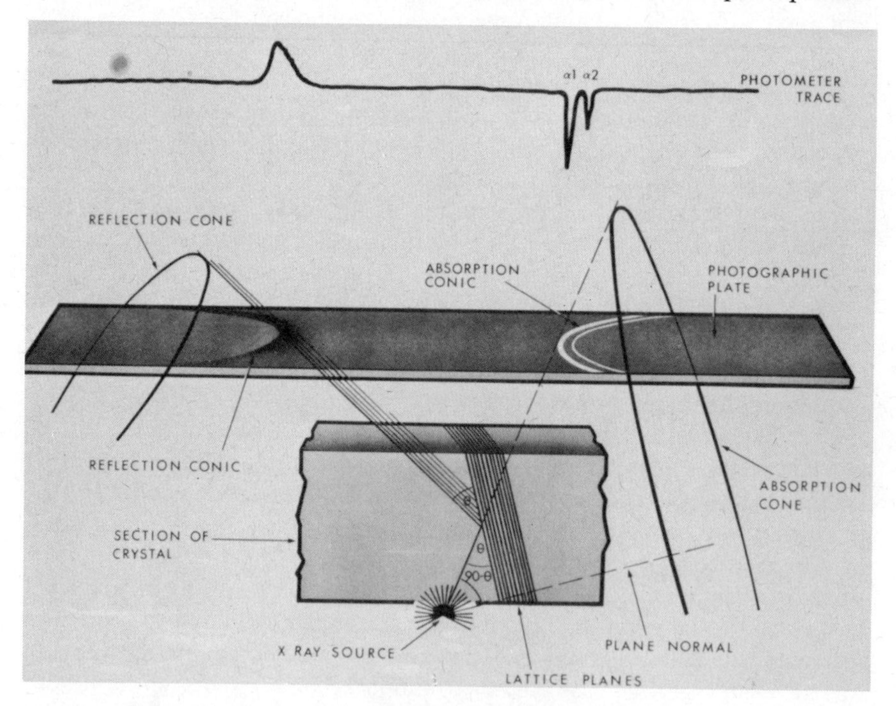

Fig. 8.24 The formation of absorption and reflection conics and their geometrical relationships. Courtesy R. S. Sharpe, A.E.R.E., Harwell, England.

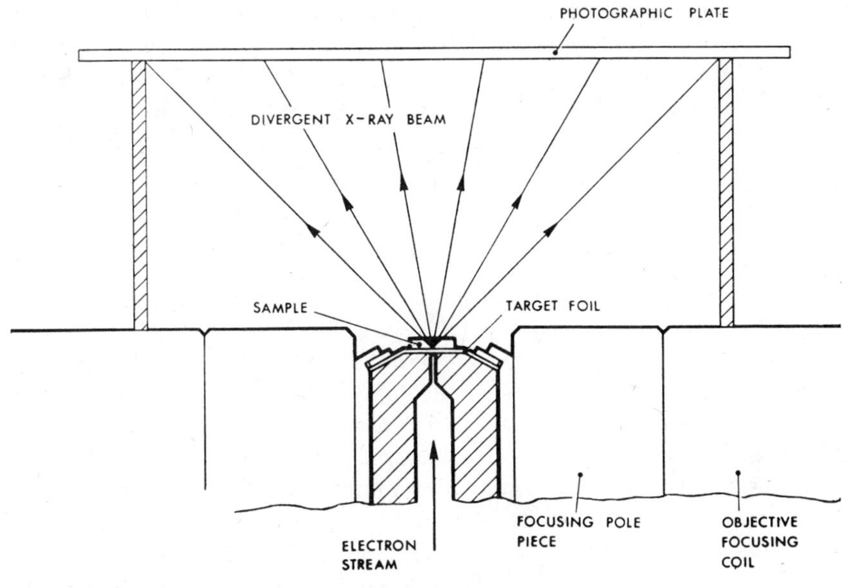

Fig. 8.25

to that of the crystal surface enabled the distribution of local stress to be measured two-dimensionally. Fe and Ge single crystals about 0.1 mm thick, with targets of corresponding materials a few microns thick, were used in these experiments. The combination of crystal and target of the same material enhanced the Kossel line contrast by absorption edge effect. Variations of strain and lattice spacing of the order of 10^{-4}, with a resolution of 1 μ, were reported.

For high-accuracy preparation of theoretical Kossel pattern projections Frazer and Arrhenius [215] developed a program for computing and automatically plotting the stereographic projection of absorption conics. The input included wavelengths, lattice parameters, and indices of the conics. In a paper on the use of Miller indices and the reciprocal lattice concept in the interpretation of Kossel patterns, Mackay [216] referred to the usual methods as complicated by spherical trigonometry and put forward techniques designed to simplify the procedure. In his method the patterns are visualized primarily as a family of circles inscribed on a sphere surrounding the specimen and centered on the x-ray point-source. An equation is derived expressing the lattice parameters as a function of the wavelength and the Miller indices of the plane. From the reciprocal lattice, to which the planes are directly related, a Kossel diagram is constructed.

The possibility that diffraction effects may cause confusion of detail in microradiographs produced by both the contact and projection methods was pointed out by Sharpe [217]. In the former method a diffraction mottling can be produced by the near-parallel radiation conditions; in the latter short sections of Kossel diffraction conics can be produced by the divergent beam if the individual grains are so orientated as to present this effect. Spurious contrast effects can also be produced at edges by total external reflection of the x-ray beam and can result in difficulties in interpretation or in measurement.

Botanical

Many botanical specimens are particularly suitable for examination by x-ray microscopy since leaves, petals, seeds, and so forth, are normally thin and transparent to long-wavelength radiations. Stereomicroradiography is also especially applicable and, with image intensification techniques, *in vivo* series exposures can be made with minimal doserates.

When sectioning is necessary, for example, for woods, much information can be obtained from sections cut with a simple microtome to a thickness of about 1 mm and contrast can be increased by injection, staining, or dehydration. Röntgen, and other early workers, envisaged the possible application of x-rays to botanical research but, apart from the early pioneering work of Goby [12], very little action took place until Lamarque and Turchini [18] produced papers on the historadiography and structure of chromatic reticulum and prochromosome nuclei of cells. They also investigated elder tree pith and onion bulb scales, with attempts at iodine impregnation.

Several years later Baranov et al. [218] published their work on the application of microradiography to the nature of the distribution of radioactive elements in natural objects. Leaflets of Lelodea were cultivated in water to which a substantial amount of Ra had been added. The broad field covered by Clark et al. [219] included the effects of heat, pressure, and swelling by pyridine and morpholine in relation to the strength and warping of swollen compressed woods.

The white spots in radiographs of rose and other plant leaves excited the interest of Barclay and Leatherdale [220] who used the contact method previously demonstrated by Barclay [53] to show, either in or closely related to the veins of leaves, bodies possessing a much higher atomic weight than that of the leaf structure. To obtain the closest possible contact between specimen and emulsion, for maximum resolution, the leaves were pressed and dried and it was not possible to detect in the microradiographs any difference between these and fresh leaves, apart from resolution. Rose, *Berberis*, *Prunus*, *Rhus cotinus*, tomato, and potato were among the many specimens examined and metallic salts, such as lead acetate in saturated solution, were also used by various injection methods to assist in the delineation of vessels. Illustrations are shown at magnifications up to $\times 240$, and include one of a rose leaf with veins filled with lead acetate. The results indicated the possibility of using the same technique to locate the site at which a drug is laid down in and around cells in animal tissue.

Mitchell [221] freeze-dried specimens of galls using a vacuum cassette. In the same year Legrand and Salmon [222] pointed the way to an adequate appreciation of the relative merits of light and x-rays in botany. Manigault and Salmon [223], in three papers, studied the induction of experimental tumors in plant tissues.

Projection x-ray microscopy, with one of the first Cosslett and Nixon type instruments, was used by Jackson [224] in his examination of obeche, mahogany, and elm burrs.

Radial and transverse sections, about 1 mm thick, indicated the relationship of the large conducting cells to the parenchyma cells and calcium oxalate crystals. A section of pine showed the tree rings resulting from different rates of seasonal growth.

In a major contribution Ong [225] includes illustrations of ashwood transverse section $\times 400$ and the stigma of a leaf $\times 520$, recorded on ultrafine-grain film with an initial x-ray magnification of $\times 4$. Tissue paper, treated with alcoholic iodine solution, is compared with gold shadowing. The latter is carried out in the same manner as for electron microscopy and is a particularly useful technique for high-magnification stereoscopic exposures. Dietrich [226] applied contact microradiography to the study of components of the plant cell and, after staining, compared each element with its corresponding radiograph. He concluded that mitosis and cytoplasm can be followed more satisfactorily by this method than by normal techniques.

Salmon [227], in a comprehensive review of plant microradiography, included a number of excellent illustrations, for example, Polargonium infection by Agrobacfaciens and a comparison of different microscopic techniques on cross sections of Nerium leaf. In a paper on recent developments in wood science, Gibson [228] refers to a microradiographic study on the location of an inorganic preservative within the cell and states that no preservative was found in the lumen.

Further progress in the study of woods was made by Jongebloed and Jutte [229] and Jongebloed [230] who used the Delft x-ray projection microscope to examine wood species in which silica aggregates and other crystals were present in the cell lumen. The radial sections employed were about 30 μ in thickness and were not pretreated. A number of illustrations indicated the scope of the work carried out and suggested future possibilities. Cellular division has been well covered by Dietrich [231] using very long wavelengths (800 V) and his paper is also distinguished by the number and quality of the illustrations. Plants that can tolerate freezing and thawing have been studied by Idle [232] with the twofold objective of providing possible clues to methods of protection from frost to expand the growing season and to help plant breeders in their search for hardy strains of valuable food plants. Microradiography was used to indicate the movement of water between the vacular system and adjacent tissue during freezing.

From this brief survey, and the accompanying illustrations, it is clear that a number of experienced botanists have appreciated how rapidly microradiography can provide valuable information, some of which cannot be obtained by other means.

General

Fabrics and Fibers

Studies of complex weaves in cloth were made at an early date by Sherwood [233] who employed stereoscopic methods for spatial analysis. Much later Newman [234] pointed out that, even with treatment by selective salts or by metallic evaporation, the results of all previous work had been limited to information on the distribution of fiber elements. Using a point projection instrument fitted with an aluminum target, he showed that it is possible to obtain contrast sufficient for structural deformation and stress distribution in a wide range of fabrics.

Paints and Paintings

The homogeneity of the pigment grinding and dispersion in paints was considered by Trillat [235] as a subject eminently suitable for investigation by x-ray methods he had adopted earlier in researches on metal alloys. Using copper radiation he was able to illustrate, at magnifications up to ×300, the grain size and dispersion of antimony oxide in a varnish.

The long-term interest in the examination of paintings is indicated by the contributions of Rawlins [236–238], who referred to the uncertainties and difficulties accompanying the early application of x-rays to works of art and the importance they could assume in the functioning of any great collection of pictures, for example, how much of the original of a "restored" painting actually existed, and whether it was safe to do further work.

Abrasion, flaking, overpainting, and inpainting were detected when using a tube with a Be window at 7.5 kV, and the various methods by which the old masters worked were demonstrated.

Kozlowski [239] constructed equipment for the production of stereomicroradiographs of large paintings. An x-ray tube was mounted on a carriage running on a semicircular rail fixed in a vertical plane. The painting was placed on a movable horizontal frame above the rail so that any desired area could be positioned at the center of the arc through which the tube could be made to move. Angular separation between two exposures could be as large as 140°.

Another form of construction is described in the subsection on stereo techniques, at the end of Section 3.

Paleontology and Fossils

Over half a century ago Barnard [240] produced radiographic enlargements of foraminifora and nearly 40 years later Schmidt [241] noted the value of microradiographic techniques as developed in medicine. These he applied to calcareous, siliceous, and pyritic microfossils to show the arrangement of internal chanbers (×36). Hedley [242] followed with a paper setting out suitable procedures and methods of specimen mounting. The results obtained in the investigation of thin wall structures by polarized light were discussed and related to the advantages offered by x-ray methods. These are enlarged upon by Hooper [243] who made quantitative population studies and considered them to be faster and more accurate than techniques necessitating sectioning of the specimens. Other examples of projection microradiographs in this field are given in a paper by Jongebloed [244].

Human teeth from the neolithic and bronze ages were found by Held, Band, and Fiore [245] to have defects in mineralization in the dentine and enamel. Using the contact method with copper radiation they found that areas of microhypoplasia were very similar to those shown in the teeth of modern man.

Ancient dentine was also studied by Wyckoff [246] with interesting results. Fossil reptilian and amphibian teeth revealed excellently preserved microstructure with little mineralization other than in the region of obvious fractures. He reported that a combination of microradiography and spectroscopy was a profitable way of studying fine structures in fossilized hard tissues.

Paper

To demonstrate the felting of paper fibers Pelgroms [247] found that impregnation with an iodine solution greatly improved contrast. The formula was as under

Iodine	5 g
Potassium iodide	20 g
Ethyl alcohol	100 ml
Until saturation.	

Adequate soaking, with prolonged drying (about 2 hr), was necessary to prevent crystallization of the potassium iodide.

Isings and co-workers [248] also used the iodine technique to study a large number of papers and concluded that three-dimensional presentation was particularly valuable in interpreting the results. This work was further developed by Ong [249] and the distribution of fillers was investigated at length by Newman and Fletcher [250], who found that the point projection method greatly simplified ease of observation and specimen preparation.

Plastics

Investigation by light microscopy of the microstructural characteristics of laminated, foamed, and loaded plastics has always presented considerable difficulty. Ascertainment of uniformity of fillings is typical of the problems encountered, and Barlow, Distler, and Hile [251] developed a method whereby microradiographs of thin sections, enlarged optically 20 to 60 times, were scanned by an electromechanical device. Shadow widths were detected by scanning the negative with a light beam collimated to a width considerably smaller than the minimum shadow to be counted. The output pulses from a photomultiplier were then sorted into groups and stored in ranges representing different orders of particle size.

Newman et al. [252] employed the projection technique to study polyurethene, polystyrene, and glass foams with aluminum radiation at 10 kV. He found that the method did not readily lend itself to precise measurement of linear dimensions but was extremely useful in the delineation of shapes of structures and their distribution in a matrix. Laminates and asphaltic materials were also selected, for example, a glass-fiber-reinforced paper tape was repeatedly flexed and examined radiographically in stages for damage to the fibers.

Another paper by the same author [234] relates to the formation of moiré fringes. It was suggested that the transverse banding noted in microradiographs of leaf fibers could be produced by a grating system formed by adjacent fibers.

Coatings on wires or perforated metals can be well defined at high magnification, as shown in a paper by Ely [253].

Rubber

An early detailed study by Trillat [254] indicated the possibility of improving production control in vulcanization, particle size, and particle distribution. Undispersed pigment agglomerates can result in weak spots in the bulk material, with a lowering in tensile strength and abrasion resistance. The contact method at 3.5 kV was used by Hess [255] to study dispersion and a special device to stress and relax specimens was developed.

5 FUTURE PROSPECTS

X-ray microscopes of the Cosslett and Nixon type should now develop further following improvements in the design of lenses, the stabilizing of high voltage and filament supply units, and in methods of collimating and focusing the electron beam. At low-energy levels it is still difficult to obtain optimum focusing conditions by optical means because image brightness on the fluorescent screen is so restricted, but new small intensifiers are now in production and critical focusing should become a simple and rapid procedure. They will also be most useful as a focusing aid for higher energy x-ray generators of the ultrafine-focus type, such as that shown in Fig. 8.7.

For magnified image television presentation, or for fluoroscopy, intensifiers with much higher resolution and sensitivity will greatly increase the possible range of dynamic studies in which the object must be at some distance from the recording medium and, in the case of living tissues, much lower dose-rates can be anticipated. The importance of primary x-ray magnification must be expected to increase with the current developments of more powerful x-ray generators with focal spots less than 0.1-mm diameter.

References

1. F. Jentsch, *Phys. Z.*, **30,** 268 (1929).
2. P. Kirkpatrick and A. V. Baez, *Phys. Rev.*, **71,** 521–529 (1947).
3. W. Ehrenberg, *J. Opt. Soc. Am.*, **39,** 741–746 (1949).
4. W. Ehrenberg, *J. Opt. Soc. Am.*, **39,** 746–751 (1949).
5. J. Dyson, *Proc. Phys. Soc. London*, **65B,** 580–589 (1952).
6. P. Kirkpatrick and H. H. Pattee, *Advances in Biological and Medical Physics*, Vol. 3, Academic, New York, 1953, pp. 247–283.
7. H. H. Pattee, *1st X-ray Conference Cambridge England*, Academic, New York, 1957, pp. 135–150.
8. M. Montel, *1st X-ray Conference Cambridge, England*, Academic, New York, 1957, pp. 177–185.
9. M. Montel, *2nd X-ray Conference Stockholm*, Elsevier, New York, 1960, pp. 129–132.
10. J. F. McGee, *Phys. Rev.*, **98,** 282 (1955).
11. J. F. McGee and J. W. Milton *2nd X-ray Conference Stockholm*, Elsevier, New York, 1960, pp. 118–128.
12. P. Goby, *Compt. Rend.*, **156,** 686–688 (1913).
13. A. Dauvillier, *Compt. Rend.*, **185,** 1460–1462 (1927).
14. A. Dauvillier, *Compt. Rend.*, **190,** 1287 (1930).
15. H. Frickie, *Radiogr. Clin. Photogr.*, **8,** 12–13 (1932).
16. H. F. Sherwood, *Radiogr. Clin. Photogr.*, **10,** 10–13 (1934).
17. P. Lamarque, *Compt. Rend.*, **202,** 684–687 (1936).
18. P. Lamarque and J. Turchini, *Compt. Rend. Ass. Anat.*, **31,** 341 (1936).
19. B. Combée, *Philips Tech. Rev.*, **17,** 45–46 (1955).
20. W. Ehrenberg and W. E. Spear, *Proc. Phys. Soc., London*, **64B,** 67–75 (1951).
21. R. V. Ely, *1st X-ray Conference Cambridge, England*, Academic, New York, 1957, pp. 59–63.

22. R. V. Ely, *2nd X-ray Conference Stockholm*, Elsevier, New York, 1960, pp. 47–50.
23. R. C. Greulich, *2nd X-ray Conference Stockholm*, Elsevier, New York, 1960, pp. 273–287.
24. B. L. Henke, *Tech. Rept. No. 2 Ultrasoft X-ray Phys.*, U.S.A.F. Off. Sci. Res., Pomona College, Cal., 1958, pp. 1–59.
25. S. K. Asunmaa, *2nd X-ray Conference Stockholm*, Elsevier, New York, 1960, pp. 66–78.
26. S. K. Asunmaa, *3rd X-ray Conference Stanford, California*, Academic, New York, 1963, pp. 35–51.
27. W. Ehrenberg and M. White, *1st X-ray Conference Cambridge, England*, Academic, New York, 1957, pp. 213–216.
28. W. Ehrenberg and M. White, *Brit. J. Radiol.*, **32,** 558 (1959).
29. E. G. Bendit, *Brit. J. Appl. Phys.*, **9,** 312–317 (1958).
30. U. W. Arndt and D. P. Riley, *Proc. Phys. Soc. London*, **65A,** 74–75 (1952).
31. V. E. Cosslett, *2nd X-ray Conference Stockholm*, Elsevier, New York, 1960, pp. 346–350.
32. N. A. Dyson, *Brit. J. Appl. Phys.*, **10,** 505–507 (1959).
33. G. L. Clark, *Applied X-rays*, McGraw-Hill, New York, 1955, pp. 238–262.
34. V. E. Cosslett and W. Nixon, *X-ray Microscopy*, Cambridge University Press, New York, 1960, 406 pp.
35. A. Engström, *X-ray Microanalysis in Biology and Medicine*, Elsevier, New York, 1962, 92 pp.
36. B. Lindström, *2nd X-ray Conference Stockholm*, Elsevier, New York, 1960, pp. 288–289.
37. V. E. Cosslett and W. Nixon, *Nature*, **168,** 24–25 (1951).
38. G. Liebmann and E. M. Grad, *Proc. Phys. Soc. London*, **64B,** 956–971 (1951).
39. G. Liebmann, *Proc. Phys. Soc. London*, **68B,** 737–745 (1955).
40. T. Mulvey, *Proc. Phys. Soc. London*, **66B,** 441–447 (1953).
41. A. C. van Dorsten and J. B. Le Poole, *Philips Tech. Rev.*, **17,** 47 (1955).
42. M. E. Haine, *Proc. Intern. Conf. Electron Microscopy*, London, 1954, pp. 92–97. (*Roy. Microsc. Soc.*, London, 1956.)
43. H. Anderton, *Sci. Progr.*, **55,** 337–356 (1967).
44. H. Anderton and K. C. A. Smith, *4th Intern. Congr. X-ray Optics and Microanalysis*, Hermann, Paris, 1966, pp. 426–431.
45. R. Halmshaw, in *Research Techniques in Nondestructive Testing*, R. S. Sharpe, Ed., Academic, New York, 1970, 492 pp.
46. E. Applebaum, *J. Dent. Res.*, **12,** 619–627 (1932).
47. E. Applebaum, F. Hollander, and C. F. Bödecker, *Trans. Dent. Soc.*, New York, 1933, pp. 77–85.
48. E. Applebaum, *Dent. Cosmos.*, **78,** 969–980 (1936).
49. H. F. Sherwood, *J. Biol. Photogr. Ass.*, **5,** 138–145 (1936).
50. P. Lamarque, J. Turchini, and P. Castel, *Arch. Soc. Sci.*, **18,** 27–29 (1937).
51. G. L. Clark, *Colloid Chem.*, **5,** 146–151 (1944).
52. K. M. Endicott and H. Yagoda, *Proc. Soc. Exp. Biol.*, New York **64,** 170–172, (1947).

53. A. E. Barclay, *Microarteriography*, Blackwell Scientific Publications, Oxford, England, 1951, 102 pp.
54. A. Engström, *Acta Radiol.*, **31,** 503–521 (1949).
55. A. Engström, *Progress in Biophysics*, Butterworth, London, 1950, pp. 164–193.
56. A. Engström, M. D. Hamberger, and S. Welin, *Brit. J. Radiol.*, **22,** 309–324 (1949).
57. S. Bellman, *Acta Radiol.*, *Suppl.*, 102 (1953), 104 pp.
58. G. A. G. Mitchell, *Brit. J. Radiol.*, **24,** 110–118 (1951).
59. J. Jowsey, M. Owen, and J. Vaughan, *Brit. J. exp. Pathol.*, **34,** 661–667 (1953).
60. H. Röckert, *Acta Odontol. Scand.*, **13,** 271–275 (1955).
61. R. L. de C. H. Saunders, *1st X-ray Conference Cambridge*, *England*, Academic, New York, 1957, pp. 561–571.
62. G. Bergman and B. Engfeldt, *Acta Odontol. Scand.*, **12,** 99–132 (1954).
63. G. Bergman and B. Engfeldt, *Acta Odontol. Scand.*, **12,** 133–144 (1954).
64. G. Bergman and B. Engfeldt, *Acta Odontol. Scand.*, **12,** 193–200 (1954).
65. F. Bohatirchuk, *Virchows Arch.*, **313,** 216–228 (1944)
66. F. Bohatirchuk, *Am. J. Roentgenol.*, **70,** 119–125 (1953).
67. F. Bohatirchuk, *Brit. J. Radiol.*, **28,** 389–404 (1955).
68. G. J. Cunningham, *Am. J. Clin. Pathol.*, **25,** 253–260 (1955).
69. G. Hammar, *Acta Oto-Laryngol.*, *Suppl.*, **127,** 1–64 (1956).
70. I. Meschan, C. S. Pool. A. Nettleship, M. Winer, and W. Zeman, *Radiology*, **65,** 770–778 (1955).
71. I. Orlandini and A. V. Bo, *Ateneo Parmense*, **26,** 530–545 (1955).
72. I. Orlandini and G. C. Vidoni, *Ateneo Parmense*, **26,** 461–477 (1955).
73. M. Owen, J. Jowsey, and J. Vaughan, *J. Bone Joint Surg.*, **37B,** 324–342 (1955).
74. F. Bohatirchuk, *1st X-ray Conference Cambridge*, *England*, Academic, New York, 1957, pp. 473–474.
75. A. Recourt, *1st X-ray Conference Cambridge*, *England*, Academic, New York, 1957, pp. 475–483.
76. P. J. Fitzgerald, *Ann. N.Y. Acad. Sci.*, **63,** 1141–1176 (1956).
77. P. J. Fitzgerald, T. G. Li, and V. Yermakov, *1st X-ray Conference Cambridge*, *England*, Academic, New York, 1957, pp. 520–530.
78. R. J. Carr, *1st X-ray Conference Cambridge*, *England*, Academic, New York, 1957, pp. 551–560.
79. M. Juster, H. Fischgold, and J. Ecoiffier, *1st X-ray Conference Cambridge*, *England*, Academic, New York, 1957, pp. 572–577.
80. A. Engström and B. Lundberg, *Exp. Cell Res.*, **12,** 198–200 (1957).
81. A. J. Hale, *J. Physiol. London*, **141,** 193–197 (1958).
82. A. G. Gustafson, *Odont. Tidskr.*, **67,** 361–472 (1959).
83. A. G. Gustafson, *Arch. Oral Biol.*, **4,** 67–69 (1961).
84. G. Gustafson and A. G. Gustafson, *Acta odontol. Scand.*, **19,** 259–287 (1961).
85. J. K. Avery, R. L. Visser, and D. E. Knapp, *J. Dent. Res.*, **40,** 1004–1019 (1961).
86. A. C. N. Hutchinson, R. E. Rowland, and L. S. Fosdick, *J. Dent. Res.*, **42,** 1040 (1963).
87. D. Weber and J. A. Yaeger, *J. Dent. Res.*, **43,** 50–56 (1963).

88. G. van Huyson, *Norelco Reptr.*, **8**, 95–97, 100 (1961).
89. G. van Huysen, *Norelco Reptr.* **8**, 98–100 (1961).
90. H. F. Atkinson and J. K. Harcourt, *Australian Dent. J.*, **6**, 194–197 (1961).
91. H. Röckert and A. Ohman, *Acta Odontol. Scand.*, **20**, 165–187 (1962).
92. T. Minura, A. Asoda, and T. Kaneko, *J. Dent. Res.*, **42**, 754 (1963).
93. D. F. Weber, *Calc. Tissue Res.*, **1**, 319–323 (1968).
94. R. L. de C. H. Saunders, *Oral Surg., Oral Med., Oral Pathol.*, **22** (4), 503–518 (1966).
95. M. J. Glimsher, U.A. Friberg, and P. T. Levine, *J. Ultrastruct. Res.*, **10**, 76–78 (1964).
96. A. W. Park, *Brit. J. Radiol.*, **39**, 597–601 (1966).
97. A. W. Park, *Brit. J. Radiol.*, **40**, 535–537 (1967).
98. F. Bohatirchuk, *Anat. Record*, **133**, 203–218 (1959).
99. J. Jowsey, *Clin. Orthop.*, **17**, 210–218 (1960).
100. R. C. Greulich, *Arch. Oral Biol.*, **3**, 137–142 (1960–61).
101. H. A. Sissons, J. Jowsey, and L. Stewart, *2nd X-ray Conference Stockholm*, Elsevier, New York, 1960, pp. 199–205.
102. R. E. Rowland, *Med. Phys.*, **3**, 525–528 (1960).
103. J. D. Manson and R. B. Lucas, *Arch. Oral Biol.*, **7**, 651–769 (1962).
104. M. C. Hall, *Clin. Orthop.*, **25**, 171–174 (1962).
105. B. R. Bhussry and J. Parikh, *Norelco Reptr.*, **9**, 83–87 (1962).
106. T. A. Mjôr, *Acta Anat.*, **53**, 259–267 (1963).
107. B. Lindström and K. A. Omnel, *Encycl. X-rays*, Reinhold, New York, 1963, pp. 606–607.
108. A. I. D. Prentice, *Nature*, **206**, 1167 (1965).
109. J. Jowsey, P. J. Kelly, B. L. Riggs, A. J. Bianco, D. A. Scholz, and J. Gershon-Cohen, *J. Bone Joint Surg.*, **47A**, 785–806 (1965).
110. R. E. Rowland, J. H. Marshall, and J. Jowsey, *Radiation Research No. 10*, Sid. 323, Ar. **10**, 323–324 (1959).
111. J. E. Stone and H. W. Ferguson, *Proc. Roy. Soc. Med.*, **60**, 849–850 (1967).
112. J. Jowsey, in *Bone Biodynamics*, Churchill, London, 1963, p. 461.
113. U. Nilsonne, *Acta Orthop. Scand., Suppl.*, **37**, 5–81 (1959).
114. M. Juster, H. Fischgold, and J. Metzger, *2nd X-ray Conference Stockholm*, Elsevier, New York, 1960, pp. 191–198.
115. H. Röckert and H. Sunzel, *Experienta*, 155–156 (1960).
116. F. Bohatirchuk, *J. Gerontol.*, **15**, 142–148 (1960).
117. H. Bohr and E. Dollerup, *2nd X-ray Conference Stockholm*, Elsevier, New York, 1960, pp. 184–190.
118. E. Lloyd, *Nature*, **34**, 521–528 (1961).
119. G. P. Vose, *Clin. Orthop.*, **24**, 206–212 (1962).
120. M. R. Urist, P. S. Zaccalini, N. S. MacDonacd, and W. A. Skoog, *J. Bone Joint Surg.*, **44B**, 464–484 (1962).
121. W. H. J. Summerskill and P. J. Kelly, *Proc. Staff Meetings Mayo Clinic*, **38**, 162–174 (1963).
122. G. Hammer, C. Radberg, and H. Röckert, *Acta Oto-Laryngol., Suppl.*, **188**, 119–124 (1963).

123. S. S. Kornblum and P. J. Kelly, *J. Bone Joint Surg.*, **46A,** 797–810 (1964).
124. U. Breine, B. Johansson, and H. Röckert, *Acta Chir. Scand.*, **129,** 250–256 (1965).
125. N. J. D. Smith and M. H. Hobdell, *5th Congr. X-ray Optics and Microanalysis*, Tubingen, Springer-Verlag, 1969.
126. M. H. Hobdell, *J. Anat.*, **106,** 165–166 (1970).
127. M. H. Hobdell and A. Boyde. *Z. Zellforsch.*, **94,** 487–494 (1969).
128. P. F. Harris, G. Haigh and B. Watson, *Acta Haematol.*, **26,** 154–168 (1961).
129. P. I. Branemark, U. Breine, B. Johansson, P. J. Roylance, H. Röckert, and J. M. Yoffey, *Acta Anat.*, **59,** 1–46 (1964).
130. J. A. Clarke, *J. Laryngol. Otol.*, **78,** 415–425 (1964).
131. H. Röckert, A. Engström, O. Hallén, G. Herberts, G. Lidén, B. Nordlund, and J. J. Shea, *J. Laryngol. Otol.*, **79,** 305–313 (1965).
132. J. Vincent and A. Dhem, *Acta Anat.*, **40,** 121–129 (1960).
133. R. L. de C. H. Saunders, *Encycl. Microsc.*, Reinhold, New York, 1961, pp. 582–586.
134. F. Bohatirchuk, *Am. J. Anat.*, **117,** 287–310 (1965).
135. D. S. Howell and L. Carlson, *4th Congr. X-ray Optics*, Hermann, Paris, 1966, pp. 664–668.
136. J. A. Veiga Pires, *Brit. J. Radiol.*, **33,** 491–495 (1960).
137. R. L. de C. H. Saunders and R. James, *Anat. Record*, **139,** 319 (1961).
138. R. L. de C. H. Saunders, *Blood Vessels and Circulation*, Pergamon, New York, 1961, pp. 38–56.
139. P. Soila, *3rd X-ray Conference Stanford, California*, Academic, New York, 1963, pp. 123–126.
140. L. F. A. Peterson and P. J. Kelly, *Encycl. X-rays*, Reinhold, New York, 1963, pp. 595–597.
141. R. L. de C. H. Saunders, *J. Roy. Microse. Soc.*, **83,** 55–62 (1964).
142. M. Niemi and M. Kormano, *Acta Pathol. Microbiol. Scand.*, **63,** 513–521 (1965).
143. C. P. Oderr, *2nd X-ray Conference Stockholm*, Elsevier, New York, 1960, pp. 290–292.
144. C. Oderr, M. Dauzat, and E. Montamat, *Med. Radiogr. Photogr.*, **39,** 38–47 (1963).
145. G. F. Leroux, J. Francois, J. M. Collette, and A. Neetens, *2nd X-ray Conference Stockholm*, Elsevier, New York, 1960, pp. 233–238.
146. G. F. Leroux, J. Francois, J. M. Collette, and A. Neetens, *2nd X-ray Conference Stockholm*, Elsevier, New York, 1960, pp. 239–243.
147. J. R. Blickman, P. J. Klopper, and A. Recourt, *2nd X-ray Conference Stockholm*, Elsevier, New York, 1960, pp. 224–225.
148. J. R. Blickman, P. J. Klopper, and A. Recourt, *2nd X-ray Conference Stockholm*, Elsevier, New York, 1960, pp. 226–232.
149. J. Salmon and P. F. Ceccaldi, *Ann. Histochim., Suppl.*, **7,** 29–33 (1962).
150. S. B. Andrus, S. N. Gershoff, F. F. Faragalla, and E. L. Prien, *Lab. Invest.*, **9,** 7–27 (1960).

151. F. Fournier, *2nd X-ray Conference Stockholm*, Elsevier, New York, 1960, pp. 518–524.
152. S. Bengmark, R. Romanus, and H. Röckert, *Urol. Intern.*, **16,** 148–157 (1963).
153. O. Hallén and H. Röckert, *2nd X-ray Conference Stockholm*, Elsevier, New York, 1960, pp. 169–176.
154. H. Bogren, *Acta Radiol.*, *Suppl.*, **226** (1964), 75 pp.
155. A. J. Hale, *J. Cell Biol.*, **15,** 417–425 (1962).
157. A. L. Ljungqvist, *Acta Med. Scand.*, *Suppl.*, **401** (1963), 38 pp.
158. C. Lagergren and A. Ljungqvist, *11th Congr. Radiol. Excerpta Med. Found.*, 1965, pp. 701–704.
159. H. Hydén, *Proc. 2nd Intern. Meeting of Neurobiologists, Amsterdam*, Elsevier, New York, 1960, pp. 348–357.
160. W. Grampp, O. Hallén, and B. Rosengren, *Exp. Cell Res.*, **19,** 437–442 (1960).
161. R. L. de C. H. Saunders, *2nd X-ray Conference Stockholm*, Elsevier, New York, 1960, pp. 244–256.
162. C. A. Perez, F. J. Hodges, and A. R. Margulis, *Radiology*, **82,** 529–535 (1964).
163. R. L. de C. H. Saunders, W. H. Feindel, and V. Carvalho, *Med. Biol. Illus.*, **15,** Pt. 1, 108–122; Pt. II, 236–246 (1965).
164. R. L. de C. H. Saunders and R. V. Ely, *4th Congr. X-ray Optics*, Hermann, Paris, 1966, pp. 642–649.
165. R. L. de C. H. Saunders, M. A. Bell, and V. R. Carvalho, *5th Congr. X-ray Optics and Microanalysis*, Tubingen, Springer-Verlag, 1969.
166. F. Fournier, *Rev. Met. Paris*, **35,** 349–355 (1938).
167. J. J. Trillat, *Rev. Sci. Paris*, **78,** 212–221 (1940).
168. H. Kirchberg and H. Moller, *Mitt. K-Wilh-Inst. Eisenforsch*, **23,** 309–314 (1941).
169. D. W. Smith and F. Keller, *J. Am. Weld. Soc.*, *Suppl.*, **21,** 573S–583S (1942)
170. S. T. Gross and G. L. Clark, *Ind. Radiogr.*, **1,** 21–25, (1942).
171. L. W. Ball, *Metal Ind. London*, **67,** 130–134 (1945).
172. L. W. Ball, *Metal Ind. London*, **67,** 210–213 (1945).
173. W. Betteridge and R. S. Sharpe, *J. Iron Steel Inst. London*, **158,** 185–191 (1948).
174. J. D. Lavender and F. W. Jones, *J. Iron Steel Instr. London*, **163,** 14–17 (1949).
175. A. Taylor, *X-ray Metallography*, Chapman and Hall, London, 1949, pp. 323–325.
176. A. Berghézan, P. Lacombe, and G. Chaudron, *Compt. Rend.*, **231,** 576–578 (1950).
177. E. Cohen, *Metallurgia*, **41,** 227–233 (1950).
178. G. A. Homès and J. Gouzou, *Rev. Met. Paris*, **48,** 251–261 (1951).
179. R. W. K. Honeycombe, *J. Inst. Metals*, **80,** 39–45 (1951).
180. H. J. Goldschmidt, *Iron & Steel* (*London*), **67,** 215 (1953).
181. S. Goldspiel and F. Bernstein, Nondestructive Testing, **11,** 15–20 (1953).
182. G. L. Clark, *Applied X-rays*, McGraw-Hill, New York, 1955, pp. 238–262.
183. W. C. Nixon and V. E. Cosslett, *Brit. J. Radiol.*, **28,** 532–536 (1955).
184. W. C. Nixon, *Proc. 3rd Intern. Conf. Electron Microscopy*, London, 1954, pp. 307–310. (Roy. Microsc. Soc., London, 1956.)
185. H. Neff, *Z. Metallk.*, **46,** 614–615 (1955).

186. R. Urlocker, J. W. Rutter, and W. C. Winegard, "Metallurgy and Microradiography," *Exp. Can. Metals*, **18,** 20, 21, 23, (1955).
187. J. M. Bermudez de Castro, J. M. Fuster Casas, and J. A. Boned Sopena, *Inst. Hierro Acero*, **9,** 75–91 (1956).
188. E. Votava, A. Berghézan, and R. H. Gillette, *Proc. 1st Intern. X-ray Conf. Cambridge, England*, 603–616 (1956).
189. R. Smoluchowski, *Compt. Rend.*, **222,** 1496–1497 (1946).
190. K. W. Andrews and W. Johnson, *Proc. 1st Intern. X-ray Conf. Cambridge, England*, 581–589 (1956).
191. R. S. Sharpe, *Proc. 1st Intern. X-ray Conf. Cambridge, England*, 590–602 (1956).
192. C. K. Jackson, *Proc. 1st Intern. X-ray Conf. Cambridge, England*, 623–627 (1956).
193. S. H. U. Bowie, *Bull. Geol. Survey G. Brit.*, No. 10, 45–57 (1955).
194. F. Erdmann-Jesnitzer and F. Gunther, *Z. Metallk.*, **49,** 9–16 (1958).
195. J. Thewlis and R. T. P. Derbyshire, *Nondestructive Testing*, **16,** 154–157 (1958).
196. S. Yamagouchi, *Ann. Chem.*, **29,** 1387 (1957).
197. R. T. Foley and S. P. Newberry, *J. Phys. Chem.*, **62,** 1184–1188 (1958).
198. W. Johnson and K. W. Andrews, *Iron & Steel* (*London*), **31,** 437–444 (1958).
199. A. W. Ruff, Jr., and L. M. Kushner, *2nd X-ray Conference Stockholm*, Elsevier, New York, 1960, pp. 153–161.
200. R. E. Spear and G. R. Gardner, *Mod. Castings*, **37,** 36–44 (1960).
201. R. S. Sharpe, *Some Applications of Microradiography to Fuel Element Inspection*, Fuel Element Symposium, Vienna, Academic, New York, 1960, **2,** pp. 45–54.
202. G. Shinoda, T. Amano, T. Tomura, and R. Shimizu, *2nd X-ray Conference Stockholm*, Elsevier, New York, 1960, pp. 162–168.
203. W. M. Williams, *Lab. Methods*, **63,** 95–101 (1961).
204. A. Taylor, *X-ray Metallography*, Wiley, New York and London, 1961, 1000 pp.
205. W. Johnson, *Encycl. Microsc.*, Reinhold, New York, 1961, pp. 575–581.
206. R. W. Andrews and W. Johnson, *Encycl. X-rays*, Reinhold, New York, 1963, pp. 618–621.
207. R. S. Sharpe, *Reactor Sci.; Reactor Technol.* (*See J. Nucl. Energy, Parts A/B*), **17,** 505–510 (1963).
208. J. J. Trillat, *Rev. Aluminium*, No. 306, 197–207 (1963).
209. R. S. Sharpe, *High Definition Radiography*, A.E.R.E. Publication R.5095, December 1965, 13 pp. (unclassified).
210. R. A. Morris and R. D. Strong, *Microradiography of Pyrocarbon-coated Microspheres to Determine Coat Thickness*, U.S. AEO Reprint LA-3814, 1967.
211. R. W. McClung, *Nondestructive Testing of Nuclear Graphite*, ASTM-STP439, 97–108 (1968).
212. R. Blue, *Tool Mfg. Eng.*, **59,** 62–64 (1967).
213. R. S. Sharpe, *Divergent Beam X-ray Diffraction and Its Application to Irradiated MgO*, UKAEA. Res. Rept. AERE-R4452, Harwell, 1963, 21 pp.

214. T. Ichinokawa and S. Shirai, *4th Congr. X-ray Optics*, Hermann, Paris, 1966, pp. 561–565.
215. J. Frazer and G. Arrhenius, *4th Congr. X-ray Optics*, Hermann, Paris, 1966, pp. 516–533.
216. K. J. H. Mackay, *4th Congr. X-ray Optics*, Hermann, Paris, 1966, pp. 544–554.
217. R. S. Sharpe, *J. Roy. Microsc. Soc.*, **86,** 271–284 (1967).
218. V. I. Baranov, A. P. Zhdanov, and M. Deizenret-Mysovskaya, *Bull. Acad. Sci. U.S.S.R.*, **1,** 20–28 (1944).
219. G. L. Clark and J. A. Howsman, *Ind. Eng. Chem.*, **38,** 1257–1262 (1946).
220. A. E. Barclay and D. Leatherdale, *Brit. J. Radiol.*, **22,** 62 (1948).
221. G. A. G. Mitchell, *J. Photogr. Sci.*, **2,** 113 (1954).
222. C. Legrand and J. Salmon, *Bull. Microsc. Appl.*, **4,** 9–17 (1954).
223. P. Manigault and J. Salmon, *Bull. Microsc. Appl.*, **8,** 14–20 (1958).
224. C. K. Jackson, *1st X-ray Conference Cambridge, England*, Academic, New York, 1957, pp. 487–491.
225. S. P. Ong. *10th Ann. Conf. X-ray Anal. Denver*, Plenum Press, New York, 1961, pp. 324–334.
226. J. Dietrich, *2nd X-ray Conference Stockholm*, Elsevier, New York, 1960, pp. 306–310.
227. J. Salmon, *Encycl. Microsc.*, Reingold, New York, 1961, pp. 636–647.
228. E. J. Gibson, *Research (London)*, **15,** 323–328 (1962).
229. W. L. Jongebloed and S. M. Jutte, *3rd European Conf. Electron Micros.*, London, 1965, p. 173.
230. W. L. Jongebloed, *Norelco Reptr.*, **12**(3), 93–97 (1965).
231. J. Dietrich, *4th Intern. Congr. X-ray Optics and Microanalysis*, Hermann, Paris, 1966, pp. 658–663.
232. B. Idle, *Sci. J.*, **4,** 59–63 (1968).
233. H. F. Sherwood, *Rayon Text. Mon.*, **18,** 53–54 (1937).
234. S. B. Newman, *Mod. Plastics*, **40,** 165–179 (1963).
235. J. J. Trillat, *Peintures, Pigments, Vernis*, **21,** 178–180 (1945).
236. F. I. G. Rawlins, *J. Sci. Instrum.*, **19,** 17–22 (1942).
237. F. I. G. Rawlins, *Photogr. J.*, **89,** 104–105 (1949).
238. F. I. G. Rawlins, *Studies Conserv.*, **1,** 135–138 (1954).
239. R. Kozlowski, Ibid **5,** 89–101 (1960).
240. J. E. Barnard, *J. Roy. Microsc. Soc.*, 1–7 (Feb. 1915).
241. R. A. M. Schmidt, *Science*, **115,** 94–95 (1952).
242. R. H. Hedley, *Micropaleontology*, **3,** 19–28 (1957).
243. K. Hooper, *2nd X-ray Conference Stockholm*, Elsevier, New York, 1960, pp. 216–223.
244. W. L. Jongebloed, Norelco Reptr., **12**(3), 93–97 (1965).
245. H. Held, C. A. Band, and G. Fiore, *Arch. Oral. Biol.* (ORCA Suppl), **4,** 1–5 (1961).
246. R. W. G. Wyckoff, V. J. Hoffman, and P. Matter, *Science*, **140,** 78–80 (1963).
247. J. D. Pelgroms, *Paper Trade J.*, **134,** 25–32 (1952).
248. J. Isings, S. P. Ong, J. B. Le Poole, and G. Van Nederveen, *Proc. Intern. Conf. Electron Microscopy*, Stockholm, Almqvist and Wiksells, 1957, pp. 282–283.

249. S. P. Ong, *Microprojection with X-rays*, Martinus Nijhoff—The Hague, 1959, 132 pp.
250. S. B. Newman and D. Fletcher, *Tappi*, **47,** 177–180 (1964).
251. O. M. Barlow, W. B. Distler, and J. L. Hile, *Nondestructive Testing*, **18,** 341–343 (1960).
252. S. B. Newman et al., *J. Res. Nat. Bur. Std.*, **67A,** 153–155 (1963).
253. R. V. Ely, *X-ray Focus*, **8**(1), 14–21 (1967).
254. J. J. Trillat, *Rev. Gen. Caoutchouc*, **22,** 63–66 (1945).
255. W. M. Hess, *Encycl. X-rays and Gamma Rays*, Reinhold, New York, 1963, pp. 947–951.
256. S. H. Thorpe and D. W. Davison, *Engineering*, **157,** 241–243 (1944).
257. A. Frantzell, *Acta Radiol.*, **33,** 83–103 (1950).
258. G. Spiegler and P. Giles, *Brit. J. Radiol.*, **26,** 130–135 (1953).
259. R. V. Ely and J. G. Hampshire, *3rd X-ray Conference Stanford, California*, Academic, New York, 1963, pp. 63–71.
260. G. Bergendahl and B. Engfeldt, *Acta Pathol. Microbiol., Scand.*, **49,** 30–38 (1960).
261. K. Cotterell and R. S. Sharpe, *Nondestructive Testing*, 234–237 (July–Aug. 1962).
262. I. Molenaar, *2nd X-ray Conference Stockholm*, Elsevier, New York, 1960, pp. 177–183.
263. H. F. Sherwood, *Tech. Studies*, **6,** 277–280 (1938).
264. R. Berthold, *Metallwirtschaft, Berlin*, **20,** 694–697 (1941).
265. A. Engström, *Acta Radiol.*, **36,** 305–310 (1951).
266. S. Bellman, *Acta Radiol., Suppl.*, 102 (1953), 104 pp.
267. R. L. de C. H. Saunders, J. Lawrence, and D. A. Maciver, *1st X-ray Conference Cambridge. England*, Academic, New York, 1957, pp. 539–550.
268. B. Odén, S. Bellman, and B. Fries, *Brit. J. Radiol.*, **31,** 70–80 (1958).
269. J. B. Nelson, *X-ray Stereo-Microradiography of Carbons*, 5th Biennial Conference on Carbon, 1961.
270. W. M. Williams, *Lab. Methods*, **63,** 95–101 (1961).
271. E. von Witte and H. Jütting, *Atompraxis*, **5,** 300–304 (1959).
272. R. V. Latham, *J. Roy. Microsc. Soc.*, **85**(3), 255–282 (1966).
273. R. V. Latham, *J. Roy. Microsc. Soc.*, **88,** 183–187 (1968).

INDEX

Abbé, 126
  condenser, 136, 261
Aberration, astigmatism, 132
  axial, 132
  distortion, 132
  chromatic, 132
  coma, 132
  curvature of field, 132
  field, 132
  monochromatic, 132
  of lenses, 131
  spherical, 132
Abrasives, 156
Absorption, 129
  atomic, 231
  atomic-molecular, 231
  coefficient, 225
  figure, 222, 238
  index, 219, 225
  of light, 130
  measurement of, 225
  molecular, 231
Achromatic, condenser, 136
  objective, 133
Acidic dyes, 235
Acute bisectrix, 171, 178, 196, 217
Addition compounds, 153
Adsorbed molecules, nmr spectra, 687
Airy's disk, 125, 261
Airy's formula, 33
Alignment of microscope, 130, 142
Allochromatism, 231
Amplifying lenses, 134
Analysis, qualitative, 160
  quantitative, 161
  x-ray qualitative, 725, 730
Analyzer, 166, 179
Angular aperture, 125, 139, 141, 159, 193
Anistropic crystals, 173
Anomalous interference colors, 182, 186
Antireflection coating, 129, 141
Aperture stop, 139, 141, 159
Aplanatic condenser, 136
Apochromatic objective, 133
Area, measurement of, 161
Areal analysis, 165, 254
Arm of microscope, 143
Aspherical lens, 137
Astigamatism, 166
Asymmetric top, 130
  near prolate, 131
  selection rules, 130
Auxiliary condenser, 137, 138
Average particle size, 164

Back focal plane, 142
Base of microscope, 143
Basic dyes, 234
Becke, lens, 169
  line, 201
Becke-Klein magnifier, 194
Beer's law, 503
Berek compensator, 212
Bertrand lens, 143, 166, 193
  diaphragm, 193
Biaxial crystals, 173, 178
  sign of, 178
Binocular body tubes, 146
Biot quartz wedge, 211
Birefringence, 195
  depth of, 210
  measurement of, 213
Bisectrix, 178
Bloch equations, 504 ff., 570, 577
Body tube, 143
Boiling point, 242
  determination of, 243
Breit-Rabi equation, 513
Bright field, 92
  microscopy, 130, 158

reflected light, 263
Brightness, 130
Brush, 297
Bulk magnetic susceptibility correction, 633

Cabannes depolarization factors, 81
Calibration of eyepiece micrometers, 162
Cardioid condenser, 261
Carr-Purcell train, 695
Cassette, vacuum for microradiography, 720, 723
Cavities, microwave, 533, 546 ff.
Cells, counting, 165
Cement clinkers, 160
Centrifugal distortion, 134
Centrifugation, 149
Ceramics, 157
Chemical exchange, 641, 659, 664, 697
Chemical microscopes, 169
Chemical shift, 614
anisotropy, 658, 684, 693
^{13}C, 671, 674
contact, 664
diamagnetic anisotropy effect on, 618
electron density effect on, 716
^{19}F, 670
measurement in solids, 684, 701
Ring-Current effect on, 621
theory, 617, 669, 674
scales, 615
solvent effects on, 632
Circularly polarized light, 175
^{13}C nmr spectra, 651, 667, 669, 673, 700
Coarse adjustment knob, 143
Coating, antireflection, 141
Coherence, length, 13
spatial, 14
temporal, 13
theory of, 12
Cold stages, 239
Collimation, 14
Colloids, 262
Color, and constitution, 231
of platinocyanide crystals, 233
Colored crystals, conoscopic examination of, 237
microscopy of, 219
orthoscopic examination of, 237
preparation of, 234
Compensating eyepiece, 134
Compensators, 167, 195, 198, 210, 281
Condenser, 127, 135
Abbe, 261
aplanatic achromatic, 261
bicentric, 262
bispheric, 262
darkfield, 261
Conditions for interference, 5
Conformational analysis, 135
Conformational equilibrium, 640, 644, 673, 703
Conical refraction, 177
Conoscope, 191, 193, 194
Conoscopic, absorption figures, 238
examination, 191, 193
of colored crystals, 237
field, 197
observation, 166
Contrast, 122, 139, 141, 142, 158, 270, 274
control of, 147
image, 130, 317
in microradiographs, 717, 719, 730
luminance, 260
Corrosion products, 154
Counting cells, 165
Coupling constants, ^{13}C-H, 674
fluorine-fluorine, 672
proton-proton, 629
protons with other isotopes, 630
spin-spin, 624
solvent-dependence, 638
sigh of, 631, 652, 690
Cover glass, 131, 133
Critical illumination, 130, 138, 261, 263
Critical solution temperature, 242
determination of, 243
Crossed axial plane dispersion, 183, 185
Crossed polarizers, 180
Crosshairs, 166
Crystal, geometry, 189
growth rate, 242
determination of, 244
habit, 234
morphology, 244
optics, 171
systems, 171
Crystallization, 149
rate, 254
Crystallographic axes, 171, 178

Crystallography, 171
Crystals, biaxial, 259
  colored, preparation of, 235
  isotropic, 259
  orientation of, 201
  preparation of, 149
  transparent, 189
  uniaxial, 259
Cubic system, 171
Cyclic twinning, 172

Dark field, 93
  microscopy, 163, 260
  reflected light, 263
  vertical illumination, 136
Debye scattering, 81
Deceptively simple spectra, 628
Decomposition, 245
  thermal, 242
Delay, 12, 16
Depth of focus, 142, 145, 163
Diamagnetic anisotropy, 618, 636
Diaphragms, 139
Dichroism, 219
  measurement of, 227
Dichroscope, 170, 221
Dielectric constant, 525, 548
Diffraction, 125, 126
  Fresnel, 729, 744
  gratings, pitch in Moiré interferometry, 59
Direction cosines, 133
Dislocations, 262
Dispersion, 171, 182, 183, 237
  crossed, 184
  of birefringence, 183, 212
    depth of, 210
  of optic axes, 183
  of refractive index, 183, 188
Dispersion, horizontal, 184
  inclined, 185
  monoclinic, 184
  orthorhombic, 184
  staining, 131, 164, 255
  triclinic, 185
Distortion, crystal, 171
Dipole moment and color, 232
Double-quantum transitions, 632
Double reflection, 228
Double resonance, 644, 648
  modulation, 132
Dovetail slide, 143
Draw tube of microscope, 144
Drawing camera, 165
Dynamic isomerism, 245, 249

Einstein-Smoluchowski relation, 95; *see also* Light scattering
Electric dipole moment, 133
Electric field, microwave, 525 ff., 533, 548 ff., 552, 555
Electric quadrupole moment, 138
  molecular, 138
  nuclear, 141
Electron beams, for x-ray microscopy, 735
  bias control of, 723, 726, 736
  focusing of, 723, 726, 735
Electron diffraction, 93
  and probe analysis, 124
  application to, dispersions, 101
    fibers, 100
    metals, 101
    photography, 100
    polymers, 100
  of dispersions, 96
  of internal structure, 97
  of surfaces, 98
  resolving power, 414
Electron micrography, 91
Electron microscopy, 129
Electron guns, for microradiography, 723, 725, 735
Electron magnetic resonance, 500; *see also* electron spin resonance
Electron microprobe, 114
  and transmission electron microscopy, 124
  application to, qualitative analysis, 124
    quantitative analysis, 124
    thin films, 123
  development, 114
  methods of analysis. 117
  production of X-rays, 116
  proportional counters, 118
  resolution, 116, 118
  specimen requirements, 116
  spectrometer geometry, 117
  techniques, 119
Electron microscopy, 87
Electron paramagnetic resonance, 500; *see also* spin resonance

Electron spin resonance, 500
  absorption, 502 ff., 542, 548, 551, 578
  dispersion, 506 ff., 551, 578
  saturation, 503 ff., 508, 542
  spin-lattice relaxation time, 503 ff., 542 ff.
  transverse relaxation time, 504 ff., 542, 560 ff., 570 ff.
Electron spin resonance spectra, computer simulation of, 523 ff., 586 ff.
  effect on, of, chemical parameters, 587
    field modulation amplitude, 570 ff.
    field modulation frequency, 576 ff., 589
    filtering, 558 ff., 588 ff.
    microwave power, 542 ff.
    sample geometry, 551 ff.
    time-averaging, 568 ff.
  interpretation of, isotropic, 510 ff.
    polycrystalline, 521 ff.
    single crystal, 519 ff.
Electron spin resonance spectrometers commercially available, 538 ff.
  homodyne, 534 ff., 538
  superheterodyne, 536 ff.
Element detection, by microradiography, 730
Elongation, sign of, 191
Embedding, 157, 165
Emulsion, graininess in microradiographs, 717, 729
  processing of, 731, 743
  selection, 729, 743
Enantiotropism, 247
Energy levels, electron spin, 502, 510 ff., 545
Enlargement, by direct x-ray projection, 717, 735
  in microradiography, optical, 731
Etalon, 37
Etching, 156, 157, 164
Etendue, 15
Eulerian angles, 129
Eutetic, 153, 251, 281
  composition of, 253
Eyepiece, 122, 134, 170
  crosshaired, 170
  see also Ocular
Eyepoint of eyepieces, 142
External conical refraction, 177
Extinction, 181
  angle, 190
  coefficient, 232
  dispersed, 190
  modulus, 225
  oblique, 190
  parallel, 190
  symmetrical, 190
Extraction replica, 154
Extraordinary ray, 175, 191

Feret's diameter, 164
Fermi contact interaction, 624, 663
Fiber optic fringe splitter, 38
Field diaphragm, 261
Field-frequency stabilization, 611
Field gradient, 141
Field stop, 134, 137, 139, 159
  diaphragm, 138
Filaments, electron gun, 723, 726, 736
Filling factor, microwave cavity, 549 ff.
Filter, 265
  barrier, 269
  interference, 6
  neutral density, 137
  polarizing, 169
  ultraviolet, 269
Fine adjustment knob, 143
Finesse, 38
First-order red compensator, 167
Fisher refractometer, 210
Fizeau-type fringes, 25
Flash interference figure, 196
Flat-field objective, 133
Fluorescence, 131
  microscopy, 268
Fluorescent screens, 265
Fluorite objective, 133
Fluorochromes, 268
^{19}F nmr spectra, 667, 669
Focal length, 124
Focus, depth of, 163
Focusing the microscope, 139, 143
Foot of microscope, 143
Form, 172
Fourier transform nmr spectra, 673, 698, 705
Fractography, 155
Fracture, 155
Free induction decay, 695
Frequency stabilization, 132

Fresnel, biprism, 271
  mirror, 271
Fringes, equal inclination, 38
  Haidinger, 37
  splitter, 38
  splitting, 47
  visibility, 8

Gaussian lineshape, 524
General references, 127
Glass formation, 249
Golay coils, 611
Gradient fusion preparations, 153
Grandjeans terraces, 250
Greenough binocular microscope, 149, 264
Grinding, 156
Growth rate, determination of, 244
  of crystals, 242

Habit, crystal, 171
Haidinger fringes, 37
Hamiltonian, internal rotation, 136
  pure rotational, 128
  spin, 503, 509 ff.
Heavy liquid separation, 149
Hemacytometer, 165
Hexagonal system, 171
High eyepoint eyepiece, 134
High-pressure microscopy, 239, 242
High resolution scanning electron microscopy, 113
Histogram, 164
Halographic interferometry, 51
  double exposure, 54
  real time, 55
  stroke interferometer, 59
  time average, 55
  see also Interferometry, halographic
Homal projection lens, 134
Hot stage, 151, 239
Huygenian eyepiece, 134
Hydrates, 245
Hydrogen bonding, 622, 645
Hydrogen-bond shift, 637
Hyperfine interaction, electron-nuclear, 509 ff., 545

Idiochromatism, 231
Illuminance, 130
Illuminating system, 130, 135
Illumination, 129, 130, 158
  critical, 130, 261
  inclined, 126
  Kohler, 261
  vertical, for examination of colored crystals, 238
Illuminators, 136, 142
  dark field, 261
  for fluorescence microscopy, 268
  infrared, 265
  ultraviolet, 265
  vertical, 263
Image, 122
  contrast, 109, 130
  converter tubes, 265
  display and recording, 107
  intensity, 90, 129
  real, 123
  transfer lens, 241
  virtual, 124
Immersion, 128
  liquids, 203, 205
Inclination joint, 143
Indicatrix, 174
Infrared microscopy, 128, 264
Integrating eyepiece, 165
Integrating stage, 166
Intensifiers, x-ray image, 739, 771
Interaction of electrons with matter, 107
Interference, 153
  colors, 42, 179, 182
  figure, 151, 166, 171, 192
    biaxial, 196
    flash, 196
    of colored crystals, 237
    uniaxial, 195
  fringes, 182, 275, 281
  microscope, 130
    for reflected light, 280
  microscopy, 270
    for reflected light, 264
Interferometers, mock or Moiré, 61
  multiple-beam Fabry-Perot, 36
  optical, nomenclature and geometry, 15
  Strobe, 59
  two beam, Jamin, 20
    Mack Zehnder, 21
    Michelson, 23
    polarizing, 28
    Rayleigh, 17

Twyman-Green, 25
x-ray, Bragg-type, 62
von Laue-type, 62
Interferometry, 153, 162, 262, 270
conditions for interference, 5
halographic, theory of, 52
light sources for, 9
microwave, theory of, 64
Moiré, theory of, 59
multiple beam, evaluation of results, 49
particle, neutron and electron, 66
radio, 65
sonic, 65
summary of uses, 66
theory of, multiple beam, 32
theory of, visible, two-beam, 4
using non-electromagentic radiation, 65
x-ray, 62
Internal conical refraction, 177
Internal rotation, 136
Isogyre, 197, 217
Isomorphism, 153
Isotopic substitution, 133
Isotropic crystals, 173

Johannsen, 217
lens, 193
wedge, 212

Kellner eyepiece, 134
Klystrons, 530, 534 ff.
Kofler, 151
Kohler illumination, 138, 159, 261, 263
Kossel lines, in x-ray diffraction, 764

Larmor precession frequency, 606
Lateral chromatic aberration, 135
Lattice, crystal, 171
Leitz-Jelley microrefractometer, 209
Length, measurement of, 161
Lens, 123
antireflection coating, 129
condenser, 127
immersion, 128
Lenses, 131
aberrations of, 131
aspherical, 132
magnetic, for x-ray microscopes, 736
Light scattering, 75-117
absolute turbidity, 107
angular scattering pattern, 84
dissymmetry, 84
gyration, radii of, 86
handling of data, 95
Zimm plot method, 96
instruments, 102
angular light scattering photometers, 103
Aminco, 104
Brice-phoenix, 104
forward-angle photometers, 108
Sofica, 105
isotropic and anisotropic particles, 81
laser scattering, 100, 110
repulsion diameter of particles, 93
see also Debye scattering; Mie scattering; and Rayleigh scattering
Light sources, 136, 158, 159
Line shapes, nmr, 682, 704
Line widths, nmr, 609, 643, 647, 655, 661, 675
Linear, counting, 165
measurement, 162
molecules, 129
Linnik interference microscope, 27; *see also* Interference, microscope
Liquid crystals, 249
Lorentz-Lorentz Law, 2
Lorentzian line shape, 508, 522, 524, 542, 558, 562, 567, 579
Luminance, 130
Lyotropic mesomorphism, 249

Magnetic dipole moment, 137
Magnetic equivalence, 628
Magnetic field, 502, 506, 509 ff., 513, 520, 522, 569
microwave, 503 ff., 527 ff., 533, 548 ff., 552, 555
modulation, 534 ff., 570 ff., 589
rotating, 507
Magnetic moment, nuclear, 602
Magnetic susceptibility, 137, 138, 505 ff.
complex, 506 ff., 548 ff., 570, 578
determination, 634
diamagnetic, 137, 138
paramagnetic, 137, 138
Magnetogyric ratio, 604, 668
Magnification, 122, 129; *see also* Enlargement

Magnifier, 124
Mallard's constant, 216
Martin's diameter, 163
Maximum horizontal intercept, 164
Maxwell's equations, 525, 527
Measurements, microscopical, 161
Mechanical stage, 166, 170
Mechanical tubelength, 144
Melatope, 196
Melt, crystallization from, 151
Melting point, 242
Mercury vapor lamp, 170
Mesomorphism, 249
Metallic reflection, 228
Michel Lévy, 217
Micromanipulator, 149
Micrometer disk for eyepieces, 170
Micrometer eyepiece, 162, 217
Micrometer stage, 162
Microphotometer, 226
Microphotometry, 266
Microprojection, 135
Microradiographs, contrast control of, 732
  photographic, enlargement of, 731
    processing of, 731, 743
Microradiography, 719
  applications, botanical, 766
    general, 768
    medical, 746
    metallurgical, 757
  stereographic, 718, 727, 744, 746
Microrefractometers, 207
Microscope, electron, 129
  mechanical parts of, 143
  optics of, 122
  stand, 143
  types of, 158
Microscopical techniques, 158
Microscopy, dark field, 260
  high pressure, 239, 242
  high temperature, 239
  see also Microradiography
Microspectrograph, 212, 238
Microspectrometry, 264
Microspectrophotometry, 266
Microtomy, 157
Microwave absorption, ESR, 503, 506, 508, 536 ff., 548, 551, 578
  cavities, 533, 546 ff.
  components, 528 ff.
  detectors, 534, 536 ff.
  double-resonance, 132
  electric field, 525 ff., 533, 548 ff., 552, 555
  frequency, 530, 537, 551, 567
  interferometry, 64; *see also* Interferometry, microwave
  klystrons, 530, 534 ff.
  magnetic field, 503 ff., 527 ff., 533, 548 ff., 552, 555
  power, 508, 530, 536 ff., 542, 548
  skin depth, 525
  spectroscopy, 128
  spectrum, 129, 130, 131
  waveguides, 526 ff., 546 ff.
  wavelength, 526, 552, 555
Mie scattering, 88
Mirror of microscope, 143
Mirror systems, in x-ray microscopy, 717
Mixed fusion test, 251
Mixed melt preparations, 153
Moiré interferometry, 59
  mechanical interference theory, 60
  obstruction theory, 60
Molal refraction, 3
Molecular motions in solids, 686
Molecular q-value, 137
Molecular structure, 133, 188
Molecular weight, 242
  determination of, 243
Moment of inertia, 128, 129
Monochromator, 171, 215, 265, 266, 269
Monotropism, 247
Morphological analysis, 160
Mounting, 147
  media, 131, 147
    for dispersion staining, 257
Multiple-beam interferometry, 277
Multiplets, first-order, 625

Natural broadening, decrease in coherence length, 3
Negative biaxial crystals, 178
  eyepiece, 133, 134
  projection lens, 135
  uniaxial crystals, 178
Nematic state, 250
Neumatic liquid crystals, 688
Newton's rings, 214
  series of interference colors, 182

Nikitin compensator, 212
Noise decoupling, 673
Nosepieces, centerable, 167
Nuclear quadrupole interaction, 141
Nucleic acids, nmr spectra, 681
Numerical aperture, 128, 142, 166, 195, 197, 261, 262, 265

Objective, choice of, 142
correction for coverglass, 133
lens, 122
numerical aperture of, 128
oil immersion, 262
Objectives, 133
conoscopic field of, 197
for polarizing microscopes, 170
infrared, 265
numerical apertures of, 170
reflecting, 265, 266
ultraviolet, 265
working distance of, 170
Obtuse bisectrix, 178, 196
Ocular, Wright universal, 212; *see also* Eyepiece
Oil immersion optics, 262
Optical activity, 175
Optical axial angle, dispersion of, 216
Optical properties, 188
of crystals, 244
Optical rotatory power, 173
Optical system, compound microscope, 139
Optic axis, 173, 175, 177
Optic axial angle, 108, 170, 178, 196, 197
measurement of, 216
Optic normal, 171
Optics, geometric, 125
microscope, 122
physical, 125
Order of interference, 5
Ordinary ray, 175
Orientation of crystals, 201
Oriented molecules, nmr spectra, 632, 688
Orthoplan®, 145
Orthorhombic system, 171
Orthoscopic, examination, 189
of colored crystals, 237
eyepiece, 134
observation, 166
Overhauser effect, nuclear, 651, 654

Paraboloid condenser, 261
Parallel polarizers, 180
Paramagnetic molecules, nmr spectra of, 665
Paramagnetic resonance, 500; *see also* Electron spin resonance
Particle interferometry, 66
Particle size, 163
distribution, 164
Particulate materials, 147, 160
Peritectic, 251
Petrographic microscopes, 166, 193
Phase, diagrams, 239, 245
binary systems, 250
ternary, 254
difference, 8
microscope, 130, 270, 277, 278
microscopy, reflected light, 264
retardation, 180
shifts, 130
Phemister, 217
Phosphorescence, 131
Photocells, 267
Photographic film, 267
Photometry, 267
Photomicrography, 135, 282
contrast control in, 732
Pillar of microscope, 143
Pinhole, diaphragm, 193
eyepiece, 143, 169
image, 122
Pitch, grating, in Moiré interferometry, 59
Planapochromatic objective, 133
Planimeter, 165
Platinocyanide crystals, color of, 233
Pleochroism, 219, 234, 237
Point counting, 165
Polarization, state of, for interference, 9
Polarized light, 172
reflection of, 229
Polarized light microscopy, 166
reflected light, 264
Polarizer, 179
Polarizing, filters, 169
microscope, 130, 166, 241, 270
Polishing, 156
Polymers, nmr spectra of, 675
Polymorphic transformation, 242
Polymorphic transitions, 153
Polymorphism, 234, 245, 247, 254

Polysynthetic twinning, 172
Positive, biaxial crystals, 178
  eyepiece, 134
  uniaxial crystals, 174
Principal, inertial axis system, 128
  moments of inertia, 129
  refractive indices, 178
  vibration directions, 178
Prisms, polarizing, 169
Propagation angles, 8
Proteins, interaction with small molecules, 659, 672, 673
  nmr spectra of, 658, 679
Pseudomorphism, 245
Pseudorotation, 135
Pulsed nmr, 694
Purity of materials, 242
Pyrometers, 241

Quadrupole, interactions in solids, 685
  moment, nuclear, 602, 631, 669
  relaxation, 647
Qualitative analysis, thermal, 254
Quality factor, 547 ff., *see also* Microwave cavities
Quantitative analysis, 161
Quarter waveplate, 167
Quartz wedge, 167, 211

Rack and pinion, 143
Radiation, *see* X-rays
Radio interferometry, 65
Ramsen, disk, 194
  eyepiece, 134
Ray velocity, 174
  surface, 177
Rayleigh, 126
  ratio in light scattering, 78
  scattering, 76
Reaction, field, 636
  rate determinations, nmr, 643, 697
Reflected, bright field illumination, 263
  dark field illumination, 263
  light microscopy, 263
Reflecting power, 6
Reflection, 141
  metallic, 228
Reflection electron microscopy, 94
Refractive index, 153, 173, 219, 242, 253, 257
  and color, 232
  determination of, 199, 200, 244
    Becke method, 201
    Clerici's method, 205
    double variation method, 204
    immersion liquids for, 203, 205
    immersion method, 200
    oblique illumination, 202
    of immersion liquids, 207
    of melts, 210
    of volatile liquids, 210
    temperature variation method, 204
    wavelength variation method, 204
  measurement of, 281
  nonprincipal, 207
  principal, 178
  standards, 244
Refractometers, 207
Refractories, 157, 160
Reinecke salt, 235
Relaxation time, 655, 661
  electron spin-lattice, 503 ff., 542 ff.
  electron transverse, 504 ff., 542 ff., 560 ff., 570 ff.
  longitudinal, 608
  measurement of, 659, 694
  transverse, 608
Replication, 154
Resinography, 156
Resistance thermometers, 241
Resolution, 110, 122, 126, 142
  enhancement, 705
  in x-ray microscopy, 716, 723, 725; *see also* Microradiography
  theory of, 126
Resolving power, 125, 126, 141, 162, 264, 265
Resonance, and color, 232
  condition, nmr, 604
  radiation, 232
Retardation, 153, 180, 192, 211, 281
  measurement of, 213
Reticule, 162
Revolving nosepiece, 144, 146, 169
Rheinberg illumination, 131
Rhombohedral system, 171
Rotamer, 135
Rotating stage, 166
Rotation stage, 201
Rotational transitions, Q-branch, 131

R-branch, 131

Sampling problem, 146
Saturation, ESR, 503 ff., 508, 542
factor, nmr, 609, 660
localized, 654
Savart plate, 29, 271
Scanning electron microscopy, 102
interaction of electrons with matter, 89
resolution, 92
Scattered light, depolarization factor, 81
Scattering, 141
of light, 130
Second moment, 683
Sectioning, 155, 156
Selenite plate, 167
Semiapochromatic objective, 133
Sensitivity, enhancement, 613, 673, 699, 705
nmr instruments, 610, 613
Separation of particulate materials, 149
Shadowing, 154, 162
Shear, angle of, 16
in interferometers, 16
lateral, 16
radial, 16
rotational, 16
Sign, determination of, 195, 198
of birefringence, 188
of uniaxial crystals, 186
Signal detectors, 105
Signal-to-noise ratio, in ESR spectra, 557, 562, 568 ff., 588; *see also* Electron spin resonance spectra
Skin depth, microwave, 525, 548
Slags, 160
Smectic state, 249
Snell's law of refraction, 175
Sodium vapor lamp, 170
Softening point, 242
Solid solutions, 251
Solids, nmr spectra, 682
Sonic interferometry, 65
Specimen, preparation, 110, 131, 146
thickness, measurement of, 281
Spectrophotometer, 226
Spectroscopy, Fourier or interference, 25, 38
Spherical aberration, 193
Spin decoupling, 648
Spin hamiltonial, 503, 509 ff.
Spin rotation interaction, 143, 658
Spin-spin interactions, 623
Spin temperature, 608
Spin tickling, 650
Split fringes, in interferometry, 47
Stage, of microscope, 143
heating, 204
high pressure, 242
micrometer, 162
rotation, 201
thermal gradient, 250, 251
universal, 201, 204
Stages, cold, 239
hot, 239
Staining, 157, 164
fluorescent, 268
Stand, microscope, 146
Stark, effect, 132
modulation, 131, 132
Stereomicroscope, 149, 264
Stereoscopy, in microradiography, 718, 727, 744, 746
Stops, 139
Strain birefringence, 241
Stray light, 141
Sublimation, 149, 151, 245
Substitutional structure, 134
Superconducting magnets, 611
Surfaces, examination of, 154
Symmetric top, 130
oblate, 130
prolate, 130
selection rules, 130
Symmetry, crystal, 171

Targets, x-ray, contamination of, 725, 739
cooling of, 725, 738
interchangeable, 722
transmission, 735, 737
Tautomerism, 249
Television systems, closed circuit, for x-ray microscopy, 740, 743
Textile fibers, 160
Thermal microscopy, 239
Thermistors, 241
Thermocouples, 241
Thermotropic mesomorphism, 249
Time-averaging, in ESR spectra, 568 ff.
Transformation temperature, 242

Transition probability, electron spin, 502
Transmission, intensity, 34
  maximum, 6
  of light, intensity, 34
Transmission electron microscopy, 88
Transmission line, microwave, 546 ff.
Transmitted illumination, 158
Trichroism, 220
Triclinic system, 171
Trigonal system, 171
"Trinocular" bocy tubes, 146
Tube length, 144
Twinning, crystal, 171

Ultramicroscope, 262
Ultramicroscopy, 163, 262
Ultramicrotomy, 157
Ultraplane projection lens, 134
Ultraviolet microscopy, 128, 264
Uniaxial crystals, 173, 174
Universal stage, 201, 204

Vacuum evaporation, 155
Van der Waals interactions, 634
Vertical illumination, 135
  for colored crystals, 238
Vertical illuminator, 263
Vibrational satellites, 130, 134
Vibration directions, principal, 178
Vibration plane, 179
Vibration-rotation interactions, 134
Virtual coupling, 629
Volume, measurement of, 161

Wave, front, 174
  normal, 174
    refractive index surface, 174-178
Waveguides, microwave, 526 ff., 546 ff.
Wedges, crystal, 182
Wollaston prism, 221, 238, 271
Working distance, 170
Wright, 217
  combination wedge, 211
  crossed-slit diaphragm, 193
  universal ocular, 212

X-ray(s), absorption in air, 719
  absorption coefficients, 730
  characteristic, 730, 738
  dose rates, 739, 771
  focusing of, 717
  interferometry, 62; *see also* Interferometry, x-ray
  microscopy, 128
  monochromatic, 731
  reflection of, 717
  spectrum, 730
  techniques, *see* Microradiography

Zeeman effect, 137
  interaction electronic, 502, 509, 520 ff.
  nuclear, 509